Fundamentals of Digital Signal Processing
using MATLAB®

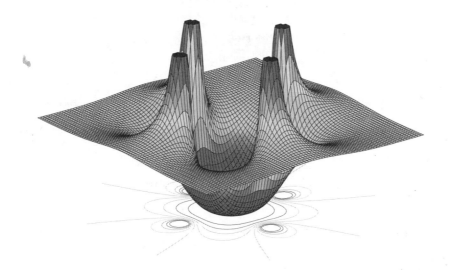

Robert J. Schilling and Sandra L. Harris

Clarkson University
Potsdam, NY

Australia   Canada   Mexico   Singapore   Spain   United Kingdom   United States

THOMSON

Fundamentals of Digital Signal Processing using MATLAB®
by Robert J. Schilling and Sandra L. Harris

**Associate Vice-President
and Editorial Director:**
Evelyn Veitch

Publisher:
Bill Stenquist

Sales and Marketing Manager:
John More

Developmental Editor:
Rose Kernan

Permissions Coordinator:
Sue Ewing

Production Services:
RPK Editorial Services

Copy Editor:
Shelly Gerger-Knechtl

Proofreader:
Jackie Twomey

Indexer:
Shelly Gerger-Knechtl

Production Manager:
Renate McCloy

Creative Director:
Angela Cluer

Interior Design:
Carmela Pereira

Cover Design:
Craig Borghesani,
Terasoft Inc.

Compositor:
PreTEX, Inc.

Printer:
Quebecor World

North America
Nelson
1120 Birchmount Road
Toronto, Ontario M1K 5G4
Canada

Asia
Thomson Learning
5 Shenton Way #01-01
UIC Building
Singapore 068808

Australia/New Zealand
Thomson Learning
102 Dodds Street
Southbank, Victoria
Australia 3006

Europe/Middle East/Africa
Thomson Learning
High Holborn House
50/51 Bedford Row
London WC1R 4LR
United Kingdom

Latin America
Thomson Learning
Seneca, 53
Colonia Polanco
11560 Mexico D.F.
Mexico

Spain
Paraninfo
Calle/Magallanes, 25
28015 Madrid, Spain

In memory of our fathers:

Edgar J. Schilling

and

George W. Harris

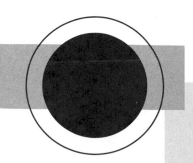

Contents

● ● ● ● ● ● ● ● ● ● ● ● ● ● ● ● ● ● ●

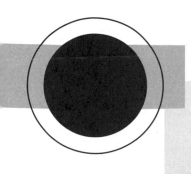

Preface

Digital signal processing (DSP) is a modern discipline built around engineering techniques that enjoy widespread application in today's world. This book focuses on the fundamentals of digital signal processing with an emphasis on practical applications. The *Fundamentals of Digital Signal Processing* includes nine chapters with the relationships between chapters shown in Figure P1.

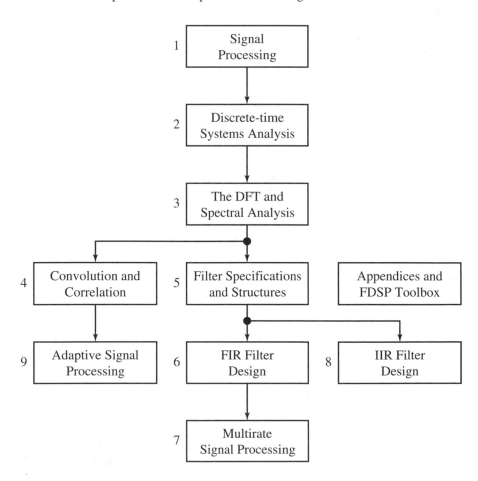

Figure P1 Chapter Dependence

● ● ● ● ● ● ● ● ● ● ● ● ● ● ●
Audience and Prerequisites

This book is targeted primarily toward second-semester juniors, seniors, and beginning graduate students in electrical and computer engineering and related fields that use modern digital signal processing. It is assumed that the students have taken a signals and systems type of course that includes exposure to the Fourier transform, the Laplace transform, and the use of transfer functions. Familiarity with the Z-transform is helpful, but it is not required. There is enough material, and sufficient flexibility in the way it can be covered, to provide for courses of different lengths without adding supplementary material. Experience with MATLAB programming is useful, but it is not essential. Indeed, a graphical user interface (GUI) module is included at the end of each chapter that allows students to investigate meaningful numerical applications without any need for programming. An FDSP toolbox and computational problems are supplied for those users who are familiar with MATLAB or are prepared to learn it, and are interested in developing their own implementations.

Courses using this book should include all or part of the first three chapters in the order indicated, because they introduce fundamental concepts and techniques that are used throughout the book. Those interested specifically in digital filter design should follow the central stem in Figure P1, but with Chapter 8 included. The stem on the left includes a number of interesting applications plus some more advanced material on adaptive signal processing. Within each chapter, sections or subsections marked with an asterisk contain more advanced or more specialized material that can be skipped without loss of continuity. This book is written in an informal style that provides motivation for each new topic, and a smooth transition between topics. However, important terms are set apart for convenient reference using Definitions, and key results are stated as Theorems in order to highlight their significance. Detailed Algorithms are also included to summarize the steps used to implement important design procedures.

In order to motivate students with examples that are meaningful and of direct interest, many of the examples feature the processing of speech and music. This *theme* is also a focus of the course software, which includes a facility for recording and playing back speech and sound on a standard PC. This way, students can experience directly the effects of various signal processing techniques. Examples are included that show how to analyze speech, synthesize sounds of stringed musical instruments, and design digital filters that produce sound effects.

● ● ● ● ● ● ● ● ● ● ● ● ● ● ●
Chapter Structure

Each chapter of this book follows the template shown in Figure P2. The chapter starts with a concise summary of student learning objectives, with each objective cross-referenced to a section or sections within the chapter. This is followed by a motivation section that introduces one or more examples of practical problems that can be solved using techniques covered in the chapter. The main body of each chapter is used to introduce a series of analysis tools and signal processing techniques applicable to the type of problems covered in the chapter. Within these sections, the analysis methods and processing techniques tend to go from the simple to the more complex. Sections marked with a * denote more advanced or specialized material that can be skipped without loss of continuity. Numerous examples are used throughout to illustrate the principles involved.

Near the end of each chapter is a software application section that introduces a GUI module designed to allow the student to explore and investigate the chapter concepts and techniques without any need for programming. The GUI modules all feature the same standard interface that is simple to use and easy to learn. Furthermore, data files created as output from one module can be imported as input into other modules. The software application section also includes a case study type of example that presents a complete solution to a practical problem in the form of a MATLAB script that is available on the distribution CD. Following the software application section is a detailed summary of the essential concepts and techniques covered in the chapter. Finally, each chapter concludes with a generous set of homework problems. There are three categories of problems, and the problems are cross-referenced to sections within the chapter. The Analysis problems can be done by hand or with a calculator. They are used to test student understanding of, and in some cases extend, the chapter material. The GUI Simulation problems allow the student to numerically investigate and compare different DSP techniques using the chapter GUI module. No programming is required for these problems. Finally, Computation problems are provided that require the user to write MATLAB scripts and functions that apply the signal processing techniques covered in the chapter. To ease the programming burden and allow the student to focus on the application, an FDSP toolbox of functions is provided that implement the more important design and analysis techniques. Solutions to selected problems, marked with the ☑ symbol, are available on the

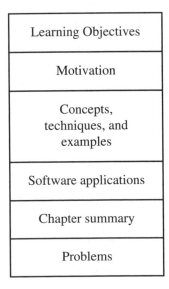

Figure P2   Chapter Template

distribution CD that comes with the text. These self-study problems are provided for the students so they can independently check their understanding of the material introduced in each chapter.

• • • • • • • • • • • • • • • • • •

FDSP Toolbox

One of the unique features of this textbook is a comprehensive integrated software package called the Fundamentals of Digital Signal Processing (FDSP) Toolbox that accompanies the text. The FDSP toolbox is contained on the distribution CD, and it can also be downloaded from an FDSP course website at:

    www.clarkson.edu/~schillin/fdsp

The directory structure of the FDSP toolbox is shown in Figure P3. The FDSP toolbox includes the chapter GUI modules, a complete library of signal processing functions, all MATLAB examples that appear in the text, the text figures, PDF file solutions to selected problems, and on-line help documentation. All of the course software can be accessed easily through a simple menu-based FDSP driver module that is executed with the following command from the MATLAB command prompt.

    f_dsp

The FDSP functions implement signal processing techniques discussed in the chapters, and their use is documented in the chapter where the technique is first introduced. The FDSP toolbox is self-contained course software designed around the student version of MATLAB. There is no need to for students to purchase additional MATLAB toolboxes such as the signal processing and filter design toolboxes. In order to maintain compatibility with those toolboxes, all FDSP toolbox functions are named using the convention f_xxx, and all chapter GUI modules are named using the convention g_xxx. Additional details on using the FDSP toolbox can be found in Chapter 1 and in Appendix B.

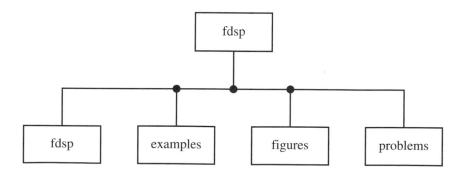

Figure P3   FDSP Toolbox Directory Structure

● ● ● ● ● ● ● ● ● ● ● ● ● ● ● ●
Support Material

Supplementary course material is available for both the student and the instructor. For the student, solutions to selected end of chapter problems, marked with a ☑, are included on the distribution CD as PDF files. Students are encouraged to use these problems as a test of their understanding of the material. For the instructor, there is an Instructors Manual, accompanied by an Instructors CD, that includes complete solutions to all of the problems that appear at the end of each chapter. Every computational example in the text can be run by the instructor or the student from the FDSP toolbox driver module, *f_dsp*. The same is true for every MATLAB figure that appears in the text. The Instructors CD also features an easy to use *Homework Builder* module. This GUI module helps the instructor create and distribute homework assignments and solutions using problems selected from the end of each chapter.

This book is also supported through the FDSP website. A self-extracting executable file containing the most recent version of the FDSP toolbox can be downloaded from this website. Additional supplementary course material, such as PowerPoint slide presentations, will be posted as they become available. Questions and comments concerning the text, the software, and the supplementary course material should be addressed to the authors using *robert.schilling@clarkson.edu*.

● ● ● ● ● ● ● ● ● ● ● ● ● ● ● ●
Acknowledgments

This project has been several years in the making and many individuals have contributed to its completion. The reviewers commissioned by Thomson made numerous thoughtful and insightful suggestions which were incorporated into the final draft. Thanks to graduate students Joe Tari and Rui Guo for helping review the FDSP toolbox software. We would also like to thank a number of individuals at Brooks/Cole who helped see this project to completion and mold the final product. Special thanks to Bill Stenquist who worked closely with us throughout, to Kamilah Reid Burell for producing the Instructors Manual and CDs, and to John More for his marketing efforts. Thanks also go to our developmental and production editor, Rose Kernan of RPK Editorial Services, and to our compositor, Paul Mailhot of PreTex, Inc. Finally, we would like to acknowledge Clarkson University for providing us with the opportunity to teach the course on which this book is based, and the Clarkson students who took that course for their valuable suggestions and their enthusiasm for learning the material.

ROBERT J. SCHILLING
SANDRA L. HARRIS
Potsdam, NY

CHAPTER 1

Signal Processing

● ● ● ● ● ● ● ● ● ● ● ● ● ● ● ●

1.1 Motivation

A signal is a physical variable whose value varies with time or space. When the value of the signal is available over a continuum of times, it is referred to as a *continuous-time* or analog signal. Everyday examples of analog signals include temperature, pressure, liquid level, chemical concentration, voltage and current, position, velocity, acceleration, force, and torque. If the value of the signal is available only at discrete instants of time, it is called a *discrete-time* signal. Although some signals, for example economic data, are inherently discrete-time signals, a more common way to produce a discrete-time signal, $x(k)$, is to take samples of an underlying analog signal, $x_a(t)$:

Samples

$$x(k) \stackrel{\Delta}{=} x_a(kT), \quad k = 0, 1, 2, \ldots$$

Here T denotes the *sampling interval* or time between samples, and $\stackrel{\Delta}{=}$ indicates equals by definition. When finite precision is used to represent the value of $x(k)$, the sequence of quantized values is then called a *digital* signal. A system or algorithm which processes one digital signal $x(k)$ as its input and produces a second digital signal $y(k)$ as its output is a digital signal processor. Digital signal processing (DSP) techniques have widespread applications, and they play an increasingly important role in the modern world. Application areas include speech recognition, detection of targets with radar and sonar, processing of music and video, seismic exploration for oil and gas deposits, medical signal processing including EEG, EKG, and ultrasound, communication channel equalization, and satellite image processing. The focus of this book is to develop, implement, and apply modern DSP techniques.

We begin this introductory chapter with a comparison of digital and analog signal processing. Next, some practical problems are posed that can be solved using DSP techniques. This is followed by characterization and classification of signals. The fundamental notion of the spectrum of a signal is then presented including the concepts of bandlimited and white noise signals. This leads naturally to the sampling process which takes a continuous-time signal and produces an equivalent discrete-time signal. Simple conditions are presented that ensure that an analog signal can be reconstructed from its samples. When these conditions are not satisfied, the phenomenon of aliasing occurs. The use of guard filters to reduce the effects of aliasing is discussed. Next DSP hardware in the form of analog-to-digital converters (ADCs) and digital-to-analog converters (DACs) is examined. The hardware discussion includes ways to model the quantization error associated with finite precision converters. A MATLAB toolbox, called FDSP, is then introduced that facilitates the development of simple DSP scripts. The FDSP toolbox also includes a number of graphical user interface (GUI) modules that can be used to browse examples and explore digital signal processing concepts without any need for programming. The GUI module *g_sample* allows the user to investigate the signal sampling process, while the companion module *g_reconstruct* allows the user to explore the signal reconstruction process. The chapter concludes with an application example that focuses on the problem of specifying and designing an anti-aliasing filter.

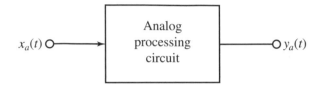

Figure 1.1   Analog
Signal Processing

1.1.1   Digital and Analog Processing

Until relatively recently, almost all signal processing was done with analog circuits as shown in Figure 1.1. For example, operational amplifiers, resistors, and capacitors are used to realize frequency selective filters.

With the advent of specialized microprocessors with built-in data conversion circuits (Papamichalis, 1990), it is now commonplace to perform signal processing digitally as shown in Figure 1.2. Digital processing of analog signals is more complex because it requires, at a minimum, the three components shown in Figure 1.2. The analog-to-digital converter or ADC at the front end converts the analog input $x_a(t)$ into an equivalent digital signal $x(k)$. The processing of $x(k)$ is then achieved with an algorithm that is implemented in software. For a filtering operation, the DSP algorithm consists of a difference equation, but other types of processing are also possible and are often used. The digital output signal $y(k)$ is then converted back to an equivalent analog signal $y_a(t)$ by the digital-to-analog converter or DAC.

Figure 1.2   Digital Signal Processing

Although the DSP approach requires more steps than analog signal processing, there are important benefits to working with signals in digital form. A comparison of the relative advantages and disadvantages of the two approaches is summarized in Table 1.1.

The primary advantages of analog signal processing are *speed* and *cost*. Digital signal processing is not as fast due to the limits on the sampling rates of the converter circuits. In addition, if substantial computations are to be performed between samples, then the clock rate of the processor can also be a limiting factor. Speed can be an issue in *real-time* applications where the kth output sample $y(k)$ must be computed and sent to the DAC as soon as possible after the kth input sample $x(k)$ is available from the ADC. However, there are also applications where the entire input signal is available ahead of time for processing off-line. For this batch mode type of processing, speed is less critical.

Feature	Analog Processing	Digital Processing
Speed	Fast	Moderate
Cost	Low	Moderate
Flexibility	Low	High
Performance	Moderate	High
Self-calibrating	No	Yes
Data-logging capability	No	Yes
Adaptive capability	No	Yes

DSP hardware is often somewhat more expensive than analog hardware because analog hardware can consist of as little as a few discrete components on a stand-alone printed circuit board. The cost of DSP hardware varies depending on the performance characteristics required. In some cases, a PC may already be available to perform other functions for a given application, and in these instances the marginal expense of adding DSP hardware is not large.

In spite of these limitations, there are great benefits to using DSP techniques. Indeed, DSP is superior to analog processing with respect to virtually all of the remaining features listed in Table 1.1. One of the most important advantages is the inherent *flexibility* available with a software implementation. Whereas an analog circuit might be tuned with a potentiometer to vary its performance over a limited range, the DSP algorithm can be completely replaced, on the fly, when circumstances warrant.

DSP also offers considerably higher *performance* than analog signal processing. For example, digital filters with arbitrary magnitude responses and linear phase responses can be designed easily whereas this is not feasible with analog filters.

A common problem that plagues analog systems is the fact that the component values tend to *drift* with age and with changes in environmental conditions such as temperature. This leads to a need for periodic calibration. With DSP there is no drift problem and therefore no need to manually calibrate.

Since data are already available in digital form in a DSP system, with little or no additional expense one can *log* the data associated with the operation of the system so that its performance can be monitored, either locally of remotely over a network connection. If an unusual operating condition is detected, its exact time and nature can be determined and a higher-level control system can be alerted. Although strip chart recorders can be added to an analog system, this substantially increases the expense thereby negating one of its potential advantages.

The flexibility inherent in software can be exploited by having the parameters of the DSP algorithm vary with time and *adapt* as the characteristics of the input signal or the processing task change. Applications like system identification and active noise control make highly effective use of adaptive signal processing, a topic that is addressed in Chapter 9.

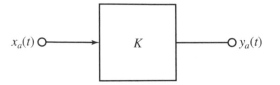

Figure 1.3 An Audio Amplifier

1.1.2 Total Harmonic Distortion (THD)

With the widespread use of digital computers, DSP applications are now common-place. As a simple initial example, consider the problem of designing an audio amplifier to boost signal strength without distorting the shape of the input signal. For the amplifier shown in Figure 1.3, suppose the input signal $x_a(t)$ is a pure sinusoidal tone of amplitude a and frequency f_0 Hz.

$$x_a(t) = a\cos(2\pi f_0 t) \tag{1.1}$$

An ideal amplifier will produce a desired output signal $y_d(t)$ that is a scaled and delayed version of the input signal. For example, if the scale factor or amplifier *gain* is K and the delay is τ, then the desired output is

$$
\begin{aligned}
y_d(t) &= K x_a(t - \tau) \\
&= K a\cos[2\pi f_0(t - \tau)]
\end{aligned}
\tag{1.2}
$$

In a practical amplifier, the relationship between the input and the output is only approximately linear, so some additional terms are present in the actual output y_a.

$$
\begin{aligned}
y_a(t) &= F[x_a(t)] \\
&\approx \frac{d_0}{2} + \sum_{i=1}^{M-1} d_i\cos(2\pi i f_0 t + \theta_i)
\end{aligned}
\tag{1.3}
$$

The presence of the additional harmonics indicates that there is *distortion* in the amplified signal due to nonlinearities within the amplifier. For example, if the amplifier is driven with an input whose amplitude a is too large, then the amplifier will saturate with the result that the output is a *clipped* sine wave that sounds distorted when played through a speaker. To quantify the amount of distortion we observe that the average power contained in the ith harmonic is $d_i^2/2$ for $i \geq 1$ and $d_i^2/4$ for $i = 0$. Thus, the *average power* of the signal $y_a(t)$ is

$$P_y = \frac{d_0^2}{4} + \frac{1}{2}\sum_{i=1}^{M-1} d_i^2 \tag{1.4}$$

The *total harmonic distortion* or THD of the output signal $y_a(t)$ is defined as the power in the spurious harmonic components, expressed as a percentage of the total power. Thus, the following measure can be used for the quality of the amplifier output.

$$\text{THD} \triangleq \frac{100(P_y - d_1^2/2)}{P_y} \% \tag{1.5}$$

For an ideal amplifier we have $d_i = 0$ for $i \neq 1$ and

$$d_1 = Ka \tag{1.6a}$$

$$\theta_1 = -2\pi f_0 \tau \tag{1.6b}$$

Consequently, for a high-quality amplifier, THD is small, and when no distortion is present THD $= 0$. Suppose the amplifier output is sampled to produce the following digital signal of length $N = 2M$.

$$y(k) = y_a(kT), \quad 0 \leq k < N \tag{1.7}$$

If the sampling interval is set to $T = 1/(Nf_0)$, then this corresponds to one period of $y_a(t)$. By processing the digital signal $x(k)$ with the discrete Fourier transform or DFT, it is possible to determine d_i and θ_i for $0 \leq i < M$. In this way the total harmonic distortion can be measured. The DFT is a key analytic tool that will be introduced in Chapter 3.

1.1.3 A Notch Filter

As a second example of a DSP application, suppose you are asked to perform some acoustic measurements in a laboratory setting using a microphone. Because of the highly sensitive nature of the measurements, any ambient background sounds in the range of frequencies of interest have the potential to corrupt your measurements with unwanted noise. After some initial measurements, you discover that the overhead fluorescent lights are emitting a 120 Hz hum which corresponds to the secondary harmonic of the 60 Hz commercial AC power used to energize them. The problem then is to remove the 120 Hz frequency component while affecting the other nearby frequency components as little as possible. That is, you want to process the acoustic data samples with a *notch* filter designed to remove the effects of the fluorescent lights. After some calculations, you arrive at the following digital filter to process the measurements $x(k)$ to produce a filtered signal $y(k)$.

$$y(k) = 1.6466y(k-1) - .9805y(k-2) + .9905x(k) -$$
$$1.6471x(k-1) + .9905x(k-2) \tag{1.8}$$

The filter in Eq. 1.8 is a notch filter with a bandwidth of 4 Hz, a notch frequency of $f_n = 120$ Hz, and a sampling frequency of $f_s = 1280$ Hz. A plot of the frequency response of this filter is shown in Figure 1.4 where a sharp notch at 120 Hz is apparent. Notice that except for frequencies near f_n, all other frequency components of $x(k)$ are passed through the filter without attenuation. The design of notch filters and other narrowband filters is discussed in Chapter 8.

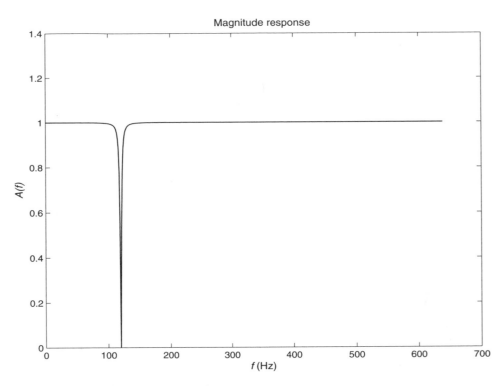

Figure 1.4 Magnitude Response of a Notch Filter with $f_n = 120$ Hz

1.1.4 Active Noise Control

An application area of DSP that makes use of adaptive signal processing is active control of acoustic noise (Kuo and Morgan, 1996). Examples include industrial noise from rotating machines, propeller and jet engine noise, road noise in an automobile, and noise caused by air flow in heating, ventilation, and air conditioning systems. As an illustration of the latter, consider the active noise control system shown in Figure 1.5 which consists of an air duct with two microphones and a speaker. The basic principle of active noise control is to inject a secondary sound into the environment so as to cancel the primary sound using destructive interference.

The purpose of the reference microphone in Figure 1.5 is to detect the primary noise $x(k)$ generated by the noise source or blower. The primary noise signal is then passed through a digital filter of the following form.

$$y(k) = \sum_{i=0}^{m} w_i(k)x(k - i) \tag{1.9}$$

The output of the filter $y(k)$ drives a speaker which creates the secondary sound sometimes called *antisound*. The error microphone, located downstream of the speaker, detects the sum of the primary and secondary sounds and produces an error signal $e(k)$. The objective of the adaptive algorithm is to take $x(k)$ and $e(k)$ as inputs and adjust the filter weights $w(k)$ so as to drive $e^2(k)$ to zero. If zero error can be achieved, then silence is observed at the error microphone. In practical systems, the error or residual sound is significantly reduced by active noise control.

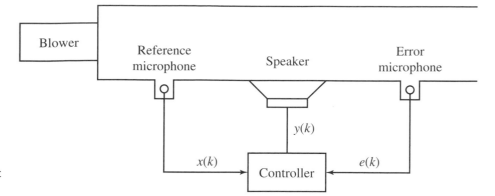

Figure 1.5 Active Control of Acoustic Noise in an Air Duct

To illustrate the operation of this adaptive DSP system, suppose the blower noise is modeled as a periodic signal with fundamental frequency f_0 and r harmonics plus some random white noise $v(k)$.

$$x(k) = \sum_{i=1}^{r} a_i \cos(2\pi i k f_0 T + \theta_i) + v(k), \quad 0 \le k < p \tag{1.10}$$

For example, suppose $f_0 = 100$ Hz and there are $r = 4$ harmonics with amplitudes $a_i = 1/i$ and random phase angles. Suppose the random white noise term is distributed uniformly over the interval $[-.5, .5]$. Let $p = 2048$ samples, and suppose the sampling interval is $T = 1/1600$ sec and the filter order is $m = 40$. The adaptive algorithm used to adjust the filter weights is called the FXLMS method, and it is discussed in detail in Chapter 9. The results of applying this algorithm are shown in Figure 1.6.

Initially the filter weights are set to $w(0) = 0$ which corresponds to no noise control at all. The adaptive algorithm is not activated until sample $k = 512$, so the first quarter of the plot in Figure 1.6 represents the ambient or primary noise detected at the error microphone. When adaptation is activated, the error begins to decrease rapidly and after a short transient period it reaches a steady-state level which, although not zero, is an order of magnitude quieter than the primary noise itself. We can quantify the noise reduction by using the following measure of overall noise cancelation.

$$E = 10 \log \left(\frac{\displaystyle\sum_{i=0}^{p/4-1} e^2(i)}{\displaystyle\sum_{i=3p/4}^{p-1} e^2(i)} \right) \text{ dB} \tag{1.11}$$

The overall noise cancelation E is the ratio of the average power of the noise during the first quarter of the samples divided by the average power of the noise during the last quarter of the samples, expressed in decibels. Using this measure, the noise cancelation observed in Figure 1.5 is $E = 37.8$ dB.

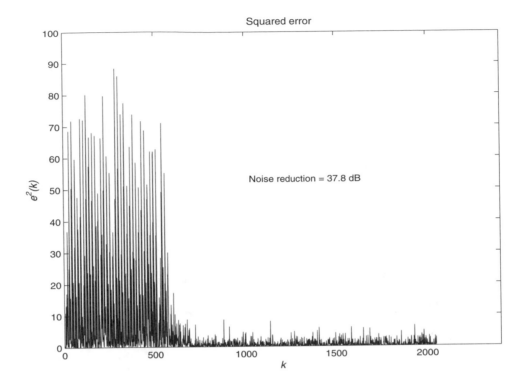

Figure 1.6 Error Signal with Active Noise Control Activated at $k = 512$

- - - - - - - - - - - - - - -

1.2 Signals and Systems

1.2.1 Signal Classification

Recall that a signal is a physical variable whose value varies with respect to time or space. To simplify the notation and terminology, we will assume that, unless noted otherwise, the independent variable denotes time. If the value of the signal (the dependent variable) is available over a continuum of times, $0 \leq t \leq \tau$, then the signal is referred to as a *continuous-time* signal. An example of a continuous-time signal, $x_a(t)$, is shown in Figure 1.7.

In many cases of practical interest, the value of the signal is only available at discrete instants of time in which case it is referred to as a *discrete-time* signal. That is, signals can be classified into continuous-time or discrete-time depending on whether the independent variable is continuous or discrete, respectively. Common everyday examples of discrete-time signals include economic statistics such as the daily balance in one's savings account, or the monthly inflation rate. In DSP applications, a more common way to produce a discrete-time signal, $x(k)$, is to sample an underlying continuous-time signal, $x_a(t)$, as follows.

$$x(k) = x_a(kT), \quad k = 0, 1, 2, \ldots \tag{1.12}$$

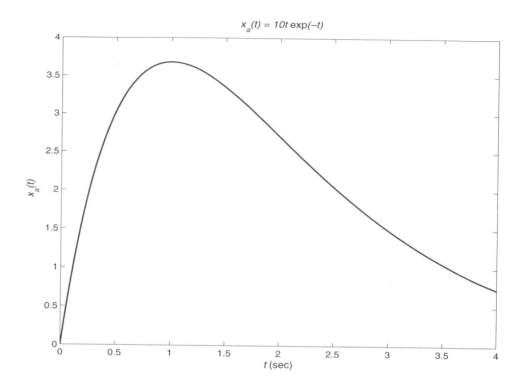

Figure 1.7 A Continuous-time Signal $x_a(t)$

Here T is the time between samples or *sampling interval* in seconds. The sample spacing can also be specified using the reciprocal of the sampling interval which is called the *sampling frequency*, f_s.

$$f_s \triangleq \frac{1}{T} \text{ Hz} \tag{1.13}$$

Here the unit of Hz is understood to mean samples/second. Notice that the integer k in Eq. 1.12 denotes *discrete time* or, more specifically, the sample number. The sampling interval T is left implicit on the left-hand side of Eq. 1.12 because this simplifies subsequent notation. In those instances where the value of T is important, it will be stated separately. An example of a discrete-time signal generated by sampling the continuous-time signal in Figure 1.7 using $T = 0.25$ seconds is shown in Figure 1.8.

Just as the independent variable can be continuous or discrete, so can the dependent variable or amplitude of the signal be continuous or discrete. If the number of bits of precision used to represent the value of $x(k)$ is finite, then we say that $x(k)$ is a *quantized* or discrete-amplitude signal. For example, if N bits are used to represent the value of $x(k)$, then there are 2^N distinct values that $x(k)$ can assume. Suppose the value of $x(k)$ ranges over the interval $[x_{\min}, x_{\max}]$. Then the *quantization level*, or spacing between adjacent discrete values of $x(k)$, is

Quantization level

$$q = \frac{x_{\max} - x_{\min}}{2^N} \tag{1.14}$$

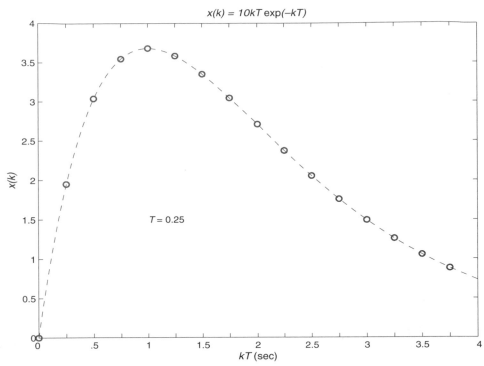

Figure 1.8 A Discrete-time Signal $x(k)$

The quantization process can be thought of as passing a signal through a piecewise-constant staircase type function. For example, if the quantization is based on rounding to the nearest N bits, then the process can be represented with the following *quantization operator*.

$$Q_N(x) \triangleq q \cdot \text{round}\left(\frac{x}{q}\right) \tag{1.15}$$

A graph of $Q_N(x)$ for x ranging over the interval $[-1, 1]$ and $N = 5$ bits is shown in Figure 1.9. A quantized discrete-time signal is called a *digital* signal. That is, a digital signal, $x_q(k)$, is discrete in both time and amplitude with

$$x_q(k) = Q_N[x_a(kT)] \tag{1.16}$$

By contrast, a signal that is continuous in both time and amplitude is called an *analog* signal. An example of a digital signal obtained by quantizing the amplitude of the discrete-time signal in Figure 1.8 is shown in Figure 1.10. In this case, the 5-bit quantizer in Figure 1.9 is used to produce $x_q(k)$. Careful inspection of Figure 1.10 reveals that at some of the samples there are noticeable differences between $x_q(k)$ and $x_a(kT)$. If rounding is used, then the magnitude of the error is, at most, $q/2$.

Most of the analysis in this book will be based on discrete-time signals rather than digital signals. That is, infinite precision is used to represent the value of the dependent variable. Finite precision, or finite word length effects, will be examined

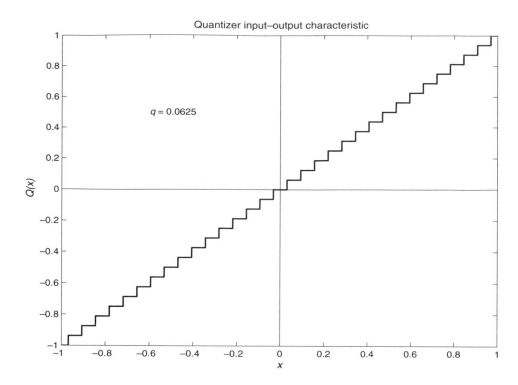

Figure 1.9 Quantization Over $[-1, 1]$ using $N = 5$ Bits

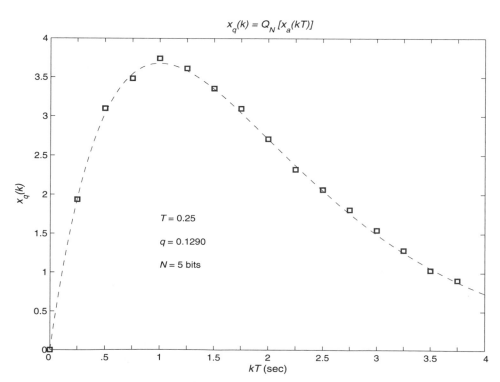

Figure 1.10 A Digital Signal $x_q(k)$

separately in Chapter 5 to see what effect they have on different DSP techniques. A digital signal $x_q(k)$ can be modeled as a discrete-time signal $x(k)$ plus random *quantization noise*, $e(k)$, as follows.

$$x_q(k) = x(k) + e(k) \tag{1.17}$$

An effective way to measure the size or strength of the quantization noise is to use average power defined as the mean, or expected value, of $e^2(k)$. Typically, $e(k)$ is modeled as a random variable uniformly distributed over the interval $[-q/2, q/2]$ with probability density $p(x) = 1/q$. In this case, the expected value of $e^2(k)$ is

$$E[e^2] = \int_{-\infty}^{\infty} p(x)x^2 dx$$

$$= \frac{1}{q} \int_{-q/2}^{q/2} x^2 dx \tag{1.18}$$

Thus, the average power of the quantization noise is proportional to the square of the quantization level with

Quantization noise power

$$E[e^2] = \frac{q^2}{12} \tag{1.19}$$

Example 1.1

Quantization Noise

Suppose the value of a discrete-time signal $x(k)$ is constrained to lie in the interval $[-10, 10]$. Let $x_q(k)$ denote a digital version of $x(k)$ using quantization level q, and consider the following problem. Suppose the average power of the quantization noise is to be less than 0.001. What is the minimum number of bits that are needed to represent the value of $x_q(k)$? The constraint on the average power of the quantization noise is

$$E[e^2] < 0.001$$

Thus from Eq. 1.14 and Eq. 1.19 we have

$$\frac{(x_{max} - x_{min})^2}{12(2^N)^2} < .001$$

Recall that the signal range is $x_{min} = -10$ and $x_{max} = 10$. Multiplying both sides by 12, taking the square root of both sides, and then solving for 2^N yields

$$2^N > \frac{20}{\sqrt{.012}}$$

Finally, taking the natural log of both sides and solving for N we have

$$N > \frac{\ln(182.5742)}{\ln(2)} = 7.5123$$

Since N must be an integer, the minimum number of bits needed to ensure that the average power of the quantization noise is less than 0.001 is $N = 8$ bits.

Signals can be further classified depending on whether or not they are nonzero for negative values of the independent variable.

DEFINITION
1.1: Causal Signal

A signal $x_a(t)$ defined for $t \in R$ is *causal* if and only if it is zero for negative t. Otherwise, the signal is *noncausal*.

$$x_a(t) = 0 \quad \text{for} \quad t < 0$$

Most of the signals that we work with will be causal signals. A simple, but important, example of a causal signal is the *unit step* which is denoted $u_a(t)$ and defined as

$$u_a(t) \triangleq \begin{cases} 0, & t < 0 \\ 1, & t \geq 0 \end{cases} \tag{1.20}$$

Note that any signal can be made into a causal signal by multiplying by the unit step. For example, $x_a(t) = \exp(-t/\tau)u_a(t)$ is a causal decaying exponential with time constant τ. Another important example of a causal signal is the *unit impulse* which is denoted $\delta_a(t)$. Strictly speaking, the unit impulse is not a function because it is not defined at $t = 0$. However, the unit impulse can be defined implicitly by the equation

$$\int_{-\infty}^{t} \delta_a(\tau)d\tau = u_a(t) \tag{1.21}$$

That is, the unit impulse $\delta_a(t)$ is a signal that, when integrated, produces the unit step $u_a(t)$. Consequently, we can loosely think of the unit impulse as the derivative of the unit step function, keeping in mind that the derivative of the unit step is not defined at $t = 0$. The two essential characteristics of the unit impulse that follow from Eq. 1.21 are

$$\delta_a(t) = 0, \quad t \neq 0 \tag{1.22a}$$

$$\int_{-\infty}^{\infty} \delta_a(t)dt = 1 \tag{1.22b}$$

A more informal way to view the unit impulse is to consider a narrow pulse of width ϵ and height $1/\epsilon$ starting at $t = 0$. The unit impulse can be thought of as the limit of this sequence of pulses as the pulse width ϵ goes to zero. By convention, we graph the unit impulse as a vertical arrow with the height of the arrow equal to the *strength*, or area, of the impulse as shown in Figure 1.11.

The unit impulse has an important property that is a direct consequence of Eq. 1.22. If $x_a(t)$ is a continuous function, then

$$\int_{-\infty}^{\infty} x_a(\tau)\delta_a(\tau - t_0)d\tau = \int_{-\infty}^{\infty} x_a(t_0)\delta_a(\tau - t_0)d\tau$$

$$= x_a(t_0) \int_{-\infty}^{\infty} \delta_a(\tau - t_0)d\tau$$

$$\left. = x_a(t_0) \int_{-\infty}^{\infty} \delta_a(\alpha)d\alpha \right\} \alpha = \tau - t_0 \tag{1.23}$$

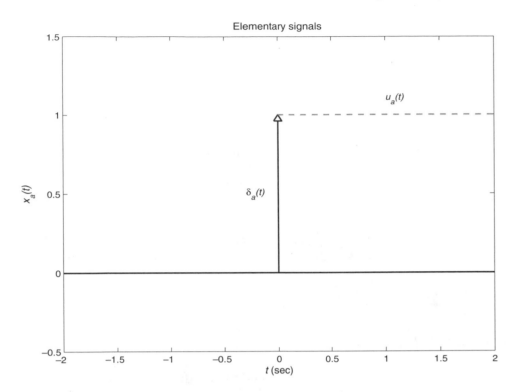

Figure 1.11 Unit Impulse, $\delta_a(t)$, and Unit Step, $u_a(t)$

Since the area under the unit impulse is one, we then have the following *sifting property* of the unit impulse.

Sifting property of impulse

$$\int_{-\infty}^{\infty} x_a(t)\delta_a(t - t_0)dt = x_a(t_0) \tag{1.24}$$

From Eq. 1.24 we see that when a continuous function of time is multiplied by an impulse and then integrated, the effect is to sift out or sample the value of the function at the time the impulse occurs.

1.2.2 System Classification

Just as signals can be classified, so can the systems that process those signals. Consider a system S with *input x* and *output y* as shown in Figure 1.12. In some instances, for

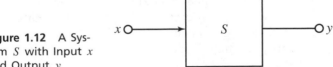

Figure 1.12 A System S with Input x and Output y

example biomedical systems, the input is referred to as the *stimulus*, and the output is referred to as the *response*. We can think of the system in Figure 1.12 as an operator S that acts on the input signal x to produce the output signal y.

$$y = Sx \qquad (1.25)$$

If the input and output are continuous-time signals, then the system S is called a *continuous-time* system. A *discrete-time* system is a system S that processes a discrete-time input $x(k)$ to produce a discrete-time output $y(k)$. There are also examples of systems that contain both continuous-time signals and discrete-time signals. These systems are referred to as *sampled-data* systems.

Almost all of the examples of systems in this book belong to an important class of systems called linear systems.

DEFINITION

1.2: Linear System

> Let x_1 and x_2 be arbitrary inputs and let a and b be arbitrary scalars. A system S is *linear* if and only if the following holds, otherwise it is a *nonlinear* system.
>
> $$S(ax_1 + bx_2) = aSx_1 + bSx_2$$

A linear system has two distinct characteristics. When $a = b = 1$, we see that the response to a sum of inputs is just the sum of the responses to the individual inputs. Similarly, when $b = 0$, we see that the response to a scaled input is just the scaled response to the original input. Examples of linear discrete-time systems include the notch filter in Eq. 1.8 and the adaptive filter in Eq. 1.9. On the other hand, if the analog audio amplifier in Figure 1.3 is over driven and its output saturates to produce harmonics as in Eq. 1.3, then this is an example of a nonlinear continuous-time system. Another important class of systems is time-invariant systems.

DEFINITION

1.3: Time-invariant System

> A system S with input $x_a(t)$ and output $y_a(t)$ is *time-invariant* if and only if whenever the input is translated in time by τ, the output is also translated in time by τ. Otherwise the system is a *time-varying* system.
>
> $$Sx_a(t - \tau) = y_a(t - \tau)$$

For a time-invariant system, delaying or advancing the input delays or advances the output by the same amount, the shape of the output signal does not change. Therefore, the results of an input-output experiment do not depend on when the experiment is initiated. Time-invariant systems described by differential or difference equations have *constant* coefficients. More generally, physical time-invariant systems have constant parameters. The notch filter in Eq. 1.8 is an example of a discrete-time system that is both linear and time-invariant. On the other hand, the adaptive digital filter in Eq. 1.9 is a time-varying system because the weights $w(k)$ are coefficients that change with time as the system adapts. The following example shows that the concepts of linearity and time-invariance can sometimes depend on how the system is characterized.

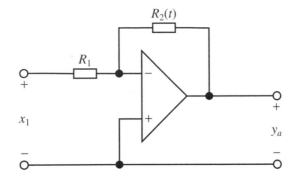

Figure 1.13 An Inverting Amplifier with a Feedback Transducer

---

Example 1.2   **System Classification**

Consider the operational amplifier circuit shown in Figure 1.13. Here input resistor R_1 is fixed, but feedback resistor R_2 represents a sensor or transducer whose resistance changes with respect to a sensed environmental variable such as temperature or pressure. For this inverting amplifier configuration, the output voltage $y_a(t)$ is

$$y_a(t) = -\left[\frac{R_2(t)}{R_1}\right] x_1(t)$$

This is an example of a linear continuous-time system that is time-varying because parameter $R_2(t)$ varies as the temperature or pressure changes. However, another way to model this system is to consider the variable resistance of the sensor as a second input $x_2(t) = R_2(t)$. Viewing the system in this way, the system output is

$$y_a(t) = -\left[\frac{x_1(t)x_2(t)}{R_1}\right]$$

This formulation of the model is a nonlinear time-invariant system, but with two inputs. Thus, by introducing a second input we have converted a single-input time-varying linear system to a two-input time-invariant nonlinear system.

Another important classification of systems focuses on the question of what happens to the signals as time increases. We say that a signal $x_a(t)$ is *bounded* if and only if there exists a $B > 0$ called a *bound* such that

$$|x_a(t)| \le B \quad \text{for} \quad t \in R \tag{1.26}$$

DEFINITION
1.4: Stable System

> A system S with input $x_a(t)$ and output $y_a(t)$ is *stable*, in a bounded-input bounded-output (BIBO) sense, if and only if every bounded input produces a bounded output. Otherwise it is an *unstable* system.

An unstable system is a system for which the magnitude of the output grows arbitrarily large for a least one bounded input.

| Example 1.3 | **Stability** |

As a simple example of a system that can be stable or unstable depending on its parameter values, consider the following first-order linear continuous-time system where $a \neq 0$.

$$\frac{dy_a(t)}{dt} + ay_a(t) = x_a(t)$$

Suppose the input is the unit step $x_a(t) = u_a(t)$ which is bounded with a bound of one. Direct substitution can be used to verify that for $t \geq 0$, the solution is

$$y_a(t) = y_a(0)\exp(-at) + \frac{1}{a}[1 - \exp(-at)]$$

If $a > 0$, then the exponential terms grow without bound which means that the bounded input $u_a(t)$ produces an unbounded output $y_a(t)$. Thus, this system is unstable, in a BIBO sense, when $a > 0$.

Just as light can be decomposed into a spectrum of colors, signals also contain energy that is distributed over a range of frequencies. To decompose a continuous-time signal $x_a(t)$ into its spectral components, we use the *Fourier transform*.

$$X_a(f) = F\{x_a(t)\} \triangleq \int_{-\infty}^{\infty} x_a(t)\exp(-j2\pi ft)dt \qquad (1.27)$$

It is assumed that the reader is familiar with the basics of the Fourier transform. Tables of common Fourier transform pairs and Fourier transform properties can be found in Appendix C. Here $f \in R$ denotes frequency in cycles/sec or Hz. In general the Fourier transform, $X_a(f)$, is complex. As such, it can be expressed in polar form in terms of its magnitude $A_a(f) = |X_a(f)|$ and phase angle $\phi_a(f) = \angle X_a(f)$ as follows.

$$X_a(f) = A_a(f)\exp[j\phi_a(f)] \qquad (1.28)$$

The real-valued function $A_a(f)$ is called the *magnitude spectrum* of $x_a(t)$, while the real-valued function $\phi_a(f)$ is called the *phase spectrum* of $x_a(t)$. More generally, $X_a(f)$ itself is called the *spectrum* of $x_a(t)$. For a real $x_a(t)$, the magnitude spectrum is an even function of f, and the phase spectrum is an odd function of f.

When a signal passes through a linear system, the shape of its spectrum changes. Systems designed to reshape the spectrum is a particular way are called *filters*. The effect that a linear system has on the spectrum of the input signal can be characterized by the frequency response.

DEFINITION

1.5: Frequency Response

Let S be a stable linear time-invariant continuous-time system with input $x_a(t)$ and output $y_a(t)$. Then the *frequency response* of the system S is denoted $H_a(f)$ and defined

$$H_a(f) \triangleq \frac{Y_a(f)}{X_a(f)}$$

The frequency response of a linear system is just the Fourier transform of the output divided by the Fourier transform of the input. Since $H_a(f)$ is complex, it can be represented by its magnitude $A_a(f) = |H_a(f)|$ and its phase angle $\phi_a(f) = \angle H_a(f)$ as follows

$$H_a(f) = A_a(f) \exp[j\phi_a(f)] \tag{1.29}$$

The function $A_a(f)$ is called the *magnitude response* of the system, while $\phi_a(f)$ is called the *phase response* of the system. The magnitude response indicates how much each frequency component of $x_a(t)$ is scaled by the system. That is, $A_a(f)$ is the *gain* of the system at frequency f. Similarly, the phase response indicates how much each frequency component of $x_a(t)$ gets advanced in phase by the system. That is, $\phi_a(f)$ is the *phase shift* of the system at frequency f. Therefore if the input to the system is a pure sinusoidal tone $x_a(t) = \sin(2\pi f_0 t)$, the steady-state output (assuming the system is stable) will be

$$y_a(t) = A_a(f_0) \sin[2\pi f_0 t + \phi_a(f_0)] \tag{1.30}$$

The magnitude response of a real system is an even function of f, while the phase response is an odd function of f. This is similar to the magnitude and phase spectra of a real signal. Indeed, there is a simple relationship between the frequency response of a system and the spectrum of a signal. To see this, consider the impulse response.

DEFINITION

1.6: Impulse Response

Suppose the initial conditions of a continuous-time system S are zero. Then the output of the system corresponding to the unit impulse input is denoted $h_a(t)$ and called the system *impulse response*.

$$h_a(t) = S\delta_a(t)$$

From the sifting property of the unit impulse in Eq. 1.24 one can show that the Fourier transform of the unit impulse is simply $\Delta_a(f) = 1$. It then follows from Definition 1.5 that when the input is the unit impulse, the Fourier transform of the system output is $Y_a(f) = H_a(f)$. That is, an alternative way to represent the frequency response is as the Fourier transform of the impulse response.

$$H_a(f) = F\{h_a(t)\} \tag{1.31}$$

In view of Eq. 1.28, the magnitude response is just the magnitude spectrum of the impulse response, and the phase response is just the phase spectrum of the impulse response. It is for this reason that the same symbol, $A_a(f)$, is used to denote both the magnitude spectrum of a signal and the magnitude response of a system. A similar remark holds for $\phi_a(f)$ which is used to denote both the phase spectrum of a signal and the phase response of a system.

Example 1.4   **Ideal Lowpass Filter**

An important example of a continuous-time system is the ideal lowpass filter. An *ideal lowpass filter* with cutoff frequency B Hz, is a system whose frequency response is the following pulse of height one and radius B centered at $f = 0$.

$$p_B(f) \triangleq \begin{cases} 1, & |f| \le B \\ 0, & |f| > B \end{cases}$$

A plot of the ideal lowpass frequency response is shown in Figure 1.14.

Recall from Definition 1.5 that $Y_a(f) = H_a(f)X_a(f)$. Consequently, the filter in Figure 1.14 passes the frequency components of $x_a(t)$ in the range $[-B, B]$ through the filter without any distortion whatsoever, not even any phase shift. Furthermore, the remaining frequency components of $x_a(t)$ outside the range $[-B, B]$ are completely eliminated by the filter. The idealized nature of the filter becomes apparent when we look at the impulse response of the filter. To compute the impulse response from the frequency response, we must apply the inverse Fourier transform. Using Table C.2 in Appendix C this yields

$$h_a(t) = 2B \cdot \text{sinc}(2\pi Bt)$$

Here the *sinc* function is a special function, that is defined as follows.

$$\text{sinc}(x) \triangleq \frac{\sin(x)}{x}$$

The sinc function is a two-sided decaying sinusoid that is confined to the envelope $1/x$. Thus, $\text{sinc}(k\pi) = 0$ for $k \ne 0$. The value of $\text{sinc}(x)$ at $x = 0$ is determined by applying L'Hospital's rule which yields $\text{sinc}(0) = 1$. The impulse response of the ideal lowpass filter is $\text{sinc}(2\pi BT)$ scaled by $2B$. A plot of the impulse response for the case $B = 100$ Hz is shown in Figure 1.15.

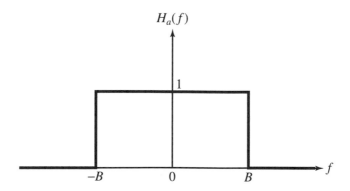

Figure 1.14 Frequency Response of Ideal Lowpass Filter

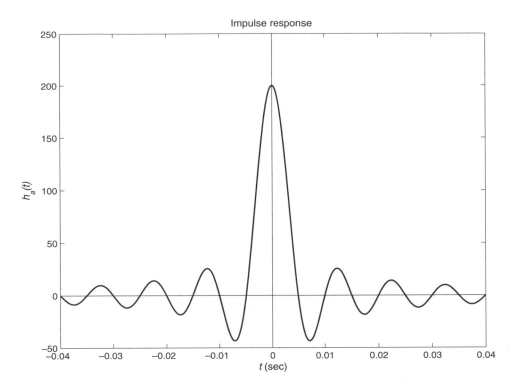

Figure 1.15 Impulse Response of Ideal Lowpass Filter when $B = 100$ Hz

Notice that the sinc function, and therefore the impulse response, is not a causal signal. But $h_a(t)$ is the filter output when a unit impulse input is applied at time $t = 0$. Consequently, for the system in Figure 1.14 we have a causal input producing a noncausal output. This is not possible for a physical system. Therefore, the frequency response in Figure 1.14 cannot be realized with physical hardware. In Section 1.4, we examine some lowpass filters that are physically realizable and can be used to closely approximate the ideal frequency response characteristic.

● ● ● ● ● ● ● ● ● ● ● ● ● ● ● ●

1.3 Signal Sampling

1.3.1 Sampling as Modulation

The process of sampling a continuous-time signal $x_a(t)$ to produce a discrete-time signal $x(k)$ can be viewed as a form of amplitude modulation. To see this, let $\delta_T(t)$ denote a periodic train of impulses of period T.

$$\delta_T(t) \triangleq \sum_{k=-\infty}^{\infty} \delta_a(t - kT) \tag{1.32}$$

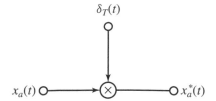

Figure 1.16 Sampling as Amplitude Modulation of an Impulse Train

Thus $\delta_T(t)$ consists of unit impulses at integer multiples of the sampling interval T. The *sampled version* of signal $x_a(t)$ is denoted $x_a^*(t)$, and is defined as the following product.

$$x_a^*(t) \overset{\Delta}{=} x_a(t)\delta_T(t) \tag{1.33}$$

Since $x_a^*(t)$ is obtained from $x_a(t)$ by multiplication by a periodic signal, this process is a form of amplitude modulation of $\delta_T(t)$. In this case, $\delta_T(t)$ plays a role similar to the high-frequency carrier wave in AM radio, and $x_a(t)$ represents the low-frequency information signal. A block diagram of the impulse model of sampling is shown in Figure 1.16.

Using the basic properties of the unit impulse in Eq. 1.22, the sampled version of $x_a(t)$ can be written as follows.

$$x_a^*(t) = x_a(t)\delta_T(t)$$

$$= x_a(t) \sum_{k=-\infty}^{\infty} \delta_a(t - kT)$$

$$= \sum_{k=-\infty}^{\infty} x_a(t)\delta_a(t - kT)$$

$$= \sum_{k=-\infty}^{\infty} x_a(kT)\delta_a(t - kT) \tag{1.34}$$

Thus, the sampled version of $x_a(t)$ is the following amplitude modulated impulse train.

Impulse sampling

$$x_a^*(t) = \sum_{k=-\infty}^{\infty} x(k)\delta_a(t - kT) \tag{1.35}$$

Whereas $\delta_T(t)$ is a constant-amplitude or uniform train of impulses, $x_a^*(t)$ is a nonuniform impulse train with the area of the kth impulse equal to sample $x(k)$. A graph illustrating the relationship between $\delta_T(t)$ and $x_a^*(t)$ for the case $x_a(t) = 10t \exp(-t)u_a(t)$ is shown in Figure 1.17.

It is useful to note from Eq. 1.35 that $x_a^*(t)$ is actually a continuous-time signal, rather than a discrete-time signal. However, it is a very special continuous-time signal in that it is zero everywhere except at the samples where it has impulses whose areas correspond to the sample values. Consequently, there is a simple one-to-one relationship between the continuous-time signal $x_a^*(t)$ and the discrete-time signal

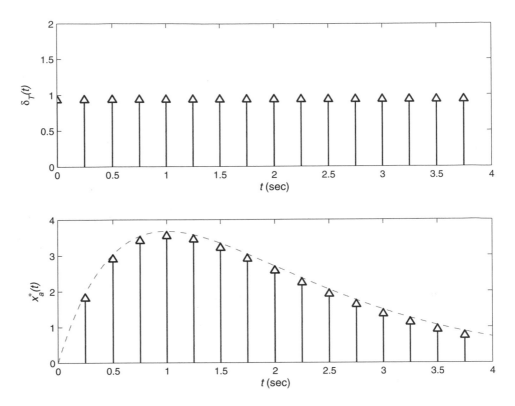

Figure 1.17 Sampled Version of $x_a(t)$ using Impulse Sampling

$x(k)$. If $x_a^*(t)$ is a causal continuous-time signal, we can apply the Laplace transform to it. The *Laplace transform* of a causal continuous-time signal $x_a(t)$ is denoted $X(s)$ and is defined

$$X(s) = L\{x_a(t)\} \triangleq \int_0^\infty x_a(t) \exp(-st)dt \qquad (1.36)$$

It is assumed that the reader is familiar with the basics of the Laplace transform. Tables of common Laplace transform pairs and Laplace transform properties can be found in Appendix C. Comparing Eq. 1.36 with Eq. 1.27 we see that for causal signals, the Fourier transform is just the Laplace transform, but with the complex variable s replaced by $j2\pi f$. Consequently, the spectrum of a causal signal can be obtained from its Laplace transform as follows.

$$X_a(f) = X(s)|_{s=j2\pi f} \qquad (1.37)$$

If the periodic impulse train $\delta_T(t)$ is expanded into a complex Fourier series, the result can be substituted into the expression for $x_a^*(t)$ in Eq. 1.33. Taking the

Laplace transform of $x_a^*(t)$ and converting the result using Eq. 1.37, we then arrive at the following expression for the spectrum of the sampled version of $x_a(t)$.

Aliasing formula

$$X_a^*(f) = \frac{1}{T} \sum_{i=-\infty}^{\infty} X_a(f - if_s) \qquad (1.38)$$

1.3.2   Aliasing

The representation of the spectrum of the sampled version of $x_a(t)$ depicted in Eq. 1.38 is called the *aliasing formula*. The aliasing formula holds the key to determining conditions under which the samples $x(k)$ contain all the information necessary to completely reconstruct or recover $x_a(t)$. To see this, we first consider the notion of a bandlimited signal.

DEFINITION

1.7: Bandlimited Signal

A continuous-time signal $x_a(t)$ is *bandlimited* to bandwidth B if and only if its magnitude spectrum satisfies

$$|X_a(f)| = 0 \quad \text{for} \quad |f| > B$$

If $x_a(t)$ is bandlimited to B, then the highest frequency component present in $x_a(t)$ is B Hz. It should be noted that some authors use a slightly different definition of the term bandlimited by replacing the strict inequality in Definition 1.7 with $|f| \geq B$.

The aliasing formula in Eq. 1.38 is quite revealing when it is applied to bandlimited signals. Notice that the aliasing formula says that the spectrum of the sampled version of a signal is just a sum of scaled and shifted spectra of the original signal with the replicated versions of $X_a(f)$ centered at integer multiples of the sampling frequency f_s. This is a characteristic of amplitude modulation in general where the unshifted spectrum ($i = 0$) is called the *baseband* and the shifted spectra ($i \neq 0$) are called *sidebands*. An illustration comparing the magnitude spectra of $x_a(t)$ and $x_a^*(t)$ is shown in Figure 1.18.

The case shown in Figure 1.18 corresponds to $f_s = 3B/2$ and is referred to as *undersampling* because $f_s \leq 2B$. The details of the shape of the even function $|X_a(f)|$ within $[-B, B]$ are not important, so for convenience a triangular spectrum is used. Note how the sidebands overlap with each other and with the baseband. This overlap is an indication of an undesirable phenomenon called *aliasing*. As a consequence of the overlap, the shape of the spectrum of $x_a^*(t)$ in $[-B, B]$ has been altered and is different from the shape of the spectrum of $x_a(t)$. The end result is that no amount of signal-independent filtering of $x_a^*(t)$ will allow us to recover the spectrum of $x_a(t)$ from the spectrum of $x_a^*(t)$. That is, the overlap or aliasing has caused the samples to be *corrupted* to the point that the original signal $x_a(t)$ can no longer be recovered from the samples. Since $x_a(t)$ is bandlimited, it is evident that there will be no aliasing if the sampling rate is sufficiently high. This fundamental result is summarized in the Shannon sampling theorem.

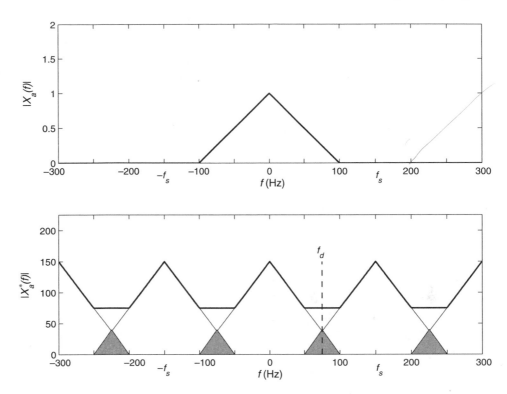

Figure 1.18 Magnitude Spectra of $x_a(t)$ and $x_a^*(t)$ when $B = 100$, and $f_s = 3B/2$

THEOREM

1.1: Sampling

Suppose a continuous-time signal $x_a(t)$ is bandlimited to B Hz. Let $x_a^*(t)$ denote the sampled version of $x_a(t)$ using impulse sampling with a sampling frequency of f_s. Then the samples $x(k)$ contain all the information necessary to recover the original signal $x_a(t)$ if

$$f_s > 2B$$

In view of the sampling theorem, it should be possible to reconstruct a continuous-time signal from its samples if the signal is bandlimited and the sampling frequency exceeds twice the bandwidth. When $f_s > 2B$, the sidebands of $X_a^*(f)$ do not overlap with each other or the baseband. By properly filtering $X_a^*(f)$, it should be possible to recover the baseband and rescale it to produce $X_a(f)$. Before we consider how to do this, it is of interest to see what happens in the time domain when aliasing occurs due to an inadequate sampling rate. If aliasing occurs, it means that there is another lower-frequency signal that will produce identical samples. Among all signals that generate a given set of samples, there is only one signal that is bandlimited to less than half the sampling rate. All other signals which generate the same samples are high-frequency imposters or *aliases*. The following example illustrates this point.

Example 1.5 | **Aliasing**

The simplest example of a bandlimited signal is a pure sinusoidal tone that has all its power concentrated at a single frequency f_0. In particular, consider the following signal where $f_0 = 90$ Hz.

$$x_a(t) = \sin(180\pi t)$$

From Table C.2 in Appendix C, the spectrum of $x_a(t)$ is

$$X_a(f) = \frac{j[\delta(f + 90) - \delta(f - 90)]}{2}$$

Thus, $x_a(t)$ is a bandlimited signal with bandwidth $B = 90$ Hz. From the sampling theorem, we need $f_s > 180$ Hz to avoid aliasing. Suppose $x_a(t)$ is sampled at the rate $f_s = 100$ Hz. In this case, $T = 0.01$ seconds, and the samples are

$$
\begin{aligned}
x(k) &= x_a(kT) \\
&= \sin(180\pi kT) \\
&= \sin(1.8\pi k) \\
&= \sin(2\pi k - .2\pi k) \\
&= \sin(2\pi k)\cos(.2\pi k) - \cos(2\pi k)\sin(.2\pi k) \\
&= -\sin(.2\pi k) \\
&= -\sin(20\pi kT)
\end{aligned}
$$

Thus, the samples of the 90 Hz signal $x_a(t) = \sin(180\pi t)$ are identical to the samples of the following lower-frequency signal which has its power concentrated at 10 Hz.

$$x_b(t) = -\sin(20\pi t)$$

A plot comparing the two signals $x_a(t)$ and $x_b(t)$ and their shared samples is shown in Figure 1.19.

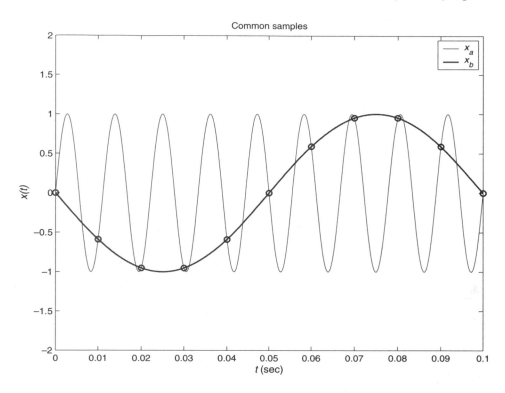

Figure 1.19 Common Samples of Two Bandlimited Signals

The existence of an equivalent low-frequency signal associated with the samples, $x(k)$, can be predicted directly from the aliasing formula. Indeed, a simple way to interpret Eq. 1.38 is to introduce something called the *folding* frequency.

Folding frequency

$$f_d \overset{\Delta}{=} \frac{f_s}{2} \tag{1.39}$$

Thus, the folding frequency is simply one half of the sampling frequency. If $x_a(t)$ has any frequency components outside of f_d, then in $x_a^*(t)$ these frequencies get reflected about f_d and folded back into the range $[-f_d, f_d]$. For the case in Example 1.5, we have $f_d = 50$ Hz. Thus, the original frequency component at $f_0 = 90$ Hz, gets reflected about f_d to produce a frequency component at 10 Hz. Notice that in Figure 1.18 the folding frequency is at the center of the first region of overlap. The part of the spectrum of $x_a(t)$ that lies outside of the folding frequency gets aliased back into the range $[-f_d, f_d]$ as a result of the overlap.

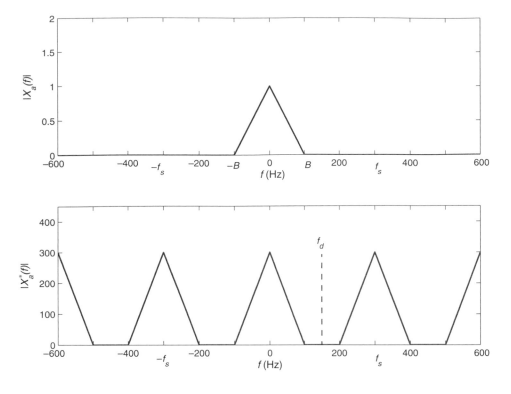

Figure 1.20 Magnitude Spectra of $x_a(t)$ and $x_a^*(t)$ when $B = 100$, and $f_s = 3B$

1.4 Signal Reconstruction

1.4.1 Reconstruction Formula

When the signal $x_a(t)$ is bandlimited and the sampling rate f_s is higher than twice the bandwidth, the samples $x(k)$ contain all the information needed to reconstruct $x_a(t)$. To illustrate the reconstruction process, consider the bandlimited signal whose magnitude spectrum is shown in Figure 1.20. In this case, we have *oversampled* by selecting a sampling frequency that is three times the bandwidth, $B = 100$ Hz. The magnitude spectrum of $x_a^*(t)$ is also shown in Figure 1.20. Note how the increase in f_s beyond $2B$ has caused the sidebands to spread out so they no longer overlap with each other or the baseband. In this case, there are no spectral components of $x_a(t)$ beyond the folding frequency $f_d = f_s/2$ to be aliased back into the range $[-f_d, f_d]$.

The problem of reconstructing $x_a(t)$ from $x_a^*(t)$ reduces to one of recovering the spectrum $X_a(f)$ from the spectrum $X_a^*(f)$. This can be achieved by passing $x_a^*(t)$ through an ideal lowpass reconstruction filter $H_{\text{ideal}}(f)$ which removes the sidebands and rescales the baseband. The required frequency response of the reconstruction filter is shown in Figure 1.21. To remove the sidebands, the cutoff frequency of the filter should be set to the folding frequency. From the aliasing formula in Eq. 1.38,

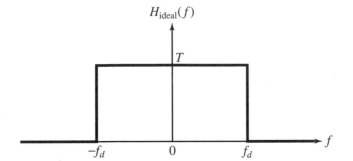

$H_{ideal}(f)$

Figure 1.21 Frequency Response of Ideal Lowpass Reconstruction Filter

the gain of the filter needed to rescale the baseband is T. Thus, the required ideal lowpass frequency response is

$$H_{ideal}(f) \triangleq \begin{cases} T, & |f| \leq f_d \\ 0, & |f| > f_d \end{cases} \tag{1.40}$$

From the aliasing formula in Eq. 1.38, one can recover the spectrum of $x_a(t)$ as follows.

$$X_a(f) = H_{ideal}(f)X_a^*(f) \tag{1.41}$$

Using Eq. 1.31 and Table C.2 in Appendix C, the impulse response of the ideal reconstruction filter is

$$\begin{aligned} h_{ideal}(t) &= F^{-1}\{H_{ideal}(f)\} \\ &= 2Tf_d \ \text{sinc}(2\pi f_d t) \\ &= \text{sinc}(\pi f_s t) \end{aligned} \tag{1.42}$$

Next, we take the inverse Fourier transform of both sides of Eq. 1.41. Using Eq. 1.35, the sifting property of the unit impulse, and the convolution property of the Fourier transform (see Table C.3), we have

$$\begin{aligned} x_a(t) &= F^{-1}\{H_{ideal}(f)X_a^*(f)\} \\ &= \int_{-\infty}^{\infty} h_{ideal}(t - \tau)x_a^*(\tau)d\tau \\ &= \int_{-\infty}^{\infty} h_{ideal}(t - \tau) \sum_{k=-\infty}^{\infty} x(k)\delta_a(\tau - kT)d\tau \\ &= \sum_{k=-\infty}^{\infty} x(k) \int_{-\infty}^{\infty} h_{ideal}(t - \tau)\delta_a(\tau - kT)d\tau \\ &= \sum_{k=-\infty}^{\infty} x(k)h_{ideal}(t - kT) \end{aligned} \tag{1.43}$$

Finally, substituting Eq. 1.42 into Eq. 1.43 yields the following formulation called the Shannon reconstruction formula.

<table>
<tr>
<td>

THEOREM

1.2: Reconstruction

</td>
<td>

Suppose a continuous-time signal $x_a(t)$ is bandlimited to B Hz. Let $x(k) = x_a(kT)$ be the kth sample of $x_a(t)$ using a sampling frequency of $f_s = 1/T$. If $f_s > 2B$, then $x_a(t)$ can be reconstructed from $x(k)$ as follows.

$$x_a(t) = \sum_{k=-\infty}^{\infty} x(k)\operatorname{sinc}[\pi f_s(t - kT)]$$

</td>
</tr>
</table>

The Shannon reconstruction formula is an elegant result that is valid as long as $x_a(t)$ is bandlimited to B and $f_s > 2B$. Note that the sinc function is used to interpolate between the samples. The importance of the reconstruction formula is that it demonstrates that all the essential information about $x_a(t)$ is contained in the samples $x(k)$ as long as the signal is bandlimited and the sampling rate exceeds twice the bandwidth.

Example 1.6   **Reconstruction**

Consider the following signal. For what range of values of the sampling interval T can this signal be reconstructed from its samples?

$$x_a(t) = \sin(5\pi t)\cos(3\pi t)$$

The signal $x_a(t)$ does not appear in Table C.2. However, using the trigonometric identities from Appendix D we have

$$x_a(t) = \frac{\sin(8\pi t) + \sin(2\pi t)}{2}$$

Since $x_a(t)$ is the sum of two sinusoids and the Fourier transform is linear, it follows that $x_a(t)$ is bandlimited with a bandwidth of $B = 4$ Hz. From the sampling theorem, we can reconstruct $x_a(t)$ from its samples if $f_s > 2B$ or $1/T > 8$. Hence, the range of sampling intervals over which $x_a(t)$ can be reconstructed from $x(k)$ is

$$0 < T < 0.125 \text{ sec}$$

1.4.2 Zero-order Hold

Exact reconstruction of $x_a(t)$ from its samples requires an ideal filter. One can approximately reconstruct $x_a(t)$ using a practical filter. We begin by noting that an effective way to characterize a linear time-invariant continuous-time system in general is in terms of its transfer function.

DEFINITION

1.8: Transfer Function

Let $x_a(t)$ be a causal nonzero input to a continuous-time system, and let $y_a(t)$ be the corresponding output assuming zero initial conditions. Then the *transfer function* of the system is defined

$$H_a(s) \triangleq \frac{Y_a(s)}{X_a(s)}$$

The transfer function is just the Laplace transform of the output divided by the Laplace transform of the input assuming zero initial conditions. A table of common Laplace transform pairs can be found in Appendix C. Note that from the sifting property of the unit impulse in Eq. 1.24 the Laplace transform of the unit impulse is $\Delta_a(s) = 1$. In view of Definition 1.8, this means that an alternative way to characterize the transfer function is to say that $H_a(s)$ is the Laplace transform of the impulse response, $h_a(t)$.

$$H_a(s) = L\{h_a(t)\} \tag{1.44}$$

Notation

At this point, a brief comment about notation is in order. Note that the same base symbol, H_a, is being used to denote both the transfer function, $H_a(s)$, in Definition 1.8, and the frequency response, $H_a(f)$, in Definition 1.5. Clearly, an alternative approach would be to introduce distinct symbols for each. However, the need for additional symbols will arise repeatedly in subsequent chapters, so using separate symbols in each case quickly leads to a proliferation of symbols that can be confusing in its own right. Instead, the notational convention adopted here is to rely on the *argument type*, a complex s or a real f, to distinguish between the two cases and dictate the meaning of H_a. The subscript a denotes a continuous-time or *analog* quantity. The less cumbersome H, without a subscript, will be reserved for discrete-time quantities introduced later.

Example 1.7   **Transportation Lag**

As an illustration of a continuous-time system and its transfer function, consider a system which delays the input by τ seconds.

$$y_a(t) = x_a(t - \tau)$$

This type of system might be used, for example, to model a transportation lag in a process control system or a signal propagation delay in a telecommunication system.

Using the definition of the Laplace transform in Eq. 1.36, and the fact that $x_a(t)$ is causal, we have

$$Y_a(s) = L\{x_a(t - \tau)\}$$

$$= \int_0^\infty x_a(t - \tau) \exp(-st)dt$$

$$= \int_{-\tau}^\infty x_a(\alpha) \exp[-s(\alpha + \tau)]d\alpha \quad \Big\} \quad \alpha = t - \tau$$

$$= \exp(-s\tau) \int_0^\infty x_a(\alpha) \exp(-s\alpha)d\alpha$$

$$= \exp(-s\tau) X_a(s)$$

It then follows from Definition 1.8 that the transfer function of a transportation lag with delay τ is

$$H_a(s) = \exp(-\tau s)$$

A block diagram of the transportation lag is shown in Figure 1.22.

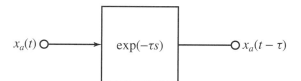

Figure 1.22 Transfer Function of Transportation Lag with a Delay of τ

$x_a(t)$ ○———→ $\exp(-\tau s)$ ———○ $x_a(t - \tau)$

The reconstruction of $x_a(t)$ in Theroem 1.2 interpolates between the samples using the sinc function. A simpler form of interpolation is to use a low degree polynomial fitted to the samples. To that end, consider the following linear system with a delay called a *zero-order hold*.

$$y_a(t) = \int_0^t [x_a(\tau) - x_a(\tau - T)]d\tau \tag{1.45}$$

Recalling that the integral of the unit impulse is the unit step, we find that the impulse response of this system is

$$h_0(t) = \int_0^t [\delta_a(\tau) - \delta_a(\tau - T)]d\tau$$

$$= \int_0^t \delta_a(\tau)d\tau - \int_0^t \delta_a(\tau - T)d\tau$$

$$= u_a(t) - u_a(t - T) \tag{1.46}$$

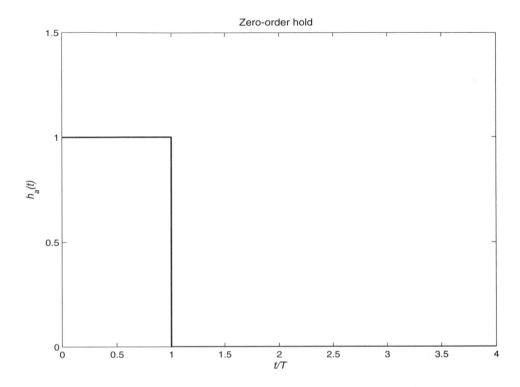

Figure 1.23 Impulse Response of a Zero-order Hold Filter

Thus, the impulse response of the zero-order hold is pulse of unit height and width T starting at $t = 0$ as shown in Figure 1.23.

Since the zero-order hold is linear and time-invariant, the response to an impulse of strength $x(k)$ at time $t = kT$ will then be a pulse of height $x(k)$ and width T starting at $t = kT$. When the input is $x_a^*(t)$, we simply add up all the responses to the scaled and shifted impulses to get a *piecewise-constant* approximation to $x_a(t)$ as shown in Figure 1.24. Notice that this is equivalent to interpolating between the samples with a polynomial of degree zero. It is for this reason that the system in Eq. 1.45 is called a zero-order hold. It holds onto the most recent sample and extrapolates to the next one using a polynomial of degree zero. Higher-degree hold filters are also possible (Proakis and Manolakis, 1992).

The zero-order hold filter also can be described in terms of its transfer function. Recall from Eq. 1.44 that the transfer function is just the Laplace transform of the impulse response in Eq. 1.46. Using the linearity of the Laplace transform, and the results of Example 1.7, we have

$$H_0(s) = L\{h_0(t)\}$$

$$= L\{u_a(t)\} - L\{u_a(t - T)\}$$

$$= [1 - \exp(-Ts)]L\{u_a(t)\} \qquad (1.47)$$

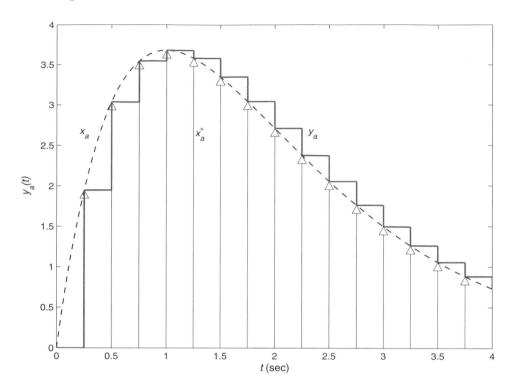

Figure 1.24 Reconstruction of $x_a(t)$ with a Zero-order Hold

Finally, from Table C.4 in Appendix C, the transfer function for a zero-order hold filter is.

Zero-order hold

$$H_0(s) = \frac{1 - \exp(-Ts)}{s} \tag{1.48}$$

A typical DSP system was described in Figure 1.2 as an analog-to-digital converter (ADC), followed by a digital signal processing program, followed by a digital-to-analog converter (DAC). We now have mathematical models available for the two converter blocks. The ADC can be modeled by an impulse sampler, while the DAC can be modeled with a zero-order hold filter as shown in Figure 1.25 which is an updated version of Figure 1.2. Note that the impulse sampler is represented symbolically with a switch that opens and closes every T seconds.

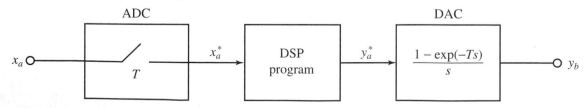

Figure 1.25 Mathematical Model of a DSP System

It should be emphasized that the formulation depicted in Figure 1.25 uses mathematical models of signal sampling and signal reconstruction. With the impulse model of sampling we are able to determine constraints on $X_a(f)$ and T that ensure that all the essential information about $x_a(t)$ is contained in the samples $x(k)$. For signal reconstruction, the piecewise-constant output from the zero-order hold filter is an effective model for the DAC output signal. To complement the mathematical models in Figure 1.25, physical circuit models of the ADC and the DAC blocks are investigated in Section 1.6.

MATLAB Toolbox

The Fundamentals of Digital Signal Processing (FDSP) toolbox contains the following function for evaluating the frequency response of a linear continuous-time system.

```
[H,f] = f_freqs (b,a,N,fmax);          % continuous-time frequency response
```

On entry to *f_freqs*, b is a vector of length $m+1$ containing the coefficients of the numerator of the transfer function, and a is a vector of length $n+1$ containing the denominator coefficients with $a(1) = 1$.

$$H_a(s) = \frac{b_0 s^m + b_1 s^{m-1} + \cdots + b_m}{s^n + a_1 s^{n-1} + \cdots + a_n} \tag{1.49}$$

Here $N \geq 2$ is the number of points at which to compute $H_a(f)$, and *fmax* is the upper limit of the range of frequencies. On exit from *f_freqs*, H is a complex vector of length N containing the frequency response at N points uniformly spaced over $[0, fmax]$, and f is vector of length N containing the evaluation frequencies. To plot the magnitude and phase responses on a single screen, one can use the following built-in MATLAB functions.

```
A = abs(H);              % magnitude response
phi = angle(H);          % phase response
subplot(2,1,1)           % top half of screen
plot (f,A)               % magnitude response plot
subplot(2,1,2)           % bottom half of screen
plot(f,phi)              % phase response plot
```

1.5 Prefilters and Postfilters

The sampling theorem tells us that to avoid aliasing during the sampling process, the signal $x_a(t)$ must be bandlimited and we must sample at a rate that is greater than twice the bandwidth. Unfortunately, a quick glance at Table C.2 in Appendix C reveals that many of those signals are, in fact, not bandlimited. The exceptions are sines

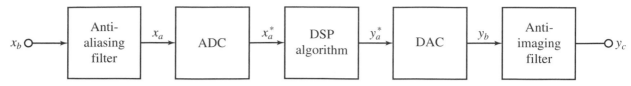

Figure 1.26 DSP System with an Analog Prefilter and Postfilter

and cosines that have all their power concentrated at a single frequency. Of course, a general periodic signal can be expressed as a Fourier series, and as long as the series is truncated to a finite number of harmonics the resulting signal will be bandlimited. For a general nonperiodic signal, if we are to avoid aliasing, we must bandlimit the signal explicitly by passing it through an analog lowpass filter as in Figure 1.26.

1.5.1 Anti-aliasing Filter

The prefilter in Figure 1.26 is called an *anti-aliasing filter* or a guard filter. Its function is to remove all frequency components outside the range $[-F_c, F_c]$ where $F_c < f_d$ so that aliasing does not occur during sampling. The optimal choice for an anti-aliasing filter is an ideal lowpass filter. Since this filter is not physically realizable, we instead approximate the ideal lowpass characteristic. For example, a widely used family of lowpass filters is the set of Butterworth filters (Ludeman, 1986). A lowpass *Butterworth filter* of order n has the following magnitude response.

$$|H_a(f)| = \frac{1}{\sqrt{1 + (f/F_c)^{2n}}}, \quad n \geq 1 \tag{1.50}$$

Notice that $|H_a(F_c)| = 1/\sqrt{2}$ where F_c is called the 3-dB *cutoff* frequency of the filter. The term arises from the fact that when the filter gain is expressed in units of decibels or dB we have

$$20 \log_{10}\{|H_a(F_c)|\} \approx -3 \text{ dB} \tag{1.51}$$

The magnitude responses of several Butterworth filters are shown in Figure 1.27. Note how as the order n increases, the magnitude response comes closer and closer to the ideal lowpass characteristic which is also shown for comparison. However, unlike the ideal lowpass filter, the Butterworth filters introduce phase shift as well.

The transfer function of a lowpass Butterworth filter of order n with radian cutoff frequency $\Omega_c = 2\pi F_c$ can be expressed as follows.

$$H_a(s) = \frac{\Omega_c^n}{s^n + \Omega_c a_1 s^{n-1} + \Omega_c^2 a_2 s^{n-2} \cdots + \Omega_c^n} \tag{1.52}$$

When $\Omega_c = 1$ rad/sec or $F_c = 1/(2\pi)$ Hz, this corresponds to a *normalized* Butterworth filter. A list of the coefficients for the first few normalized Butterworth filters is summarized in Table 1.2. Notice that the denominator polynomials in Table 1.2 are factored into quadratic factors when n is even and quadratic and linear factors when n is odd. This is done because the preferred way to realize $H_a(s)$ with a circuit is as a series connection of first and second order blocks. That way, the overall transfer function is less sensitive to the precision of the circuit elements.

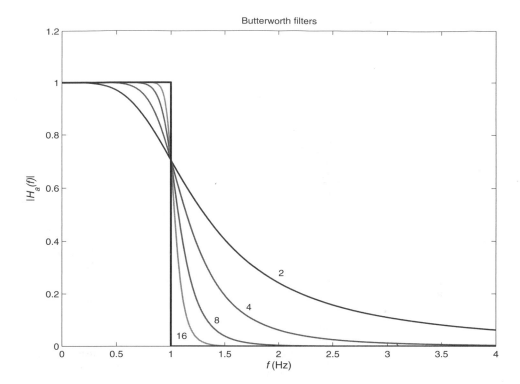

Figure 1.27 Magnitude Responses of Lowpass Butterworth Filters with $F_c = 1$

Table 1.2: ▶ Second-order Factors of Normalized Butterworth Lowpass Filters

Order	$a(s)$
1	$(s + 1)$
2	$(s + 1.5142s + 1)$
3	$(s + 1)(s^2 + s + 1)$
4	$(s^2 + 1.8478s + 1)(s^2 + 0.7654s + 1)$
5	$(s + 1)(s^2 + 1.5180s + 1)(s^2 + 0.6180s + 1)$
6	$(s^2 + 1.9318s + 1)(s^2 + 1.5142s + 1)(s^2 + 0.5176s + 1)$
7	$(s + 1)(s^2 + 1.8022s + 1)(s^2 + 1.2456s + 1)(s^2 + 0.4450s + 1)$
8	$(s^2 + 1.9622s + 1)(s^2 + 1.5630s + 1)(s^2 + 1.1110s + 1)(s^2 + 0.3986s + 1)$

Example 1.8   **First-order Filter**

Consider the problem of realizing a lowpass Butterworth filter of order $n = 1$ with a circuit. From Table 1.2, we have $a_1 = 1$. Thus from Eq. 1.52 the first-order transfer function is

$$H_1(s) = \frac{\Omega_c}{s + \Omega_c}$$

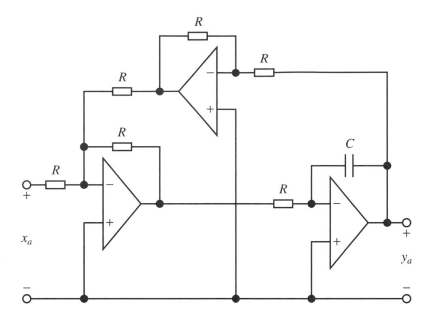

Figure 1.28 Active Circuit Realization of a First-order Lowpass Butterworth Filter Block

This transfer function can be realized with a simple RC circuit with $RC = 1/\Omega_c$. However, a passive circuit realization of $H_1(s)$ can experience electrical loading effects when it is connected in series with other blocks to form a more general filter. Therefore, consider instead the active circuit realization shown in Figure 1.28 which uses three operational amplifiers (op amps). This circuit has a high-input impedance and a low-output impedance which means it will not introduce significant loading effects when connected to other circuits. The transfer function of the circuit (Dorf and Svoboda, 2000) is

$$H_1(s) = \frac{1/(RC)}{s + 1/(RC)}$$

Thus, we require $1/(RC) = \Omega_c$. Typically, C is chosen to be a convenient value, and then R is computed using

$$R = \frac{1}{\Omega_c C}$$

As an illustration, suppose a cutoff frequency of $F_c = 1000$ Hz is desired. Then $\Omega_c = 2000\pi$. If $C = 0.01\ \mu$F, which is a common value, then the required value for R is

$$R = 15.915\ \text{k}\Omega$$

Integrated circuits typically contain up to four op amps. Therefore, the first-order filter section can be realized with a single integrated circuit and seven discrete components.

Since all of the first-order filter sections in Table 1.2 have $a_1 = 1$, the filter in Figure 1.28 can be used for a general first-order filter section. An nth-order Butterworth filter also contains second-order filter sections.

Example 1.9   Second-order Filter

Consider the problem of realizing a lowpass Butterworth filter of order $n = 2$ with a circuit. From Eq. 1.52 the general form of the second-order transfer function is

$$H_2(s) = \frac{\Omega_c^2}{s^2 + \Omega_c a_1 + \Omega_c^2}$$

Using a state-space formulation of $H_2(s)$, this second-order block can be realized with the active circuit shown in Figure 1.29 which uses four op amps. The transfer function of circuit (Dorf and Svoboda, 2000) is

$$H_2(s) = \frac{1/(R_1 C)^2}{s^2 + s/(R_2 C) + 1/(R_1 C)^2}$$

From the last term of the denominator, we require $1/(R_1 C)^2 = \Omega_c^2$. Again, typically C is chosen to be a convenient value, and then R_1 is computed using

$$R_1 = \frac{1}{\Omega_c C}$$

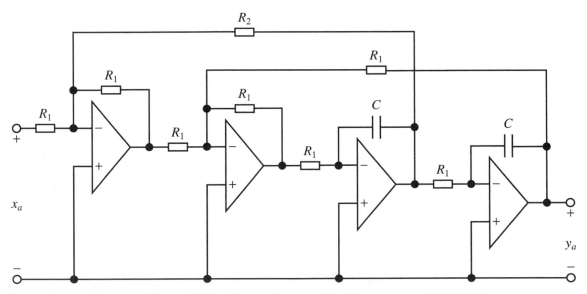

Figure 1.29   Active Circuit Realization of a Second-order Lowpass Butterworth Filter Block

Next, from the linear term of the denominator, we have $1/(R_2C) = \Omega_c a_1$. Solving for R_2 and expressing the final result in terms of R_1 yields

$$R_2 = \frac{R_1}{a_1}$$

As an illustration, suppose a cutoff frequency of $F_c = 5000$ Hz is desired. Then $\Omega_c = 10^4\pi$, and from Table 1.2 we have $a_1 = 1.5142$. If we pick $C = 0.01$ μF, then the two resistors are

$$R_1 \approx 3.183 \text{ k}\Omega$$

$$R_2 \approx 2.251 \text{ k}\Omega$$

This filter also can be realized with a single integrated circuit plus 10 discrete components.

A general lowpass Butterworth filter can be realized by using a series or cascade connection of first- and second-order blocks where the output of one block is used as the input to the next block. In this way, an anti-aliasing filter of order n can be constructed.

The Butterworth family of lowpass filters is one of several families of classical analog filters (Lam, 1979). Other classical analog filters that could be used for anti-aliasing filters include Chebyshev filters and elliptic filters. Classical analog lowpass filters are considered in detail in Chapter 8 when we investigate a digital design technique that converts an analog filter into an equivalent digital filter.

Classical analog filters are sufficiently popular, that they have been implemented as integrated circuits using switched capacitor technology (Jameco, 2004). For example, the National Semiconductor LMF6-100 is a sixth-order lowpass Butterworth filter whose cutoff frequency is tunable with an external clock signal of frequency $f_{\text{clock}} = 100F_c$. The switched capacitor technology involves internal sampling or switching at the rate f_{clock}. Therefore, if the signal $x_a(t)$ does not have any significant spectral content beyond 50 times the desired sampling rate f_s, then the switch-capacitor filter can be used as an anti-aliasing filter to remove the frequency content in the range $[f_s/2, 50f_s]$. For example, with a desired sampling rate of 2 kHz, a switched-capacitor filter will remove (i.e., significantly reduce) frequencies in the range from 1 kHz to 100 kHz.

1.5.2   Anti-imaging Filter

The postfilter in Figure 1.26 is called an *anti-imaging filter* or smoothing filter. The function of this filter is to remove the residual high-frequency components of $y_b(t)$. The zero-order hold transfer function of the DAC tends to reduce the size of the

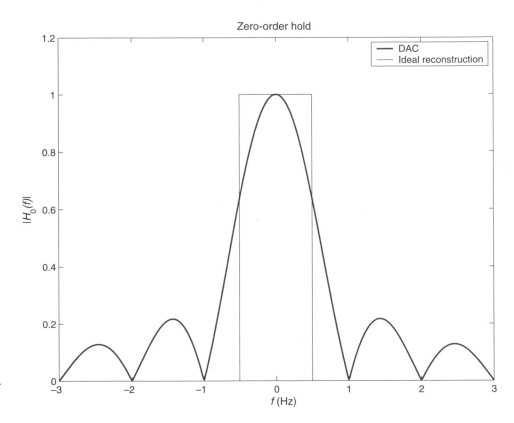

Figure 1.30 Magnitude Response of a Zero-order Hold

sidebands of $y_a^*(t)$ which can be thought of as *images* of the baseband spectrum. However, it does not eliminate them completely. The magnitude response of the zero-order hold, shown in Figure 1.30, reveals a lowpass type of characteristic that is different from the ideal reconstruction filter due to the presence of a series of lobes. For the lowpass filter in Figure 1.30, the output is the piecewise constant signal $y_b(t)$, rather than the ideally reconstructed signal $y_a(t)$.

Note from Figure 1.30 that the zero-order hold has zero gain at multiples of the sampling frequency where the images of the baseband spectrum are centered. The effect of this lowpass type of characteristics is to reduce the size of the sidebands as can be seen Figure 1.31 which shows a discrete-time DAC input with a triangular spectrum and the corresponding spectrum of the piecewise-constant DAC output.

The purpose of the anti-imaging filter is to further reduce the residual images of the baseband spectrum centered at multiples of the sampling frequency. A low-pass filter similar to the anti-aliasing filter can be used for this purpose. A careful inspection of the baseband spectrum of $y_b(t)$ in Figure 1.31 reveals that it is slightly distorted due to the nonflat passband characteristic of the zero-order hold shown in

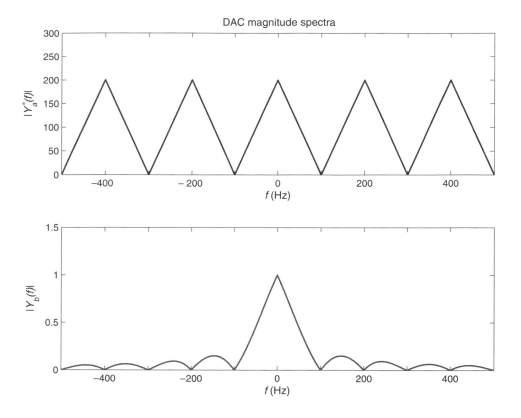

Figure 1.31 Magnitude Spectra of a DAC Input and a DAC Output when $f_s = 200$ Hz and $B = f_s/2$

Figure 1.30. Interestingly enough, this can be compensated for with a digital filter as part of the DSP algorithm by preprocessing $y_a^*(t)$ before it is sent into the DAC (see Chapter 7).

A simple way to reduce the need for an anti-imaging filter is to increase the sampling rate f_s beyond the minimum needed to avoid aliasing. As we shall see in Chapter 7, the DAC sampling rate can be increased as part of the DSP algorithm by inserting samples between those coming from the ADC, a process known as interpolation. The effect of oversampling is to spread out the images along the frequency axis which allows the zero-order hold to more effectively attenuate them. This can be seen in Figure 1.32 which is identical to Figure 1.31, except that the signals have been oversampled by a factor of two. Notice that oversampling also has the beneficial effect of reducing the distortion of the baseband spectrum that is caused by the nonideal passband characteristic of the zero-order hold.

Finally, it is worth noting that there are applications where an anti-imaging filter may not be needed at all. For example, in a digital control application the output of the DAC might be used to drive a relatively slow electro-mechanical device such as a motor. These devices already have a lowpass frequency response characteristic so it is not necessary to filter the DAC output. For some DSP applications, the desired output may be information that can be extracted directly from the discrete-time signal $y_a^*(t)$ in which case there is no need to convert from digital back to analog.

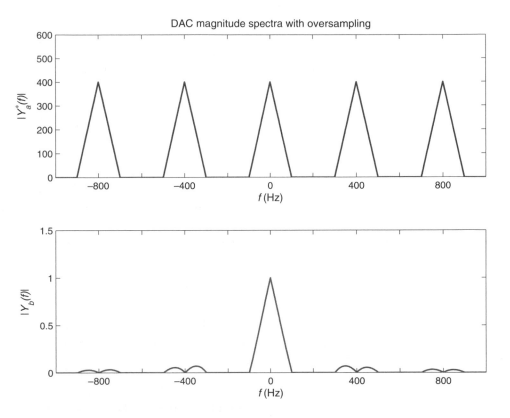

Figure 1.32 Magnitude Spectra of DAC Input and Output when $f_s = 400$ and $B = f_s/4$

*1.6 Conversion Circuits

This optional section presents circuit realizations of digital-to-analog and analog-to-digital converters. This material is included for those readers specifically interested in hardware. This section, and others like it marked with *, can be skipped without loss of continuity.

1.6.1   Digital-to-Analog Conversion (DAC)

Recall that a digital-to-analog converter or DAC can be modeled mathematically as a zero-order hold as in Eq. 1.48. A physical model of a DAC is entirely different because the input is an N-bit binary number, $b = b_{N-1}b_{N-2}\cdots b_1 b_0$, rather than an amplitude modulated impulse train. A DAC is designed to produce an analog output y_a that is proportional to the decimal equivalent of the binary input b. DAC circuits can be classified as *unipolar* if $y_a \geq 0$ or *bipolar* if y_a is both positive and negative. For the simpler case of a unipolar DAC, the decimal equivalent of the binary input b is

$$x = \sum_{k=0}^{N-1} b_k 2^k \tag{1.53}$$

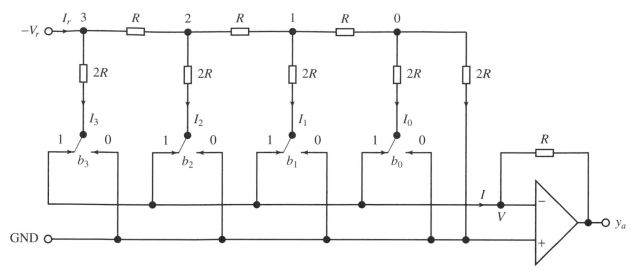

Figure 1.33 A 4-bit Unipolar DAC Using an R-$2R$ Ladder

The binary input of a bipolar DAC represents negative numbers as well using two's complement, offset binary, or a sign-magnitude format (Grover and Deller, 1999). The most common type of DAC is the R-$2R$ ladder circuit shown in Figure 1.33 for the case $N = 4$. The configuration of resistors across the top is the R-$2R$ ladder. On each rung of the ladder is a digitally controlled single pole double throw (SPDT) switch that sends current into either the inverting input (when $b_k = 1$) or the non-inverting input (when $b_k = 0$) of the operational amplifier or *op amp*.

To analyze the operation of the circuit in Figure 1.33, we begin at the end of the ladder and work backwards. First note that for an ideal op amp, the voltage at the inverting input is the same as that at the noninverting input, namely $V = 0$. Consequently, the current I_k through the kth switch does not depend on the switch position. The equivalent resistance looking into node 0 from the left is therefore $R_0 = R$ because it consists of two resistors of resistance $2R$ in parallel. Thus the current entering node 0 from the left is split in half. Next, consider node 1. The equivalent resistance to the right of node 1 is $2R$ because it consists of a series combination of R and R_0. This means that the equivalent resistance looking into node 1 from the left is again $R_1 = R$ because it consists of two resistors of resistance $2R$ in parallel. This again means that the current entering node 1 from the left is split in half. This process can be repeated as many times as needed until we conclude that the equivalent resistance looking into node $N - 1$ from the left is R. Consequently, the current drawn by the R-$2R$ ladder, regardless of the switch positions, is as follows where V_r is the *reference voltage*.

$$I_r = \frac{-V_r}{R} \qquad (1.54)$$

Since the current is split in half each time it enters another node of the ladder, the current shunted through the kth switch is $I_k = I_r/2^{N-k}$. Consequently, using

Eq. 1.53 the total current entering the op amp section is

$$I = \sum_{k=0}^{N-1} b_k I_k$$

$$= \sum_{k=0}^{N-1} b_k I_r 2^{k-N}$$

$$= \left(\frac{I_r}{2^N}\right) \sum_{k=0}^{N-1} b_k 2^k$$

$$= \left(\frac{I_r}{2^N}\right) x \tag{1.55}$$

Finally, an ideal op amp has infinite input impedance which means that the current drawn by the inverting input is zero. Therefore, using Eq. 1.54 and Eq. 1.55, the op amp output is

$$y_a = -RI$$

$$= \left(\frac{-RI_r}{2^N}\right) x$$

$$= \left(\frac{V_r}{2^N}\right) x \tag{1.56}$$

Setting $x = 1$ in Eq. 1.56 we see that the quantization level of the DAC is $q = V_r/2^N$. The range of output values for the DAC in Figure 1.33 is

$$0 \le y_a \le \left(\frac{2^N - 1}{2^N}\right) V_r \tag{1.57}$$

In view of Eq. 1.57, the DAC in Figure 1.33 is a unipolar DAC. It can be converted to a bipolar DAC with outputs in the range $-V_r \le y_a < V_r$ by replacing V_r with $2V_r$ and adding a second op amp circuit at the output that performs level shifting (see Problem P1.15). In this case, b is interpreted as an offset binary input whose decimal equivalent is as follows where x is as in Eq. 1.53.

$$x_{\text{bipolar}} = x - 2^{N-1} \tag{1.58}$$

In general, a *signal conditioning* circuit can be added to the DAC in Figure 1.33 to perform scaling, offset, and impedance matching (Dorf and Svoboda, 2000). It is useful for the DAC circuit to have a low output impedance so that the DAC output is capable of supplying adequate current to drive the desired load.

1.6.2 Analog-to-Digital Conversion (ADC)

An analog-to-digital converter or ADC must take an analog input $-V_r \le x_a < V_r$ and convert it to a binary output b whose decimal value is equivalent to x_a. The input-output characteristic of an N-bit bipolar ADC is shown in Figure 1.34 for the case $V_r = 5$ and $N = 4$.

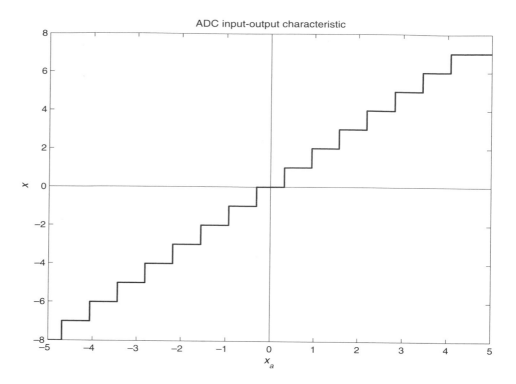

Figure 1.34 Input-output Characteristic of 4-bit ADC with $V_r = 5$

Note that the staircase is shifted to the left by half a step so that the ADC output is less sensitive to low-level noise when $x_a = 0$. The horizontal length of the step is the quantization level which, for a bipolar ADC, can be expressed as

$$q = \frac{V_r}{2^{N-1}} \tag{1.59}$$

Successive-approximation Converters

The most widely used ADC consists of a comparator circuit plus a DAC in the configuration shown in Figure 1.35. The input is the analog voltage x_a to be converted and the output is the equivalent N-bit binary number b.

A very simple form of the ADC can be realized by replacing the block labeled SAR logic with a binary counter which counts the pulses of the periodic pulse train or clock signal, f_{clock}. As the counter output b increases from 0 to $2^N - 1$, the DAC output y_a ranges from $-V_r$ to V_r. When the value of y_a becomes larger than the input x_a, the comparator output u_a switches from 1 to 0 at which time the counter is disabled. The count b is then the digital equivalent of the analog input x_a using the offset binary code in Eq. 1.58.

Although an ADC based on the use of a counter is appealing because of its conceptual simplicity, there is a significant practical drawback. The time required to perform a conversion is variable and can be quite long. For random inputs with a

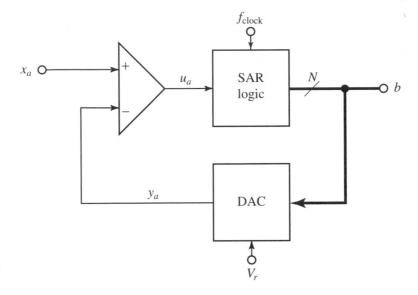

Figure 1.35 An N-bit Successive-approximation ADC

mean value of zero, it takes on the average 2^{N-1} clock pulses to perform a conversion, and it can take as long as 2^N clock pulses when $x_a = V_r$. A much more efficient way to perform a conversion is to execute a binary search for the proper value of b by using a successive approximation register (SAR). The basic idea is to start with the most significant bit, b_{N-1}, and determine if it should be 0 or 1 based on the comparator output. This cuts the range of uncertainty for the value of x_a in half and can be done in one clock pulse. Once b_{N-1} is determined, the process is then repeated for bit b_{N-2} and so on until the least significant bit, b_0, is determined. The successive approximation technique is summarized in Algorithm 1.1.

ALGORITHM 1.1:
Successive
Approximation

1. Set $b_k = 0$ for $0 \le k \le N - 1$.
2. For $k = N - 1$ down to 0 do
{

    (a) Set $b_k = 1$
    (b) If $u_a = 0$, set $b_k = 0$

}

The virtue of the binary search approach is that it takes exactly N clock pulses to perform a conversion, independent of the value of x_a. Thus, the conversion time is constant and the process is much faster. For example, for a precision of $N = 12$ bits, the conversion time in comparison with the counter method is reduced, on the average, by a factor of $2^{11}/12 = 170.7$ or two orders of magnitude.

| **Example 1.10** | **Successive Approximation** |

As an illustration of the successive approximation conversion technique, suppose the reference voltage is $V_r = 5$ volts, the converter precision is $N = 10$ bits, and the value to be converted is $x_a = 2.891$ volts. When Algorithm 1.1 is applied, the traces of $y_a(k)$ and $u_a(k)$ for $0 \leq k < N$ are as shown in Figure. 1.36. Note how the DAC output quickly adjusts to the value of x_a by cutting the interval of uncertainty in half with each clock pulse. From Eq. 1.59, the quantization level in this case is

$$q = \frac{V_r}{2^{N-1}}$$

$$= \frac{5}{512}$$

$$= 9.8 \, \text{mV}$$

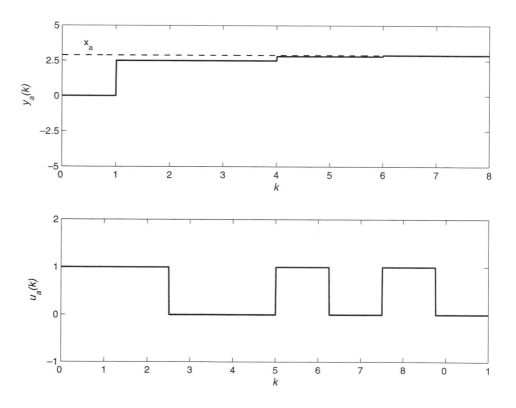

Figure 1.36 DAC and Comparator Outputs During Successive-approximation Steps

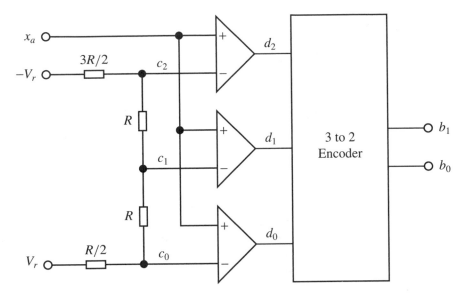

Figure 1.37   A 2-bit Flash Converter

Flash Converters

There is another type of ADC, called the *flash* converter, that is used in applications where very high-speed conversion is essential such as in a digital oscilloscope. A simple 2-bit flash converter is shown in Figure 1.37. It consists of a linear resistor array, an array of comparator circuits, and an encoder circuit.

The resistors in the resistor array are selected such that the voltage drop between successive inverting inputs to the comparators is the quantization level q in Eq. 1.59. The fractional resistances at the ends of the array cause the input-output characteristic to shift to the left by $q/2$ as shown previously in Figure 1.34. Therefore, the voltage at the inverting input of the kth comparator is

$$c_k = -V_r + \left(k + \frac{1}{2}\right)q, \quad 0 \le k < N - 1 \tag{1.60}$$

The analog input x_a is compared to each of the threshold voltages c_k. For those k for which $x_a > c_k$, the comparator outputs will be $d_k = 1$, while the remaining comparators will all have outputs of $d_k = 0$. Thus the comparator outputs d can be thought of as a bar graph code where all the bits to the right of a certain bit are turned on. The encoder circuit takes the $2^N - 1$ comparator outputs d and converts them to an N-bit binary output b. It does so by setting the decimal equivalent of b equal to i where i is the largest subscript such that $d_i = 1$. For example, for the 2-bit converter in Figure 1.37, the four possible inputs and outputs are summarized in Table 1.3.

The beauty of the flash converter is that the entire conversion can be done in a single clock pulse. More specifically, the conversion time is limited only by the

Input Range	$d = d_2 d_1 d_0$	$b = b_1 b_0$
$-1 \leq x_a/V_r < -.75$	000	00
$-.75 \leq x_a/V_r < -.25$	001	01
$-.25 \leq x_a/V_r < .25$	011	10
$.25 \leq x_a/V_r \leq 1$	111	11

settling time of the comparator circuits and the propagation delay of the encoder circuit. Unfortunately, the extremely fast conversion time is achieved at a price. The converter shown in Figure 1.37 has a precision of only 2 bits. In general, for an N-bit converter, there will be a total of $2^N - 1$ comparator circuits required. Furthermore, the encoder circuit will require $2^N - 1$ inputs. Consequently, as N increases, the flash converter becomes very hardware intensive. As a result, practical high speed flash converters are typically lower precision (6 to 8 bits) in comparison with the medium speed successive approximation converters (8 to 16 bits).

There are a number of other types of ADCs as well including the slower, but very high precision, sigma delta converters and dual integrating converters (Grover and Deller, 1999). The successive approximation converters and the flash converters can be combined to form a hybrid successive flash or *multistage* converter. For example, a 4-bit flash converter can be used reduce the interval of uncertainty of x_a by a factor of 16. During the second clock pulse, this subinterval can be reduced by another factor of 16 with a second 4-bit flash converter, etc. The multistage converters have performance characteristics that lie between those of the successive approximation converters and the flash converting in terms of speed and hardware complexity. A table which summarizes the characteristics of the different converters is shown in Table 1.4.

Converter Type	Precision N	Conversion Time	Comparators
Successive approximation	8	8	0
Multistage (2 stages)	8	2	30
Flash	8	1	255
Successive approximation	12	12	0
Multistage (3 stages)	12	3	45
Flash	12	1	4095

MATLAB Toolbox

The FDSP toolbox contains two functions for performing analog-to-digital and digital-to-analog conversions.

```
        y = f_quant (x,q,qtype);        % quantization operator
[b,d,y] = f_adc (x,N,Vr);               % analog-to-digital
        y = f_dac (b,N,Vr);             % digital-to-analog
```

Function f_quant implements the quantization operation. On entry to f_quant, x is the input to be quantized, q is the quantization level, and $qtype$ selects the quantization type, zero for rounding and nonzero for truncation. On exit from f_quant, y is the quantized version of x. Function f_adc performs a bipolar analog-to-digital conversion. On entry to f_adc, $-Vr \leq x < Vr$ is the analog value to be converted, $N \geq 1$ is the number of bits, and Vr is the reference voltage of the ADC. On exit from f_adc, b is a vector of length N containing the binary equivalent of x, d is the decimal value of b using offset binary, and y is the quantized analog output. Function f_dac performs a bipolar digital-to-analog conversion. On entry to f_dac, b is a vector of length N containing the offset binary input, $N \geq 1$ is the number of bits, and Vr is the reference voltage. On exit from f_dac, $-Vr \leq y < Vr$ is the analog output.

● ● ● ● ● ● ● ● ● ● ● ● ● ● ● ●

1.7 The FDSP Toolbox

The software distribution CD that accompanies this text includes a Fundamentals of Digital Signal Processing (FDSP) toolbox containing MATLAB software that implements the signal processing algorithms developed in the text. This software is provided as an aid to help students solve the Simulation problems and the Computation problems appearing at the end of each chapter. The FDSP toolbox also provides the user with a convenient way to run the examples and reproduce the figures that appear in the text. A novel component of the toolbox is a collection of graphical user interface (GUI) modules that allow the user to investigate, and explore, the signal processing techniques covered in each chapter without any need for programming. The FDSP toolbox is installed using MATLAB itself (Marwan, 2003). For example, if the distribution CD is in drive D, then enter the following commands in the MATLAB command window.

```
cd d:\
setup
```

It is assumed that students have access to MATLAB, Version 6.1 or later. Only the student version of MATLAB is required. The FDSP toolbox is described in more detail in Appendix B. This section provides an overview of the main features of the toolbox.

1.7.1 Driver Module: *f_dsp*

All of the software on the distribution CD can be conveniently accessed through a driver module called *f_dsp*. The driver module is launched by entering the following command from the MATLAB command prompt:

```
f_dsp
```

The startup screen for *f_dsp* is shown in Figure 1.38. Most of the options on the menu toolbar at the top of the screen produce submenus of selections. The GUI Modules option is used to run the end of chapter graphical user interface modules. Using the Examples option, all MATLAB examples appearing in the text can be executed. Similarly, the Figures option is used to recreate and display the MATLAB figures in the text. The Problems option is used to display solutions to selected end of chapter problems. They are displayed as PDF files using the Adobe Acrobat Reader. The Help option provides online help for the GUI modules and the toolbox functions. For users interested in obtaining updates and supplementary course material, the Web option connects the user the following FDSP web site:

```
www.clarkson.edu/~schillin/fdsp
```

Using this option, the user can download a self-extracting file that contains the latest version of the FDSP software. Additional supplementary support material for courses based on the *Fundamentals of Digital Signal Processing* will be posted at this site as it becomes available. The Exit option returns control to the MATLAB command window.

1.7.2 GUI Modules

When the GUI Modules option is selected from the FDSP driver module, the user is provided with the list of chapter GUI modules summarized in Table 1.5.

Table 1.5: ▶
FDSP Toolbox GUI
Modules

Module	Description	Chapter
g_sample	Signal sampling	1
g_reconstruct	Signal reconstruction	1
g_system	Discrete-time systems	2
g_spectra	Signal spectral analysis	3
g_correlate	Signal correlation and convolution	4
g_filter	Filter specifications and structures	5
g_fir	FIR filter design	6
g_multirate	Multirate signal processing	7
g_iir	IIR filter design	8
g_adapt	Adaptive signal processing	9

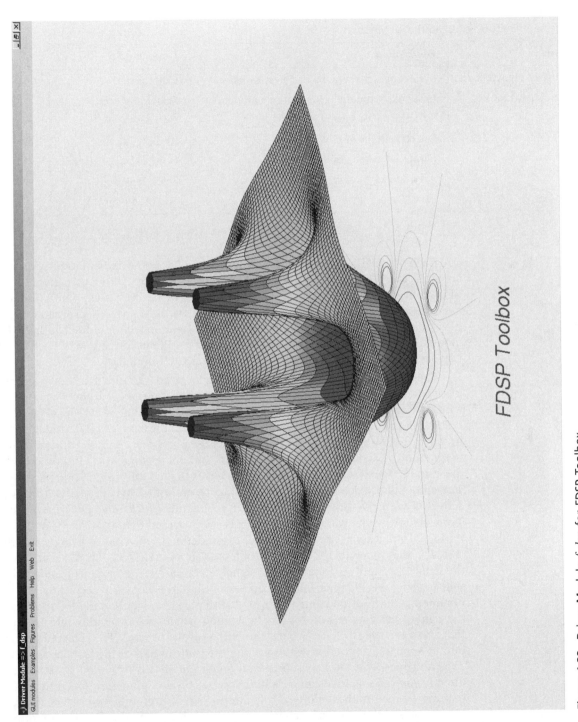

Figure 1.38 Driver Module *f_dsp* for FDSP Toolbox

Table 1.6: ▶
User Interface Features of Chapter GUI Modules

Feature	Description	Comments
Block diagram window	System or algorithm under investigation	Color coded
Parameters window	Edit boxes containing simulation parameters	Scalar and vector
Type window	Select signal or system type	Radio buttons
View window	Select contents of plot window	Radio buttons
Plot window	Display selected graphical results	Bottom half of screen
Slider bar	Adjust scalar design parameter	Fixed range
Menu bar	Save, Caliper, Print, Help, Exit	Top of screen

Each of the GUI modules is described in detail near the end of the corresponding chapter. The GUI modules are designed to provide users with a convenient means of investigating the signal processing concepts covered in that chapter without any need for programming. Users who are familiar with MATLAB programming can provide optional MAT-files and optional M-file functions that interact with the GUI modules. There is also a set of GUI Simulation problems at the end of each chapter that are designed to be solved using the chapter GUI modules.

The GUI modules in Table 1.5 have a standardized user interface that is simple to learn and easy to use. The startup screens consist of a set of tiled windows populated with GUI controls. A typical startup screen for a GUI module is shown in Figure 1.39. A summary of the common user interface features of the chapter GUI modules can be found in Table 1.6. In the upper left of the screen is a *Block diagram* window showing the system or signal processing operation under investigation. Below the block diagram is a *Parameters* window with edit boxes that contain simulation parameters. The contents of each edit box can be directly modified by the user with changes activated with the Enter key. To the right of the *Block diagram* window is the *Type* window where radio button controls are provided that allow the user to select the signal or system type. In addition to predefined types, there is a User-defined selection that prompts for the name of a user-supplied MATLAB file that contains the signal or system information. For signal selection, there is also an option to record signals using the PC microphone. To verify that an acceptable recording has been obtained, a pushbutton control is provided to play the signal on the PC speaker. To the right of the *Type* window is a *View* window which includes radio button controls that allow the user to select which simulation results to view. The selected graphical output appears in the *Plot* window along the bottom half of the screen. Below the *Type* and *View* windows is a horizontal *Slider bar* that is provided so the user can directly adjust a scalar design parameter such as the sampling frequency, the number of samples, the number of bits, or the filter order. Within the *Parameters*, *Type*, and *View* windows there are also checkbox controls that allow the user to turn optional features on and off such as dB display, signal clipping, and additive noise.

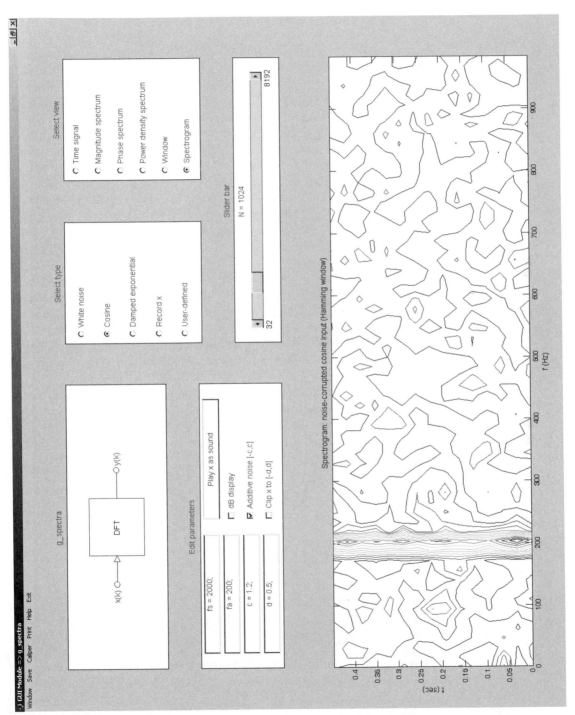

Figure 1.39 Example Screen, GUI Module *g_spectra*

The *Menu* bar at the top of the screen contains options specific to the GUI module. The common options include Save, Caliper, Print, Help, and Exit. The Caliper option is provided so the user can use crosshairs and the mouse to measure a point of interest on the current plot. The Save option allows the user to save data in a user-specified MAT-file. The contents of the MAT-file can later be loaded into the same or other GUI modules using the User-defined option in the *Type* window. In this way, results can be exported and imported between the GUI modules. MAT-files produced by GUI modules use a single common format. The Print option provides a hard copy graphical output of the current plot. The Help option is used to display directions on how to make effective use of the GUI module. Finally, the Exit option returns control to the calling program, either the FDSP driver module or the MATLAB command prompt.

1.7.3 Toolbox Functions

The use of the GUI modules is designed to be convenient, but it is not as flexible as having users write their own MATLAB scripts to perform signal processing tasks. Algorithms developed in the text are implemented as a library of FDSP toolbox functions. These functions are described in detail in the chapters, and they are summarized in Appendix B. They fall into two broad categories, main program support functions and chapter functions. The main program support functions consist of general low-level utility functions that are designed to simplify the process of writing MATLAB scripts by performing some routine tasks. These functions are summarized in Table 1.7.

The second group of toolbox functions consists of implementations of algorithms developed in the chapters. Specialized functions are developed in those instances where corresponding built-in MATLAB functions are not available as part of the student version of MATLAB. In order to minimize the expense to the student, it is

Table 1.7: ▶
FDSP Main Program
Support Functions

Name	Description
f_caliper	Measure points on plot using mouse crosshairs
f_clip	Clip value to an interval, check calling arguments
f_getsound	Record signal from PC microphone
f_labels	Label graphical output
f_prompt	Prompt for a scalar in a specified range
f_randg	Gaussian random matrix
f_randu	Uniformly distributed random matrix
f_torow	Convert vector to a row vector
f_tocol	Convert vector to a column vector
f_wait	Pause to examine displayed output

assumed that no supplementary MATLAB toolboxes (e.g., Signal Processing toolbox, Filter Design toolbox, DSP blockset) are available. Instead, these functions are provided in the FDSP toolbox as needed. However, if students do have access to a supplementary toolbox, it can be used without conflict because the FDSP functions all follow the *f_xxx* naming convention in Table 1.7. A summary of the chapter functions can be found in Table 1.8. To learn more about the usage of any of these functions simply type *helpwin* followed by the function name.

Table 1.8: ▶
FDSP Toolbox
Functions

Chapter	Functions	Description
1. Signal sampling and reconstruction	*f_freqs*	Analog frequency response
	f_quant	Quantization operator
	f_adc	Analog-to-digital conversion
	f_dac	Digital-to-analog conversion
2. Discrete-time system analysis	*f_freqz*	Frequency response
	f_impulse	Impulse response
	f_spec	Signal spectra
	f_pzplot	Pole-zero sketch
	f_pzsurf	Pole-zero surface
3. The DFT and spectral analysis	*f_pds*	Power density spectrum
	f_window	Data windows
	f_specgram	Spectrogram
	f_unscramble	Reorder FFT output
4. Convolution and correlation	*f_conv*	Fast linear and circular convolution
	f_corr	Fast linear and circular correlation
	f_blockcorr	Fast linear block correlation
5. Filter specifications and structures	*f_cascade*	Find cascade form realization
	f_parallel	Find parallel form realization
	f_lattice	Find lattice form realization
	f_filtcas	Evaluate cascade form output
	f_filtpar	Evaluate parallel form output
	f_filtlat	Evaluate lattice form output
	f_minall	Minimum-phase allpass factorization
6. FIR filter design	*f_firamp*	Frequency-selective amplitude response
	f_firideal	Design frequency-selective windowed FIR filter
	f_firwin	Design windowed FIR filter
	f_firsamp	Design frequency-sampled FIR filter
	f_firls	Design least-squares FIR filter
	f_firparks	Design optimal equiripple FIR filter
7. Multirate signal processing	*f_decimate*	Integer sampling rate decimator
	f_interpol	Integer sampling rate interpolator
	f_rateconv	Rational sampling rate converter

Table 1.8 continues on the following page

Chapter	Functions	Description
8. IIR filter design	*f_iirres*	Design IIR resonator filter
	f_iirnotch	Design IIR notch filter
	f_iircomb	Design IIR comb filter
	f_iirinv	Design IIR inverse comb filter
	f_butters	Design analog Butterworth lowpass filter
	f_cheby1s	Design analog Chebyshev-I lowpass filter
	f_cheby2s	Design analog Chebyshev-II lowpass filter
	f_elliptics	Design analog elliptic lowpass filter
	f_low2lows	Analog lowpass to lowpass transformation
	f_low2highs	Analog lowpass to highpass transformation
	f_low2bps	Analog lowpass to bandpass transformation
	f_low2bss	Analog lowpass to bandstop transformation
	f_bilin	Bilinear analog-to-digital filter transformation
	f_butterz	Design IIR Butterworth filter
	f_cheby1z	Design IIR Chebyshev-I filter
	f_cheby2z	Design IIR Chebyshev-II filter
	f_ellipticz	Design IIR elliptic filter
	f_string	Find IIR plucked-string filter output
	f_reverb	Find IIR reverb filter output
9. Adaptive signal processing	*f_lms*	Least Mean Square (LMS) method
	f_normlms	Normalized LMS method
	f_corrlms	Correlation LMS method
	f_leaklms	Leaky LMS method
	f_fxlms	Filtered-x LMS active noise control method
	f_sigsyn	Signal synthesis active noise control method
	f_rls	Recursive least-squares (RLS) method
	f_rbfw	Radial basis function (RBF) system identification
	f_rbf0	Evaluate zero-order RBF network output
	f_state	Evaluate state of nonlinear discrete-time system

1.7.4 Toolbox Help

The Help option in the driver module in Figure 1.38 provides documentation on all of the GUI modules, the main program support functions, and the chapter functions. An alternative way to obtain online documentation of the toolbox functions directly from the MATLAB command prompt is to use the MATLAB *helpwin* command. To get instructions for usage of any of the MATLAB functions or GUI modules, use the helpwin command.

```
helpwin fdsp        % Help for all FDSP toolbox functions
helpwin f_dsp       % Help for the FDSP driver module
helpwin g_xxx       % Help for chapter GUI module g_xxx
```

Once the name of a module or function is determined, more detailed help on that item can be obtained by activating that link. Alternatively, if the name of the function is known, a help window for that function can be obtained directly by using the function name as the command-line argument.

```
helpwin f_xxx            % Help for FDSP toolbox function f_xxx
```

The MATLAB *lookfor* command can also be used to locate the names of candidate functions by searching for key words in the first comment line of each function in the toolbox.

<table>
<tr><td>**Example 1.11**</td><td>**FDSP Help**</td></tr>
</table>

To illustrate the type of information provided by the help command, consider the following example which elicits help documentation on one of the toolbox functions appearing in Chapter 4.

```
helpwin f_corr
```

The resulting display, shown in Figure 1.40, follows a standard format for all toolbox functions. It includes a description of what the function does and how the function is used including definitions of all input and output arguments.

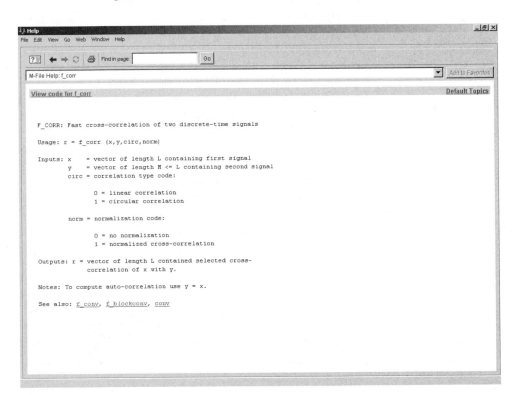

Figure 1.40   Help Window for Function *f_corr*

● ● ● ● ● ● ● ● ● ● ● ● ● ● ● ●●●

1.8 Software Applications

This section focuses on applications of signal sampling and reconstruction. Two GUI modules are presented, one for signal sampling and the other for signal reconstruction. Both allow the user to visualize and explore the concepts covered in this chapter without any need for programming. An application example is then presented and solved using a MATLAB script. Relevant MATLAB functions from the FDSP toolbox are described in Appendix B.

1.8.1 GUI Module: *g_sample*

The graphical user interface module *g_sample* is designed allow the user to interactively investigate the sampling process. GUI module *g_sample* features a display screen with tiled windows as shown in Figure 1.41.

The *Block diagram* window in the upper-left corner shows a block diagram that includes two signal processing blocks, an anti-aliasing filter and an ADC. Below each block is a pair of edit boxes that allow the user to change the parameters of the signal processing blocks. Parameter changes are activated with the Enter key. The anti-aliasing filter is a lowpass Butterworth filter with user-selectable filter order, n, and cutoff frequency, F_c. Selecting $n = 0$ removes the filtering operation completely ($x_b = x_a$). The user-selectable parameters for the analog-to-digital converter are the number of bits of precision, N, and the reference voltage, V_r. Input signals larger in magnitude than V_r get clipped, and when N is small the effects of quantization error become apparent.

The *Type* and *View* windows in the upper-right corner of the screen allow the user to select both the type of input signal x_a, and the viewing mode. The inputs include several common signals plus a user-defined input. For the latter selection, the user must provide the file name (without the .m extension) of a user-created M-file function that returns the value of the input. For example, if the following file is saved under the name *userfun1.m*, then the signal that it generates can be sampled and analyzed with the GUI module *g_sample*.

```
%------------------------------------------------------------------
% Example user function (g_sample)
%------------------------------------------------------------------
function x_a = userfun1(t)
x_a = t .* exp(-t);
%------------------------------------------------------------------
```

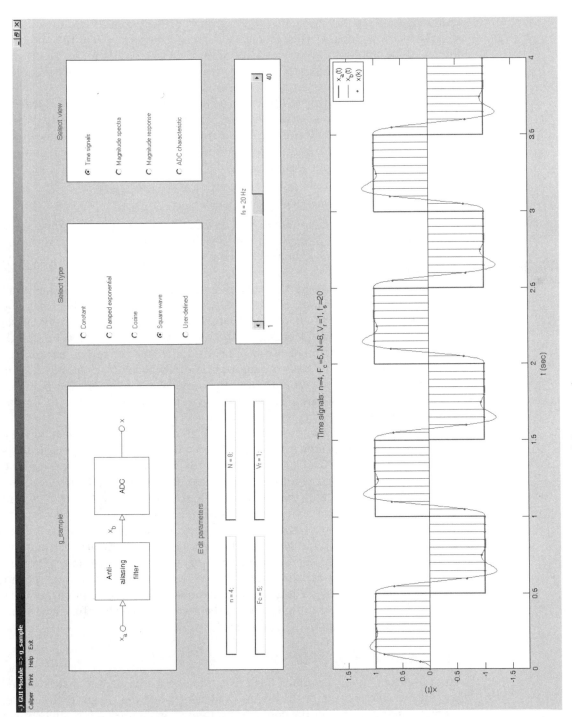

Figure 1.41 Display Screen of Chapter GUI Module *g_sample*

The *View* window options include the color-coded time signals (x_a, x_b, x), and their magnitude spectra. Other viewing options portray the characteristics of the two signal-processing blocks. They include the magnitude response of the anti-aliasing filter, and the input-output characteristic of the ADC. The *Plot* window along the bottom half of the screen shows the selected view. The sampling frequency is controlled with the slider bar.

The *Menu* bar at the top of the screen includes several menu options. The *Caliper* option allows the user to measure any point on the current plot by moving the mouse crosshairs to that point and clicking. The *Print* option prints the contents of the plot window. Finally, the *Help* option provides the user with some helpful suggestions on how to effectively use module *g_sample*.

FDSP Toolbox

1.8.2   GUI Module: *g_reconstruct*

The graphical user interface module *g_reconstruct* is a companion to module *g_sample* that allows the user to interactively investigate the signal reconstruction process. GUI module *g_reconstruct* features a display screen with tiled windows as shown in Figure 1.42.

The *Block diagram* window in the upper-left corner shows a block diagram that includes two signal processing blocks, a DAC, and an anti-imaging filter. Below each block is a pair of edit boxes that allow the user to change the parameters of the signal processing blocks. Parameter changes are activated with the Enter key. For the DAC, the user can select the number of bits of precision, N, and the reference voltage, V_r. When N is small, the effects of quantization error become apparent. The user-selectable parameters for the anti-imaging Butterworth filter are the filter order, n, and the cutoff frequency, F_c. Selecting $n = 0$ removes the filtering operation completely ($y_a = y_b$).

The *Type* and *View* windows in the upper-right corner of the screen allow the user to select both the type of input signal y, and the viewing mode. The inputs include several common signals plus a User-defined input selection where the user supplies the file name of a user-defined M-file function as was described for GUI module *g_sample*. The viewing options include the color-coded time signals (y_b, y_a, y), and their magnitude spectra. Another viewing option shows the magnitude responses of the DAC and the anti-aliasing filter. The *Plot* window along the bottom half of the screen shows the selected view. The sampling frequency is controlled with the slider bar.

The *Menu* bar at the top of the screen includes several menu options. The *Caliper* option allows the user to measure any point on the current plot by moving the mouse crosshairs to that point and clicking. The *Print* option prints the contents of the plot window. Finally, the *Help* option provides the user with some helpful suggestions on how to effectively use module *g_reconstruct*.

1.8.3   Anti-Aliasing Filter Design

An anti-aliasing filter, or guard filter, is an analog lowpass filter that is placed in front of the ADC to reduce the effects of aliasing when the signal to be sampled is not bandlimited. Consider the configuration shown in Figure 1.43 which features an nth-order lowpass Butterworth filter followed by an ADC.

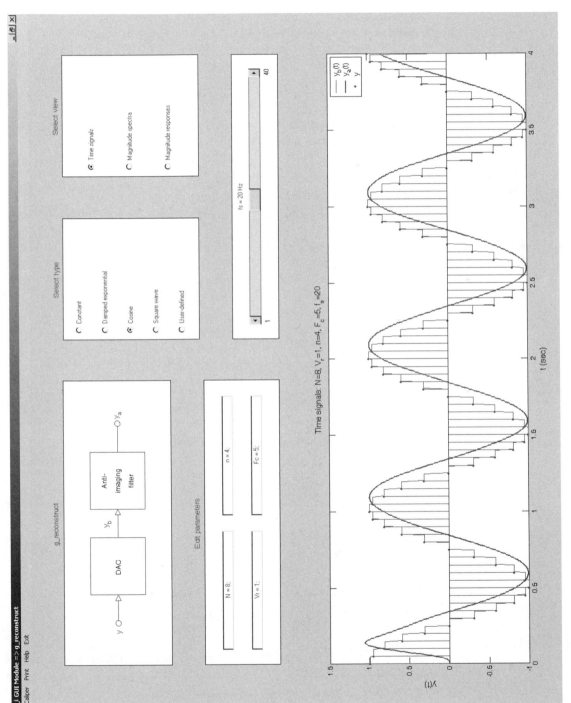

Figure 1.42 Display Screen of Chapter GUI Module *g_reconstruct*

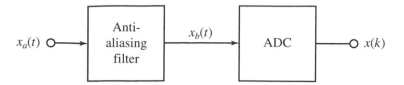

Figure 1.43 Pre-processing with an Anti-aliasing Filter

Suppose the Butterworth filter has a cutoff frequency of F_c and the ADC is a bipolar N-bit analog-to-digital converter with a reference voltage of V_r. Since the Butterworth filter is not an ideal lowpass filter, we oversample by a factor of $\alpha > 1$. That is,

$$f_s = 2\alpha F_c, \quad \alpha > 1 \tag{1.61}$$

The design task is then as follows. Find the minimum filter order, n, that will ensure that the magnitude of the aliasing error is no larger than the quantization error of the ADC.

Given the monotonically decreasing nature of the magnitude response of the Butterworth filter (see Figure 1.27), the maximum aliasing error occurs at the folding frequency $f_d = f_s/2$. The largest signal that the ADC can process has magnitude V_r. Thus from Eq. 1.50 and Eq. 1.61 the maximum aliasing error is

$$
\begin{aligned}
E_a &= V_r |H_a(f_s/2)| \\
&= V_r |H_a(\alpha F_c)| \\
&= \frac{V_r}{\sqrt{1 + \alpha^{2n}}}
\end{aligned}
\tag{1.62}
$$

Next, if the bipolar ADC input-output characteristic is offset by $q/2$ as in Figure 1.34, then the size of the quantization error is $|e_q| \le q/2$ where q is the quantization level. Thus, from Eq. 1.59, the maximum quantization error is

$$
\begin{aligned}
E_q &= \frac{q}{2} \\
&= \frac{V_r}{2^N}
\end{aligned}
\tag{1.63}
$$

Setting $E_a^2 = E_q^2$, we observe that the reference voltage, V_r drops out. Taking reciprocals then yields:

$$1 + \alpha^{2n} = 2^{2N} \tag{1.64}$$

Finally, solving for n, the required order of the anti-aliasing filter must satisfy

$$n \ge \frac{\ln(2^{2N} - 1)}{2 \ln(\alpha)} \tag{1.65}$$

Of course, the filter order n must be an integer. Consequently, the required order of n is as follows where *ceil* rounds up to the nearest integer.

Anti-aliasing filter order

$$n = \text{ceil} \left[\frac{\ln(4^N - 1)}{2 \ln(\alpha)} \right] \tag{1.66}$$

The following MATLAB script, labeled *exam1_12* on the distribution CD, computes *n* using Eq. 1.66 for oversampling rates in the range $2 \leq \alpha \leq 4$ and ADC precisions in the range $10 \leq N \leq 16$ bits.

```
function exam1_12

% Example 1.12: Anti-aliasing filter design

clear
clc
fprintf ('Example 1.12: Anti-aliasing filter design\n\n')
F_c = 1;

% Compute minimum filter order

alpha = [2 : 4]
N = [8 : 12]
r = length(N);
n = zeros(r,3);
for i = 1 : r
   for j = 1 : 3
      n(i,j) = ceil(log(4^N(i)-1)/(2*log(alpha(j))));
   end
end
n

% Display results

figure
plot (N,n,'o-','LineWidth',1.5)
f_labels ('Anti-Aliasing Filter Order, \alpha = f_s/(2F_c)','N (bits)','n (filter order)')
text (9.5,5.5,'\alpha = 4')
text (9.5,7.3,'\alpha = 3')
text (9.5,10.3,'\alpha = 2')
f_wait
```

When *exam1_12* is executed, it produces the plot shown in Figure 1.44 which displays the required anti-aliasing filter order versus the ADC precision for three different values of oversampling. As expected, as the oversampling factor α increases, the required order decreases.

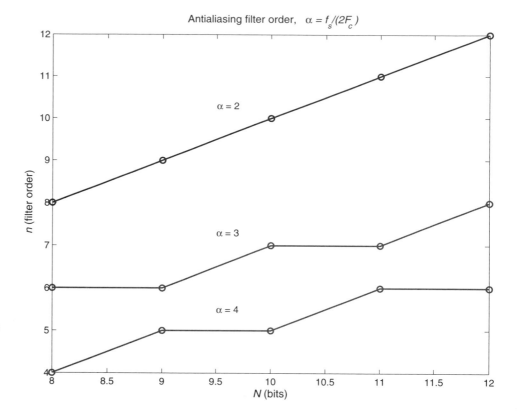

Figure 1.44 Minimum Order of Anti-aliasing Filter Needed to Ensure that the Magnitude of Aliasing Error Matches the ADC Quantization Error for Different Levels of Oversampling α

• • • • • • • • • • • •

1.9 Chapter Summary

This chapter focused on continuous-time signal analysis techniques, and sampling and reconstruction of continuous-time signals. A signal is a physical variable whose value changes with time or space. Although our focus is on time signals, the DSP techniques introduced also can be applied to two-dimensional spatial signals such as images.

Signals and Systems

A *continuous-time* signal, $x_a(t)$, is a signal whose independent variable, t, takes on values over a continuum. A *discrete-time* signal, $x(k)$, is a signal whose independent variable is available only at discrete instants of time $t = kT$ where T is the sampling interval. If $x(k)$ is the sampled version of $x_a(t)$, then

$$x(k) = x_a(kT), \quad k = 0, 1, 2, \ldots \tag{1.67}$$

Just as the independent variable can be continuous or discrete, so can the dependent variable or amplitude of the signal. When a discrete-time signal is represented with finite precision, the signal is said to be quantized because it can only take on discrete values. A quantized discrete-time signal is called a *digital* signal. That is, a digital signal is discrete both in time and in amplitude. The spacing between adjacent discrete values is called the quantization level q. For an N-bit signal with values ranging over the interval $[x_{min}, x_{max}]$, the quantization level is

$$q = \frac{x_{max} - x_{min}}{2^N} \tag{1.68}$$

Just as light can be decomposed into a spectrum of colors, a signal $x_a(t)$ contains energy that is distributed over a range of frequencies. These spectral components are obtained by applying the Fourier transform to produce a complex-valued frequency domain representation of the signal, $X_a(f)$, called the signal spectrum. The magnitude of $X_a(f)$ is called the *magnitude spectrum*, and the phase angle of $X_a(f)$ is called the *phase spectrum*. The *frequency response* of a linear continuous-time system with input $x_a(t)$ and output $y_a(t)$ is the spectrum of the output signal divided by the spectrum of the input signal.

$$H_a(f) = \frac{Y_a(f)}{X_a(f)} \tag{1.69}$$

The frequency response is complex-valued, with the magnitude, $A_a(f) = |H_a(f)|$, called the *magnitude response* of the system, and the phase angle, $\phi_a(f) = \angle H_a(f)$, called the *phase response* of the system. A linear system that is designed to reshape the spectrum of the input signal in some desired way is called a frequency-selective *filter*.

Signal Sampling

The process of creating a discrete-time signal by sampling a continuous-time signal $x_a(t)$ can be modeled mathematically as amplitude modulation of a uniform periodic impulse train, $\delta_T(t)$.

$$x_a^*(t) = x_a(t)\delta_T(t) \tag{1.70}$$

The effect of sampling is to scale the spectrum of $x_a(t)$ by $1/T$ where T is the sampling interval, and to replicate the spectrum of $x_a(t)$ at integer multiples of the sampling frequency $f_s = 1/T$. A signal $x_a(t)$ is *bandlimited* to B Hz if the spectrum is zero for $|f| > B$. A bandlimited signal of bandwidth B can be reconstructed from it samples by passing it through an ideal lowpass filter with gain T and cutoff frequency $F_c = f_s/2$ as long as the sampling frequency satisfies:

$$f_s > 2B \tag{1.71}$$

Consequently, a signal can be reconstructed from its samples if the signal is bandlimited and the sampling frequency is greater than twice the bandwidth. This fundamental result is known as the Shannon *sampling theorem*. When the signal $x_a(t)$ is not bandlimited, or the sampling rate does not exceed twice the bandwidth, the replicated spectral components of $x_a^*(t)$ overlap and it is not possible to recover $x_a(t)$ from its samples. The phenomenon of spectral overlap is known as *aliasing*.

Most signals of interest are not bandlimited. To minimize the effects of aliasing the input signal is preprocessed with a lowpass analog filter called an *anti-aliasing* filter before it is sampled with an analog-to-digital converter (ADC). A practical lowpass filter, such as a Butterworth filter, does not completely remove spectral components above the cutoff frequency F_c. However, the residual aliasing can be further reduced by *oversampling* at a rate $f_s = 2\alpha F_c$ where $\alpha > 1$ is the oversampling factor.

Signal Reconstruction

Once the digital input signal $x(k)$ from the ADC is processed with a DSP algorithm, the resulting output $y(k)$ is typically converted back to an analog signal using a digital-to-analog converter (DAC). Whereas an ADC can be modeled mathematically with an impulse sampler, a DAC is modeled as a zero-order hold filter with transfer function

$$H_0(s) = \frac{1 - \exp(-Ts)}{s} \tag{1.72}$$

The transfer function of the DAC is the Laplace transform of the output divided by the Laplace transform of the input assuming zero initial conditions. The piecewise-constant output of the DAC contains high-frequency spectral components called images centered at integer multiples of the sampling frequency. These can be reduced by postprocessing with a second lowpass analog filter called an *anti-imaging* filter.

A circuit realization of a DAC was presented that featured an R-$2R$ resistor network, an operational amplifier, and a bank of digitally-controlled analog switches. Circuit realizations of ADCs that were considered included the successive approximation ADC, the flash ADC, and hybrid designs that combine the two. The successive approximation ADC is a widely used, high precision, medium speed converter that requires N clock pulses to perform an N-bit conversion. By contrast, the flash ADC is a very fast, hardware intensive, medium precision converter that requires $2^N - 1$ comparator circuits, but performs a conversion in a single clock pulse.

The final topic covered in this chapter was the FDSP toolbox, a collection of graphical user interface (GUI) modules and MATLAB functions designed for use with the student version of MATLAB. The toolbox modules and functions are discussed in detail in Appendix B. The GUI modules are provided so the user can investigate the DSP topics covered in each chapter without any need for programming. MATLAB scripts can make use of the toolbox functions that include general support functions provided to facilitate user program development plus chapter functions that implement the DSP algorithms developed in each chapter. All of the FDSP toolbox software that is included with the distribution CD can be conveniently accessed through the driver module, *f_dsp*. This includes running the GUI modules, executing all of the MATLAB examples that appear in the text, recreating and displaying the text figures, and solving selected end-of-chapter problems. The driver module also provides toolbox help and includes an option for downloading the latest version of the FDSP toolbox from the internet.

• • • • • • • • • • • • • • •

1.10 Problems

The problems are divided into Analysis problems that can be solved by hand or with a calculator, GUI Simulation problems that are solved using the GUI modules *g_sample* and *g_reconstruct*, and Computation problems. The Computation problems require the student to write a MATLAB script using the FDSP toolbox functions summarized in Appendix B. Solutions to selected problems are available on the distribution CD. Students are encouraged to use these problems, which are identified with a ☑, as a check on their understanding of the material.

1.10.1 Analysis

Section 1.2

P1.1. Suppose the input to an amplifier is $x_a(t) = \sin(2\pi f_0 t)$ and the steady-state output is

$$y_a(t) = 100\sin(2\pi f_0 t + \phi_1) - 2\sin(4\pi f_0 t + \phi_1) + \cos(6\pi f_0 t + \phi_3)$$

(a) Is the amplifier a linear system or is it a nonlinear system?
(b) What is the gain of the amplifier?
(c) Find the average power of the output signal.
(d) What is the total harmonic distortion of the amplifier?

☑ **P1.2.** Consider the following *signum* function which returns the sign of its argument.

$$\operatorname{sgn}(t) \triangleq \begin{cases} 1, & t > 0 \\ 0, & t = 0 \\ -1, & t < 0 \end{cases}$$

(a) Find the magnitude spectrum.
(b) Find the phase spectrum.

P1.3. Parseval's relation states that a signal and its spectrum are related in the following way.

$$\int_{-\infty}^{\infty} |x_a(t)|^2 dt = \int_{-\infty}^{\infty} |X_a(f)|^2 df$$

Use Parseval's relation to compute the following integral.

$$J = \int_{-\infty}^{\infty} \operatorname{sinc}^2(2\pi Bt)dt$$

P1.4. Consider the causal exponential function.

$$x_a(t) = \exp(-ct)u_a(t)$$

(a) Find the magnitude spectrum.
(b) Find the phase spectrum.
(c) Sketch the magnitude and phase spectra when $c = 1$.

P1.5. If an analog signal $x_a(t)$ is square integrable, then the *energy* that the signal contains within the frequency band $[f_0, f_1]$ can be computed as follows.

$$E(f_0, f_1) = 2 \int_{f_0}^{f_1} |X_a(f)|^2 df$$

Consider the following double exponential signal with $c > 0$.

$$x_a(t) = \exp(-c|t|)$$

(a) Find the total energy, $E(0, \infty)$.

(b) Find the percentage of the total energy that lies in the frequency range $[0, 2]$ Hz.

P1.6. Let $x_a(t)$ be a periodic signal with period τ. The *average power* of $x_a(t)$ can be defined as follows.

$$P_x = \frac{1}{\tau} \int_0^\tau |x_a(t)|^2 dt$$

Find the average power of the following periodic continuous-time signals.

(a) $x_a(t) = \cos(2\pi f_0 t)$

(b) $x_a(t) = c$

(c) A periodic train of pulses of amplitude a, duration T, and period τ.

P1.7. Consider the following discrete-time signal where the samples are represented using N bits.

$$x(k) = \exp(-ckT)u(k)$$

(a) How many bits are needed to ensure that the quantization level is less than 0.001?

(b) Suppose $N = 8$ bits. What is the average power of the quantization noise?

P1.8. Show that the spectrum of a causal signal $x_a(t)$ can be obtained from the Laplace transform $X_a(s)$ be replacing s by $j2\pi f$. Is this also true for noncausal signals?

Section 1.3

P1.9. Consider the following periodic signal.

$$x_a(t) = 1 + \cos(10\pi t)$$

(a) Compute and sketch the magnitude spectrum of $x_a(t)$.

(b) Suppose $x_a(t)$ is sampled with a sampling frequency of $f_s = 8$ Hz. Sketch the magnitude spectrum of $x_a^*(t)$.

(c) Does aliasing occur when $x_a(t)$ is sampled at the rate $f_s = 8$ Hz? What is the folding frequency in this case?

(d) Find a range of values for the sampling interval T that ensures that aliasing will not occur.

(e) Assuming $f_s = 8$ Hz, find an alternative lower-frequency signal, $x_b(t)$, that has the same set of samples as $x_a(t)$.

☑ **P1.10.** Consider the following bandlimited signal.

$$x_a(t) = \sin(4\pi t)[1 + \cos^2(2\pi t)]$$

(a) Using the trigonometric identities in Appendix D, find the maximum frequency present in $x_a(t)$.

(b) For what range of values for the sampling interval T can this signal be reconstructed from its samples?

Section 1.4

P1.11. It is not uncommon for students to casually restate the sampling theorem in the following way: "A signal must be sampled at twice the highest frequency present to avoid aliasing." Interestingly enough, this informal formulation is not quite correct. To verify this, consider the following simple signal.

$$x_a(t) = \sin(2\pi t)$$

(a) Find and sketch the magnitude spectrum of $x_a(t)$, and verify that the highest frequency present is $f_0 = 1$ Hz.

(b) Suppose $x_a(t)$ is sampled at the rate $f_s = 2$ Hz. Sketch the magnitude spectrum of $x_a^*(t)$. Do the replicated spectra overlap?

(c) Compute the samples $x(k) = x_a(kT)$ using the sampling rate $f_s = 2$ Hz. Is it possible to reconstruct $x_a(t)$ from $x(k)$ using the reconstruction formula in Theorem 1.2 in this instance?

(d) Restate the Sampling theorem in terms of the highest frequency present, but this time correctly.

P1.12. Why is it not possible to physically construct an ideal lowpass filter? Use the impulse response, $h_a(t)$, to explain your answer.

Section 1.5

P1.13. Consider the problem of using an anti-aliasing filter as shown in Figure P1.13. Suppose the anti-aliasing filter is a lowpass Butterworth filter of order $n = 4$ with cutoff frequency $F_c = 2$ kHz.

(a) Find a lower bound f_L on the sampling frequency that ensures that the aliasing error is reduced by a factor of at least 0.005.

(b) The lower bound f_L represents oversampling by what factor?

P1.13 Preprocessing with an Anti-aliasing Filter

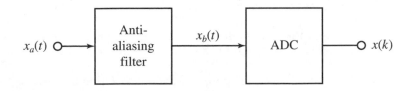

P1.14. Show that the transfer function of a linear continuous-time system is the Laplace transform of the impulse response.

Section 1.6

P1.15. A bipolar DAC can be constructed from a unipolar DAC by inserting an operational amplifier at the output as shown in Figure P1.15. Note that the unipolar N-bit DAC uses a reference voltage of $2V_R$, rather than $-V_r$ as in Figure 1.33. This means that the unipolar DAC output is $-2y_a$ where y_a is given in Eq. 1.56. Analysis of the operational amplifier section of the circuit reveals that the bipolar DAC output is then

$$z_a = 2y_a - V_r$$

(a) Find the range of values for z_a.

(b) Suppose the binary input is $b = b_{N-1}b_{N-2}\cdots b_0$. For what value of b is $z_a = 0$?

(c) What is the quantization level of this bipolar DAC?

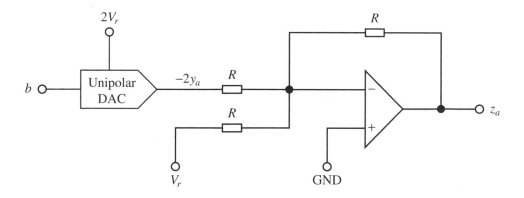

P1.15   A Bipolar
N-bit DAC

P1.16. Suppose a bipolar ADC is used with a precision of $N = 12$ bits, and a reference voltage of $V_r = 10$ volts.

(a) What is the quantization level q?

(b) What is the maximum value of the magnitude of the quantization noise assuming the ADC input-output characteristics are offset by $q/2$ as in Figure 1.34.

(c) What is the average power of the quantization noise?

P1.17. Suppose an 8-bit bipolar successive approximation ADC has reference voltage $V_r = 10$ volts.

(a) If the analog input is $x_a = -3.941$ volts, find the successive approximations by filling in the entries in Table P1.17.

(b) If the clock rate is $f_{clock} = 200$ kHz, what is the sampling rate of this ADC?

(c) Find the quantization level of this ADC.

(d) Find the average power of the quantization noise.

P1.17: ▶
Successive
Approximations

k	b_{n-k}	u_k	y_k
0			
1			
2			
3			
4			
5			
6			
7			

P1.18. An alternative to the $R - 2R$ ladder DAC is the weighted-resistor DAC shown in Figure P1.18 for the case $N = 4$. Here the switch controlled by bit b_k is open when $b_k = 0$ and closed when $b_k = 1$. Recall that the decimal equivalent of the binary input b is as follows.

$$x = \sum_{k=0}^{N-1} b_k 2^k$$

(a) Show that the current through the kth branch of an N-bit weighted-resistor DAC is

$$I_k = \frac{-V_r b_k}{2^{N-k} R}, \quad 0 \le k < N$$

(b) Show that the DAC output voltage is

$$y_a = \left(\frac{V_r}{2^N} \right) x$$

(c) Find the range of output values for this DAC.

(d) Is this DAC unipolar, or is it bipolar?

(e) Find the quantization level of this DAC.

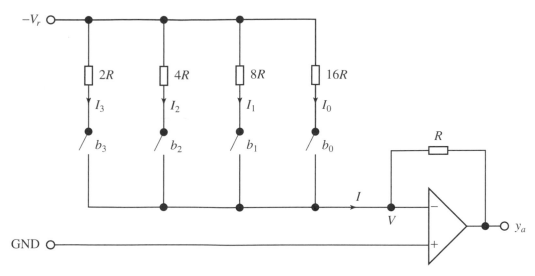

P1.18 A 4-bit Weighted-resistor DAC

1.10.2 GUI Simulation

Section 1.6

P1.19. Use GUI module *g_sample* to plot the time signals, the magnitude spectra, and the magnitude response using the following anti-aliasing filters. What is the oversampling factor, α, in each case?

(a) $n = 2$, $F_c = 1$

(b) $n = 6$, $F_c = 2$

P1.20. Use GUI module *g_sample* to plot the time signals and the ADC characteristics using the following ADCs. For what cases does the ADC output saturate? Write down the quantization level on each time plot.

(a) $N = 4$, $V_r = 1.0$

(b) $N = 8$, $V_r = 0.5$

(c) $N = 8$, $V_r = 1.0$

P1.21. Use GUI module *g_sample* to plot the time signals and magnitude spectra of the square wave using $f_s = 8$ Hz. On the magnitude spectra plot, use the Caliper option to display the amplitude and frequency of the fundamental harmonic. Are there even harmonics present in the square wave?

P1.22. Use GUI module *g_sample* to plot the magnitude spectra for the User-defined signal in the file, *u_sample1*. Do the following two cases. For which ones is there noticeable aliasing?

(a) $f_s = 2$ Hz

(b) $f_s = 10$ Hz

P1.23. Consider the following exponentially damped sine wave with $c = 1$ and $f_0 = 1$.

$$x_a(t) = \exp(-ct)\sin(2\pi f_0 t)u(t)$$

(a) Write an M-file function called *user1* that returns the value $x_a(t)$.

(b) Use the User-defined option in GUI module *g_sample* to sample this signal at $f_s = 12$ Hz. Plot the time signals.

(c) Adjust the sampling rate to $f_s = 6$ Hz and plot the magnitude spectra.

(d) Repeat part (c), but with the anti-aliasing filter removed (set $n = 0$).

P1.24. Use GUI module *g_reconstruct* to plot the magnitude responses using the following anti-imaging filters. What is the oversampling factor in each case?

(a) $n = 2$, $F_c = 1$

(b) $n = 6$, $F_c = 2$

P1.25. Use GUI module *g_reconstruct* to plot the time signals using the following DACs. What is the quantization level in each case?

(a) $N = 4$, $V_r = 0.5$

(b) $N = 12$, $V_r = 2.0$

☑ **P1.26.** Use GUI module *g_reconstruct* to plot the time signals and the magnitude spectra for the User-defined signal in the file, *u_reconstruct1* using $f_s = 12$ Hz and $V_r = 4$. On the time signals plot, use the Caliper option to display the amplitude and time of the peak value of the output.

P1.27. Consider the exponentially damped sine wave in Problem P1.23.

(a) Write an M-file function called *user1* that returns the value $x_a(t)$.

(b) Use the User-defined option in GUI module *g_reconstruct* to sample this signal at $f_s = 8$ Hz. Plot the time signals.

(c) Adjust the sampling rate to $f_s = 4$ Hz and plot the magnitude spectra.

(d) Repeat part (c), but with the anti-imaging filter removed (set $n = 0$).

P1.28. Use the GUI module *g_reconstruct* to plot the magnitude responses of a 12-bit DAC with reference voltage $V_r = 10$ volts, and a 6th order Butterworth anti-imaging filter with cutoff frequency $F_c = 2$ Hz. Use oversampling by a factor of two.

1.10.3 Computation

Section 1.4

P1.29. Write a MATLAB function called *u_sinc* that returns the value of the sinc function

$$\text{sinc}(x) = \frac{\sin(x)}{x}$$

Note that, by L'Hospital's rule, sinc(0) $= 1$. Make sure your function works properly when $x = 0$. Plot sinc($2\pi t$) for $-4 \le t \le 4$.

P1.30. The purpose of this problem is to numerically verify the signal reconstruction formula in Theorem 1.2. Consider the following bandlimited periodic signal which can be thought of as a truncated Fourier series.

$$x_a(t) = 1 - 2\sin(\pi t) + \cos(2\pi t) + 3\cos(3\pi t)$$

Write a MATLAB script which uses the function *u_sinc* from Problem P1.29 to approximately reconstruct $x_a(t)$ as follows.

$$x_p(t) = \sum_{k=-p}^{p} x_a(kT)\text{sinc}[\pi f_s(t - kT)]$$

Use a sampling rate of $f_s = 6$ Hz. Plot $x_a(t)$ and $x_p(t)$ on the same graph using 101 points equally spaced over the interval $[-2, 2]$. Prompt for the number p and do the following three cases.

(a) $p = 5$
(b) $p = 10$
(c) $p = 20$

Section 1.5

P1.31. The Butterworth filter is optimal in the sense that, for a given filter order, the magnitude response is as flat as possible in the passband. If ripples are allowed in the passband, then an analog filter with a sharper cutoff can be achieved. Consider the following Chebyshev-I filter that will be discussed in detail in Chapter 8.

$$H_a(s) = \frac{1263.7}{s^5 + 6.1s^4 + 67.8s^3 + 251.5s^2 + 934.3s + 1263.7}$$

Write a MATLAB script that uses the FDSP toolbox function *f_freqs* to compute the magnitude response of this filter. Plot it over the range $[0, 3]$ Hz. This filter is optimal in the sense that the passband ripples are all of the same size.

P1.32. Consider the following Chebyshev-II lowpass filter that will be discussed in detail in Chapter 8.

$$H_a(s) = \frac{3s^4 + 499s^2 + 15747}{s^5 + 20s^4 + 203s^3 + 1341s^2 + 5150s + 15747}$$

Write a MATLAB script that uses the FDSP toolbox function *f_freqs* to compute the magnitude response of this filter. Plot it over the range $[0, 3]$ Hz. This filter is optimal in the sense that the stopband ripples are all of the same size.

P1.33. Consider the following elliptic lowpass filter that will be discussed in detail in Chapter 8.

$$H_a(s) = \frac{2.0484s^2 + 171.6597}{s^3 + 6.2717s^2 + 50.0487s + 171.6597}$$

Write a MATLAB script that uses the FDSP toolbox function *f_freqs* to compute the magnitude response of this filter. Plot it over the range $[0, 3]$ Hz. This filter is optimal in the sense that the passband ripples and the stopband ripples are all of the same size.

Discrete-time System Analysis

Learning Objectives

- Know how to compute the Z-transform using the geometric series and the Z-transform properties (Sections 2.2–2.3)
- Be able to invert the Z-transform using synthetic division, partial fractions, and the residue method (Section 2.4)
- Be able to go back and forth between difference equations, transfer functions, impulse responses, and signal flow graphs (Sections 2.5–2.7)
- Understand the relationship between poles, zeros, and natural and forced modes of a linear system (Section 2.5)
- Know how to find the impulse response and use it to compute the zero-state response to any input (Section 2.7)
- Understand the differences between FIR systems and IIR systems, and know which system is always stable (Section 2.7)
- Appreciate the significance of the unit circle in the Z-plane in terms of stability, and know how to evaluate stability using the Jury test (Section 2.8)
- Be able to compute the frequency response from the transfer function (Section 2.9)
- Know how to compute the steady-state output of a stable discrete-time system corresponding to a periodic input (Section 2.9)
- Know how to use the GUI module *g_system* to investigate the input-output behavior of a discrete-time system (Section 2.10)

● ● ● ● ● ● ● ● ● ● ● ● ● ● ● ⋯ ⋯

2.1 Motivation

A discrete-time system is a system that processes a discrete-time input signal, $x(k)$, to produce a discrete-time output signal, $y(k)$. Recall from Chapter 1 that if the signals $x(k)$ and $y(k)$ are discrete in amplitude, as well as in time, then they are digital signals and the associated system is a digital signal processor or digital filter. In this chapter, we focus our attention on analyzing the input-output behavior of linear time-invariant discrete-time systems. This material lays a mathematical foundation for subsequent chapters where we design digital filters and develop digital signal processing (DSP) algorithms.

An essential tool for the analysis of discrete-time systems is the Z-transform, a transformation which maps or transforms a causal discrete-time signal $x(k)$ into a function $X(z)$ of a complex variable z.

Z-transform

$$X(z) = Z\{x(k)\}$$

With the help of the Z-transform, a difference-equation description of a discrete-time system can be converted to a simple algebraic equation which is readily solved for the Z-transform of the output $Y(z)$. Applying the inverse Z-transform to $Y(z)$ then produces the system output $y(k)$. Important qualitative features of discrete-time systems can also be obtained with the help of the Z-transform. For example, a discrete-time system is said to be *stable* if, and only if, every bounded input signal is guaranteed to produce a bounded output signal. Stability is an essential characteristic of practical digital filters, and the easiest way to establish stability is with the Z-transform.

We begin this chapter by examining a number of practical problems that can be modeled using discrete-time systems and solved using the Z-transform. Next, the Z-transform is defined and a table of Z-transform pairs is developed. The size of this table is then expanded by introducing a number of important Z-transform properties. The problem of inverting the Z-transform is then addressed using the synthetic division, partial fraction, and residue methods. Next, four alternative representations of a discrete-time system are introduced. The first is the difference equation. The second is the transfer function representation which includes a discussion of poles, zeros, and modes. The third is a compact, graphical representation called the signal flow graph. Finally, the fourth is a time-domain representation based on the impulse response and convolution. Using the impulse response, systems are classified into finite impulse response (FIR) and infinite impulse response (IIR) systems. Next, criteria to determine when a system is bounded-input bounded-out (BIBO) stable are presented and the Jury stability test is introduced. The frequency response is then introduced, and interpretations of it are provided in terms of the transfer function and the steady-state response to periodic inputs. Finally, a GUI module called *g_system* is presented that allows the user the investigate the input-output behavior of discrete-time systems without any need for programming. The chapter concludes with some application examples, and a summary of discrete-time system analysis techniques. The selected examples include an analysis of a home mortgage loan, satellite attitude control, and the Fibonacci sequence.

2.1.1 Home Mortgage

As a simple illustration of a discrete-time system that affects many families, consider the problem of purchasing a home with a mortgage loan. Suppose a fixed-rate mortgage is taken out at an annual interest rate of r percent compounded monthly. Let $y(k)$ denote the balance owed to the lending agency at the end of month k, and let $x(k)$ denote the monthly mortgage payment. The balance owed at the end of month k is the balance owed at the end of month $k - 1$, plus the monthly interest on that balance, and minus the monthly payment.

$$y(k) = y(k - 1) + \left(\frac{r}{120}\right) y(k - 1) - x(k) \tag{2.1}$$

Here the initial condition, $y(-1)$ is the size of the mortgage. Given the *difference-equation* representation in Eq. 2.1, there are a number of practical questions that a potential home buyer might want to ask. For example, if the duration of the mortgage is d years, what is the required monthly mortgage payment? Another question is how many months does it take before the majority of the monthly payment goes toward reducing principal rather than paying interest? These questions can be easily answered once we develop the tools for solving this system. As we shall see, the mortgage loan system is an example of an unstable discrete-time system.

2.1.2 Satellite Attitude Control

As a second example of a discrete-time system, consider the problem of finding a discrete-equivalent of a sampled-data system. Recall from Chapter 1 that a *sampled-data* system is system that contains both continuous-time signals and discrete-time signals. Consequently, any system that has a digital-to-analog converter (DAC) or an analog-to-digital converter (ADC) is a sampled-data system. For example, consider the feedback control system shown in Figure 2.1 which contains both.

The task of this feedback system is to control the angular position of a satellite that is spinning about one axis in space, as shown in Figure 2.2. Here $r(k)$ is the desired angular position of the satellite at discrete time k, and $y(k)$ is the actual angular position. Since there is no friction, the motion of the satellite can be modeled using Newton's second law as follows:

$$J \frac{d^2 y_a(t)}{dt^2} = \tau_a(t) \tag{2.2}$$

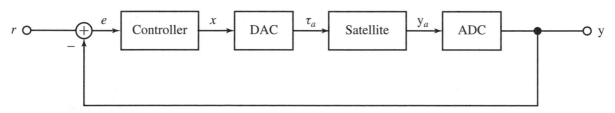

Figure 2.1 Single-axis Satellite Attitude Control System

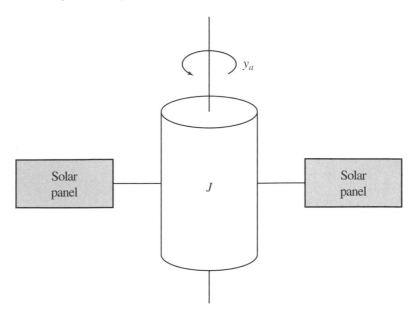

Figure 2.2 A Spinning Satellite

From Figure 2.1, the input $\tau_a(t)$ is the torque generated by the thrusters, and J is the moment of inertia about the axis of rotation. The digital controller acts on the *error* signal input, $e(k) = r(k) - y(k)$ to produce a control signal $x(k)$. For example, consider the following difference equation used to implement a simple controller.

$$x(k) = c[e(k) - e(k-1)] \tag{2.3}$$

Here the *controller gain*, c, is an engineering design parameter whose value can be chosen to satisfy some performance specification. The controller in Eq. 2.3 is an example of a finite impulse response or FIR discrete-time system.

Next, suppose the DAC is modeled as a zero-order hold as in Chapter 1. Then the discrete equivalent of the DAC and the satellite can be modeled with the following hold-equivalent discrete-time system (Franklin *et al.*, 1990) where $y(k) = y_a(kT)$.

$$y(k) = 2y(k-1) - y(k-2) + \left(\frac{T^2}{2J}\right)[x(k-1) + x(k-2)] \tag{2.4}$$

Substituting Eq. 2.3 into Eq. 2.4, and recalling that $e(k) = r(k) - y(k)$, the entire feedback system then can be modeled by the following discrete-equivalent system whose input is the desired angle $r(k)$ and output is the actual angle $y(k)$.

$$y(k) = (d-1)y(k-1) + dy(k-2) + d[r(k-1) + r(k-2)] \tag{2.5a}$$

$$d = \frac{cT^2}{2J} \tag{2.5b}$$

Thus, the sampled-data control system consists of three separate discrete-time systems, the controller in Eq. 2.3, the hold equivalent of the DAC and the satellite in

Eq. 2.4, and the overall closed-loop discrete-equivalent system in Eq. 2.5. Later in this chapter, we will see that this control system is stable, and therefore operates successfully for the following values of the controller gain.

$$0 < c < \frac{2J}{T^2} \tag{2.6}$$

2.1.3 Running Average Filter

As a third illustration of a discrete-time system, suppose you are given the task of analyzing customer evaluation data for a large automobile dealership. One of the key statistics is the ratio of the overall customer evaluation score for a given model to the average of the overall scores for that model over the past several years. If $x(k)$ denotes the overall customer evaluation score for the kth year, then the running average of the last $m + 1$ evaluation scores is

$$y(k) = \frac{1}{m+1} \sum_{i=0}^{m} x(k - i) \tag{2.7}$$

The discrete-time system in Eq. 2.7 is called a *running average* filter of order m. One measure of the computational complexity of a signal-processing algorithm is the required number of floating-point arithmetic operations or FLOPs. Note that the running average filter in Eq. 2.7 requires $m + 1$ additions and one division. We can develop a more efficient representation of the running average filter for large values of m by rewriting the right-hand side of Eq. 2.7 in terms of $y(k - 1)$ plus a correction term. Note that $y(k - 1)$ can be written as follows where we use a change of variable $j = i + 1$.

$$y(k - 1) = \frac{1}{m+1} \sum_{i=0}^{m} x(k - 1 - i)$$

$$= \frac{1}{m+1} \sum_{j=1}^{m+1} x(k - j) \quad \left. \right\} j = i + 1$$

$$= \frac{1}{m+1} \left[\sum_{j=0}^{m} x(k - j) - x(k) + x(k - m - 1) \right]$$

$$= \frac{1}{m+1} \sum_{j=0}^{m} x(k - j) - \frac{x(k) - x(k - m - 1)}{m+1}$$

$$= y(k) - \frac{x(k) - x(k - m - 1)}{m+1} \tag{2.8}$$

Solving Eq. 2.8 for $y(k)$ then yields the following more efficient implementation of the running average filter.

$$y(k) = y(k-1) + \frac{x(k) - x(k-m-1)}{m+1} \tag{2.9}$$

In this case, the number of floating-point additions has been reduced from m to three. Thus, for $m > 3$, the implementation in Eq. 2.9 is more efficient than the implementation in Eq. 2.7.

● ● ● ● ● ● ● ● ● ● ● ● ● ● ● ●

2.2 Z-transform Pairs

The Z-transform is a powerful tool that is useful for analyzing and solving linear discrete-time systems.

DEFINITION

2.1: Z-transform

The *Z-transform* of a causal discrete-time signal $x(k)$ is a function $X(z)$ of a complex variable z defined as

$$X(z) \overset{\Delta}{=} \sum_{k=0}^{\infty} x(k)z^{-k}$$

From Definition 2.1, we see that the Z-transform is a power series in the variable z^{-1}. The operation of performing a Z-transform can also be represented with the Z-transform *operator*, Z, as follows

$$Z\{x(k)\} = X(z) \tag{2.10}$$

Note that, by convention, the corresponding upper-case letter is used to denote the Z-transform of a signal. For most practical signals, the Z-transform can be expressed as a ratio of two polynomials.

$$X(z) = \frac{b_0(z - z_1)(z - z_2) \cdots (z - z_m)}{(z - p_1)(z - p_2) \cdots (z - p_n)} \tag{2.11}$$

Here the roots of the numerator polynomial are called the *zeros* of $X(z)$, and the roots of the denominator polynomial are called the *poles* of $X(z)$.

The Z-transform in Definition 2.1 converges to a finite value, $X(z)$, for certain values of z in the complex plane. The set of values over which the power series in z^{-1} converges is called the *region of convergence* or ROC. For causal signals, it can be

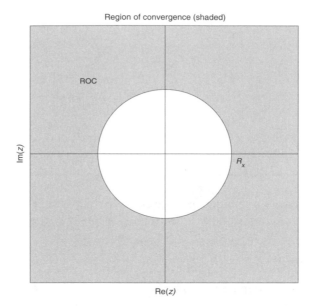

Region of convergence (shaded)

Figure 2.3 Typical Region of Convergence (ROC) of the Z-transform for a Causal Signal

shown that the ROC is the part of the complex plane that lies outside the outer-most pole of $X(z)$. That is,

$$\text{ROC} = \{z \in C \mid |z| > R_x\} \tag{2.12a}$$

$$R_x \stackrel{\Delta}{=} \max_{i=1}^{n}\{|p_i|\} \tag{2.12b}$$

A sketch of a typical region of convergence is shown in Figure 2.3. For causal signals, the ROC is the entire complex plane except for the smallest disk, centered at the origin, that includes all of the poles. It should be pointed out that the one-sided version of the Z-transform depicted in Definition 2.1 can be extended to a two-sided Z-transform that includes noncausal signals as well. For the more general two-sided Z-transform, where the summation index k goes from $-\infty$ to ∞, the ROC is an annular or donut-shaped region centered at the origin of the complex plane. All of the signals we use in this chapter are causal signals, and therefore, we restrict our attention to the simpler one-sided Z-transform in Definition 2.1.

Example 2.1   **Finite Sequence**

As a simple initial example of a Z-transform, consider the following signal which has a finite number of nonzero samples.

$$x = \{2, -3, 7, 4, 0, 0, \cdots\}$$

Applying Definition 2.1 yields:

$$X(z) = 2 - 3z^{-1} + 7z^{-2} + 4z^{-3}$$

$$= \frac{2z^3 - 3z^2 + 7z + 4}{z^3}, \quad |z| > 0$$

Comparing with Eq. 2.12 we see that $m = n = 3$, and the radius of the region of convergence is $R_x = 0$. Thus, the ROC is the entire complex plane except the origin.

There is one finite sequence, presented in the next example, that is of particular interest because it is useful as a test signal for linear systems.

Example 2.2 **Unit Impulse**

The *unit impulse*, denoted $\delta(k)$, is a signal that is zero everywhere except at $k = 0$, where it takes on the value one.

Unit Impulse

$$\delta(k) \triangleq \begin{cases} 1, & k = 0 \\ 0, & k \neq 0 \end{cases}$$

Using Definition 2.1, we find that the only nonzero term in the series is the constant term whose coefficient is unity. Consequently,

$$Z\{\delta(k)\} = 1$$

Since the Z-transform of a unit impulse is constant, it has no poles, which means that its ROC is the entire complex plane, C.

Most practical signals have an infinite number of nonzero samples. In these cases, when computing the Z-transform, it is helpful to make use of the following fundamental result. When $m = 0$ and $f(z) = z$, this is called the *geometric series*.

Geometric series

$$\sum_{k=m}^{\infty} f^k(z) = \frac{f^m(z)}{1 - f(z)}, \quad |f(z)| < 1, \ m \geq 0 \tag{2.13}$$

The simplest instance of a signal with an infinite number of nonzero samples is the unit step function. The following example illustrates how to use the geometric series to find its Z-transform.

Example 2.3   **Unit Step**

The *unit step*, denoted $u(k)$, is defined as follows:

Unit step

$$u(k) \triangleq \begin{cases} 1, & k \geq 0 \\ 0, & k < 0 \end{cases}$$

Applying Definition 2.1, and using the generalized geometric series in Eq. 2.13 with $m = 0$ and $f(z) = z^{-1}$, we have

$$U(z) = \sum_{k=0}^{\infty} z^{-k}$$

$$= \sum_{k=0}^{\infty} (z^{-1})^k$$

$$= \frac{1}{1 - z^{-1}}, \quad |z^{-1}| < 1$$

Negative powers of z can be cleared by multiplying the numerator and the denominator by z which then yields

$$Z\{u(k)\} = \frac{z}{z - 1}, \quad |z| > 1$$

Note that the Z-transform has a zero at $z = 0$ and a pole at $z = 1$. Plots of the signal $u(k)$ and the pole-zero pattern of its Z-transform are shown in Figure 2.4.

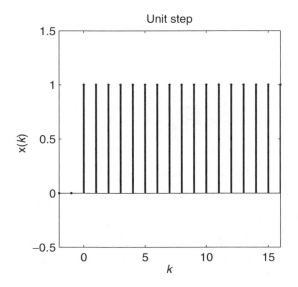

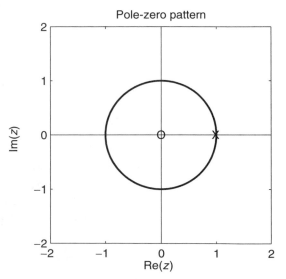

Figure 2.4   Time Plot and Pole-zero Pattern of a Unit Step Signal $u(k)$ for Example 2.3

The unit step function $u(k)$ is useful because, if we multiply another signal by $u(k)$, the result is a causal signal. By generalizing the unit step function slightly, we generate the most important Z-transform pair.

Example 2.4 **Causal Exponential**

Let a be real and consider the following *causal exponential* signal generated by taking powers of a.

$$x(k) = a^k u(k)$$

Using Definition 2.1 and the generalized geometric series in Eq. 2.13 with $m = 0$ and $f(z) = az^{-1}$, we have

$$X(z) = \sum_{k=0}^{\infty} a^k z^{-k}$$

$$= \sum_{k=0}^{\infty} (az^{-1})^k$$

$$= \frac{1}{1 - az^{-1}}, \quad |az^{-1}| < 1$$

Clearing z^{-1} by multiplying the numerator and denominator by z, the Z-transform of the causal exponential is

Causal exponential

$$Z\{a^k u(k)\} = \frac{z}{z - a}, \quad |z| > a$$

Plots of the causal exponential signal and its pole-zero pattern are shown in Figure 2.5. The first case, $a = 0.8$, corresponds to a damped exponential, and the second, $a = 1.1$, to a growing exponential. If the pole at $z = a$ is negative, then the time signal oscillates between positive and negative values as it decays or grows. When $a = 1$, the causal exponential reduces to the unit step in Example 2.3.

The causal exponential has a single real pole. Another important case corresponds to a complex conjugate pair of poles. The associated signal is referred to as an exponentially damped sinusoid.

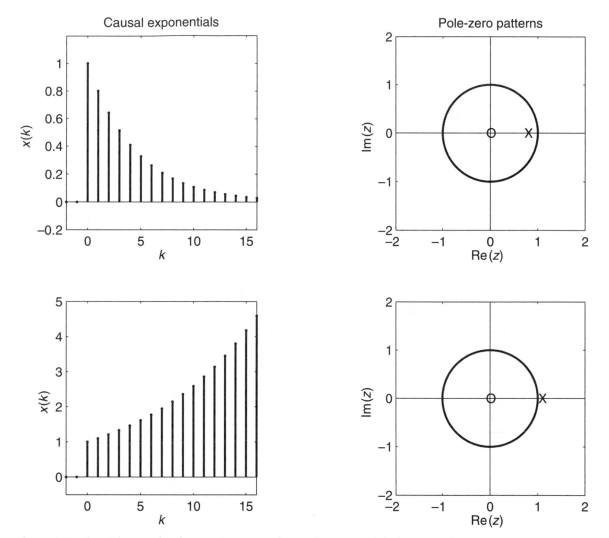

Figure 2.5 Time Plots and Pole-zero Patterns of Causal Exponentials for Example 2.4 with $a = 0.8$ (Bounded) and $a = 1.1$ (Unbounded)

Example 2.5 **Exponentially Damped Sine**

Let a and b be real, and consider the following causal exponentially damped sine wave.

$$x(k) = a^k \sin(bk)u(k)$$

Using Euler's identity (in Appendix D), we can express $\sin(bk)$ as:

$$\sin(bk) = \frac{\exp(jbk) - \exp(-jbk)}{j2}$$

Thus, the signal $x(k)$ can be represented as

$$x(k) = \frac{a^k[\exp(jbk) - \exp(-jbk)]u(k)}{j2}$$

$$= \frac{[a\exp(jb)]^k u(k) - [a\exp(-jb)]^k u(k)}{j2}$$

It is clear from Definition 2.1 that the Z-transform operator is a *linear* operator. That is, the transform of the sum of two signals is the sum of the transforms, and the transform of a scaled signal is just the scaled transform. Using this property, Euler's identity, and the results of Example 2.4 we have

$$X(z) = Z\left\{\frac{[a\exp(jb)]^k u(k) - [a\exp(-jb)]^k u(k)}{j2}\right\}$$

$$= \frac{Z\{[a\exp(jb)]^k u(k)\}}{j2} - \frac{Z\{[a\exp(-jb)]^k u(k)\}}{j2}$$

$$= \frac{1}{j2}\left\{\frac{z}{z - a\exp(jb)} - \frac{z}{z - a\exp(-jb)}\right\}, \quad |z| > a$$

$$= \frac{-az[\exp(-jb) - \exp(jb)]}{j2[z - a\exp(jb)][z - a\exp(-jb)]}, \quad |z| > a$$

$$= \frac{a\sin(b)z}{(z^2 - a[\exp(jb) + \exp(-jb)] + a^2)}, \quad |z| > a$$

Finally, applying Euler's identity to the denominator, the Z-transform of the causal exponentially damped sine wave is

$$Z\{a^k \sin(bk)u(k)\} = \frac{a\sin(b)z}{z^2 - 2a\cos(b)z + a^2}, \quad |z| > a$$

Plots of the exponentially damped sine wave and its pole-zero pattern are shown in Figure 2.6. Note that the *magnitude* of the poles, a, determines whether, and how fast, the signal decays to zero, while the *angle* of the poles, b, determines the frequency of the oscillation.

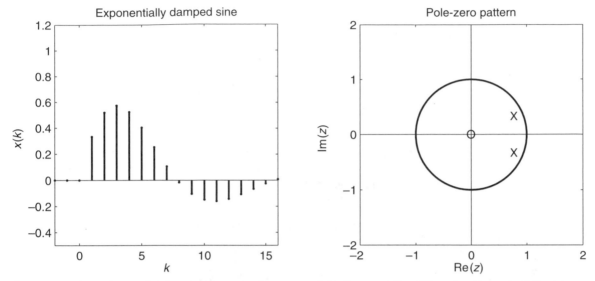

Figure 2.6   Time Plot and Pole-zero Pattern of Exponentially Damped Sine Wave for Example 2.5 with $a = 0.8$ and $b = 0.6$

A brief summary of the most important Z-transform pairs is shown in Table 2.1. Note that the unit step function is a special case of the causal exponential function with $a = 1$. For all but the first entry, the region of convergence is $|z| > a$. A more complete table of Z-transform pairs can be found in Appendix C.

Table 2.1: ▶
Basic Z-transform
Pairs

Signal	Z-transform	Poles
$\delta(k)$	1	none
$a^k u(k)$	$\dfrac{z}{z-a}$	$z = a$
$a^k \sin(bk)u(k)$	$\dfrac{a\sin(b)z}{z^2 - 2a\cos(b)z + a^2}$	$z = a\exp(\pm jb)$
$a^k \cos(bk)u(k)$	$\dfrac{[z - a\cos(b)]z}{z^2 - 2a\cos(b)z + a^2}$	$z = a\exp(\pm jb)$

● ● ● ● ● ● ● ● ● ● ● ● ● ● ● ●

2.3 Z-transform Properties

The effective size of Table 2.1 can be increased significantly by judiciously applying a number of the properties of the Z-transform.

2.3.1 Linearity Property

The most fundamental property of the Z-transform is the notion that the Z-transform operator is a *linear* operator. In particular, if $x(k)$ and $y(k)$ are two signals and a and b are two constants, then from Definition 2.1 we have

$$Z\{ax(k) + by(k)\} = \sum_{k=0}^{\infty} [ax(k) + by(k)]z^{-k}$$

$$= a \sum_{k=0}^{\infty} x(k)z^{-k} + b \sum_{k=0}^{\infty} y(k)z^{-k}$$

$$= aX(z) + bY(z) \tag{2.14}$$

Thus, the Z-transform of the sum of two signals is just the sum of the Z-transforms of the signals. Similarly, the Z-transform of a scaled signal is just the scaled Z-transform of the signal.

Linearity property

$$Z\{ax(k) + by(k)\} = aX(z) + bY(z) \tag{2.15}$$

Recall that Example 2.5 was an example that made use of the linearity property to compute a Z-transform.

2.3.2 Delay Property

Perhaps the most widely used property, particularly for the analysis of linear discrete-time systems, is the *delay* property. Let $M \geq 0$ denote the number of samples by which a causal discrete-time signal $x(k)$ is delayed. Applying Definition 2.1, and using the change of variable $i = k - M$, we have

$$Z\{x(k - M)\} = \sum_{k=0}^{\infty} x(k - M)z^{-k}$$

$$= \sum_{i=-M}^{\infty} x(i)z^{-(i+M)} \quad \Big\} \, i = k - M$$

$$= z^{-M} \sum_{i=0}^{\infty} x(i)z^{-i}$$

$$= z^{-M} X(z) \tag{2.16}$$

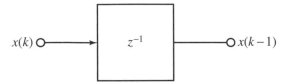

Figure 2.7 The Unit Delay Operator

Consequently, delaying a signal by M samples is equivalent to multiplying its Z-transform by z^{-M}.

Delay property

$$Z\{x(k - M)\} = z^{-M} X(z) \qquad (2.17)$$

In view of the delay property, we can regard z^{-1} as a *unit delay* operator as shown in Figure 2.7. That is, processing by the operator z^{-1} delays a signal by one sample.

Example 2.6   **Pulse**

As a simple illustration of the linearity and delay properties, consider the problem of finding the Z-transform of a pulse of height a and duration M. This signal can be written in terms of steps as a step up of amplitude a at time $k = 0$ followed by a step down of amplitude a at time $k = M$.

$$x(k) = a[u(k) - u(k - M)]$$

Applying the linearity and the delay properties, and using Table 2.1, we have

$$X(z) = a[U(z) - z^{-M} U(z)]$$

$$= a(1 - z^{-M})U(z)$$

$$= \frac{a(1 - z^{-M})z}{z - 1}$$

$$= \frac{a(z^M - 1)}{z^{M-1}(z - 1)}$$

Note that when $a = 1$ and $M = 1$, this reduces to $X(z) = 1$, which is the Z-transform of the unit impulse.

2.3.3   Z-scale Property

Another property that can be used to produce new Z-transform pairs is obtained by multiplying a discrete-time signal by a damped exponential. In particular, let a be constant and consider the signal $a^k x(k)$. Here,

$$Z\{a^k x(k)\} = \sum_{k=0}^{\infty} a^k x(k) z^{-k}$$

$$= \sum_{k=0}^{\infty} x(k) \left(\frac{z}{a}\right)^{-k}$$

$$= X\left(\frac{z}{a}\right) \qquad (2.18)$$

That is, multiplying a time signal by a^k is equivalent to scaling the Z-transform variable z by $1/a$.

Z-scale property

$$Z\{a^k x(k)\} = X\left(\frac{z}{a}\right) \qquad (2.19)$$

A simple illustration of the *z-scale* property can be found in Table 2.1. Notice that the Z-transform of the causal exponential signal can be obtained from the Z-transform of the unit step $U(z) = z/(z-1)$ by simply replacing z with z/a.

2.3.4 Time Multiplication Property

We can add an important family of entries to Table 2.1 by considering what happens when we take the derivative of the Z-transform of a signal.

$$\frac{dX(z)}{dz} = \frac{d}{dz} \sum_{k=0}^{\infty} x(k)z^{-k}$$

$$= -\sum_{k=0}^{\infty} kx(k)z^{-(k+1)}$$

$$= -z^{-1} \sum_{k=0}^{\infty} kx(k)z^{-k} \qquad (2.20)$$

Since the summation in Eq. 2.20 is just the Z-transform of $kx(k)$, we can rewrite Eq. 2.20 as

Time multiplication property

$$Z\{kx(k)\} = -z\frac{dX(z)}{dz} \qquad (2.21)$$

Thus, multiplying a discrete-time signal by the time variable k is equivalent to taking the derivative of the Z-transform and scaling by $-z$.

Example 2.7 **Unit Ramp**

In addition to the unit pulse and the unit step, another useful test signal is the *unit ramp* which is defined as

$$r(k) \triangleq ku(k)$$

Applying the time multiplication property, we have

$$R(z) = -z\frac{dU(z)}{dz}$$

$$= -z\frac{d}{dz}\left\{\frac{z}{z-1}\right\}$$

$$= \frac{z}{(z-1)^2}$$

Thus, the following entry can be added to the Z-transform pairs in Table 2.1.

$$Z\{ku(k)\} = \frac{z}{(z-1)^2}$$

The time multiplication property can be applied repeatedly to obtain the Z-transform of additional signals such as $k^m u(k)$ for $m \geq 0$. The results for $0 \leq m \leq 2$ are summarized in Appendix C.

2.3.5   Initial and Final Value Theorems

The Z-transform properties considered thus far all help expand the effective size of the table of Z-transform pairs. There are also some properties that serve as partial checks on the validity of a computed Z-transform. Note from Definition 2.1 that as we let $z \to \infty$, all of the terms go to zero except for the zeroth term which represents the initial value of the causal signal $x(k)$. Thus, we have the *initial value* theorem.

Initial value theorem

$$x(0) = \lim_{z \to \infty} X(z) \qquad (2.22)$$

The initial value $x(0)$ can be thought of as the value of $x(k)$ and one end of the time range. The value of $x(k)$ at the other end of the range, as $k \to \infty$, can also be obtained from $X(z)$ if certain conditions are satisfied. In particular, if the poles of $(z-1)X(z)$ are all strictly inside the unit circle of the complex plane, then it can be shown that the final, or steady-state, value of $x(k)$ can be obtained from its Z-transform as follows. This is called the *final value* theorem.

Final value theorem

$$x(\infty) = \lim_{z \to 1}(z-1)X(z) \qquad (2.23)$$

It is important to emphasize that the final value theorem is applicable *only* if the region of convergence of $(z-1)X(z)$ includes the unit circle.

Example 2.8   **Initial and Final Value Theorems**

Consider the Z-transform of a pulse of amplitude a and duration M that was previously computed in Example 2.6.

$$x(k) = a[u(k) - u(k-M)]$$

$$X(z) = \frac{a(z^M - 1)}{z^{M-1}(z-1)}$$

Using Eq. 2.22, the initial value of $x(k)$ is

$$x(0) = \lim_{z \to \infty} X(z)$$

$$= \lim_{z \to \infty} \frac{a(z^M - 1)}{z^{M-1}(z - 1)}$$

$$= a$$

Similarly, using Eq. 2.23 and noting that all of the poles of $(z - 1)X(z)$ are inside the unit circle, the final value of $x(k)$ is

$$x(\infty) = \lim_{z \to 1}(z - 1)X(z)$$

$$= \lim_{z \to 1} \frac{a(z^M - 1)}{z^{M-1}}$$

$$= 0$$

These values are consistent with $x(k)$ being a pulse of amplitude a and duration M.

A summary of the basic properties of the Z-transform can be found in Table 2.2. The last property, called the *convolution* property, is sufficiently important that it is covered separately in Section 2.7. A more complete table of Z-transform properties can be found in Appendix C.

Table 2.2: ▶
Basic Z-transform
Properties

Property	Description
Linearity	$Z\{ax(k) + by(k)\} = aX(z) + bY(z)$
Delay	$Z\{x(k - M)\} = z^{-M}X(z)$
Z-scale	$Z\{a^k x(k)\} = X\left(\dfrac{z}{a}\right)$
Time multiplication	$Z\{kx(k)\} = -z\dfrac{dX(z)}{dz}$
Initial value	$x(0) = \lim_{z \to \infty} X(z)$
Final value	$x(\infty) = \lim_{z \to 1}(z - 1)X(z)$
Convolution	$Z\{h(k) \star x(k)\} = H(z)X(z)$

● ● ● ● ● ● ● ● ● ● ● ● ● ● ● ● ●

2.4 Inverse Z-transform

There are a number of application areas where it is relatively easy to find the Z-transform of the solution to a problem. To find the actual solution, $x(k)$, we must then go backward by computing the *inverse* Z-transform of $X(z)$.

$$x(k) = Z^{-1}\{X(z)\} \tag{2.24}$$

Recall that $X(z)$ can be represented as a ratio of two polynomials in z. By convention, the denominator polynomial is always *normalized* to make its leading coefficient one.

$$X(z) = \frac{b_0 z^m + b_1 z^{m-1} + \cdots + b_m}{z^n + a_1 z^{n-1} + \cdots + a_n} \tag{2.25}$$

If $x(k)$ is a causal signal, then $X(z)$ will be a *proper* rational polynomial with $m \leq n$. That is, the number of zeros of $X(z)$ will be less than or equal to the number of poles.

2.4.1 Synthetic Division Method

If only a finite number of samples of $x(k)$ are required, then the synthetic division method can be used to invert $X(z)$. First, we rewrite Eq. 2.25 in terms of negative powers of z. Recalling that $m \leq n$, one can multiply the numerator and denominator by z^{-n} which yields

$$X(z) = \frac{z^{-r}(b_0 + b_1 z^{-1} + \cdots b_m z^{-m})}{1 + a_1 z^{-1} + \cdots + a_n z^{-n}} \tag{2.26}$$

In this case, $r = n - m$ is the difference between the number of poles and the number of zeros. The basic idea behind the synthetic division method is to perform long division of the numerator polynomial $b(z^{-1})$ by the denominator polynomial $a(z^{-1})$ to produce a quotient polynomial $q(z^{-1})$.

$$X(z) = z^{-r}[q(0) + q(1)z^{-1} + q(2)z^{-2} + \cdots] \tag{2.27}$$

For the special case when $r = 0$, we recognize the right-hand side of Eq. 2.27 as the power series definition of $X(z)$. That is, $x(k) = q(k)$ when $r = 0$. For the more general case when $r \geq 0$, $x(k)$ is $q(k)$ delayed by r samples. That is,

Synthetic division
method

$$x(k) = \begin{cases} 0, & 0 \leq k < r \\ q(k-r), & r \leq k < \infty \end{cases} \tag{2.28}$$

Example 2.9   **Synthetic Division**

As an illustration of the synthetic division method, consider the following Z-transform.

$$X(z) = \frac{z+1}{z^2 - 2z + 3}$$

Thus, there are $n = 2$ poles and $m = 1$ zero. Multiplying the top and bottom by z^{-2} yields

$$X(z) = \frac{z^{-1}(1 + z^{-1})}{1 - 2z^{-1} + 3z^{-2}}$$

Performing long division yields

$$
\begin{array}{r}
1 + 3z^{-1} + 3z^{-2} - 3z^{-3} - 15z^{-4} + \cdots \\
\hline
1 - 2z^{-1} + 3z^{-2} \enclose{longdiv}{1 + z^{-1}} \\
\end{array}
$$

$$
\begin{aligned}
& \underline{1 - 2z^{-1} + 3z^{-2}} \\
& \quad 3z^{-1} - 3z^{-2} \\
& \quad \underline{3z^{-1} - 6z^{-2} + 9z^{-3}} \\
& \qquad\quad 3z^{-2} - 9z^{-3} \\
& \qquad\quad \underline{3z^{-1} - 6z^{-2} + 9z^{-3}} \\
& \qquad\qquad\quad - 3z^{-3} - 9z^{-3} \\
& \qquad\qquad\quad \underline{-3z^{-3} + 6z^{-4} - 9z^{-4}} \\
& \qquad\qquad\qquad\quad - 15z^{-4} + 9z^{-5}
\end{aligned}
$$

Here $r = 1$. Thus, from Eq. 2.28, the inverse Z-transform of $X(z)$ is

$$x(k) = \{0, 1, 3, 3, -3, -15, \cdots\}$$

One advantage of the synthetic division method is that it can be automated (Ifeachor and Jervis, 2002). Algorithm 2.1 computes the first p samples of $x(k)$ from $X(z)$.

ALGORITHM 2.1:
Synthetic Division
Method

1. Write $X(z)$ as a normalized rational polynomial in z^{-1} as in Eq. 2.26
2. Set $q(0) = b_0$
3. For $k = 1$ to p compute
{

$$q(k) = \begin{cases} \dfrac{1}{b_0}\left[a_k - \displaystyle\sum_{i=1}^{k} q(k-i)b_i\right], & 1 \le k \le n \\[2em] -\dfrac{1}{b_0}\displaystyle\sum_{i=1}^{n} q(k-i)b_i, & n < k \le p \end{cases}$$

}
4. Compute $x(k)$ as in Eq. 2.28

The synthetic division method for inverting the Z-transform can be easily implemented in MATLAB. A drawback of the method is that it does not produce a closed-form expression for $x(k)$ as a function of k. Instead, it produces numerical values for a finite number of samples of $x(k)$. If the total number of samples p is large, then care must be taken to minimize the effects of accumulated roundoff error by using high-precision arithmetic.

2.4.2   Partial Fraction Method

If it is important to find a closed-form expression for the inverse Z-transform of $X(z)$, then the method of partial fraction expansion can be used. The easiest way to find the inverse Z-transform of $X(z)$ is to locate $X(z)$ in a table of Z-transform pairs and read across the row to find the corresponding $x(k)$. Unfortunately, for most problems of interest, $X(z)$ does not appear in the table. The basic idea behind partial fractions is to write $X(z)$ as a sum of terms, each of which is in a Z-transform table. To simplify the discussion, suppose that $X(z)$ has n distinct poles. Then $X(z)$ in Eq. 2.25 can be written in factored form as

$$X(z) = \frac{b(z)}{(z - p_1)(z - p_2)\cdots(z - p_n)} \tag{2.29}$$

Next, express $X(z)/z$ using partial fractions.

$$\frac{X(z)}{z} = \sum_{i=1}^{n} \frac{R_i}{z - p_i} \tag{2.30}$$

Here the coefficient R_i is called the *residue* of $X(z)/z$ at pole $z = p_i$. Multiplying both sides of Eq. 2.30 by $(z - p_k)$ and evaluating the result at $z = p_k$, we find that

Partial fraction
coefficient

$$R_i = \left.\frac{(z - p_i)X(z)}{z}\right|_{z=p_i}, \quad 1 \le i \le n \tag{2.31}$$

Thus, the residue is what *remains* of $X(z)/z$ at $z = p_i$, after the pole at $z = p_i$ has been removed. Once the partial fraction representation in Eq. 2.30 is obtained, then it is a simple matter to find the inverse Z-transform. First, multiply both sides of Eq. 2.30 by z. Then, using the linearity property and entry two of Table 2.1, the inverse Z-transform of $X(z)$ is

Partial fraction
method

$$x(k) = \sum_{i=1}^{n} R_i\, p_i^k u(k) \tag{2.32}$$

Example 2.10   **Partial Fractions**

As an illustration of the partial fraction expansion method, consider the following Z-transform.

$$X(z) = \frac{10z(z^2 + 4)}{(z - 2)(z + 1)(z - 3)}$$

Thus, $X(z)$ has simple real poles at $p_1 = 2$, $p_2 = -1$, and $p_3 = 3$. Using Eq. 2.31, the residues are

$$R_1 = \left. \frac{(z - p_1)X(z)}{z} \right|_{z=p_1}$$

$$= \left. \frac{10(z^2 + 4)}{(z + 1)(z - 3)} \right|_{z=2}$$

$$= \frac{-80}{3}$$

$$R_2 = \left. \frac{(z - p_2)X(z)}{z} \right|_{z=p_2}$$

$$= \left. \frac{10(z^2 + 4)}{(z - 2)(z - 3)} \right|_{z=-1}$$

$$= \frac{25}{6}$$

$$R_3 = \left. \frac{(z - p_3)X(z)}{z} \right|_{z=p_3}$$

$$= \left. \frac{10(z^2 + 4)}{(z - 2)(z + 1)} \right|_{z=3}$$

$$= \frac{65}{2}$$

It then follows from Eq. 2.32 that a closed-form expression for the inverse Z-transform of $X(z)$ is

$$x(k) = \left[\frac{-80(2)^k}{3} + \frac{25(-1)^k}{6} + \frac{65(3)^k}{2} \right] u(k)$$

The partial fraction method is quite effective for simple real poles. However, it can become cumbersome for complex poles and multiple poles. Complex poles will occur in conjugate pairs, and their corresponding residues will be complex conjugate pairs as well. When the two complex terms are combined, the result will be real because $x(k)$ is real. For the case of multiple poles, the expansion in Eq. 2.30 is no longer valid. Instead, a pole that is repeated q times will generate q terms. For the more general cases of complex poles and multiple poles, the following method is recommended.

2.4.3   Residue Method

There is an elegant alternative technique for inverting $X(z)$ that does not rely on the use of a table at all. Instead, it uses the Z-transform properties and Cauchy's residue theorem (Proakis and Manolakis, 1992). To formulate the approach, suppose the denominator polynomial of $X(z)$ in Eq. 2.25 is factored as follows.

$$X(z) = \frac{b(z)}{z^r (z - p_1)^{m_1} (z - p_2)^{m_2} \cdots (z - p_q)^{m_q}} \tag{2.33}$$

The r poles at $z = 0$ are treated separately because they correspond to a delay of r samples. We say that p_k is a pole of *multiplicity* m_k for $1 \leq k \leq q$. A pole of multiplicity one is referred to as a *simple* pole. It is assumed that the degree of the numerator polynomial $b(z)$ is, at most, $m_T = m_1 + m_2 + \ldots + m_q$. To simplify the initial discussion, suppose the number of poles at the origin is $r = 0$. The residue theorem says that $x(k)$ is the sum of the residues of $X(z)z^{k-1}$ evaluated at the poles of $X(z)z^{k-1}$.

Residue method

$$x(k) = \sum_{i=1}^{q} \text{Res}(p_i) \tag{2.34}$$

Here $\text{Res}(p_i)$ denotes the *residue* of $X(z)z^{k-1}$ at pole p_i. For the case of a simple pole, the residue can be thought of as what *remains* of $X(z)z^{k-1}$ at the pole after the pole has been removed. That is,

Simple residue

$$\text{Res}(p_i) = (z - p_i)X(z)z^{k-1}\Big|_{z=p_i} \quad \text{if} \quad m_i = 1 \tag{2.35}$$

Comparing Eq. 2.35 with Eq. 2.31, we find that when p_i is a simple pole, the residue is just the partial fraction coefficient $R_i(k)$ of $X(z)z^{k-1}$. Consequently, for simple poles, the residue method is the same amount of work as the partial fraction expansion method, but there is no need to use a table.

When p_i is a pole of multiplicity $m_i > 1$, the expression for the residue in Eq. 2.35 has to be generalized as follows.

Multiple residue

$$\text{Res}(p_i) = \frac{1}{(m_i - 1)!} \frac{d^{m_i-1}}{dz^{m_i-1}} \left\{ (z - p_i)^{m_i} X(z)z^{k-1} \right\}\Big|_{z=p_i} \quad \text{for} \quad m_i \geq 1 \tag{2.36}$$

Thus, the pole is again removed, but before the result is evaluated at the pole it is differentiated $m_i - 1$ times and scaled by $1/(m_i - 1)!$. Note that the expression for a multiple-pole residue in Eq. 2.36 reduces to the simpler expression for a simple-pole residue in Eq. 2.35 when $m_i = 1$.

The residue method requires the same amount of computational effort as the partial fraction method for the case of simple poles, and it requires less computational effort for multiple poles because there is only a single term associated with a multiple pole. The residue method also does not require the use of a table. However, the residue method does suffer from a drawback. Because $X(z)z^{k-1}$ depends on k, there can be some values of k that cause a pole at $z = 0$ to appear, and when this happens its residue must be included in Eq. 2.34 as well. For example, if $X(0) \neq 0$, then $X(z)z^{k-1}$ will have a pole at $z = 0$ when $k = 0$, but the pole disappears for $k \geq 1$. Fortunately, this problem can be easily resolved by separating $x(k)$ into two cases, $k = 0$ and $k \geq 1$. The initial value theorem can be used to compute $x(0)$. Finally, if $X(z)$ itself has r poles at $z = 0$, as in Eq. 2.33, these poles are removed first. Using the delay property, the final result is then delayed by r samples. This modified version of the basic residue method is summarized in Algorithm 2.2.

ALGORITHM 2.2:
Residue Method

1. Factor the denominator polynomial of $X(z)$ as in Eq. 2.33
2. Remove any poles at $z = 0$ by setting $V(z) = z^r X(z)$
3. Compute the initial value:

$$v(0) = \lim_{z \to \infty} V(z)$$

4. For $i = 1$ to q do
   {
         If $m_i = 1$ then p_i is a simple pole and

   $$\text{Res}(p_i) = (z - p_i)V(z)z^{k-1}\Big|_{z=p_i}$$

         else p_i is a multiple pole and

   $$\text{Res}(p_i) = \frac{1}{(m_i - 1)!} \frac{d^{m_i - 1}}{dz^{m_i - 1}} \left\{ (z - p_i)^{m_i} V(z) z^{k-1} \right\}\Big|_{z=p_i}$$

   }
5. Set

$$v(k) = v(0)\delta(k) + \sum_{i=1}^{q} \text{Res}(p_i)u(k-1)$$

6. Using the delay property, set $x(k) = v(k - r)$

In most cases, $r = 0$, which means that $V(z) = X(z)$ and steps 2 and 6 can be skipped. The following examples illustrate the use of Algorithm 2.2 to find inverse Z-transforms using the modified residue method.

Example 2.11   **Simple Real Poles**

As a simple initial example, consider a Z-transform with two real nonzero poles:

$$X(z) = \frac{z^2}{(z-a)(z-b)}$$

Thus, in this case, $r = 0$ and $V(z) = X(z)$. The initial value of $x(k)$ is

$$x(0) = \lim_{z \to \infty} X(z) = 1$$

The two residues are

$$\text{Res}(a) = \left. \frac{z^{k+1}}{z-b} \right|_{z=a} = \frac{a^{k+1}}{a-b}$$

$$\text{Res}(b) = \left. \frac{z^{k+1}}{z-a} \right|_{z=b} = \frac{b^{k+1}}{b-a}$$

Thus,

$$x(k) = x(0)\delta(k) + [\text{Res}(a) + \text{Res}(b)]u(k-1)$$

$$= \delta(k) + \left(\frac{a^{k+1} - b^{k+1}}{a-b} \right) u(k-1)$$

$$= \left(\frac{a^{k+1} - b^{k+1}}{a-b} \right) u(k)$$

Example 2.12   **Multiple Poles**

Next, consider a case that has multiple poles.

$$X(z) = \frac{1}{z^3(z-a)^2}$$

Here $r = 3$ and the undelayed Z-transform is

$$V(z) = z^3 X(z) = \frac{1}{(z-a)^2}$$

The initial value of $v(k)$ is

$$v(0) = \lim_{z \to \infty} V(z) = 0$$

The residue of the multiple pole is

$$\text{Res}(a) = \frac{d}{dz}\left\{z^{k-1}\right\}\Big|_{z=a}$$

$$= (k-1)z^{k-2}\Big|_{z=a}$$

$$= (k-1)a^{k-2}$$

The time signal $v(k)$ is then

$$v(k) = v(0)\delta(k) + \text{Res}(a)u(k-1)$$

$$= (k-1)a^{k-2}u(k-1)$$

Using the delay property:

$$x(k) = v(k-3)$$

$$= (k-4)a^{k-5}u(k-4)$$

Example 2.13   Complex Poles

Finally, consider an example that has a complex conjugate pair of poles:

$$X(z) = \frac{z}{z^2 + a^2} = \frac{z}{(z - ja)(z + ja)}$$

Here, $r = 0$ and $V(z) = X(z)$. The initial value of $x(k)$ is

$$x(0) = \lim_{z \to \infty} X(z) = 0$$

The two residues are

$$\text{Res}(ja) = \frac{z^k}{z + ja}\Big|_{z=ja} = \frac{(ja)^k}{j2a}$$

$$\text{Res}(-ja) = \frac{z^k}{z - ja}\Big|_{z=-ja} = \frac{(-ja)^k}{-j2a}$$

Finally, using Euler's identity from Appendix D, the time signal $x(k)$ is

$$x(k) = x(0)\delta(k) + [\text{Res}(ja) + \text{Res}(-ja)]u(k-1)$$

$$= \left[\frac{(ja)^k - (-ja)^k]}{j2a}\right]u(k-1)$$

$$= \left\{\frac{[a\exp(j\pi/2)]^k - [a\exp(-j\pi/2)]^k}{j2a}\right\}u(k-1)$$

$$= a^{k-1}\left[\frac{\exp(jk\pi/2) - \exp(-jk\pi/2)}{j2}\right]u(k-1)$$

$$= a^{k-1}\sin(k\pi/2)u(k-1)$$

$$= a^{k-1}\sin(k\pi/2)u(k)$$

Comparing this result with Table 2.1, we see that it corresponds to $b = \pi/2$.

2.5 Transfer Functions

2.5.1 The Transfer Function

The examples of discrete-time systems introduced in Section 2.1 are all special cases of the following generic linear time-invariant difference equation.

Difference equation

$$y(k) + \sum_{i=1}^{n} a_i y(k-i) = \sum_{i=0}^{m} b_i x(k-i) \tag{2.37}$$

Note that, by convention, the coefficient of the current output, $y(k)$, has been normalized to one by dividing both sides of the equation, if needed, by a_0. The complete solution to Eq. 2.37 depends on both the causal input $x(k)$ and the *initial condition*, $\{y(-1), y(-2), \ldots, y(-n)\}$. That is, in general, the output $y(k)$ can be decomposed into the sum of two parts.

$$y(k) = \underbrace{y_0(k)}_{zero-input} + \underbrace{y_1(k)}_{zero-state} \tag{2.38}$$

The first term, $y_0(k)$, is called the *zero-input* response. It is the part of the output that is generated by the initial condition. When the input is zero, $y(k) = y_0(k)$. For a stable system:

$$y_0(k) \rightarrow 0 \quad \text{as} \quad k \rightarrow \infty \tag{2.39}$$

The term $y_1(k)$ is called the *zero-state* response. The zero-state response is the part of the output that is generated by the input $x(k)$. When the initial condition is zero, $y(k) = y_1(k)$. In view of Eq. 2.39, one can measure the zero-state response of a stable system by first waiting for $y_0(k)$ to die out. Then excite the system with the input and measure the output.

    The difference equation in Eq. 2.37 is merely one way to represent a discrete-time system. The following alternative representation, based on the Z-transform, is a more compact algebraic characterization.

DEFINITION

2.2: Transfer Function

Let $x(k)$ be a nonzero input to a linear time-invariant discrete-time system, and let $y(k)$ be the output assuming a zero initial condition. Then the *transfer function* of the system is defined:

$$H(z) \triangleq \frac{Y(z)}{X(z)}$$

    The transfer function is simply the Z-transform of the zero-state response divided by the Z-transform of the input. The transfer function, $H(z)$, is a concise algebraic representation of the system. By multiplying both sides of the expression for $H(z)$ by $X(z)$, we get the frequency-domain input-output representation

Input-output
representation

$$Y(z) = H(z)X(z) \tag{2.40}$$

A block diagram which shows the relationship between the input, the output, and the transfer function of a discrete-time system is shown in Figure 2.8.

    The transfer function of the system in Eq. 2.37 can be determined using the delay property of the Z-transform. Taking the Z-transform of both sides of Eq. 2.37 yields

$$Y(z) + \sum_{i=1}^{n} a_i z^{-i} Y(z) = \sum_{i=0}^{m} b_i z^{-i} X(i) \tag{2.41}$$

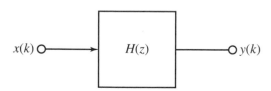

Figure 2.8 The Transfer Function Representation

Factoring $Y(z)$ from the left-hand side, and $X(z)$ from the right-hand side then yields

$$\left(1 + \sum_{i=1}^{n} a_i z^{-i}\right) Y(z) = \left(\sum_{i=0}^{m} b_i z^{-i}\right) X(z) \tag{2.42}$$

Finally, solving for $Y(z)/X(z)$, this produces the following transfer function for the discrete-time system in Eq. 2.37.

Transfer function

$$H(z) = \frac{b_0 + b_1 z^{-1} + \cdots + b_m z^{-m}}{1 + a_1 z^{-1} + \cdots + a_n z^{-n}} \tag{2.43}$$

Comparing Eq. 2.43 with Eq. 2.37, it is apparent that the transfer function in Eq. 2.43 can be written down directly from *inspection* of the difference equation. The transfer function representation in Eq. 2.43 is in terms of z^{-1}. To convert it to a ratio of two polynomials in z, we multiply the top and bottom by z^r where $r = \max\{m, n\}$.

Example 2.14 **Transfer Function**

As a simple illustration of computing the transfer function, consider the following discrete-time system.

$$y(k) = 1.2y(k-1) - .32y(k-2) + 10x(k-1) + 6x(k-2)$$

Using Eq. 2.43, we see from inspection that:

$$H(z) = \frac{10z^{-1} + 6z^{-2}}{1 - 1.2z^{-1} + .32z^{-2}}$$

Note that when the terms corresponding to delayed outputs are on the right-hand side of the difference equation, the signs of the coefficients must be reversed in the denominator of $H(z)$. To reformulate $H(z)$ in terms of positive powers of z, we multiply the numerator and the denominator by z^2, which yields the positive-power form of the transfer function.

$$H(z) = \frac{10z + 6}{z^2 - 1.2z + .32}$$

2.5.2 Zero-state Response

One of the useful features of the transfer function representation is that it provides us with a simple means of computing the zero-state response of a discrete-time system for an arbitrary input $x(k)$. Taking the inverse Z-transform of both sides of Eq. 2.40

yields the following formulation of the zero-state response.

Zero-state response

$$y(k) = Z^{-1}\{H(z)X(z)\} \tag{2.44}$$

Consequently, if the initial condition is zero, the output of the system is just the inverse Z-transform of the product of the transfer function and the Z-transform of the input. The following example illustrates this technique for computing the output.

Example 2.15   **Zero-state Response**

Consider the discrete-time system from Example 2.14. Suppose the input is the unit step $x(k) = u(k)$, and the initial condition is zero. Then the Z-transform of the output is

$$Y(z) = H(z)X(z)$$

$$= \left(\frac{10z + 6}{z^2 - 1.2z + .32}\right)\frac{z}{z - 1}$$

$$= \frac{(10z + 6)z}{(z^2 - 1.2z + .32)(z - 1)}$$

To find $y(k)$, we invert $Y(z)$ using the residue method in Algorithm 2.2. The denominator of $Y(z)$ is already partially factored. Applying the quadratic formula, the remaining two roots are

$$p_{1,2} = \frac{1.2 \pm \sqrt{1.44 - 1.28}}{2}$$

$$= \frac{1.2 \pm .4}{2}$$

$$= \{.8, .4\}$$

Since there are no poles at $z = 0$, $V(z) = Y(z)$. From the initial value theorem:

$$y(0) = \lim_{z \to \infty} Y(z) = 0$$

The residues at the three poles are

$$\text{Res}(.8) = \left.\frac{(10z + 6)z^k}{(z - .4)(z - 1)}\right|_{z=.8} = \frac{14(.8)^k}{.4(-.2)} = -175(.8)^k$$

$$\text{Res}(.4) = \left.\frac{(10z + 6)z^k}{(z - .8)(z - 1)}\right|_{z=.4} = \frac{10(.4)^k}{-.4(-.6)} = 41.7(.4)^k$$

$$\text{Res}(1) = \left.\frac{(10z + 6)z^k}{(z - .8)(z - .4)}\right|_{z=1} = \frac{16}{.2(.6)} = 133.3$$

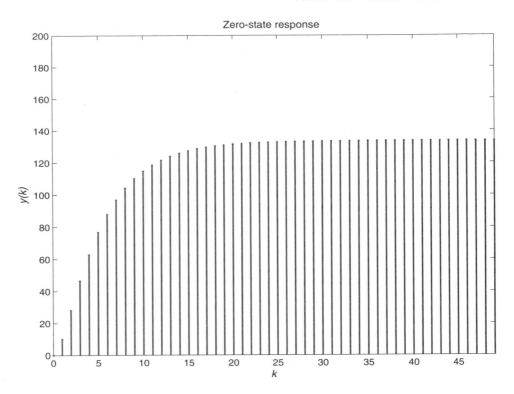

Figure 2.9 Zero-state Response for Example 2.15

Finally, the zero-state response is

$$y(k) = y(0)\delta(k) + [\text{Res}(.8) + \text{Res}(.6) + \text{Res}(1)]u(k-1)$$

$$= [133.3 - 175(.8)^k + 41.7(.4)^k]u(k-1)$$

$$= [133.3 - 175(.8)^k + 41.7(.4)^k]u(k)$$

A plot of the step response is shown in Figure 2.9. It was generated by running the script *exam2_15* on the distribution CD.

2.5.3   Poles, Zeros, and Modes

Recall that the poles and zeros of a signal were defined earlier in Section 2.2. This notion can also be applied to discrete-time systems using the transfer function. If we multiply the numerator and the denominator of $H(z)$ in Eq. 2.43 by z^m, and then factor the numerator and denominator, this results in the following *factored form* of the transfer function.

Factored form

$$H(z) = \frac{b_0(z - z_1)(z - z_2)\cdots(z - z_m)}{z^{m-n}(z - p_1)(z - p_2)\cdots(z - p_n)} \qquad (2.45)$$

The roots of the numerator polynomial are called the *zeros* of the discrete-tir system, and the roots of the denominator polynomial are called the *poles* of system. Recall that the *multiplicity* of a pole or zero corresponds to the numb times it appears. If $m \neq n$, then $H(z)$ will have either some poles $(m > n)$ or zeros $(m < n)$ at $z = 0$. Each pole at $z = 0$ corresponds to a delay in $y(k)$ sample.

Example 2.16 | **Poles and Zeros**

Consider the home mortgage discrete-time system introduced in Section 2.1.1. Recan from Eq. 2.1 that the home mortgage difference equation is

$$y(k) = y(k-1) + ry(k-1) - x(k)$$

Thus, the transfer function of this system is

$$H(z) = \frac{-1}{1 - (1+r)z^{-1}} = \frac{-z}{z - (1+r)}$$

The home mortgage system has one zero at $z = 0$ and one pole at $z = 1 + r$, where r is the annual interest rate expressed as a fraction.

Poles and zeros have simple interpretations in the time domain as well. Suppose $Y(z) = H(z)X(z)$. Then the output $y(k)$ can be decomposed into the sum of two types of terms called *modes*.

$$y(k) = \text{natural modes} + \text{forced modes} \tag{2.46}$$

Each *natural-mode* term is generated by a pole of $H(z)$, while each *forced-mode* term is generated by a pole of the input or forcing function $X(z)$. For a simple pole at $z = p$, the corresponding mode is of the form $c(p)^k$ for some constant c. More generally, if $z = p$ is a multiple pole of multiplicity r, then the mode is of the form

System mode

$$\text{mode} = (c_0 + c_1 k + \cdots + c_{r-1}k^{r-1})p^k, \quad r \geq 1 \tag{2.47}$$

Thus, a multiple pole is similar to a simple pole, except that the coefficient is a polynomial in k of degree $r - 1$. For a simple pole, $r = 1$, and the coefficient reduces to a polynomial of degree zero, a constant. Interpretation of poles of $Y(z)$ as modes of $y(k)$ is useful because it allows us to write down the *form* of $y(k)$, showing the dependence of k, directly from inspection of $Y(z)$.

The zeros of $H(z)$ and $X(z)$ also have simple interpretations in terms of $y(k)$. Since $Y(z) = H(z)X(z)$, there is a possibility of pole-zero cancellation between $H(z)$ and $X(z)$. For example, if $H(z)$ has a zero at $z = a$ and $X(z)$ has a pole at $z = a$, then the pole at $z = a$ will not appear in $Y(z)$, which means that there is no corresponding forced-mode term in $y(k)$. That is, the zeros of $H(z)$ can *suppress* certain forced-mode terms and prevent them from appearing in $y(k)$. Similarly, with a judicious choice for $x(k)$, one can suppress natural-mode terms in $y(k)$. In particular, if $H(z)$ has a pole at $z = a$ and $X(z)$ has a zero at $z = a$, then the natural-mode term generated by the pole of $H(z)$ will not appear in the output $y(k)$, assuming the initial condition is zero. We illustrate these ideas with the following example.

Example 2.17   **Modes**

Consider the discrete-time system introduced in Example 2.15. The factored form of its transfer function is

$$H(z) = \frac{10(z + .6)}{(z - .8)(z - .4)}$$

Thus, there are typically two natural-mode terms in $y(k)$. Next, consider the following input signal:

$$x(k) = 10(-.6)^k u(k) - 4(-.6)^{k-1} u(k - 1)$$

Using the properties of the Z-transform and row two of Table 2.1, the Z-transform is

$$X(z) = 10 \left(\frac{z}{z + .6} \right) - 4z^{-1} \left(\frac{z}{z + .6} \right)$$

$$= \frac{10z - 4}{z + .6}$$

$$= \frac{10(z - .4)}{z + .6}$$

Consequently, there is potentially one forced-mode term in $y(k)$. Using Eq. 2.44, the zero-state response is

$$y(k) = Z^{-1}\{H(z)X(z)\}$$

$$= Z^{-1} \left\{ \frac{100}{z - .8} \right\}$$

$$= y(0)\delta(k) + \text{Res}(.8)u(k - 1)$$

$$= 100(.8)^{k-1} u(k - 1)$$

For this particular input, neither the forced mode generated by the pole at $z = -0.6$, nor the natural mode generated by the pole at $z = 0.4$, survives and appears in the zero-state output $y(k)$.

2.5.4   DC Gain

When the poles of the system all lie strictly inside the unit circle, it is clear from Eq. 2.47 that all of the natural-mode terms decay to zero with increasing time. In this case, we say the system is *stable*. For a stable system, it is possible to determine the steady-state response of the system to a step input directly from $H(z)$. Suppose the input is a step of amplitude a. Then, using Eq. 2.40 and the final value theorem,

$$\lim_{k \to \infty} y(k) = \lim_{z \to 1}(z - 1)Y(z)$$

$$= \lim_{z \to 1}(z - 1)H(z)X(z)$$

$$= \lim_{z \to 1}(z - 1)H(z)\left(\frac{az}{z - 1}\right)$$

$$= H(1)a \tag{2.48}$$

Thus, the steady-state response to a step of amplitude a is the constant $H(1)a$. A unit step input can be regarded as a cosine input $x(k) = \cos(2\pi f_0 k)u(k)$, whose frequency happens to be $f_0 = 0$ Hz. That is, a step or constant input is actually a DC input. The amount by which the amplitude of the DC input is scaled to produce the steady-state output is called the *DC gain* of the system. From Eq. 2.48, it is evident that

DC gain

$$\text{DC gain} = H(1) \tag{2.49}$$

That is, the DC gain of a stable system with transfer function $H(z)$ is simply $H(1)$. In Section 2.9, we will see that the gain of the system at other frequencies can also be obtained by evaluating $H(z)$ at appropriate values of z.

Example 2.18   **Comb Filter**

As an illustration of poles, zeros, and DC gain, consider the following transfer function where $0 < r < 1$.

$$H(z) = \frac{z^n}{z^n - r^n}$$

This is the transfer function of a *comb filter*. Note that $H(z)$ has n poles and n zeros. The zeros are all at the origin, while the n poles are equally spaced around a circle of radius r. A plot of $|H(z)|$ for the case $N = 6$ and $r = 0.9$ is shown in Figure 2.10. Note the placement of the six poles (the plot is clipped at $|H(z)| \le 2$) and the multiple zero at the origin. The DC gain of the comb filter is

$$H(1) = \frac{1}{1 - r^n} = 2.134$$

Magnitude of transfer function

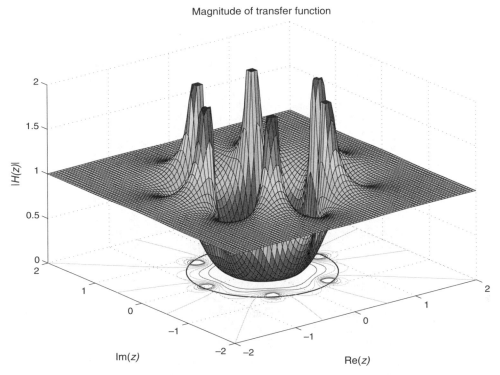

Figure 2.10   Magnitude of Comb Filter Transfer Function with $n = 6$ and $r = 0.9$

MATLAB Toolbox

MATLAB has a number of built-in functions that are useful for analyzing and computing the responses of linear discrete-time systems.

```
y = filter (b,a,x);          % zero-state response to x
stem (y)                     % stem plot of signal y
pole = roots (a);            % poles of H(z)
zero = roots (b);            % zeros of H(z)
a = poly (pole);             % construct polynomial from roots
p = polyval (a,z);           % evaluate polynomial at z
[R,pole] = residue (b,a);    % partial fraction coefficients
```

Function *filter* is used to evaluate the zero-state response of a linear discrete-time system. On entry to *filter*, b is a vector of length $m + 1$ containing the coefficients of the numerator of the transfer function, and a is a vector of length $n + 1$ containing the denominator coefficients with $a(1) = 1$.

$$H(z) = \frac{b_0 + b_1 z^{-1} + \cdots + b_m z^{-m}}{1 + a_1 z^{-1} + \cdots + a_n z^{-n}} \tag{2.50}$$

Argument x is a vector of length N containing the samples of the input signal. On exit from *filter*, y is a vector of length N containing the samples of the zero-state response to x. The coefficient vectors can be defined as follows prior to calling *filter*.

```
a = [1 a_1 ... a_n];        % denominator coefficients
b = [b_0 b_1 ... b_m];      % numerator coefficients
```

MATLAB function *stem* is used to plot a discrete-time signal. On entry to *stem*, y is a vector containing the samples of the signal to be plotted. MATLAB function *roots* is used to find the roots of a polynomial. On entry to *roots*, a is a vector of length $n + 1$ containing the coefficients of the polynomial, and on exit, *pole* is a complex vector of length n containing the roots. MATLAB function *poly* is used to reconstruct the coefficient vector of a polynomial from the roots. Thus, except for scale factor, *poly* is the inverse of *roots*. MATLAB function *polyval* is used to evaluate a polynomial. On entry to *polyval*, a is the vector of length $n + 1$ containing the polynomial coefficients, and z is a scalar or vector containing the points at which the polynomial is to be evaluated. On exit from *polyval*, p is a vector of the same length as z containing the evaluated polynomial. Finally, *residue* computes the partial fraction expansion coefficients of $H(z)$. An example of using *roots*, *filter*, and *stem* to compute the zeros, poles, and zero-state response of a linear system can be found in script *exam2_15* on the distribution CD.

● ● ● ● ● ● ● ● ● ● ● ● ● ● ● ● ●

2.6 Signal Flow Graphs

Engineers often use diagrams to show relationships between variables and subsystems. Discrete-time systems can be represented efficiently using a special type of diagram called a *signal flow graph*. A signal flow graph is a collection of *nodes* interconnected by *arcs*. Each node is represented with a dot and each arc is represented with a directed line segment. An arc is an input arc if it enters a node and an output arc if it leaves a node. Each arc transmits or carries a signal in the direction indicated. When two or more input arcs enter a node, their signals are *added* together. That is, a node can be thought of as a summing junction with respect to its inputs. An output arc carries the value of the signal leaving the node. When an arc is labeled, the labeling indicates what type of operation is performed on the signal as it traverses the arc. For example, a signal might be scaled by a constant or, more generally, acted on by a transfer function. As a simple illustration, consider the signal flow graph shown in Figure 2.11 that consists of four nodes and three arcs.

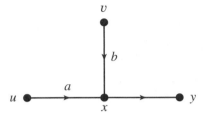

Figure 2.11 A Simple Signal Flow Graph

Observe that the node labeled x has two input arcs and one output arc. Thus, the value of the signal leaving the node is $x = au + bv$. When an arc is not labeled, the default gain is assumed to be one. Therefore, the value of the signal at the output node is

$$y = x$$
$$= au + bv \tag{2.51}$$

In order to develop a signal flow graph for the discrete-time system in Eq. 2.43, we first factor the transfer function $H(z)$ into a product of two transfer functions as follows.

$$H(z) = \frac{b_0 + b_1 z^{-1} + \cdots b_m z^{-m}}{1 + a_1 z^{-1} + \cdots + a_n z^{-n}}$$

$$= \left(\frac{b_0 + b_1 z^{-1} + \cdots b_m z^{-m}}{1}\right)\left(\frac{1}{1 + a_1 z^{-1} + \cdots + a_n z^{-n}}\right)$$

$$= H_b(z) H_a(z) \tag{2.52}$$

Thus, $H(z)$ consists of two subsystems in series, one associated with the numerator polynomial, and the other with the denominator polynomial as shown in Figure 2.12. Here the intermediate variable, $V(z) = H_a(z) X(z)$, is the output from the first subsystem and the input to the second subsystem.

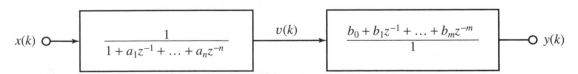

Figure 2.12 Decomposition of Transfer Function $H(z)$

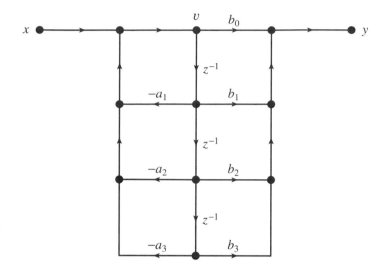

Figure 2.13 Signal Flow Graph of a Discrete-time System, $m = n = 3$

Subsystem $H_a(z)$ processes input $x(k)$ to produce $v(k)$, and then subsystem $H_b(z)$ acts on $v(k)$ to produce the output $y(k)$. The decomposition in Figure 2.12 can be represented by the following pair of difference equations.

$$v(k) = x(k) - \sum_{i=1}^{n} a_i v(k - i) \qquad (2.53a)$$

$$y(k) = \sum_{i=0}^{m} b_i v(k - i) \qquad (2.53b)$$

Given the decomposition in Eq. 2.53, the entire system can be represented with a signal flow graph as shown in Figure 2.13, which illustrates the case $m = n = 3$.

The output of the top center node is $v(k)$, and nodes below it produce delayed versions of $v(k)$. The left-hand side of the signal flow graph ladder is a feedback system that implements Eq. 2.53a to produce $v(k)$, while the right-hand side is a feed-forward system that implements Eq. 2.53b to produce $y(k)$. In the general case, the order of the signal flow graph will be $r = \max\{m, n\}$. If $m \neq n$, then some of the arcs will be missing or labeled with gains of zero.

Example 2.19   **Signal Flow Graph**

Consider a discrete-time system with the following transfer function.

$$H(z) = \frac{2.4z^{-1} + 1.6z^{-2}}{1 - 1.8z^{-1} + .9z^{-2}}$$

By inspection of Figure 2.13, it is evident that the signal flow graph of this system is as shown in Figure 2.14. Note how the gain of one of the arcs on the right-hand side

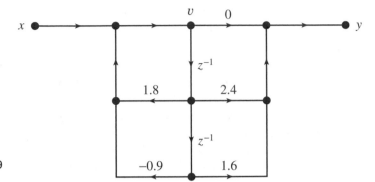

Figure 2.14 Signal Flow Graph of System in Example 2.19

is zero due to a missing term in the numerator. Given $H(z)$, the difference equation of this system is

$$y(k) = 1.8y(k-1) - .9y(k-2) + 2.4x(k-1) + 1.6x(k-2)$$

Thus, there are three distinct ways to represent linear time-invariant discrete-time systems as shown in Figure 2.15. With care, it should be possible to go directly from one representation to another by inspection.

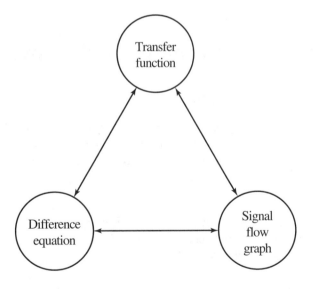

Figure 2.15 Alternative Ways to Represent a Discrete-time System

ARMA Models

The transfer function of the first subsystem in Figure 2.12 is a special case of the following special structure.

Auto-regressive model

$$H_{\text{ar}}(z) = \frac{b_0}{1 + a_1 z^{-1} + \cdots + a_n z^{-n}} \tag{2.54}$$

The system $H_{ar}(z)$, whose numerator polynomial is constant, is called an *auto-regressive* or AR model. Similarly, the second subsystem in Figure 2.12 has the following special structure.

Moving average model

$$H_{ma}(z) = b_0 + b_1 z^{-1} + \cdots + b_m z^{-m} \tag{2.55}$$

System $H_{ma}(z)$, whose denominator polynomial is one, is called a *moving average* or MA model. The name "moving average" arises from the fact that, if $b_k = 1/(m+1)$, then the output is a running average of the last $m + 1$ samples. More generally, if the b_k are not equal, then $H_{ma}(z)$ represents a weighted moving average of the last $m + 1$ samples.

The general case of a transfer function $H(z)$ in Eq. 2.43 is called an *auto-regressive moving average* or ARMA model. It is evident from Figure 2.12 that a general ARMA model can always be factored into a product of an AR model and a MA model.

$$H_{arma}(z) = H_{ma}(z) H_{ar}(z) \tag{2.56}$$

For the signal flow graph representation in Figure 2.13, the left side of the ladder is the AR part and the right side of the ladder is the MA part.

- - - - - - - - - - - - - - - -

2.7 The Impulse Response and Convolution

2.7.1 Impulse Response

At this point, we have three ways of representing a discrete-time system: the difference equation, the transfer function, and the signal flow graph. There is a fourth and final representation that is closely tied to the transfer function, called the impulse response.

DEFINITION
2.3: Impulse Response

The *impulse response* of a discrete-time system is denoted $h(k)$, and defined as the system output produced by the unit impulse input $x(k) = \delta(k)$, assuming the initial condition is zero.

The impulse response is the zero-state response to a unit impulse. Recall from Eq. 2.40 that the Z-transform of the zero-state response is $Y(z) = H(z)X(z)$. But if $x(k) = \delta(k)$, then $X(z) = 1$. Consequently, the impulse response can be expressed in terms of the transfer function as follows.

Impulse response

$$h(k) = Z^{-1}\{H(z)\} \tag{2.57}$$

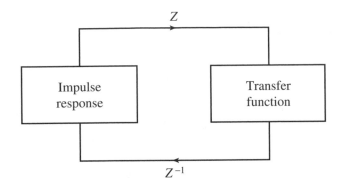

Figure 2.16 Impulse Response and Transfer Function Representations

In view of Eq. 2.57, an alternative way to define the transfer function is to say that $H(z)$ is the Z-transform of the impulse response. See Figure 2.16 for an illustration of the relationship between the two equivalent representations of a discrete-time system.

Example 2.20   **Impulse Response**

As an illustration of using Eq. 2.57 to compute the impulse response, consider the following discrete-time system.

$$y(k) = 1.6y(k-1) - 0.64y(k-2) + 3x(k)$$

Using Eq. 2.42, the factored form of the transfer function is

$$H(z) = \frac{3}{1 - 1.6z^{-1} + .64z^{-2}}$$

$$= \frac{3z^2}{z^2 - 1.6z + .64}$$

$$= \frac{3z^2}{(z - .8)^2}$$

From Eq. 2.57 and Algorithm 2.2, the impulse response of this system is

$$h(k) = Z^{-1}\{H(z)\}$$

$$= h(0)\delta(k) + \text{Res}(.8)u(k-1)$$

From the initial value theorem, $h(0) = 3$. The residue at $z = .8$ is

$$\text{Res}(.8) = \frac{d}{dz}\left\{3z^{k+1}\right\}\Big|_{z=.8}$$

$$= 3(k+1)(.8)^k$$

Thus, the impulse response of this system is

$$h(k) = 3\delta(k) + 3(k+1)(.8)^k u(k-1)$$
$$= 3(k+1)(.8)^k u(k)$$

A plot of the impulse response, obtained by running script *exam2_20* on the distribution CD, is shown in Figure 2.17.

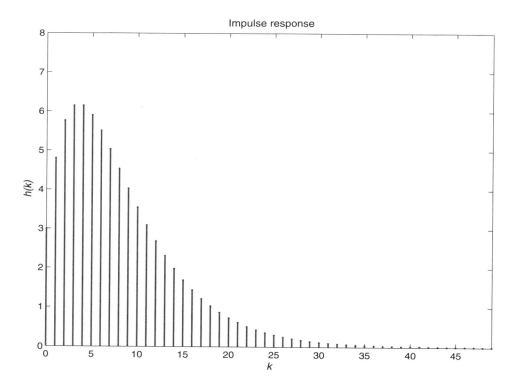

Figure 2.17 Impulse Response of System in Example 2.20

2.7.2 Convolution

Once the impulse response is known, the zero-state response to an arbitrary input $x(k)$ can be computed from it. To see this, first note that $x(k)$ can be written as a weighted sum of impulses as follows.

$$x(k) = \sum_{i=0}^{k} x(i)\delta(k-i), \quad k \geq 0 \tag{2.58}$$

Here the ith term is an impulse of amplitude $x(i)$ occurring at time $k = i$. Recall from Definition 2.3 that the response to an impulse of unit amplitude occurring at $k = 0$ is $h(k)$. Since the system $H(z)$ is linear and time-invariant, when we scale the input by $x(i)$ and delay it by i samples, the corresponding output is also scaled by $x(i)$ and delayed by i samples. That is, the output corresponding to the ith term in Eq. 2.58 is simply $x(i)h(k - i)$. Finally, because $H(z)$ is a linear system, the response to the sum of inputs in Eq. 2.58 is just the sum of the responses to the individual inputs. Thus, the zero-state output produced by $x(k)$ in Eq. 2.58 can be written as

Convolution
$$y(k) = \sum_{i=0}^{k} h(k - i)x(i) \qquad (2.59)$$

The formulation on the right-hand side of Eq. 2.59 is referred to as the convolution of the signal $h(k)$ with the signal $x(k)$. We define discrete convolution more formally as follows.

DEFINITION
2.4: Convolution

Let $h(k)$ and $x(k)$ be causal discrete-time signals. Then the *convolution* of $h(k)$ with $x(k)$ is denoted $h(k) \star x(k)$ and defined

$$h(k) \star x(k) \triangleq \sum_{i=0}^{k} h(k - i)x(i)$$

The operator $\star$ is called the *convolution operator*. Comparing Definition 2.4 with Eq. 2.59, we see that the zero-state response of a discrete-time system can be expressed in terms of convolution as

$$y(k) = h(k) \star x(k) \qquad (2.60)$$

That is, the zero-state response is the convolution of the impulse response with the input. Recall from Eq. 2.40 and Eq. 2.57 that $Y(z) = H(z)X(z)$ where $H(z) = Z\{h(k)\}$. Taking the Z-transform of both sides of Eq. 2.60, it then follows that:

$$Z\{h(k) \star x(k)\} = H(z)X(z) \qquad (2.61)$$

Consequently, the *convolution* operator, in the time-domain, maps into the *multiplication* operator in the Z-transform domain. This completes the Z-transform properties listed in Table 2.2.

If we interchange the roles of $h(k)$ and $x(k)$ in Eq. 2.61, we find that it does not affect the result because $H(z)X(z) = X(z)H(z)$. This means that the convolution operator is a *commutative* operator. That is,

$$h(k) \star x(k) = x(k) \star h(k) \tag{2.62}$$

In view of Eq. 2.62 and Eq. 2.60, the following is an alternative way to write the output of a discrete-time system.

Alternative convolution

$$y(k) = \sum_{i=0}^{k} h(i)x(k-i) \tag{2.63}$$

Example 2.21   **Convolution**

Suppose the input to the system in Example 2.20 is the following causal sine wave.

$$x(k) = 10\sin\left(\frac{\pi k}{10}\right)u(k)$$

If the initial condition is zero, the output of this system using the convolution representation is

$$y(k) = h(k) \star x(k)$$

$$= \sum_{i=0}^{k} h(k-i)x(i)$$

$$= \sum_{i=0}^{k} 3(k+1-i)(.8)^{k-i}u(k-i)10\sin\left(\frac{\pi i}{10}\right)u(i)$$

$$= 30\sum_{i=0}^{k}(k+1-i)(.8)^{k-i}\sin\left(\frac{\pi i}{10}\right), \quad k \geq 0$$

A plot of the impulse response, obtained by running script *exam2_21* on the distribution CD, is shown in Figure 2.18.

The convolution operation is an example of a DSP technique that processes a pair of signals to produce a third signal. Another example of an operation of this type is the correlation operation. The topics of convolution and correlation are considered in detail in Chapter 4.

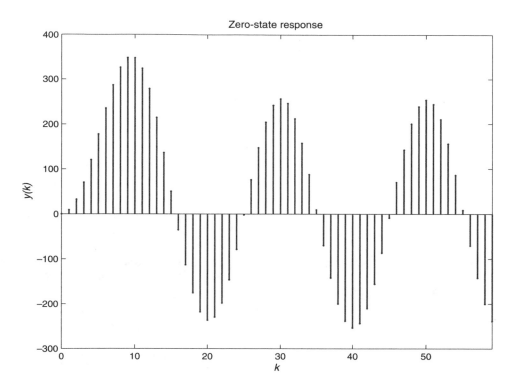

Figure 2.18 Zero-state Response for Example 2.21

2.7.3   FIR and IIR Systems

Discrete-time systems can be classified into two broad categories based on the *duration* of the impulse response.

DEFINITION
2.5: FIR and IIR

> A discrete-time system $H(z)$ is a *finite-impulse response* (FIR) system if and only if the impulse response $h(k)$ has a finite number of nonzero samples. Otherwise, it is an *infinite-impulse response* (IIR) system.

As an illustration of an FIR system, consider the MA model in Eq. 2.55, whose transfer function is

$$H(z) = b_0 + b_1 z^{-1} + \cdots + b_m z^{-m} \tag{2.64}$$

Using the delay property, the difference equation representation of the MA model is

$$y(k) = \sum_{i=0}^{m} b_i x(k - i) \tag{2.65}$$

Thus, the output is simply a *weighted sum* of the past inputs. When we substitute

$x(k) = \delta(k)$ into Eq. 2.65, the resulting impulse response of the MA model is

$$h(k) = \sum_{i=0}^{m} b_i \delta(k - i) \tag{2.66}$$

Consequently, $h(k) = 0$ for $k > m$. That is, the impulse response of a MA model has a *finite* number of nonzero samples. An alternative way to write Eq. 2.66, that makes this more explicit, is

FIR impulse response

$$h(k) = \begin{cases} b_k, & 0 \le k \le m \\ 0, & m < k < \infty \end{cases} \tag{2.67}$$

It follows that a MA model is an FIR system. Often the terms MA and FIR are used interchangeably. It is important to emphasize that the simple expression for the impulse response in Eq. 2.67 holds only for a MA model where the denominator polynomial is one. For an AR model and the more general ARMA model, the impulse response typically has an *infinite* number of nonzero terms (barring complete pole-zero cancelation) which makes these IIR systems.

If the transfer function of an FIR system is expressed in terms of positive powers of z by multiplying the numerator and denominator of Eq. 2.64 by z^m, this yields

FIR transfer function

$$H(z) = \frac{b_0 z^m + b_1 z^{m-1} \cdots + b_{m-1} z + b_m}{z^m} \tag{2.68}$$

Consequently, every FIR system has the same number of poles and zeros. Furthermore, the poles of an FIR system are all located at the origin.

Example 2.22   FIR System

As an illustration of an FIR system, consider the running average filter introduced in Section 2.1.3 Recall from Eq. 2.7 that the difference equation for an mth-order running average filter is

$$y(k) = \frac{1}{m+1} \sum_{i=0}^{m} x(k - i)$$

Thus, the impulse response of the running average filter is a pulse of amplitude $1/(m+1)$ and duration $m + 1$.

$$h(k) = \begin{cases} \dfrac{1}{m+1}, & 0 \le k \le m \\ 0, & m < k < \infty \end{cases}$$

The transfer function of the mth-order running average filter is

$$H(z) = \frac{1 + z^{-1} + \cdots + z^{-m}}{m + 1}$$

MATLAB Toolbox

The FDSP toolbox contains the following functions for computing the impulse response of a linear discrete-time system and viewing the pole-zero pattern of the transfer function.

```
[h,k] = f_impulse (b,a,N);      % impulse response
f_pzplot (b,a,title)            % pole-zero pattern
f_pzsurf (b,a,title)            % pole-zero surface
```

Function *f_impulse* computes the impulse response. On entry to *f_impulse*, b is a vector of length $m + 1$ containing the coefficients of the numerator of the transfer function, and a is a vector of length $n + 1$ containing the denominator coefficients with $a(1) = 1$.

$$H(z) = \frac{b_0 + b_1 z^{-1} + \cdots + b_m z^{-m}}{1 + a_1 z^{-1} + \cdots + a_n z^{-n}} \tag{2.69}$$

Argument N is an integer specifying the number of points at which the impulse response is to be evaluated. On exit from *f_impulse*, h is a vector of length N containing the impulse response and k is an optional vector of length N containing the discrete times at which h is evaluated. Function *f_pzplot* is used to create a sketch showing the poles and zeros of $H(z)$. The calling arguments for *f_pzplot* are the same as for *f_impulse* except for the string *title* which contains the title of the pole-zero sketch. Function *f_pzsurf* creates a surface plot showing the magnitude of the transfer function, $|H(z)|$, over a section of the complex plane that includes the poles and zeros. The surface is clipped at the poles. The calling arguments for *f_pzsurf* are identical to those for *f_pzplot*. An example of a pole-zero surface plot created by a call to *f_pzsurf* can be seen in Figure 2.10, which shows the magnitude of the transfer function of a comb filter.

● ● ● ● ● ● ● ● ● ● ● ● ● ● ● ● ●

2.8 Stability

2.8.1 BIBO Stable

Practical discrete-time systems, particularly digital filters, tend to have one qualitative feature in common: they are stable. To define what is meant by a stable system, we first consider the notion of a bounded signal.

DEFINITION
2.6: Bounded Signal

A discrete-time signal $x(k)$ is *bounded* if and only if there exist a $B_x > 0$, called a *bound*, such that

$$|x(k)| \leq B_x, \qquad 0 \leq k < \infty$$

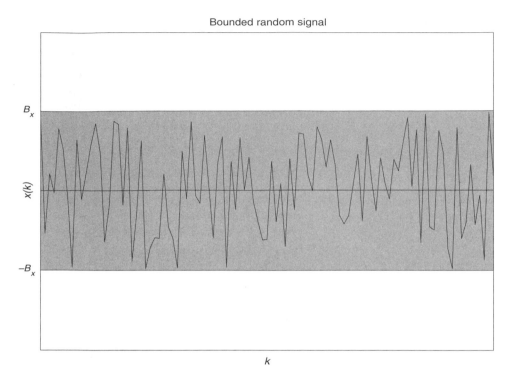

Bounded random signal

Figure 2.19 A Bounded Signal $x(k)$

A bounded signal is a signal whose value is constrained to lie inside a strip of radius B_x centered along the time axis. An example of a random bounded signal is shown in Figure 2.19.

Examples of bounded signals include the unit impulse, the unit step, sines and cosines, and signals such as damped exponentials that go to zero as $k \to \infty$.

DEFINITION

2.7: Stable System

A discrete-time system is bounded-input bounded-output (BIBO) *stable* if and only if every bounded input produces a bounded output. Otherwise the system is *unstable*.

The stability of a discrete-time system can be determined from either the impulse response $h(k)$, or the transfer function, $H(z)$. First, consider the impulse response. Suppose the input $x(k)$ is bounded by a bound B_x as in Figure 2.19. Recalling that the output is the convolution of the input with the impulse response, from Eq. 2.63, we have

$$|y(k)| = |\sum_{i=0}^{k} h(i)x(k-i)|$$

$$\leq \sum_{i=0}^{k} |h(i)x(k-i)|$$

$$= \sum_{i=0}^{k} |h(i)| \cdot |x(k-i)|$$

$$\leq B_x \sum_{i=0}^{k} |h(i)|$$

$$\leq B_x \sum_{i=0}^{\infty} |h(i)| \qquad (2.70)$$

It follows from Eq. 2.70 that the system is BIBO stable if the infinite series on the right-hand side converges to a finite value. This condition is both necessary and sufficient for stability. Consequently, a discrete-time system with impulse response $h(k)$ is BIBO stable if and only if

Time-domain stability constraint

$$S_h = \sum_{k=0}^{\infty} |h(k)| < \infty \qquad (2.71)$$

The stability criterion in Eq. 2.71 can be summarized in words by saying that the system is BIBO stable if and only if the impulse response is *absolutely summable*.

Example 2.23 **Stable Impulse Response**

Consider a first order discrete-time system with the following transfer function.

$$H(z) = \frac{z}{z-a}$$

From Table 2.1, this is the Z-transform of a causal exponential. Therefore, the impulse response is

$$h(k) = a^k u(k)$$

Recalling the geometric series in Eq. 2.13, we have

$$S_h = \sum_{k=0}^{\infty} |h(k)|$$

$$= \sum_{k=0}^{\infty} |a^k u(k)|$$

$$= \sum_{k=0}^{\infty} |a|^k$$

$$= \frac{1}{1-|a|}, \quad |a| < 1$$

Thus, the impulse response is absolutely summable if and only if $|a| < 1$. Consequently, this system is BIBO stable if and only if $|a| < 1$.

The impulse response stability criterion in Eq. 2.71 can be used to establish the stability of an entire class of discrete-time systems. Recall that FIR systems, by definition, have an impulse response that is of finite duration. In particular, from Eq. 2.67, we have the following for an FIR system:

$$S_h = \sum_{k=0}^{m} |b_k| \tag{2.72}$$

Consequently, *all* FIR systems are BIBO stable. This is an important qualitative feature of FIR systems that sets them apart from IIR systems and makes them particularly attractive for use as digital filters.

Whereas an FIR system is always stable, the more general IIR system can be stable or unstable depending on the denominator of the transfer function. Recall from Example 2.23 that the first order system is stable if and only if the pole at $z = a$ is located inside the unit circle $|z| = 1$. As it turns out, this simple criterion generalizes to systems with multiple poles. This leads to the following fundamental result given in Theorem 2.1.

THEOREM

2.1: Stability

> A linear time-invariant discrete-time system with transfer function $H(z)$ is BIBO stable if and only if the poles of $H(z)$ satisfy
>
> $$|p_k| < 1, \quad 1 \le k \le n$$

The poles of a stable system, whether simple or multiple, must all lie strictly within the unit circle as shown in Figure 2.20. Recall from Eq. 2.47 that this is equivalent to saying that all of the natural modes of the system must decay to zero. From Eq. 2.68, the transfer function of an FIR system is

$$H(z) = \frac{b_0 z^m + b_1 z^{m-1} + \cdots + b_m}{z^m} \tag{2.73}$$

In this case, all m poles are at $z = 0$, well inside the unit circle. This confirms the previous time-domain analysis in Eq. 2.72 that FIR systems are always stable.

Example 2.24   **Unstable Transfer Function**

Consider a discrete-time system with the following transfer function.

$$H(z) = \frac{10}{z+1}$$

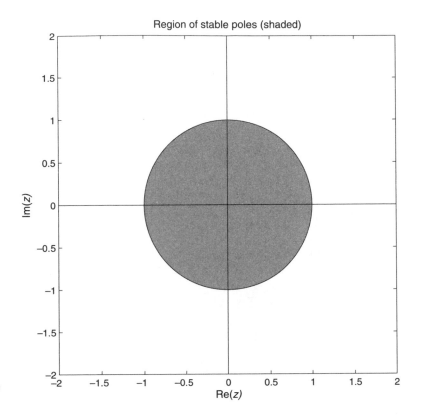

Figure 2.20   Region of Stable Poles

This system has a single pole on the unit circle at $z = -1$. Since it is not strictly inside as in Theorem 2.1, this is an example of an unstable system. When a discrete-time system has one or more simple poles on the unit circle, and the rest of its poles are inside the unit circle, it is sometimes referred to as a *marginally* unstable system. For a marginally unstable system there are natural modes that neither grow without bound nor decay to zero. From Algorithm 2.2, the impulse response of the system $H(z)$ is

$$h(k) = h(0)\delta(k) + \text{Res}(-1)u(k - 1)$$

$$= 10(-1)^{k-1}u(k - 1)$$

Thus, the pole at $z = -1$ produces a term that oscillates about zero. Since this system is unstable, there must be at least one bounded input that produces an unbounded output. Consider the following input that is bounded by $B_x = 1$.

$$x(k) = (-1)^k u(k)$$

The Z-transform of the zero-state response is

$$Y(z) = H(z)X(z)$$

$$= \frac{10z}{(z + 1)^2}$$

Applying Algorithm 2.2, we have

$$y(k) = y(0)\delta(k) + \text{Res}(-1)u(k-1)$$

$$= 10k(-1)^{k-1}u(k-1)$$

$$= 10k(-1)^{k-1}u(k)$$

Clearly, $y(k)$ is not bounded. Note that the forced mode of $X(z)$ at $z = -1$ exactly matches the marginally unstable natural mode of $H(z)$ at $z = -1$. This causes $Y(z)$ to have a double pole at $z = -1$, thereby generating a mode that grows. This phenomenon of driving a system with one of its natural modes is sometimes known as *harmonic forcing*. It tends to elicit a large response because the input reinforces the natural motion of the system.

---

*2.8.2 The Jury Test

The stability criterion in Theorem 2.1 provides us with a simple and easy way to test for stability. In particular, if $a = [1, a_1, a_2, \ldots, a_n]^T$ denotes the $(n+1) \times 1$ coefficient vector of the denominator polynomial of $H(z)$, then the following MATLAB command returns the radius of the largest pole.

```
r = max(abs(roots(a)))
```

In this case, the system is stable if, and only if, $r < 1$. There are instances, when a discrete-time system has one or more unspecified design parameters and we want to determine a *range* of values for the parameters over which the system is stable. To see how this can be done, let $a(z)$ denote the denominator polynomial of the transfer function $H(z)$ where it is assumed, for convenience, that $a_0 > 0$.

$$a(z) = a_0 z^n + a_1 z^{n-1} + \cdots + a_n \tag{2.74}$$

The objective is to determine if $a(z)$ is a *stable* polynomial: a polynomial whose roots lie strictly inside the unit circle. There are two simple, necessary conditions that a stable polynomial must satisfy, namely:

$$a(1) > 0 \tag{2.75a}$$

$$(-1)^n a(-1) > 0 \tag{2.75b}$$

If the polynomial in question violates either Eq. 2.75a or Eq. 2.75b, then there is no point in testing further because it is known to be *unstable*.

Row	Coefficients			
1	$\mathbf{a_0}$	a_1	a_2	a_3
2	a_3	a_2	a_1	a_0
3	$\mathbf{b_0}$	b_1	b_2	
4	b_2	b_1	b_0	
5	$\mathbf{c_0}$	c_1		
6	c_1	c_0		
7	$\mathbf{d_0}$			
8	d_0			

Suppose $a(z)$ passes the necessary conditions in Eq. 2.75, which means that it is a viable candidate for a stable polynomial. We then construct a table from the coefficients of $a(z)$ known as a *Jury table*. The case $n = 3$ is shown in Table 2.3.

Observe that the first two rows of the Jury table are obtained directly from inspection of the coefficients of $a(z)$. The rows appear in pairs with the even rows being reversed versions of the odd rows immediately above them. The odd rows are constructed using 2×2 determinants as follows.

$$b_0 = \frac{1}{a_0} \begin{vmatrix} a_0 & a_3 \\ a_3 & a_0 \end{vmatrix}, \quad b_1 = \frac{1}{a_0} \begin{vmatrix} a_0 & a_2 \\ a_3 & a_1 \end{vmatrix}, \quad b_2 = \frac{1}{a_0} \begin{vmatrix} a_0 & a_1 \\ a_3 & a_2 \end{vmatrix} \qquad (2.76a)$$

$$c_0 = \frac{1}{b_0} \begin{vmatrix} b_0 & b_2 \\ b_2 & b_0 \end{vmatrix}, \quad c_1 = \frac{1}{b_0} \begin{vmatrix} b_0 & b_1 \\ b_2 & b_1 \end{vmatrix} \qquad (2.76b)$$

$$d_0 = \frac{1}{c_0} \begin{vmatrix} c_0 & c_1 \\ c_1 & c_0 \end{vmatrix} \qquad (2.76c)$$

As a matter of convenience, any odd row can be scaled by a positive value. Once the Jury table is constructed, it is a simple matter to determine if the polynomial is stable. The polynomial $a(z)$ is stable if the first entry in each *odd* row of the Jury table is positive.

Stability condition

$$a_0 > 0, \ b_0 > 0, \ c_0 > 0, \ \cdots \qquad (2.77)$$

Example 2.25   **Jury Test**

Consider the following general second-order discrete-time system with design parameters a_1 and a_2.

$$H(z) = \frac{b(z)}{z^2 + a_1 z + a_2}$$

For this system,

$$a(z) = z^2 + a_1 z + a_2$$

The first two rows of the Jury table are

$$J_2 = \begin{bmatrix} 1 & a_1 & a_2 \\ a_2 & a_1 & 1 \end{bmatrix}$$

Thus, the elements in the third row are

$$b_0 = \begin{vmatrix} 1 & a_2 \\ a_2 & 1 \end{vmatrix} = 1 - a_2^2$$

$$b_1 = \begin{vmatrix} 1 & a_1 \\ a_2 & a_1 \end{vmatrix} = a_1(1 - a_2)$$

From the stability condition $b_0 > 0$, we have

$$|a_2| < 1$$

The first four rows of the Jury table are

$$J_4 = \begin{bmatrix} 1 & a_1 & a_2 \\ a_2 & a_1 & 1 \\ 1 - a_2^2 & a_1(1 - a_2) & \\ a_1(1 - a_2) & a - a_2^2 & \end{bmatrix}$$

Thus, the element in the fifth row of the Jury table is

$$c_0 = \frac{1}{1 - a_2^2} \begin{vmatrix} 1 - a_2^2 & a_1(1 - a_2) \\ a_1(1 - a_2) & 1 - a_2^2 \end{vmatrix}$$

$$= \frac{(1 - a_2^2)^2 - a_1^2(1 - a_2)^2}{1 - a_2^2}$$

$$= \frac{(1 - a_2)^2[(1 + a_2)^2 - a_1^2]}{1 - a_2^2}$$

Since $|a_2| < 1$, the denominator is positive and the first factor in the numerator is positive. Thus, the condition $c_0 > 0$ reduces to

$$(1 + a_2)^2 > a_1^2$$

We can break this inequality into two cases. If $a_1 \geq 0$, then $(1 + a_2) > a_1$ or

$$a_2 > a_1 - 1, \quad a_1 \geq 0$$

Thus, in the a_2 versus a_1 plane, a_2 must be above a line of slope 1 and intercept -1 when $a_1 \geq 0$. If $a_1 < 0$, then $1 + a_2 > -a_1$ or

$$a_2 > -a_1 - 1, \quad a_1 < 0$$

Thus, a_2 must be above a line of slope -1 and intercept -1 when $a_1 < 0$. Adding the row three constraint, $|a_2| < 1$, this generates the stable region in parameter space shown in Figure 2.21. As long as the two coefficients lie in the shaded triangular region, the second-order system will be BIBO stable.

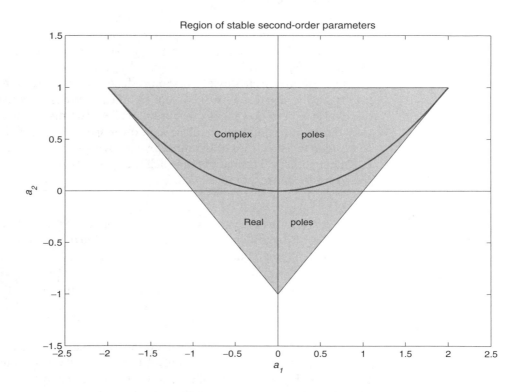

Figure 2.21 Stable Parameter Region for a Second-order System

The parabolic line in Figure 2.21 that separates real poles from complex poles is obtained by factoring the second-order denominator polynomial $a(z) = z^2 + a_1 z + a_2$, which yields poles at

$$p_{1,2} = \frac{-a_1 \pm \sqrt{a_1^2 - 4a_2}}{2} \tag{2.78}$$

When $4a_2 > a_1^2$, the poles go from real to complex. Thus, the parabola $a_2 = a_1^2/4$ separates the two regions in Figure 2.21. Recall that stable real poles generate natural modes in the form of damped exponentials. A stable complex conjugate pair of poles generates a natural mode that is an exponentially damped sinusoid.

Figure 2.22 Digital Filter with Transfer Function $H(z)$

• • • • • • • • • • • • • •

2.9 Frequency Response

The spectral characteristics of a signal $x(k)$ can be modified in a desired manner by passing $x(k)$ through a linear discrete-time system to produce a second signal $y(k)$ as shown in Figure 2.22. In this case, we refer to $x(k)$ as the input, $y(k)$ as the output, and the system that processes $x(k)$ to produce $y(k)$ as a digital filter.

Typically, signal $x(k)$ will contain significant power at certain frequencies and less power (or no power) at other frequencies. The distribution of average power over different frequencies is called the power spectrum. A digital filter is designed to *reshape* the spectrum of the input by removing certain spectral components and enhancing others. The manner in which the spectrum of $x(k)$ is reshaped is specified by the frequency response of the filter.

DEFINITION

2.8: Frequency Response

Let $H(z)$ be the transfer function of a stable linear discrete-time system, and let T be the sampling interval. Then the *frequency response* of the system is denoted $H(f)$ and defined

$$H(f) \triangleq H(z)|_{z=\exp(j2\pi fT)}, \qquad 0 \le f \le \frac{f_s}{2}$$

As the *frequency* f ranges over the interval $[0, f_s/2]$, the argument $z = \exp(j2\pi fT)$ traces out the top half of the unit circle in a counterclockwise sense as shown in Figure 2.23. In other words, the frequency response is just the transfer function $H(z)$ evaluated along the top half of the unit circle. Recall that the upper-frequency limit, $f_s/2$, is the highest frequency that a sampled signal can contain without aliasing. In Chapter 1, we referred to $f_s/2$ as the folding frequency.

Notation

A comment about notation is in order. Note that the same base symbol, H, is being used to denote both the transfer function, $H(z)$, and the frequency response, $H(f)$: two quantities that are distinct but related. An alternative approach would be to introduce separate symbols for each. However, the need for separate symbols arises repeatedly, and using distinct symbols in each instance quickly leads to a proliferation of symbols that can be confusing in its own right. Instead, we adopt the convention that the *argument type*, a complex z or a real f, is used to dictate the meaning of H.

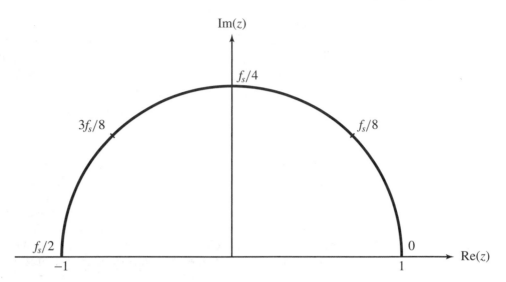

Figure 2.23 Evaluation of Frequency Response

The frequency response can be regarded as being defined over the larger frequency range, $[-f_s/2, f_s/2]$. However, if $H(z)$ is generated by a difference equation with real coefficients, then all of the information about $H(f)$ is contained in the positive frequency range $[0, f_s/2]$. More specifically, if the coefficients of $H(z)$ are real, then the frequency response satisfies the following *symmetry* property.

$$H(-f) = H^*(f), \quad 0 \le f \le f_s/2 \tag{2.79}$$

Since $H(f)$ is complex, it can be written in polar form as $H(f) = A(f) \exp[\phi(f)]$. The magnitude of $H(f)$ is called the *magnitude response* of the filter.

Magnitude response

$$A(f) \stackrel{\Delta}{=} |H(f)|, \quad 0 \le f \le f_s/2 \tag{2.80}$$

In view of the symmetry property in Eq. 2.79, the magnitude response of a real filter is an *even* function of f. Similarly, the phase angle $\phi(f)$ is called the *phase response* of the filter. From Eq. 2.79, the phase response of a real filter is an *odd* function of f.

Phase response

$$\phi(f) \stackrel{\Delta}{=} \tan^{-1}\left\{\frac{\text{Im}[H(f)]}{\text{Re}[H(f)]}\right\}, \quad 0 \le f \le f_s/2 \tag{2.81}$$

Just as the impulse response $h(k)$ can be recovered from the transfer function $H(z)$ by applying the inverse Z-transform, the impulse response also can be obtained from the frequency response $H(f)$. To see this, note from Definition 2.8 that $H(f)$ can be expressed as

$$H(f) = \sum_{k=0}^{\infty} h(k) \exp(-jk2\pi fT) \qquad (2.82)$$

Recall that for a physically realizable system, the impulse response is causal with $h(k) = 0$ for $k < 0$. Thus, the summation limits in Eq. 2.82 can be extended from 0 to ∞ to $-\infty$ to ∞ without changing the result. When this is done, Eq. 2.82 can be regarded as a complex Fourier series representation of a periodic function $H(f)$. A direct evaluation using Eq. 2.82 reveals that $H(f + f_s) = H(f)$. Thus, $H(f)$ is periodic with period f_s. Given that Eq. 2.82 is a complex Fourier series expansion of $H(f)$, from Appendix C the kth complex Fourier coefficient can be computed as

Impulse response

$$h(k) = T \int_{-f_s/2}^{f_s/2} H(f) \exp(jk2\pi fT) df \qquad (2.83)$$

Thus, the impulse response can be recovered from the frequency response using Eq. 2.83. Recall from Eq. 2.79 that if the coefficients of the transfer function $H(z)$ are real, then $H^*(f) = H(-f)$. Consequently, using Eq. 2.83 the complex conjugate of $h(k)$ is

$$
\begin{aligned}
h^*(k) &= T \int_{-f_s/2}^{f_s/2} H^*(f) \exp(-jk2\pi fT) df \\
&= T \int_{-f_s/2}^{f_s/2} H(-f) \exp(-jk2\pi fT) df \\
&= -T \int_{f_s/2}^{-f_s/2} H(F) \exp(jk2\pi FT) dF, \quad F = -f \\
&= T \int_{-f_s/2}^{f_s/2} H(F) \exp(jk2\pi FT) dF \\
&= h(k) \qquad (2.84)
\end{aligned}
$$

It follows that if $H(f)$ is the frequency response of a filter with real coefficients, then the complex Fourier series coefficients in Eq. 2.83 are real.

2.9.1   Sinusoidal Inputs

There is a simple physical interpretation of the frequency response. Suppose we excite the digital filter with a pure tone: a sinusoidal input with frequency f_a Hz where $0 \le f_a \le f_s/2$.

$$x(k) = \sin(2\pi f_a kT) \tag{2.85}$$

The zero-state response or output of the filter will consist of a natural response due to the poles of the transfer function $H(z)$, plus a forced response due to the poles of the forcing function $X(z)$. To simplify the notation, let

$$\theta \overset{\Delta}{=} 2\pi f_a T \tag{2.86}$$

Then the input is $x(k) = \sin(k\theta)$. From Table 2.1, the Z-transform of the zero-state response is

$$
\begin{aligned}
Y(z) &= H(z)X(z) \\
&= \frac{H(z)\sin(\theta)z}{z^2 - 2\cos(\theta)z + 1} \\
&= \frac{H(z)\sin(\theta)z}{[z - \exp(j\theta)][z - \exp(-j\theta)]}
\end{aligned}
\tag{2.87}
$$

We can obtain the output $y(k)$ by inverting $Y(z)$ in Eq. 2.87 using the residue method. Suppose $H(z)$ is stable. Then the poles of $H(z)$ all have magnitudes less than one. As a result, the natural-mode terms in $y(k)$ will all decay to zero as $k \to \infty$. Therefore, the steady-state response of the filter consists of the forced modes associated with the poles of $X(z)$. Using the residue method, the steady-state response is then

$$y_{ss}(k) = \mathrm{Res}[\exp(j\theta)] + \mathrm{Res}[\exp(-j\theta)] \tag{2.88}$$

From Eq. 2.87, Definition 2.8, and Euler's identity, the residue of $Y(z)z^{k-1}$ at $z = \exp(j\theta)$ is

$$
\begin{aligned}
\mathrm{Res}[\exp(j\theta)] &= \frac{H[\exp(j\theta)]\sin(\theta)\exp(jk\theta)}{\exp(j\theta) - \exp(-j\theta)} \\
&= \frac{H(f_a)\sin(\theta)\exp(jk\theta)}{\exp(j\theta) - \exp(-j\theta)} \\
&= \frac{H(f_a)\exp(jk\theta)}{j2}
\end{aligned}
\tag{2.89}
$$

Since the poles of $X(z)$ form a complex-conjugate pair, so do their residues. Thus, from Eq. 2.79 and Eq. 2.89, the residue of $H(z)z^{k-1}$ at $z = \exp(-j\theta)$ is

$$\mathrm{Res}[\exp(-j\theta)] = \frac{H(-f_a)\exp(-jk\theta)}{-j2} \tag{2.90}$$

Finally, substituting Eq. 2.89 and Eq. 2.90 into Eq. 2.88 and simplifying using Eq. 2.79 and trigonometric identities from Appendix D, we arrive at the following steady-state response

$$
\begin{aligned}
y_{ss}(k) &= \frac{H(f_a)\exp(jk\theta)}{j2} + \frac{H(-f_a)\exp(-jk\theta)}{-j2} \\
&= \frac{H(f_a)\exp(jk\theta) - H(-f_a)\exp(-jk\theta)}{j2} \\
&= \frac{H(f_a)\exp(jk\theta) - H^*(f_a)\exp(-jk\theta)}{j2} \\
&= \frac{H(f_a)\exp(jk\theta) - [H(f_a)\exp(jk\theta)]^*}{j2} \\
&= \text{Im}\{H(f_a)\exp(jk\theta)\} \tag{2.91}
\end{aligned}
$$

Recall that the frequency response can be represented in polar form as $H(f_a) = A(f_a)\exp[j\phi(f_a)]$. Thus, Eq. 2.91 can be simplified further as

$$
\begin{aligned}
y_{ss}(k) &= \text{Im}\{A(f_a)\exp[\phi(f_a)]\exp(jk\theta)\} \\
&= A(f_a)\text{Im}\{\exp[j\phi(f_a)]\exp(jk\theta)\} \\
&= A(f_a)\text{Im}\{\exp[jk\theta + j\phi(f_a)]\} \\
&= A(f_a)\sin[k\theta + \phi(f_a)] \tag{2.92}
\end{aligned}
$$

Consequently, for a stable digital filter with frequency response $H(f)$, the steady-state response to the sinusoidal input of frequency f_a in Eq. 2.85 is

Sinusoidal
steady-state
response

$$
y_{ss}(k) = A(f_a)\sin[2\pi f_a kT + \phi(f_a)] \tag{2.93}
$$

It follows that the magnitude response and the phase response have simple interpretations which allow them to be measured directly. The magnitude response $A(f_a)$ specifies the *gain* of the system at frequency f_a: the amount by which the sinusoidal input $x(k)$ is amplified or attenuated. The phase response $\phi(f_a)$ specifies the *phase shift* of the system at frequency f_a: the number of radians by which the sinusoidal input $x(k)$ is advanced if positive or delayed if negative. Recall that a cosine is just a sign shifted by $\pi/2$ radians. Therefore, if the sine in Eq. 2.85 is replaced by a cosine, the steady-state response in Eq. 2.93 continues to hold, but with the sine replaced by the cosine.

By designing a filter with a prescribed magnitude response, certain frequencies can be removed and others enhanced. The design of digital FIR filters is the focus of Chapter 6, and the design of digital IIR filters is the focus of Chapter 8.

Example 2.26   **Frequency Response**

As an example of computing the frequency response of a discrete-time system, consider a second-order digital filter with the following transfer function.

$$H(z) = \frac{z + 1}{z^2 + .64}$$

Let $\theta = 2\pi f T$. Then, using Definition 2.8 and Euler's identity, the frequency response is

$$H(f) = \left. \left(\frac{z + 1}{z^2 - .64} \right) \right|_{z = \exp(j\theta)}$$

$$= \frac{\exp(j\theta) + 1}{\exp(j2\theta) - .64}$$

$$= \frac{\cos(\theta) + 1 + j \sin(\theta)}{\cos(2\theta) - .64 + j \sin(2\theta)}$$

Thus, from Eq. 2.80 and Eq. 2.86, the magnitude response of the filter is

$$A(f) = |H(f)|$$

$$= \frac{\sqrt{[\cos(2\pi f T) + 1]^2 + \sin^2(2\pi f T)}}{\sqrt{[\cos(4\pi f T) - .64]^2 + \sin^2(4\pi f T)}}$$

Similarly, from Eq. 2.81 and Eq. 2.86, the phase response of the filter is

$$\phi(f) = \angle H(f)$$

$$= \tan^{-1} \left[\frac{\sin(2\pi f T)}{\cos(2\pi f T) + 1} \right] - \tan^{-1} \left[\frac{\sin(4\pi f T)}{\cos(4\pi f T) - .64} \right]$$

Plots of the magnitude response and the phase response for the case $f_s = 2000$ Hz are shown in Figure 2.24. Notice that frequencies near $f = 500$ Hz are enhanced by the filter, whereas the frequency component at $f = 1000$ Hz is eliminated.

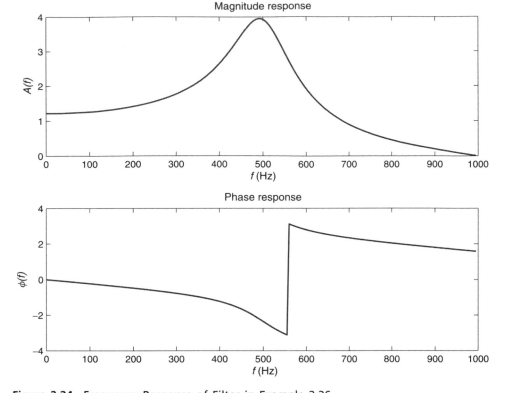

Figure 2.24 Frequency Response of Filter in Example 2.26

2.9.2 Periodic Inputs

The expression for the sinusoidal steady-state output in Eq. 2.93 can be generalized to periodic inputs. Suppose $x_a(t)$ is periodic with period τ and fundamental frequency $f_0 = 1/\tau$. The signal $x_a(t)$ can be approximated with a truncated Fourier series as follows.

$$x_a(t) \approx \frac{d_0}{2} + \sum_{i=1}^{M} d_i \cos(2\pi i f_0 t + \theta_i) \tag{2.94}$$

Let c_i denote the ith complex Fourier coefficient of $x_a(t)$. That is,

$$c_i = \frac{1}{\tau} \int_{-\tau/2}^{\tau/2} x_a(t) \exp(-j2\pi i f_0 t) dt, \quad i \geq 0 \tag{2.95}$$

From Appendix C, the magnitude d_i and phase angle θ_i of the ith harmonic of $x_a(t)$ can then be obtained from c_i as $d_i = 2|c_i|$ and $\theta_i = \angle c_i$, respectively. Next, let the samples $x(k) = x_a(kT)$ be a discrete-time input to a stable linear system with transfer function $H(z)$.

$$x(k) = \frac{d_0}{2} + \sum_{i=1}^{M} d_i \cos(2\pi i f_0 kT + \theta_i) \tag{2.96}$$

Since the system $H(z)$ is linear, it follows that the steady-state response to a sum of inputs is just the sum of the steady-state responses to the individual inputs. Setting $f_a = i f_0$ in Eq. 2.93, we find that the steady-state response to $\cos(2\pi i f_0 kT + \theta_i)$ is scaled by $A(i f_0)$ and shifted in phase by $\phi(i f_0)$. Thus, the steady-state response to the sampled periodic input $x(k)$ is

Periodic steady-state response

$$y_{ss}(k) = \frac{A(0)d_0}{2} + \sum_{i=1}^{M} A(i f_0) d_i \cos[2\pi i f_0 kT + \theta_i + \phi(i f_0)] \tag{2.97}$$

It should be pointed out that there is a practical upper limit on the number of harmonics M. The sampling process will introduce aliasing if there are harmonics located above the folding frequency, $f_s/2$. Therefore, for Eq. 2.97 to be valid, the number of harmonics must satisfy

$$M < \frac{f_s}{2 f_0} \tag{2.98}$$

Example 2.27 **Steady-state Response**

As an illustration of using Eq. 2.97 to compute the steady-state response, consider the following stable first-order filter.

$$H(z) = \frac{0.2z}{z - 0.8}$$

Let $\theta = 2\pi f T$. Then, from Definition 2.8 and Euler's identity, the frequency response of this filter is

$$H(f) = \left(\frac{0.2z}{z - 0.8} \right) \Big|_{z = \exp(j\theta)}$$

$$= \frac{0.2 \exp(j\theta)}{\exp(j\theta) - 0.8}$$

$$= \frac{0.2 \exp(j\theta)}{\cos(\theta) - 0.8 + j \sin(\theta)}$$

Thus, from Eq. 2.80, the filter magnitude response is

$$A(f) = |H(f)|$$

$$= \frac{0.2}{\sqrt{[\cos(2\pi fT) - 0.8]^2 + \sin^2(2\pi fT)}}$$

Similarly, from Eq. 2.81, the filter phase response is

$$\phi(f) = \angle H(f)$$

$$= 2\pi fT - \tan^{-1}\left[\frac{\sin(2\pi fT)}{\cos(2\pi fT) - 0.8}\right]$$

Next, suppose the input $x_a(t)$ is an even periodic-pulse train of period τ where the pulses are of unit amplitude and radius $0 \le a \le \tau/2$. From Appendix C, the truncated Fourier series of $x_a(t)$ is

$$x_a(t) \approx \frac{2a}{\tau} + \frac{4a}{\tau}\sum_{i=1}^{M}\text{sinc}(2\pi i f_0 a)\cos(2\pi i f_0 t)$$

Recall from Example 1.4 that $\text{sinc}(x) = \sin(x)/x$. Suppose the period is $\tau = 0.01$ sec and $a = \tau/4$, which corresponds to a square wave with fundamental frequency $f_0 = 100$ Hz. If we sample at $f_s = 2000$ Hz, then, to avoid aliasing, the M harmonics must be below 1000 Hz. Thus, from Eq. 2.98 the maximum number of harmonics used to approximate $x_a(t)$ should be $M = 9$. The sampled version of $x_a(t)$ is then

$$x(k) = \frac{1}{2} + \sum_{i=1}^{9}\text{sinc}\left(\frac{\pi i}{2}\right)\cos(0.1\pi i k)$$

Finally, from Eq. 2.97, the steady-state response to the periodic input $x(k)$ is

$$y_{ss}(k) = \frac{1}{2} + \sum_{i=1}^{9}A(100i)\text{sinc}\left(\frac{\pi i}{2}\right)\cos[0.1\pi i k + \phi(100i)]$$

Plots of $x(k)$ and $y_{ss}(k)$, obtained by running script *exam2_27* on the distribution CD, are shown in Figure 2.25. Note the small oscillations or ringing in the approximation to the square wave due to the fact that only a finite number of harmonics are used (Gibb's phenomenon). The steady-state output is much smoother than $x(k)$ due to the lowpass nature of the filter $H(z)$.

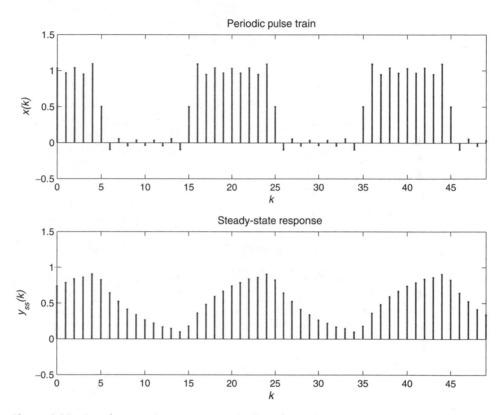

Figure 2.25 Steady-state Response to Periodic Pulse Train

MATLAB Toolbox

The FDSP toolbox contains the following function for finding the frequency response of a stable linear discrete-time system.

```
[H,f] = f_freq (b,a,N,fs);    % discrete-time frequency response
```

On entry to *f_freq*, b is a vector of length $m+1$ containing the coefficients of the numerator polynomial of the transfer function, and a is a vector of length $n+1$ containing the denominator coefficients with $a(1) = 1$.

$$H(z) = \frac{b_0 + b_1 z^{-1} + \cdots + b_m z^{-m}}{1 + a_1 z^{-1} + \cdots + a_n z^{-n}} \tag{2.99}$$

Here N is the number of points at which the frequency response is to be evaluated, and fs is an optional scalar specifying the sampling frequency. The default value is $fs = 1$ Hz. On exit from *f_freq*, H is a complex vector of length N containing the frequency response evaluated using Definition 2.8, and f is a vector of length N containing the frequencies at which H is evaluated. The frequency response is evaluated for non-negative frequencies along the top half of the unit circle. That is, $f(i) = (i-1)f_s/(2N)$ for $1 \le i \le N$. To compute and plot the magnitude response and phase response, the following standard MATLAB functions can be used.

```
A = abs(H);              % magnitude response
plot (f,A)               % magnitude response plot
phi = angle(H);          % phase response
plot (f,phi)             % phase response plot
```

An example that uses *f_freq* to compute and plot the magnitude and phase responses of a digital filter can be found in script *exam2_26* on the distribution CD. Note that *f_freq* is the discrete-time version of the continuous-time function *f_freqs* introduced in Chapter 1.

● ● ● ● ● ● ● ● ● ● ● ● ● ● · · · · ·

2.10 Software Applications

This section focuses on applications of discrete-time systems. A graphical user interface module called *g_system* is introduced that allows users to explore the input-output behavior of linear discrete-time systems without any need for programming. Application examples are then presented and solved using MATLAB scripts. The relevant MATLAB functions from the FDSP toolbox are described in Appendix B.

2.10.1 GUI Module: *g_system*

The graphical user interface module *g_system* allows the user to investigate the input-output behavior of linear discrete-time systems. GUI module *g_system* features a display screen with tiled windows as shown in Figure 2.26.

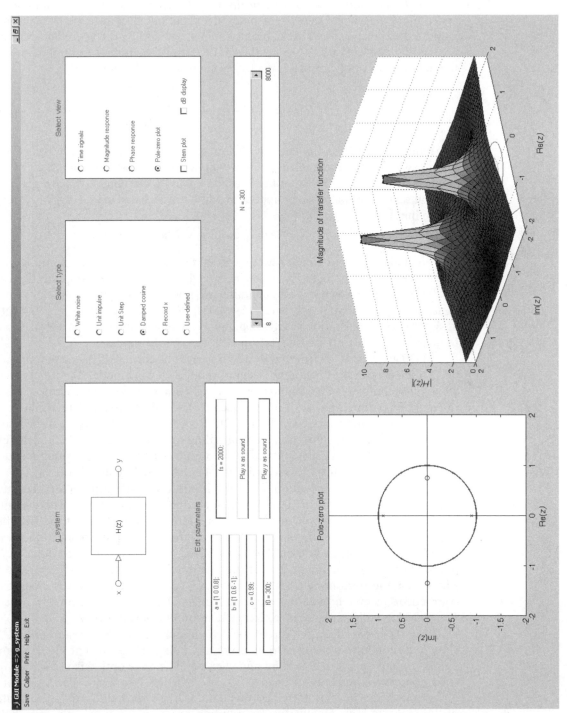

Figure 2.26 Display Screen of Chapter GUI Module *g_system*

The *Block diagram* window in the upper-left corner of the screen contains a color-coded block diagram of the system under investigation. Below the block diagram are a number of edit boxes whose contents can be modified by the user. The edit boxes for a and b allow the user to select the coefficients of the numerator polynomial and the denominator polynomial of the following transfer function.

$$H(z) = \frac{b_0 + b_1 z^{-1} + \cdots + b_m z^{-m}}{1 + a_1 z^{-1} + \cdots + a_n z^{-n}} \qquad (2.100)$$

The numerator and denominator coefficient vectors can be edited directly by clicking on the shaded area and entering in new values. Any MATLAB statement defining a and b can be used. The Enter key is used to activate a change to a parameter. Additional scalar parameters that appear in edit boxes are associated with the damped cosine input.

$$x(k) = c^k \cos(2\pi f_0 kT) u(k) \qquad (2.101)$$

They include the input frequency $0 \le f0 \le fs/2$, the exponential damping factor c, which is constrained to lie in the interval $[-1, 1]$, and the sampling frequency fs. The *Parameters* window also contains two pushbutton controls. The pushbutton controls play the signals $x(k)$ and $y(k)$ on the PC speaker using a sampling rate of $f_s = 8000$ Hz. This option is active on any PC with a sound card. It allows the user to hear the filtering effects of $H(z)$ on various types of inputs.

The *Type* and *View* windows in the upper-right corner of the screen allow the user to select both the type of input signal and the viewing mode. The inputs include white noise uniformly distributed over $[-1, 1]$, a unit impulse input, a unit step input, the damped cosine input in Eq. 2.101, recorded sounds from a PC microphone, and user-defined inputs from a MAT-file. The Recorded sound option can be used to record up to one second of sound at a sampling rate of $f_s = 8000$ Hz. For the User-defined option, a MAT-file containing the input vector x, the sampling frequency fs, and the coefficient vectors a and b must be supplied by the user.

The *View* options include plots of the input $x(k)$ and output $y(k)$, the magnitude response $A(f)$, the phase response $\phi(f)$, and a pole-zero plot. The magnitude response is either linear or logarithmic depending on the status of the dB checkbox control. Similarly, the plots of the input and the output use continuous time or discrete time, depending on the status of the Stem plot checkbox control. The pole-zero plot also includes a plot of the transfer function surface $|H(z)|$. The *Plot* window on the bottom half of the screen shows the selected view. The curves are color-coded to match the block diagram labels. The slider bar below the *Type* and *View* windows allows the user to change the number of samples N.

The *Menu* bar along the top of the screen includes several menu options. The *Caliper* option allows the user to measure any point on the current plot by moving the mouse crosshairs to that point and clicking. The *Save* option is used to save the current x, fs, a, and b in a user-specified MAT-file for future use. The User-defined input option can be used to reload this data. The *Print* option prints the contents of the plot window. Finally, the *Help* option provides the user with some helpful suggestions on how to effectively use module *g_system*.

2.10.2 Home Mortgage

Recall from Section 2.1.1 that the following discrete-time system can be used to model a home mortgage loan.

$$y(k) = y(k-1) + \left(\frac{r}{12}\right) y(k-1) - x(k) \tag{2.102}$$

Here $y(k)$ is the balance owed at the end of month k, r is the annual interest rate expressed as a fraction, and $x(k)$ is the monthly mortgage payment. One of the questions posed in Section 2.1.1 was the following. If the size of the mortgage is b dollars, and the duration of the mortgage is d years, what is the required monthly payment? Now that we have the necessary tools in place, we can answer this and related questions. To streamline the notation, let

$$a = 1 + \frac{r}{12} \tag{2.103}$$

Then from Example 2.16, the transfer function of the home mortgage system is

$$H(z) = \frac{-z}{z-a} \tag{2.104}$$

This system is different from the examples we have considered thus far in that it has a nonzero initial condition $y(-1) = b$ where b is the size of the mortgage. Consequently, the system output has two components: the zero-input response $y_0(k)$ and the zero-state response $y_1(k)$.

$$y(k) = y_0(k) + y_1(k) \tag{2.105}$$

Since $H(z)$ has a simple pole at $z = a$, the zero-input response consists of a single natural-mode term of the form $y_0(k) = c(a)^k$. Setting $y_0(-1) = b$ and solving for the constant c, we get $c = ba$ and the following zero-input response:

$$y_0(k) = b(a)^{k+1} \tag{2.106}$$

Note from Eq. 2.103 that $a > 1$. Consequently, in the absence of monthly payments, the amount owed to the lending agency grows with time. If the monthly payments are constant, $x(k) = p$, then this input can be modeled as a step of amplitude p. Thus, the Z-transform of the zero-state response is

$$Y_1(z) = H(z)X(z)$$

$$= \left(\frac{-z}{z-a}\right) \frac{pz}{z-1}$$

$$= \frac{-pz^2}{(z-a)(z-1)} \tag{2.107}$$

Next we apply Algorithm 2.2 to obtain $y_1(k)$. From the initial value theorem, $y_1(0) = -p$. The residues at the two poles are

$$\text{Res}(a) = \left.\frac{-pz^{k+1}}{(z-1)}\right|_{z=a} = \frac{-p(a)^{k+1}}{a-1} \tag{2.108a}$$

$$\text{Res}(1) = \left.\frac{-pz^{k+1}}{(z-a)}\right|_{z=1} = \frac{p}{a-1} \tag{2.108b}$$

Thus, the zero-state response is

$$
\begin{aligned}
y_1(k) &= y_1(0)\delta(k) + [\text{Res}(a) + \text{Res}(1)]u(k-1) \\
&= -p\delta(k) - p\left(\frac{a^{k+1}-1}{a-1}\right)u(k-1) \\
&= -p\left(\frac{a^{k+1}-1}{a-1}\right)u(k) \tag{2.109}
\end{aligned}
$$

If the duration of the mortgage is d years, then the loan will be paid off after $12d$ monthly payments, which means $y(12d) = 0$ in Eq. 2.105. From Eq. 2.106 and Eq. 2.109, this constraint yields

$$b(a)^{12d+1} = \frac{p(a^{12d+1}-1)}{a-1} \tag{2.110}$$

Finally, solving Eq. 2.110 for p, we get the following expression for the required monthly payment.

$$p = \frac{b(a)^{12d+1}(a-1)}{a^{12d+1}-1} \tag{2.111}$$

A family of curves which shows the required monthly payment as a function of the size of the mortgage for different length mortgages is shown in Figure 2.27. These curves, which correspond to an interest rate of 7.5 percent, were generated by the first half of the following MATLAB script, labeled *exam2_28* on the distribution CD.

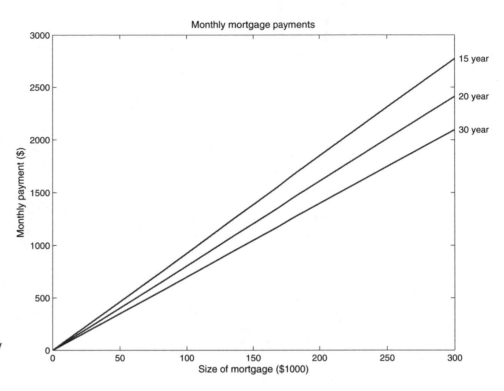

Monthly mortgage payments

Figure 2.27   Monthly Mortgage Payments for 7.5% Interest

```
function exam2_28

% Example 2.28: Home mortgage

clear
clc
fprintf('Example 2.28: Home mortgage\n')
n = 31;
b = linspace(0,300,n)';          % size of mortgage
d = [15 20 30];                  % duration in years
p = zeros(n,3);                  % monthly payments

% Compute monthly payments

r = f_prompt('Enter interest rate in percent:',0,20,7.5)/100;
a = 1 + r/12;
c = a.^(12*d+1);
for k = 1 : n
   for j = 1 : length(d)
      p(k,j) = 1000*b(k)*c(j)*(a-1)/(c(j)-1);
   end
```

```
end
figure
plot(b,p,'LineWidth',1.5)
f_labels ('Monthly mortgage payments','Size of mortgage ($1000)',...
          'Monthly payment ($)')
hold on
for j = 1 : length(d)
    duration = sprintf ('%d year',d(j));
    text (b(n)+3,p(n,j),duration)
end
f_wait

% Compute balance vs time

q = 12*d(3)+1;                     % number of months
k = [0 : q-1]';                    % discrete times
y0 = b(21)*a.^(k+1);               % zero-input response
num = [-1 0];                      % numerator coefficients
den = [1 -a];                      % denominator coefficients
x = p(21,3)*ones(q,1);            % step of amplitude p
y1 = filter(num,den,x)/1000;       % zero-state response
y(:,1) = y0 + y1;
y(:,2) = (6*p(21,3)/(1000*r))*ones(q,1);
figure
h = plot (k,y(:,1),k,y(:,2));
set (h(1),'LineWidth',1.5)
f_labels ('Balance due on 30 year $200,000 mortgage','Month','Balance ($1000)')

% Find crossover month

text (k(q)+5,y(q,2),'6\it{p/r}')
crossover = min(find(y(:,1) < y(:,2)))
hold on
plot ([k(crossover),k(crossover)],[0,y(crossover,2)],'r--')
plot (k(crossover),y(crossover),'.')
axis ([0 400 0 200])
f_wait
```

Another question that was asked in Section 2.1.1 was: How many months does it take before more than half of the monthly payment goes toward reducing the principal, rather than paying interest? The monthly interest is $(r/12)y(k)$. Thus we require $ry(k)/12 < p/2$ or

$$y(k) < \frac{6p}{r} \tag{2.112}$$

The smallest k that satisfies Eq. 2.112 is the *crossover* month. The second half of script *exam2_28* computes the zero-input response directly and uses the standard

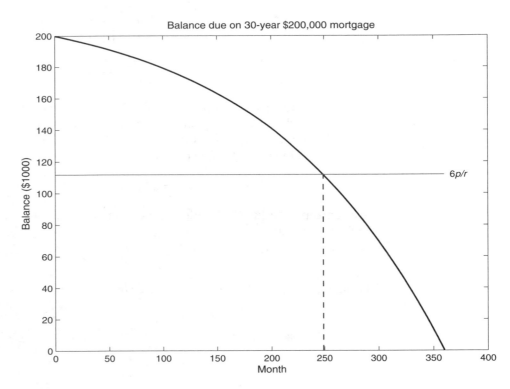

Figure 2.28 Balance Owed over the Duration of the Mortgage

MATLAB function *filter* to compute the zero-state response corresponding to a $200,000 mortgage over 30 years. The resulting plot of the balance owed is shown in Figure 2.28. Note that the crossover month does not occur until month 250, well beyond the midpoint of the 30 year loan.

2.10.3   Satellite Attitude Control

In Section 2.1.2, the following discrete-time model was introduced for a single-axis satellite attitude control system.

$$y(k) = (1 - d)y(k - 1) - dy(k - 2) + d[r(k - 1) + r(k - 2)] \qquad (2.113a)$$

$$d = \frac{cT^2}{2J} \qquad (2.113b)$$

Here $r(k)$ is the desired angular position of the satellite at the kth sampling time, and $y(k)$ is the actual angular position. The constants c, T, and J denote the controller gain, the sampling interval, and the satellite moment of inertia, respectively. By inspection of Eqs. 2.113, the transfer function of this control system is

$$H(z) = \frac{Y(z)}{R(z)}$$

$$= \frac{d(z^{-1} - z^{-2})}{1 - (1 - d)z^{-1} + dz^{-2}}$$

$$= \frac{d(z + 1)}{z^2 + (d - 1)z + d} \tag{2.114}$$

The controller gain c appearing in Eq. 2.113b is an engineering design parameter that must be chosen to satisfy some performance specification. The most fundamental performance constraint is that the control system be stable. The denominator polynomial of the transfer function is

$$a(z) = z^2 + (d - 1)z + d \tag{2.115}$$

We can apply the Jury test to determine a stable range for c. The first two rows of the Jury table are

$$J_2 = \begin{bmatrix} 1 & d - 1 & d \\ d & d - 1 & 1 \end{bmatrix} \tag{2.116}$$

Using Eq. 2.76, the elements of the third row are

$$b_0 = \begin{vmatrix} 1 & d \\ d & 1 \end{vmatrix} = 1 - d^2 \tag{2.117a}$$

$$b_1 = \begin{vmatrix} 1 & d - 1 \\ d & d - 1 \end{vmatrix} = -(1 - d)^2 \tag{2.117b}$$

From the stability condition $b_0 > 0$, we have

$$|d| < 1 \tag{2.118}$$

The first four rows of the Jury table are

$$J_4 = \begin{bmatrix} 1 & d - 1 & d \\ d & d - 1 & 1 \\ 1 - d^2 & -(1 - d)^2 & \\ -(1 - d)^2 & 1 - d^2 & \end{bmatrix} \tag{2.119}$$

Thus, the element in the fifth row of the Jury table is

$$c_0 = \frac{1}{1 - d^2} \begin{vmatrix} 1 - d^2 & -(1 - d)^2 \\ -(1 - d)^2 & 1 - d^2 \end{vmatrix}$$

$$= \frac{(1 - d^2)^2 - (1 - d)^4}{1 - d^2}$$

$$= \frac{(1 - d)^2[(1 + d)^2 - (1 - d)^2]}{1 - d^2}$$

$$= \frac{4d(1 - d)^2}{1 - d^2} \tag{2.120}$$

Since $|d| < 1$, the denominator is positive and the second factor in the numerator is positive. Thus, the stability condition $c_0 > 0$ reduces to $d > 0$. Together with Eq. 2.118, this yields $0 < d < 1$. From the definition of d in Eq. 2.113b, we conclude that the control system is BIBO stable for the following range of controller gains.

Stable range

$$0 < c < \frac{2J}{T^2} \qquad (2.121)$$

To test the effectiveness of the control system, suppose $T = 0.1$ sec. and $J = 5$ N-m-sec^2. Then $0 < c < 1000$ is the stable range. Suppose a ground station issues a command to rotate the satellite one quarter of a turn. Then the desired angular position signal is

$$r(k) = \left(\frac{\pi}{2}\right) u(k) \qquad (2.122)$$

The following MATLAB script, labeled *exam2_29* on the distribution CD, computes the resulting zero-state response for three different values of the controller gain c.

```
function exam2_29

% Example 2.29: Satellite attitude control

clear
clc
fprintf('Example 2.29: Satellite attitude control\n')
n = 21;
T = .1;                             % sampling interval
J = 5;                              % moment of inertia
c = [.1 3-sqrt(8) .5]*(2*J/T^2)     % controller gains
d = (T^2/(2*J))*c;
m = length(c);

% Compute step response

r = (pi/2)*ones(n,1);
for i = 1 : m
    a = [1 d(i)-1 d(i)];
    b = [0 d(i) d(i)];
    pole = roots(a)
    y(:,i) = (180/pi)*filter(b,a,r);
end

% Plot curves

figure
k = [0 : n-1];
```

```
for i = 1 : 3
    subplot (3,1,i);/
    hp = stem (k,y(:,i),'filled','.')
    set (hp, 'LineWidth', 1.5)
    axis ([k(1) k(n) 0 150])
    switch i
    case 1,
        title ('Satellite step response')
        text (10,120,'Overdamped','HorizontalAlignment','center')
    case 2,
        ylabel ('{\ity(k)} (deg)')
        text (10,120,'Critically damped','HorizontalAlignment','center')
    case 3,
        xlabel ('\it{k}')
        text (10,120,'Underdamped','HorizontalAlignment','center')
    end
    hold on
    plot(k,(180/pi)*r,'r')
end
f_wait
```

When script *exam2_29* is executed, it produces the plot shown in Figure 2.29. Note that for all three controller gains, the satellite turns to the desired 90 degree orientation. The control system pole locations for the three controller gains are summarized in Table 2.4. When $c = 100$, there are two distinct real natural modes and sluggish *overdamped* response results. When $c = 171.6$, there is a double real natural mode which results in a *critically damped* response. This is the fastest possible step response among those that do not overshoot the final position. When $c = 500$, there is a pair of complex conjugate natural modes which generates an oscillatory *underdamped* response.

Table 2.4: ▶
Control System Pole
Locations for
Different Controller
Gains

c	**Poles**	**Case**
100	$p_{1,2} = .770, .130$	Overdamped
171.6	$p_{1,2} = .414, .414$	Critically damped
500	$p_{1,2} = .25 \pm .661j$	Underdamped

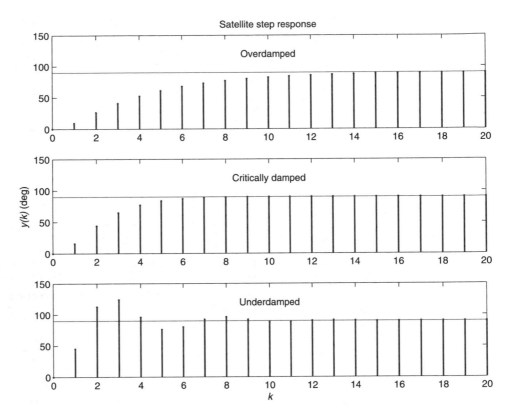

Figure 2.29 Response of Satellite to a 90 Degree Step Input using Different Controller Gains

2.10.4   The Fibonacci Sequence and the Golden Ratio

There is a simple discrete-time system that can be used to produce a well-known sequence of numbers called the *Fibonacci sequence*.

$$y(k) = y(k-1) + y(k-2) + x(k) \tag{2.123}$$

The impulse response of this system is the Fibonacci sequence. Note that with $x(k) = \delta(k)$ and a zero initial condition, we have $y(0) = 1$. For $k > 0$, the next number in the sequence is just the sum of the previous two. This yields the following impulse response.

$$h(k) = \{1, 1, 2, 3, 5, 8, 13, 21, 34, 55, 89, \ldots\} \tag{2.124}$$

Fibonacci introduced this model in 1202 to describe how fast rabbits could breed under ideal circumstances (Cook, 1979). He started with one male–female pair and assumed that at the end of each month they breed. One month later the female produces another male–female pair and the process continues. The number of pairs at the end of each month then follows the pattern shown in Eq. 2.124. The Fibonacci numbers occur in a surprising number of places in nature. For example, the number of petals in flowers is often a Fibonacci number as can be seen from Table 2.5.

Flower	Number of petals
Iris	3
Wild rose	5
Delphinium	8
Corn marigold	13
Aster	21
Pyrethrum	34
Michelmas daisy	55

The system used to generate the Fibonacci sequence is an unstable system with $h(k)$ growing without bound as $k \rightarrow \infty$. However, it is of interest to investigate the *ratio* of successive samples of the impulse response. This ratio converges to a special number called the *golden ratio*.

$$\gamma \triangleq \lim_{k \to \infty} \left\{ \frac{h(k)}{h(k-1)} \right\} \approx 1.618 \qquad (2.125)$$

The golden ratio is noteworthy in that it has been used in Greek architecture as far back as 430 BC in the construction of the Parthenon: a temple to the goddess Athena. For example, the ratio of the width to the height of the front of the temple is γ.

Consider the problem of finding the exact value of the golden ratio. From Eq. 2.123, the transfer function of the Fibonacci system is

$$H(z) = \frac{1}{1 - z^{-1} - z^{-2}}$$

$$= \frac{z^2}{z^2 - z - 1} \qquad (2.126)$$

Factoring the denominator, this system has poles at

$$p_{1,2} = \frac{1 \pm \sqrt{5}}{2} \qquad (2.127)$$

Using Algorithm 2.2, the initial value is $h(0) = 1$, and the residues at the two poles are

$$\text{Res}(p_1) = \frac{p_1^{k+1}}{p_1 - p_2} \qquad (2.128a)$$

$$\text{Res}(p_2) = \frac{p_2^{k+1}}{p_2 - p_1} \qquad (2.128b)$$

Thus, the closed-form expression for the impulse response is

$$h(k) = h(0)\delta(k) + [\text{Res}(p_1) + \text{Res}(p_2)]u(k-1)$$

$$= \delta(k) + \left(\frac{p_1^{k+1} - p_2^{k+1}}{p_1 - p_2}\right)u(k-1)$$

$$= \left(\frac{p_1^{k+1} - p_2^{k+1}}{p_1 - p_2}\right)u(k) \tag{2.129}$$

Note from Eq. 2.127 that $|p_1| > 1$ and $|p_2| < 1$. Consequently, $p_2^{k+1} \to 0$ as $k \to \infty$. Thus, from Eq. 2.125 and Eq. 2.129, we have

$$\gamma = \lim_{k\to\infty}\left\{\frac{p_1^{k+1}}{p_1^k}\right\} = p_1 \tag{2.130}$$

Consequently, the golden ratio is

Golden ratio

$$\gamma = \frac{1 + \sqrt{5}}{2} = 1.6180339\cdots \tag{2.131}$$

The following MATLAB script, labeled *exam2_30* on the distribution CD, computes the impulse response of the system in Eq. 2.123 to produce the Fibonacci sequence. It also computes the golden ratio, both directly as in Eq. 2.131 and indirectly as in Eq. 2.125.

```
function exam2_30

% Example 2.30: The Fibonacci sequence and the golden ratio

clear
clc
fprintf('Example 2.30: The Fibonacci sequence and the golden ratio\n')
N = 21;
gamma = (1 + sqrt(5))/2;        % golden ratio
g = zeros(N,1);                 % estimates of gamma
a = [1 -1 -1];                  % denominator coefficients
b = 1;                          % numerator coefficients

% Estimate golden ratio with pulse response

[h,k] = f_impulse (b,a,N);
h
for i = 2 : N
   g(i) = h(i)/h(i-1);
end
```

```
figure
stem (k(2:N),g(2:N),'filled','.')
hp = /;/
set (hp, 'LineWidth', 1.5)
f_labels ('The golden ratio','\it{k}','\it{h(k)/h(k-1)}')
axis ([k(1) k(N) 0 3])
hold on
plot (k,gamma*ones(N),'r')
golden = sprintf ('\\gamma = %.6f',gamma);
text (10,2.4,golden,'HorizontalAlignment','center')
f_wait
```

When script *exam2_30* is executed, it produces the plot shown in Figure 2.30 which graphs $g(k) = h(k)/h(k-1)$. It is evident that $g(k)$ rapidly converges to the Golden ratio, γ.

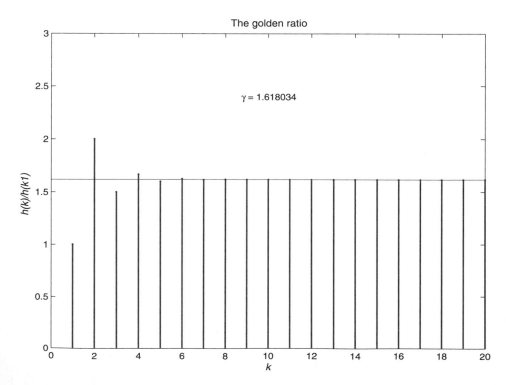

Figure 2.30 Numerical Approximation of the Golden Ratio

● ● ● ● ● ● ● ● ● ● ● ● ● ● ● ●
2.11 Chapter Summary

The Z-transform

This chapter focused on the analysis of linear time-invariant discrete-time systems using the Z-transform. The Z-transform is an essential analytical tool for digital signal processing. It is a transformation which maps a causal sequence $x(k)$ into an algebraic function $X(z)$ of a complex variable z. Often the Z-transform can be computed by consulting a table of Z-transform pairs. The size of the table is effectively enlarged by using the Z-transform properties which include the linearity, delay, z-scale, and differentiation properties. The initial and final value theorems allow us to recover the initial and final values of values of the signal $x(k)$ directly from its Z-transform $X(z)$. More generally, a finite number of samples of $x(k)$ can be obtained by inverting the Z-transform using the synthetic division method. If a closed-form expression for $x(k)$ is desired, then the inverse Z-transform should be computed using either the partial fraction method or the residue method. The residue method is the method of choice for the case of multiple poles because it requires less computational effort than the partial fraction method. Furthermore, unlike the partial fraction method, the residue method does not require the use of a table of Z-transform pairs.

The Transfer Function

The *transfer function* of a discrete-time system is a compact algebraic representation of the system defined as the Z-transform of the output divided by the Z-transform of the input assuming the initial condition is zero.

$$H(z) = \frac{Y(z)}{X(z)} \tag{2.132}$$

Using the linearity and delay properties, the transfer function can be obtained directly from inspection of the difference equation representation of a discrete-time system. A third representation of a discrete-time system is the signal flow graph which is a concise graphical description.

The transfer function can be used to compute the zero-state response of the system, the part of the output that is generated by the input. The complete response also includes the zero-input response which is the part of the output that is generated by the initial condition. Typically, the transfer function is a ratio of two polynomials in z. The roots of the numerator polynomial are called the *zeros* of the system, and the roots of the denominator polynomial are called the *poles* of the system. Each pole of $H(z)$ generates a term in the output $y(k)$ called a *natural mode*. If the magnitude of the pole is less than one, then the natural mode will decay to zero with increasing time.

Impulse Response

The time-domain equivalent of the transfer function is the impulse response. The *impulse response* is the zero-state output of the system when the input is the unit impulse $\delta(k)$. The easiest way to compute the impulse response $h(k)$ is to take the inverse Z-transform of the transfer function $H(z)$. The zero-state response of the system to any input can be obtained from the impulse response using convolution.

$$y(k) = \sum_{i=0}^{k} h(i)x(k-i) \tag{2.133}$$

Convolution, in the time domain, maps into multiplication in the Z-transform domain. Therefore, the Z-transform of Eq. 2.133 is simply $Y(z) = H(z)X(z)$. If a system has an impulse response that is zero after a finite number of samples, then it is called an FIR system. Otherwise, it is an IIR system. The impulse response of an FIR system can be obtained directly from inspection of the transfer function, the difference equation, or the signal flow graph.

A system is *BIBO stable* if, and only if, every bounded input is guaranteed to produced a bounded output. Otherwise, the system is *unstable*. For stable systems, the natural modes all decay to zero. Therefore, a system is stable if, and only if, all of the poles lie strictly inside the unit circle of the complex plane. The Jury test is a tabular stability test that can be used to determine ranges for the system parameters over which a system is stable. All FIR systems are stable, but IIR systems may or may not be stable.

The Frequency Response

The frequency response of a discrete-time system is just the transfer function evaluated along the top half of the unit circle. If f_s is the sampling frequency and $T = 1/f_s$ is the sampling interval, then the frequency response is

$$H(f) = H(z)|_{z=\exp(j2\pi fT)}, \quad 0 \le f \le \frac{f_s}{2} \tag{2.134}$$

A digital filter is a discrete-time system that is designed to have a prescribed frequency response. The magnitude $A(f) = |H(f)|$ is called the *magnitude response* of the filter, and the phase angle $\phi(f) = \angle H(f)$ is called the *phase response* of the filter. When a stable discrete-time system is driven with a sinusoidal input with frequency $0 \le f_a \le f_s/2$, the steady-state output is a sinusoid of frequency f_a whose amplitude is scaled by $A(f_a)$ and whose phase is shifted by $\phi(f_a)$. By designing a digital filter with a prescribed magnitude response, certain frequencies can be removed from the input signal, and other frequencies can be enhanced. The impulse response can be recovered from the frequency response.

The FDSP toolbox includes a GUI module called *g_system* that allows the user to investigate the input-output behavior of a discrete-time system. A variety of systems and signals can be examined graphically without any programming required.

2.12 Problems

The problems are divided into Analysis problems that can be solved by hand or with a calculator, GUI Simulation problems that are solved using GUI module *g_system*, and Computation problems. The Computation problems require the student to write a MATLAB script using the FDSP toolbox functions summarized in Appendix B. Solutions to selected problems are available on the distribution CD. Students are encouraged to use these problems, which are identified with a ☑, as a check on their understanding of the material.

2.12.1 Analysis

Section 2.2

P2.1. Consider the following finite discrete-time signal.

$$x = \{8, -6, 4, -2, 0, 0, \cdots\}$$

(a) Find the Z-transform $X(z)$, and express it as a ratio of two polynomials in z.

(b) What is the region of convergence of $X(z)$?

P2.2. Find the Z-transform of the following signal.

$$x(k) = k^3 u(k)$$

P2.3. Consider the following discrete-time signal.

$$x(k) = a^k \sin(bk + \theta)u(k)$$

(a) Use Table 2.1 and the trigonometric identities in Appendix D to find $X(z)$.

(b) Verify that $X(z)$ reduces to entry three of Table 2.1 when $\theta = 0$.

(c) Verify that $X(z)$ reduces to entry four of Table 2.1 when $\theta = \pi/2$.

P2.4. Show that the formulation of the geometric series in Eq. 2.13 can be generalized as follows where $m \geq 0$ and $n \geq m$. Verify that this reduces to the classic geometric series when $m = 0$, $n = \infty$, and $f(z) = z$. *Hint*: Write the finite sum as a difference of two infinite series.

$$\sum_{k=m}^{n} f^k(z) = \frac{f^m(z) - f^{n+1}(z)}{1 - f(z)}, \quad |f(z)| < 1$$

Section 2.3

P2.5. Consider the following signal.

$$x(k) = \begin{cases} 10, & 0 \le k < 4 \\ -2, & 4 \le k < \infty \end{cases}$$

(a) Write $x(k)$ as a difference of two step signals.

(b) Use the delay property to find $X(z)$. Express your final answer as a ratio of two polynomials in z.

(c) Find the region of convergence of $X(z)$.

P2.6. Consider the following signal.

$$x(k) = \begin{cases} 2k, & 0 \le k < 9 \\ 18, & 9 \le k < \infty \end{cases}$$

(a) Write $x(k)$ as a difference of two ramp signals.

(b) Use the delay property and Example 2.7 to find $X(z)$. Express your final answer as a ratio of two polynomials in z.

(c) Find the region of convergence of $X(z)$.

P2.7. Consider the following Z-transform.

$$X(z) = \frac{10(z-2)^2(z+1)^3}{(z-.8)^2(z-1)(z-.2)^2}$$

(a) Find $x(0)$ without inverting $X(z)$.

(b) Find $x(\infty)$ without inverting $X(z)$.

(c) Write down the *form* of $x(k)$ from inspection of $X(z)$. You can leave the coefficients of each term of $X(z)$ unspecified.

P2.8. A student attempts to apply the final value theorem to the following Z-transform and gets the steady-state value $x(\infty) = -5$. Is this correct? If not, what is the value of $x(k)$ as $k \to \infty$? Explain your answer.

$$X(z) = \frac{10z^3}{(z^2 - z - 2)(z - 1)}, \quad |z| > 2$$

Section 2.4

P2.9. Consider the following Z-transform.

$$X(z) = \frac{z^4 + 2z^3 + 3z^2 + 2z + 1}{z^4}, \quad |z| > 0$$

(a) Rewrite $X(z)$ in terms negative powers of z.

(b) Use Definition 2.1 to find $x(k)$.

(c) Check to see that $x(k)$ is consistent with the initial value theorem.

(d) Check to see that $x(k)$ is consistent with the final value theorem.

P2.10. Consider the following Z-transform.

$$X(z) = \frac{2z}{z^2 - 1}, \quad |z| > 1$$

(a) Find $x(k)$ for $0 \le k \le 5$ using the synthetic division method.
(b) Find $x(k)$ using the partial fraction method.
(c) Find $x(k)$ using the residue method.

P2.11. Consider the following Z-transform. Find $x(k)$ using the residue method.

$$X(z) = \frac{100}{z^2(z - .5)^3}, \quad |z| > .5$$

P2.12. Consider the following Z-transform. Use Algorithm 2.2 to find $x(k)$. Express your final answer as a real signal.

$$X(z) = \frac{1}{z^2 + 1}, \quad |z| > 1$$

P2.13. Repeat Problem P2.12, but use Table 2.1 and the Z-transform properties.

☑ **P2.14.** Consider the following Z-transform. Find $x(k)$.

$$X(z) = \frac{5z^3}{(z^2 - z + .25)(z + 1)}, \quad |z| > 1$$

Section 2.5

P2.15. Consider the running average filter of order m from Section 2.1.3.

$$y(k) = \frac{1}{m + 1} \sum_{i=0}^{m} x(k - i)$$

(a) Find the transfer function $H(z)$. Express it as a ratio of two polynomials in z.
(b) Use the geometric series in Eq. 2.13 to show that an alternative form of the transfer function is as follows. *Hint*: Express $y(k)$ as a difference of two sums.

$$H(z) = \frac{z^{m+1} - 1}{(m + 1)z^m(z - 1)}$$

(c) Convert the transfer function in part (b) to a difference equation. Compare your result with Eq. 2.9.

P2.16. Consider a discrete-time system described by the following difference equation.

$$y(k) = y(k - 1) - .24y(k - 2) + 2x(k - 1) - 1.6x(k - 2)$$

(a) Write down the form of the natural-mode terms of this system.
(b) Find the zero-state response to the step input $x(k) = 10u(k)$.
(c) Find the zero-state response to the causal exponential input $x(k) = (.8)^k u(k)$. Does a forced-mode term appear in $y(k)$? If not, why not?
(d) Find the zero-state response to the causal exponential input $x(k) = (.4)^k u(k)$. Is this an example of harmonic forcing? Why or why not?

Section 2.6

P2.17. Find the overall transfer function of the signal flow graph in Figure P2.17. This is called a *cascade* configuration of $H_1(z)$ and $H_2(z)$.

P2.17 Signal Flow Graph of a Cascade Configuration

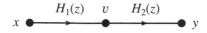

P2.18. Find the overall transfer function of the signal flow graph in Figure P2.18. This is called a *parallel* configuration of $H_1(z)$ and $H_2(z)$.

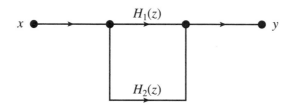

P2.18 Signal Flow Graph of a Parallel Configuration

P2.19. Find the overall transfer function of the signal flow graph in Figure P2.19. This is called a *feedback* configuration of $H_1(z)$ and $H_2(z)$.

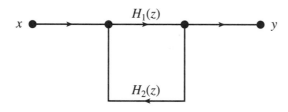

P2.19 Signal Flow Graph of a Feedback Configuration

Section 2.7

P2.20. Consider a discrete-time system described by the following difference equation.

$$y(k) = 0.6y(k-1) + 0.16y(k-2) + 10x(k-1) + 5x(k-2)$$

(a) Find the transfer function $H(z)$.
(b) Find the impulse response $h(k)$.
(c) Sketch the signal flow graph.

P2.21. Consider a discrete-time system described by the following transfer function.

$$H(z) = \frac{4z^2 + 1}{z^2 - 1.8z + 0.81}$$

(a) Find the difference equation.
(b) Find the impulse response $h(k)$.
(c) Sketch the signal flow graph.

☑ **P2.22.** Consider a discrete-time system described by the following impulse response.

$$h(k) = [2 - (.5)^k + (.2)^{k-1}]u(k)$$

(a) Find the transfer function $H(z)$.

(b) Find the difference equation.

(c) Sketch the signal flow graph.

P2.23. Consider a discrete-time described by the signal flow graph shown in Figure P2.23.

(a) Find the transfer function $H(z)$.

(b) Find the impulse response $h(k)$.

(c) Find the difference equation.

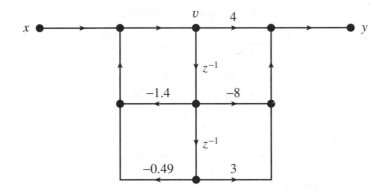

P2.23 Signal Flow Graph of System in Problem P2.23

P2.24. A discrete-time system has poles at $z = \pm 0.5$ and zeros at $z = \pm j2$. The system has a DC gain of 20.

(a) Find the transfer function $H(z)$.

(b) Find the impulse response $h(k)$.

(c) Find the difference equation.

(d) Sketch the signal flow graph.

Section 2.8

P2.25. For each of the following signals, determine whether or not it is bounded. For the bounded signals, find a bound.

(a) $x(k) = [1 + \sin(5\pi k)]u(k)$

(b) $x(k) = k(.5)^k u(k)$

(c) $x(k) = \left[\dfrac{(1+k)\sin(10k)}{1+(.5)^k} \right] u(k)$

(d) $x(k) = [1 + (-1)^k]\cos(10k)u(k)$

P2.26. Is the following system BIBO stable? Show your work.

$$H(z) = \frac{5z^2(z+1)}{(z-0.8)(z^2+0.2z-0.8)}$$

P2.27. Use the Jury test to find a range of values for the parameter α over which the following system is BIBO stable. Is your answer consistent with Figure 2.21?

$$H(z) = \frac{z^2}{(z-.8)(z^2-z+\alpha)}$$

P2.28. Use the Jury test to find a range of values for the parameter β over which the following system is BIBO stable. Is your answer consistent with Figure 2.21?

$$H(z) = \frac{z^2}{(z+0.7)(z^2+\beta z+0.5)}$$

P2.29. Consider the following discrete-time system.

$$H(z) = \frac{10z}{z^2-1.5z+0.5}$$

(a) Find the poles and zeros of $H(z)$.

(b) Show that this system is BIBO unstable.

(c) Find a bounded input $x(k)$ that will produce an unbounded output. Show that $x(k)$ is bounded. *Hint*: Use harmonic forcing.

(d) Find the zero-state response produced by the input in part (c) and show that it is unbounded.

Section 2.9

P2.30. Consider the following first-order IIR digital filter, which is called a trapezoid rule integrator.

$$H(z) = \frac{T(z+1)}{2(z-1)}$$

(a) Find the frequency response $H(f)$.

(b) Find and sketch the magnitude response $A(f)$.

(c) Find and sketch the phase response $\phi(f)$.

P2.31. Consider the following first-order FIR digital filter, which is called a backwards Euler differentiator.

$$H(z) = \frac{z-1}{Tz}$$

(a) Find the frequency response $H(f)$.

(b) Find and sketch the magnitude response $A(f)$.

(c) Find and sketch the phase response $\phi(f)$.

(d) Find the steady-state response to the following periodic input.

$$x(k) = 2\cos(0.8\pi k) - \sin(0.5\pi k)$$

☑ **P2.32.** Consider the following second-order digital filter.

$$H(z) = \frac{3(z+1)}{z^2 - 0.81}$$

(a) Find the frequency response $H(f)$.

(b) Find and sketch the magnitude response $A(f)$.

(c) Find and sketch the phase response $\phi(f)$.

(d) Find the steady-state response to the following periodic input.

$$x(k) = 10\cos(0.6\pi k)$$

2.12.2   GUI Simulation

Section 2.10

P2.33. Consider the system in Problem P2.21. Use GUI module *g_system* to plot the pole-zero pattern and the impulse response. Estimate the peak of the impulse response using the *Caliper* option. Is this system BIBO stable?

P2.34. Consider the system in Problem P2.21. Use GUI module *g_system* to plot the step response. Estimate the DC gain from the step response using the *Caliper* option.

☑ **P2.35.** Consider the following linear discrete-time system.

$$H(z) = \frac{5z^{-2} + 4.5z^{-4}}{1 - 1.8z^{-2} + 0.81z^{-4}}$$

Use GUI module *g_system* to plot the following damped cosine input and the zero-state response to it.

$$x(k) = (0.96)^k \cos(0.4\pi k)$$

P2.36. Consider the following linear discrete-time system.

$$H(z) = \frac{6 - 7.7z^{-1} + 2.5z^{-2}}{1 - 1.7z^{-1} + 0.8z^{-2} - 0.1z^{-3}}$$

Create a MAT-file called *prob2_36.mat* that contains $fs = 100$, the appropriate coefficient vectors a and b, and the following input samples where $v(k)$ is white noise uniformly distributed over $[-0.5, 0.5]$.

$$x(k) = k\exp(-k/50) + v(k), \quad 0 \le k < 500$$

Use GUI module *g_system* and the User-defined option to plot this input and the zero-state response to this input.

P2.37. Consider the following linear discrete-time system. Suppose the sampling frequency is $f_s = 1000$ Hz. Use GUI module *g_system* to plot the magnitude response and the phase response.

$$H(z) = \frac{10(z^2 + 0.8)}{(z^2 + 0.9)(z^2 + 0.7)}$$

☑ **P2.38.** Consider the following linear discrete-time system. Use GUI module *g_system* to plot the magnitude response and the phase response. Use $f_s = 100$ Hz, and use the dB scale for the magnitude response.

$$H(z) = \frac{5(z^2 + 0.9)}{(z^2 - 0.9)^2}$$

P2.39. Consider the following discrete-time system, which is a narrow band *bandpass* filter with sampling frequency $f_s = 800$ Hz.

$$H(z) = \frac{.141(z^2 - 1)}{z^2 - .704z + .723}$$

Use GUI module *g_system* to find the output corresponding to the sinusoidal input $x(k) = \cos(2\pi f_0 kT)u(k)$. Do the following cases.

(a) Plot the magnitude response.

(b) Plot the output when $f_0 = 0$ Hz.

(c) Plot the output when $f_0 = 150$ Hz.

P2.40. Consider the following discrete-time system, which is a *notch* filter with sampling interval $T = 1/360$ sec.

$$H(z) = \frac{z^2 - z + 1}{z^2 - .956z + .914}$$

Use GUI module *g_system* to find the output corresponding to the sinusoidal input $x(k) = \cos(2\pi f_0 kT)u(k)$. Do the following cases.

(a) Plot the magnitude response.

(b) Plot the output when $f_0 = 60$ Hz.

(c) Plot the output when $f_0 = 120$ Hz.

2.12.3   Computation

Section 2.5

P2.41. Consider the following discrete-time system.

$$H(z) = \frac{0.5z^4 - 0.75z^3 - 1.2z^2 + 0.4z - 1.2}{z^4 - 0.95z^3 - 0.035z^2 + 0.462z - 0.351}$$

Write a MATLAB script that uses *filter* and *plot* to compute and plot the zero-state response of this system to the following input. Plot both the input and the output on the same graph.

$$x(k) = (k + 1)^2(.8)^k u(k), \quad 0 \le k \le 100$$

P2.42. Write a MATLAB script that computes and displays the poles, zeros, and DC gain of the following discrete-time system. Is this system stable? Plot the poles and zeros using the FDSP toolbox function *f_pzplot*. Plot the transfer function surface using *f_pzsurf*.

$$H(z) = \frac{2z^5 + 0.25z^4 - 0.8z^3 - 1.4z^2 + 0.6z - 0.9}{z^5 + 0.055z^4 - 0.85z^3 - 0.04z^2 + 0.49z - 0.32}$$

☑ **P2.43.** Consider the following running average filter.

$$h(k) = \frac{1}{10} \sum_{i=0}^{9} x(k - i), \quad 0 \le k \le 100$$

Write a MATLAB script that uses *filter* and *plot* to compute and plot the zero-state response to the following input where $v(k)$ is a random white noise uniformly distributed over $[-0.1, 0.1]$. Plot $x(k)$ and $y(k)$ on the same graph.

$$x(k) = \exp(-k/20) \cos(\pi k/10)u(k) + v(k)$$

P2.44. Consider the following discrete-time system.

$$H(z) = \frac{10(z + 0.5)^3}{z^4 - 0.3z^3 - 0.57z^2 + 0.115z + 0.0168}$$

This system has four simple nonzero poles. Therefore, the zero-input response consists of a sum of the following four natural-mode terms.

$$y_0(k) = c_1 p_1^k + c_2 p_2^k + c_3 p_3^k + c_4 p_4^k$$

The coefficients can be determined from the initial condition $\{y(-1), y(-2), y(-3), y(-4)\}$. Setting $y_0(k) = y(k)$ for $-4 \le k \le -1$ yields the following linear algebraic system in the coefficient vector $c = [c_1, c_2, c_3, c_4]^T$.

$$\begin{bmatrix} p_1^{-1} & p_2^{-1} & p_3^{-1} & p_4^{-1} \\ p_1^{-2} & p_2^{-2} & p_3^{-2} & p_4^{-2} \\ p_1^{-3} & p_2^{-3} & p_3^{-3} & p_4^{-3} \\ p_1^{-4} & p_2^{-4} & p_3^{-4} & p_4^{-4} \end{bmatrix} \begin{bmatrix} c_1 \\ c_2 \\ c_3 \\ c_4 \end{bmatrix} = \begin{bmatrix} y(-1) \\ y(-2) \\ y(-3) \\ y(-4) \end{bmatrix}$$

Write a MATLAB script that uses *roots* to find the poles and then solves this linear algebraic system for the coefficient vector c using the MATLAB left divisor or \ operator when the initial condition is $[y(-1), y(-2), y(-3), y(-4)] = [2, -1, 0, 3]$. Print the poles and the coefficient vector c. Use *stem* to plot the zero-input response $y_0(k)$ for $0 \le k \le 40$.

The DFT and Spectral Analysis

Learning Objectives

- Know how to compute the spectra of discrete-time signals using the discrete-time Fourier transform or DTFT (Section 3.2)

- Know how to use the discrete Fourier transform or DFT, and its properties, to find the magnitude, phase, and power density spectra of finite signals (Sections 3.3–3.4)

- Know how to compute and evaluate the speed of the fast Fourier transform or FFT (Section 3.5)

- Understand how to characterize white noise and why this random signal is useful for signal modeling and system testing (Section 3.6)

- Know how to compute the frequency response of a stable linear discrete-time system using the DFT (Section 3.7)

- Understand how zero padding can be used to interpolate between discrete frequencies to better estimate the spectral content of a signal (Section 3.8)

- Understand how to estimate the power density spectrum of a signal using Bartlett's and Welch's methods (Section 3.9)

- Understand what a spectrogram is and how it is used to characterize signals whose spectral characteristics change with time (Section 3.10)

- Know how to use GUI module *g_spectra* to perform a spectral analysis of discrete-time signals and systems (Section 3.11)

3.1 Motivation

Recall that the Z-transform starts with a discrete-time signal $x(k)$ and transforms it into a function $X(z)$ of a complex variable z. In this chapter, we restrict the Z-transform in two important and practical ways. First, we focus our attention primarily on time signals that are of finite duration, $x(0), x(1), \ldots, x(N-1)$. The second restriction is that we evaluate $X(z)$ at a finite number of points $z_0, z_1, \ldots, z_{N-1}$: points equally spaced around the unit circle of the complex plane. The resulting transformation from $x(k)$ to $X(i)$ is called the discrete Fourier transform or DFT.

DFT

$$X(i) = \text{DFT}\{x(k)\}, \quad 0 \le i < N$$

If the number of samples N is chosen to be a power of two, then the computational speed can be increased dramatically, and this leads to the well known fast Fourier transform or FFT. The FFT is a highly efficient implementation of the DFT that is a widely used computational tool in digital signal processing. In this chapter, we use the DFT to compute the spectra of discrete-time signals and the frequency responses of discrete-time systems. Practical applications of the DFT involving convolution and correlation of pairs of signals are treated separately in Chapter 4.

We begin this chapter by introducing a number of examples where the DFT and spectral analysis can be put to use. As a lead-in to the DFT, the discrete-time Fourier transform (DTFT) is first introduced. The DTFT is used to compute the spectra of discrete-time signals of infinite duration. Next, the finite discrete Fourier transform and its inverse are defined. We then develop some useful properties of the DFT based on symmetry. The highly efficient FFT implementation of the DFT is then derived using the decimation in time approach. This is followed by a comparison of the relative computational effort required by the FFT and the DFT in terms of the number of floating-point operations or FLOPs. The use of the DFT to compute the magnitude, phase, and power density spectra of finite discrete-time signals is then presented. This is followed by an introduction to white noise. White noise is an important type of random signal that is useful for signal modeling and system testing. The use of the DFT to approximate the frequency response of a discrete-time system or digital filter is then examined. This is followed by an examination of techniques that are used to estimate the underlying continuous-time power density spectrum, including Bartlett's method and Welch's method. The use of zero padding to interpolate between discrete frequencies is then presented. Finally, a GUI module called *g_spectra* is introduced that allows the user to perform a spectral analysis of discrete-time signals and systems without any need for programming. Using *g_spectra*, the user can examine a variety of signals including user-defined signals stored in MAT files and audio signals recorded live from a microphone. The chapter concludes with some application examples and a summary of the DFT and spectral analysis techniques. The selected examples include the detection of unknown sinusoidal components in signals that are corrupted with noise, and an analysis of the total harmonic distortion of an overdriven audio amplifier.

3.1.1 Fourier Series

Consider a periodic continuous-time signal $x_a(t)$ with period τ. For example, $x_a(t)$ might represent the hum or whine of a rotating machine where the fundamental frequency $f_0 = 1/\tau$ changes with the speed of rotation. Since $x_a(t)$ is periodic, it can be approximated by a truncated Fourier series using M harmonics.

$$x_a(t) \approx \sum_{i=-(M-1)}^{M-1} c_i \exp(ji2\pi f_0 t) \tag{3.1}$$

This is the complex form of the Fourier series with the ith Fourier coefficient being

$$c_i = \frac{1}{\tau} \int_0^\tau x_a(t) \exp(-ji2\pi f_0 t) dt \tag{3.2}$$

Fourier series coefficients of common periodic waveforms can be found in Appendix C. Next, suppose we convert $x_a(t)$ to a discrete-time signal $x(k)$ by sampling it at $N = 2M$ points using a sampling frequency of $f_s = N f_0$. Thus, the sampling interval is $T = \tau/N$ and

$$x(k) = x_a(kT), \quad 0 \leq k < N \tag{3.3}$$

Note that the N samples of $x_a(t)$ span one period because $NT = \tau$. To compute the ith Fourier coefficient of $x_a(t)$, we approximate the integral in Eq. 3.2 with a sum. Recalling that $f_0 = f_s/N$ and $T = \tau/N$ we have

$$c_i = \frac{1}{\tau} \int_0^\tau x_a(t) \exp(-ji2\pi f_0 t)$$

$$\approx \frac{1}{\tau} \sum_{k=0}^{N-1} x_a(kT) \exp(-ji2\pi f_0 kT)T, \quad N \gg 1$$

$$= \frac{T}{\tau} \sum_{k=0}^{N-1} x(k) \exp(-ji2\pi f_s kT/N)$$

$$= \frac{1}{N} \sum_{k=0}^{N-1} x(k) \exp(-jik2\pi/N), \quad -N/2 < i < N/2 \tag{3.4}$$

Thus, the Fourier coefficients of the periodic signal $x_a(t)$ can be approximated using the samples $x(k)$. Interestingly enough, the summation part of Eq. 3.4 is the discrete Fourier transform or DFT of the samples. Let $X(i) = \mathrm{DFT}\{x(k)\}$. Then the Fourier coefficients of $x_a(t)$ can be approximated as follows.

$$c_i \approx \frac{X(i)}{N}, \quad 0 \leq i < N/2 \tag{3.5}$$

The DFT produces $M = N/2$ complex Fourier coefficients $\{c_0, c_1, \ldots, c_{M-1}\}$. To obtain the remaining coefficients in Eq. 3.1, observe from Eq. 3.2 that for real $x_a(t)$

the coefficients corresponding to negative values of i are the *complex conjugates* of the coefficients corresponding to positive values of i. Thus,

$$c_{-i} \approx \frac{X^*(i)}{N}, \quad 0 \le i < N/2 \tag{3.6}$$

3.1.2 DC Wall Transformer

Many electronic items receive power from batteries or a DC wall transformer. A DC wall transformer is an inexpensive power supply that approximates the constant voltage produced by batteries. A block diagram of a typical DC wall transformer is shown in Figure 3.1.

The transformer block steps the 120-volt 60-Hz sinusoidal AC input signal, x_a, down to a lower-voltage AC signal, u_a. The full wave bridge rectifier consists of four diodes configured to take the absolute value of the AC signal, u_a. Thus, the signals $x_a(t)$, $u_a(t)$, and $v_a(t)$ can be modeled as follows, where $0 < \alpha < 1$ depends on the desired DC voltage.

$$x_a(t) = 120\sqrt{2}\sin(120\pi t) \tag{3.7a}$$

$$u_a(t) = \alpha x_a(t) \tag{3.7b}$$

$$v_a(t) = |u_a(t)| \tag{3.7c}$$

The bridge rectifier output v_a has a DC component or average value, $d_0/2$, plus a periodic component that has a fundamental frequency of $f_0 = 120$ Hz. To produce a pure DC output similar to that of a battery, the nonconstant part of v_a must be filtered out with the lowpass RC filter section whose transfer function is

$$H_a(s) = \frac{1}{RCs + 1} \tag{3.8}$$

Of course, the lowpass filter is not an ideal filter, so some of the nonconstant part of v_a survives in the wall transformer output y_a in the form of a small AC *ripple* voltage. The end result is that a DC wall transformer output can be modeled as a DC component, $d_0/2$, plus a small periodic ripple component with a fundamental frequency of 120 Hz.

$$y_a(t) = \frac{d_0}{2} + \sum_{i=1}^{M-1} d_i \cos(240i\pi t + \theta_i) \tag{3.9}$$

Figure 3.1 DC Wall Transformer

For an ideal wall transformer, there is no AC ripple, so $d_i = 0$ for $i > 0$. Thus, we can measure the quality of the wall transformer (or any other DC power supply for that matter) by using the total harmonic distortion, THD, as follows.

$$P_y = \frac{d_0^2}{4} + \frac{1}{2} \sum_{i=1}^{M-1} d_i^2 \tag{3.10a}$$

$$\text{THD} = \frac{100(P_y - d_0^2/4)}{P_y} \% \tag{3.10b}$$

Notice that this definition of total harmonic distortion is similar to that used in Chapter 1 for an ideal amplifier, except that in Eqs 3.10, the term $d_0^2/4$ is removed from the numerator, whereas in Eq. 1.5, the term $d_1^2/2$ was removed from the numerator. This is because, for an ideal DC power supply, the output should be the zeroth harmonic, while for an ideal amplifier the output should be the first harmonic. The term $d_0^2/4$ represents the power of the DC term or zeroth harmonic, while $d_i^2/2$ represents the power of the i th harmonic for $i > 0$.

From Appendix C, the cosine form coefficients, d_i and θ_i, are related to the complex Fourier series coefficients in Eq. 3.2 as

$$d_i = 2|c_i|, \quad 0 \leq i < M \tag{3.11a}$$

$$\theta_i = \tan^{-1}\left(\frac{-\text{Im}\{c_i\}}{\text{Re}\{c_i\}}\right), \quad 0 \leq i < M \tag{3.11b}$$

To measure the harmonic distortion THD, we sample the output at $N = 2M$ points using a sampling frequency of $f_s = Nf_0$. If $Y(i) = \text{DFT}\{y_a(kT)\}$, then from Eq. 3.5 and Eq. 3.11 we have

$$d_i = \frac{2|Y(i)|}{N}, \quad 0 \leq i < M \tag{3.12a}$$

$$\theta_i = \tan^{-1}\left(\frac{-\text{Im}\{Y(i)\}}{\text{Re}\{Y(i)\}}\right), \quad 0 \leq i < M \tag{3.12b}$$

3.1.3   Frequency Response

The DFT can also be used to characterize a linear discrete-time system or digital filter. Consider the discrete-time system shown in Figure 3.2. Recall from Chapter 2

Figure 3.2 A Linear Discrete-time System with Transfer Function $H(z)$

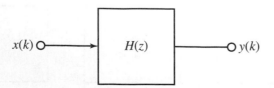

$x(k)$ → $H(z)$ → $y(k)$

that the system transfer function is the Z-transform of the zero-state output divided by the Z-transform of the input.

$$H(z) = \frac{Y(z)}{X(z)} \tag{3.13}$$

Also recall from Chapter 2 that, if the system is stable and we evaluate the transfer function along the unit circle using $z = \exp(j2\pi fT)$, the resulting function of f is called the *frequency response*. The samples of the frequency response can be easily approximated using the DFT. In particular, if $X(i) = \text{DFT}\{x(k)\}$ and $Y(i) = \text{DFT}\{y(k)\}$, then the approximate frequency response evaluated at discrete frequency $f_i = if_s/N$ Hz is as follows, where the accuracy of the approximation improves as N increases.

$$H(i) = \frac{Y(i)}{X(i)}, \quad 0 \le i < N \tag{3.14}$$

The frequency response specifies how much each sinusoidal input gets scaled in magnitude and shifted in phase as it passes through the system. For the DFT method of evaluating the frequency response in Eq. 3.14, the input signal $x(k)$ should be chosen such that it has power at all frequencies of interest so as to avoid division by zero. For example, we can let $x(k)$ be a unit impulse or a random white-noise signal. Since $H(k)$ is complex, it can be expressed in polar form in terms of its magnitude and phase angle.

$$H(i) = A(i)\exp[j\phi(i)], \quad 0 \le i < N \tag{3.15}$$

Once $H(i)$ is known, the steady-state response of the system to a sinusoidal input at a discrete frequency can be obtained from inspection. For example, suppose $x(k) = \alpha \sin(2\pi f_n kT)$ for some $0 \le n < N$. Then the steady-state output generated by this input is scaled in amplitude by $A(n)$ and shifted in phase by $\phi(n)$.

$$y_{ss}(k) = A(n)\alpha \sin[2\pi f_n kT + \phi(n)] \tag{3.16}$$

As an illustration, consider a stable second-order discrete-time system characterized by the following transfer function.

$$H(z) = \frac{10 + 20z^{-1} - 5z^{-2}}{1 - 0.2z^{-1} - 0.63z^{-2}} \tag{3.17}$$

This system is stable with poles at $z = 0.5$ and $z = -0.9$. Suppose the sampling frequency is $f_s = 100$ Hz and the DFT is evaluated using $N = 256$ points. The resulting magnitude response $A(i)$ and phase response $\phi(i)$ for $0 \le i \le N/2$ are shown in Figure 3.3. Note that only the positive frequencies are plotted. Values of i in the range $N/2 < i < N$ correspond to the values of $z = \exp(j2\pi f_i T)$ that are in the lower half of the unit circle and therefore represent negative frequencies.

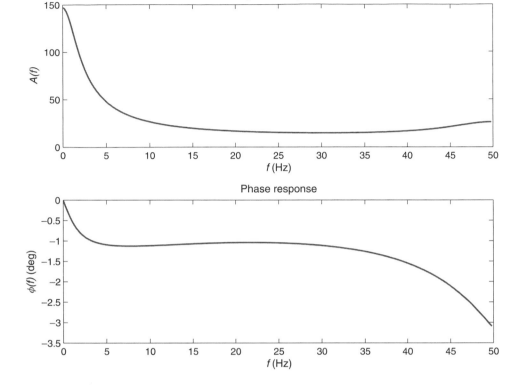

Figure 3.3 Magnitude Response and Phase Response of System in Eq. 3.17 using $f_s = 100$ Hz and $N = 256$

3.2 The Discrete-time Fourier Transform (DTFT)

Recall from Chapter 2 that the transfer function of a discrete-time system, $H(z)$, is the Z-transform of the impulse response. If the system is stable, then the frequency response, $H(f)$, is just the transfer function evaluated along the unit circle of the complex plane. Since the Z-transform can be applied to any discrete-time signal, we can define the *spectrum* of a discrete-time signal $x(k)$ as its Z-transform $X(z)$ evaluated along the unit circle. That is,

$$X(f) \triangleq X(z)|_{z=\exp(j2\pi fT)}, \quad -f_s/2 < f \le f_s/2 \tag{3.18}$$

Since $X(f)$ is complex, it can be represented in polar form as $X(f) = A(f) \exp[j\phi(f)]$ where

$$A(f) = |X(f)| \tag{3.19a}$$

$$\phi(f) = \angle X(f) \tag{3.19b}$$

In this case, $A(f)$ is called the *magnitude spectrum* of $x(k)$, and $\phi(f)$ is called the *phase spectrum* of $x(k)$. It is of interest to see what happens when the unit circle substitution for z in Eq. 3.18 is made using the definition of the Z-transform. Here

$$X(f) = \sum_{k=0}^{\infty} x(k)z^{-k}|_{z=\exp(j2\pi fT)}$$

$$= \sum_{k=0}^{\infty} x(k)\exp(-jk2\pi fT) \tag{3.20}$$

The transformation from discrete-time, $x(k)$, to continuous-frequency, $X(f)$, depicted in Eq. 3.20 is referred to as the discrete-time Fourier transform.

DEFINITION
3.1: DTFT

> The *discrete-time Fourier transform* or DTFT of a causal discrete-time signal, $x(k)$, is denoted $X(f) = \text{DTFT}\{x(k)\}$ and defined
>
> $$X(f) = \sum_{k=0}^{\infty} x(k)\exp(-jk2\pi fT), \quad -f_s/2 < f \le f_s/2$$

Like the Z-transform, the DTFT can be generalized to noncausal signals by changing the lower limit of the summation to start at $k = -\infty$ rather than $k = 0$. The DTFT has several properties analogous to those of the Z-transform and the continuous-time Fourier transform. In the remainder of this section, we examine a few of the fundamental ones and some examples. First, consider the question of when the summation in Definition 3.1 actually converges. Since the DTFT is the Z-transform evaluated along the unit circle, the infinite series in Definition 3.1 will converge if the region of convergence of $X(z)$ includes the unit circle. Recall that the region of convergence of the Z-transform of a causal signal is the part of the complex plane that lies outside the outer-most pole. Thus, $X(f)$ converges as long as the poles of $X(z)$ all lie strictly inside the unit circle. Using the stability analysis from Chapter 2, this constraint on the poles of $X(z)$ is equivalent to $x(k)$ being absolutely summable. That is, $X(f)$ converges for signals satisfying

$$\sum_{k=0}^{\infty} |x(k)| < \infty \tag{3.21}$$

The time signal $x(k)$ can be recovered from its spectrum $X(f)$ using the inverse transform. To determine the inverse DTFT, consider multiplying $X(f)$ by a complex conjugate exponential, $\exp(ji2\pi fT)$, and integrating over one period.

$$\int_{-f_s/2}^{f_s/2} X(f)\exp(ji2\pi fT)df = \int_{-f_s/2}^{f_s/2} \sum_{k=0}^{\infty} x(k)\exp(jk2\pi fT)\exp(ji2\pi fT)df$$

$$= \sum_{k=0}^{\infty} x(k)\int_{-f_s/2}^{f_s/2} \exp[j(i-k)2\pi fT]df$$

$$= \sum_{k=0}^{\infty} x(k)f_s\delta(i-k)$$

$$= x(i)f_s \tag{3.22}$$

In Eq. 3.22, it is valid to interchange the order of the integral and the sum because $x(k)$ is absolutely summable. Solving Eq. 3.22 for $x(i)$, we then find that the *inverse DTFT* is

Inverse DTFT
$$x(k) = \frac{1}{f_s}\int_{-f_s/2}^{f_s/2} X(f)\exp(jk2\pi fT)df, \quad k \geq 0 \tag{3.23}$$

The DTFT is sometimes called the *analysis* equation because it decomposes a signal into its spectral components. The inverse DTFT is then called the *synthesis* equation because it reconstructs or synthesizes the signal from its spectral components.

Among the absolutely summable signals satisfying Eq. 3.21, there are signals that are also square summable. A signal is square summable if Eq. 3.21 continues to hold when $|x(k)|$ is replaced by $|x(k)|^2$. Square summable signals are also called energy signals. For energy signals, there is a simple relationship between the time signal and its spectrum called *Parseval's theorem*.

Parseval's theorem
$$\sum_{k=0}^{\infty} |x(k)|^2 = \frac{1}{f_s}\int_{-f_s/2}^{f_s/2} |X(f)|^2 df \tag{3.24}$$

Next, note from Euler's identity in Appendix D that $\exp(j2\pi fT)$ is periodic with period f_s. It then follows from Eq. 3.18 that the spectrum of $x(k)$ is also periodic with period f_s.

$$X(f+f_s) = X(f) \tag{3.25}$$

It is for this reason that frequency typically is restricted to the interval $-f_s/2 < f \leq f_s/2$. If the time signal $x(k)$ is real, then the frequency interval can be restricted still further. This is a consequence of the *symmetry* property of the DTFT. When $x(k)$ is real, it follows from Definition 3.1 that:

$$X(-f) = X^*(f) \tag{3.26}$$

Thus, the spectrum of $x(k)$ at negative frequencies is just the complex conjugate of the spectrum of $x(k)$ at positive frequencies. Using the symmetry property, we find that the magnitude spectrum of a real signal is an *even* function of f, and the phase spectrum of a real signal is an *odd* function of f.

$$A(-f) = A(f) \tag{3.27a}$$

$$\phi(-f) = -\phi(f) \tag{3.27b}$$

Consequently, for real signals, all of the essential information about the spectrum is contained in the nonnegative frequency range $[0, f_s/2]$.

Signals can be categorized or classified based on the part of the overall frequency spectrum that they occupy (Proakis and Manolakis, 1992). Table 3.1 summarizes some practical signals that include examples taken from biomedical, geological, and communications applications. Note the impressive range of frequencies (26 orders of magnitude) going from circadian rhythms that oscillate at approximately one cycle per day to lethal gamma rays, which are faster than a billion billion Hz.

Table 3.1: ▶
Frequency Ranges of
Some Practical
Signals

Signal Type	Frequency Range (Hz)
Circadian rhythm	$1.1 \times 10^{-5} - 1.2 \times 10^{-5}$
Earthquake	$10^{-2} - 10^{1}$
Electrocardiogram (ECG)	$0 - 10^{2}$
Electroencephalogram (EEG)	$0 - 10^{2}$
AC power	$5 \times 10^{2} - 6 \times 10^{2}$
Wind	$10^{2} - 10^{3}$
Speech	$10^{2} - 4 \times 10^{3}$
Audio	$2 \times 10^{1} - 2 \times 10^{4}$
AM radio	$5.4 \times 10^{5} - 1.6 \times 10^{6}$
FM radio	$8.8 \times 10^{7} - 1.08 \times 10^{8}$
Cell phone	$8.1 \times 10^{8} - 9 \times 10^{8}$
TV	$3 \times 10^{8} - 9.7 \times 10^{8}$
GPS	$1.52 \times 10^{9} - 1.66 \times 10^{9}$
Shortwave radio	$3 \times 10^{6} - 3 \times 10^{9}$
Radar, microwave	$3 \times 10^{9} - 3 \times 10^{12}$
Infrared light	$3 \times 10^{12} - 4.3 \times 10^{14}$
Visible light	$4.3 \times 10^{14} - 7.5 \times 10^{14}$
Ultraviolet light	$7.5 \times 10^{14} - 3 \times 10^{17}$
X-ray	$3 \times 10^{17} - 3 \times 10^{19}$
Gamma ray	$5 \times 10^{19} - 10^{21}$

Example 3.1 **Spectrum of Causal Exponential**

As an illustration of using the DTFT to analyze a discrete-time signal, consider the following causal exponential.

$$x(k) = a^k u(k)$$

From Table 2.1, the Z-transform of this signal is

$$X(z) = \frac{z}{z - a}$$

$$= \frac{1}{1 - az^{-1}}$$

Here $X(z)$ has a pole at $z = a$. Thus, for the DTFT to converge, it is necessary that $|a| < 1$. From Eq. 3.18, the spectrum of $x(k)$ is

$$X(f) = \frac{1}{1 - a\exp(-j2\pi fT)}$$

Using Euler's identity, the magnitude spectrum of the causal exponential is

$$A(f) = |X(f)|$$

$$= \frac{1}{\sqrt{[1 - a\cos(2\pi fT)]^2 + a^2 \sin^2(2\pi fT)}}$$

$$= \frac{1}{\sqrt{2[1 - a\cos(2\pi fT)]}}$$

Similarly, the phase spectrum is

$$\phi(f) = \angle X(f)$$

$$= -\tan^{-1}\left[\frac{a\sin(2\pi fT)}{1 - a\cos(2\pi fT)}\right]$$

Plots of the magnitude spectra and the phase spectra, obtained by running script *exam3_1* on the distribution CD, are shown in Figure 3.4 for the case $a = 0.8$. Note how the magnitude spectrum is even and the phase spectrum is odd. Also observe that *normalized frequency* is plotted along the abscissa because $x(k)$ may or may not have been obtained by sampling an underlying continuous-time signal. Thus, the independent variable is

$$\hat{f} \triangleq \frac{f}{f_s}$$

Using normalized frequency, $-0.5 \le \hat{f} \le 0.5$ is equivalent to setting the sampling interval to $T = 1$ sec.

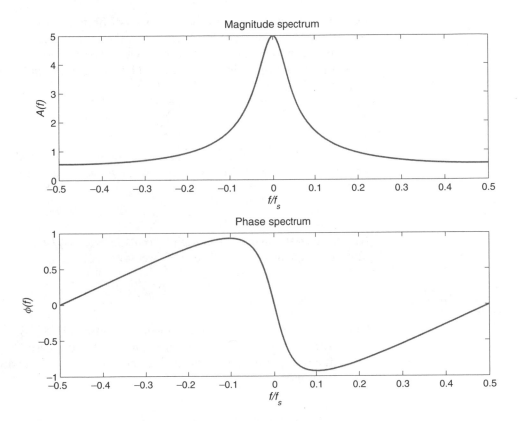

Figure 3.4 Magnitude and Phase Spectra of Causal Exponential with $a = 0.8$

• • • • • • • • • • • • • •

3.3 The Discrete Fourier Transform (DFT)

The discrete Fourier transform can be regarded as a special case of the discrete-time Fourier transform or DTFT. Recall from Definition 3.1 that the DTFT of a causal signal $x(k)$ is defined as follows.

$$X(f) = \sum_{k=0}^{\infty} x(k) \exp(-jk2\pi fT), \quad -f_s/2 < f \leq f_s/2 \qquad (3.28)$$

Although the DTFT is an important *theoretical* tool for analyzing discrete-time signals and systems, it does suffer from practical limitations when used as a *computational* tool. One drawback is that a direct evaluation of $X(f)$ using Eq. 3.28 requires an infinite number of floating-point operations or FLOPs. This is compounded by the second computational drawback; namely, that the transform itself must be evaluated at an infinite number of frequencies, f. The first limitation can be removed by restricting our consideration to signals of finite duration. Recall from Eq. 3.21 that the DTFT converges for time signals that are absolutely summable. If $x(k)$ is absolutely summable, then it is necessary that $|x(k)| \to 0$ as $k \to \infty$. Consequently, for sufficiently large values of N, the DTFT in Eq. 3.28 can be approximated by the following finite sum.

$$X(f) \approx \sum_{k=0}^{N-1} x(k) \exp(-jk2\pi fT) \tag{3.29}$$

To address the second limitation, we evaluate $X(f)$ at N discrete values of f. In particular, consider the following *discrete frequencies* equally spaced over one period of $X(f)$.

$$f_i = \frac{if_s}{N}, \quad 0 \le i < N \tag{3.30}$$

It is of interest to view the complex points $z_i = \exp(j2\pi f_i T)$ corresponding to the discrete frequencies. Using Eq. 3.30, we have

$$z_i = \exp\left(\frac{ji2\pi}{N}\right) \tag{3.31}$$

Notice that $|z_i| = 1$. Thus, the N evaluation points in Eq. 3.31 are equally spaced around the unit circle, as shown in Figure 3.5 for the case $N = 8$. Observe that the unit circle is traversed in the counterclockwise direction with $z_0 = 1$, $z_{N/4} = j$, $z_{N/2} = -1$ and $z_{3N/4} = -j$.

3.3.1 Roots of Unity

The formulation of the discrete Fourier transform can be simplified if we introduce the following factor which corresponds to z_i for $i = -1$.

*N*th root of unity

$$W_N \triangleq \exp\left(\frac{-j2\pi}{N}\right) \tag{3.32}$$

By using Euler's identity from Appendix D, we find that $W_N^N = 1$. Consequently, the factor W_N can be thought of as the Nth *root of unity*. More generally, we can express W_N^k as follows using Euler's identity.

$$W_N^k = \cos\left(\frac{2\pi k}{N}\right) - j \sin\left(\frac{2\pi k}{N}\right) \tag{3.33}$$

The Nth root of unity has a number of interesting symmetry properties that are useful for subsequent computations. For convenient reference, they are summarized in Table 3.2. Each entry can be verified using Eq. 3.33. Note that the first four properties verify that W_N^k moves clockwise around the unit circle as k ranges from 0 to N. Property 5 verifies that W_N^k is a periodic function of k with period N. The remaining properties are useful for the derivation of the fast Fourier transform: a highly efficient implementation of the discrete Fourier transform.

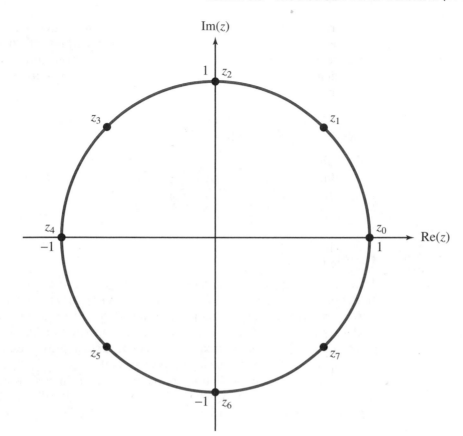

Figure 3.5 Evaluation Points of DFT when $N = 8$

Table 3.2: ▶
Basic Properties of W_N^k

Property	Description
1	$W_N^{N/4} = -j$
2	$W_N^{N/2} = -1$
3	$W_N^{3N/4} = j$
4	$W_N^N = 1$
5	$W_N^{k+N} = W_N^k$
6	$W_N^{k+N/2} = -W_N^k$
7	$W_N^{2k} = W_{N/2}^k$
8	$W_N^* = W_N^{-1}$

The discrete values of z in Eq. 3.31 can be reformulated in terms of the Nth root of unity as $z_i = W_N^{-i}$. If this value for z_i is substituted into the truncated expression for the DTFT in Eq. 3.29, the resulting transformation from discrete-time $x(k)$ to discrete-frequency $X(i)$ is called the discrete Fourier transform.

DEFINITION

3.2: DFT

Let $x(k)$ be an N-point signal, and let W_N be the Nth root of unity in Eq. 3.32. Then the *discrete Fourier transform* of $x(k)$ is denoted $X(i) = \text{DFT}\{x(k)\}$ and defined

$$X(i) \triangleq \sum_{k=0}^{N-1} x(k) W_N^{ik}, \quad 0 \leq i < N$$

Notation

In terms of notation, it should pointed out that the same base symbol is being used to denote the Z-transform, $X(z)$; the DTFT, $X(f)$; and the DFT, $X(i)$. This is consistent with the convention adopted earlier where the argument type, a complex z, a real f, or an integer i, are used to distinguish between the different cases and dictate the meaning of X. This approach is used in order to avoid a proliferation of different symbols, which can be confusing in its own right.

The continuous-time Fourier transform has an inverse whose form is almost identical to the original transform. The same is true of the discrete Fourier transform. The inverse of the DFT, which is denoted $x(k) = \text{IDFT}\{X(i)\}$, is computed as follows.

$$x(k) = \frac{1}{N} \sum_{i=0}^{N-1} X(i) W_N^{-ki}, \quad 0 \leq k < N \tag{3.34}$$

Thus, the IDFT is identical to the DFT except that W_N has been replaced by its complex conjugate $W_N^* = W_N^{-1}$, and the final result is normalized by N. In practical terms, this means that any algorithm devised to compute the DFT can also be used, with only minor modification, to compute the IDFT.

3.3.2   Matrix Formulation

The DFT can be interpreted as a transformation (or mapping) from a vector of input samples x to a vector of output samples X.

$$x = [x(0), x(1), \ldots, x(N-1)]^T \tag{3.35a}$$

$$X = [X(0), X(1), \ldots, X(N-1)]^T \tag{3.35b}$$

Here x and X are $N \times 1$ column vectors, but they are written as transposed row vectors to conserve space. Since the DFT is a linear transformation from x to X, it can be represented as an $N \times N$ matrix. Consider, in particular, the matrix $W_{ik} = W_N^{ik}$ where the row index i and column index k are assumed to start from zero. For example, for the case $N = 5$, we have

$$
W = \begin{bmatrix}
W_N^0 & W_N^0 & W_N^0 & W_N^0 & W_N^0 \\
W_N^0 & W_N^1 & W_N^2 & W_N^3 & W_N^4 \\
W_N^0 & W_N^2 & W_N^4 & W_N^6 & W_N^8 \\
W_N^0 & W_N^3 & W_N^6 & W_N^9 & W_N^{12} \\
W_N^0 & W_N^4 & W_N^8 & W_N^{12} & W_N^{16}
\end{bmatrix}
\tag{3.36}
$$

Comparison of Eq. 3.35 and Eq. 3.36 with Definition 3.2 reveals that the DFT can be expressed in matrix form as

Matrix form of DFT

$$
X = Wx
\tag{3.37}
$$

For small values of N, Eq. 3.37 can be used to compute the DFT. As we shall see, for moderate to large values of N, a much more efficient FFT implementation of the DFT is the method of choice.

The matrix form of the DFT can also be used to compute the IDFT. Multiplying both sides of Eq. 3.37 on the left by W^{-1} yields $x = W^{-1}X$. If we compare this equation with Eq. 3.34, we find that there is no need to explicitly invert the matrix W because $W^{-1} = W^*/N$. Thus, the matrix form for the IDFT is

Matrix form of IDFT

$$
x = \frac{W^*X}{N}
\tag{3.38}
$$

The following examples illustrate computation of the DFT and the IDFT for small values of N.

Example 3.2   **DFT**

As an example of computing a DFT using the matrix form, suppose the input samples are as follows.

$$
x = [3, -1, 0, 2]^T
$$

Thus, $N = 4$ and from Eq. 3.33

$$
W_4 = \cos\left(\frac{2\pi}{4}\right) - j\sin\left(\frac{2\pi}{4}\right)
$$

$$
= -j
$$

Next, from Eq. 3.36 and Eq. 3.37, the DFT of x is

$$X = Wx$$

$$= \begin{bmatrix} W_4^0 & W_4^0 & W_4^0 & W_4^0 \\ W_4^0 & W_4^1 & W_4^2 & W_4^3 \\ W_4^0 & W_4^2 & W_4^4 & W_4^6 \\ W_4^0 & W_4^3 & W_4^6 & W_4^9 \end{bmatrix} x$$

$$= \begin{bmatrix} 1 & 1 & 1 & 1 \\ 1 & -j & -1 & j \\ 1 & -1 & 1 & -1 \\ 1 & j & -1 & -j \end{bmatrix} \begin{bmatrix} 3 \\ -1 \\ 0 \\ 2 \end{bmatrix}$$

$$= \begin{bmatrix} 4 \\ 3 + j3 \\ 2 \\ 3 - j3 \end{bmatrix}$$

Note that even though the signal $x(k)$ is real, its DFT $X(i)$ is complex.

Example 3.3   **IDFT**

As a numerical check, suppose we compute the IDFT of the results from Example 3.2.

$$X = [4, 3 + j3, 2, 3 - j3]^T$$

Using $N = 4$, the matrix W from Example 3.2, and Eq. 3.38, we have

$$x = \frac{W^*X}{N}$$

$$= \frac{1}{4} \begin{bmatrix} 1 & 1 & 1 & 1 \\ 1 & -j & -1 & j \\ 1 & -1 & 1 & -1 \\ 1 & j & -1 & -j \end{bmatrix}^* \begin{bmatrix} 4 \\ 3 + j3 \\ 2 \\ 3 - j3 \end{bmatrix}$$

$$= \frac{1}{4} \begin{bmatrix} 1 & 1 & 1 & 1 \\ 1 & j & -1 & -j \\ 1 & -1 & 1 & -1 \\ 1 & -j & -1 & j \end{bmatrix} \begin{bmatrix} 4 \\ 3 + j3 \\ 2 \\ 3 - j3 \end{bmatrix}$$

$$= \frac{1}{4} \begin{bmatrix} 12 \\ -4 \\ 0 \\ 8 \end{bmatrix} = \begin{bmatrix} 3 \\ -1 \\ 0 \\ 2 \end{bmatrix} \checkmark$$

3.3.3 Signal Spectra

To interpret the meaning of $X(i)$, suppose $x_a(t)$ is a periodic continuous-time signal with period τ and fundamental frequency $f_0 = 1/\tau$. Next, let $x(k)$ be the kth sample of $x_a(t)$ using a sampling interval of $T = \tau/N$.

$$x(k) = x_a(kT), \quad 0 \le k < N \tag{3.39}$$

In this case, the N samples $x(k)$ span one period of $x_a(t)$ using a sampling frequency of $f_s = Nf_0$. Since $x_a(t)$ is periodic, it can be approximated with a complex Fourier series using N harmonics.

$$x_a(t) \approx \sum_{i=-(N-1)}^{N-1} c_i \exp(ji2\pi f_0 t) \tag{3.40}$$

Recall from Eq. 3.4 that the ith Fourier coefficient can be approximated as follows when N is large.

$$c_i \approx \frac{1}{N} \sum_{k=0}^{N-1} x(k) \exp(-jik2\pi/N)$$

$$= \frac{1}{N} \sum_{k=0}^{N-1} x(k) W_N^{ik}$$

$$= \frac{X(i)}{N}, \quad 0 \le i < N/2 \tag{3.41}$$

Thus, the Fourier coefficients of $x_a(t)$ can be obtained from the DFT of the samples of $x_a(t)$. The ith Fourier coefficient specifies the magnitude and phase angle of the ith harmonic, the spectral component of $x_a(t)$ at frequency

$$f_i = \frac{if_s}{N}, \quad 0 \le i < N/2 \tag{3.42}$$

From Eq. 3.41, we see that DFT sample $X(i)$ specifies the magnitude and phase angle of the ith spectral component of $x(k)$. The DFT can be written in polar form as $X(i) = A(i) \exp[j\phi(i)]$. In this case, the *magnitude spectrum*, $A(i)$, and *phase spectrum*, $\phi(i)$, are defined as follows for $0 \le i < N$.

$$A(i) \triangleq |X(i)| \tag{3.43a}$$

$$\phi(i) \triangleq \angle X(i) \tag{3.43b}$$

The amount of power that $x(k)$ contains at frequency, f_i, can be determined from the *power density spectrum*, which is just the square of the magnitude spectrum normalized by N.

$$S_N(i) \triangleq \frac{|X(i)|^2}{N}, \quad 0 \le i < N \tag{3.44}$$

Example 3.4   **Spectra**

As a very simple example of a discrete-time signal and its spectra, consider the unit impulse $x(k) = \delta(k)$. Using Definition 3.2, we have

$$
\begin{aligned}
X(i) &= \sum_{k=0}^{N-1} x(k) W_N^{ik} \\
&= \sum_{k=0}^{N-1} \delta(k) W_N^{ik} \\
&= 1, \quad 0 \le i < N
\end{aligned}
$$

It then follows from Eq. 3.43 and Eq. 3.44 that the magnitude, phase, and power density spectra of the unit impulse for $0 \le i < N$ are

$$
A(i) = 1
$$

$$
\phi(i) = 0
$$

$$
S_N(i) = \frac{1}{N}
$$

The fact that $S_N(i)$ is a constant nonzero value for all i means that the unit impulse has its power distributed evenly over all N discrete frequencies.

At this point, we have introduced three tools for computing the spectra of signals. For comparison, they are summarized in Table 3.3. They differ from one another in the types of independent variables: continuous or discrete. Later in this chapter, we will introduce a highly efficient implementation of the discrete Fourier transform called the fast Fourier transform or FFT.

Table 3.3: ▶
Comparison of
Transforms for
Computing Signal
Spectra

Transform	Symbol	Time	Frequency
Fourier transform	FT	Continuous	Continuous
Discrete-time Fourier transform	DTFT	Discrete	Continuous
Discrete Fourier transform	DTF	Discrete	Discrete

MATLAB Toolbox

The FDSP toolbox contains the following function for evaluating the magnitude, phase, and power density spectra of a finite discrete-time signal.

```
[A,phi,S,f] = f_spec (x,N,fs);          % compute spectra
```

On entry to *f_spec*, x is a vector of length M containing the samples, N is the number of samples to use, and fs is an optional argument specifying the sampling frequency. The default value is $fs = 1$. If $N > M$, then x is first padded with $N - M$ zeros. On exit from *f_spec*, A, *phi*, and S are vectors of length N containing the magnitude spectrum, phase spectrum, and power density spectrum, respectively. Output f is a vector of length N containing the discrete frequencies at which the spectra are evaluated where $f(i) = (i - 1)f_s/N$.

● ● ● ● ● ● ● ● ● ● ● ● ● ● ● ● ●

3.4 DFT Properties

Like the Z-transform, the DFT has a number of useful qualitative properties. With the help of these properties, we can more effectively apply the DFT.

3.4.1 Linearity Property

The DFT is an operator that maps one N-point sequence into another. This operator is a *linear* operator. That is, if $x(k)$ and $y(k)$ are signals and a and b are constants, then from Definition 3.2 we have

$$\text{DFT}\{ax(k) + by(k)\} = \sum_{k=0}^{N-1}[ax(k) + by(k)]W_N^{ik}$$

$$= a\sum_{k=0}^{N-1}x(k)W_N^{ik} + b\sum_{k=0}^{N-1}y(k)W_N^{ik}$$

$$= a\text{DFT}\{x(k)\} + b\text{DFT}\{y(k)\} \tag{3.45}$$

Thus, the DFT of the sum of two signals is just the sum of the DFTs of the signals. Similarly, the DFT of a scaled signal is just the scaled DFT of the signal.

Linearity property

$$\text{DFT}\{ax(k) + by(k)\} = aX(i) + bY(i) \tag{3.46}$$

3.4.2 Periodic Property

Note from Definition 3.2 that the dependence of $X(i)$ on i occurs only in the exponent of W_N. As a result, $X(i)$ can be regarded as a function that is defined for *all* integer values of i, not just for $0 \leq i < N$. It is of interest to examine what happens when we go outside the range from 0 to $N - 1$. First, note from entry four of Table 3.2 that $W_N^{Nk} = 1$ for every integer k. We can use this result to then demonstrate that $X(i)$ is periodic.

$$X(i + N) = \sum_{k=0}^{N-1} x(k) W_N^{(i+N)k}$$

$$= \sum_{k=0}^{N-1} x(k) W_N^{ik} W_N^{Nk}$$

$$= \sum_{k=0}^{N-1} x(k) W_N^{ik} \tag{3.47}$$

The right-hand side of Eq. 3.47 is just $X(i)$. Thus, the DFT is periodic with a period of N.

Periodic property

$$\boxed{X(i + N) = X(i)} \tag{3.48}$$

3.4.3 Symmetry Property

If the signal $x(k)$ is real, then all of the information needed to reconstruct the N points of $x(k)$ is contained in the first $N/2$ complex points of $X(i)$. To see this, we compute the DFT by starting at the end and working backwards. In particular, from Definition 3.2 and Eq. 3.47, we have

$$X(N - i) = \sum_{k=0}^{N-1} x(k) W_N^{(N-i)k}$$

$$= \sum_{k=0}^{N-1} x(k) W_N^{-ik} W_N^{Nk}$$

$$= \sum_{k=0}^{N-1} x(k) W_N^{-ik} \tag{3.49}$$

Recall from Eq. 3.32 that the complex conjugate of W_N is $W_N^* = W_N^{-1}$. Consequently, if $x(k)$ is real, then the right-hand side of Eq. 3.49 is just the complex conjugate of $X(i)$. This yields the symmetry property of the DFT, which is valid for real $x(k)$.

Symmetry property

$$X(N - i) = X^*(i) \qquad (3.50)$$

There are a number of consequences of the symmetry property. First, note that when $i = 0$, it says $X(N) = X^*(0)$. But from the periodic property in Eq. 3.48, we have $X(N) = X(0)$. Therefore, $X^*(0) = X(0)$, which means that $X(0)$ must be real. Indeed, from Definition 3.2, we see that $X(0)$ is just the mean of $x(k)$ scaled by N.

$$X(0) = \sum_{k=0}^{N-1} x(k) \qquad (3.51)$$

The most important consequence of Eq. 3.50 lies in the observation that, for a real signal $x(k)$, the DFT contains redundant information. Recall that $X(i)$ can be expressed in polar form as $X(i) = A(i) \exp[j\phi(i)]$ where $A(i)$ is the magnitude and $\phi(i)$ is the phase angle. In most cases of practical interest, N will be a power of two, and therefore, N will be even. When N is even, it follows from Eq. 3.50 that the magnitude and phase spectra exhibit the following symmetry about the midpoint, $X(N/2)$.

$$A\left(\frac{N}{2} + i\right) = A\left(\frac{N}{2} - i\right), \quad 0 \le i < N/2 \qquad (3.52a)$$

$$\phi\left(\frac{N}{2} + i\right) = -\phi\left(\frac{N}{2} - i\right), \quad 0 \le i < N/2 \qquad (3.52b)$$

Thus, the magnitude spectrum $A(i)$ exhibits *even* symmetry about the midpoint, while the phase spectrum $\phi(i)$ exhibits *odd* symmetry about the midpoint. The following example illustrates this important observation.

Example 3.5 **DFT Symmetry**

As an example of computing a DFT of a real signal, suppose $x(k)$ is as follows.

$$x(k) = (0.8)^k - (-0.9)^k, \quad 0 \le k < 256$$

Here, $N = 256$ points in this case. The DFT of this signal can be found by running the script *exam3_5* on the distribution CD. Plots of the resulting magnitude spectrum $A(i)$ and phase spectrum $\phi(i)$ for $0 \le i < N$ are shown in Figure 3.6. Observe the even symmetry of $A(i)$ about $i = 128$ and the odd symmetry of $\phi(i)$ about $i = 128$. In this case, all of the information about the 256 points in $x(k)$ is encoded in the first 128 complex points of $X(i)$.

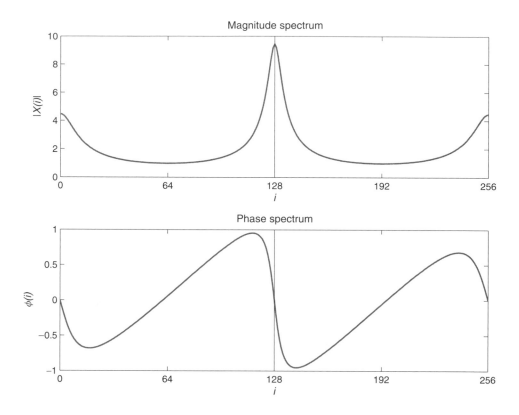

Figure 3.6 Magnitude and Phase Spectra of Signal in Example 3.5

3.4.4 Circular Shift Property

The next property requires that we work with a periodic extension of $x(k)$ to a signal $x_p(k)$ that is defined for all integers k. To concisely formulate $x_p(k)$, it is helpful to first define the *modulo* function as follows.

$$\text{mod}(k, N) \triangleq k - N \cdot \text{floor}\left(\frac{k}{N}\right) \tag{3.53}$$

Here *floor* is a function that rounds down toward $-\infty$, which means that $\text{mod}(k, N)$ represents the *remainder* after dividing k by N. The expression $\text{mod}(k, N)$ is often pronounced k modulo N. Notice that $\text{mod}(k, N) = k$ for $0 \leq k < N$. More generally, $r(k) = \text{mod}(k, N)$ is a periodic ramp-type function with period N as shown in Figure 3.7 for the case $N = 8$.

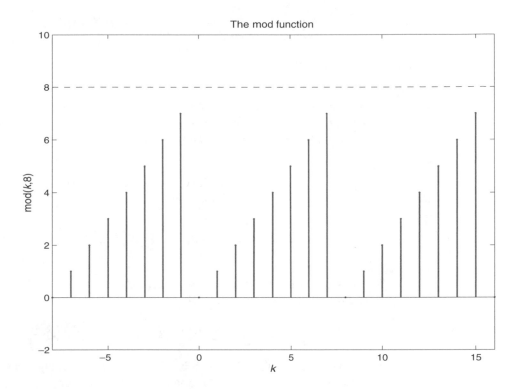

Figure 3.7 The Periodic Function $r(k) = \text{mod}(k, N)$ with $N = 8$

Given Eq. 3.53, the *periodic extension* of an N-point signal $x(k)$ is a signal $x_p(k)$ defined for all integers k as follows.

$$x_p(k) \triangleq x[\text{mod}(k, N)] \tag{3.54}$$

Therefore, $x_p(k) = x(k)$ for $0 \leq k < N$ and $x_p(k)$ extends $x(k)$ periodically in both positive and negative directions. Next, suppose we shift $x_p(k)$ by some number of samples M. Then, using the change of variable $q = k - M$, we have

$$
\begin{aligned}
\text{DFT}\{x_p(k - M)\} &= \sum_{k=0}^{N-1} x_p(k - M) W_N^{ik} \\
&= \left. \sum_{q=-M}^{N-1-M} x_p(q) W_N^{i(q+M)} \right\} q = k - M \\
&= W_N^{iM} \sum_{q=-M}^{N-1-M} x_p(q) W_N^{iq} \tag{3.55}
\end{aligned}
$$

Just as $x_p(q)$ is a periodic function of q with period N, the factor W_N^{iq} is also a periodic function of q of period N. In particular, from Eq. 3.32:

$$W_N^{i(q+N)} = \exp[-j2\pi i(q + N)/N]$$

$$= \exp(-j2\pi iq/N)\exp(-j2\pi i)$$

$$= \exp(-j2\pi iq/N)$$

$$= W_N^{iq} \tag{3.56}$$

It then follows that the term $x_p(q)W_N^{iq}$ in Eq. 3.55 is periodic with period N. As a result, the summation over one period can start at $q = 0$ rather than $q = -M$ without affecting the result. Thus, using Eq. 3.54 we have

$$\text{DFT}\{x_p(k)\} = W_N^{iM}\sum_{q=0}^{N-1} x_p(q)W_N^{iq}$$

$$= W_N^{iM}\sum_{q=0}^{N-1} x(q)W_N^{iq} \tag{3.57}$$

The sum on the right-hand side of Eq. 3.57 is just $X(i)$. This yields the following *circular shift* property of the DFT.

Circular shift property

$$\boxed{\text{DFT}\{x_p(k - M)\} = W_N^{iM} X(i)} \tag{3.58}$$

The time shift in Eq. 3.58 is referred to as a circular shift because $x_p(k)$ is a periodic extension of $x(k)$. One can think of the N samples of $x(k)$ as being *wrapped around* the unit circle of the complex plane shown previously in Figure 3.5. The signal $x_p(k - M)$ then represents a counterclockwise shift of $x(k)$ by M samples.

3.4.5 Reflection Property

Another property that makes use of the periodic extension $x_p(k)$ is the reflection property. Since the periodic extension is defined for all integers k, we can compute the DFT of $x_p(-k)$. Using the fact that $x_p(k)$ is periodic with period N and the change of variable $m = N - k$, we have

$$\text{DFT}\{x_p(-k)\} = \sum_{k=0}^{N-1} x_p(-k)W_N^{ik}$$

$$= \left.\sum_{k=0}^{N-1} x_p(N - k)W_N^{ik}\right\} \quad m = N - k$$

$$= \sum_{m=N}^{1} x_p(m)W_N^{i(N-m)}$$

$$= \sum_{m=1}^{N} x_p(m)W_N^{-im} \tag{3.59}$$

Here we have made use of the identity $W_N^N = 1$. Next, note that $x_p(m) = x(m)$ for $1 \le m < N$. Furthermore, $x_p(N) = x(0)$. Thus, the DFT can be rewritten as

$$\text{DFT}\{x_p(-k)\} = \sum_{m=1}^{N-1} x_p(m) W_N^{-im} + x_p(N) W_N^{-mN}$$

$$= \sum_{m=1}^{N-1} x(m) W_N^{-im} + x(0)(W_N^N)^{-m}$$

$$= \sum_{m=0}^{N-1} x(m) W_N^{-im} \tag{3.60}$$

Finally, recall from Table 3.2 that $W_N^{-im} = (W_N^*)^{im}$. Consequently, if $x(k)$ is real, the right-hand side of Eq. 3.60 is just the complex conjugate of $X(i)$. Thus, we have the following reflection property of the DFT valid for real $x(k)$.

Reflection property

$$\text{DFT}\{x_p(-k)\} = X^*(i) \tag{3.61}$$

3.4.6 Parseval's Theorem

Parseval's theorem is a simple and elegant relationship between a time signal and its transform. There are several versions of Parseval's theorem depending on the independent variable (continuous or discrete) and the signal duration (finite or infinite). For example, the DTFT version of Parseval's theorem is given in Eq. 3.24. The DFT version of Parseval's theorem is very similar and is as follows.

Parseval's theorem

$$\sum_{k=0}^{N-1} |x(k)|^2 = \frac{1}{N} \sum_{i=0}^{N-1} |X(i)|^2 \tag{3.62}$$

As an illustration of how we can apply Parseval's theorem, suppose $x_p(k)$ is a periodic extension of the N-point signal $x(k)$. The *average power* of $x_p(k)$ can be computed from one period as

$$P_x \triangleq \frac{1}{N} \sum_{k=0}^{N-1} |x(k)|^2 \tag{3.63}$$

Comparing Eq. 3.63 with Eq. 3.62, we see that Parseval's theorem provides us with an alternative frequency-domain version of average power. In particular, recalling from Eq. 3.44 that $S_N(i) = |X(i)|^2/N$ is the power density spectrum, one can show that

$$P_x = \frac{1}{N} \sum_{i=0}^{N-1} S_N(i) \tag{3.64}$$

Thus, the average power is just the average of the power density spectrum of $x(k)$; hence the name power density spectrum.

Example 3.6   **Parseval's Theorem**

To verify Parseval's theorem using a specific signal, consider the signal $x(k)$ from Example 3.2.

$$x = [3, -1, 0, 2]^T$$

In this case, $N = 4$. A direct time-domain computation of the average power of the periodic extension $x_p(k)$ using Eq. 3.63 yields

$$P_x = \frac{1}{N} \sum_{k=0}^{N-1} |x(k)|^2$$

$$= \frac{1}{4} (9 + 1 + 0 + 4)$$

$$= 3.5$$

From Example 3.2, the DFT of $x(k)$ is

$$X = [4, 3 + j3, 2, 3 - j3]^T$$

Thus, from Eq. 3.44, the power density spectrum is

$$S_N = \left(\frac{1}{4}\right) [16, 18, 4, 18]^T$$

$$= [4, 4.5, 1, 4.5]^T$$

Finally, from Eq. 3.62, the frequency-domain method of determining the average power is

$$P_x = \frac{1}{N} \sum_{i=0}^{N-1} S_N(i)$$

$$= \left(\frac{1}{4}\right) (4 + 4.5 + 1 + 4.5)$$

$$= \frac{14}{4}$$

$$= 3.5 \checkmark$$

Property	Description	Comments
Linearity	$\text{DFT}\{ax(k) + by(k)\} = aX(i) + bY(i)$	General
Periodic	$X(i + N) = X(i)$	General
Symmetry	$X(N - i) = X^*(i)$	Real x
Circular shift	$\text{DFT}\{x_p(k - M)\} = W_N^{iM} X(i)$	General
Reflection	$\text{DFT}\{x_p(-k)\} = X^*(i)$	Real x
Parseval's theorem	$\displaystyle\sum_{k=0}^{N-1} |x(k)|^2 = \frac{1}{N} \sum_{i=0}^{N-1} |X(i)|^2$	General
Circular convolution	$\text{DFT}\{x(k) \circ y(k)\} = X(i)Y(i)$	General
Circular correlation	$C_{xy}(k) = X(i)Y^*(i)$	General

A summary of the basic properties of the DFT can be found in Table 3.4. There are a number of other properties that could be added (Ingle and Proakis, 2000). The circular convolution and circular correlation properties listed at the bottom of Table 3.4 are sufficiently important that they are investigated separately in Chapter 4.

● ● ● ● ● ● ● ● ● ● ● ● ● ● ● ●

3.5 The Fast Fourier Transform (FFT)

The discrete Fourier transform or DFT is an important computational tool that is widely used in DSP. As with any computational method, it can be rated in terms of how the computational effort grows as the size of the problem increases. Recall that the DFT of an N-point signal $x(k)$ is

$$X(i) = \sum_{k=0}^{N-1} x(k) W_N^{ik}, \quad 0 \le i < N \tag{3.65}$$

The N^2 values of W_N^{ik} do not depend on $x(k)$, so they can be precomputed once and stored in an $N \times N$ matrix W. Each point $X(i)$ then requires N complex multiplications. Since there are N values of i, the total number of complex *floating-point operations* or FLOPs required to compute the entire DFT is

$$n_{\text{DFT}} = N^2 \ \text{FLOPs} \tag{3.66}$$

Thus, the computational effort, measured in complex multiplications, grows as the square of the size of the problem, N. In this case, we say that the DFT computation is of order $O(N^2)$. More generally, a computational algorithm is of *order* $O(N^p)$ if, for some constant α, the number of computations n satisfies

$$n \approx \alpha N^p, \quad N \gg 1 \tag{3.67}$$

3.5.1 Decimation in Time FFT

The popularity of the DFT arises from the fact that there is an implementation of it that dramatically decreases the computational time, particularly for large values of N. To see how the improvement in speed is achieved, suppose that the number of points N is a power of two. That is, $N = 2^r$ for some integer r where

$$r = \log_2(N) \tag{3.68}$$

Next, suppose we *decimate* the N-point signal $x(k)$ into two $N/2$-point signals $x_e(k)$ and $x_o(k)$ corresponding to the even and odd indices or subscripts of x, respectively.

$$x_e \overset{\Delta}{=} [x(0), x(2), \ldots, x(N/2 - 2)]^T \tag{3.69a}$$

$$x_o \overset{\Delta}{=} [x(1), x(3), \ldots, x(N/2 - 1)]^T \tag{3.69b}$$

The DFT in Eq. 3.65 can then be recast as two sums: one corresponding to the even values of k, and the other corresponding to the odd values of k. Recalling from Table 3.2 that $W_N^{2k} = W_{N/2}^k$, we have

$$
\begin{aligned}
X(i) &= \sum_{k=0}^{N/2-1} x(2k) W_N^{2ki} + \sum_{k=0}^{N/2-1} x(2k+1) W_N^{(2k+1)i} \\
&= \sum_{k=0}^{N/2-1} x_e(k) W_N^{2ki} + W_N^i \sum_{k=0}^{N/2-1} x_o(k) W_N^{2ki} \\
&= \sum_{k=0}^{N/2-1} x_e(k) W_{N/2}^{ki} + W_N^i \sum_{k=0}^{N/2-1} x_o(k) W_{N/2}^{ki} \\
&= X_e(i) + W_N^i X_o(i), \quad 0 \le i < N
\end{aligned}
\tag{3.70}
$$

Here $X_e(i) = \text{DFT}\{x_e(k)\}$ and $X_o(i) = \text{DFT}\{x_o(k)\}$ are $N/2$-point transforms of the even and odd parts of $x(k)$, respectively. Thus, Eq. 3.69 decompose the original N-point problem into two $N/2$-point subproblems with N complex multiplications and N complex additions required to merge the solutions. Each $N/2$-point DFT requires $N^2/4$ FLOPs. Thus, for large values of N, the number of floating-point operations required to implement the even–odd decomposition in Eq. 3.70 is

$$n_{\text{eo}} \approx \frac{N^2}{2} \text{ FLOPs}, \quad N \gg 1 \tag{3.71}$$

Comparing Eq. 3.71 with Eq. 3.66, we see that the computational effort has been reduced by a factor of two for large values of N. It is helpful to break the *merging formula* in Eq. 3.70 into two cases where $0 \le i < N/2$.

$$X(i) = X_e(i) + W_N^i X_o(i) \tag{3.72a}$$

$$X(i + N/2) = X_e(i + N/2) + W_N^{i+N/2} X_o(i + N/2) \tag{3.72b}$$

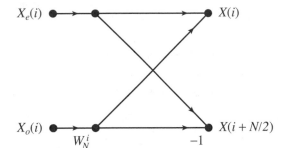

Figure 3.8 Signal Flow Graph of ith Order Butterfly Computation

Since $X_e(i)$ and $X_o(i)$ are $N/2$-point transforms, we know from the periodic property in Table 3.4 that $X_e(i + N/2) = X_e(i)$ and similarly for $X_o(i)$. Furthermore, from Table 3.2 we have $W_N^{i+N/2} = -W_N^i$. Thus, Eq. 3.72 can be simplified as follows where $0 \le i < N/2$.

$$Y(i) = W_N^i X_o(i) \tag{3.73a}$$

$$X(i) = X_e(i) + Y(i) \tag{3.73b}$$

$$X(i + N/2) = X_e(i) - Y(i) \tag{3.73c}$$

The merging formula is written as three equations using a temporary variable $Y(i)$. This increases storage by one complex scalar, but it reduces the number of complex multiplications from two to one. The computations in Eq. 3.73 are referred to as an ith order *butterfly*. The name arises from the signal flow graph representation of Eq. 3.73 shown in Figure 3.8. With a little imagination, the "wings" of the butterfly are apparent.

The even–odd decomposition of the DFT in Eq. 3.73 consists of two $N/2$-point DFTs plus $N/2$ interleaved butterfly computations. A block diagram that illustrates the case $N = 8$ is shown in Figure 3.9.

The beauty of the even–odd decomposition technique in Figure 3.9 is that there is nothing to prevent us from applying it again! For example, the even samples $x_e(k)$ and the odd samples $x_o(k)$ can each be decimated into even and odd parts. In this way, we can decompose each $N/2$-point transform into a pair of $N/4$-point transforms. Since $N = 2^r$, this process can be continued a total of r times. In the end, we are then left with a collection of elementary two-point DFTs. Interestingly enough, the two-point DFT is simply a zeroth order butterfly. The resulting algorithm is called the decimation in time *fast Fourier transform* or FFT (Cooley and Tukey, 1965). The signal flow graph for the case $N = 8$ is shown in Figure 3.10. Note how there are r iterations, and each iteration consists of $N/2$ butterfly computations.

Observe from Figure 3.10 that the input to the first iteration has been *scrambled* due to repeated decimation into even and odd subsequences. As it turns out, there is a simple numerical relationship between the original DFT order and the scrambled FFT order. This becomes apparent when we look at the binary representations of the indices, as shown in Table 3.5 for the case $N = 8$. Note how the scrambled FFT indices are obtained by taking the normal DFT indices, converting to binary,

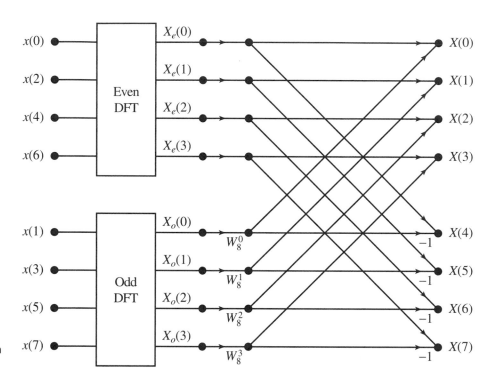

Figure 3.9 Even–Odd Decomposition of DFT with $N = 8$

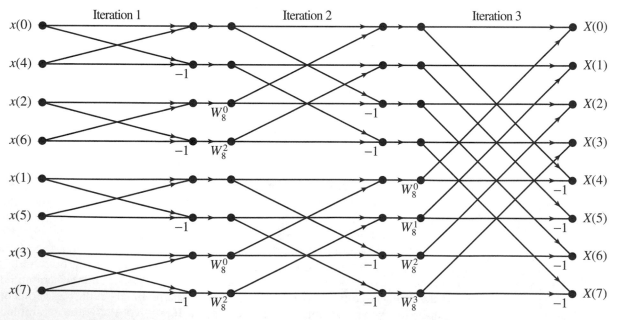

Figure 3.10 Signal Flow Graph of Decimation in Time FFT with $N = 8$

Table 3.5: ▶
Scrambled FFT Order
using Bit Reversal

DFT Order	Forward Binary	Reverse Binary	FFT Order
0	000	000	0
1	001	100	4
2	010	010	2
3	011	110	6
4	100	001	1
5	101	101	5
6	110	011	3
7	111	111	7

reversing the bits, and then converting back to decimal. The following algorithm is designed to interchange the elements of the input vector x using the bit-reversal process shown in Table 3.5.

ALGORITHM 3.1:

Bit Reversal

1. For $k = 0$ to $N - 1$ do
{

    (a) Set $m = 0$ and $d = k$

    (b) For $i = 0$ to $r - 1$ do
      {

$$b = \text{mod}(d, 2)$$

$$d = \text{floor}(d/2)$$

$$m = m + b2^{r-1-i}$$

      }

    (c) Set $y(m) = x(k)$

}

2. For $k = 0$ to $N - 1$ set $x(k) = y(k)$.

On entry to Algorithm 3.1, the vector x contains the input samples in the original DFT order. On exit, the elements of x have been reordered to the bit-reversed FFT order. The function $\text{mod}(d, 2)$ in Algorithm 3.1 returns the remainder after dividing d by 2, while the function floor$(d/2)$ rounds $d/2$ toward minus infinity. The integer m is the bit-reversed version of the index k. If Algorithm 3.1 is applied twice, the original ordering of $x(k)$ is restored.

The overall decimation in time FFT is summarized in Algorithm 3.2. The input vector is scrambled in step 1 with a call to Algorithm 3.1. The $rN/2$ butterfly computations are performed in step 2, and the final result is copied from x to X in step 3. To analyze step 2, it is helpful to refer to Figure 3.10. Step 2 consists of three loops. The outer loop goes through the r iterations moving from left to right in Figure 3.10. Notice that for each iteration there are groups of butterflies. In step 2, s is the spacing between groups, g is the number of groups, and b is the number of butterflies per group. The parameter b is also the wingspan of the butterfly. The second loop goes through the g groups associated with the current iteration, while the third and innermost loop goes through the b butterflies of each group. The first three equations in the innermost loop compute the weight w and the butterfly location n, while the last three equations compute the butterfly outputs as in Eq. 3.73.

ALGORITHM 3.2:
FFT

1. Call Algorithm 3.1 to scramble the input vector x.
2. For $i = 1$ to r do   % iteration
   {

   (a) Compute

   $$s = 2^i \qquad \text{\% group spacing}$$

   $$g = N/s \qquad \text{\% number of groups}$$

   $$b = s/2 \qquad \text{\% butterflies/group}$$

   (b) For $k = 0$ to $g - 1$ do
       {

       For $m = 0$ to $b - 1$ do
           {

           $$\theta = -2\pi mg/N$$
           $$w = \cos(\theta) + j\sin(\theta)$$
           $$n = ks + m$$
           $$y = wx(n + b)$$
           $$x(n) = x(n) + y$$
           $$x(n + b) = x(n) - y$$

           }

       }

   }
3. For $k = 0$ to $N - 1$ set $X(k) = x(k)$

3.5.2   FFT Computational Effort

Although the derivation of the FFT requires some attention to detail, the payoff is worthwhile because the end result is an algorithm that is dramatically faster than the DFT for practical values of N. To see how much faster, note from Figure 3.10 that there are r iterations and $N/2$ butterflies per iteration. Given the weight W_N^i, the butterfly computation in Eq. 3.73 requires one complex multiplication. Thus, there are $rN/2$ complex multiplications required for the FFT. It then follows from Eq. 3.68 that the computational effort of the FFT is

$$n_{\text{FFT}} = \frac{N \log_2(N)}{2} \text{ FLOPs} \tag{3.74}$$

Thus, the FFT algorithm is of order $O[N \log_2(N)]$, while the DFT is of order $O(N^2)$. A graphical comparison of the number of FLOPs required by the DFT and the FFT is shown in Figure 3.11 for $1 \leq N \leq 256$. Even over this modest range of values for N, the improvement in the speed of the FFT in comparison with the DFT is quite dramatic. For many practical problems, larger values of N in the range 1024 to 8192 are often used. For the case $N = 1024$, the DFT requires 1.049×10^6 FLOPs, while the FFT requires only 5.12×10^3 FLOPs. In this instance, a speed improvement by a factor of 204.8, or more than two orders of magnitude, is achieved.

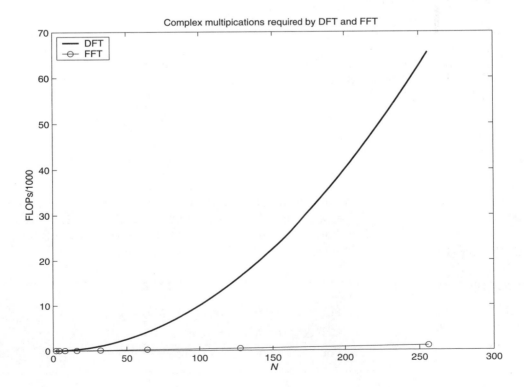

Figure 3.11   Computational Effort of DFT and FFT for $1 \leq N \leq 256$

Given the highly efficient FFT, there are a number of DSP applications where the best way to solve a time-domain problem is to use the following steps.

1. Transform to the frequency domain using the FFT.

2. Solve the problem in the frequency domain.

3. Transform back using the IFFT to recover the time-domain solution.

Recall that the formulation for the inverse discrete Fourier transform or IDFT is very similar to that of the DFT, namely

$$x(k) = \frac{1}{N} \sum_{i=0}^{N-1} X(i) W_N^{-ki} \tag{3.75}$$

One way to devise an algorithm for the IDFT is to modify Algorithm 3.2 by adding an input parameter which specifies which direction we want to transform: forward or reverse. Still another approach is to use Algorithm 3.2 itself, without modification. Recall that the complex conjugate of W_N is just $W_N^* = W_N^{-1}$. This being the case, we can reformulate Eq. 3.75 in terms of the FFT as follows.

$$x(k) = \frac{1}{N} \sum_{i=0}^{N-1} X(i) W_N^{-ki}$$

$$= \frac{1}{N} \sum_{i=0}^{N-1} X(i)(W_N^{ki})^*$$

$$= \frac{1}{N} \left(\sum_{i=0}^{N-1} X^*(i) W_N^{ki} \right)^*$$

$$= \left(\frac{1}{N} \right) \text{FFT}^*\{X^*(i)\} \tag{3.76}$$

Thus, we can implement an IFFT of $X(i)$ by taking the complex conjugate of the FFT of $X^*(i)$ and then normalizing the result by N. These steps are summarized in the following algorithm. Note that this approach takes advantage of the fact that the FFT can be applied to either a real or a complex signal.

ALGORITHM 3.3:
IFFT

1. For $k = 0$ to $N - 1$ set $x(k) = X^*(k)$
2. Call Algorithm 3.2 to compute $X(i) = \text{FFT}\{x(k)\}$
3. For $k = 0$ to $N - 1$ set $x(k) = X^*(k)/N$

3.5.3   Alternative FFT Implementations

There is one drawback to the FFT, as implemented, that is evident from Figure 3.11. For the FFT, the number of points N must be a power of two, while the DFT is defined for all $N \geq 1$. Often this is not a serious limitation because the user may be at liberty to choose a value for N, say in collecting data samples for an experiment. In still other cases, it may be possible to *pad* the signal $x(k)$ with a sufficient number of zeros to make N a power of two. Because N is a power of two, the formulation of the FFT in Algorithm 3.2 is called a *radix-two* version of the FFT. It is also possible to factor N in other ways and achieve alternative versions of the FFT (Ingle and Proakis, 2000). The alternative versions have computational speeds that lie between the DFT and the radix-two FFT shown in Figure 3.11.

There is also an alternative to the decimation in time method summarized in Algorithm 3.2 called the decimation in frequency method (see, for example, Schilling and Harris, 2000). The decimation in frequency method starts by decomposing $x(k)$ into a first half, $0 \leq k < N/2$, and a second half, $N/2 \leq k < N$. The merging formula for the decimation in frequency method is $X(2i) = \text{DFT}\{a(k)\}$ and $X(2i + 1) = \text{DFT}\{b(k)\}$ where

$$a(k) = x(k) + x(k + N/2) \tag{3.77a}$$

$$b(k) = [x(k) - x(k + N/2)]W_N^k \tag{3.77b}$$

The process proceeds in a manner generally similar to the decimation in time approach and again leads to $rN/2$ butterfly computations. In this case, it is the output vector $X(i)$ that ends up being scrambled, hence the name decimation in frequency.

MATLAB Toolbox

There are a number of a built-in MATLAB functions for computing the FFT and signal spectra that are very simple to use.

```
X   = fft (x,N);                  % compute N-point FFT
A   = abs (X);                    % magnitude spectrum
phi = angle(X);                   % phase spectrum
S   = A.^2/N;                     % power density spectrum
```

MATLAB function *fft* computes the FFT, when possible, and the DFT, when needed, depending on the value of N. One entry to *fft*, x is a real or complex vector of length M containing the samples to be transformed, and N is an optional integer argument specifying how many samples of x to use. The default value is $N = M$. If $N > M$, then x is padded with $N - M$ zeros. This has the effect of decreasing the spacing between discrete frequencies, which is given by

$$\Delta f = \frac{f_s}{N} \tag{3.78}$$

MATLAB functions *abs* and *angle* compute the magnitude and phase angle of X, respectively. Recall that the FDSP toolbox function *f_spec*, described at the end of Section 3.3, can also be used to compute the magnitude, phase, and power density spectra of a finite discrete-time signal. Often it is helpful to be able to determine the coordinates of critical points on a graph, such as peaks in a magnitude spectrum plot. This can be done interactively with the following FDSP toolbox function.

```
[x,y] = f_caliper (n);              % record coordinates of n points
```

When *f_caliper* is called, it displays a pair of crosshairs on the displayed graph. The crosshairs can be positioned anywhere on the graph using the mouse. When the left mouse button is clicked, the x and y coordinates of the current crosshair position are displayed on the graph. In addition, the coordinates are saved as the output variables x and y. The input argument n specifies the number of points to measure. Thus, x and y are returned as vectors of length n. When $n > 1$, the outputs from *caliper* can be used to compute the distance between critical points on a graph. For example, if the graph is a periodic waveform and matching points on the waveform are measured, then the period can be estimated using $x(2) - x(1)$.

3.6 White Noise

In this section, we examine the spectral characteristics of an important class of random signals called white noise. White noise signals are useful because they provide an effective way to model physical signals, signals that are often corrupted with additive noise. For example, the quantization error associated with finite precision analog-to-digital conversion can be modeled with white noise. Another important application area is in the modeling or identification of linear discrete-time systems, a topic that is treated in Chapter 9. White-noise input signals are particularly suitable as test signals for system identification because they contain power at all frequencies and therefore excite all of the natural modes of the system under investigation.

3.6.1 Uniform White Noise

Let x be a random variable in the interval $[a, b]$. If each value of x is equally likely to occur, then we say that x is *uniformly distributed* over the interval $[a, b]$. A uniformly distributed random variable can be characterized by the following *probability density function*.

$$p(x) = \begin{cases} \dfrac{1}{b - a}, & a \leq x \leq b \\ 0, & \text{otherwise} \end{cases} \qquad (3.79)$$

A graph of the uniform probability density function is shown in Figure 3.12. Note that the area under the probability density curve is always one. For any probability

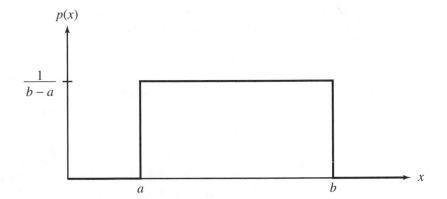

Figure 3.12 Uniform Probability Density Function

density function, the probability that a random variable x lies in an interval $[\alpha, \beta]$ can be computed as

$$P_{[\alpha,\beta]} = \int_{\alpha}^{\beta} p(x)dx \tag{3.80}$$

Random numbers can be characterized by their statistical properties. To define the statistical characteristics, it is helpful to introduce the expected value operator.

DEFINITION

3.3: Expected Value

> Let x be a random variable with probability density $p(x)$ and let $f(x)$ be a function of x. Then the *expected value* of $f(x)$ is denoted $E[f(x)]$ and defined
>
> $$E[f(x)] \triangleq \int_{-\infty}^{\infty} f(x)p(x)dx$$

The kth statistical *moment* of x is defined as $E[x^k]$ for $k \geq 1$. Thus, the kth moment is just the expected value of x^k. The most fundamental moment is the first moment which is called the *mean* of x.

$$\mu \triangleq E[x] \tag{3.81}$$

The mean, μ, is the average value about which the random variables are distributed. For random variables uniformly distributed over the interval $[a, b]$, a direct application of Definition 3.3 reveals that the mean is $\mu = (a+b)/2$.

Once the mean is determined, a second set of moments, called the central moments, can be computed. The kth *central moment* is defined as $E[(x - \mu)^k]$. The central moments specify the distribution of x about the mean. The first nonzero central moment is the second central moment, which is called the *variance* of x.

$$\sigma^2 \triangleq E[(x - \mu)^2] \tag{3.82}$$

The variance σ^2 is a measure of the spread of the random variables about the mean. The square root of the variance, σ, is called the *standard deviation* of x.

Most programming languages provide a facility for generating uniformly distributed random numbers. In the case of MATLAB, successive calls to the function *rand*, with no calling arguments, will produce a sequence of uniformly distributed random values in the interval $[0, 1]$. More generally, *rand(m,n)* returns a random $m \times n$ matrix. Suppose $r(k)$ for $0 \le k < N$ is a random sequence uniformly distributed over the interval $[0, 1]$. To obtain random values uniformly distributed over an arbitrary interval $[a, b]$, we use the following offset and scale factor.

$$x(k) = a + (b - a)r(k), \quad 0 \le k < N \tag{3.83}$$

The sequence of values in Eq. 3.83 is referred to as *uniform white noise*. More specifically, it is white noise that is uniformly distributed over the interval $[a, b]$. The reason for the term *white* arises from the fact that, just as white light contains all colors, the random noise signal in Eq. 3.83 contains power at all frequencies.

The average power of a random variable x is the second moment $E[x^2]$. For a uniformly distributed random variable, the average power can be computed using Definition 3.3 and the probability density function in Eq. 3.79 as follows.

$$P_x = E[x^2]$$

$$= \int_{-\infty}^{\infty} x^2 p(x) dx$$

$$= \frac{1}{b - a} \int_a^b x^2 dx$$

$$= \left. \frac{x^3}{3(b - a)} \right|_a^b \tag{3.84}$$

Consequently, for large N, the average power of the uniformly distributed white noise signal in Eq. 3.83 can be approximated as

Power of uniform white noise

$$P_u = \frac{b^3 - a^3}{3(b - a)} \tag{3.85}$$

Recall from Eq. 3.44 that the power density spectrum, $S_N(i) = |X(i)|^2/N$, specifies the amount of power at frequency $f_i = i f_s/N$. For a white noise signal, the power density spectrum is flat, as can be seen from the following example.

Example 3.7   **Uniform White Noise**

Suppose $a = -5$, $b = 5$, and $N = 512$. Then from Eq. 3.85 the average power of the white noise signal $x(k)$ in Eq. 3.83 is

$$P_u = \frac{5^3 - (-5)^3}{3(5 - (-5))}$$

$$= \frac{250}{30}$$

$$= 8.333$$

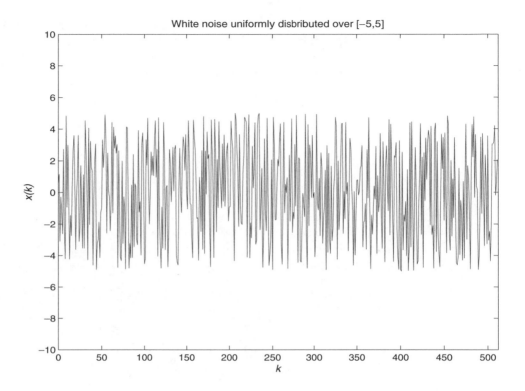

Figure 3.13 White Noise Uniformly Distributed Over $[-5, 5]$ with $N = 512$

Plots of a uniform white noise signal and its power density spectrum can be obtained by running the script *exam3_7* on the distribution CD. The time signal $x(k)$ for $0 \leq k < N$ is shown in Figure 3.13, and its power density spectrum $S_N(i)$ for $0 \leq i < N/2$ is shown in Figure 3.14. Note how the power density spectrum is, at least roughly, flat and nonzero, indicating that there is power at all $N/2$ discrete frequencies. The uneven nature of the computed power density spectrum in Figure 3.13 will be addressed in Section 3.8, where we will develop more sophisticated ways of estimating the power at each frequency.

The horizontal line in Figure 3.14 specifies the theoretical average power P_u from Eq. 3.85. The exact value of the average power of $x(k)$ can be computed directly for comparison. Applying the time-domain formula in Eq. 3.63 yields

$$P_x = \frac{1}{N} \sum_{i=0}^{N-1} x^2(k)$$

$$= 8.781$$

As the signal length N increases, the difference between the actual average power, P_x, and the theoretical average power, P_u, decreases.

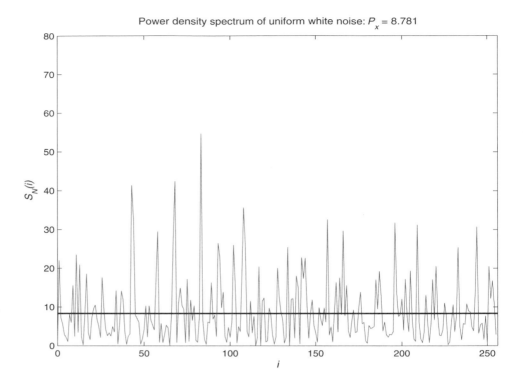

Figure 3.14 Power Density Spectrum of Uniformly Distributed White Noise

3.6.2 Gaussian White Noise

Uniformly distributed white noise is appropriate in those instances where the value of the signal is constrained to lie within a fixed interval. For example, a signal produced by a bipolar DAC is constrained to lie in the interval $[-V_r/2, V_r/2]$, where V_r is the reference voltage. However, there are other cases where it is more natural to model the noise using a *Gaussian* or normal probability density function as follows.

$$p(x) = \frac{1}{\sigma\sqrt{2\pi}} \exp\left[\frac{-(x-\mu)^2}{\sigma^2}\right] \tag{3.86}$$

Here the two parameters are the mean, μ, and the standard deviation, σ. The mean specifies where the random values are centered, while the standard deviation specifies the spread of the random values about the mean. A plot of the bell-shaped normal or Gaussian probability density function is shown in Figure 3.15 for the case where $\mu = 0$ and $\sigma = 1$.

Random numbers with a Gaussian distribution can be generated using the MAT-LAB function *randn*. The usage of the function *randn* is identical to that of the uniform random number generator *rand*. It produces random numbers with a Gaussian distribution with mean $\mu = 0$ and standard deviation $\sigma = 1$. Suppose $r(k)$

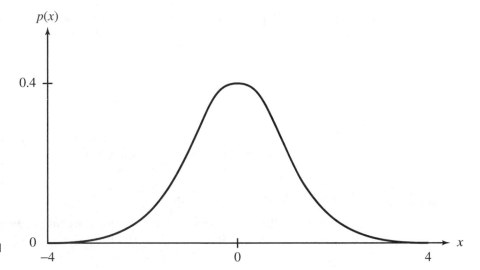

Figure 3.15 Gaussian (Normal) Probability Density Function with $\mu = 0$ and $\sigma = 1$

for $0 \le k < N$ is a Gaussian random sequence with zero mean and unit standard deviation. To produce a sequence with mean μ and standard deviation σ, we then use the following offset and scale factor.

$$x(k) = \mu + \sigma r(k), \quad 0 \le k < N \tag{3.87}$$

The random sequence in Eq. 3.87 is referred to as *Gaussian white noise* with mean μ and standard deviation σ. Again, it is called white noise because $x(k)$ contains power at all frequencies. Later, we will see how to filter it to produce *colored* noise.

For a Gaussian random variable, the average power can be computed as $E[x^2]$ using the probability density function in Eq. 3.86. To simplify the final result, we restrict our attention to the important special case of zero-mean noise. Using a table of integrals (Dwight, 1961), we have

$$
\begin{aligned}
P_x &= E[x^2] \\
&= \int_{-\infty}^{\infty} x^2 p(x) dx \\
&= \frac{2}{\sigma \sqrt{2\pi}} \int_0^{\infty} x^2 \exp\left(\frac{-x^2}{\sigma^2}\right) dx \quad \Big\} \quad \mu = 0 \\
&= \frac{2}{\sigma \sqrt{2\pi}} \left(\frac{\sqrt{\pi}\sigma^3}{4}\right)
\end{aligned}
\tag{3.88}
$$

Consequently, for large N, the average power of zero-mean Gaussian white noise with standard deviation σ can be approximated as

Power of zero-mean
Gaussian noise

$$P_g = \frac{\sigma^2}{2\sqrt{2}} \tag{3.89}$$

It is apparent that the average power of zero-mean Gaussian white noise is proportional to the square of the standard deviation, which is the variance.

Example 3.8   **Gaussian White Noise**

Consider the following continuous-time periodic signal produced by an AM mixer with frequencies $f_1 = 300$ Hz and $f_2 = 100$ Hz.

$$x_a(t) = \sin(2\pi f_1 t)\cos(2\pi f_2 t)$$

Recall from Appendix D, that the product of two sinusoids produces sum and difference frequencies. Suppose $x_a(t)$ is corrupted with additive Gaussian noise, $v(k)$, with zero mean and standard deviation $\sigma = 0.8$. If the result is then sampled at $f_s = 1$ kHz using $N = 1024$ samples, the corresponding discrete-time signal is

$$x(k) = \sin(.3\pi k)\cos(.1\pi k) + v(k), \quad 0 \le k < N$$

Using Eq. 3.89, the average power of the noise term is

$$P_v = \frac{\sigma^2}{2\sqrt{2}}$$

$$= \frac{.64}{2\sqrt{2}}$$

$$= 0.2263$$

The signal $x(k)$ and its power density spectrum $S_N(i)$ can be obtained by running the script *exam3_8* on the distribution CD. The first quarter of the noise-corrupted time-signal is shown in Figure 3.16, and its power density spectrum is shown in Figure 3.17. Note how it is difficult to tell from the time plot in Figure 3.16 that $x(k)$ contains a periodic component due to the presence of the noise. However, the two spectral components at the sum frequency $f_1 + f_2 = 400$ Hz and the difference frequency $f_1 - f_2 = 200$ Hz are evident from the power density spectrum plot in Figure 3.17, where two distinct spikes are present. The Gaussian white noise is also apparent in Figure 3.17 as low-level power that is distributed over all frequencies.

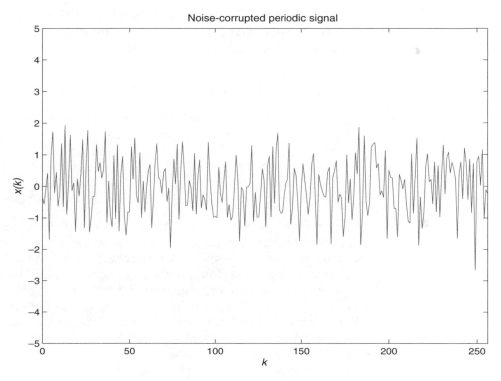

Figure 3.16 Periodic Signal Corrupted with Zero-mean Gaussian White Noise

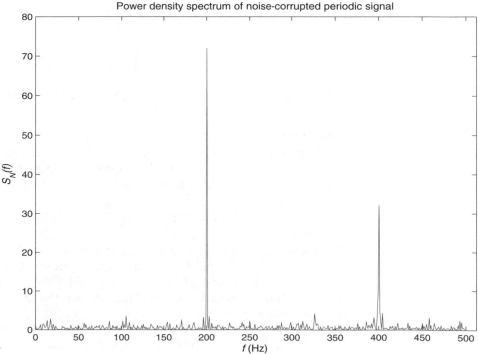

Figure 3.17 Power Density Spectrum of Noise-corrupted Periodic Signal

The power density spectrum plot in Figure 3.17 uses the independent variable $f = if_s/N$, rather than i, in order to facilitate interpretation of the frequency. Thus, it is a plot of $S_N(f)$ versus f rather than $S_N(i)$ versus i.

MATLAB Toolbox

As a convenience to the user, the FDSP toolbox contains the following functions for generating white noise signals.

```
rand ('state',s)              % Initialize random numbers using seed s
A = f_randu (m,n,a,b);        % Uniform white noise
A = f_randg (m,n,mu,sigma);   % Gaussian white noise
```

The built-in MATLAB function *rand* can be used to initialize the random number generator using an integer seed, s. Otherwise, the default seed is used. Each $s \geq 0$ produces a different pseudo-random sequence. FDSP toolbox function *f_randu* is used to generate a matrix containing uniformly distributed white noise, while function *f_randg* produces a matrix of Gaussian white noise. On entry to *f_randu*, $m \geq 1$ is the number of rows, $n \geq 1$ is the number of columns, a is the lower limit of the random values, and b is the upper limit. On exit from *f_randu*, A is an $m \times n$ matrix whose elements are real random values uniformly distributed over $[a, b]$. The calling arguments for *f_randg* are identical to those for *f_randu*, except that the lower limit, a, is replaced with the mean, mu, and the upper limit, b, is replaced with the standard deviation, $sigma$, of the Gaussian distribution.

3.7 Discrete-time Frequency Response

Recall that the frequency response of a discrete-time system can be obtained from the discrete-time Fourier transform or DTFT by evaluating the transfer function along the unit circle. If f_s is the sampling frequency and $T = 1/f_s$ is the sampling interval, then the frequency response is

$$H(f) = H(z)|_{z=\exp(j2\pi fT)}, \quad -f_s/2 < f \leq f_s/2 \tag{3.90}$$

The magnitude $A(f) = |H(f)|$ is the magnitude response, and the phase angle $\phi(f) = \angle H(f)$ is the phase response of the system. There is a practical way to approximate the frequency response using the DFT. Moreover, in some instances, the approximation is exact. Using Definition 3.1, the DTFT of the system impulse response $h(k)$ is

$$H(f) = \sum_{k=0}^{\infty} h(k) \exp(-jk2\pi fT) \tag{3.91}$$

To develop an efficient way to approximate the frequency response, let $H(i)$ denote the DFT of the first N samples of $h(k)$.

$$H(i) = \text{DFT}\{h(k)\}, \quad 0 \leq i < N \tag{3.92}$$

Next, suppose we evaluate the frequency response at the ith discrete frequency, $f_i = if_s/N$. Recalling from Eq. 3.32 that $W_N = \exp(-j2\pi/N)$ and using Eq. 3.91, we have

$$
\begin{aligned}
H(f_i) &= \sum_{k=0}^{\infty} h(k) \exp(-jki2\pi/N) \\
&= \sum_{k=0}^{\infty} h(k) W_N^{ik} \\
&= \sum_{k=0}^{N-1} h(k) W_N^{ik} + \sum_{k=N}^{\infty} h(k) W_N^{ik} \\
&= H(i) + \sum_{k=N}^{\infty} h(k) W_N^{ik}
\end{aligned}
\tag{3.93}
$$

Thus, the difference between $H(f_i)$ and $H(i)$ is represented by the *tail* of the DTFT of the impulse response $h(k)$. The magnitude of this difference can be bounded in the following way.

$$
\begin{aligned}
|H(f_i) - H(i)| &= \left| \sum_{k=N}^{\infty} h(k) W_N^{ik} \right| \\
&\leq \sum_{k=N}^{\infty} |h(k) W_N^{ik}| \\
&= \sum_{k=N}^{\infty} |h(k)| \cdot |W_N^{ik}| \\
&= \sum_{k=N}^{\infty} |h(k)|
\end{aligned}
\tag{3.94}
$$

Notice that the upper bound on the right-hand side of Eq. 3.94 no longer depends on i. For a stable filter, the impulse response is absolutely summable, so the right-hand side is finite and goes to zero as $N \to \infty$. Consequently, the DFT of the impulse response can be used to approximate the discrete-time frequency response as long as the number of samples N is sufficiently large.

Discrete-time
frequency response

$$H(i) \approx H(f_i), \quad 0 \le i \le \frac{N}{2} \tag{3.95}$$

In Eq. 3.95, we restrict i to the interval $0 \le i \le N/2$ because, for a system with real coefficients, $H(i)$ exhibits symmetry about the midpoint $i = N/2$, as in Eq. 3.50. Thus, all of the essential information is contained in the subinterval $0 \le i \le N/2$, which corresponds to positive frequencies.

For an important class of digital filters, the approximation in Eq. 3.95 is in fact exact. To see this, recall that if $H(z)$ is an FIR filter of order m, then $h(k) = 0$ for $k > m$. It then follows from Eq. 3.94 that the upper bound on the error is zero for $N > m$. That is, if $H(z)$ is an FIR filter of order m and $N > m$, then $H(i)$ in Eq. 3.92 is an exact representation of the N samples of the frequency response. The frequency response, $H(i)$, can be expressed in polar form as $H(i) = A(i) \exp[j\phi(i)]$, where the $A(i)$ is the *magnitude response* and $\phi(i)$ is the *phase response*.

$$A(i) \overset{\Delta}{=} |H(i)| \tag{3.96a}$$

$$\phi(i) \overset{\Delta}{=} \angle H(i) \tag{3.96b}$$

Example 3.9 **Discrete-time Frequency Response**

As an example of using Eq. 3.92 to find the frequency response, consider a running average filter of order m.

$$y(k) = \frac{1}{m+1} \sum_{i=0}^{m} x(k-i)$$

The impulse response of this FIR filter is a pulse of amplitude, $1/(m+1)$, and duration, $m+1$, samples starting at $k = 0$.

$$h(k) = \frac{1}{m+1} \sum_{i=0}^{m} \delta(k-i)$$

The frequency response $H(i) = \text{DFT}\{h(k)\}$ for the case $m = 8$, $N = 1024$, and $f_s = 200$ Hz can be obtained by running the script *exam3_9* on the distribution CD. Note that since $N = 2^{10}$, the FFT implementation can be used. The resulting magnitude response, $A(f)$, and phase response, $\phi(f)$, are shown in Figure 3.18. There are $m/2$ lobes in the magnitude response, and the phase response has jump discontinuities between the lobes but is otherwise linear. The independent variable used in Figure 3.18 is $f = if_s/N$.

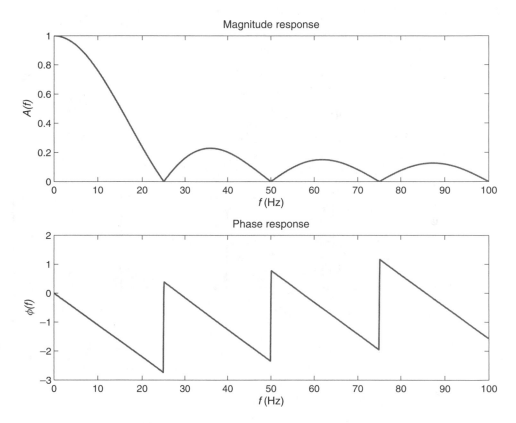

Figure 3.18 Frequency Response of Running Average Filter with $m = 8$, $N = 1024$, and $f_s = 200$ Hz

---

Decibel Scale (dB)

The plot of the magnitude response in Figure 3.18 is referred to as a *linear* plot because both axes use linear scales. If there is a large frequency range or a large range in the values of the magnitude response, then logarithmic units are often used. When plotting the magnitude response of a filter or the magnitude spectrum of a signal, the logarithmic unit that is typically used is the decibel which is abbreviated dB. A *decibel* is defined as 10 times the logarithm (base 10) of the signal power. Since power corresponds to $|H(i)|^2$, this translates to 20 times $\log_{10}$ of $|H(i)|$.

Gain in decibels

$$A(i) = 20 \log_{10}(|H(i)|) \ \ \text{dB} \tag{3.97}$$

One of the useful characteristics of the dB scale is that it allows us to better quantify how close to zero the magnitude response gets. Notice from Eq. 3.97 that a

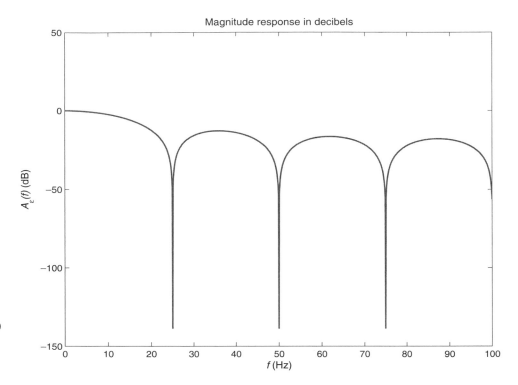

Figure 3.19 Magnitude Response of Running Average Filter in Example 3.9 using the Logarithmic Decibel Scale

gain of one corresponds to zero dB, and that every reduction in $|H(i)|$ by a factor of 10 corresponds to a change of -20 dB. For an ideal filter, the signal attenuation in the stopband is complete, so the magnitude response is zero. However, in practice, Wiener and Paley (1934) showed that the magnitude response of a causal system can not be identically zero over a continuum of frequencies, it can only be zero at isolated frequencies. By using decibels, we can see how close to zero the magnitude response gets. However, there is one problem that arises with the use of decibels. If the magnitude response does go to zero at certain isolated frequencies as in Figure 3.18, then from Eq. 3.97 this corresponds to a decibel value of $-\infty$. To accommodate this situation, we usually replace $|H(i)|$ in Eq. 3.97 as follows

$$A_\epsilon(i) = 20 \log_{10}(\max\{|H(i)|, \epsilon\}) \ \ \text{dB} \qquad (3.98)$$

Here $\epsilon > 0$ is taken to be a very small number. For example, the single-precision machine epsilon, $\epsilon_M = 1.19 \times 10^{-7}$, might be used. A plot of the magnitude response of the filter in Example 3.9 using $\epsilon = \epsilon_M$ and the logarithmic dB scale is shown in Figure 3.19.

MATLAB Toolbox

The FDSP toolbox contains the following function for computing the frequency response of a linear discrete-time system.

```
[H,f] = f_freqz (b,a,N,fs);
A = abs(H);
phi = angle(H);
```

On entry to *f_freqz*, b and a are vectors of length $m+1$ and $n+1$. They contain the numerator polynomial coefficients and the denominator polynomial coefficients, respectively, of the following system transfer function where $a(1) = 1$.

$$H(z) = \frac{b_0 + b_1 z^{-1} + \cdots + b_m z^{-m}}{1 + a_1 z^{-1} + \cdots + a_n z^{-n}} \tag{3.99}$$

Input argument N is the number of samples of the impulse response to use in evaluating the frequency response, and fs is the sampling frequency in Hz. Argument fs is optional with a default value of $fs = 1$. On exit from *f_freqz*, H is a complex vector of length N containing the samples of the frequency response, and f is a vector of length N containing the frequencies at which the frequency response is evaluated with $f(i) = (i-1)fs/(2N)$. Here the frequency response is obtained by computing the DFT of the impulse response. The vectors A and *phi* contain the magnitude response and the phase response, respectively. Function *f_freqz* has some optional calling arguments that will be discussed in Chapter 5.

• • • • • • • • • • • • • • • •

3.8 Zero Padding

There is a simple way to interpolate between the discrete points of signal spectra without increasing the sampling rate. The length of the signal can be increased from N to M by padding $x(k)$ with $M - N$ zeros as follows.

$$x_z(k) \triangleq \begin{cases} x(k), & 0 \le k < N \\ 0, & N \le k < M \end{cases} \tag{3.100}$$

Here $x_z(k)$ denotes the *zero-padded* version for $x(k)$. Suppose $S_M(i)$ is the power density spectrum of $x_z(k)$ but scaled by M/N because no new power was added by the zero padding. That is,

$$S_M(i) \triangleq \frac{|X_z(i)|^2}{N}, \quad 0 \le i < M \tag{3.101}$$

where $X_z(i) = \text{DFT}\{x_z(k)\}$. Then the *frequency increment*, or space between adjacent discrete frequencies, of $S_M(i)$ is

$$\Delta f = \frac{f_s}{M} \tag{3.102}$$

The effect of zero padding is to interpolate between the original points of the spectrum. In particular, if $M = qN$ for some integer q, then there will be $q - 1$ new points between each of the original N points. The original points will be unchanged.

$$X_z(qi) = X(i), \quad 0 \le i < N \tag{3.103}$$

By decreasing Δf through zero padding, the location of an individual sinusoidal spectral component can be more precisely determined, as can be seen from the following example.

Example 3.10   **Zero Padding**

As an illustration of how zero padding can be put to effective use, suppose $f_s = 1024$ Hz and $N = 256$. Consider a noise-free signal with a single sinusoidal spectral component at $f_a = 330.5$ Hz.

$$x(k) = \cos(2\pi f_a kT), \quad 0 \le k < N$$

If we zero pad $x(k)$ by a factor of eight, then $M = 8N = 2048$. Thus, the frequency increment before and after zero padding is

$$\Delta f_N = \frac{f_s}{N} = 4 \text{ Hz}$$

$$\Delta f_M = \frac{f_s}{M} = 0.5 \text{ Hz}$$

The corresponding discrete-frequency indices for the two cases are

$$i_N = \frac{f_a}{\Delta f_N} = 82.625$$

$$i_M = \frac{f_a}{\Delta f_M} = 661$$

It then follows that the peaks in the power density spectra will occur at

$$f_N = 83\Delta f_N = 332 \text{ Hz}$$

$$f_M = 661\Delta f_M = 330.5 \text{ Hz}$$

Consequently, with a frequency increment of $\Delta f_M = 0.5$ Hz, the zero-padded power density spectrum should be able to locate the sinusoidal frequency exactly. The two power density spectra can be computed by running the script *exam3_10* on the distribution CD. The resulting plot is shown in Figure 3.20. To clarify the display, only the frequency range $300 \le f \le 360$ Hz is plotted. The low-precision power density spectrum, $S_N(f)$, is specified with isolated points, while the high-precision power density spectrum, $S_M(f)$, is plotted as points connected by straight lines. Note how $S_M(f)$ interpolates between the points of $S_N(f)$. In this case, $M/N = 8$, so there are seven points of $S_M(f)$ between each pair of points in $S_N(f)$.

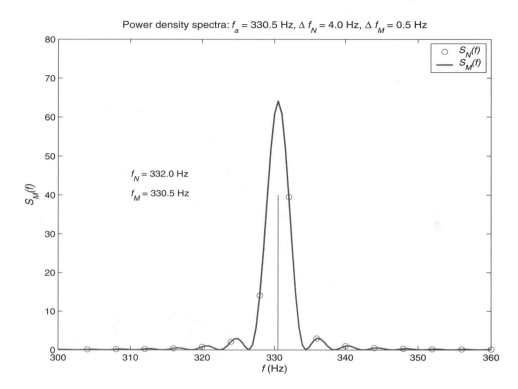

Power density spectra: $f_a = 330.5$ Hz, $\Delta f_N = 4.0$ Hz, $\Delta f_M = 0.5$ Hz

$f_N = 332.0$ Hz

$f_M = 330.5$ Hz

Figure 3.20   Power Density Spectra using Zero Padding with $N = 256$ and $M = 2048$

The ringing in $S_M(f)$ arises because we are multiplying $x(k)$ by a rectangular window to obtain $x_z(k)$. As we shall see, we can reduced the side lobes, at the expense of broadening the main lobe, by multiplying by a different window.

Although zero padding allows us to identify, more precisely, an isolated sinusoidal frequency component, it is less helpful when it comes to trying to *resolve* the difference between two closely spaced sinusoidal components. When two sinusoidal components are too close together, their peaks in the power density spectrum tend to merge into a single, somewhat broader, peak. The basic problem with zero padding is that it does not add any new information to the signal. The *frequency resolution*, or smallest frequency difference that we can detect, is limited by the sampling frequency and the number of nonzero samples. The following ratio, called the *Rayleigh limit*, represents the value of the frequency resolution.

Frequency resolution

$$\Delta F = \frac{f_s}{N} \tag{3.104}$$

Thus, there is a difference between the frequency increment, which is the spacing between discrete frequencies of the DFT, and the frequency resolution, which is the smallest difference in frequencies that can be reliably detected. Zero padding can be helpful in distinguishing between two spectral components, but we cannot go below the limit in Eq. 3.104. Observe that since $f_s = 1/T$, the Rayleigh limit is really just the reciprocal of the length of the original signal, $\tau = NT$. Thus, for a given sampling frequency, to improve the frequency resolution we must take more samples.

Example 3.11   **Frequency Resolution**

As an illustration of the detection of two closely spaced sinusoidal components, suppose $f_s = 1024$ Hz and $N = 1024$. Consider the following noise-free signal with spectral components at $f_a = 330$ Hz and $f_b = 331$ Hz.

$$x(k) = \sin(2\pi f_a kT) + \cos(2\pi f_b kT), \quad 0 \le k < N$$

If we zero pad $x(k)$ by a factor of two, then $M = 2N = 2048$. Thus, the frequency increment before and after zero padding is

$$\Delta f_N = \frac{f_s}{N} = 1.0 \text{ Hz}$$

$$\Delta f_M = \frac{f_s}{M} = 0.5 \text{ Hz}$$

The two power density spectra can be computed by running the script *exam3_11* on the distribution CD. The resulting plots are shown in Figures 3.21 and 3.22, respectively. To clarify the display, the independent variable f is used instead of i,

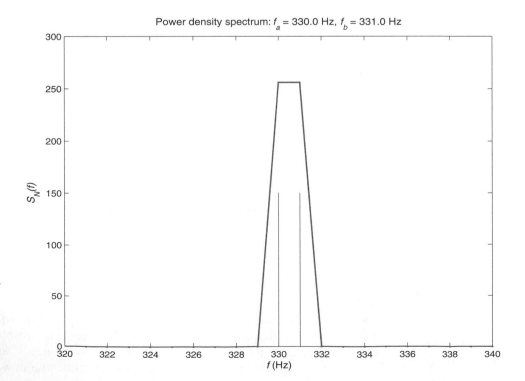

Figure 3.21   Power Density Spectrum of Two Closely Spaced Sinusoidal Components with $f_s = 1024$ Hz and $N = 1024$

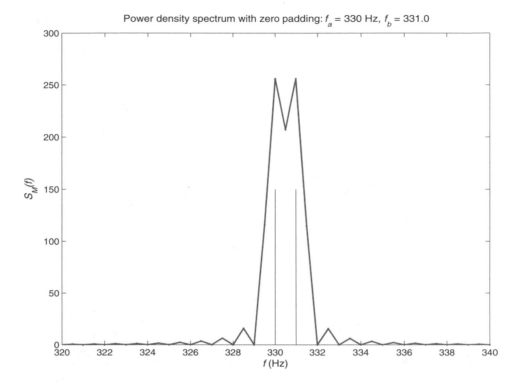

Figure 3.22 Spectral Resolution of Two Closely Spaced Sinusoidal Components using Zero Padding with $f_s = 1024$ Hz, $N = 1024$, and $M = 2048$

and only the frequency range of $320 \leq f \leq 340$ Hz is plotted in each case. The low-precision power density spectrum in Figure 3.22 shows a single broad peak centered at the midpoint,

$$f_p = \frac{f_a + f_b}{2} = 330.5 \text{ Hz}$$

By zero padding, we can decrease the frequency increment from 1 Hz to 0.5 Hz. This results in the plot shown in Figure 3.22, which features two distinct, but overlapping, peaks centered at f_a and f_b, respectively. The power density spectrum between the peaks decreases sufficiently to make the two peaks discernible, but it does not go to zero between the peaks. From Eq. 3.104, the Rayleigh limit is

$$\Delta F = \frac{f_s}{N} = 1 \text{ Hz}$$

Interestingly enough, increasing M beyond 2048 does *not* make the two peaks more discernible because the separation between f_a and f_b is the Rayleigh limit.

$$|f_b - f_a| = \Delta F$$

Of course, we can decrease the Rayleigh limit by adding more samples. If the number of samples is doubled to $N = 2048$, then the resulting power density spectrum, without any zero padding, is shown in Figure 3.23. Here it is clear that to two peaks are clearly separated.

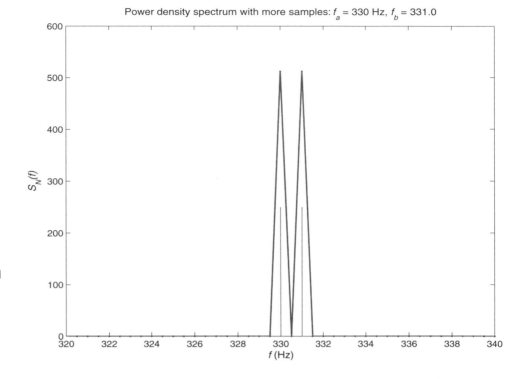

Figure 3.23 Spectral Resolution of Two Closely Spaced Sinusoidal Components using Improved Frequency Resolution with $f_s = 1024$ Hz and $N = 2048$

Zero Padding and the FFT

The concept of zero padding is useful even when the motivation is not to reduce the frequency increment. If the number of samples N is not a power of two, then a radix-two FFT cannot be used, instead a DFT or perhaps a less efficient alternative form of the FFT is needed. This problem can be circumvented by simply padding $x(k)$ with enough zeros to make N the next power of two. For example, if $N = 1000$, then padding $x(k)$ with 24 zeros will increase the number of samples to $N = 2^{10} = 1024$. This increases storage requirements by 2.4 percent, but in exchange, it significantly decreases the computational time. Recall that if a 1000-point DFT is computed, this requires approximately $N^2 = 10^6$ FLOPs. However, a 1024-point FFT requires only $N \log_2(N) = 1.02 \times 10^4$ FLOPs. Thus, at a cost of 2.4 percent in storage space, we have purchased an increase in speed by a factor of 97.7: A real bargain!

● ● ● ● ● ● ● ● ● ● ● ● ● ● ● ●

*3.9 Power Density Spectrum Estimation

In this section, we develop techniques for estimating the underlying continuous-time power density spectrum (PDS) of a signal when the length of the data sequence N is large. In addition, we focus on the practical problem of detecting the presence and location of one or more sinusoidal components buried in a signal that is corrupted with noise. The problem is made more challenging when the frequencies of the unknown sinusoidal components do not correspond to any of the discrete frequencies $f_i = if_s/N$, which are sometimes called the *bin frequencies*.

The topic of power density spectrum estimation is one that has been studied in some depth (Proakis and Manolakis, 1992; Ifeachor and Jervis, 2002; Prat, 1997). The treatment presented here includes a basic presentation of the classical nonparametric estimation methods. Recall from Eq. 3.44 that the power density spectrum of an N-point signal is given by

$$S_N(i) = \frac{|X(i)|^2}{N}, \quad 0 \le i < N \tag{3.105}$$

Here $S_N(i)$ specifies the average power contained in the ith harmonic of the periodic extension, $x_p(k)$, of the N-point signal $x(k)$. Recall from Eq. 3.64 that the average power of $x_p(k)$ is the average of the power density spectrum.

$$P_x = \frac{1}{N} \sum_{i=0}^{N-1} S_N(i) \tag{3.106}$$

3.9.1   Bartlett's Method

The formulation of the power density spectrum in Eq. 3.105 is sometimes called a *periodogram*. If the sequence $x(k)$ is long, then a more reliable way to estimate the power density spectrum is to partition $x(k)$ into a number of subsignals. Suppose $N = LM$ for a pair of positive integers, L and M. Then $x(k)$ can be partitioned into M subsignals, each of length L, as shown in Figure 3.24.

If the subsignals are denoted $x_m(k)$ for $0 \le m < M$, then the mth subsignal can be extracted from the original signal $x(k)$ as

$$x_m(k) \overset{\Delta}{=} x(mL + k), \quad 0 \le k < L \tag{3.107}$$

Next, let $X_m(i) = \text{DFT}\{x_m(k)\}$ for $0 \le m < M$. The estimated power density spectrum of $x(k)$ is obtained by taking the average of the M individual power density spectra.

Bartlett's method

$$S_B(i) = \frac{1}{LM} \sum_{m=0}^{M-1} |X_m(i)|^2, \quad 0 \le i < L \tag{3.108}$$

The power density spectrum estimate $S_B(i)$ is called an *average periodogram*. It was first introduced by Bartlett (1948). Note that one of the consequences of

Figure 3.24 Partition of $x(k)$ into M Subsignals of Length L

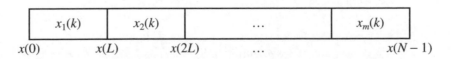

Bartlett's method is that it changes the frequency increment of the power density spectrum. If f_s is the sampling frequency, the frequency increment, or space between adjacent discrete frequencies, is

$$\Delta f = \frac{f_s}{L} \tag{3.109}$$

This is in contrast to the periodogram method in Eq. 3.105 where the frequency increment is $\Delta f = f_s/N$. Since $N = LM$, the frequency increment of Bartlett's method is larger by a factor of M. This loss of precision is offset by the observation that the variance in the estimated power density spectrum is reduced by the same factor. We can approximate the variance of the estimate as follows.

$$\sigma_B^2 = \frac{1}{L} \sum_{i=0}^{L-1} [S_B(i) - P_x]^2 \tag{3.110}$$

Ideally, the power density spectrum of white noise should be a flat line at $S_N(i) = P_x$. By reducing the variance of the computed power density spectrum, Bartlett's method comes closer to this ideal as can be seen in the following example.

Example 3.12 **Bartlett's Method: White Noise**

As an illustration of how the average periodogram can improve the estimate of the power density spectrum, let $x(k)$ be white noise uniformly distributed over the interval $[-5, 5]$. Then from Eq. 3.85 the average power of the white noise signal is

$$P_u = \frac{5^3 - (-5)^3}{3[5 - (-5)]}$$

$$= \frac{250}{30}$$

$$= 8.333$$

Suppose the length of $x(k)$ is $N = 4096$. One way to factor N is $L = 512$ and $M = N/L = 8$. In this case, we compute eight periodograms, each of length 512. The resulting average periodogram can be obtained by running script *exam3_12* on the distribution CD. A plot of the estimated power density spectrum, $S_B(i)$, is shown in Figure 3.25. This uniform white noise signal was previously analyzed in Example 3.7. Comparing Figure 3.25 with Figure 3.14 from Example 3.7, we see that the average periodogram estimate is noticeably flatter and closer to the ideal characteristic, $P_u = 8.333$. The average power of $x(k)$ is also closer to the theoretical value of P_u as a result of working with the longer signal of length, $N = 4096$. In this case, we have

$$P_x = \frac{1}{N} \sum_{k=0}^{N-1} x^2(k)$$

$$= 8.588$$

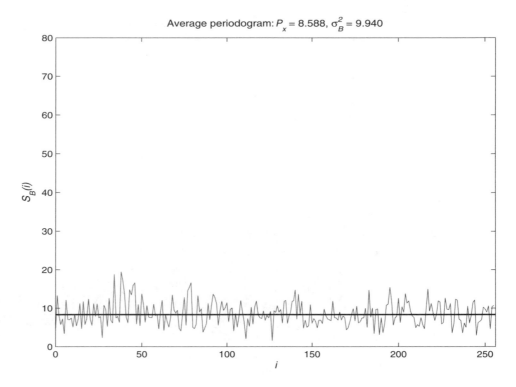

Average periodogram: $P_x = 8.588$, $\sigma_B^2 = 9.940$

Figure 3.25 Average Periodogram of White Noise Uniformly Distributed over $[-5, 5]$ with $N = 4096$, $L = 512$, and $M = 8$

The reduction in the variance of the estimate is a consequence of the averaging of M periodograms. Using Eq. 3.110, the variance of the estimate in this case is

$$\sigma_B^2 = \frac{1}{L} \sum_{i=0}^{L-1} [S_B(i) - P_x]^2$$

$$= 9.940$$

The curious reader might wonder if perhaps the same improvement in the estimate of the power density spectrum might be achieved with a single periodogram by simply increasing N from 512 in Example 3.7 to the 4096 samples used in Example 3.12. This is an intuitively appealing conjecture, but unfortunately, it does not hold, as can be seen in Figure 3.26. The power density spectrum shown in Figure 3.26 was obtained by running script *exam3_12*, but using $L = 4096$ and $M = 1$. Although the average power P_x does come closer to P_u as N increases, it is apparent that the variance in the power density spectrum curve does *not* decrease in this case. In particular, by using Eq. 3.110, the variance using a single periodogram of length $N = 4096$ is

$$\sigma_N^2 = 76.225$$

If we take the ratio of the two variances, we find that $\sigma_N^2 / \sigma_B^2 = 7.669$. Thus, the improvement achieved by Bartlett's average periodogram method was by a factor of approximately $M = 8$.

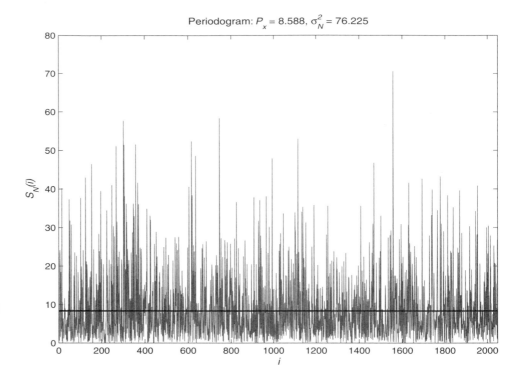

Periodogram: $P_x = 8.588$, $\sigma_N^2 = 76.225$

Figure 3.26 Single Periodogram of White Noise Uniformly Distributed over $[-5, 5]$ with $N = 4096$

Next, consider the practical problem of detecting the presence and precise location of unknown sinusoidal components in a signal $x(k)$ that may also contain noise.

Example 3.13 **Bartlett's Method: Periodic Input**

Suppose the sampling rate is $f_s = 1024$ Hz, and the number of samples is $N = 512$. If Bartlett's method is used with $L = 128$ and $M = 4$, then from Eq. 3.109 the frequency increment is

$$\Delta f = \frac{f_s}{L}$$

$$= 8 \text{ Hz}$$

Consider a signal $x(k)$ that contains two sinusoidal components, one at frequency $f_a = 200$ Hz, and the other at frequency $f_b = 331$ Hz.

$$x(k) = \sin(2\pi f_a kT) - \sqrt{2}\cos(2\pi f_b kT)$$

Observe that f_a is an integer multiple of Δf, so f_a corresponds to discrete frequency i_a with

$$i_a = \frac{f_a}{\Delta f}$$

$$= 25$$

However, f_b is not an integer multiple of the frequency increment Δf. Therefore, f_b falls between two discrete frequencies at

$$i_b = \frac{f_b}{\Delta f}$$

$$= 41.375$$

The average periodogram estimate of the power density spectrum of $x(k)$ can be obtained by running script *exam3_13* on the distribution CD. The resulting plot of $S_B(f)$ is shown in Figure 3.27. Here we have used the independent variable $f = if_s/L$ to facilitate interpretation of the frequencies. Note that there are two spikes indicating the presence of two sinusoidal components.

The first spike is centered at $f_a = 200$ Hz as expected. This spike is symmetric about $f = f_a$, and it is relatively narrow with a width of ± 8 Hz which corresponds to the frequency increment Δf. Next, consider the spectral spike associated with f_b.

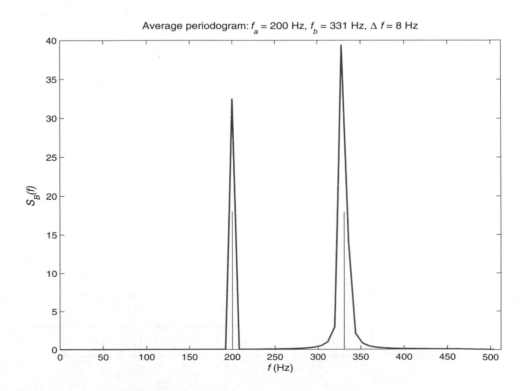

Figure 3.27 Average Periodogram of Signal in Example 3.13 with $N = 512$, $L = 128$, and $M = 4$

Unfortunately, this spike is not as narrow as the low-frequency spike, particularly near the bottom. Furthermore, the high-frequency spike is also not centered at $f_b = 331$ Hz. Instead, the peak of the second spike occurs below f_b at discrete frequency:

$$f_{41} = \frac{41 f_s}{L}$$

$$= 328 \text{ Hz}$$

The estimate of the location of the second spectral component is low by 0.91 percent. Thus, the estimated power density spectrum does an effective job in detecting sinusoidal components located at the discrete frequencies, but the performance is less impressive for frequencies located between the discrete frequencies.

3.9.2   Welch's Method

The broad spike in Figure 3.27 associated with a spectral component that is not aligned with one of the discrete frequencies, $f_i = i f_s / L$, is a manifestation of something called the *leakage phenomenon*. Basically, the spectral power that should be concentrated at a single frequency has spread out or *leaked* into adjacent frequencies. To reduce the leakage Welch (1967) proposed modifying the average periodogram approach in two ways. First, instead of partitioning the signal into M distinct subsignals, as in Figure 3.24, we instead decompose $x(k)$ into a number of overlapping subsignals. If $N = LM$, then a total of $2M - 1$ subsignals of length L can be constructed using an overlap of $L/2$ as shown in Figure 3.28. If the subsignals are denoted $x_m(k)$ for $0 \leq m < 2M - 1$, then mth subsignal can be extracted from the original signal $x(k)$ as

$$x_m(k) \overset{\Delta}{=} x(mL/2 + k), \quad 0 \leq k < L \tag{3.111}$$

The extraction of $x_0(k)$ from $x(k)$ can be regarded as a multiplication of $x(k)$ by the following *rectangular window* of width L.

$$w_R(k) = \begin{cases} 1, & 0 \leq k < L \\ 0, & \text{otherwise} \end{cases} \tag{3.112}$$

Figure 3.28   Decomposition of $x(k)$ into $2M - 1$ Overlapping Subsignals of Length L

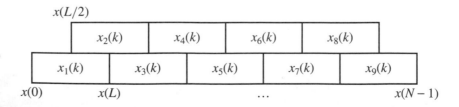

The other subsignals can be extracted in a similar manner with a shifted rectangular window. When the DFT of subsignal $x_m(k)$ is computed, it can be thought of as finding the Fourier coefficients of the periodic extension of $x_m(k)$. However, if subsignal $x_m(k)$ does not represent an integer number of cycles of a given sinusoidal component, then the periodic extension of $x_m(k)$ will have a jump discontinuity. This discontinuity leads to the well known Gibb's phenomenon of Fourier series, a phenomenon that causes ringing or oscillations near the discontinuity when the signal is reconstructed. This ringing manifests itself in the frequency domain as the leakage observed in the second spectral spike in Figure 3.27.

If T is the sampling interval, then the length of subsignal $x_m(k)$ is $\tau = LT$. Thus, for a sinusoidal component at frequency f_a, the number of cycles per subsignal is

$$i_a = f_a \tau$$

$$= \frac{L f_a}{f_s} \tag{3.113}$$

Since $f_i = i f_s / L$, it follows that discrete frequency f_i contains exactly i cycles per subsignal. For the first spectral component in Example 3.13 at $f_a = 200$ Hz, the number of cycles per subsignal was $i_a = 25$. However, for the troublesome spectral component at $f_b = 331$ Hz, the number of cycles per subsignal was $i_b = f_b \tau = 41.375$. Thus, there was a jump discontinuity associated with the second spectral component.

In order to reduce the effects of jump discontinuities in the periodic extensions of the subsignals, Welch proposed that the subsignals be multiplied by a window other than the rectangular window. If the window goes to zero gradually, rather than abruptly as in Eq. 3.112, then the jump discontinuities in the periodic extensions of the subsignals can be eliminated. A number of popular candidates for data window functions are summarized in Table 3.6. Plots of the window functions in Table 3.6 are shown in Figure 3.29 for the case $L = 256$.

Table 3.6: ▶
Window Functions

Number	Name	$w(k), \ 0 \leq k < L$
0	Rectangular	1
1	Hanning	$0.5 - 0.5 \cos\left(\dfrac{2\pi k}{L}\right)$
2	Hamming	$0.54 - 0.46 \cos\left(\dfrac{2\pi k}{L}\right)$
3	Blackman	$0.42 - 0.5 \cos\left(\dfrac{2\pi k}{L}\right) + 0.08 \sin\left(\dfrac{4\pi k}{L}\right)$

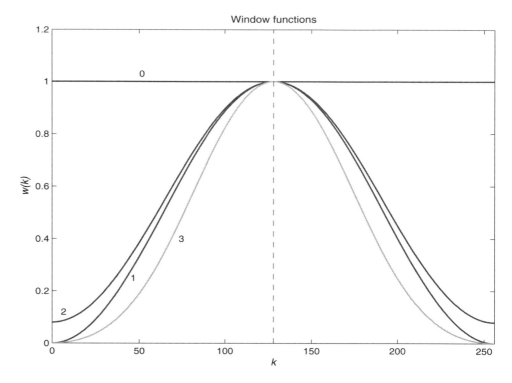

Figure 3.29 Window Functions for $L = 256$: 0 = Rectangular, 1 = Hanning, 2 = Hamming, and 3 = Blackman

Given a window $w(k)$ of width L, we can compute the DFT of the mth windowed subsequence of $x(k)$ as

$$Y_m(i) = \text{DFT}\{w(k)x_m(k)\}, \quad 0 \le m < 2M - 1 \tag{3.114}$$

An estimate of the power density spectrum is then obtained by averaging the $2M - 1$ overlapping windowed periodograms.

Welch's method

$$S_W(i) = \frac{1}{L(2M - 1)} \sum_{m=0}^{2M-2} |Y_m(i)|^2, \quad 0 \le i < L \tag{3.115}$$

The power density spectrum estimate $S_W(i)$ is called a *modified average periodogram* or Welch's method. The overlap between subsignals could be increased or decreased. However, the 50 percent overlap has been shown to improve certain statistical properties of the estimate (Welch, 1967). For the Hanning window, the 50 percent overlap means that each sample is equally weighted (see Problem P3.17).

Example 3.14 **Welch Method: Noisy Periodic Input**

As an illustration of Welch's method, consider a noise-corrupted version of the signal from Example 3.13. That is, suppose $f_a = 200$ Hz, $f_b = 331$ Hz, and $v(k)$ is white noise uniformly distributed over the interval $[-3, 3]$.

$$x(k) = \sin(2\pi f_a kT) - \sqrt{2}\cos(2\pi f_b kT) + v(k), \quad 0 \le k < N$$

Suppose $N = 1024$, and we factor N as $L = 256$, $M = N/L = 4$. Thus, Welch's method uses $2M - 1 = 7$ subsignals, each of length $L = 256$. From Eq. 3.109, the frequency increment is

$$\Delta f = \frac{f_s}{L}$$

$$= 4 \text{ Hz}$$

In this case, the discrete frequency indices of f_a and f_b are

$$i_a = \frac{f_a}{\Delta f} = 50$$

$$i_b = \frac{f_b}{\Delta f} = 82.75$$

An estimate of the power density spectrum of $x(k)$ using Welch's method with the Hanning window can be obtained by running script *exam3_14* on the distribution CD. The resulting plot of $S_W(f)$ is shown in Figure 3.30. Here we have used the independent variable $f = if_s/L$ to facilitate interpretation of the frequencies. Note

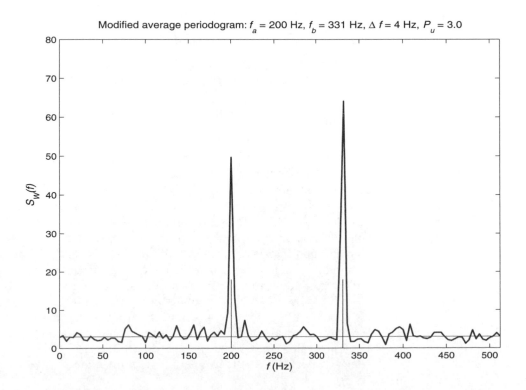

Modified average periodogram: f_a = 200 Hz, f_b = 331 Hz, Δf = 4 Hz, P_u = 3.0

Figure 3.30 Modified Average Periodogram of Noisy Periodic Signal

that there are two spikes indicating the presence of two sinusoidal components. In addition, there is power uniformly distributed over all frequencies which represents the white noise term $v(k)$. Using Eq. 3.85, the average power of the white noise is $P_u = 3$, which is indicated by the horizontal line. The two peaks occur at

$$f_{50} = 50\Delta f = 200 \text{ Hz}$$

$$f_{83} = 83\Delta f = 332 \text{ Hz}$$

In this case, there is improvement in the estimate of f_b in comparison with Example 3.13, but only because L has been doubled, which reduces the frequency precision from 8 Hz to 4 Hz. The important difference between Figure 3.30 and Figure 3.27 lies in the observation that the width of the second peak is now comparable to the width of the first peak. This reduction in width is a result of the windowing which reduces the leakage into nearby frequencies. The peaks in Figure 3.30 are actually somewhat wider than the frequency precision of ± 4 Hz. This is the price that is paid for the leakage reduction through windowing. We will study windows in more detail, including their spectral characteristics, in Chapter 6.

MATLAB Toolbox

The FDSP toolbox contains the following function for estimating the power density spectrum using both Bartlett's method and Welch's method.

```
[S,f,Px] = f_pds (x,N,L,fs,win,meth);          % estimate PDS of x
```

On entry to *f_pds*, x is a vector of length n containing the samples, and N is the number of samples to use. If $N > n$, then x is padded with $N - n$ zeros. Calling argument L is the length of the subsequence to use. It must be an integer submultiple of N. That is, $N = LM$ for a pair of integers L and M. Next, *fs* is the sampling frequency, and *win* is an integer specifying the number of the data window to use from Table 3.6. Finally, *meth* is an integer specifying the estimation method to use. When *meth* is zero and *win* is zero, Bartlett's average periodogram is computed using a rectangular window and no overlap of the subsignals. If *meth* is nonzero, then Welch's modified average periodogram is computed using the selected window and a 50 percent overlap of the subsignals. On exit from *f_pds*, S is a vector of length L containing the estimated power density spectrum, f is a vector of length L containing the frequencies at which the power density spectrum is evaluated, and Px is a real scalar containing the average power of x. The evaluation frequencies are $f(i) = (i - 1)f_s/L$. For a real x, the power density spectrum $S(i)$ is symmetric about the midpoint $i = L/2$.

● ● ● ● ● ● ● ● ● ● ● ● ● ● ● ● ●

*3.10 The Spectrogram

Many signals of practical interest are sufficiently long that their spectral characteristics can be thought of as changing with time. For example, a segment of recorded
speech can be broken down into more basic units such as words, syllables, or phonemes.
Each basic unit will have its own distinct spectral characteristics. One way to cap-
ture the fact that the spectral characteristics are changing with time is to compute a
sequence of short overlapping DFTs. For example, suppose $x(k)$ is a long N-point
signal with $N = LM$ for integers L and M. Then $x(k)$ can be partitioned into $2M - 1$
overlapping subsignals of length L as shown previously in Figure 3.28. If the subsignals are denoted $x_m(k)$ for $0 \leq m < 2M - 1$, then the mth subsignal can be extracted
from the original signal $x(k)$ as

$$x_m(k) \triangleq x(mL/2 + k), \quad 0 \leq m < 2M - 1, \ 0 \leq k < L \qquad (3.116)$$

Thus, the subsignals overlap one another by $L/2$ samples. If we take the DFTs
of $x_m(k)$ for $0 \leq m < 2M - 1$ and arrange them into a matrix, this leads to the
spectrogram.

DEFINITION

3.4: Spectrogram

> Let $x(k)$ be an N-point signal partitioned into $2M - 1$ overlapping subsignals
> $x_m(k)$ of length L as in Eq. 3.116, and let $w(k)$ be a window function from
> Table 3.6. The *spectrogram* of $x(k)$ is a $(2M - 1) \times L$ matrix denoted $G(m, i)$
> and defined
>
> $$G(m, i) \triangleq |\text{DFT}\{w(k)x_m(k)\}|, \quad 0 \leq m < 2M - 1, \ 0 \leq i < L$$

The spectrogram $G(m, i)$ is a collection of short DFTs parameterized by the
starting time. The first independent variable m specifies the starting time, in increments of $L/2$, while the second variable i specifies the frequency, in increments of
f_s/L. The purpose of the window function $w(k)$ is to reduce the effects of the leakage
phenomenon. When no window is used, the values of $G(m, i)$ tend to get smeared
or spread out along the frequency dimension.

The spectrogram is typically displayed as a two-dimensional contour plot. Note
that $G(m, i) \geq 0$. The two-dimensional display uses different colors or shades of a
color to denote the values of $G(m, i)$ within certain bands or levels. Thus, a spectrogram is a contour plot of slices or level surfaces of the windowed magnitude spectra.
Note that if $x(k)$ is real, then $G(m, i)$ will exhibit even symmetry about the line
$i = L/2$. Consequently, only the $(2M - 1) \times L/2$ submatrix needs to be plotted for
real $x(k)$.

Example 3.15

Spectrogram

Consider the signal $x(k)$ shown in Figure 3.31, which consists of a recording of four seconds of the vowels, {A, E, I, O, U}, spoken in succession. In this case, the sampling rate was $f_s = 8000$ Hz and the number of samples was $N = 32,000$. If we choose $L = 256$, then $M = 125$, then $G(m, i)$ is a 250×256 matrix. Letting $T = 1/f_s$, the time and frequency increments are as follows.

$$\Delta t = \frac{LT}{2} = 16 \text{ msec}$$

$$\Delta f = \frac{f_s}{L} = 31.25 \text{ Hz}$$

Script *exam3_15* on the distribution CD can be used to compute $G(m, i)$. The resulting spectrogram using $p = 12$ levels and a rectangular window is shown in Figure 3.32. Note how each vowel has its own distinct set of contour lines. For example, the vowel "I" has significant power between zero and 1500 Hz, while the vowel "A" has power between zero and about 600 Hz, with some additional power centered around 2500 Hz. The use of a rectangular window (i.e. no tapering) tends to smear the spectrogram features parallel to the frequency axis due to the leakage phenomenon. A somewhat cleaner spectrogram can be obtained by using the Hamming window, as shown in Figure 3.33. Here the islands associated with peaks in the magnitude response are more isolated from one another.

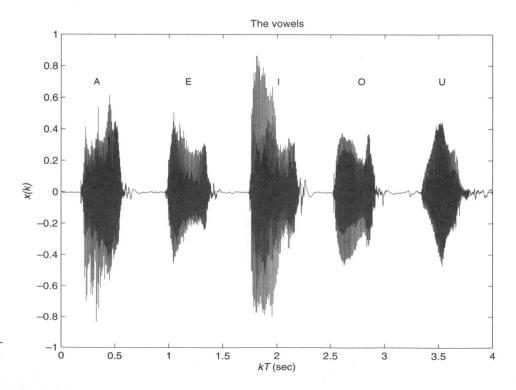

Figure 3.31 Recording of Vowels at $f_s = 8000$ Hz

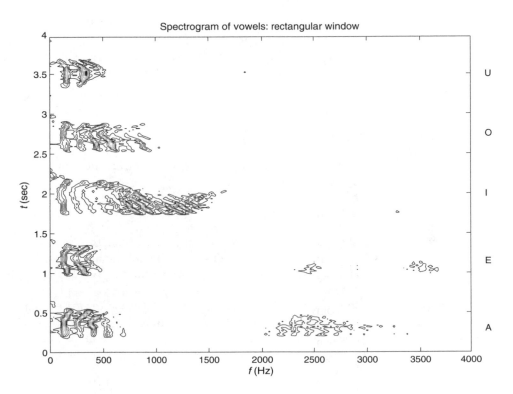

Figure 3.32 Spectrogram of Vowels using a Rectangular Window and 20 Levels

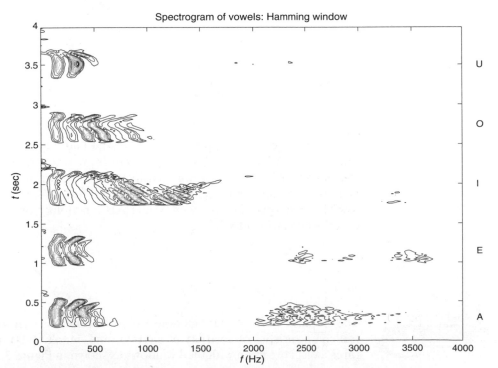

Figure 3.33 Spectrogram of Vowels using a Hamming Window and 20 Levels

It should be pointed out that overlaps other than $L/2$ can be used in Eq. 3.116 to compute the spectrogram. A still more flexible approach is to decompose the signal $x(k)$ into low-frequency and high-frequency parts using the wavelet transform. For a discussion of wavelets, the interested reader is referred, for example, to Burrus and Guo (1997).

MATLAB Toolbox

The FDSP toolbox contains the following function for computing the spectrogram of a discrete-time signal.

```
[G,f,t] = f_specgram (x,L,fs,win);
```

On entry to *f_specgram*, x is a vector of length N containing the samples of the signal, $L < N$ is the length of the subsignal, *fs* is the sampling frequency in Hz, and *win* is an integer specifying the number of the window to be used from Table 3.6. On exit from *f_specgram*, G is a $(2M - 1) \times L$ matrix containing the spectrogram, f is a vector of length L containing the evaluation frequencies, and t is a vector of length $2M - 1$ containing the starting times. The subsignal length L does not have to be an integer submultiple of N because the value for M is computed using $M = \text{floor}(N/L)$. See script *exam3_15* for an example of using *f_specgram*.

• • • • • • • • • • • • • • ••••

3.11 Software Applications

This section focuses on applications of the DFT and the spectral analysis of discrete-time signals. A GUI module called *g_spectra* is presented that allows user to examine the magnitude, phase, and power density spectra of discrete-time signals without any need for programming. Application examples are then presented and solved using MATLAB scripts. The relevant MATLAB functions from the FDSP toolbox are described in Appendix B.

3.11.1   GUI Module: *g_spectra*

The graphical user interface module *g_spectra* allows the user to investigate the spectral characteristics of a variety of discrete-time signals. GUI module *g_spectra* features a display screen with tiled windows as shown in Figure 3.34.

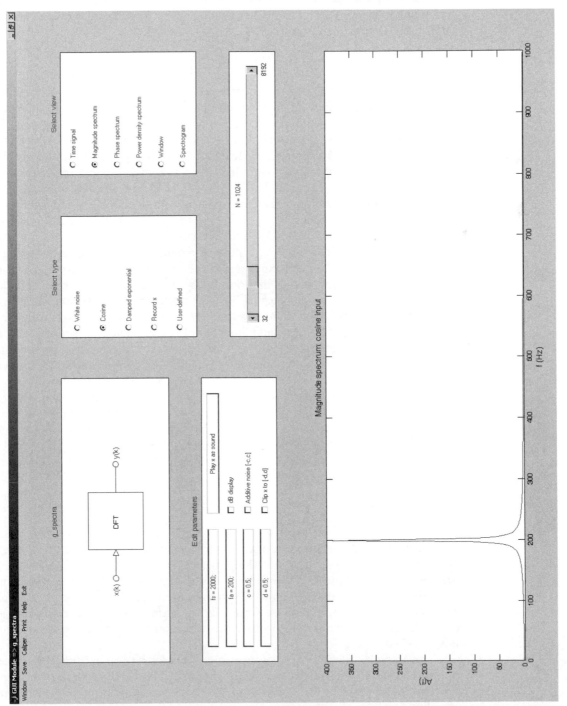

Figure 3.34 Display Screen of Chapter GUI Module *g_spectra*

The *Block diagram* window in the upper-left corner of the screen contains a color-coded block diagram of the operation being performed. This module computes the DFT of an N-point signal $x(k)$.

$$X(i) = \text{DFT}\{x(k)\}, \quad 0 \leq i < N \tag{3.117}$$

Below the block diagram are a number of edit boxes containing parameters that can be modified by the user. The parameters include the sampling frequency fs, the frequency of a cosine input fa, the magnitude c of additive zero-mean uniform white noise, and a clipping threshold d. Changes to the parameters are activated with the Enter key. To the right of the edit boxes are pushbutton and checkbox controls. The Play x as sound pushbutton plays the signal $x(k)$ as sound on the PC speaker. The dB Display checkbox control toggles the graphical display between a linear scale and a logarithmic dB scale. The Add noise checkbox adds white noise uniformly distributed over $[-c, c]$ to the input $x(k)$. Finally, the Clip checkbox activates clipping of the input $x(k)$ to the interval $[-d, d]$.

The *Type* and *View* windows in the upper-right corner of the screen allow the user to select both the type of input signal $x(k)$ and the viewing mode. The inputs include several common signals that optionally can include white noise, depending on the status of the Add noise checkbox. There are two inputs that the user can customize. The Record sound input allows the user to record one second of sound from the PC microphone at a sampling rate of $fs = 8000$ Hz. Proper recording can be verified by using the Play x as sound pushbutton. The User-defined input prompts for the name of a user-supplied MAT-file that contains up to 8192 samples of $x(k)$ plus the sampling frequency fs. If no fs is present in the MAT-file, then the current value for the sampling frequency is used. Near the bottom of the *Type* and *View* windows is a *Slider* bar that allows the user to control the number of samples N. The *Slider* bar is active for all inputs except the recorded and user-defined inputs.

The viewing options include the time signal $x(k)$, the magnitude spectrum $A(f)$, the phase spectrum $\phi(f)$, and the estimated power density spectrum $S_W(f)$, where

$$f_i = \frac{i f_s}{N}, \quad 0 \leq i \leq N/2 \tag{3.118}$$

The power density spectrum estimate uses Welch's modified average periodogram method with $L = N/4$. There are also viewing options for displaying the data window and the spectrogram of x using overlapping subsignals of length $L = N/8$. The *Plot* window along the bottom half of the screen shows the selected view. The curves are color-coded to match the block diagram in the upper-left corner of the screen. The magnitude response and power density spectrum plots can be displayed using either the linear scale or the logarithmic dB scale, depending on the status of the dB Display checkbox.

The *Menu* bar along the top of the screen includes several options. The *Window* option allows the user to choose a data window to be used for the power density spectrum estimate. The *Save* option is used to save the current x and fs in a user-specified MAT-file for future use. The *Caliper* option allows the user to measure any point on the current plot by moving the mouse crosshairs to that point and clicking.

Files created in this manner can be reloaded with the User-defined input option. The *Print* option prints the contents of the plot window. Finally, the *Help* option provides the user with some helpful suggestions on how to effectively use module *g_spectra*.

3.11.2　Signal Detection

One of the application areas of spectral analysis is the detection of signals buried in noise. Suppose the sampling rate is f_s, and the signal to be analyzed includes M sinusoidal components with frequencies $F_1, F_2, \ldots, F_M$, where $0 \le F_i \le f_s/2$ are *unknown* and must be determined from a spectral analysis of the following signal

$$x(k) = \sum_{i=1}^{M} \sin(2\pi F_i kT) + v(k), \quad 0 \le k < N \tag{3.119}$$

Here $v(k)$ is additive white noise uniformly distributed over the interval $[-b, b]$. For example, $v(k)$ might represent measurement noise or noise picked up in a communication channel over which $x(k)$ is transmitted.

The unknown frequencies can be detected and identified by executing the following MATLAB script, labeled *exam3_16* on the distribution CD, which uses $N = 1024$, $f_s = 2000$ Hz, and $b = 2$.

```
function exam3_16

% Example 3.16: Signal detection

clear
clc
fprintf('Example 3.16: Signal detection\n')

% Prompt for simulation parameters

seed = f_prompt ('Enter seed for random number generator',0,10000,3000);
M = f_prompt ('Enter number of random sinusoidal terms',0,10,3);
rand ('state',seed);
N = 1024;
k = 0 : N-1;
b = 2;
x = -b + 2*b*rand(1,N);
fs = 2000;
T = 1/fs;
F = (fs/2)*rand(1,M);
for i = 1 : M
   x = x + sin(2*pi*F(i)*k*T);
end
save u_spectra2 x fs                          % MAT file
```

```
% Plot portion of signal

figure
t = k*T;
plot (t(1:N/8),x(1:N/8))
axis([t(1),t(N/8),-5,5])
f_labels ('Noisy time signal with sinusoidal components','{\itt} (sec)','\it{x(t)}')
f_wait


% Compute and plot power density spectrum

figure
A = abs(fft(x));
S_N = A.^2/N;
f = linspace (0,(N-1)*fs/N,N);
plot (f(1:N/2),S_N(1:N/2))
set(gca,'Xlim',[0,fs/2])
f_labels ('Power density spectrum','{\itf} (Hz)','\it{S_N(f)}')

% Find frequencies with user-supplied threshold

S_max = max(S_N);
s = f_prompt ('Enter threshold for locating peaks',0,S_max,.7*S_max);
ipeak = S_N > s;
for i = 1 : N/2
   if ipeak(i) == 1
      fprintf ('f = %.0f Hz\n',f(i))
   end
end
f_wait
```

The first part of the script prompts the user for an integer seed for the random number generator and for the number of unknown frequencies, M. Each seed generates a different set of M random frequencies. Once the signal x is constructed, it is saved in the MAT-file *u_spectra2* so that it can also be analyzed with GUI module *g_spectra* as a user-defined input.

A plot of the first $N/8$ samples of the signal $x(k)$ generated using the default responses is shown in Figure 3.35. Note that because of the additive white noise, it is not clear from direct inspection of $x(k)$ whether it has *any* sinusoidal components, much less how many there are and where they are located. However, the existence of sinusoidal spectral components is apparent when we examine the power density spectrum plot, $S_N(f)$, shown in Figure 3.36.

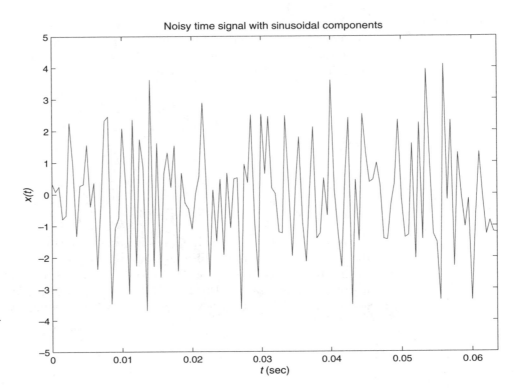

Figure 3.35 A Noise-corrupted Signal with Unknown Sinusoidal Components, $N = 1024$ and $f_s = 2000$ Hz

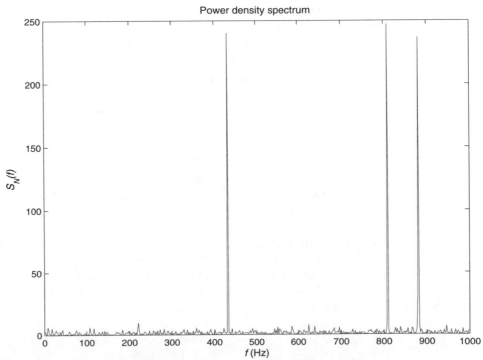

Figure 3.36 Power Density Spectrum of the Signal in Figure 3.35

In this case, there are three sinusoidal components corresponding to the three distinct peaks. To determine the locations of these peaks, the user is prompted for a threshold value, s, so that all f for which $S_N(f) > s$ can be displayed. Using the default threshold value, the three unknown frequencies are identified to be

$$F_1 = 432 \text{ Hz} \tag{3.120a}$$

$$F_2 = 809 \text{ Hz} \tag{3.120b}$$

$$F_3 = 881 \text{ Hz} \tag{3.120c}$$

3.11.3   Distortion Due to Clipping

Many a parent has had the experience of having a child turn the music up so loud that the sound begins to distort. The distortion is caused by the fact that the amplifier or the speakers are being overdriven to the point that they are no longer operating in their linear range. This type of distortion can be modeled with a saturation nonlinearity where the output is *clipped* at lower and upper limits.

$$y = \text{clip}(x, a, b) \triangleq \begin{cases} a, & -\infty < x < a \\ x, & a \le x \le b \\ b, & b < x < \infty \end{cases} \tag{3.121}$$

When the input x is in the interval $[a, b]$, there is a simple linear relationship, $y = x$. Outside this range, y saturates to a for $x < a$ or b for $x > b$. A graph of the clipping nonlinearity for the case $[a, b] = [-1, 1]$ is shown in Figure 3.37.

In a sound system, clipping typically occurs because the magnitude of the amplified output signal exceeds the DC power supply level of the amplifier. To illustrate this phenomenon, consider a cosine input of unit amplitude and frequency $f_a = 625$ Hz. Using a sampling frequency of $f_s = 20$ kHz and $N = 32$ yields one period of $x(k)$.

$$x(k) = \cos(1300\pi kT), \quad 0 \le k < N \tag{3.122}$$

Suppose this signal is sent through a saturation nonlinearity with a clipping interval $[-c, c]$:

$$y(k) = \text{clip}[x(k), -c, c], \quad 0 \le k < N \tag{3.123}$$

A plot of the resulting input and output for the case $c = 0.7$ is shown in Figure 3.38. We can quantify the amount of distortion caused by clipping by computing the total harmonic distortion, THD. Output $y(k)$ is a sampled version of an underlying periodic signal $y_a(t)$ that can be approximated by the following truncated Fourier series.

$$y_a(t) = \frac{d_0}{2} + \sum_{i=1}^{N/2-1} d_i \cos(2\pi f_a t + \theta_i) \tag{3.124}$$

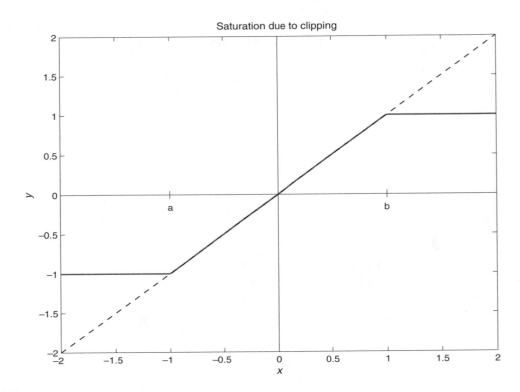

Figure 3.37 Saturation of x due to Clipping to the Interval $[-1, 1]$

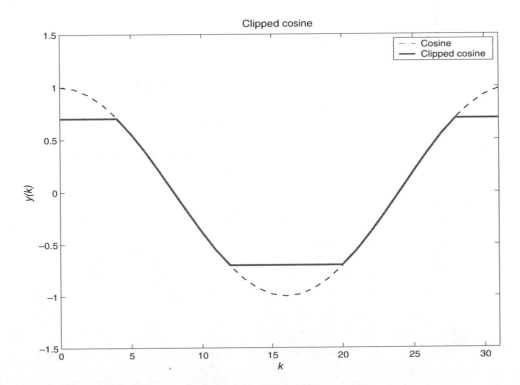

Figure 3.38 Clipping of $x(k)$ to Interval $[-c, c]$ with $c = 0.7$

Recall from Eq. 3.12 that the Fourier series coefficients can be obtained directly from the DFT of $y(k)$. In particular, if $A(i)$ is the magnitude spectrum of $y(k)$, then

$$d_i = \frac{2A(i)}{N}, \quad 0 \le i < N/2 \tag{3.125}$$

Each term of the Fourier series has an average power associated with it. From Eq. 3.10a, the total average power of $y_a(t)$ is

$$P_y = \frac{d_0^2}{4} + \frac{1}{2} \sum_{i=1}^{N/2} d_i^2 \tag{3.126}$$

The total harmonic distortion, THD, is the average power of the unwanted harmonics expressed as a percentage of the total average power. Thus, from Eq. 3.10b we have

$$\text{THD} = \frac{100(P_y - d_1^2/2)}{P_y} \% \tag{3.127}$$

The total harmonic distortion can be obtained by executing the following script, labeled *exam3_17* on the distribution CD.

```
function exam3_17

% Example 3.17: Distortion due to clipping

clear
clc
fprintf('Example 3.17: Distortion due to clipping\n')

% Construct input and output

N = 32;
k = 0 : N-1;
fs = 20000;
T = 1/fs;
fa = fs/N;
c = 0.70;
x = cos(2*pi*fa*k*T);
y = f_clip(x,-c,c);
```

```
% Plot clipped signal

figure
hp = plot (k,x,'--',k,y);
set (hp(1),'LineWidth',1.5)
axis([k(1),k(N),-1.5,1.5])
legend ('Cosine', 'Clipped cosine')
f_labels ('Clipped cosine','\it{k}','\it{y(k)}')
f_wait

% Compute total harmonic distortion

A = abs(fft(y));
d = 2*A/N;
Delta_f = fs/N;
i = round(fa/Delta_f) + 1;
P_y = d(1)^2/4 + (1/2)*sum(d(2:N/2).^2)
D = 100*(P_y - (d(i)^2)/2)/P_y

% Compute and plot magnitude spectrum

figure
i = 1 : N/2;
hp = stem (i-1,A(i),'filled','.');
set (hp,'LineWidth',1.5)
f_labels('Magnitude spectrum','\it{i}','\it{A(i)}')
set(gca,'Xlim',[0,N/2])
f_wait
```

When script *exam3_17* is run, it generates the clipping plot in Figure 3.38 and the magnitude spectrum plot shown in Figure 3.39. Note the presence of power at the third, fifth, and other odd harmonics due to the clipping operation. Using a clipping threshold of $c = 0.7$, the total harmonic distortion in this case is found to be

$$\text{THD} = 1.93\% \tag{3.128}$$

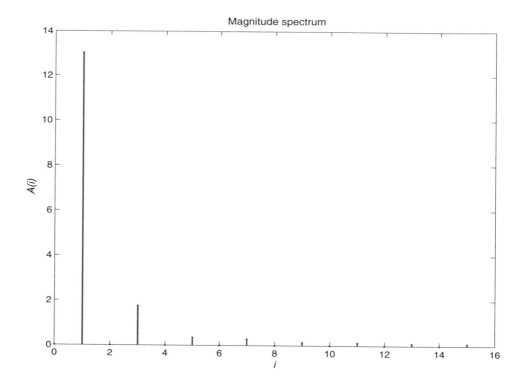

Figure 3.39 Magnitude Spectrum of Clipped Cosine Signal, THD $= 1.93\%$

● ● ● ● ● ● ● ● ● ● ● ● ● ●

3.12 Chapter Summary

The Discrete Fourier Transform

This chapter focused on the discrete Fourier transform (DFT) and the spectral analysis of discrete-time signals. The DFT is a special case of the discrete-time Fourier transform (DTFT) which is obtained by evaluating the Z-transform of a signal $x(k)$ along the unit circle of the complex plane.

$$X(f) = X(z)|_{z=\exp(j2\pi fT)}, \quad -f_s/2 \leq f \leq f_s/2 \tag{3.129}$$

The resulting function of f is the *spectrum* of $x(k)$. The magnitude $A(f) = |X(f)|$ is called the *magnitude spectrum*, and the phase angle $\phi(f) = \angle X(f)$ is called the *phase spectrum*. When the DTFT is applied to an N-point signal, and $X(f)$ is evaluated at N discrete frequencies equally spaced around the unit circle, the resulting transform is called the *discrete-Fourier transform* or DFT. Let W_N denote the Nth root of unity:

$$W_N = \exp\left(\frac{-j2\pi}{N}\right) \tag{3.130}$$

Then the DFT, which is denoted $X(i) = DFT\{x(k)\}$, is defined in terms of W_N as

$$X(i) = \sum_{k=0}^{N-1} x(k) W_N^{ik}, \quad 0 \leq i < N \tag{3.131}$$

Since W_N is complex, $X(i)$ will be complex. We can express the DFT in polar form as $X(i) = A(i) \exp[j\phi(i)]$ where

$$A(i) = |X(i)| \tag{3.132a}$$

$$\phi(i) = \angle X(i) \tag{3.132b}$$

The magnitude $A(i)$ is called the *magnitude spectrum* of $x(k)$, and the phase angle $\phi(i)$ is called the *phase spectrum* of $x(k)$. The *power density spectrum* of $x(k)$ is defined as

$$S_N(i) = \frac{|X(i)|^2}{N} \tag{3.133}$$

For a real signal, the magnitude and power density spectra exhibit even symmetry about the midpoint $i = N/2$, and the phase spectrum exhibits odd symmetry about the midpoint. Consequently, for real signals, all of the essential information is contained in the range $0 \leq i \leq N/2$, which corresponds to the positive discrete frequencies.

$$f_i = \frac{if_s}{N}, \quad 0 \leq i \leq \frac{N}{2} \tag{3.134}$$

Here $f_s = 1/T$ is the sampling frequency. The highest discrete frequency, $f_{N/2}$, is the folding frequency $f_s/2$.

Just as light can be decomposed into different colors, signals have power that is distributed over different frequencies. If $x_p(k)$ denotes the periodic extension of $x(k)$, then the power of $x_p(k)$ at discrete frequency f_i is given by $P_i = S_N(i)$. The total average power of $x(k)$ is

$$P_x = \frac{1}{N} \sum_{i=0}^{N-1} S_N(i) \tag{3.135}$$

The Fast Fourier Transform

When the number of samples N is an integer power of two, a highly efficient implementation of the DFT called the radix-two *fast Fourier transform* or FFT is available. For large values of N, the number of complex floating-point operations or FLOPs required to perform a DFT is approximately N^2, while the number of FLOPs required for an FFT is only $(N/2) \log_2(N)$. Thus, the FFT is r times faster than the DFT where

$$r = \frac{N}{2 \log_2(N)} \tag{3.136}$$

For $N = 1024$, the FFT is about 200 times faster, and for $N = 8192$, it is more than 1200 times faster than the DFT.

The frequency response of a stable linear discrete-time system with transfer function $H(z)$ can be approximated at discrete frequencies $f_i = i f_s / N$ using the DFT. If $H(i) = \mathrm{DFT}\{h(k)\}$ where $h(k)$ is the system impulse response, then

$$H(i) \approx H(f_i), \quad 0 \le i < N \tag{3.137}$$

For an IIR system, the approximation in Eq. 3.137 becomes increasingly accurate as the number of samples N increases. For an FIR system of order m, the approximation in Eq. 3.137 is exact when $N > m$. The frequency response approximation can be expressed in polar form as $H(i) = A(i) \exp[j\phi(i)]$, where $A(i) = |H(i)|$ is the *magnitude response* and $\phi(i) = \angle H(i)$ is the *phase response* of the system. Thus, the magnitude response is just the magnitude spectrum of the impulse response, and the phase response is the phase spectrum of the impulse response. The magnitude response can be expressed using either a linear scale or a logarithmic dB scale.

The spacing between discrete frequencies, $\Delta f = f_s / N$, is called the *frequency increment*. It can be decreased to $\Delta f = f_s / M$ by padding $x(k)$ with $M - N$ zeros. In this way, isolated sinusoidal spectral components of $x(k)$ can be more accurately detected and located, even when $x(k)$ is corrupted with white noise. White noise is a random signal whose power density spectrum is nonzero and flat, indicating that the power is uniformly distributed over all frequencies. The smallest frequency difference that can be detected is called the *frequency resolution*. The frequency resolution ΔF is the reciprocal of the signal duration and is referred to as the *Rayleigh limit*

$$\Delta F = \frac{f_s}{N} \tag{3.138}$$

The Power Density Spectrum Estimation

A number of techniques have been proposed for obtaining improved estimates of the underlying continuous-time power density spectrum of a signal. The basic definition in Eq. 3.133 is called the *periodogram* method. Improved estimates can be obtained using Bartlett's *average periodogram* method, and Welch's *modified average periodogram* method. Average periodograms are computed by partitioning a long signal $x(k)$ into subsignals of length L. For Bartlett's method, the subsignals do not overlap, while for Welch's method they overlap by $L/2$ samples. Welch's method also multiplies the subsignals by data windows which taper gradually to zero at each end. The use of data windows tends to reduce the effects of the *leakage phenomenon*: a computational artifact that causes overly broad spikes in the estimated power density spectrum.

Some practical signals, such as speech and music, are sufficiently long that their spectral characteristics can be thought of as changing with time. A long signal $x(k)$ can be partitioned into $2M - 1$ overlapping subsignals of length L, and these subsignals can be windowed and then transformed with L-point DFTs. The resulting magnitude spectra can then be arranged as the rows of a $(2M - 1) \times L$ matrix called the *spectrogram* of $x(k)$.

$$G(m, i) = |\mathrm{DFT}\{w(k)x(mL/2 + k)\}| \tag{3.139}$$

The spectrogram shows how the spectrum changes with time. The first independent variable m specifies the starting time in increments of $L/2$ samples, while the second independent variable i specifies the discrete frequency in increments of f_s/L.

The FDSP toolbox includes a GUI module called *g_spectra* that allows the user to perform a spectral analysis of discrete-time signals without any need for programming. Several common signals are included, plus signals recorded from a PC microphone and user-defined signals saved in MAT-files. The signals can be noise-corrupted or noise-free and clipped or unclipped. Viewing options include the magnitude spectrum, the phase spectrum, the estimated power density spectrum, and the spectrogram.

3.13　Problems

The problems are divided into Analysis problems that can be solved by hand or with a calculator, GUI Simulation problems that are solved using GUI module *g_spectra*, and Computation problems. The Computation problems require the student to write a MATLAB script using the FDSP toolbox functions summarized in Appendix C. Solutions to selected problems are available on the distribution CD. Students are encouraged to use these problems, which are identified with a ☑, as a check on their understanding of the material.

3.13.1　Analysis

Section 3.2

P3.1. Find the discrete-time Fourier transform (DTFT) of the following signals.

(a) $x(k) = \delta(k)$

(b) $x(k) = a^k u(k), \quad |a| < 1$

(c) $x(k) = a^k \cos(2\pi b k T), \quad |a| < 1$

P3.2. Let $x_a(t)$ be periodic with period τ, and let $x(k)$ be a sampled version of $x_a(t)$ using sampling interval T.

(a) For what values of T is $x(k)$ periodic? Give an example.

(b) For what values of T is $x(k)$ not periodic? Give an example.

Section 3.3

P3.3. Find the value of the following complex expression. *Hint*: Use Euler's identity. The answer is real.

$$\alpha = j^j$$

P3.4. Consider the following discrete-time signal.

$$x = [2, -1, 3]^T$$

(a) Find the third root of unity, W_3.

(b) Find the 3×3 DFT transformation matrix W.

(c) Use W to find the DFT of x.

(d) Find the inverse DFT transformation matrix W^{-1}.

(e) Find the discrete-time signal x whose DFT is given by

$$X = [3, -j, j]^T$$

P3.5. Verify the following basic properties of W_N in Table 3.2.

(a) $W_N^{k+N} = W_N^k$

(b) $W_N^{k+N/2} = -W_N^k$

(c) $W_N^{2k} = W_{N/2}^k$

(d) $W_N^* = W_N^{-1}$

P3.6. Compete the following DFT pairs for N-point signals.

(a) $x(k) = \delta(k) \implies X(i) = ?$

(b) $X(i) = \delta(i) \implies x(k) = ?$

✓ **P3.7.** Consider the following discrete-time signal.

$$x = [1, 2, 1, 0]^T$$

(a) Find $X(i) = \text{DFT}\{x(k)\}$.

(b) Compute and sketch the magnitude spectrum $A(i)$.

(c) Compute and sketch the phase spectrum $\phi(i)$.

(d) Compute and sketch the power density spectrum $S_N(i)$.

Section 3.4

P3.8. Let $x_p(k)$ be the periodic extension of an N-point signal $x(k)$.

(a) Find the average power P_x of $x_p(k)$. This is the average power of $x(k)$.

(b) Let $S_N(i)$ be the power density spectrum of $x(k)$. Use Parseval's theorem (Eq. 3.62) to show that the average power of $x(k)$ is the average of the power density spectrum.

$$P_x = \frac{1}{N} \sum_{i=0}^{N-1} S_N(i)$$

P3.9. Consider the following discrete-time signal.

$$x = [-1, 2, 2, 1]^T$$

(a) Find the average power P_x.

(b) Find the DFT of x.

(c) Verify Parseval's theorem in this case.

P3.10. Consider the following discrete-time signal, where $|a| < 1$.

$$x(k) = a^k, \quad 0 \le k < N$$

(a) Find $X(i)$.

(b) Use the geometric series to simplify $X(i)$ as much as possible. Express your final answer as a ratio of two polynomials in a.

P3.11. Recall from Eq. 3.58 that a circular shift of $x(k)$ by M samples is equivalent to multiplying $X(i)$ by W_N^{iM}. This is sometimes called the circular time shift property of the DFT. Show that the following *circular frequency shift* property also holds.

$$\text{DFT}\{W_N^{-kM} x(k)\} = X(i - M)$$

Section 3.5

P3.12. Consider an N-point signal $x(k)$. Find the smallest integer N such that a radix-two FFT of $x(k)$ is at least 100 times as fast as the DFT of $x(k)$ when speed is measured in FLOPs.

Section 3.6

P3.13. Let x be a random variable that is uniformly distributed over the interval $[a, b]$.

(a) Find the mth statistical moment, $E[x^m]$, for $m \ge 0$.

(b) Verify that $E[x^m] = P_u$ in Eq. 3.85 when $m = 2$.

P3.14. Let x be a random variable whose probability density function is given in Figure P3.14.

(a) What is the probability that $-.5 \le x \le .5$?

(b) Find $E[x^2]$.

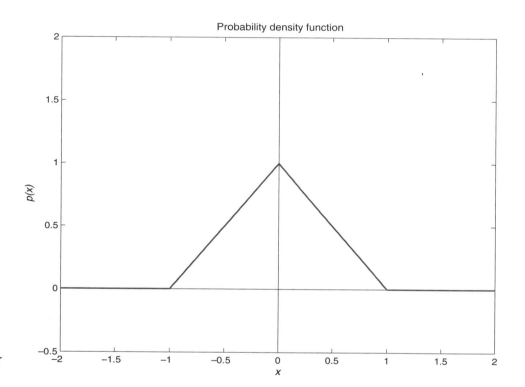

P3.14 Probability Density Function for Problem P3.14

Section 3.7

P3.15. Consider the following digital filter where $|a| < 1$.

$$H(z) = \frac{1}{1 - az^{-1}}$$

(a) Find the impulse response $h(k)$.

(b) Find the frequency response $H(f)$.

(c) Let $H(i)$ be the N-point DFT of $h(k)$, and let $f_i = if_s/N$. Given an arbitrary $\epsilon > 0$, use Eq. 3.94 to find a lower-bound n such that for $N \ge n$,

$$|H(i) - H(f_i)| \le \epsilon \quad \text{for} \quad 0 \le i \le \frac{N}{2}$$

Section 3.8

P3.16. Suppose a signal $x(t)$ is sampled at $N = 300$ points using a sampling rate of $f_s = 1600$ Hz. Let $x_z(k)$ be a zero-padded version of $x(k)$ using $M - N$ zeros. Suppose a radix-two FFT is used to find $X_z(i)$.

(a) Find a lower bound on M that ensures that the frequency increment of $X_z(i)$ is no larger than 2 Hz.

(b) How much faster or slower is the FFT of $x_z(k)$ in comparison with the DFT of $x(k)$? Express your answer as a ratio of the computational effort of the FFT to the computational effort of the DFT.

Section 3.9

P3.17. One of the problems with using data windows to reduce the Gibb's phenomenon in the periodic extension of an N-point signal $x(k)$ is that the samples are no longer weighted equally when computing an estimate of the power density spectrum. This is particularly the case when no overlap of subsignals is used.

(a) Use the trigonometric identities in Appendix D to show that the Hanning window in Table 3.6 can be expressed as

$$w(k) = 0.5 + 0.5 \cos\left[\frac{2\pi(k - L/2)}{L}\right]$$

(b) If a 50 percent overlap of subsignals is used for the power density spectrum estimate, then each overlapped sample gets counted twice: once with weight $w(k)$ and once with weight $w(k + L/2)$. Show that if the Hanning window is used, the overlapped samples are weighted equally. Find the total weight for each overlapped sample.

(c) Are there any other windows in Table 3.6 for which the total weighting of the overlapped samples is uniform when a 50 percent overlap is used? If so, which ones?

3.13.2   GUI Simulation

Section 3.11

✅ **P3.18.** Use the GUI module *g_spectra* to plot the spectrogram of the following signals. Use $f_s = 3000$ Hz and $N = 2048$ samples for each.

(a) Cosine of unit amplitude and frequency $f_a = 400$ Hz.

(b) Cosine of unit amplitude and frequency $f_s = 400$ Hz, clipped to $[-0.5, 0.5]$.

(c) Cosine of unit amplitude and frequency $f_a = 400$ Hz, plus white noise uniformly distributed over $[-1.5, 1.5]$.

P3.19. Use the GUI module *g_spectra* to plot the power density spectrum of a noise-free cosine input using the default parameter values. Use the dB scale and do the following cases.

(a) Rectangular window.

(b) Hamming window.

(c) Hanning window.

(d) Blackman window.

P3.20. Use the GUI module *g_spectra* to plot the following characteristics of a noise-corrupted damped exponential input using the default parameter values. Use the linear scale.

(a) Time signal.

(b) Magnitude spectrum.

(c) Power density spectrum (Blackman window).

(d) Blackman window.

☑ **P3.21.** Using the GUI module *g_spectra*, record the word HELLO. Play it back to make sure it is recorded properly. Save it in a MAT-file called *hello.mat*. Then reload it as a User-defined input. Plot the following spectral characteristics.

(a) Magnitude spectrum.

(b) Power density spectrum (Hamming window).

(c) Spectrogram.

P3.22. Consider the signal shown in Figure P3.22 which contains one or more sinusoidal components corrupted with white noise. The complete signal $x(k)$ and the sampling frequency f_s are stored in the file *prob3_22.mat* on the distribution CD. Use the GUI module *g_spectra* to plot the following spectral characteristics.

(a) The power density spectrum (Hamming window). Use the *Caliper* option to estimate the frequencies of the sinusoidal components.

(b) The spectrogram (Hamming window).

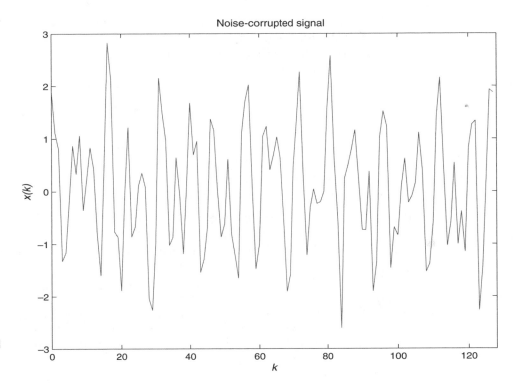

P3.22 Noise-Corrupted Signal with Unknown Sinusoidal Components (Samples 0 to $N/8$)

P3.23. Consider the following noise-corrupted periodic signal with a sampling frequency of $f_s = 1600$ Hz and $N = 1024$. Here $v(k)$ is white noise uniformly distributed over $[-1, 1]$.

$$x(k) = \sin(600\pi kT)\cos^2(200\pi kT) + v(k), \quad 0 \le k < N$$

Create a MAT-file called *prob3_23* containing x and f_s. Then use *g_spectra* to plot the following.

(a) Magnitude spectrum.

(b) Power density spectrum using Welch's method (rectangular window).

(c) Power density spectrum using Welch's method (Blackman window).

P3.24. Use the GUI module *g_spectra* to perform the following analysis of the vowels. Play back the sound in each case to make sure you have a good recording.

(a) Record one second of the vowel "A" and plot the time signal.

(b) Record one second of the vowel "E" and plot the time signal.

(c) Record one second of the vowel "I" and plot the time signal.

(d) Record one second of the vowel "O" and plot the time signal.

(e) Record one second of the vowel "U" and plot the time signal.

P3.25. A signal stored in *prob3_25.mat* on the distribution CD contains white noise plus a single sinusoidal component whose frequency does not correspond to any of the discrete frequencies. Use GUI module *g_spectra* to plot the following spectral characteristics.

(a) The magnitude spectrum of $x(k)$ using the linear scale.

(b) The power density spectrum of $x(k)$ using the Blackman window. Use the *Caliper* option to estimate the frequency of the sinusoidal component.

3.13.3 Computation

Section 3.1

P3.26. Let $x_a(t)$ be a periodic pulse train of period T_0. Suppose the pulse amplitude is $a = 10$, and the pulse duration is $\tau = T_0/5$, as shown in Figure P3.26 for the case $T_0 = 1$. This signal can be represented by the following cosine form Fourier series.

$$x_a(t) = \frac{d_0}{2} + \sum_{i=1}^{\infty} d_i \cos\left(\frac{2\pi it}{T_0} + \theta_i\right)$$

Write a MATLAB script that uses the Eq. 3.12 and the DFT to compute coefficients d_0 and (d_i, θ_i) for $1 \le i < 32$. Plot d_i and θ_i using the MATLAB function *stem*.

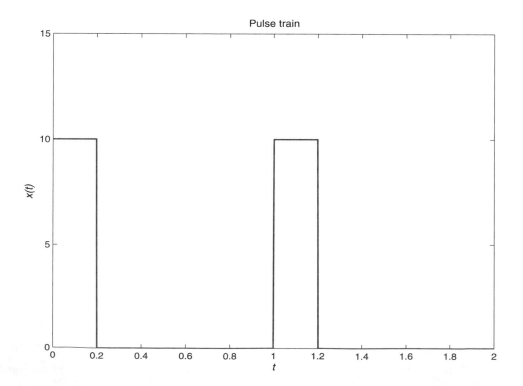

P3.26 Periodic Pulse Train with $a = 10$ and $T_0 = 1$

☑ **P3.27.** In addition to saturation due to clipping, another common type of nonlinearity is the *dead-zone* nonlinearity shown in Figure P3.27. The algebraic representation of a dead zone of radius a is as follows.

$$F(x, a) \triangleq \begin{cases} 0, & 0 \le |x| \le a \\ x, & a < |x| < \infty \end{cases}$$

Suppose $f_s = 2000$ Hz, and $N = 100$. Consider the following input signal where $0 \le k < N$ corresponds to one cycle.

$$x(k) = \cos(40\pi kT), \quad 0 \le k < N$$

Let the dead-zone radius be $a = 0.25$. Write a MATLAB script that does the following.

(a) Compute and plot $y(k) = F[x(k), a]$ versus k.

(b) Compute and plot the magnitude spectrum of $y(k)$.

(c) Compute and print the total harmonic distortion of $y(k)$ caused by the dead zone. Here, if d_i and θ_i for $0 \le i < M$ are the cosine form Fourier coefficients of $y(k)$ with $M = N/2$, then

$$\text{THD} = \frac{100(P_y - d_1^2/2)}{P_y}\%$$

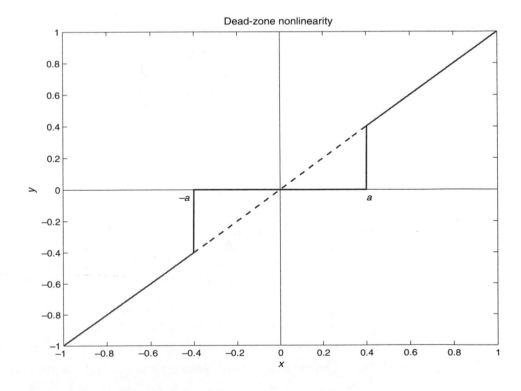

P3.27 Dead-Zone Nonlinearity of Radius a

P3.28. Repeat Problem P3.27, but using $f_s = 1000$ Hz, $N = 50$ samples, and the cubic nonlinearity

$$F(x) = x^3$$

Section 3.5

P3.29. Consider the following digital filter of order $m = 2p$ where $p = 20$.

$$H(z) = \sum_{i=0}^{2p} b_i z^{-i}$$

$$b_p = 0.5$$

$$b_i = \frac{[0.54 - 0.46\cos(\pi i/p)]\{\sin[0.75\pi(i-p)] - \sin[0.25\pi(i-p)]\}}{\pi(i-p)}, i \neq p$$

Suppose $f_s = 200$ Hz. Write a script that uses *filter* to do the following.

(a) Compute and plot the impulse response $h(k)$ for $0 \le k < N$ where $N = 64$.

(b) Compute and plot the magnitude response $A(f)$ for $0 \le f \le f_s/2$.

(c) What type of filter is this, FIR or IIR? What range of frequencies gets passed by this filter?

P3.30. Consider the following digital filter of order n where $n = 11$ and $r = 0.98$.

$$H(z) = \frac{(1 + r^n)(1 - z^{-n})}{2(1 - r^n z^{-n})}$$

Suppose $f_s = 2200$ Hz. Write a script that uses *filter* to do the following.

(a) Compute and plot the impulse response $h(k)$ for $0 \le k < N$ where $N = 1001$.

(b) Compute and plot the magnitude response $A(f)$ for $0 \le f \le f_s/2$.

(c) What type of filter is this, FIR or IIR? Which frequencies get rejected by this filter?

Section 3.9

P3.31. Consider the following noise-corrupted periodic signal with a sampling frequency of $f_s = 1600$ Hz and $N = 1024$.

$$x(k) = \sin^2(400\pi kT)\cos^2(300\pi kT) + v(k), \quad 0 \le k < N$$

Here $v(k)$ is zero-mean Gaussian white noise with a standard deviation of $\sigma = 1/\sqrt{2}$. Write a script which does the following.

(a) Compute and plot the power density spectrum $S_N(f)$ for $0 \le f \le f_s/2$.

(b) Compute and print the average power of $x(k)$ and the average power of $v(k)$.

P3.32. Write a script which creates a 1×2048 vector x of white noise uniformly distributed over $[-0.5, 0.5]$. The script should then compute and display the following.

(a) The average power P_x, the predicted average power P_u, and the percent error in P_x.

(b) Plot the estimated power density spectrum using Bartlett's method with $L = 512$. Use a y-axis range of $[0, 1]$. In the plot title, print L and the estimated variance σ_B^2 of the power density spectrum.

(c) Repeat part (b), but use $L = 32$.

Convolution and Correlation

Learning Objectives

- Know how to compute the zero-state response of a linear discrete-time system using linear convolution (Section 4.2)

- Understand the relationship between linear convolution, circular convolution, and zero padding (Section 4.2)

- Know how to recover the impulse response from the input and the output using deconvolution (Section 4.2)

- Know how to implement fast linear convolution and fast block convolution using the FFT (Section 4.3)

- Be able to characterize and interpret cross-correlation and auto-correlation (Section 4.4)

- Know how to implement fast cross-correlation and auto-correlation using the FFT (Section 4.5)

- Understand the relationship between auto-correlation, average power, and the power density spectrum (Section 4.6)

- Know how to characterize the auto-correlation of white noise (Section 4.6)

- Be able to extract a periodic signal from noise using correlation techniques (Section 4.7)

- Know how to use the GUI module *g_correlate* to perform convolutions and correlations without any programming (Section 4.8)

• • • • • • • • • • • • • • • •

4.1 Motivation

There are a number of important DSP applications that require the processing of *pairs* of signals. Recall from Chapter 2 that if $h(k)$ is the impulse response of a linear discrete-time system and $x(k)$ is an input signal, then the zero-state response of the system is the *convolution* of the impulse response with the input.

Convolution

$$y(k) = \sum_{i=0}^{k} h(i)x(k-i), \quad k \geq 0$$

Thus, the convolution operation is a filtering operation that allows us to compute the zero-state output corresponding to an arbitrary input. There is a second operation involving a pair of signals that closely resembles convolution. To measure the degree to which the pattern of an L-point signal $h(k)$ is similar to the pattern of an M-point signal $x(k)$ where $M \leq L$, we compute the *cross-correlation* of h with x.

Cross-correlation

$$r_{hx}(k) = \frac{1}{L} \sum_{i=0}^{L-1} h(i)x(i-k), \quad 0 \leq k < L$$

Comparing the cross-correlation with the convolution, we see that, apart from scaling, the essential difference is in the sign of the independent variable of the second signal. The cross-correlation operation has a number of important applications. For example, in radar or sonar processing, the cross-correlation can be used to determine if a received signal contains an echo of the transmitted signal. Another use of correlation is to extract a periodic signal from noise. When the signals are finite, convolution and correlation have highly efficient implementations, called fast convolution and fast correlation, that are based on the use of the FFT.

We begin this chapter by introducing a number of examples of applications where convolution and correlation occur. A finite N-point version of convolution, called circular convolution, is then introduced. Next, zero padding is used to convert circular convolution into linear convolution. Using the FFT, a highly efficient implementation of convolution, called fast convolution, is presented. This technique is then extended to block convolution, a method that is useful when one of the signals is very long. The operations of cross-correlation and normalized cross-correlation are then defined, and an efficient implementation of cross-correlation based on the FFT called fast correlation is presented. An important special case of cross-correlation, called auto-correlation, is the cross-correlation of a signal with itself. Simple interpretations of auto-correlation in terms of average power and the power density spectrum are developed. Next, correlation techniques are put to use to extract periodic signals from noise. Finally, a GUI module called *g_correlate* is introduced that allows the user to perform convolutions and correlations without any need for programming. The chapter concludes with a presentation of application examples, and a summary of the convolution and correlation operations. The selected examples include echo detection in radar and the analysis of speech.

4.1.1 Modeling the Vocal Tract

One of the exciting application areas of DSP is the analysis and synthesis of human speech. Both speech recognition and speech synthesis are becoming increasingly commonplace as a user interface between humans and machines. An effective technique for generating synthetic speech is to use a discrete-time signal to represent the output from the vocal chords and a slowly varying digital filter to model the vocal tract as illustrated by the block diagram in Figure 4.1 (Rabiner and Schafer, 1978; Markel and Gray, 1976).

Speech sounds can be decomposed into fundamental units called *phonemes* that are either voiced or unvoiced. Unvoiced phonemes are associated with turbulence in the vocal tract and are therefore modeled using filtered white noise. Unvoiced phonemes include the fricatives, such as the s, sh, and f sounds, and the terminal sounds p, t, and k. Voiced phonemes are associated with periodic excitation of the vocal chords. They include the vowels, nasal sounds, and transient terminal sounds such as b, d, and g. Voiced phonemes can be modeled as the response of a digital filter to a periodic impulse train with period p.

$$x(k) = \sum_{i=0}^{\infty} \delta(k - ip) \tag{4.1}$$

If T is the sampling interval, then the period of the impulse train in seconds is $T_0 = pT$. The fundamental frequency or *pitch* of the speaker is then

$$f_0 = \frac{1}{pT} \tag{4.2}$$

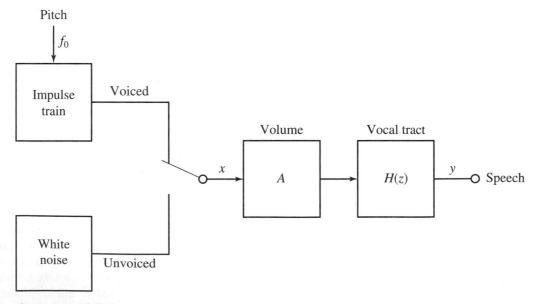

Figure 4.1 Digital Speech Synthesis

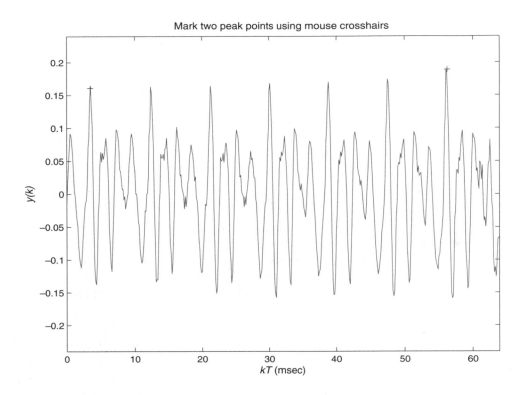

Figure 4.2 Segment of Recorded Vowel "O"

Speaker pitch typically ranges from about 50 Hz to 400 Hz, with male speakers having a lower pitch, on the average, than female speakers. An illustration of a short segment of the vowel "O" is shown in Figure 4.2.

To estimate the pitch of this speaker, we examine the distance between peaks in Figure 4.2. Using the FDSP toolbox function *f_caliper*, the period of the signal in Figure 4.2 is about $T_0 = 8.4$ msec, which corresponds to a pitch of

$$f_0 \approx 113.6 \text{ Hz} \tag{4.3}$$

The linear system $H(z)$ in Figure 4.1 models the vocal-tract cavity including the throat, mouth, and lips. An effective model for most sounds is an nth order auto-regressive (AR) all pole model.

$$H(z) = \frac{1}{1 + a_1 z^{-1} + \cdots + a_n z^{-n}} \tag{4.4}$$

The speech output, $y(k)$, can be represented as the convolution of $h(k)$ with $x(k)$ where the filter impulse response $h(k)$ is the inverse Z-transform of the transfer function.

$$h(k) = Z^{-1}\{H(z)\} \tag{4.5}$$

Given $x(k)$ and a $y(k)$ associated with a specific sound, the problem is then to find $h(k)$ so that the sound can be reproduced artificially using convolution as

$$y(k) = \sum_{i=0}^{k} h(i)x(k-i) \tag{4.6}$$

Note that the limits of the convolution summation in Eq. 4.6 are finite because both the input $x(k)$ and the impulse response $h(k)$ are causal signals. Finding the vocal tract impulse response $h(k)$ from input $x(k)$ and output $y(k)$ is an example of system identification, a topic that is treated in detail in Chapter 9.

4.1.2 Range Measurement with Radar

An application where correlation arises is in the measurement of range to a target using radar. Consider the radar installation shown in Figure 4.3. Here the radar antenna transmits an electromagnetic wave $y(k)$ into space. When a target is illuminated by the radar, some of the signal energy is reflected back and returns to the radar receiver. The received signal $x(k)$ can be modeled as follows.

$$x(k) = ay(k-d) + v(k) \tag{4.7}$$

The first term in Eq. 4.7 represents the *echo* of the transmitted signal. Typically the echo is very faint ($0 < a \ll 1$) because only a small fraction of the original signal is reflected back with most of the signal energy dissipated through dispersion into space. The delay of d samples accounts for the propagation time required for the signal to travel from the transmitter to the target and back again. Thus, d is proportional to the time of flight. The second term in Eq. 4.6 accounts for random atmospheric measurement noise, $v(k)$, that is picked up and amplified by the receiver.

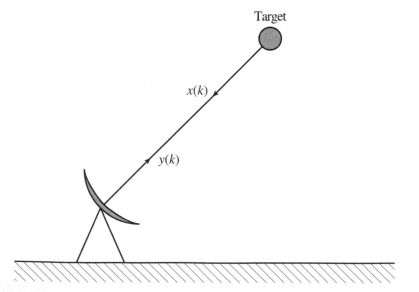

Figure 4.3 Radar Il-luminating a Target

The objective in processing the received signal $x(k)$ is to determine whether or not an echo is present. If an echo is detected, then the distance to the target can be obtained from the delay d. If T is the sampling interval, then the time of flight in seconds is $\tau = dT$. Suppose c denotes the signal propagation speed. For radar, this is the speed of light. Then the distance to the target is computed as follows where the factor two arises because the signal must make a round trip.

$$r = \frac{cdT}{2} \tag{4.8}$$

To detect if an echo is present in the noise-corrupted received signal, $x(k)$, we compute the normalized cross-correlation of $x(k)$ with $y(k)$. If the measurement noise $v(k)$ is not present, the normalized cross-correlation will have a spike of unit amplitude at $i = d$. When $v(k) \neq 0$, the height of the spike will be reduced.

As an illustration, suppose the transmitted signal $y(k)$ consists of $M = 512$ points of white noise uniformly distributed over $[-1, 1]$. Other broadband signals, for example a multifrequency chirp, might also be used. Next, let the received signal $x(k)$ consist of $N = 2048$ points. Suppose the attenuation factor is $a = 0.05$. Let $v(k)$ consist of zero-mean Gaussian white noise with a standard deviation of $\sigma = 0.05$. A plot of the received signal is shown in Figure 4.4.

Note that from a direct inspection of Figure 4.4 it is not at all apparent whether $x(k)$ contains a delayed and attenuated echo of $y(k)$, much less where it is located.

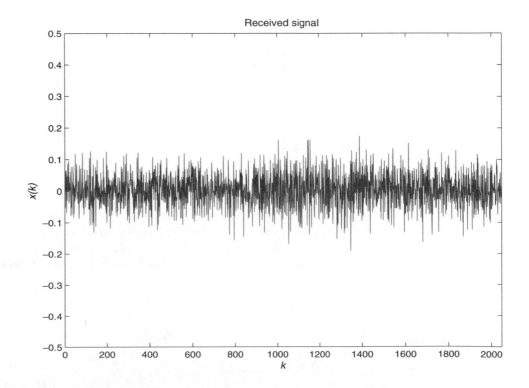

Received signal

Figure 4.4 Signal Received at Radar Station

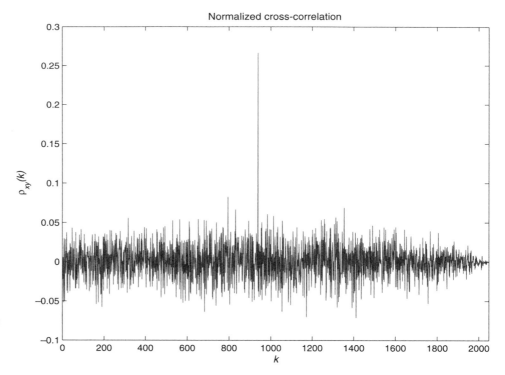

Figure 4.5 Normalized Cross-correlation of Transmitted Signal with Received Signal

However, if we compute the normalized cross-correlation of $x(k)$ with $y(k)$, then the presence and location of an echo are evident as can be seen in Figure 4.5. Using the MATLAB *max* function, we find that the time of flight in this case is $d = 939$ samples. The range to the target then can be found using Eq. 4.8.

4.2 Convolution

4.2.1 Linear Convolution

Consider a linear discrete-time system with input $x(k)$ and output $y(k)$, as shown in Figure 4.6. Here $H(z)$ is the system transfer function and $h(k) = Z^{-1}\{H(z)\}$ is the system impulse response. Recall from Chapter 2 that the zero-state response of the system in Figure 4.6 can be expressed in terms of the convolution operation.

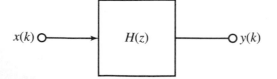

Figure 4.6 A Linear Discrete-time System

DEFINITION

4.1: Linear
Convolution

Let $h(k)$ and $x(k)$ be two causal discrete-time signals. Then the *linear convolution* of $h(k)$ with $x(k)$ is denoted $h(k) \star x(k)$ and defined as

$$h(k) \star x(k) \overset{\Delta}{=} \sum_{i=0}^{k} h(i)x(k-i), \quad k \geq 0$$

The zero-state output $y(k)$ is the linear convolution of the impulse response $h(k)$ with the input $x(k)$. Consequently, if the initial conditions are zero, the output produced by an arbitrary input can be expressed as

$$y(k) = h(k) \star x(k) \tag{4.9}$$

In general, the limits of the convolution summation in Definition 4.1 go from $-\infty$ to ∞. However, because $h(k)$ is causal, the lower limit can be set to zero, and because $x(k)$ is causal, the upper limit can be set to k. As we shall see, there are several forms of convolution, so to distinguish them from one another we refer to the formulation in Definition 4.1 as *linear convolution*.

Linear convolution has a number of useful properties. For example, the convolution operator is a *commutative* operator which means that the order of the two operands can be interchanged without affecting the result.

$$h(k) \star x(k) = x(k) \star h(k) \tag{4.10}$$

Another fundamental property has to do with the frequency-domain interpretation of convolution. Recall from Table 2.2 that linear convolution in the time domain maps into multiplication in the frequency, or Z-transform, domain. Consequently, if we take the Z-transform of Eq. 4.9, the result is the familiar transfer function representation.

$$Y(z) = H(z)X(z) \tag{4.11}$$

Example 4.1   **Linear Convolution**

As a simple illustration of using convolution to compute the zero-state output, consider the following first-order digital IIR filter.

$$H(z) = \frac{bz}{z - a}$$

Taking the inverse Z-transform of $H(z)$, we arrive at the impulse response:

$$h(k) = Z^{-1}\{H(z)\}$$

$$= b(a)^k u(k)$$

Using Eq. 4.9, the zero-state response to an arbitrary input $x(k)$ can be expressed in terms of convolution as

$$y(k) = h(k) \star x(k)$$

$$= \sum_{i=0}^{k} h(i)x(k-i)$$

$$= b \sum_{i=0}^{k} a^i x(k-i)$$

The expression for the zero-state response of a filter using linear convolution becomes particularly simple when the filter is an FIR filter. Recall that an FIR filter is a filter whose output is a finite weighted sum of the past inputs.

$$y(k) = \sum_{i=0}^{m} b_i x(k-i) \tag{4.12}$$

If we replace $x(k)$ in Eq. 4.12 with the unit impulse $\delta(k)$, we find that the impulse response of the FIR filter is simply

$$h(k) = \{b_0, b_1, \ldots, b_m, 0, 0, \ldots\} \tag{4.13}$$

Thus, the impulse response of an FIR system can be obtained directly from an inspection of the difference equation in Eq. 4.12. Using Eq. 4.13, the b_i in Eq. 4.12 can be replaced by $h(i)$. Furthermore, the upper limit of the sum in Eq. 4.12 can be changed from m to k because $x(k-i) = 0$ for $i > k$. Making these two changes, and comparing the result with Definition 4.1, we see that the summation in Eq. 4.12 is, in fact, the linear convolution of $h(k)$ with $x(k)$. Therefore, for an FIR filter of order m, the zero-state output can be expressed using the following formulation of convolution.

FIR convolution

$$y(k) = \sum_{i=0}^{m} h(i)x(k-i) \tag{4.14}$$

It should be emphasized that replacing the variable upper limit k in Definition 4.1 with a fixed upper limit m in Eq. 4.14 is valid *only* for an FIR filter. For an IIR filter, the impulse response is not finite, so either the variable limit k or a fixed limit of ∞ must be used.

4.2.2   Circular Convolution

For a numerical implementation of the convolution operation to be practical, the signals $h(k)$ and $x(k)$ must be finite. Suppose $h(k)$ is defined for all k, but nonzero only for $0 \le k < L$. Similarly, let $x(k)$ be a signal that is nonzero for $0 \le k < M$. Then the linear convolution in Definition 4.1 can be expressed as

$$y(k) = \sum_{i=0}^{L-1} h(i)x(k-i), \quad 0 \le k < L + M \tag{4.15}$$

The upper limit on the summation has been changed from k in Definition 4.1 to $L-1$ because $h(i) = 0$ for $k \ge L$. Observe that when $i = L - 1$, we have $x(k-i) \ne 0$ for $L - 1 \le k < L + M$. Consequently, the linear convolution of an L-point signal with an M-point signal is a signal of length $L + M$.

There is an alternative way to define convolution where the length of the result is the same as the lengths of the two operands.

DEFINITION

4.2: Circular
Convolution

Let $h(k)$ and $x(k)$ be N-point signals, and let $x_p(k)$ be the periodic extension of $x(k)$ as previously defined in Eq. 3.54. Then the *circular convolution* of $h(k)$ with $x(k)$ is denoted $y_c(k) = h(k) \circ x(k)$ and defined

$$h(k) \circ x(k) \overset{\Delta}{=} \sum_{i=0}^{N-1} h(i)x_p(k-i), \quad 0 \le k < N$$

Evaluating the periodic extension, $x_p(k)$, at $k - i$ is equivalent to a counterclockwise circular shift of $x(-i)$ by k samples as illustrated in Figure 4.7. It is for

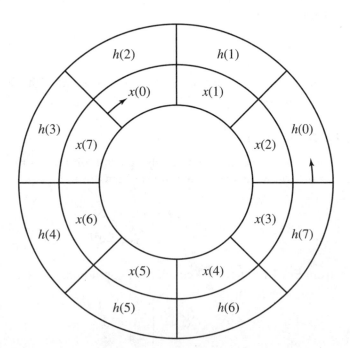

Figure 4.7 Counterclockwise Circular Shift of $x(-i)$ by $k = 2$ with $N = 8$

this reason that the operation in Definition 4.2 is referred to as *circular convolution*. Note how $h(i)$ is distributed counterclockwise around the outer ring, while $x(-i)$ is distributed clockwise around the inner ring. For different values of k, signal $x(-i)$ gets rotated k samples counterclockwise. The circular convolution is just the sum of the products of the N points distributed around the circle.

Just as was the case with the DFT, there is a matrix formulation of circular convolution. Let h and y_c be $N \times 1$ column vectors containing the samples of $h(k)$ and $y_c(k) = h(k) \circ x(k)$.

$$h = [h(0), h(1), \ldots, h(N-1)]^T \tag{4.16a}$$

$$y_c = [y_c(0), y_c(1), \ldots, y_c(N-1)]^T \tag{4.16b}$$

Since circular convolution is a linear transformation from h to y_c, it can be represented by an $N \times N$ matrix $C(x)$. Consider, in particular, the following matrix which corresponds to the case $N = 5$.

$$C(x) = \begin{bmatrix} x(0) & x(4) & x(3) & x(2) & x(1) \\ x(1) & x(0) & x(4) & x(3) & x(2) \\ x(2) & x(1) & x(0) & x(4) & x(3) \\ x(3) & x(2) & x(1) & x(0) & x(4) \\ x(4) & x(3) & x(2) & x(1) & x(0) \end{bmatrix} \tag{4.17}$$

Observe that the rows of the circular convolution matrix $C(x)$ are just rotated versions of $x(-i)$. Using Eqs. 4.16 and Eq. 4.17, the circular convolution in Definition 4.2 can be expressed in matrix form as

$$y_c = C(x)h \tag{4.18}$$

Note from Eq. 4.17 that if the input $x(k)$ is selected with some care, the circular convolution matrix $C(x)$ will be nonsingular.

| Example 4.2 | **Circular Convolution** |

To illustrate how to compute a circular convolution using the matrix formulation, consider the following two finite signals of length $N = 3$.

$$h = [2, -1, 6]^T$$

$$x = [5, 3, -4]^T$$

Thus, from Eq. 4.17, the 3×3 circular convolution matrix $C(x)$ is

$$C(x) = \begin{bmatrix} 5 & -4 & 3 \\ 3 & 5 & -4 \\ -4 & 3 & 5 \end{bmatrix}$$

Using Eq. 4.18, if $y_c(k) = h(k) \circ x(k)$, then

$$y_c = C(x)h$$

$$= \begin{bmatrix} 5 & -4 & 3 \\ 3 & 5 & -4 \\ -4 & 3 & 5 \end{bmatrix} \begin{bmatrix} 2 \\ -1 \\ 6 \end{bmatrix}$$

$$= [32, -23, 19]^T$$

4.2.3   Zero Padding

Circular convolution is more efficient than linear convolution. However, circular convolution *cannot* be directly used to compute the zero-state response of the linear discrete-time system in Figure 4.6, even when it is an FIR system with a finite input. To see this, note from the Definition 4.2 that if we evaluate $y_c(k)$ for $k \geq N$ we find that circular convolution is periodic with period N.

$$y_c(k + N) = y_c(k) \tag{4.19}$$

Since the zero-state response is not, in general, periodic, it follows that circular convolution produces a different response than linear convolution. Fortunately, there is a simple preprocessing step that can be performed that allows us to achieve linear convolution using circular convolution. To keep the formulation general, let $h(k)$ be an L-point signal, and let $x(k)$ be an M-point signal. Suppose we construct two new signals, each of length $N = L + M - 1$, by using zero padding. In particular, we can pad $h(k)$ with $M - 1$ zeros and $x(k)$ with $L - 1$ zeros.

$$h_z = [h(0), h(1), \ldots, h(L-1), \overbrace{0, \ldots, 0}^{M-1}]^T \tag{4.20a}$$

$$x_z = [x(0), x(1), \ldots, x(M-1), \underbrace{0, \ldots, 0}_{L-1}]^T \tag{4.20b}$$

Thus, h_z and x_z are zero-padded vectors of length $N = L + M - 1$. Next, consider the circular convolution of $h_z(k)$ with $x_z(k)$. If $x_{zp}(k)$ is the periodic extension of $x_z(k)$, then

$$y_c(k) = h_z(k) \circ x_z(k)$$

$$= \sum_{i=0}^{N-1} h_z(i) x_{zp}(k - i)$$

$$= \sum_{i=0}^{L-1} h_z(i) x_{zp}(k - i), \quad 0 \leq k < N \tag{4.21}$$

Since $0 \leq i < L$ and $0 \leq k < N$, the minimum value for $k - i$ is $-(L - 1)$. But $x_z(i)$ has $L - 1$ zeros padded to the end of it. Therefore, $x_{zp}(-i) = 0$ for $0 \leq i < L$. This means that $x_{zp}(k - i)$ in Eq. 4.21 can be replaced by $x_z(k - i)$. The result is then the linear convolution of $h_z(k)$ with $x_z(k)$. Thus, from Eq. 4.9 the zero-state response of the system in Figure 4.6 can be computed using circular convolution as follows.

Zero-state response

$$y(k) = h_z(k) \circ x_z(k), \quad 0 \le k < N \tag{4.22}$$

In summary, linear convolution can be achieved by computing the circular convolution of zero-padded versions of the two signals. The following example illustrates this technique.

Example 4.3   **Zero Padding**

Consider the signals from Example 4.2. In this case, $L = 3$, $M = 3$, and $N = L + M - 1 = 5$. The zero-padded signals are

$$h_z = [2, -1, 6, 0, 0]^T$$

$$x_z = [5, 3, -4, 0, 0]^T$$

From Eq. 4.17, the 5×5 circular convolution matrix $C(x)$ is

$$C(x) = \begin{bmatrix} 5 & 0 & 0 & -4 & 3 \\ 3 & 5 & 0 & 0 & -4 \\ -4 & 3 & 5 & 0 & 0 \\ 0 & -4 & 3 & 5 & 0 \\ 0 & 0 & -4 & 3 & 5 \end{bmatrix}$$

Thus, from Eq. 4.18 and Eq. 4.22, the linear convolution of $h(k)$ with $x(k)$ is

$$y = C(x_z)h_z$$

$$= \begin{bmatrix} 5 & 0 & 0 & -4 & 3 \\ 3 & 5 & 0 & 0 & -4 \\ -4 & 3 & 5 & 0 & 0 \\ 0 & -4 & 3 & 5 & 0 \\ 0 & 0 & -4 & 3 & 5 \end{bmatrix} \begin{bmatrix} 2 \\ -1 \\ 6 \\ 0 \\ 0 \end{bmatrix}$$

$$= [10, 1, 19, 22, -24]^T$$

Using the MATLAB convolution function, *conv*, one can verify that $h_z(k) \circ x_z(k) = h(k) \star x(k)$ in this case.

4.2.4 Deconvolution

Recall from Section 4.1 that speech synthesis is achieved by identifying a model for the human vocal tract and then driving this model with either white noise (unvoiced phonemes) or a periodic impulse train (voiced phonemes). The vocal-tract cavity is modeled with a digital filter that can be characterized by the following convolution operation.

$$y(k) = \sum_{i=0}^{k} h(i)x(k-i), \quad k \geq 0 \tag{4.23}$$

The process of finding the input $x(k)$ given the impulse response $h(k)$ and the output $y(k)$ is referred to as *deconvolution*. Deconvolution also includes finding the impulse response $h(k)$, given the input $x(k)$ and the output $y(k)$. This is a special case of a more general problem called system identification that is treated in detail in Chapter 9. When $h(k)$ and $x(k)$ are both causal noise-free signals as in Eq. 4.23, recovery of $h(k)$ is relatively simple. Suppose the input $x(k)$ is chosen such that $x(0) \neq 0$. Evaluating Eq. 4.23 at $k = 0$ then yields $y(0) = h(0)x(0)$ or

$$h(0) = \frac{y(0)}{x(0)} \tag{4.24}$$

Once $h(0)$ is known, the remaining samples of $h(k)$ can be obtained recursively. For example, evaluating Eq. 4.23 at $k = 1$ yields

$$y(1) = h(0)x(1) + h(1)x(0) \tag{4.25}$$

Solving Eq. 4.25 for $h(1)$, we then have

$$h(1) = \frac{y(1) - h(0)x(1)}{x(0)} \tag{4.26}$$

This process can be repeated for $2 \leq k < N$. The expression for the general case is obtained by decomposing the sum in Eq. 4.23 into the $i < k$ terms and the $i = k$ term. Solving for $h(k)$, we then arrive at the following recursive reconstruction of the impulse response.

Deconvolution

$$h(k) = \frac{1}{x(0)} \left\{ y(k) - \sum_{i=0}^{k-1} h(i)x(k-i) \right\}, \quad k \geq 1 \tag{4.27}$$

With the initialization in Eq. 4.24, the formulation in Eq. 4.27 solves the deconvolution problem, thereby reconstructing the impulse response from the input and the output. The following example illustrates this technique.

Example 4.4   **Deconvolution**

Consider the signals from Example 4.2. In this case, both $h(k)$ and $x(k)$ are of length $N = 3$ with

$$h = [2, -1, 6]^T$$

$$x = [5, 3, -4]^T$$

Let $y(k) = h(k) \star x(k)$. Circular convolution with zero padding was used in Example 4.3 to find the following zero-state response.

$$y = [10, 1, 19, 22, -24]^T$$

To recover $h(k)$ from $x(k)$ and $y(k)$ using deconvolution, we start with Eq. 4.24. The first sample of the impulse response is

$$h(0) = \frac{y(0)}{x(0)}$$

$$= \frac{10}{5}$$

$$= 2$$

Next, applying Eq. 4.27 with $k = 1$ yields

$$h(1) = \frac{y(1) - h(0)x(1)}{x(0)}$$

$$= \frac{1 - 2(3)}{5}$$

$$= -1$$

Finally, applying Eq. 4.27 with $k = 2$ then yields

$$h(2) = \frac{y(2) - h(0)x(2) - h(1)x(1)}{x(0)}$$

$$= \frac{19 - 2(-4) - (-1)3}{5}$$

$$= 6$$

Thus, the impulse response vector is

$$h = [2, -1, 6]^T \; \checkmark$$

4.2.5 Polynomial Arithmetic

Linear convolution has a simple and useful interpretation in terms of polynomial arithmetic. Suppose $A(z)$ and $B(z)$ are polynomials of degree L and M, respectively.

$$A(z) = a(0)z^L + a(1)z^{L-1} + \cdots + a(L) \tag{4.28a}$$

$$B(z) = b(0)z^M + b(1)z^{M-1} + \cdots + b(M) \tag{4.28b}$$

Thus, the coefficient vectors a and b are of length $L + 1$ and $M + 1$, respectively. Next, let $C(z)$ be the following product polynomial.

$$C(z) = A(z)B(z) \tag{4.29}$$

In this case, $C(z)$ is of degree $N = L + M$, and its coefficient vector c is of length $L + M + 1$. That is,

$$C(z) = c(0)z^{L+M} + c(1)z^{L+M-1} + \cdots + c(L + M) \qquad (4.30)$$

The coefficient vector of $C(z)$ can be obtained directly from the coefficient vectors of $A(z)$ and $B(z)$ using linear convolution as follows.

Polynomial
multiplication

$$c(k) = a(k) \star b(k), \quad 0 \le k < L + M + 2 \qquad (4.31)$$

Thus, linear convolution is equivalent to *polynomial multiplication*. Since deconvolution allows us to recover $a(k)$ from $b(k)$ and $c(k)$, deconvolution is equivalent to *polynomial division*.

Example 4.5 **Polynomial Multiplication**

To illustrate the relationship between linear convolution and polynomial arithmetic, consider the following two polynomials.

$$A(z) = 2z^2 - z + 6$$
$$B(z) = 5z^2 + 3z - 4$$

In this case, the coefficient vectors are $a = [2, -1, 6]^T$ and $b = [5, 3, -4]^T$. Suppose $c(k) = a(k) \star b(k)$. Then from Example 4.3:

$$c = [10, 1, 19, 22, -24]^T$$

Thus, the product of $A(z)$ with $B(z)$ is the polynomial of degree four whose coefficient vector is given by c. That is,

$$C(z) = A(z)B(z)$$
$$= 10z^4 + z^3 + 19z^2 + 22z - 24$$

Using direct multiplication of $A(z)$ times $B(z)$, we can verify this result.

$$C(z) = \left\{ \begin{array}{lllll} 10z^4 & -5z^3 & +30z^2 & & \\ & 6z^3 & -3z^2 & +18z & \\ & & -8z^2 & +4z & -24 \\ \hline 10z^4 & +z^3 & +19z^2 & +22z & -24 \end{array} \right\}$$

MATLAB Toolbox

MATLAB has two built-in functions for performing linear convolution and de-convolution.

```
y    = conv (h,x);          % linear convolution
[h,r] = deconv (y,x)        % linear deconvolution
```

The function *conv* performs convolution. On entry to *conv*, h is a vector of length L containing the samples of the first signal, and x is a vector of length M containing the samples of the second signal. On exit from *conv*, y is vector of length $L + M - 1$ containing $y(k) = h(k) \star x(k)$. The function *deconv* performs deconvolution. On entry to *deconv*, y is a vector of length N containing the output signal, and x is a vector of length $M < N$ containing the input signal. If y is the zero-state output of an FIR system produced by input x, then on exit from *deconv*, h is a vector of length N containing the impulse response, and r is zero. More generally, if y and x contain the coefficients of polynomials, then on exit from *deconv*, h is a vector containing the coefficients of the quotient polynomial, and r is a vector containing the coefficients of the remainder polynomial. Here:

$$y(z) = h(z)x(z) + r(z) \tag{4.32}$$

4.3 Fast Convolution

4.3.1 Fast Linear Convolution

The most important characteristic of linear convolution is the fact that convolution in the time domain maps into multiplication in the frequency domain using the Z-transform. Therefore, an effective way to compute the linear convolution of $h(k)$ with $x(k)$ is

$$h(k) \star x(k) = Z^{-1}\{H(z)X(z)\}, \quad k \geq 0 \tag{4.33}$$

In order to develop a relationship analogous to Eq. 4.33 that is valid for finite signals, we must substitute circular convolution for linear convolution and the DFT for the Z-transform. To verify this, we make use of the linearity property and the circular shift property of the DFT from Table 3.2. Taking the DFT of the circular convolution of $h(k)$ with $x(k)$ using Definition 4.2, we have

$$\text{DFT}\{h(k) \circ x(k)\} = \text{DFT}\left\{\sum_{m=0}^{N-1} h(m)x_p(k-m)\right\}$$

$$= \sum_{m=0}^{N-1} h(m)\text{DFT}\{x_p(k-m)\}$$

$$= \sum_{m=0}^{N-1} h(m) W_N^{im} X(m)$$

$$= H(i) X(i) \qquad (4.34)$$

Consequently, circular convolution in the time domain maps into multiplication in the frequency domain using the DFT. Therefore, an effective way to perform circular convolution is

$$h(k) \circ x(k) = \text{IDFT}\{H(i)X(i)\}, \quad 0 \le k < N \qquad (4.35)$$

In view of Eq. 4.35, we now have in place all of the tools needed to perform a practical, highly efficient, linear convolution of two finite signals. Suppose $h(k)$ is an L-point signal, and $x(k)$ is an M-point signal. Recall from Eq. 4.22 that linear convolution can be achieved with circular convolution using zero padding. In particular, let $h_z(k)$ be the zero-padded version of $h(k)$ using $M + p$ zeros, and let $x_z(k)$ be the zero-padded version of $x(k)$ using $L + p$ zeros. Thus, the common length of $h_z(k)$ and $x_z(k)$ is $N = L + M + p$. In developing the relationship in Eq. 4.22, we used $p = -1$, which corresponds to the smallest value for N that ensures that Eq. 4.22 holds. However, any $p \ge -1$ will work. Suppose we pick p such that $N = L + M + p$ is a power of two. This can be achieved by using the following value for N where *ceil* rounds up to the next integer.

$$N = 2^{\text{ceil}[\log_2(L+M-1)]} \qquad (4.36)$$

Selecting N as in Eq. 4.36 ensures that N is the smallest power of two such that $N \ge L + M - 1$. The highly efficient radix-two FFT then can be used to compute the DFTs of $h_z(k)$ and $x_z(k)$. This results in the following version of linear convolution which is called *fast convolution*.

Fast convolution

$$h(k) \star x(k) = \text{IFFT}\{H_z(i)X_z(i)\}, \quad 0 \le k < L + M - 1 \qquad (4.37)$$

A block diagram of the fast convolution operation is shown in Figure 4.8. Even though fast convolution involves quite a few steps, there is a value for N beyond which fast convolution is more efficient than the direct computation of linear convolution in Eq. 4.15.

To simplify the analysis of the computational effort of fast convolution, suppose the signals $h(k)$ and $x(k)$ are both of length L where L is a power of two. In this case, zero padding to length $N = 2L$ is sufficient. Thus, the two FFTs in Figure 4.8 each require $(N/2) \log_2(N)$ complex floating-point operations or FLOPs, while the inverse FFT requires $(N/2) \log_2(N) + 1$ FLOPs. The multiplication of $H_z(i)$ times $X_z(i)$ for $0 \le i < N$ requires an additional N FLOPs. Thus, the total number of complex FLOPs is $(3N/2) \log_2(N) + N + 1$. The direct computation of linear convolution does not involve any complex arithmetic, so to provide a fair comparison, we should count the number of real multiplications. The product of two complex numbers can

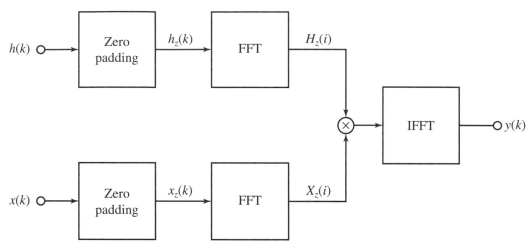

Figure 4.8   Fast Linear Convolution

be expressed as follows.

$$(a + jb)(c + jd) = ac - bd + j(ad + bc) \tag{4.38}$$

Thus, each complex multiplication requires four real multiplications. Recalling that $N = 2L$, the total number of real FLOPs required to perform a fast linear convolution of two L-point signals is then

$$n_{\text{fast}} = 12L \log_2(2L) + 8L + 4 \ \text{FLOPs} \tag{4.39}$$

Next, consider the number of real multiplications required to implement linear convolution directly. Setting $M = L$ in Eq. 4.15, we find that the number of real multiplications or FLOPs is

$$n_{\text{dir}} = 2L^2 \ \text{FLOPs} \tag{4.40}$$

Comparing Eq. 4.40 with Eq. 4.39, we see that for small values of L a direct computation of linear convolution will be faster. However, the $2L^2$ term grows faster than the $L \log_2(2L)$ term, so eventually the fast convolution will outperform direct convolution. A plot of the number of real FLOPs required by the two methods for signal lengths in the range $2 \leq L \leq 1024$ is shown in Figure 4.9. The two methods require roughly the same number of FLOPs for $L \leq 32$. However, fast linear convolution is superior to direct linear convolution for signal lengths in the range $L \geq 64$, and as L increases it becomes significantly faster.

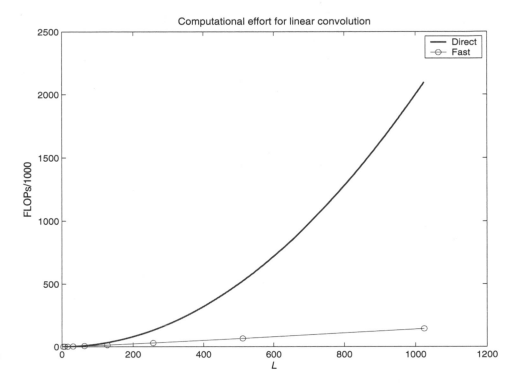

Figure 4.9 Comparison of Computational Effort Required for Linear Convolution of Two L-point Signals

---

Example 4.6   **Fast Convolution**

To illustrate the use of fast convolution, consider a linear discrete-time system with the following transfer function.

$$H(z) = \frac{0.98 \sin(\pi/24)z}{z^2 - 1.96 \cos(\pi/24)z + 0.9604}$$

Using the Z-transform pairs in Table 2.1, the impulse response of this system is

$$h(k) = 0.98^k \sin(\pi k/24)u(k)$$

Next, suppose this system is driven with the following exponentially damped sinusoidal input.

$$x(k) = k^2(0.99)^k \cos(\pi k/48), \quad k \geq 0$$

Since both $h(k)$ and $x(k)$ decay to zero, we can approximate them as L-point signals when L is sufficiently large. The zero-state response of the system can be obtained by running script *exam4_6* on the distribution CD. This script computes the linear convolution of $h(k)$ with $x(k)$ using fast convolution with $L = 512$. Plots of the input $x(k)$ and zero-state output $y(k)$ are shown in Figure 4.10. In this case, the total number of real FLOPs was $n_{\text{fast}} = 4.67 \times 10^4$. This is in contrast to the direct method which would require $n_{\text{dir}} = 5.24 \times 10^5$ FLOPs.

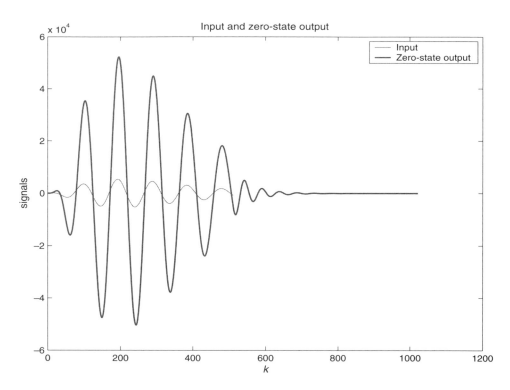

Figure 4.10 Zero-state Response to $x(k)$ using Fast Linear Convolution

---

*4.3.2 Fast Block Convolution

The efficient formulation of convolution based on the FFT assumes that the two signals are of finite duration. There are some applications where the input $x(k)$ is available continuously and is of indefinite duration. For example, the input might represent a long speech signal obtained from a microphone. In cases like these, the number of input samples M will be very large, and computation of an FFT of length $N = L + M + p$ may not be practical. Another potential drawback is that none of the samples of the filtered output are available until the entire N points have been processed.

These difficulties associated with very long inputs can be addressed by using a technique known as *block convolution*. Suppose the impulse response consists of L samples, and the input contains M samples where $M \gg L$. The basic idea is to break up the input signal into blocks or sections of length L. Each of these blocks is convolved with the L-point impulse response $h(k)$. If the results are combined in the proper way, the original $(L + M - 1)$-point convolution can be recovered.

To see how this is achieved, first note that the $x(k)$ can be padded with up to $L-1$ zeros, if needed, such that $M = QL$ for some integer $Q > 1$. The zero-padded input signal, $x_z(k)$, then can be expressed as a sum of Q blocks of length L as follows.

$$x_z(k) = \sum_{i=0}^{P-1} x_i(k - iL), \quad 0 \le k < M \tag{4.41}$$

Here the Q blocks, or subsignals of length L, are extracted from the original signal $x(k)$ using a window of length L as follows.

$$x_i(k) \triangleq \begin{cases} x(k + iL), & 0 \le k < L \\ 0, & \text{otherwise} \end{cases} \tag{4.42}$$

Subsignal $x_i(k)$ is the segment of $x(k)$ starting at $k = Li$, but it has been shifted so that it starts at $k = 0$. Using Eq. 4.41 and Definition 4.1, the linear convolution of $h(k)$ with $x(k)$ is then

$$h(k) \star x(k) = \sum_{i=0}^{Q-1} h(k) \star x_i(k - iL)$$

$$= \sum_{i=0}^{Q-1} y_i(k - iL) \tag{4.43}$$

Here $y_i(k)$ is the convolution of $h(k)$ with the ith subsignal $x_i(k)$. That is,

$$y_i(k) = h(k) \star x_i(i), \quad 0 \le i < Q \tag{4.44}$$

The block convolutions in Eq. 4.44 are between two L-point signals. If these signals are padded with $L + p$ zeros such that $p \ge -1$ and $2L + p$ is a power of two, then a radix-two FFT can be used. The results must then be shifted and added as in Eq. 4.43. The resulting procedure, known as the *overlap-add method* of block convolution, is summarized in the following algorithm.

ALGORITHM 4.1:
Fast Block
Convolution

1. Compute

$$M_{save} = M$$

$$r = L - \text{mod}(M, L)$$

$$M = M + r$$

$$x_z = [x(0), \ldots, x(M_{save} - 1), 0, \ldots, 0]^T \in R^M$$

$$Q = \frac{M}{L}$$

$$N = 2^{\text{ceil}[\log_2(2L-1)]}$$

$$h_z = [h(0), \ldots, h(L - 1), 0, \ldots, 0]^T \in R^N$$

$$H_z = \text{FFT}\{h_z(k)\}$$

$$y_0 = [0, \ldots, 0]^T \in R^{L(Q-1)+N}$$

2. For $i = 0$ to $Q - 1$ compute

{

$$x_i(k) = x_z(k + iL), \quad 0 \le k < L$$

$$x_{iz}(k) = [x_i(0), \ldots, x_i(L - 1), 0, \ldots, 0]^T \in R^N$$

$$X_{iz}(i) = \text{FFT}\{x_{iz}(k)\}$$

$$y_i(k) = \text{IFFT}\{H_z(i)X_{iz}(i)\}$$

$$y_0(k) = y_0(k) + y_i(k - Li), \quad Li \le k < Li + 2N - 1$$

}

3. Set

$$y(k) = y_0(k), \quad 0 \le k < L + M_{save} - 1$$

In step 1 of Algorithm 4.1, the original number of input samples is saved in M_{save}. If $\text{mod}(M, L) > 0$, then M is not an integer multiple of L. In this case, r zeros are padded to the end of x so that the length of $x_z(k)$ is $M = QL$, where Q is an integer representing the number of blocks. Next N is computed so that the zero-padded length of h satisfies $N \ge 2L - 1$ and is a power of two. Since H_z only has to be computed once, it is also computed in step 1. The Q block convolutions are then performed in step 2, and the results are overlapped and added using y_0 for storage. Finally, the relevant part of $y_0(k)$ is extracted in step 3. There is an alternative to the overlap-add method summarized in Algorithm 4.1 called the overlap-save method of block convolution. The interested reader is referred to (Oppenheim *et al.*, 1999).

| Example 4.7 | **Fast Block Convolution** |

As an illustration of the fast block convolution technique, suppose the impulse response is

$$h(k) = (0.8)^k \sin\left(\frac{\pi k}{4}\right), \quad 0 \le k < 12$$

Thus, $L = 12$. Next, let the input consist of white noise uniformly distributed over $[-1, 1]$.

$$x(k) = v(k), \quad 0 \le k < 70$$

In this case, $M_{\text{save}} = 70$. The number of samples in the zero-padded version of x is

$$
\begin{aligned}
M &= M + [L - \text{mod}(M, L)] \\
&= 70 + 12 - \text{mod}(70, 12)) \\
&= 82 - 10 \\
&= 72
\end{aligned}
$$

Thus, $Q = 72/12$ and there are exactly $Q = 6$ blocks of length 12 in $x_z(k)$. Since both h and x_i are of length L, the minimum length of the zero-padded versions of h and x_i will be $L + L + p$ for $p \ge -1$. We can choose $N = 2L + p$ to be a power of two as follows.

$$
\begin{aligned}
N &= 2^{\text{ceil}[\log_2(2L-1)]} \\
&= 2^{\text{ceil}(4.5236)} \\
&= 2^5 \\
&= 32
\end{aligned}
$$

Plots of the impulse response $h(k)$, the zero-padded input $x_z(k)$, and the output $y(k)$ are shown in Figure 4.11. These we generated by running script *exam4_7* on the distribution CD. The zero-padded version of the input is sectioned into Q blocks, each of length L. Script *exam4_7* also computes the convolution in the direct manner. Both outputs are plotted in Figure 4.11 where it can be seen that they are identical. For this example, a modest value for M was used so that the results are easier to visualize.

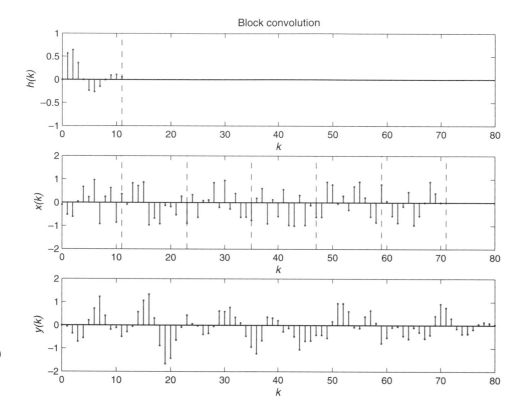

Figure 4.11 Block Convolution of $h(k)$ with $x(k)$, where $L = 12$, $M = 72$, $Q = 6$, and $N = 32$

MATLAB Toolbox

The FDSP toolbox contains the following functions for performing fast convolutions and fast block convolutions.

```
y = f_conv (h,x,circ);          % fast convolution
y = f_blockconv (h,x);          % fast block convolution
```

Function *f_conv* performs a fast convolution. On entry to *f_conv*, h is a vector of length L containing the samples of the first signal, x is a vector of length M containing the samples of the second signal, and *circ* is an integer that specifies the type of convolution to perform. If *circ* is zero, then fast linear convolution is performed. Otherwise, fast circular convolution is performed where it is assumed that $M = L$. Function *f_blockconv* performs a fast block convolution. The input and output arguments for *f_blockconv* are identical to the corresponding arguments for *f_conv*, except that it is assumed that $M > L$.

• • • • • • • • • • • • • •

4.4 Cross-correlation

Next we turn our attention to an operation that is related to convolution, called correlation.

DEFINITION

4.3: Linear
Cross-correlation

> Let $x(k)$ be an L-point signal, and let $y(k)$ be an M-point signal where $M \leq L$. Then the *linear cross-correlation* of $x(k)$ with $y(k)$ is denoted $r_{xy}(i)$ and defined
>
> $$r_{xy}(k) \triangleq \frac{1}{L} \sum_{i=0}^{L-1} x(i) y(i-k), \quad 0 \leq k < L$$

Since $y(k)$ is causal, the lower limit of the sum in Definition 4.3 can be set to $i = k$. The variable k is sometimes call the *lag* variable because it represents the number of samples that $y(i)$ is shifted to the right, or delayed, before the sum of products is computed.

Definition 4.3 indicates how to compute the cross-correlation of two deterministic discrete-time signals. Linear cross-correlation is sometimes defined without the scale factor, $1/L$. A scale factor is included in Definition 4.3, because this way Definition 4.3 matches more closely with an alternative statistical formulation of the cross-correlation of two random signals. The cross-correlation of a pair of random signals is introduced in Chapter 9 and uses the expected value operator.

Just as was the case with convolution, there is a matrix formulation of cross-correlation. Let x and r_{xy} be $L \times 1$ column vectors containing the samples of $x(k)$ and $r_{xy}(k)$, respectively.

$$x = [x(0), x(1), \ldots, x(L-1)]^T \tag{4.45a}$$

$$r_{xy} = [r_{xy}(0), r_{xy}(1), \ldots, r_{xy}(L-1)]^T \tag{4.45b}$$

Cross-correlation is a linear transformation from x to r_{xy}. Consequently, it can be represented by an $L \times L$ matrix $D(y)$. Consider, in particular, the following matrix which corresponds to the case where $L = 5$ and $M = 3$.

$$D(y) = \frac{1}{5} \begin{bmatrix} y(0) & y(1) & y(2) & 0 & 0 \\ 0 & y(0) & y(1) & y(2) & 0 \\ 0 & 0 & y(0) & y(1) & y(2) \\ 0 & 0 & 0 & y(0) & y(1) \\ 0 & 0 & 0 & 0 & y(0) \end{bmatrix} \tag{4.46}$$

Note how the rows of $D(y)$ are constructed by shifting $y(k)$ to the right. Unlike the circular convolution matrix in Eq. 4.17, when the samples of y get shifted off the right end of $D(y)$, they do not wrap around and reappear on the left end. Using

Eqs. 4.45, Eq. 4.46, and Definition 4.3, the linear cross-correlation of $x(k)$ with $y(k)$ can be expressed in matrix form as

$$r_{xy} = D(y)x \qquad (4.47)$$

Observe from Eq. 4.46 that if $y(0) \neq 0$, then the cross-correlation matrix $D(y)$ is nonsingular, which means signal $x(k)$ can be recovered from the cross-correlation using $x = D^{-1}(y)r_{xy}$.

Linear cross-correlation can be used to measure the degree to which the shape of one signal is *similar* to the shape of another signal. The following example illustrates this point.

Example 4.8   **Linear Cross-correlation**

To demonstrate how linear cross-correlation can be computed using the matrix formulation, consider the following pair of discrete-time signals.

$$x = [4, -1, -2, 0, 4, 2, 4, 2, 0, -2, 2]^T$$

$$y = [0, 2, 1, 2, 0]^T;$$

Here $L = 10$ and $M = 5$. A plot of the two signals $x(k)$ and $y(k)$ is shown in Figure 4.12. Note how the graph of $y(k)$ is a flattened "M" shaped signal. Furthermore,

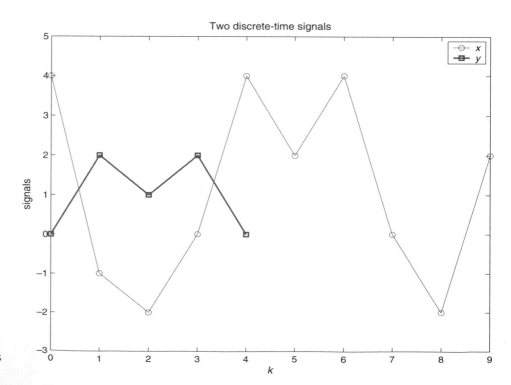

Figure 4.12   Two Discrete-time Signals

the longer signal $x(k)$ contains a scaled and translated version of $y(k)$ with a larger "M" starting at $k = 3$. To verify that $x(k)$ contains a scaled and shifted version of $y(k)$, we compute the linear cross-correlation. From Eq. 4.46 and Eq. 4.47, the linear cross-correlation of $x(k)$ with $y(k)$ is

$$r_{xy} = D(y)x$$

$$= \frac{1}{10}
\begin{bmatrix}
0 & 2 & 1 & 2 & 0 & 0 & 0 & 0 & 0 & 0 \\
0 & 0 & 2 & 1 & 2 & 0 & 0 & 0 & 0 & 0 \\
0 & 0 & 0 & 2 & 1 & 2 & 0 & 0 & 0 & 0 \\
0 & 0 & 0 & 0 & 2 & 1 & 2 & 0 & 0 & 0 \\
0 & 0 & 0 & 0 & 0 & 2 & 1 & 2 & 0 & 0 \\
0 & 0 & 0 & 0 & 0 & 0 & 2 & 1 & 2 & 0 \\
0 & 0 & 0 & 0 & 0 & 0 & 0 & 2 & 1 & 2 \\
0 & 0 & 0 & 0 & 0 & 0 & 0 & 0 & 2 & 1 \\
0 & 0 & 0 & 0 & 0 & 0 & 0 & 0 & 0 & 2 \\
0 & 0 & 0 & 0 & 0 & 0 & 0 & 0 & 0 & 0
\end{bmatrix}
\begin{bmatrix}
4 \\ -1 \\ -2 \\ 0 \\ 4 \\ 2 \\ 4 \\ 0 \\ -2 \\ 2
\end{bmatrix}$$

$$= [-0.4, 0.4, 0.8, 1.8, 0.8, 0.4, 0.2, -0.2, 0.4, 0.0]^T$$

In this case, $r_{xy}(k)$ reaches its maximum value at a lag of $k = 3$. This is evident from the plot of $r_{xy}(k)$ shown in Figure 4.13. The fact that $r_{xy}(k)$ has a clear peak at $k = 3$

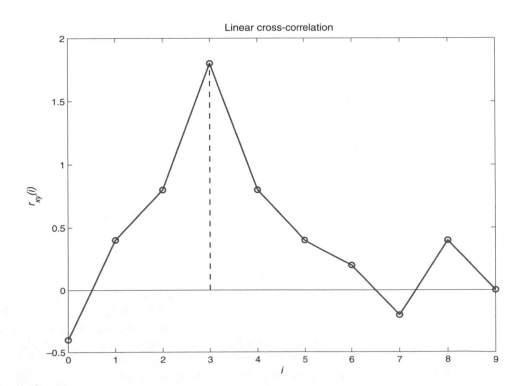

Figure 4.13 Linear Cross-correlation of the Signals in Figure 4.12

indicates there is a strong positive correlation between $x(k)$ and $y(k-3)$. Indeed, except for scaling, the two signals exactly match in this case.

$$x(k) = 2y(k-3), \quad 3 \leq k < 3 + M$$

Thus, a dominant peak in $r_{xy}(k)$ at $k = p$ is an indication that a signal similar to $y(k)$ is present in signal $x(k)$ starting at $k = p$.

Although the cross-correlation in Definition 4.3 can be used to detect the presence of one signal in another signal, it does suffer from a practical drawback. When $x(k)$ contains a scaled and shifted version of $y(k)$, the cross-correlation will contain a distinct peak. However, the height of the peak will depend on the data in $x(k)$ and $y(k)$. For example, scaling $x(k)$ or $y(k)$ by α will scale the height of the peak by α. Consequently, one is left with the question, how high does the peak have to be for a significant correlation to exist? A way to simplify this problem is to develop a normalized version of cross-correlation. It can be shown (Proakis and Manolakis, 1992), that the square of the cross-correlation is bounded from above as follows.

$$r_{xy}^2(k) \leq \left(\frac{M}{L}\right) r_{xx}(0) r_{yy}(0), \quad 0 \leq k < L \tag{4.48}$$

By making use of Eq. 4.48, we can introduce the following scaled version of cross-correlation, called the *normalized linear cross-correlation* of $x(k)$ with $y(k)$.

Normalized linear cross-correlation

$$\rho_{xy}(k) \triangleq \frac{r_{xy}(k)}{\sqrt{(M/L)r_{xx}(0)r_{yy}(0)}}, \quad 0 \leq k < L \tag{4.49}$$

By construction, the magnitude of the normalized cross-correlation is bounded by one. That is, the normalized cross-correlation is guaranteed to lie within the following interval.

$$-1 \leq \rho_{xy}(k) \leq 1, \quad 0 \leq k < L \tag{4.50}$$

If the normalized cross-correlation is used, then any peak that begins to approach the maximum value of one indicates a very strong positive correlation between $x(k)$ and $y(k)$, regardless of whether one of the signals is very small or very large in comparison with the other. For the cross-correlation example shown in Figure 4.13, the peak of the normalized cross-correlation is $\rho_{xy}(3) = 0.744$, indicating there is a strong positive correlation between $x(k)$ and $y(k-3)$.

● ● ● ● ● ● ● ● ● ● ● ● ● ●

4.5 **Fast Correlation**

Practical cross-correlations often involve long signals, so it is important to develop a numerical implementation of linear cross-correlation that is more efficient than the direct method in Definition 4.3.

4.5.1　Circular Cross-correlation

Recall that fast linear convolution was achieved by using circular convolution with zero padding. A similar approach can be used to develop a fast version of linear cross-correlation.

DEFINITION

4.4: Circular
Cross-correlation

Let $x(k)$ and $y(k)$ be N-point signals, and let $y_p(k)$ be the periodic extension of $y(k)$. The *circular cross-correlation* of $x(k)$ with $y(k)$ is denoted $c_{xy}(k)$ and defined

$$c_{xy}(k) \triangleq \frac{1}{N} \sum_{i=0}^{N-1} x(i) y_p(i - k), \quad 0 \le k < N$$

Circular cross-correlation operates on two signals of the same length. Comparing Definition 4.4 with Definition 4.3, we see that for circular cross-correlation $y(k)$ is replaced by its periodic extension $y_p(k)$. The effect of this change is to replace the linear shift of $y(k)$ with a circular shift or rotation of $y(k)$, hence the name circular cross-correlation. A diagram illustrating circular cross-correlation for the case $N = 8$ and $k = 2$ is shown in Figure 4.14. Note that evaluating $y_p(i)$ at $i - k$ is equivalent to a clockwise circular shift of $y(i)$ by k samples, as shown in Figure 4.14. Observe how $x(i)$ is distributed counterclockwise around the outer ring, while $y(i)$ is distributed counterclockwise around the inner ring. For different values of the lag k, the signal $y_p(i - k)$ gets rotated k samples clockwise. The circular cross-correlation is just the sum of the products of the N points distributed around the circle.

Just as was the case with linear cross-correlation, we can scale $c_{xy}(k)$ to produce the following *normalized circular cross-correlation*, whose value is restricted to the interval $[-1, 1]$.

Normalized circular
cross-correlation

$$\sigma_{xy}(k) = \frac{c_{xy}(k)}{\sqrt{c_{xx}(0)c_{yy}(0)}} \tag{4.51}$$

There are a number of useful properties of circular cross-correlation. For example, consider the effect of interchanging the roles of $x(k)$ and $y(k)$. Note from

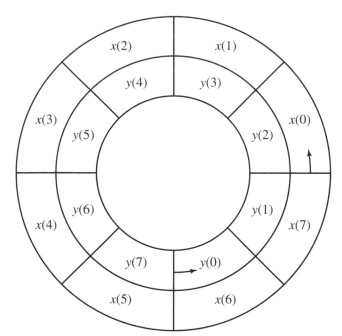

Figure 4.14 Clockwise Circular Shift of $y(i)$ by $k = 2$ with $N = 8$

Definition 4.4 that $x(k)$ can be replaced with its periodic extension $x_p(k)$ without affecting the result. Consequently, using the change of variable $q = i - k$ we have

$$c_{yx}(k) = \frac{1}{N} \sum_{i=0}^{N-1} y(i)x_p(i - k)$$

$$= \frac{1}{N} \sum_{i=0}^{N-1} y_p(i)x_p(i - k)$$

$$= \frac{1}{N} \sum_{q=-k}^{N-1-k} y_p(q + k)x_p(q) \left. \right\} \quad q = i - k$$

$$= \frac{1}{N} \sum_{q=0}^{N-1} x_p(q)y_p(q + k)$$

$$= \frac{1}{N} \sum_{q=0}^{N-1} x(q)y_p(q + k) \tag{4.52}$$

Observe that the summations in Eq. 4.52 all extend over one period. Consequently, the starting sample of the sum can be changed from $q = -k$ to $q = 0$ without affecting the result. From Definition 4.4, the last line of Eq. 4.52 is $c_{xy}(-k)$. Thus, we have the following *symmetry property* of circular cross-correlation, which says that changing the order of x and y is equivalent to changing the sign of the lag variable.

$$c_{yx}(k) = c_{xy}(-k), \quad 0 \le k < N \tag{4.53}$$

There is a simple and elegant relationship between circular cross-correlation and circular convolution. Comparing the expression in Definition 4.4 with that in Definition 4.2 observe that the circular cross-correlation of $x(k)$ with $y(k)$ is just a scaled version of the circular convolution of $x(k)$ with $y(-k)$. That is,

Correlation using convolution

$$c_{xy}(k) = \frac{x(k) \circ y(-k)}{N}, \quad 0 \le k < N \tag{4.54}$$

This simple relationship between circular correlation and circular convolution can now be put to use to develop a fast version of correlation. Recall from Eq. 4.34 that circular convolution in the time domain maps into multiplication in the frequency domain using the DFT. Furthermore, the circular convolution of $x(k)$ with $y(-k)$ is the same as the circular convolution of $x(k)$ with $y_p(-k)$, because computing a circular convolution requires replacing $y(k)$ with $y_p(k)$. Thus, from Eq. 4.54 the DFT of the circular correlation of $x(k)$ with $y(k)$ is

$$
\begin{aligned}
C_{xy}(i) &= \text{DFT}\left\{ \frac{x(k) \circ y(-k)}{N} \right\} \\
&= \frac{\text{DFT}\{x(k) \circ y_p(-k)\}}{N} \\
&= \frac{X(i)\text{DFT}\{y_p(-k)\}}{N} \tag{4.55}
\end{aligned}
$$

Using the reflection property of the DFT from Table 3.2, the expression for the DFT of the circular correlation of $x(k)$ with $y(k)$ then reduces to the following, where $Y^*(i)$ denotes the complex conjugate of $Y(i)$.

Circular cross-correlation using DFT

$$C_{xy}(i) = \frac{X(i)Y^*(i)}{N}, \quad 0 \le i < N \tag{4.56}$$

4.5.2   Fast Linear Cross-correlation

With the use of zero padding, linear cross-correlation can be achieved using circular cross-correlation. Suppose $x(k)$ is an L-point signal and $y(k)$ is an M-point signal with $M \le L$. Let $x_z(k)$ be a zero-padded version of $x(k)$ with $M + p$ zeros appended where $p \ge -1$. Similarly, let $y_z(k)$ be a zero-padded version of $y(k)$ with $L + p$ zeros. Therefore, x_z and y_z are both signals of length $N = L + M + p$.

$$x_z = [x(0), \cdots, x(L-1), \overbrace{0, \cdots, 0}^{M+p}]^T \tag{4.57a}$$

$$y_z = [y(0), \cdots, y(M-1), \underbrace{0, \cdots, 0}_{L+p}]^T \tag{4.57b}$$

Next, let y_{zp} be the periodic extension of $y_z(k)$, as in Eq. 3.54, and consider the circular cross-correlation of $x_z(k)$ with $y_z(k)$.

$$c_{x_z y_z}(k) = \frac{1}{N} \sum_{i=0}^{N-1} x_z(i) y_{zp}(i-k), \quad 0 \le k < N \tag{4.58}$$

If we restrict $c_{x_z y_z}(k)$ to $0 \le k < L$, it can be shown to be proportional to the linear cross-correlation in Definition 4.3. In particular, recalling that $x(k)$ is an L-point signal, we have

$$c_{x_z y_z}(k) = \frac{1}{N} \sum_{i=0}^{L-1} x_z(i) y_{zp}(i-k), \quad 0 \le k < L \tag{4.59}$$

Since $0 \le k < L$, the minimum value for $i - k$ is $-(L-1)$. But $y_z(k)$ has $L + p$ zeros padded to the end of it. Therefore, $y_{zp}(-k) = 0$ for $0 \le k \le L + p$. It follows that $y_{zp}(i - k)$ in Eq. 4.59 can be replaced by $y_z(i - k)$ as long as $p \ge -1$. The result is then the linear cross-correlation of $x_z(k)$ with $y_z(k)$. But for $0 \le k < L$, the linear cross-correlation of $x_z(k)$ with $y_z(k)$ is identical to the linear cross-correlation of $x(k)$ with $y(k)$, except for a scale factor of L/N. Consequently,

$$c_{x_z y_z}(k) = \left(\frac{L}{N}\right) r_{xy}(k), \quad 0 \le k < L \tag{4.60}$$

Thus, linear cross-correlation can be achieved with circular cross-correlation using zero padding. Suppose we pick $p \ge -1$, such that $N = L + M + p$ is a power of two. That is,

$$N = 2^{\text{ceil}[\log_2(L+M-1)]} \tag{4.61}$$

This means that a radix-two FFT can be used in place of the DFT. Using Eq. 4.56 and Eq. 4.60, we then arrive at the following highly efficient version of linear cross-correlation, called *fast linear cross-correlation*.

Fast linear
cross-correlation

$$r_{xy}(k) = \frac{\text{IFFT}\{X_z(i)Y_z^*(i)\}}{L}, \quad 0 \le k < L \tag{4.62}$$

Note the strong similarity between fast convolution in Eq. 4.37 and fast correlation in Eq. 4.62. The only differences are that $Y_z(i)$ is replaced by its complex conjugate, $Y_z^*(i)$, and the final result is scaled by N/L and evaluated only over $0 \le k < L$.

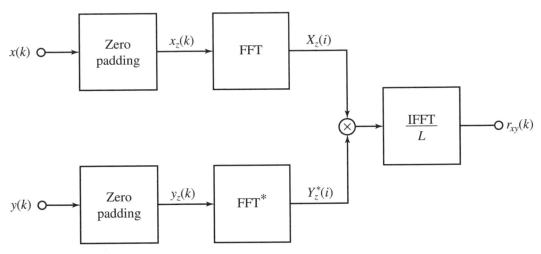

Figure 4.15 Fast Linear Cross-correlation

A block diagram of the fast correlation operation is shown in Figure 4.15. Just as with fast convolution, there is a value for L beyond which fast correlation is more efficient than the direct computation of cross-correlation using Definition 4.3.

The analysis of the computational effort for fast cross-correlation is similar to that for fast convolution. For simplicity, suppose the signals $x(k)$ and $y(k)$ are both of length L, where L is a power of two. Then from Eq. 4.61, zero padding to length $N = 2L$ is sufficient. Proceeding as was done in Section 4.3, the number of real FLOPs required to perform a fast linear cross-correlation of two L-point signals is then as follows.

$$n_{\text{fast}} = 12L \log_2(2L) + 8L + 6 \text{ FLOPs} \tag{4.63}$$

Next, consider the number of real multiplications required to implement linear cross-correlation directly. If the matrix formulation in Eq. 4.47 is used, then the number of real multiplication is L^2. However, if Definition 4.3 is used, then the lower limit on the sum can be replaced by $i = k$, because $y(k)$ is causal. This reduces the number of real multiplications by a factor of two.

$$n_{\text{dir}} = \frac{L^2}{2} + 1 \text{ FLOPs} \tag{4.64}$$

Comparing Eq. 4.64 with Eq. 4.63, we again see that for small values of L, a direct computation of linear cross-correlation will be faster. However, the L^2 term grows faster than the $L \log_2(2L)$ term, so eventually fast cross-correlation will outperform direct cross-correlation. A plot of the number of real FLOPs required by the two methods for signal lengths in the range $2 \leq L \leq 2048$ is shown in Figure 4.16. The two methods require roughly the same number of FLOPs for $L = 256$. However, fast linear cross-correlation is superior to direct linear cross-correlation for signal lengths in the range $L \geq 512$, and as L increases, it becomes significantly faster.

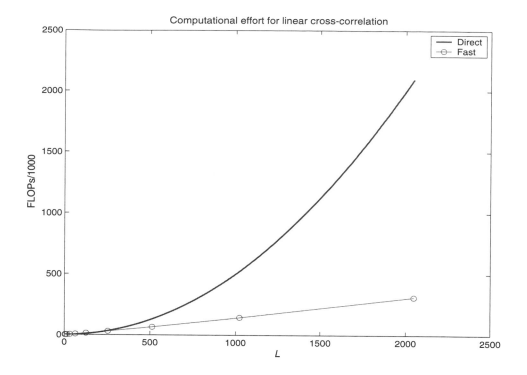

Figure 4.16 Comparison of Computational Effort for Linear Crosscorrelation of Two L-point Signals

---

Example 4.9   **Fast Linear Correlation**

To illustrate the use of fast correlation, let $L = 1024$ and $M = 512$, and consider the following pair of signals where $v(k)$ is white noise uniformly distributed over the interval $[-1, 1]$.

$$y(k) = \frac{3k}{M} \exp\left(\frac{-4k}{M}\right) \sin\left(\frac{5\pi k^2}{M}\right), \quad 0 \le k < M$$

$$x(k) = y_z(k - p) + v(k), \quad 0 \le k < L$$

We refer to $y(k)$ as a multi-frequency *chirp* signal because it contains a range of frequencies due to the k^2 factor in the sin. Here $y_z(k)$ denotes the zero-padded extension of $y(k)$. Thus, $y_z(k-p)$ is just $y(k)$ shifted to the right by p samples. For this example, $p = 279$. A plot of these two signals, obtained by running script *exam4_9* on the distribution CD, is shown in Figure 4.17. Also computed is the normalized linear cross-correlation of $x(k)$ with $y(k)$, as shown in Figure 4.18. Observe that the peak correlation occurs at $\rho_{xy}(279) = .173$, as expected. The number of real FLOPs required by fast cross-correlation in this case was $n_{\text{fast}} = 1.02 \times 10^5$. This is in contrast to the direct computation using Definition 4.3, which requires $n_{\text{dir}} = 5.24 \times 10^5$ real FLOPs.

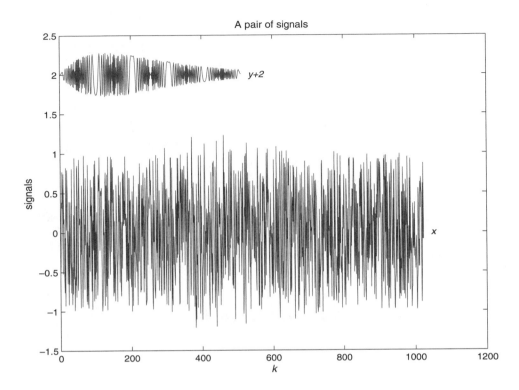

A pair of signals

Figure 4.17 A Pair of Discrete-time Signals

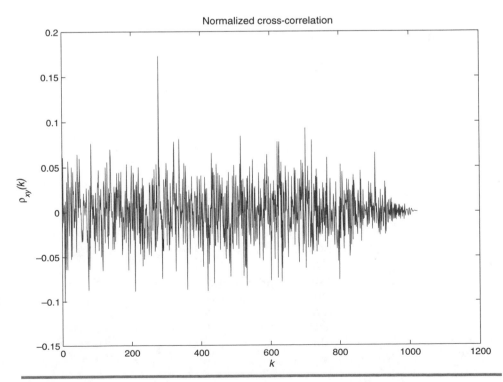

Normalized cross-correlation

Figure 4.18 Normalized Cross-correlation of the Signals in Figure 4.17

MATLAB Toolbox

The FDSP toolbox contains the following function for computing fast linear and circular cross-correlations.

```
r = f_corr (x,y,circ,norm);      % fast correlation
```

On entry to *f_corr*, x is a vector of length L containing the samples of the first signal, y is a vector of length $M \leq L$ containing the samples of the second signal, and *circ* and *norm* are integers which select the type of cross-correlation to perform. If *circ* is nonzero, then a fast circular cross-correlation is performed, otherwise a fast linear cross-correlation is performed. When *circ* is nonzero, the two signals should be of the same length: $M = L$. If *norm* is nonzero, the selected type of cross-correlation, circular or linear, is normalized. On exit from *f_corr*, r is a vector of length L containing the selected cross-correlation.

4.6 Auto-correlation

There is an important special case of cross-correlation that arises when one takes the cross-correlation of a signal with itself.

DEFINITION

4.5: Linear Auto-correlation

Let $x(k)$ be an N-point signal. Then the *linear auto-correlation* of $x(k)$ is denoted $r_{xx}(k)$ and defined

$$r_{xx}(k) \overset{\Delta}{=} \frac{1}{N} \sum_{i=k}^{N-1} x(i)x(i-k), \quad 0 \leq k < N$$

As the notation in Definition 4.5 suggests, auto-correlation is a special case of cross-correlation with $y = x$. Circular auto-correlation is denoted $c_{xx}(k)$ and is as defined in Definition 4.5, but with $y = x$.

There is a simple interpretation of auto-correlation in terms of the average power. It is apparent from Definition 4.5 that evaluating $r_{xx}(k)$ at a lag of $k = 0$ yields the average power P_x. Thus, the average power can be expressed in terms of linear auto-correlation and circular auto-correlation as follows.

Average power

$$P_x = r_{xx}(0) = c_{xx}(0) \tag{4.65}$$

Auto-correlation can be normalized just as cross-correlation. Using Eq. 4.49 and Eq. 4.65, the *normalized linear auto-correlation*, which is denoted $\rho_{xx}(k)$, can be expressed as

Normalized auto-correlation

$$\rho_{xx}(k) = \frac{r_{xx}(k)}{P_x}, \quad 0 \leq k < N \tag{4.66}$$

In view of Eq. 4.65, it follows that $\rho_{xx}(0) = 1$. Since $|\rho_{xx}(k)| \leq 1$, this means that the normalized linear auto-correlation always reaches its peak value of one at a lag of $k = 0$. Normalized circular auto-correlation is denoted $\sigma_{xx}(k)$ and defined as in Eq. 4.51, but with $y = x$.

4.6.1 Auto-correlation of White Noise

White noise has a particularly simple auto-correlation. To see this, let $v(k)$ be a random white-noise signal of length N with a mean value of $\mu = 0$. Recall that the mean of a random signal is the expected value of the signal, $E[v(k)]$. If a random signal has the property that it is *ergodic*, then the expected value of $f\{v(k)\}$ for a function f can be approximated by replacing the ensemble average (which depends on the probability density) with the simpler time average as follows.

Time average

$$E[f\{v(k)\}] \approx \frac{1}{N} \sum_{k=0}^{N-1} f\{v(k)\} \qquad (4.67)$$

Using Eq. 4.67, the auto-correlation of $v(k)$ can be expressed in terms of expected values as follows.

$$r_{vv}(k) = \frac{1}{N} \sum_{i=0}^{N-1} v(i)v(i-k)$$

$$\approx E[v(i)v(i-k)] \qquad (4.68)$$

Typically, the samples of a random white-noise signal are *statistically independent* from one another. For statistically independent or uncorrelated random variables, the expected value of the product is equal to the product of the expected values. Since the signal $v(i)$ has zero mean, this means that for $k \neq 0$ we have

$$r_{vv}(k) \approx E[v(i)v(i-k)]$$

$$= E[v(i)]E[v(i-k)]$$

$$= 0, \quad k \neq 0 \qquad (4.69)$$

For the case $k = 0$, using Eq. 4.65 yields $r_{vv}(0) = P_v$ where P_v is the average power of $v(k)$. Combining the two cases, we conclude that the linear auto-correlation of a zero-mean white-noise signal with average power P_v can be expressed as

Zero-mean white noise

$$r_{vv}(k) \approx P_v\delta(k), \quad 0 \leq k < N \qquad (4.70)$$

Thus, the linear auto-correlation of zero-mean white noise is simply an impulse of strength P_v at $k = 0$. As the value of N increases, the approximation in Eq. 4.70 becomes more accurate. The same analysis that was used to develop the approximation in Eq. 4.70 can be applied using circular auto-correlation, and the result is

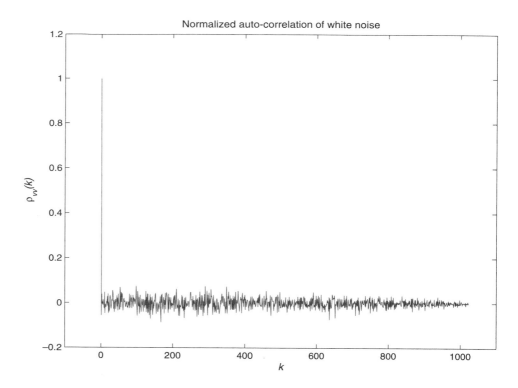

Figure 4.19 Normalized Linear Auto-correlation of Zero-mean Gaussian White Noise

identical. Thus, for $N \gg 1$, the circular auto-correlation of zero-mean white noise can be approximated as follows.

$$c_{vv}(k) \approx P_v \delta(k), \quad 0 \leq k < N \tag{4.71}$$

To numerically illustrate Eq. 4.70, suppose $v(k)$ is Gaussian white noise with mean $\mu = 0$ and standard deviation $\sigma = 1$. The normalized linear auto-correlation for the case $N = 1024$ is shown in Figure 4.19. Since the auto-correlation is normalized, the theoretical result should be $\rho_{vv}(k) = \delta(k)$. When circular auto-correlation is used, the only difference is that there is no narrowing of the tail of $\sigma_{vv}(k)$ for large values of k.

4.6.2 Power Density Spectrum

There is also a simple and elegant relationship between the power density spectrum and circular auto-correlation. Recall that the power density spectrum specifies the distribution of power over the discrete frequencies. For an N-point signal $x(k)$, the power density spectrum is as follows where $X(i)$ is the DFT of $x(k)$.

$$S_N(i) \triangleq \frac{|X(i)|^2}{N}, \quad 0 \leq i < N \tag{4.72}$$

To determine the relationship between $S_N(i)$ and $c_{xx}(k)$, we use Eq. 4.56 with $y = x$. Computing the DFT of the auto-correlation, we find

$$C_{xx}(i) = \text{DFT}\{c_{xx}(k)\}$$

$$= \frac{X(i)X^*(i)}{N}$$

$$= \frac{|X(i)|^2}{N}, \quad 0 \le i < N \tag{4.73}$$

Comparing Eq. 4.73 with Eq. 4.72 then yields the following alternative formulation of the power density spectrum.

Power density spectrum

$$\boxed{S_N(i) = C_{xx}(i), \quad 0 \le i < N} \tag{4.74}$$

Thus, the DFT of the circular auto-correlation of $x(k)$ is the power density spectrum. A block diagram illustrating the steps involved in this formulation of the power density spectrum is shown in Figure 4.20.

It is instructive to apply Eq. 4.74 to white noise. Suppose $v(k)$ is zero-mean white noise with average power P_v. Then recalling Eq. 4.71 and using Eq. 4.74, we have

$$S_N(i) = C_{vv}(i)$$

$$\approx \text{DFT}\{P_v \delta(k)\}$$

$$= P_v \tag{4.75}$$

Thus, zero-mean white noise with average power P_v has a power density spectrum that is *flat* and equal to P_v. It is for this reason that we refer to the noise as *white* because it contains power at all frequencies just as white light contains all colors.

Example 4.10

Power Density Spectrum

To illustrate the use of Eq. 4.74 to compute the power density spectrum, let $N = 512$, and consider a signal $x(k)$ that consists of a double pulse of width $M = 4$ centered at $k = N/2$.

$$x(k) = \begin{cases} 0, & 0 \le k < N/2 - M \\ 1, & N/2 - M \le k < N/2 \\ -1, & N/2 \le k < N/2 + M \\ 0, & N/2 + M \le k < N \end{cases}$$

Script *exam4_10* on the distribution CD computes the power density spectrum, $S_N(i)$, by finding the DFT of the circular auto-correlation as in Figure 4.20. The resulting plots of $x(k)$ and its power density spectrum, $S_N(f)$, are shown in Figure 4.21.

Figure 4.20 Power Density Spectrum using Circular Auto-correlation

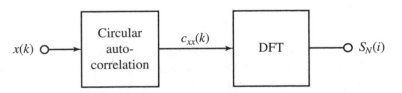

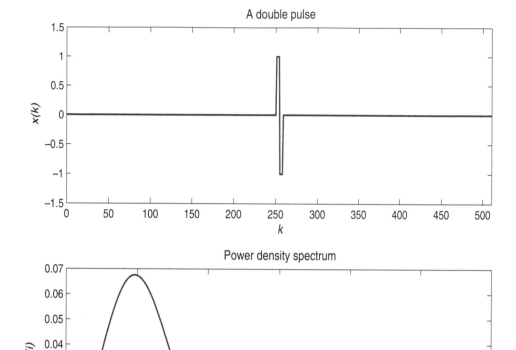

Figure 4.21 Power Density Spectrum of Double Pulse of Width $M = 4$ using Circular Auto-correlation

4.7 Extracting Periodic Signals from Noise

Practical signals are often corrupted with noise. Suppose $x(k)$ is a periodic signal with period M. We can model a noisy version of $x(k)$ as follows.

$$y(k) = x(k) + v(k), \quad 0 \le k < N \tag{4.76}$$

Here $v(k)$ represents additive zero-mean white noise that may arise, for example, from the measurement process or perhaps because $x(k)$ is transmitted over a noisy communication channel.

4.7.1 Estimating the Period of a Noisy Periodic Signal

Our initial objective is to estimate the period of $x(k)$ given $y(k)$. Since $v(k)$ contains power at all frequencies, completely removing $v(k)$ with a filtering operation is not an option. Instead, we use correlation techniques. We begin by examining the circular auto-correlation of the noisy signal $y(k)$. Let $y_p(k)$ and $v_p(k)$ be the periodic extensions of the N-point signals $y(k)$ and $v(k)$, respectively. Using Eq. 4.68, Definition 4.4, and the fact that the expected value operator is linear, we have

$$
\begin{aligned}
c_{yy}(k) &\approx E[y(i)y_p(i - k)] \\
&= E[\{x(i) + v(i)\}\{x_p(i - k) + v_p(i - k)\}] \\
&= E[x(i)x_p(i - k) + x(i)v_p(i - k) + v(i)x_p(i - k) + v(i)v_p(i - k)] \\
&= E[x(i)x_p(i - k)] + E[x(i)v_p(i - k)] + E[v(i)x_p(i - k)] + E[v(i)v_p(i - k)] \\
&\approx c_{xx}(k) + c_{xv}(k) + c_{vx}(k) + c_{vv}(k)
\end{aligned}
\tag{4.77}
$$

Typically, the noise $v(k)$ is statistically independent of the signal $x(k)$. Since $E[v(k)] = 0$, this means that the circular cross-correlation terms, $c_{xv}(k)$ and $c_{vx}(k)$, are both zero. Then using Eq. 4.71, the circular auto-correlation of the noisy signal $y(k)$ simplifies to

$$
c_{yy}(k) \approx c_{xx}(k) + P_v\delta(k)
\tag{4.78}
$$

Consequently, $c_{yy}(0) = P_x + P_v$, where P_x is the average power of the signal, and P_v is the average power of the noise. For nonzero values of k, we have $c_{yy}(k) \approx c_{xx}(k)$. That is, the effect of auto-correlation is to average out or reduce the noise. The circular auto-correlation of $y(k)$ is less noisy than $y(k)$ itself. Not only is the circular auto-correlation less sensitive to noise, but it is also periodic with the same period as $x(k)$. In particular, using $x(k + M) = x(k)$, we have

$$
\begin{aligned}
c_{xx}(k + M) &= \frac{1}{N} \sum_{i=0}^{N-1} x(i)x_p(i - k - M) \\
&= \frac{1}{N} \sum_{i=0}^{N-1} x(i)x_p(i - k) \\
&= c_{xx}(k)
\end{aligned}
\tag{4.79}
$$

Thus, the circular auto-correlation of a periodic signal is itself periodic with the same period.

Periodic circular
auto-correlation

$$
c_{xx}(k + M) = c_{xx}(k), \quad 0 \le k < N - M
\tag{4.80}
$$

Since $c_{xx}(k)$ is periodic with period M, and $c_{yy}(k) \approx c_{xx}(k)$ for $k > 0$, it follows that the auto-correlation of $y(k)$ will be periodic with period M. We can estimate the period of $x(k)$ by examining peaks or valleys in $c_{yy}(k)$.

Example 4.11   **Period Estimation**

Suppose $N = 256$. Consider the following periodic signal which includes two sinusoidal components.

$$x(k) = \cos\left(\frac{32\pi k}{N}\right) + \sin\left(\frac{48\pi k}{N}\right)$$

Note that the cos term has period $N/16$ and the sin term has period $N/24$. Thus, the period of $x(k)$ is the common period

$$M = \frac{N}{8}$$

$$= 32$$

Suppose $y(k)$ is a noisy version of $x(k)$ as in Eq. 4.76, where $v(k)$ is white noise uniformly distributed over the interval $[-0.5, 0.5]$. A plot of $y(k)$, obtained by running script *exam4_11* on the distribution CD, is shown in Figure 4.22. Note that the periodic nature of the underlying signal $x(k)$ is apparent, but it is difficult to precisely estimate the period due to the presence of the noise. By contrast, the circular auto-correlation of $y(k)$ is much less noisy, as can be seen from the plot in Figure 4.23. Using the FDSP function *f_caliper* to measure several cycles, we can estimate the period of $c_{yy}(k)$ to be approximately 31.91, which rounds to $M = 32$.

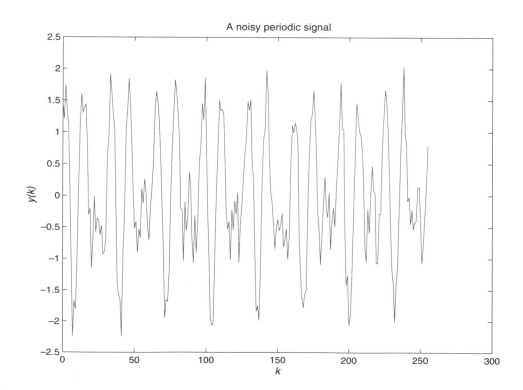

Figure 4.22 A Noisy Periodic Signal

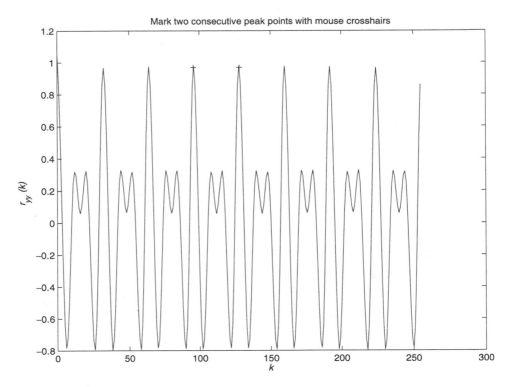

Figure 4.23 Circular Auto-correlation of Noisy Periodic Signal in Figure 4.22

4.7.2 Extracting a Periodic Signal from Noise

Once the period of a noise-corrupted periodic signal has been determined, we can use this information to extract the signal itself from the noise. Suppose $x(k)$ is an N-point signal that is periodic with period $M \ll N$. Then the number of complete cycles of $x(k)$ in $y(k)$ is

$$L = \text{floor}\left(\frac{N}{M}\right) \tag{4.81}$$

Next, let $\delta_M(k)$ be an N-point periodic impulse train with period M. Then we can represent $\delta_M(k)$ as follows.

$$\delta_M(k) = \sum_{i=0}^{L-1} \delta(k - iM), \quad 0 \le k < N \tag{4.82}$$

Suppose $y(k)$ is a noisy version of $x(k)$ that has been corrupted by zero-mean white noise $v(k)$ as in Eq. 4.76. The underlying periodic signal $x(k)$ can be extracted from the noisy signal $y(k)$ by cross-correlating with $\delta_M(k)$. To see this, let $y_p(k)$ be the periodic extension of $y(k)$. Recalling the symmetry property of circular cross-correlation in Eq. 4.53 and using Eq. 4.82, the circular cross-correlation of $y(k)$ with

$\delta_M(k)$ is

$$c_{y\delta_M}(k) = c_{\delta_M y}(-k)$$

$$= \frac{1}{N} \sum_{i=0}^{N-1} \delta_M(i) y_p(i+k)$$

$$= \frac{1}{N} \sum_{i=0}^{N-1} \left[\sum_{q=0}^{L-1} \delta(i - qM) \right] y_p(i+k)$$

$$= \frac{1}{N} \sum_{q=0}^{L-1} y_p(qM + k) \tag{4.83}$$

Next the expression for $y(k)$ in Eq. 4.76 can be substituted in Eq. 4.83. Recalling that $x(k)$ is periodic with period M, we then have

$$c_{y\delta_M}(k) = \frac{1}{N} \sum_{q=0}^{L-1} x_p(qM + k) + \frac{1}{N} \sum_{q=0}^{L-1} v_p(qM + k)$$

$$= \frac{1}{N} \sum_{q=0}^{L-1} x_p(k) + \frac{1}{N} \sum_{q=0}^{L-1} v_p(qM + k)$$

$$= \frac{Lx(k)}{N} + \frac{1}{N} \sum_{q=0}^{L-1} v_p(qM + k), \quad 0 \le k < N \tag{4.84}$$

For each k, the last term in Eq. 4.84 represents a sum of L noise terms. Since $v(k)$ is zero-mean white noise, this means that, for $L \gg 1$, the last term is approximately zero. Thus, for $N \gg M$, we have the following method of extracting a periodic signal from noise. The underlying periodic signal $x(k)$ can be approximated using $\hat{x}(k)$ where

Periodic signal estimate

$$\hat{x}(k) = \left(\frac{N}{L} \right) c_{y\delta_M}(k), \quad 0 \le k < N \tag{4.85}$$

A block diagram summarizing the steps required to extract a periodic signal from noise using circular cross-correlation is shown in Figure 4.24.

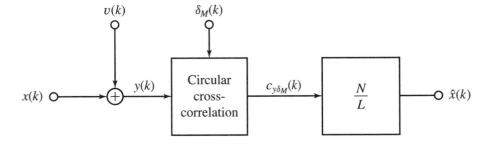

Figure 4.24 Extracting a Periodic Signal of Period M from Noise using Circular Cross-correlation

Example 4.12   **Extracting a Periodic Signal from Noise**

To illustrate the use of Eq. 4.85 to extract a periodic signal from noise, let $N = 256$, and consider the following noisy periodic signal.

$$x(k) = \cos\left(\frac{32\pi k}{N}\right) + \sin\left(\frac{48\pi k}{N}\right)$$

$$y(k) = x(k) + v(k)$$

Suppose $v(k)$ is white noise uniformly distributed over $[-.5, .5]$. The signal $y(k)$ was considered previously in Example 4.11 and is shown in Figure 4.22. The analysis of the auto-correlation of $y(k)$ in Example 4.11 revealed the period of $x(k)$ to be $M = 32$. Therefore, the number of complete cycles of $x(k)$ in $y(k)$ is

$$L = \text{floor}\left(\frac{N}{M}\right)$$

$$= 8$$

Extraction of $x(k)$ from the noise-corrupted $y(k)$ using Eq. 4.85 can be performed by running the script *exam4_12* on the distribution CD. The estimate of $x(k)$ using circular cross-correlation is

$$\hat{x}(k) = \left(\frac{N}{L}\right) c_{y\delta_M}(k)$$

A plot comparing the first two periods of $x(k)$ with the estimate, $\hat{x}(k)$, is shown in Figure 4.25. The reconstruction, in this case, is quite reasonable, but it is not exact given that the noise term in Eq. 4.84 is only approximately zero.

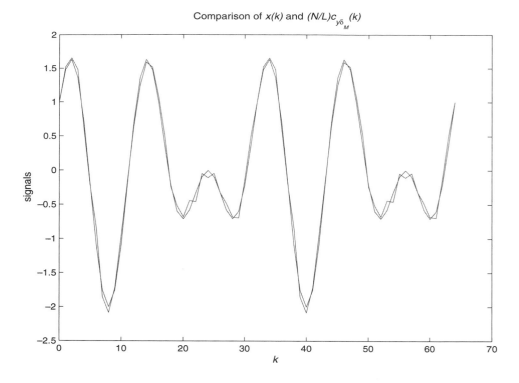

Comparison of $x(k)$ and $(N/L)c_{y\delta_M}(k)$

Figure 4.25 Comparison of Periodic Signal $x(k)$ and Estimate $\hat{x}(k)$ Extracted from $y(k)$ using Circular Cross-correlation

● ● ● ● ● ● ● ● ● ● ● ● ●
4.8 Software Applications

This section focuses on applications of convolution and correlation. A graphical user interface module called *g_correlate* is presented that allows the user to examine linear and circular convolution, cross-correlation, and auto-correlation. Application examples are then presented and solved using MATLAB scripts. The relevant MATLAB functions from the FDSP toolbox are described in Appendix B.

4.8.1 GUI Module: *g_correlate*

The graphical user interface module *g_correlate* is designed to allow the user to investigate convolutions and cross-correlations of pairs of signals and auto-correlations of individual signals. GUI module *g_correlate* features a display screen with tiled windows as shown in Figure 4.26.

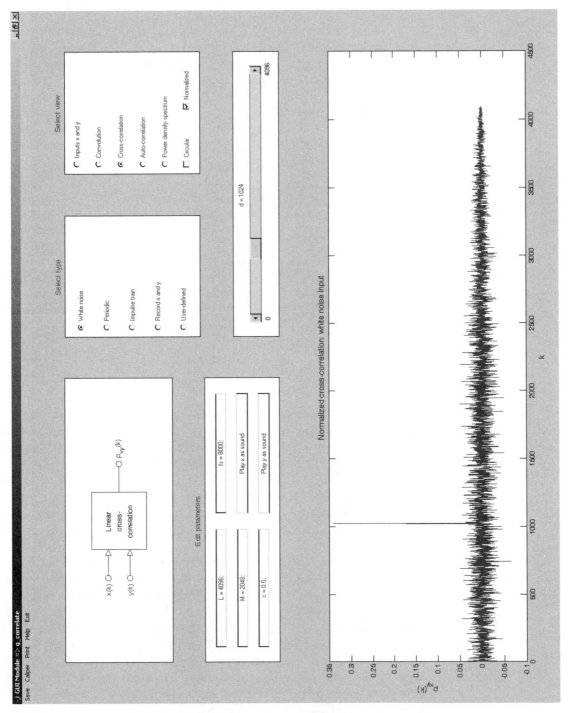

Figure 4.26 Display Screen of Chapter GUI Module *g_correlate*

The *Block diagram* window in the upper-left corner of the screen contains a block diagram that specifies the operation being performed. The signals are color-coded, and the labels change depending on the signal processing operation selected. The *Parameters* window below the block diagram contains four edit boxes. Parameters L, M, f_s and c can be edited directly by the user. Parameters L and M are the lengths of the input signals $x(k)$ and $y(k)$, respectively, with $M \leq L$, while f_s is the sampling frequency. Parameter c represents a scale factor that is used to add a scaled and shifted version of $y(k)$ to $x(k)$, so that its presence can be detected using cross-correlation. Changes to parameter values are activated with the Enter key. The *Parameters* window also includes two pushbutton controls. The pushbutton controls play the signals $x(k)$ and $y(k)$ on the PC speaker using a sampling rate of $f_s = 8192$ Hz.

The *Type* and *View* windows in the upper-right corner of the screen allow the user to select both the type of input signal and the viewing mode. The inputs include white noise inputs, periodic inputs, a periodic x, with a synchronized impulse train for y, inputs recorded from a PC microphone, and user-defined inputs stored in a MAT-file. For the recorded inputs, up to two seconds for x and 0.5 seconds for y can be recorded at $f_s = 8192$ Hz. The user-defined inputs are specified in a user-supplied MAT-file containing the vectors x, y, and f_s. For the white noise inputs, the signal $x(k)$ contains a scaled and delayed version of the signal $y(k)$. That is, $x(k)$ is computed as follows.

$$x(k) = v(k) + cy_z(k - d), \quad 0 \leq k < L \tag{4.86}$$

Here $y_z(k)$ is a zero-extended version of $y(k)$. Consequently, $x(k)$ contains a scaled and delayed version of $y(k)$. The scale factor c in the *Parameters* window can be modified by the user. The delay d can be set anywhere between 0 and L using the horizontal slider bar that appears below the *Type* and *View* windows.

The *View* options include plots of $x(k)$ and $y(k)$, the convolution of $x(k)$ with $y(k)$, the cross-correlation of $x(k)$ with $y(k)$, the auto-correlation of $x(k)$, and the power density spectrum of $x(k)$. The power density spectrum is computed by taking the DFT of the circular auto-correlation. Below the View options are two checkbox controls. The first checkbox allows the user to toggle between linear correlation or convolution and circular correlation or convolution. The second checkbox allows the user to choose between regular and normalized cross-correlations and auto-correlations. The *Plot* window on the bottom half of the screen shows the selected view. The curves are color-coded to match the block diagram labels.

The *Menu* bar at the top of the screen includes several menu options. The *Save* option is used to save the current x, y, and f_s in a user-specified MAT-file for future use. Files created in this manner can be loaded with the User-defined input option. The *Caliper* option allows the user to measure any point on the current plot by moving the mouse crosshairs to that point and clicking. The *Print* option prints the contents of the *Plot* window. Finally, the *Help* option provides the user with some helpful suggestions on how to effectively use module g_correlate.

4.8.2 Echo Detection

Recall from Section 4.1 that one of the applications of cross-correlation is in radar processing, as shown previously in Figure 4.3. Here $y(k)$ is the transmitted signal, and $x(k)$ is the received signal. First, consider the transmitted signal. Suppose the sampling frequency is $f_s = 1$ MHz, and the number of transmitted samples is $M = 512$. One possible choice for the transmitted signal is a multi-frequency *chirp*: a sinusoidal signal whose frequency varies with time. For example, let $T = 1/f_s$, $f_0 = f_s/8$, and $f_1 = f_s/4$, and consider the following chirp signal with variable frequency $f(k)$.

$$f(k) = f_0 + \frac{k(f_1 - f_0)}{M - 1}, \quad 0 \le k < M \tag{4.87a}$$

$$y(k) = \sin[2\pi f(k)kT], \quad 0 \le k < M \tag{4.87b}$$

The received signal $x(k)$ includes a scaled and delayed version of the transmitted signal plus measurement noise. Suppose the received signal consists of $L = 2048$ samples. If $y_z(k)$ denotes the transmitted signal, zero-extended to L points, then the received signal can be expressed as follows.

$$x(k) = ay_z(k - d) + v(k), \quad 0 \le k < L \tag{4.88}$$

The first term in $x(k)$ represents the echo of the transmitted signal that is reflected back from the illuminated target. Typically, the echo will be attenuated due to the dispersion, so $a \ll 1$. In addition, the echo will be delayed by d samples due to the time it takes for the transmitted signal to travel to the target, bounce off, and return. The second term in $x(k)$ represents random atmospheric measurement noise picked up by the receiver. For example, suppose $v(k)$ is white noise uniformly distributed over the interval $[-0.1, 0.1]$.

To determine the range to the target, let c be the propagation speed of the transmitted signal. For radar applications, this corresponds to the speed of light or $c = 1.86 \times 10^8$ miles/sec. The time of flight of the signal is then $\tau = dT$ sec. Multiplying τ by the signal propagation speed, and dividing by two for the round trip, we then arrive at the following expression for the *range* to the target.

$$r = \frac{cdT}{2} \tag{4.89}$$

Thus, the key to finding the distance to the target is to detect the presence, and location, of the echo of $y(k)$ in the received signal $x(k)$. This can be achieved by running the following MATLAB script, labeled *exam4_13* on the distribution CD.

```
function exam4_13

% Example 4.13: Echo detection

clear
clc
fprintf('Example 4.13: Echo detection\n')
rand('state',1000)

% Construct chirp signal y and its power density spectrum

M = 512;
fs = 1.e7;
T = 1/fs;
f0 = fs/8;
f1 = fs/4;
m = 0 : M-1;
freq = f0 + (f1-f0)*m/(M-1);
y = sin(2*pi*freq.*m*T);
[A,phi,S,f] = f_spec (y,M,fs);
figure
i = 1 : M/2+1;
plot (f(i),S(i))
f_labels ('Power density spectrum of chirp signal','{\itf} (Hz)','\it{S_N(f)}')
f_wait

% Construct received signal x

L = 2048;
a = 0.03;
d = 1304;
x = f_randu(1,L,-0.1,0.1);
x(d+1:d+M) = x(d+1:d+M) + a*y;
k = 0 : L-1;
figure
plot (k,x)
f_labels ('Received signal','\it{k}','\it{x(k)}')
f_wait

% Locate echo and compute range

rho = f_corr (x,y,0,1);
figure
plot (k,rho)
f_labels ('Normalized linear cross-correlation','\it{k}','\it{\rho_{xy}(k)}')
[rmax,kmax] = max(rho)
```

```
delay = kmax - 1
c = 1.86e5;
r = c*delay*T/2
precision = c*T/2
f_wait
```

When script *exam4_13* is run, it first constructs the chirp signal, $y(k)$, and computes its power density spectrum, $S_N(f)$, using the FDSP toolbox function, *f_spec*. The resulting plot is shown in Figure 4.27. Note that power is distributed approximately uniformly over a range of frequencies. Script *exam4_13* then constructs the received signal $x(k)$ and computes the normalized linear cross-correlation of $x(k)$ with $y(k)$. The echo of $y(k)$ buried in $x(k)$ becomes evident when we examine the cross-correlation plot shown in Figure 4.28. For this example, $a = 0.03$, and $d = 1304$. Although the peak at $\rho_{xy}(d) = 0.2115$ is not large due to the noise; it is distinct. Using the MATLAB function *max* to locate the peak, the value reported for the range to the target is

$$r = \frac{cdT}{2}$$

$$= 12.13 \text{ miles} \tag{4.90}$$

The precision with which the range can be measured depends on both the sampling

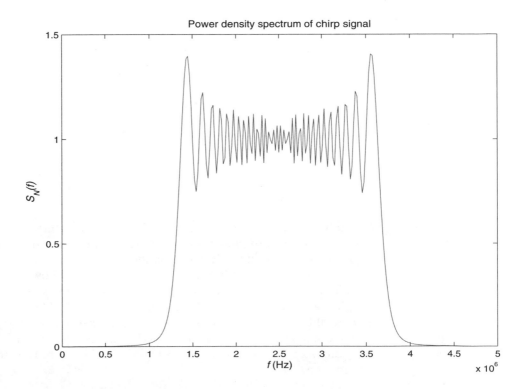

Figure 4.27 Power Density Spectrum of Multi-frequency Chirp Signal $y(k)$

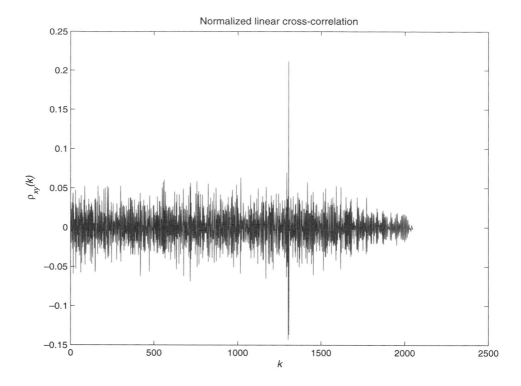

Normalized linear cross-correlation

Figure 4.28 Normalized Linear Cross-correlation of Received Signal $x(k)$ with Transmitted Signal $y(k)$.

frequency f_s and the propagation speed c. The smallest increment for r corresponds to a delay of $d = 1$ sample. Consequently, from Eq. 4.89

$$\Delta r = \frac{cT}{2}$$

$$= 0.0093 \text{ miles} \tag{4.91}$$

4.8.3 Speech Analysis and Pitch

Recall from Section 4.1 that synthetic speech can be generated by passing a periodic impulse train (voiced phonemes) or white noise (unvoiced phonemes) through an appropriate digital filter. System identification techniques for determining the coefficients of the filter are introduced in Chapter 9. For now, we will use correlation techniques to investigate the problem of speech analysis. Consider, in particular, the plot of a recorded vowel "E" shown in Figure 4.29. Here the sampling rate was $f_s = 8000$ Hz. The data for this plot was generated using the FDSP toolbox function *f_getsound*, and then stored in a MAT-file named *vowel_e*. The plot in Figure 4.29 was generated by running the following MATLAB script labeled *exam4_14* on the distribution CD.

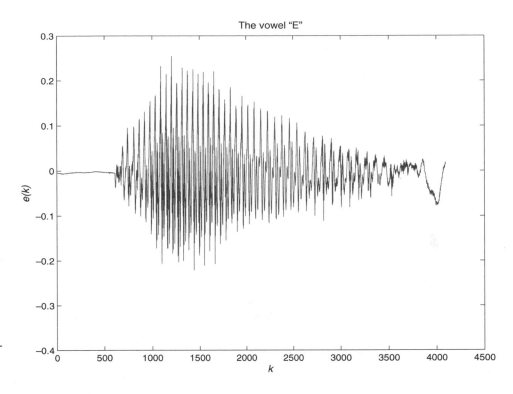

Figure 4.29 Recording of the Vowel "E"

```
function exam4_14

% Example 4.14: Speech analysis and pitch

clear
clc
fprintf('Example 4.14: Speech analysis and pitch\n')

% Load sound data and plot it

load vowel_e
N = length(e);
k = 0 : N-1;
fs = 8000;
T = 1/fs;
figure
plot (k,e)
f_labels ('The vowel E','\it{k}','\it{e(k)}')
f_wait ('Press any key to play sound')
soundsc (e)
```

```
% Estimate pitch of speaker

M = 512;
m = 1201:1200+M;
r = f_corr (e(m),e(m),1,0);
figure
plot (0:511,r)
f_labels ('Circular cross-correlation of segment of vowel "E"','\it{k}','\it{c_{ee}(k)}')
f_wait

% Compute pitch from density spectrum

S = real(fft(r));
[A,phi,R,f] = f_spec (e(m),M,fs);
figure
n = 1 : M/8;
plot (f(n),S(n),f(n),R(n))
f_labels ('Power density spectrum of segment of vowel "E"','{\itf (Hz)}','{\itS_N(f)}')
[Smax,imax] = max(S(n));
pitch = f(imax);
hold on
plot (pitch,Smax,'rs')
fprintf ('\npitch = %.0f Hz\n',pitch)
f_wait
x = cos(2*pi*pitch*k*T);
soundsc (x,fs)
f_wait
```

Consider the problem of estimating the pitch of the speaker. Since the vowel has transient segments at both the beginning and end, suppose we apply a $M = 512$ point window and compute the circular auto-correlation of the segment beginning at sample $L = 1200$.

$$c_{ee}(k) = \frac{1}{M} \sum_{i=L}^{L+M-1} e(i)e_p(i-k), \quad 0 \le k < M \tag{4.92}$$

A plot of the circular auto-correlation is shown in Figure 4.30. Note that it is less noisy than the original recording and approximately periodic. Using the FDSP function *f_caliper* in *exam4_14*, we can mark the first and the second-to-last peaks, thus selecting eight cycles. Dividing the difference in x coordinates by eight, and multiplying by the sampling period $T = 1/f_s$, we get the following estimate of the speaker pitch.

$$\text{pitch} \approx 140.3 \text{ Hz} \tag{4.93}$$

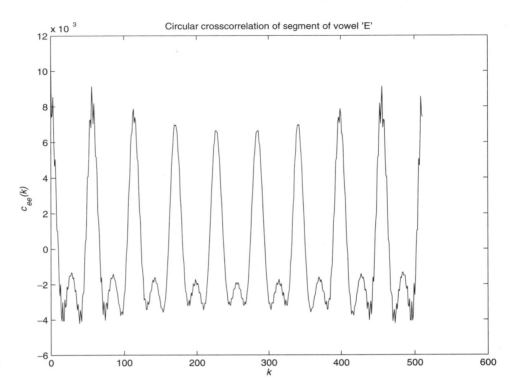

Figure 4.30 Circular Auto-correlation of the Vowel "E" Starting at $k = 1200$

An alternative way to estimate the pitch is to compute the power density spectrum of $e(k)$. This can be done either directly using $S_N(i) = |E(i)|^2/N$ or indirectly by computing the FFT of the circular auto-correlation of $e(k)$. Both methods are used in *exam4_14*, and it is seen from the plot in Figure 4.31 that they produce identical results. For clarity, only the frequency range from 0 to 1000 Hz is plotted. The peaks in the power density spectrum of a speech signal are sometimes referred to as *formants* (Roads *et al.*, 1997). In this case, there are two clear formants in the displayed range. Using the *f_caliper* function, the locations of the formants can be approximated as follows.

$$f_1 \approx 140.6 \text{ Hz} \tag{4.94a}$$

$$f_2 \approx 281.1 \text{ Hz} \tag{4.94b}$$

The first formant corresponds to the speaker pitch. Note that it is within about 0.2 percent of the pitch estimate obtained from circular auto-correlation in Eq. 4.93. In general, the distribution of formants can be used as a distinguishing characteristic of a speech fragment.

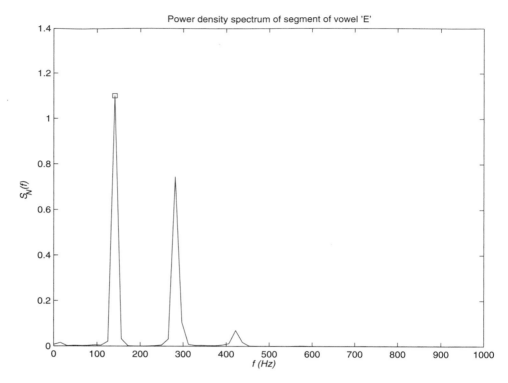

Figure 4.31 Power Density Spectrum of Vowel "E"

● ● ● ● ● ● ● ● ● ● ● ● ● ● ●

4.9 Chapter Summary

This chapter focused on convolution, cross-correlation, and auto-correlation of discrete-time signals.

Convolution

Given an L-point signal $h(k)$ and an M-point signal $x(k)$, the *linear convolution* of $h(k)$ with $x(k)$ is defined

$$h(k) \star x(k) = \sum_{i=0}^{k} h(i)x(k-i), \quad 0 \leq k < L + M \tag{4.95}$$

If $h(k)$ is the impulse response of a linear discrete-time system and $x(k)$ is the system input, then the zero-state response of the system is the linear convolution of $h(k)$ with $x(k)$. That is,

$$y(k) = h(k) \star x(k) \tag{4.96}$$

The convolution operation is commutative, so $h(k) \star x(k) = x(k) \star h(k)$. In addition, if $x(0) \neq 0$, then $h(k)$ can be recovered from $x(k)$ and $y(k)$ using a process known as *deconvolution*. From Eq. 4.95 and Eq. 4.96, $h(0) = y(0)/x(0)$. The remaining samples of the impulse response can then be obtained recursively using

$$h(k) = \frac{1}{x(0)} \left\{ y(k) - \sum_{i=0}^{k-1} h(i)x(k-i) \right\}, \quad k \geq 1 \tag{4.97}$$

Suppose $h(k)$ and $x(k)$ are both signals of length N, and $x(k)$ in Eq. 4.95 is replaced by its periodic extension, $x_p(k)$. This results in an operation called the *circular convolution* of $h(k)$ with $x(k)$.

$$h(k) \circ x(k) = \sum_{i=0}^{N-1} h(i)x_p(k-i), \quad 0 \leq k < N \tag{4.98}$$

The most important property of circular convolution is that it maps into multiplication in the frequency domain under the DFT transformation. More specifically, taking the DFT of both sides of Eq. 4.98 yields

$$\text{DFT}\{h(k) \circ x(k)\} = H(i)X(i), \quad 0 \leq i < N \tag{4.99}$$

The circular convolution operation can be used to achieve linear convolution. Suppose $h_z(k)$ is a zero-padded version of $h(k)$ using $M + p$ zeros, and $x_z(k)$ is a zero-padded version of $x(k)$ using $L + p$ zeros. Then $h_z(k)$ and $x_z(k)$ are both of length $N = L + M + p$. If $p \geq -1$, then the circular convolution of the zero-padded signals is the same as the linear convolution of the original signals. Suppose p is chosen, such that $N = L + M + p$ is a power of two using

$$N = 2^{\text{ceil}[\log_2(L+M-1)]} \tag{4.100}$$

Then a radix-two FFT can be used in place of the DFT in Eq. 4.99. This leads to the following highly efficient way to compute linear convolution called *fast linear convolution*.

$$h(k) \star x(k) = \text{IFFT}\{H_z(i)X_z(i)\}, \quad 0 \leq k < N \tag{4.101}$$

Suppose $L = M$. Then using Eq. 4.39 and Eq. 4.40, the ratio of the computational effort for fast convolution to the computational effort for direct linear convolution, measured in real FLOPs, is

$$r \approx \frac{6\log_2(2L)}{L}, \quad L \gg 1 \tag{4.102}$$

This ratio is less than one for $L \geq 64$ and goes to zero as $L \to \infty$. Thus, fast convolution is superior to direct linear convolution for signals of length $L \geq 64$.

Cross-correlation

An operation that is closely related to convolution is correlation. The *linear cross-correlation* of an L-point signal $x(k)$ with an M-point signal $y(k)$ is defined as follows, where it is assumed that $M \leq L$.

$$r_{xy}(k) = \frac{1}{L} \sum_{i=0}^{L-1} x(i)y(i-k), \quad 0 \leq k < L \tag{4.103}$$

Here k is referred to as the *lag* variable because it represents the amount by which the second signal is delayed before the sum of products is computed. Linear cross-correlation can be used to measure the degree to which the shape of a segment of signal $y(k)$ is similar to the shape of a segment of signal $x(k)$. In particular, if $x(k)$ contains a scaled version of $y(k)$ delayed by d samples, then $r_{xy}(k)$ will exhibit a peak at $k = d$. The following normalized version of linear cross-correlation takes on values in the interval $[-1, 1]$.

$$\rho_{xy}(k) = \frac{r_{xy}(k)}{\sqrt{(M/L)r_{xx}(0)r_{yy}(0)}}, \quad 0 \leq k < L \tag{4.104}$$

Just as there is a circular version of convolution, there is also a circular version of cross-correlation. Suppose $x(k)$ and $y(k)$ are both of length N. If $y(k)$ in Eq. 4.103 is replaced by its periodic extension $y_p(k)$, then this results in the following formulation of cross-correlation called *circular cross-correlation*.

$$c_{xy}(k) = \frac{1}{N} \sum_{i=0}^{N-1} x(i)y_p(i-k), \quad 0 \leq k < N \tag{4.105}$$

The following normalized version of circular cross-correlation takes on values in the interval $[-1, 1]$.

$$\sigma_{xy}(k) = \frac{c_{xy}(k)}{\sqrt{c_{xx}(0)c_{yy}(0)}}, \quad 0 \leq k < N \tag{4.106}$$

There is a simple relationship between circular correlation and circular convolution. Comparing Eq. 4.105 with Eq. 4.98 we find that circular cross-correlation of $x(k)$ with $y(k)$ is just a scaled version of circular convolution of $x(k)$ with $y(-k)$.

$$c_{xy}(k) = \frac{x(k) \circ y(-k)}{N}, \quad 0 \leq k < N \tag{4.107}$$

The most important property of circular cross-correlation is that it maps into conjugate multiplication in the frequency domain under the DFT transformation. In particular, using Eq. 4.99, Eq. 4.107, and the properties of the DFT, we have the following expression for the DFT of $c_{xy}(k)$ where $Y^*(i)$ denotes the complex conjugate of the $Y(i)$.

$$C_{xy}(i) = \frac{X(i)Y^*(i)}{N}, \quad 0 \leq i < N \tag{4.108}$$

Circular cross-correlation can be used to achieve linear cross-correlation using zero padding. Suppose $x_z(k)$ is a zero-padded version of $x(k)$ using $M + p$ zeros, and $y_z(k)$ is a zero-padded version of $y(k)$ using $L + p$ zeros. Then $x_z(k)$ and $y_z(k)$ are both of length $N = L + M + p$. If $p \geq -1$, then the circular cross-correlation of the zero-padded signals is the same as the linear cross-correlation of the original signals for $0 \leq k < L$. Suppose p is chosen, such that $N = L + M + p$ is a power of two using Eq. 4.100. Then a radix-two FFT can be used in place of the DFT in Eq. 4.108, and this leads to the following highly efficient way to compute linear cross-correlations, called *fast linear cross-correlation*.

$$r_{xy}(k) = \left(\frac{N}{L}\right) \text{IFFT}\{X_z(i)Y_z^*(i)\}, \quad 0 \leq k < L \tag{4.109}$$

For convenience, suppose $L = M$. Then using Eq. 4.63 and Eq. 4.64, the ratio of the computational effort for fast linear cross-correlation to the computational effort for direct linear cross-correlation, measured in real FLOPs, is

$$r \approx \frac{24 \log_2(2L)}{L}, \quad L \gg 1 \tag{4.110}$$

This ratio is less than one for $L \geq 256$ and goes to zero as $L \to \infty$. Thus, fast convolution is superior to direct linear convolutions for signals of length $L \geq 256$.

Auto-correlation

An important special case of cross-correlation arises when one takes the cross-correlation of a signal with itself. This is referred to as *auto-correlation*. Auto-correlation can be linear, as in $r_{xx}(k)$, or circular, as in $c_{xx}(k)$. When the auto-correlation of $x(k)$ is evaluated at a lag of $k = 0$, this yields the average power P_x. Thus, the average power can be expressed as

$$P_x = r_{xx}(0) = c_{xx}(0) \tag{4.111}$$

There is also a simple and elegant relationship between circular auto-correlation and the power density spectrum. The DFT of the circular auto-correlation of $x(k)$ is just the power density spectrum of $x(k)$.

$$S_N(i) = C_{xx}(i) \tag{4.112}$$

White noise has a particularly simple auto-correlation. If $v(k)$ is a zero-mean white-noise signal with average power P_v, then the circular auto-correlation of $v(k)$ can be approximated as a impulse of strength P_v at $k = 0$.

$$c_{vv}(k) \approx P_v \delta(k), \quad 0 \leq k < N \tag{4.113}$$

The approximation in Eq. 4.113 also holds when linear auto-correlation is used. If we take the DFT of both sides of Eq. 4.113 and use Eq. 4.112, we find that white noise has a *flat* power density spectrum

$$S_N(i) \approx P_v \tag{4.114}$$

Circular correlation can also be used to extract periodic signals from noise. Suppose $x(k)$ is an N-point signal that is periodic with period $M \ll N$. We can then represent a noisy version of $x(k)$ as follows, where $v(k)$ is zero-mean white noise.

$$y(k) = x(k) + v(k) \tag{4.115}$$

The circular auto-correlation of a periodic signal is itself periodic with the same period. Since auto-correlation tends to average out the effects of zero-mean noise, this means that the auto-correlation of the noisy signal $y(k)$ can be used to estimate the period of $x(k)$. Once the period M is known, the periodic signal itself can be extracted from the noise by using a circular cross-correlation with a periodic impulse train $\delta_M(k)$ of period M. If L denotes the number of complete cycles of $x(k)$ in the N-point signal $y(k)$, then $x(k)$ can be approximated by the estimate $\hat{x}(k)$ where

$$\hat{x}(k) = \left(\frac{N}{L}\right) c_{y\delta_M}(k), \quad 0 \le k < N \tag{4.116}$$

• • • • • • • • • • • • • •

4.10 Problems

The problems are divided into Analysis problems that can be solved by hand or with a calculator, GUI Simulation problems that are solved using GUI module *g_correlate*, and Computation problems. The Computation problems require the student to write a MATLAB script using the FDSP toolbox functions summarized in Appendix B. Solutions to selected problems are available on the distribution CD. Students are encouraged to use these problems, which are identified with a ☑, as a check on their understanding of the material.

4.10.1 Analysis

Section 4.2

P4.1. Using the definition of linear convolution show that

$$h(k) \star \delta(k) = h(k)$$

P4.2. Suppose $h(k)$ and $x(k)$ are defined as follows.

$$h = [2, -1, 0, 4]^T$$
$$x = [5, 3, -7, 6]^T$$

(a) Let $y_c(k) = h(k) \circ x(k)$. Find the circular convolution matrix $C(x)$ such that $y_c = C(x)h$.

(b) Use $C(x)$ to find $y_c(k)$.

P4.3. Suppose $h(k)$ and $x(k)$ are the following signals of length L and M, respectively.

$$h = [3, 6, -1]^T$$

$$x = [2, 0, -4, 5]^T$$

(a) Let h_z and x_z be zero-padded versions of $h(k)$ and $x(k)$ of length $N = L + M - 1$. Construct h_z and x_z.

(b) Let $y_c(k) = h_z(k) \circ x_z(k)$. Find the circular convolution matrix $C(x_z)$, such that $y_c = C(x_z)h_z$.

(c) Use $C(x_z)$ to find $y_c(k)$.

(d) Use $y_c(k)$ to find the linear convolution $y(k) = h(k) \star x(k)$ for $0 \le k < N$.

P4.4. Consider the linear discrete-time system shown in Figure P4.4. Here $H(z)$ is driven by an input $x(k)$ for $0 \le k < L$ and produces a zero-state output $y(k)$. Use deconvolution to find the impulse response $h(k)$ for $0 \le k < L$ if $x(k)$ and $y(k)$ are as follows.

$$x = [2, 0, -1, 4]^T$$

$$y = [6, 1, -4, 3]^T$$

P4.4 A Linear Discrete-time System

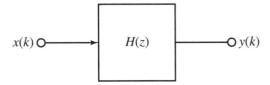

$x(k) \circ\!\!\longrightarrow\quad H(z) \quad\longrightarrow\!\!\circ\, y(k)$

Section 4.3

☑ **P4.5.** Suppose $h(k)$ and $x(k)$ are both of length $L = 2048$.

(a) Find the number of real FLOPs for a fast linear convolution of $h(k)$ with $x(k)$.

(b) Find the number of real FLOPs for a direct linear convolution of $h(k)$ with $x(k)$.

(c) Express the answer to (a) as a percentage of the answer to (b).

P4.6. Suppose $h(k)$ is of length L, and $x(k)$ is of length M. Let L and M be powers of two with $M \ge L$.

(a) Find the number of real FLOPs for a fast linear convolution of $h(k)$ with $x(k)$. Does your answer agree with Eq. 4.39 when $M = L$?

(b) Using Eq. 4.15, find the number of real FLOPs for a direct linear convolution of $h(k)$ with $x(k)$. Does your answer agree with Eq. 4.40 when $M = L$?

P4.7. Suppose L is a power of two, and $M = QL$ for some positive integer Q. Let n_{block} be the number of real FLOPs needed to compute a fast block convolution of an L-point signal $h(k)$ with an M-point signal $x(k)$. Find n_{block}.

Section 4.4

P4.8. Some books use the following alternative way to define the linear cross-correlation of an L-point signal $x(k)$ with an M-point signal $y(k)$. Using a change of variable, show that this is equivalent to Definition 4.3.

$$r_{xy}(k) = \frac{1}{L} \sum_{n=0}^{L-1-k} x(n+k)y(n)$$

P4.9. Suppose $x(k)$ and $y(k)$ are defined as follows.

$$x = [1, 0, -2, 3, 5, 4]^T$$
$$y = [3, -11]^T$$

(a) Find the linear cross-correlation matrix $D(y)$, such that $r_{xy} = D(y)x$.

(b) Use $D(y)$ to find the linear cross-correlation $r_{xy}(k)$.

(c) Find the normalized linear cross-correlation $\rho_{xy}(k)$.

☑ **P4.10.** Suppose $x(k)$ is as follows.

$$x = [5, 7, -2, 4, 8, 6, 1]^T$$

(a) Construct a 3-point signal $y(k)$, such that $r_{xy}(k)$ reaches its peak positive value at $k = 3$ and $|y(0)| = 1$.

(b) Construct a 4-point signal $y(k)$, such that $r_{xy}(k)$ reaches its peak negative value at $k = 4$ and $|y(0)| = 1$.

Section 4.5

P4.11. Suppose $x(k)$ and $y(k)$ are defined as follows.

$$x = [2, 3, 1, -1]^T$$
$$y = [4, 0, -12, 8]^T$$

(a) Find a circular cross-correlation matrix $E(y)$, such that $c_{xy} = E(y)x$.

(b) Use $E(y)$ to find the circular cross-correlation $c_{xy}(k)$.

(c) Find the normalized circular cross-correlation $\sigma_{xy}(k)$.

P4.12. Suppose $x(k)$ is as follows.

$$x = [8, 2, -3, 4, 5, 7]^T$$

(a) Construct a 6-point signal $y(k)$, such that $\sigma_{xy}(2) = 1$ and $|y(0)| = 6$.

(b) Construct a 6-point signal $y(k)$, such that $\sigma_{xy}(3) = -1$ and $|y(0)| = 12$.

P4.13. Use the DFT to solve the following.

    (a) Recover $x(k)$ from $c_{xy}(k)$ and $y(k)$.

    (b) Recover $y(k)$ from $c_{xy}(k)$ and $x(k)$.

P4.14. Suppose $x(k)$ and $y(k)$ are both of length $L = 4096$.

    (a) Find the number of real FLOPs for a fast linear cross-correlation of $x(k)$ with $y(k)$.

    (b) Find the number of real FLOPs for a direct linear cross-correlation of $x(k)$ with $y(k)$.

    (c) Express the answer to (a) as a percentage of the answer to (b).

P4.15. Suppose $x(k)$ is of length L, and $y(k)$ is of length $M \leq L$.

    (a) Find the number of real FLOPs for a fast linear cross-correlation of $x(k)$ with $y(k)$. Does your answer agree with Eq. 4.63 when $M = L$?

    (b) Find the number of real FLOPs for a direct linear cross-correlation of $x(k)$ with $y(k)$. Does your answer agree with Eq. 4.64 when $M = L$?

Section 4.6

P4.16. Consider the following discrete-time signal.

$$x = [10, -5, 20, 0, 15]^T$$

    (a) Find a linear auto-correlation matrix $D(x)$, such that $r_{xx} = D(x)x$.

    (b) Use $D(x)$ to find the linear auto-correlation $r_{xx}(k)$.

    (c) Find the normalized linear auto-correlation $\rho_{xx}(k)$.

    (d) Find the average power P_x.

P4.17. Consider the following discrete-time signal.

$$x = [12, 4, -8, 16]^T$$

    (a) Find a circular auto-correlation matrix $E(x)$, such that $c_{xx} = E(x)x$.

    (b) Use $E(x)$ to find the circular auto-correlation $c_{xx}(k)$.

    (c) Find the normalized circular auto-correlation $\sigma_{xx}(k)$.

P4.18. A white-noise signal $v(k)$ is uniformly distributed over the interval $[-a, a]$. Suppose $v(k)$ has the following circular auto-correlation.

$$c_{vv}(k) = 8\delta(k), \quad 0 \leq k < 1024$$

    (a) Find the interval bound a.

    (b) Sketch the power density spectrum of $v(k)$.

4.10.2 GUI Simulation

Section 4.8

☑ **P4.19.** Use the GUI module *f_correlate* to record the sequence of vowels "A","E","I","O","U" in x. Play x to make sure you have a good recording of all five vowels. Then record the vowel "O" in y. Play y back to make sure you have a good recording of "O" that sounds similar to the "O" in x. Save this data in a MAT-file named *my_vowels*.

(a) Plot the inputs x and y showing the vowels.

(b) Plot the normalized cross-correlation of x with y using the *Caliper* option to mark the peak which should show the location of y in x.

(c) Based on the plots in (a), estimate the lag d_1 that would be required to get the "O" in y to align with the "O" in x. Compare this with the peak location d_2 in (b). Find the percent error relative to the estimated lag d_1. There will be some error due to the overlap of y with adjacent vowels and coarticulation effects in creating x.

P4.20. Use the GUI module *f_correlate* and the User-defined option to load the MAT-file *prob4_20*.

(a) What sentence does $x(k)$ contain?

(b) Plot $x(k)$ and $y(k)$.

(c) Plot the normalized linear cross-correlation of $x(k)$ with $y(k)$.

(d) Plot the power density spectrum of $x(k)$ by taking the DFT of $c_{xx}(k)$. The peaks in $S_L(f)$ are called *formants*.

P4.21. Using the GUI module *f_correlate*, select the periodic input.

(a) Plot $x(k)$ and $y(k)$.

(b) Plot the normalized circular auto-correlation, $\sigma_{xx}(k)$. Notice how the noise has been reduced.

(c) Estimate the period of $x(k)$ in seconds by using the *Caliper* option to estimate the period of σ_{xx}.

☑ **P4.22.** Using the GUI module *f_correlate*, select the white-noise input. Set the scale factor for y to $c = 0$.

(a) Plot $x(k)$ and $y(k)$. What is the range of values over which the uniform white noise is distributed?

(b) Verify that $r_{xx}(k) \approx P_x \delta(k)$ by plotting the auto-correlation of $x(k)$.

(c) Use the *Caliper* option to estimate P_x.

(d) Verify that this estimate of P_x is consistent with the theoretical value, P_u, in Eq. 3.85.

P4.23. The file *prob4_23.mat* contains two signals, x and y, and their sampling frequency f_s. Use the GUI module *g_correlate* to load x, y, and f_s.

(a) Plot $x(k)$ and $y(k)$.

(b) Plot the normalized linear cross-correlation $\rho_{xy}(k)$. Does $x(k)$ contain any scaled and shifted versions of $y(k)$? Determine how many, and use the *Caliper* option to estimate the locations of $y(k)$ within $x(k)$.

P4.24. Using the GUI module *f_correlate*, select the impulse-train input. This sets $x(k)$ to a periodic input, and $y(k)$ to an impulse train whose period matches the period of $x(k)$. Set $L = 4096$ and $M = 4096$.

(a) Plot the noise-corrupted periodic input $x(k)$, and the impulse train input $y(k)$.

(b) Plot the normalized circular auto-correlation of $x(k)$.

(c) Plot the normalized circular cross-correlation $\sigma_{xy}(k)$. This should be proportional to $x(k)$, but with the noise reduced.

P4.25. The file *prob4_25.mat* contains two signals, x and y, and their sampling frequency f_s. Use the GUI module *g_correlate* to load x, y, and f_s.

(a) The signal $x(k)$ is the input to a discrete-time system, and the signal $y(k)$ is the impulse response of the system. Plot $x(k)$ and $y(k)$.

(b) Plot the zero-state response of the discrete-time system to the input $x(k)$.

4.10.3   Computation

Section 4.2

P4.26. Consider the following FIR filter. Write a MATLAB script that performs the following tasks.

$$H(z) = \sum_{i=0}^{20} \frac{(-1)^i z^{-i}}{10 + i^2}$$

(a) Use the function *filter* to compute and plot the impulse response $h(k)$ for $0 \le k < N$, where $N = 50$.

(b) Compute and plot the following periodic input.

$$x(k) = \sin(0.1\pi k) - 2\cos(0.2\pi k) + 3\sin(0.3\pi k), \quad 0 \le k < N$$

(c) Use the FDSP toolbox function *f_conv* to compute the zero-state response to the input $x(k)$ using convolution. Also compute the zero-state response to $x(k)$ using *filter*. Plot both responses on the same graph using a legend.

P4.27. Consider the following pair of signals.

$$h = [1, 2, 3, 4, 5, 4, 3, 2, 1]^T$$

$$x = [2, -1, 3, 4, -5, 0, 7, 9, -6]^T$$

Verify that linear convolution and circular convolution produce different results by writing a script that uses the FDSP function f_conv to compute the linear convolution $y(k) = h(k) \star x(k)$ and the circular convolution $y_c(k) = h(k) \circ x(k)$. Plot $y(k)$ and $y_c(k)$ below one another on the same screen.

P4.28. Consider the following pair of signals.

$$h = [1, 2, 4, 8, 16, 8, 4, 2, 1]^T$$

$$x = [2, -1, -4, -4, -1, 2]^T$$

Verify that linear convolution can be achieved by zero padding and circular convolution by writing a MATLAB script that pads these signals with an appropriate number of zeros and uses the FDSP toolbox function f_conv to compare the linear convolution $y(k) = h(k) \star x(k)$ with the circular convolution $y_{zc}(k) = h_z(k) \circ x_z(k)$. Plot the following.

(a) The zero-padded signals $h_z(k)$ and $x_z(k)$ on the same graph using a legend.

(b) The linear convolution $y(k) = h(k) \star x(k)$.

(c) The zero-padded circular convolution $y_{zc}(k) = h_z(k) \circ x_z(k)$.

P4.29. Consider the following two polynomials.

$$A(z) = z^4 + 4z^3 + 2z^2 - z + 3$$

$$B(z) = z^3 - 3z^2 + 4z - 1$$

Let $a = [1, 4, 2, -1, 3]^T$ and $b = [1, -3, 4, -1]^T$ be the coefficient vectors of the polynomials $A(z)$ and $B(z)$, respectively. The coefficients of the *product* polynomial $C(z) = A(z)B(z)$ can be obtained by computing the linear convolution of the two coefficient vectors.

$$c(k) = a(k) \star b(k)$$

Once $c(k)$ is known, the coefficient vector a can be recovered from b and c using deconvolution. Thus, convolution performs polynomial multiplication, and deconvolution performs polynomial division.

(a) Find the coefficient vector of $C(z) = A(z)B(z)$ by direct multiplication by hand.

(b) Write a MATLAB script that uses *conv* to find the coefficient vector of $C(z)$ by computing c as the linear convolution of a with b.

(c) In the script, show that a can be recovered from b and c by using the MATLAB function *deconv* to perform deconvolution.

Section 4.3

☑ **P4.30.** Let $h(k)$ and $x(k)$ be two N-point white-noise signals uniformly distributed over $[-1, 1]$. Recall that the MATLAB function *conv* can be used to compute direct linear convolution. Write a MATLAB script which uses *tic* and *toc* to compute the computational time, t_{dir}, of *conv* and the computational time, t_{fast}, of the FDSP toolbox function *f_conv* for the cases $N = 4096$, $N = 8192$, and $N = 16384$.

    (a) Print the two computational times t_{dir} and t_{fast} for $N = 4096, 8192$, and 16384.

    (b) Plot t_{dir} versus $N/1024$ and t_{fast} versus $N/1024$ on the same graph and include a legend.

P4.31. Consider the following linear discrete-time system. Write a MATLAB script that performs the following tasks.

$$H(z) = \frac{z}{z^2 - 1.4z + 0.98}$$

    (a) Compute and plot the impulse response $h(k)$ for $0 \le k < L - 1$, where $L = 500$.

    (b) Construct an M-point white-noise input $x(k)$ that is distributed uniformly over $[-5, 5]$, where $M = 10000$. Use the FDSP toolbox function *f_blockconv* to compute the zero-state response $y(k)$ to the input $x(k)$ using block convolution. Plot $y(k)$ for $9500 \le k < 10000$.

    (c) Print the number of FFTs and the lengths of the FFTs used to perform the block convolution.

Section 4.5

P4.32. Consider the following pair of signals.

$$x = [3, 2, 1, 0, -1, -2, -3, -2, -1, 0, 1, 2]^T$$
$$y = [2, -4, 3, 7, 6, 1, 9, 4, -3, 2, 7, 8]^T$$

Verify that linear cross-correlation and circular cross-correlation produce different results by writing a MATLAB script that uses the FDSP function *f_corr* to compute the linear cross-correlation, $r_{xy}(k)$ and the circular cross-correlation, $c_{xy}(k)$. Plot $r_{xy}(k)$ and $c_{xy}(k)$ below one another on the same screen.

☑ **P4.33.** Consider the following pair of signals.

$$x = [1, 2, 4, 8, 16, 8, 4, 2, 1]^T$$
$$y = [2, -1, -4, -4, -1, 2]^T$$

Verify that linear cross-correlation can be achieved by zero-padding and circular cross-correlation by writing a MATLAB script that pads these signals with an appropriate number of zeros, and uses the FDSP toolbox function f_corr to compute the linear cross-correlation $r_{xy}(k)$ and the circular cross-correlation $c_{x_z y_z}(k)$. Plot the following.

(a) The zero-padded signals $x_z(k)$ and $y_z(k)$ on the same graph using a legend.

(b) The linear cross-correlation $r_{xy}(k)$ and the scaled zero-padded circular cross-correlation $(N/L)c_{x_z y_z}(k)$ on the same graph using a legend.

P4.34. Consider the following pair of signals of length $N = 8$.

$$x = [3, 1, -5, 2, 4, 9, 7, 0]^T$$
$$y = [2, -4, 7, 3, 8, -6, 5, 1]^T$$

Write a MATLAB script that performs the following tasks.

(a) Use the FDSP toolbox function f_corr to compute and plot the circular cross-correlation, $c_{xy}(k)$.

(b) Compute and print $v(k) = y(-k)$ using the periodic extension, $y_p(k)$.

(c) Verify that $c_{xy}(k) = [x(k) \circ y(-k)]/N$ by using the FDSP toolbox function f_conv to compute and plot the scaled circular convolution, $w(k) = [x(k) \circ v(k)]/N$. Plot $c_{xy}(k)$ and $w(k)$ on separate graphs.

Section 4.6

P4.35. Let $x(k)$ be an N-point white-noise signal uniformly distributed over $[-1, 1]$ where $N = 4096$. Write a script that performs the following operations.

(a) Create $x(k)$, and then compute and plot the normalized circular auto-correlation, $\sigma_{xx}(k)$.

(b) Compute $c_{xx}(k)$, and use the result to compute and plot the power density spectrum of $x(k)$.

(c) Compute and print the average power P_x.

Section 4.7

P4.36. Consider the following N-point periodic signal of period M. Suppose $M = 128$ and $N = 1024$.

$$x(k) = 1 + 3\cos\left(\frac{2\pi k}{M}\right) - 2\sin\left(\frac{4\pi k}{M}\right), \quad 0 \le k < N$$

Let $y(k)$ be a noise-corrupted version of $x(k)$, where $v(k)$ is white noise uniformly distributed over $[-.5, .5]$.

$$y(k) = x(k) + v(k), \quad 0 \le k < N$$

The objective of this problem is to study how *sensitive* the periodic signal extraction technique is to the estimate of the period M.

$$\hat{x}_m(k) = \left(\frac{N}{L}\right) c_{y\delta_m}(k)$$

Write a script which performs the following tasks.

(a) Compute and plot the noise-corrupted periodic signal $y(k)$.

(b) Compute and plot on the same graph $x(k)$ and $\hat{x}_m(k)$ for $m = M - 5$ using a legend.

(c) Compute and plot on the same graph $x(k)$ and $\hat{x}_m(k)$ for $m = M$ using a legend.

(d) Compute and plot on the same graph $x(k)$ and $\hat{x}_m(k)$ for $m = M + 5$ using a legend.

CHAPTER 5

Filter Specifications and Structures

Learning Objectives

- Know how to specify the design characteristics of frequency-selective filters (Section 5.2)

- Understand what a linear-phase filter is and how to construct FIR linear-phase filters (Section 5.3)

- Know how to decompose a general transfer function into its minimum-phase and allpass factors (Section 5.4)

- Know how to find direct, cascade, parallel, and lattice-form realizations of FIR and IIR filters (Sections 5.5–5.6)

- Be able to estimate the finite word length effects of FIR and IIR filters, and know which filter structures are less sensitive to these effects (Sections 5.7–5.8)

- Know how to model signal and coefficient quantization error using white noise (Sections 5.7–5.8)

- Know how to use the GUI module *g_filters* to examine design specifications, compute filter realization structures, and evaluate finite word length effects (Section 5.9)

● ● ● ● ● ● ● ● ● ● ● ● ● ● ● ●

5.1 Motivation

The remaining chapters focus primarily on the design and application of various types of digital filters. To lay a foundation for filter design, it is helpful to first consider certain fundamental characteristics that filters have in common. One characteristic of frequency-selective filters is that they are constructed to meet certain *design specifications*. For example, the design specifications dictate which frequencies or spectral components of the input are passed by the filter, which are rejected, and the degree to which rejected frequencies are blocked by the filter. In addition, each filter can be realized physically in hardware, or mathematically in software, using any one of several *filter structures*. For example, both FIR and IIR transfer functions can be realized with the following cascade structure, where the M blocks are second-order subsystems.

Cascade-form realization

$$H(z) = b_0 H_1(z) H_2(z) \cdots H_M(z)$$

There are several other filter realization structures, including direct forms, the parallel form, and the lattice form. When infinite-precision arithmetic is used, all of the different filter structures are equivalent in terms of their input-output characteristics. However, when finite-precision arithmetic is used to implement the filter, some of the structures are superior to others in terms of their sensitivity to detrimental finite-word length effects. These effects include ADC and coefficient quantization error, roundoff error, numerical overflow, and limit-cycle oscillations.

We begin this chapter by introducing examples of filter specifications and structures. The filter design problem is then formulated by presenting a set of filter design specifications, both linear and logarithmic. Next the notion of a linear-phase filter is introduced, and four types of FIR linear-phase filters are presented. This is followed by an investigation that reveals that a general filter can be decomposed into a minimum-phase part, whose phase lag is as small as possible, and an allpass part, whose magnitude response is constant. The discussion then turns to alternative structures for realizing digital FIR filters. Several direct-form structures are presented, which use parameters that are obtained directly from inspection of the transfer function. This is followed by some indirect structures, including the cascade-form and lattice-form realizations, both of which are based on factorization. Filter realization structures for IIR filters are then investigated, including direct forms, the parallel form based on partial fractions, and the cascade form. Once the different filter realization structures are in place, the effects of finite-precision arithmetic can be investigated. Finite word length effects on both FIR and IIR filters are examined, including ADC quantization error, coefficient quantization error, roundoff error, scaling to avoid overflow, and limit-cycle oscillations. It is shown that some filter realization structures are less sensitive to finite word length effects than others. Finally, a GUI module called *g_filters* is introduced that allows the user to construct filters from design specifications and examine finite word length effects, all without any need for programming. The chapter concludes with a presentation of an application example, and a summary of filter design specifications, realization structures, and finite word length effects. The selected application example examines the effects of finite word length on an elliptic highpass filter.

5.1.1 Filter Design Specifications

Perhaps the most common type of digital filter is a lowpass filter. A digital lowpass filter is a filter that removes the higher frequencies but passes the lower frequencies. An example of a magnitude response of a digital lowpass filter is shown in Figure 5.1. This is a fourth-order Chebyshev-I filter. The design of FIR filters is discussed in Chapter 6, and the design of IIR filters, including Chebyshev filters, is discussed in Chapter 8.

Passband

The shaded areas in Figure 5.1 represent the design specifications. The shaded region in the upper-left corner represents the filter passband, and the shaded region in the lower-right corner represents the filter stopband. Notice that the passband has width F_p and height δ_p. That is, the desired magnitude response must meet or exceed the following *passband specification*.

$$1 - \delta_p \leq A(f) \leq 1, \quad 0 \leq f \leq F_p \tag{5.1}$$

Here $0 < F_p < f_s/2$ is the *passband cutoff frequency*, and $\delta_p > 0$ is the *passband ripple*. The passband ripple can be made small, but must be positive for a physically

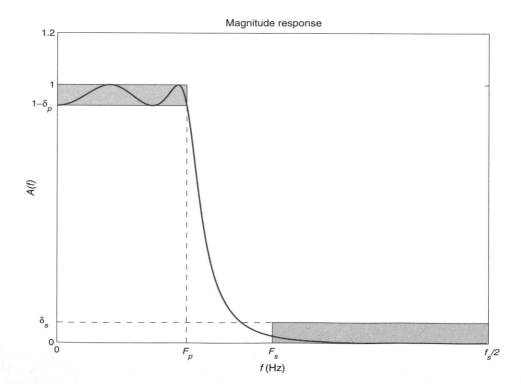

Figure 5.1 Magnitude Response of a Lowpass Chebyshev-I Filter of Order $n = 4$

realizable filter. It is called a ripple factor because the magnitude response sometimes oscillates within the passband, as shown in Figure 5.1. However, for some filters such as Butterworth filters and Chebyshev-II filters, the magnitude response decreases monotonically within the passband. For the filter in Figure 5.1, the passband cutoff frequency is $F_p/f_s = 0.15$, and the passband ripple is $\delta_p = 0.08$.

Stopband

Similar to the passband, the shaded stopband region in the lower-right corner of Figure 5.1 has width $f_s/2 - F_s$ and height δ_s. Thus, the desired magnitude response must meet or exceed the following *stopband specification*.

$$0 \leq A(f) \leq \delta_s, \quad F_s \leq f \leq f_s/2 \tag{5.2}$$

It is evident from Figure 5.1 that the passband specification is met exactly, whereas the stopband specification is exceeded in this case. Here $F_p < F_s < f_s/2$ is the *stopband cutoff frequency*, and $\delta_s > 0$ is the *stopband attenuation*. Again, the stopband attenuation can be made small, but must be positive for a physically realizable filter. For the filter in Figure 5.1, the stopband cutoff frequency is $F_s/f_s = 0.25$, and the stopband attenuation is $\delta_s = 0.08$.

Transition Band

Notice that there is a significant part of the spectrum that is left unspecified. The frequency band, $[F_p, F_s]$ between the passband and the stopband is called the *transition band*. The width of the transition band can be made small, but it must be positive for a physically realizable filter. Indeed, as the passband ripple, the stopband attenuation, and the transition bandwidth all approach zero, the required order of the filter approaches infinity. The limiting special case of the filter with $\delta_p = 0$, $\delta_s = 0$, and $F_s = F_p$ is an ideal lowpass filter.

5.1.2 Filter Realization Structures

Each digital filter has a number of alternative realizations depending on which filter structure is used. The different filter realizations are equivalent to one another as long as infinite-precision arithmetic is used. To illustrate some filter realization structures, consider the fourth-order lowpass filter in Figure 5.1. Using design techniques covered in Chapter 8, the transfer function of this Chebyshev-I lowpass filter is

$$H(z) = \frac{0.0095 + 0.0379z^{-1} + 0.0569z^{-2} + 0.0379z^{-3} + 0.0095z^{-4}}{1 - 2.2870z^{-1} + 2.5479z^{-2} - 1.4656z^{-3} + 0.3696z^{-4}} \tag{5.3}$$

Direct Form I

Filter realization structures can be represented graphically using signal flow graphs as discussed in Chapter 2. For example, a direct form I realization of $H(z)$ is shown in Figure 5.2. Recall that the nodes are summing junctions, and arcs without labels have a default gain of unity. Notice that the gains of the branches correspond directly to the coefficients of the numerator and denominator polynomials of $H(z)$. This is a characteristic of direct-form realizations that sets them apart from the indirect forms.

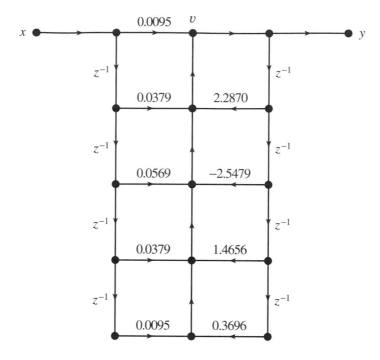

Figure 5.2 Signal Flow Graph of a Direct Form I Realization of the Fourth-order Chebyshev-I Filter

Filter realizations can also be represented mathematically using the difference equations associated with the signal flow graph. The direct form I filter realization shown in Figure 5.2 can be implemented with the following pair of difference equations, where $v(k)$ is an intermediate variable.

$$v(k) = 0.0095x(k) + 0.0379x(k-1) + 0.0569x(k-2)+$$

$$0.0379x(k-3) + 0.0095x(k-4) \tag{5.4a}$$

$$y(k) = v(k) + 2.2870y(k-1) - 2.5479y(k-2)+$$

$$1.4656y(k-3) - 0.3696y(k-4) \tag{5.4b}$$

Cascade Form

To develop an alternative to the direct form I realization in Figure 5.2, we first recast the transfer function in 5.43 in terms of positive powers of z which yields

$$H(z) = \frac{0.0095z^4 + 0.0379z^3 + 0.0569z^2 + 0.0379z + 0.0095}{z^4 - 2.2870z^3 + 2.5479z^2 - 1.4656z + 0.3696} \tag{5.5}$$

The factor 0.0095 can be removed from the numerator terms. The resulting numerator polynomial and denominator polynomial can then be factored into zeros and poles. Suppose complex-conjugate pairs of zeros are grouped together (similarly for complex-conjugate pairs of poles). The transfer function can then be written as a product of two second-order transfer functions, each with real coefficients. This is called a cascade-form realization.

$$H(z) = 0.0095\,H_1(z)H_2(z) \tag{5.6}$$

There are several possible formulations of the two second-order blocks depending on how the zeroes and poles are ordered and grouped together. Using techniques covered later in this chapter, one such ordering is

$$H_1(z) = \frac{1 + 2z^{-1} + z^{-2}}{1 - 1.0328z^{-1} + 0.7766z^{-2}} \tag{5.7}$$

$$H_2(z) = \frac{1 + 2z^{-1} + z^{-2}}{1 - 1.2542z^{-1} + 0.4759z^{-2}} \tag{5.8}$$

Each of the second-order blocks can be realized using one of the direct forms. For example, a signal flow graph which uses direct form II realizations for the two blocks is shown in Figure 5.3.

As with the direct-form realization, the signal flow graph of the cascade realization can be implemented with a system of difference equations as follows.

$$w_0(k) = 0.0095x(k) \tag{5.9a}$$

$$w_1(k) = w_0(k) + 2w_0(k-1) - w_0(k-2) + 1.0328w_1(k-1) - 0.7766w_1(k-2) \tag{5.9b}$$

$$w_2(k) = w_1(k) + 2w_1(k-1) - w_1(k-2) + 1.2542w_2(k-1) - 0.4749w_2(k-2) \tag{5.9c}$$

$$y(k) = w_2(k) \tag{5.9d}$$

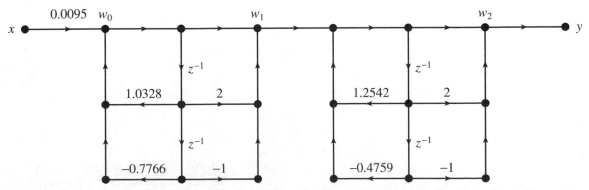

Figure 5.3 Signal Flow Graph of a Cascade-form Realization of the Fourth-order Chebyshev-I Filter using Direct Form II Realizations for the Second-order Blocks

Quantization Error

When a filter is implemented in software using MATLAB, double-precision floating-point arithmetic is used for all calculations. This typically involves 64 bits of precision, which corresponds to about 16 decimal digits for the mantissa or factional part and the remaining bits used to represent the exponent. For convenience of display, only four decimal places are shown in Figures 5.2 and 5.3. In most instances, double-precision arithmetic is a good approximation to infinite-precision arithmetic, so no significant finite word length effects are apparent. However, if a filter is implemented on specialized DSP hardware, or if storage space or speed requirements dictate the need to use single-precision floating-point arithmetic or integer fixed-point arithmetic, then finite word length effects can begin to manifest themselves.

To illustrate the detrimental effects that limited precision can have, suppose the coefficients of the fourth-order Chebyshev-I lowpass filter in Figure 5.1 are represented using N bits. The resulting magnitude responses for three cases are shown in Figure 5.4. Comparing Figure 5.4 with Figure 5.1, we see that the case using $N = 12$ bits is essentially correct, but the lower-precision cases, $N = 6$ and $N = 9$, have magnitude responses that differ significantly from the double-precision version shown in Figure 5.1. Interestingly enough, if the precision is lowered still further to $N = 4$ bits, the coefficient quantization error becomes so large that the poles of the filter migrate outside the unit circle, at which point the implementation becomes unstable.

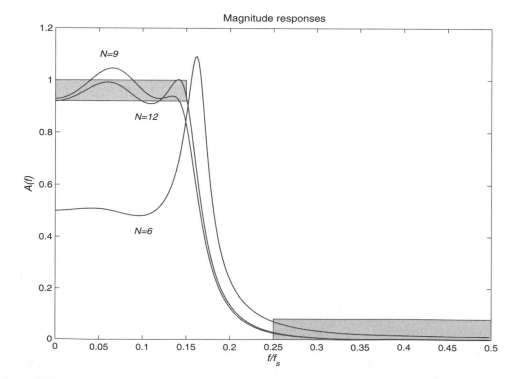

Figure 5.4 Magnitude Responses of Fourth-order Chebyshev-I Lowpass Filter using N Bits of Precision to Represent the Coefficients

- - - - - - - - - - - - - - -

5.2 Filter Design Specifications

A digital filter is a discrete-time system that reshapes the spectrum of the input signal to produce desired spectral characteristics in the output signal. Recall from Definition 2.8 that a filter with transfer function, $H(z)$, and sampling frequency, $f_s = 1/T$, has the following *frequency response*.

Frequency response

$$H(f) \triangleq H(z)|_{z=\exp(j2\pi fT)}, \quad 0 \le f \le \frac{f_s}{2} \tag{5.10}$$

Thus, the frequency response is just the transfer function evaluated along the top half of the unit circle. The complex-valued function $H(f)$ can be expressed in polar form as $H(f) = A(f)\exp[j\phi(f)]$, where $A(f)$ denotes the *magnitude response* and $\phi(f)$ denotes the *phase response* of the filter.

$$A(f) \triangleq |H(f)|, \quad 0 \le f \le \frac{f_s}{2} \tag{5.11a}$$

$$\phi(f) \triangleq \angle H(f), \quad 0 \le f \le \frac{f_s}{2} \tag{5.11b}$$

In Chapter 2, it was shown that if the system $H(z)$ is stable, then the steady-state response to the sinusoidal input $x(k) = \sin(2\pi f_a kT)$ is

$$y(k) = A(f_a)\sin[2\pi f_a kT + \phi(f_a)] \tag{5.12}$$

Thus, the magnitude response $A(f_a)$ can be interpreted as the *gain* of the filter at frequency f_a. It specifies the amount by which a sinusoidal signal of frequency f_a is scaled as it passes through the filter. Similarly, the phase response $\phi(f_a)$ can be interpreted as the *phase shift* of the filter at frequency f_a. It specifies the number of radians by which a sinusoidal signal of frequency f_a gets advanced as it passes through the filter.

By designing a filter with a specified $A(f)$ or $\phi(f)$, we can control the spectral characteristics of the output signal $y(k)$. Most digital filters are designed to produce a desired magnitude response $A(f)$. However, there are specialized filters, such as allpass filters, that are designed to produce a desired phase response. A particularly useful phase response is a *linear* phase response of the form:

$$\phi(f) = -2\pi\tau f \tag{5.13}$$

Linear-phase filters have the property that each spectral component gets delayed by the same amount, namely, τ seconds. Consequently, the spectral components of the input signal that survive at the filter output are not otherwise distorted. Although it is possible to approximate linear-phase filters in the passband with IIR filters (e.g., with Bessel filters), it turns out that it is much simpler to use FIR filters to design linear-phase filters. Linear-phase FIR filters are discussed in detail in the next section.

5.2.1 Linear Design Specifications

There are many specialized frequency-selective filters that one can consider. However, the most common filters fall into four basic categories: lowpass, highpass, bandpass, and bandstop. The magnitude responses of *ideal* versions of the four basic filter types are shown in Figure 5.5.

Note that in each case, the upper-frequency limit is $f = f_s/2$ because this is the highest frequency that a digital filter can process. Recall from Chapter 1 that analog signals at higher frequencies get aliased back into the range $[0, f_s/2]$ during the sampling process. The range of frequencies over which $A(f) = 1$ is called the *passband*, and the range of frequencies over which $A(f) = 0$ is called the *stopband*. One of the advantages of digital filters is that the passband gain can be set to a value greater than one, if desired, in which case the signal is amplified in the passband. Analog filters can also have passbands with gains greater than one, but in this case they must be implemented as active filters, rather than passive filters.

The filters in Figure 5.5 are idealized in several respects. For a practical filter, it is not possible to go abruptly from the passband to the stopband without a *transition band* separating the two. In addition, it is not possible for a physically realizable filter to completely attenuate a signal throughout the stopband (Weiner and Payley,

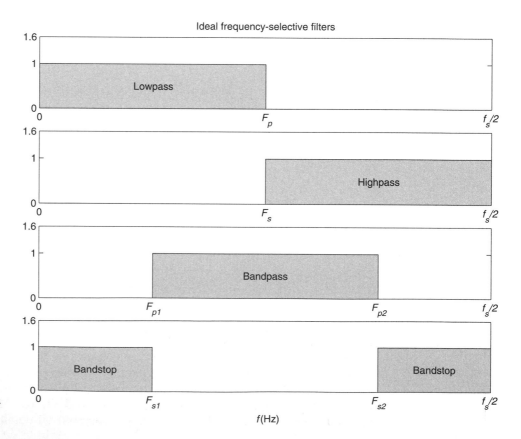

Figure 5.5 Ideal Magnitude Responses of the Four Basic Filter Types

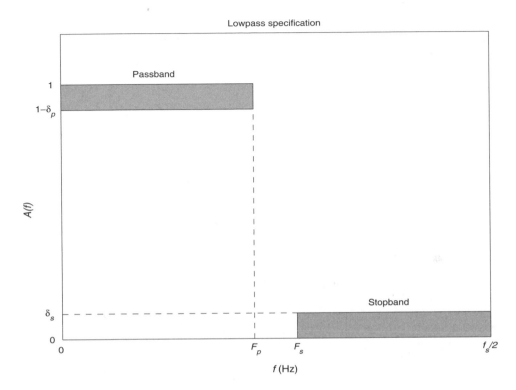

Figure 5.6 Linear Magnitude Response Specifications for a Lowpass Filter

1934). Indeed, the idealized filters in Figure 5.5 all have impulse responses $h(k)$ that are noncausal.

A more realistic design specification for the magnitude response of a lowpass filter is shown in Figure 5.6. There are two differences worth noting. First, both the passband and the stopband are specified by a *range* of acceptable values for the desired magnitude response. The parameter δ_p is called the *passband ripple* because the magnitude response often oscillates within the passband. Similarly, δ_s is called the *stopband attenuation*. The passband ripple and stopband attenuation can be made small, but not zero. The second difference is that there is a transition band of width $F_s - F_p$ between the passband and the stopband. Again, the width of the transition band can be made small (at the expense of the filter order), but not zero.

The desired magnitude response must fall within the shaded area in Figure 5.6. Note how the ideal cutoff F_p in Figure 5.5 has been split into two cutoff frequencies, F_p and F_s, to create a transition band in Figure 5.6. A practical design specification for the magnitude response of a highpass filter is shown in Figure 5.7. Again, a transition band has been created by splitting the ideal cutoff frequency F_s into two cutoff frequencies, F_s and F_p.

The design specification for the magnitude response of a bandpass filter is a bit more involved because there are two transition bands bracketing the passband, as can be seen in Figure 5.8. Thus, there are four cutoff frequencies plus a passband ripple δ_p and a stopband attenuation δ_s. The design specification for the magnitude

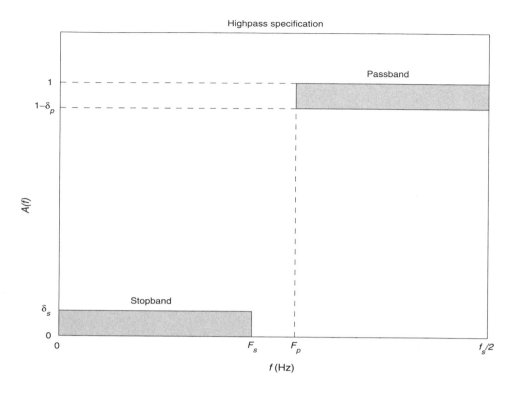

Figure 5.7 Linear Magnitude Response Specifications for a Highpass Filter

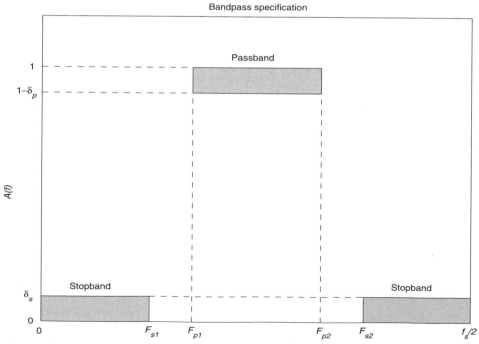

Figure 5.8 Linear Magnitude Response Specifications for a Bandpass Filter

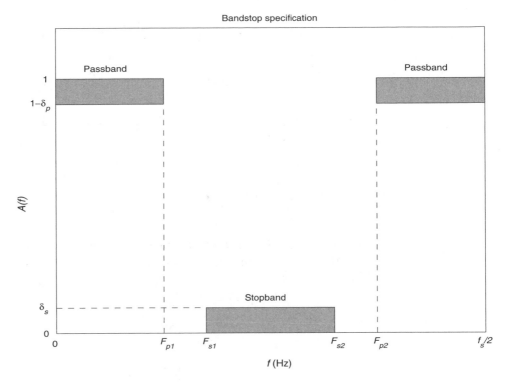

Figure 5.9 Linear Magnitude Response Specifications for a Bandstop Filter

response of a stopband filter also has two transition bands; this time bracketing the stopband, as can be seen in Figure 5.9.

Example 5.1   **Linear Design Specifications**

As a simple illustration of design specifications for a desired magnitude response, consider the following first-order IIR filter.

$$H(z) = \frac{0.5(1-c)(1+z^{-1})}{1-cz^{-1}}$$

Recasting $H(z)$ in terms of positive powers of z, we have

$$H(z) = \frac{0.5(1-c)(z+1)}{z-c}$$

Thus, $H(z)$ has a zero at $z = -1$ and a pole at $z = c$. For the filter to be stable, it is necessary that the pole satisfy $|c| < 1$. Before we compute the complete frequency response, we can evaluate the filter gain at the two ends of the spectrum. Setting $f = 0$ in $z = \exp(2\pi f T)$ yields $z = 1$. Thus, the low-frequency or DC gain of the filter is

$$A(0) = |H(z)|_{z=1}$$

$$= \frac{0.5(1-c)2}{1-c}$$

$$= 1$$

Next, setting $f = f_s/2$ in $z = \exp(2\pi f T)$ yields $z = -1$. Thus, the high-frequency gain of the filter is

$$A(f_s/2) = |H(z)|_{z=-1}$$

$$= \frac{0.5(1-c)0}{-1-c}$$

$$= 0$$

It follows that $H(z)$ is a lowpass filter with a passband gain of one. To make the example specific, suppose $c = 0.5$. Then from Eq. 5.10, the frequency response of this IIR filter is

$$H(f) = H(z)|_{z=\exp(j2\pi f T)}$$

$$= \frac{0.25[\exp(j2\pi f T) + 1]}{\exp(j2\pi f T) - 0.5}$$

$$= \frac{0.25[(\cos(2\pi f T) + 1) + j\sin(2\pi f T)]}{(\cos(2\pi f T) - 0.5) + j\sin(2\pi f T)}$$

The magnitude response of the lowpass filter is then

$$A(f) = |H(f)|$$

$$= \frac{0.25\sqrt{[\cos(2\pi f T) + 1]^2 + \sin^2(2\pi f T)}}{\sqrt{[\cos(2\pi f T) - 0.5]^2 + \sin^2(2\pi f T)}}$$

For this simple first-order filter, suppose the cutoff frequencies for the transition band are taken to be

$$F_p = 0.1 f_s$$

$$F_s = 0.4 f_s$$

The magnitude response decreases monotonically in this case. Consequently, the passband ripple satisfies $1 - \delta_p = A(F_p)$ or

$$\delta_p = 1 - A(F_p)$$

$$= 1 - \frac{0.25\sqrt{[\cos(0.2\pi) + 1]^2 + \sin^2(0.2\pi)}}{\sqrt{[\cos(0.2\pi) - 0.5]^2 + \sin^2(0.2\pi)}}$$

$$= 0.2839$$

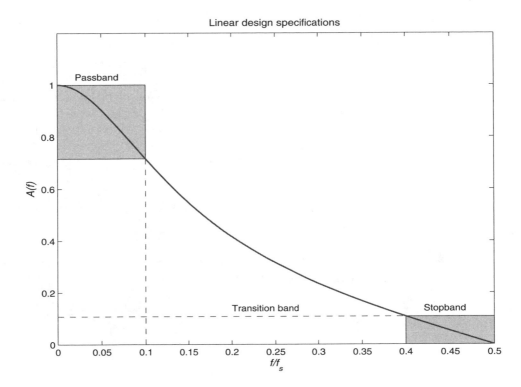

Figure 5.10 Magnitude Response of a First-order IIR Lowpass Filter

Similarly, from Figure 5.6, the stopband attenuation satisfies

$$\delta_s = A(F_s)$$

$$= \frac{0.25\sqrt{[\cos(0.8\pi)+1]^2 + \sin^2(0.8\pi)}}{\sqrt{[\cos(0.8\pi)-0.5]^2 + \sin^2(0.8\pi)}}$$

$$= 0.1077$$

A plot of the magnitude response of the first-order filter is shown in Figure 5.10. For convenience, *normalized frequency*, $\hat{f} = f/f_s$, is used as the independent variable in this case. For this filter, the passband ripple, the stopband attenuation, and the transition bandwidth are all relatively large because this is the lowest-order IIR filter possible.

5.2.2 Logarithmic Design Specifications (dB)

The filter specifications in Figures 5.6 through 5.9 are referred to as *linear* specifications because they are applied to the actual value of $A(f)$. It is also common to use a *logarithmic* specification that represents the value of the magnitude response using the *decibel* or dB scale.

$$A(f) \triangleq 10 \log_{10}\{|H(f)|^2\} \text{ dB} \tag{5.14}$$

A logarithmic design specification for the magnitude response of a lowpass filter is shown in Figure 5.11. Note that the passband ripple in dB is A_p, and stopband attenuation in dB is A_s. The dB scale is useful to show the degree of attenuation in the stopband. Of course, the lowpass specifications in Figures 5.6 and 5.11 are equivalent. Using Eq. 5.14, we find the logarithmic specifications can be expressed in terms of the linear specifications as

Logarithmic specifications

$$A_p = -20 \log_{10}(1 - \delta_p) \text{ dB} \tag{5.15a}$$

$$A_s = -20 \log_{10}(\delta_s) \text{ dB} \tag{5.15b}$$

Similarly, solving Eq. 5.15 for δ_p and δ_s, the linear specifications can be expressed in terms of the logarithmic specifications as

Linear specifications

$$\delta_p = 1 - 10^{-A_p/20} \tag{5.16a}$$

$$\delta_s = 10^{-A_s/20} \tag{5.16b}$$

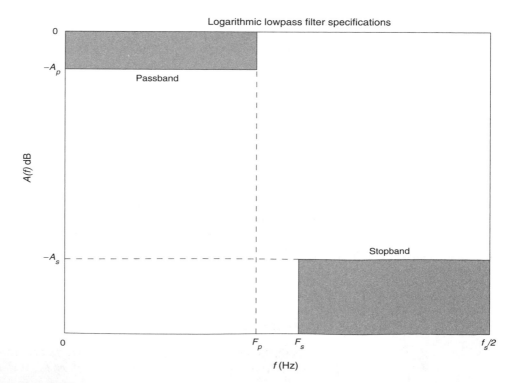

Figure 5.11 Logarithmic Magnitude Response Specifications for a Lowpass Filter

Example 5.2   **Logarithmic Design Specifications**

To facilitate a comparison of the two types of filter design specifications, consider the following first-order IIR system.

$$H(z) = \frac{0.25(1 + z^{-1})}{1 - 0.5z^{-1}}$$

This filter, which was considered in Example 5.1, has a passband cutoff frequency of $F_p = 0.1 f_s$ and a stopband cutoff frequency of $F_s = 0.4 f_s$. Using Eq. 5.15a and the linear passband ripple from Example 5.1, we arrive at the following equivalent passband ripple in dB.

$$A_p = -20 \log_{10}(1 - 0.2839)$$

$$= 2.901 \text{ dB}$$

Next, using Eq. 5.15b and the linear stopband attenuation from Example 5.1, the equivalent logarithmic stopband attenuation is

$$A_s = -20 \log_{10}(0.1077)$$

$$= 9.358 \text{ dB}$$

A plot of the magnitude response in dB can be obtained by running script *exam5_2* on the distribution CD, with the results shown in Figure 5.12. Note that the display range had to be clipped (at -40 dB) because $A(f_s/2) = 0$, which means that $A(f_s/2) = -\infty$ dB.

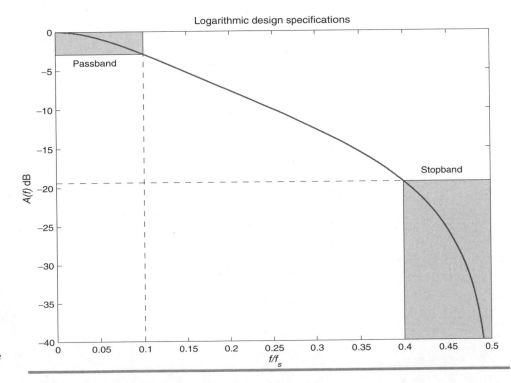

Figure 5.12   Magnitude Response of First-order IIR Lowpass Filter using the dB Scale

• • • • • • • • • • • • • • • •

5.3 Linear-phase Filters

5.3.1 Group Delay

The filter design specifications discussed thus far are specifications on the desired magnitude response of the filter. It is also possible to design filters with prescribed phase responses. To illustrate the type of phase responses that are desirable and achievable, consider the following analog system.

$$H_a(s) = \exp(-\tau s) \tag{5.17}$$

Recall from the properties of the Laplace transform (see Appendix C), that $H_a(s)$ represents an ideal *delay line* of delay τ. Thus, the input is delayed by τ but is not otherwise distorted by the system. The frequency response of the delay line is $H_a(f) = \exp(-j2\pi\tau f)$. Consequently, if we regard the delay line as a filter, it is an allpass filter with $A(f) = 1$ and with a phase response of

$$\phi_a(f) = -2\pi\tau f \tag{5.18}$$

The linear-phase response in Eq. 5.18 represents a pure delay. To interpret the meaning of a nonlinear-phase response, it is helpful to introduce the concept of group delay.

DEFINITION

5.1: Group Delay

Let $\phi(f)$ be the phase response of a linear system. The *group delay* of the system is denoted, $D(f)$, and defined

$$D(f) \triangleq \left(\frac{-1}{2\pi}\right)\frac{d\phi(f)}{df}$$

The group delay is the negative of the slope of the phase response scaled by $1/(2\pi)$. Observe from Eq. 5.18 that the group delay of a delay line is simply $D(f) = \tau$. That is, for a pure delay line the group delay specifies the amount by which the signal is delayed as it is processed by the filter. For most filters, the group delay $D(f)$ is not constant. In these cases, we can interpret $D(f)$ as the amount by which the spectral component at frequency f gets delayed as it is processed by the filter. Given the notion of group delay, we are now in a position to define what is meant by a linear-phase digital filter.

DEFINITION

5.2: Linear-phase Filter

Let F_z denote the set of frequencies at which the magnitude response is $A(f) = 0$. A digital filter $H(z)$ is a *generalized linear-phase filter* if and only if there exists a constant τ such that

$$D(f) = \tau, \quad f \notin F_z$$

A digital filter is a linear-phase filter if the group delay is constant, except possibly at frequencies at which the magnitude response is zero. These spectral components do not appear in the filter output, so the group delay is not meaningful for $f \in F_z$. Typically, F_z is a finite, isolated set of frequencies, and often, F_z is the empty set. Although Definition 5.2 applies to generalized linear-phase filters, for convenience, we will simply refer to them as *linear-phase* filters. Note from Definition 5.1 that Definition 5.2 implies the following form for the phase response.

$$\phi(f) = \alpha - 2\pi\tau f + \beta(f) \qquad (5.19)$$

Here α is a constant and $\beta(f)$ is piecewise constant with jump discontinuities permitted at the frequencies, F_z, at which $A(f) = 0$. An alternative way to characterize a linear-phase filter is in terms of the frequency response which must be of the following general form.

Linear-phase frequency response

$$H(f) = A_r(f) \exp[j(\alpha - 2\pi\tau f)] \qquad (5.20)$$

Here the factor $A_r(f)$ is *real*, but it may change signs and is referred to as the *amplitude response* of $H(z)$. This is to distinguish it from the magnitude response $A(f)$, which is never negative. Taking the magnitudes of both sides of Eq. 5.20, we see that the amplitude response and the magnitude response are related to one another as follows.

$$|A_r(f)| = A(f) \qquad (5.21)$$

The points at which $A_r(f) = 0$ are the points, F_z, where the phase can abruptly change by π. Thus, the piecewise-constant function $\beta(f)$ in Eq. 5.19 jumps between zero and π each time the amplitude response, $A_r(f)$, changes sign.

A linear-phase characteristic in the passband can be achieved by an analog IIR Bessel filter (Proakis and Manolakis, 1992). However, the linear-phase feature does not survive the analog-to-digital transformation. Consequently, it is better to start with a digital FIR filter as follows.

$$H(z) = \sum_{i=0}^{m} b_i z^{-i} \qquad (5.22)$$

For an FIR filter, there is a simple symmetry condition on the coefficients that guarantees a linear phase response. First, we illustrate this with a special case.

Example 5.3   **Even Symmetry**

Consider an FIR filter of order $m = 4$ having the following transfer function.

$$H(z) = c_0 + c_1 z^{-1} + c_2 z^{-2} + c_1 z^{-3} + c_0 z^{-4}$$

Recall from Eq. 4.13 that $h(k) = b_k$. Thus, the impulse response, $h = [c_0, c_1, c_2, c_1, c_0]^T$, exhibits even symmetry about the midpoint $k = m/2$. The frequency response of this filter, in terms of $\theta = 2\pi f T$, is

$$H(f) = H(z)|_{z=\exp(j\theta)}$$

$$= c_0 + c_1 \exp(-j\theta) + c_2 \exp(-j2\theta) + c_1 \exp(-j3\theta) + c_0 \exp(-j4\theta)$$

$$= \exp(-j2\theta)[c_0 \exp(j2\theta) + c_1 \exp(j\theta) + c_2 + c_1 \exp(-j\theta) + c_0 \exp(-2j\theta)]$$

Combining terms with identical coefficients and using Euler's identity, we have

$$H(f) = \exp(-j2\theta)\{c_0[\exp(j2\theta) + \exp(-j2\theta)] + c_1[\exp(j\theta) + \exp(-j\theta)] + c_2\}$$

$$= \exp(-j2\theta)\{2c_0\cos(2\theta) + 2c_1\cos(\theta) + c_2\}$$

$$= \exp(-j4\pi fT)A_r(f)$$

Comparing with Eq. 5.20, we see that this is a linear-phase system with $\alpha = 0$ and $\tau = 2T$. In this instance, the amplitude response, $A_r(f)$, is the following real *even* function that may be positive or negative.

$$A_r(f) = 2c_0\cos(4\pi fT) + 2c_1\cos(2\pi fT) + c_2$$

The even symmetry of $h(k)$ about the midpoint $k = m/2$ is one way to obtain a linear-phase filter. Another approach is to use odd symmetry of $h(k)$ about $k = m/2$, as can be seen from the following example.

Example 5.4   **Odd Symmetry**

Consider an FIR filter of order $m = 4$ having the following transfer function.

$$H(z) = c_0 + c_1 z^{-1} - c_1 z^{-3} - c_0 z^{-4}$$

Recalling that $h(k) = b_k$, we see that for this filter the impulse response, $h = [c_0, c_1, 0, -c_1, -c_0]^T$, exhibits odd symmetry about the midpoint $k = m/2$. The frequency response of this filter, in terms of $\theta = 2\pi fT$, is

$$H(f) = H(z)|_{z=\exp(j\theta)}$$

$$= c_0 + c_1\exp(-j\theta) - c_1\exp(-j3\theta) - c_0\exp(-j4\theta)$$

$$= \exp(-j2\theta)[c_0\exp(j2\theta) + c_1\exp(j\theta) - c_1\exp(-j\theta) - c_0\exp(-2j\theta)]$$

Combining terms with identical coefficients and using Euler's identity, we have

$$H(f) = \exp(-j2\theta)\{c_0[\exp(j2\theta) - \exp(-j2\theta)] + c_1[\exp(j\theta) - \exp(-j\theta)]\}$$

$$= j\exp(-j2\theta)\{2c_0\sin(2\theta) + 2c_1\sin(\theta)\}$$

$$= \exp[j(\pi/2 - 4\pi fT)]A_r(f)$$

Comparing with Eq. 5.20, this is a linear-phase system with $\alpha = \pi/2$ and $\tau = 2T$. In this instance, the amplitude response, $A_r(f)$, is the following real *odd* function that may be positive or negative.

$$A_r(f) = 2c_0\sin(4\pi fT) + 2c_1\sin(2\pi fT)$$

Table 5.1: ▶
Types of Linear-phase FIR Filters, $\tau = mT/2$

Filter Type	Impulse Response Symmetry	Filter Order m	Phase Offset α	Amplitude Response $A_r(f)$	End-point Zeros	Candidate Filters
1	Even	Even	0	Even	None	All
2	Even	Odd	0	Even	$z = -1$	Lowpass, Bandpass
3	Odd	Even	$\pi/2$	Odd	$z = \pm 1$	Bandpass
4	Odd	Odd	$\pi/2$	Odd	$z = 1$	Highpass, Bandpass

The results of Examples 5.3 and 5.4 can be generalized. In particular, it is possible to show that an FIR filter of order m is a linear-phase filter with constant group delay $D(f) = mT/2$ if the impulse response satisfies the following *symmetry condition*.

Linear-phase symmetry condition

$$h(k) = \pm h(m - k), \quad 0 \le k \le m \tag{5.23}$$

The symmetry condition in Eq. 5.23 says that the impulse response must exhibit *even symmetry* about the midpoint, $k = m/2$, if the plus sign is used, or *odd symmetry* about the midpoint if the minus sign is used. If we further decompose Eq. 5.23 into cases where the order m is even or odd, then there are a total of four linear-phase filter types as summarized in Table 5.1.

Observe that when the filter order m is even, there is a middle sample, $h(m/2)$, about which the impulse response exhibits either even or odd symmetry. When m is odd, the symmetry is about a point that is midway between a pair of samples. The impulse responses of the four linear-phase filter types are summarized in Figure 5.13 for the case $h(k) = k^2$ for $0 \le k \le \text{floor}(m/2)$. Note that the middle sample of the type 3 filter must satisfy $h(m/2) = -h(m/2)$, which means that $h(m/2) = 0$ in this case. For filters with even symmetry, the phase offset is $\alpha = 0$ and the amplitude response is an even function in Eq. 5.20. For filters with odd symmetry, $\alpha = \pi/2$, and the amplitude response is an odd function. The latter can be useful in certain special applications such as the design of differentiators and Hilbert transformers. Note that for the type 1 and type 2 filters, the impulse responses are the numerical equivalents of a *palindrome*: a word that is the same whether it is spelled forwards or backwards.

5.3.2 Linear-phase Zeros

The symmetry condition that guarantees a linear phase response also imposes certain constraints on the zeros of an FIR filter. To see this, we start with Eq. 5.22 and use Eq. 5.23 and the change of variable $k = m - i$.

$$H(z) = \sum_{i=0}^{m} h(i) z^{-i}$$

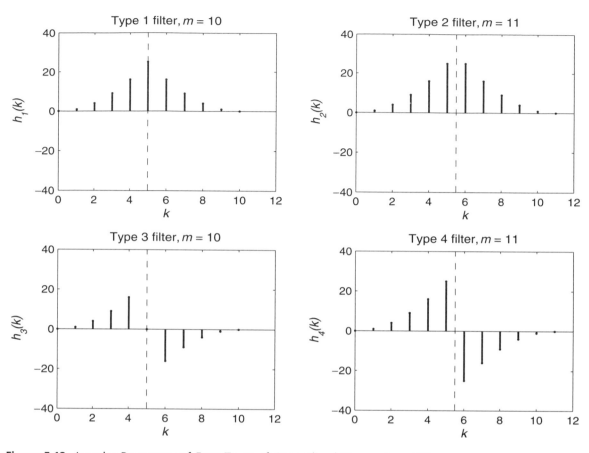

Figure 5.13 Impulse Responses of Four Types of Generalized Linear-phase Filters

$$= \pm \sum_{i=0}^{m} h(m - i)z^{-i}$$

$$= \pm \sum_{k=m}^{0} h(k)z^{-(m-k)} \left.\right\} k = m - i$$

$$= \pm z^{-m} \sum_{k=0}^{m} h(k)z^{k} \qquad (5.24)$$

The final summation on the right-hand side of Eq. 5.24 is simply $H(z^{-1})$. Thus, the linear-phase symmetry condition, expressed in terms of the transfer function, is

Linear-phase
constraint

$$H(z) = \pm z^{-m} H(z^{-1}) \qquad (5.25)$$

There are a number of consequences of Eq. 5.25. First, consider a type 2 filter with even symmetry and odd order. Evaluating Eq. 5.25 at $z = -1$ yields

$H(-1) = -H(-1)$. Therefore, every type 2 linear-phase filter has a zero at $z = -1$. Consequently, type 2 filters should not be used to design highpass or bandstop filters.

Next, consider a type 3 filter with odd symmetry and even order. Again, evaluating Eq. 5.25 at $z = -1$ yields $H(-1) = -H(-1)$, which means that $z = -1$ is also a zero of the type 3 filter. Evaluating Eq. 5.25 at $z = 1$ yields $H(1) = -H(1)$. Hence, a type 3 filter has zeros at $z = \pm 1$. It follows that a type 3 filter might be used as a bandpass filter, but it should not be used for a lowpass, highpass, or bandstop filter.

Finally, consider a type 4 filter where both the symmetry and the order are odd. Evaluating Eq. 5.25 at $z = 1$ yields $H(1) = -H(1)$, which means that $z = 1$ is a zero of $H(z)$. Thus, a type 4 filter should not be used as a lowpass or bandstop filter. The end-point constraints placed by the zeros at $f = 0$ and $f = f_s/2$ are summarized in the last two columns of Table 5.1. It is apparent that the type 1 filter (even symmetry, even order) is the most general in that it does not suffer from either end-point constraint.

In addition to the zeros at the two ends of the spectrum, the constraint on $H(z)$ in Eq. 5.25 also implies that complex zeros must appear in certain patterns. Let $z = r \exp(j\phi)$ be a complex zero of $H(z)$ with $r > 0$. Then, from Eq. 5.25, $z^{-1} = r^{-1} \exp(-j\phi)$ must also be a zero of $H(z)$. Furthermore, if the coefficient vector b is real, then zeros must occur in complex-conjugate pairs. Hence, for $r \neq 1$, the zeros will appear in groups of four and satisfy the following *reciprocal symmetry*.

$$Q = \{r \exp(\pm j\phi), r^{-1} \exp(\mp j\phi)\} \tag{5.26}$$

When the zeros are real, we have $\phi = 0$ or $\phi = \pi$, and the set Q in Eq. 5.26 reduces to a pair of reciprocal real zeros. A pole-zero plot for a type 1 FIR linear-phase filter showing a typical arrangement of both real and complex zeros is shown in Figure 5.14.

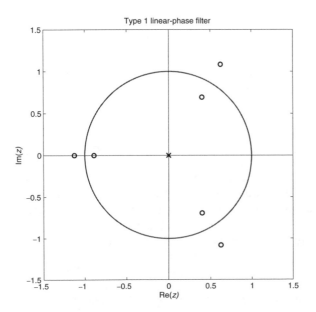

Figure 5.14 Poles and Zeros of a Type 1 FIR Linear-phase Filter of Order $m = 6$

• • • • • • • • • • • • • • • • •

5.4 Minimum-phase and Allpass Filters

Every digital filter with a rational transfer function can be expressed as a product of two specialized filters. The first is a minimum-phase filter, and the second is an allpass filter.

5.4.1 Minimum-phase Filters

The magnitude response, by itself, does not provide enough information to completely specify a filter. Indeed, among IIR filters having m zeros, there are up to 2^m distinct filters each having an identical magnitude response $A(f)$. To see this, recall that a rational IIR transfer function can be written as a ratio of two polynomials in z.

$$H(z) = \frac{b(z)}{a(z)} \tag{5.27}$$

Since the polynomial $b(z)$ has real coefficients, the complex conjugate $b^*(z)$ can be obtained by replacing z by z^*. On the unit circle, $z = \exp(j2\pi fT)$, which means that $z^* = z^{-1}$. Thus, the square of the magnitude response can be expressed as follows.

$$
\begin{aligned}
A^2(f) &= \left. \frac{|b(z)|^2}{|a(z)|^2} \right|_{z=\exp(j2\pi fT)} \\[2mm]
&= \left. \frac{b(z)b^*(z)}{|a(z)|^2} \right|_{z=\exp(j2\pi fT)} \\[2mm]
&= \left. \frac{b(z)b(z^{-1})}{|a(z)|^2} \right|_{z=\exp(j2\pi fT)}
\end{aligned}
\tag{5.28}
$$

Next, suppose $H(z)$ has a zero at $z = c$ with $c \neq 0$. Then $b(z)$ and $b(z^{-1})$ can be written in partially factored form as

$$b(z) = (z - c)b_0(z) \tag{5.29a}$$

$$b(z^{-1}) = (z^{-1} - c)b_0(z^{-1}) \tag{5.29b}$$

If the factor, $z - c$, in $b(z)$ is interchanged with the corresponding factor, $z^{-1} - c$, in $b(z^{-1})$, then the product $b(z)b(z^{-1})$ does not change, which means $A^2(f)$ in Eq. 5.28 does not change. Replacing the factor, $z - c$, with the factor, $z^{-1} - c$, is equivalent to replacing the zero at $z = c$ with a zero at its reciprocal, $z = c^{-1}$, and scaling by a constant. To determine the constant, first note that

$$
\begin{aligned}
z^{-1} - c &= z^{-1}(1 - cz) \\
&= -cz^{-1}(z - c^{-1})
\end{aligned}
\tag{5.30}
$$

When the magnitude response of $H(z)$ is computed, we evaluate $H(z)$ along the unit circle, which means $|z^{-1}| = 1$. Thus, a new numerator polynomial which does not change the magnitude response of $H(z)$ is

$$b_1(z) = -c(z - c^{-1})b_0(z) \qquad (5.31)$$

Since this can be done with any of the m zeros of $b(z)$, there are up to 2^m distinct combinations of zeros of $H(z)$ that all yield filters with identical magnitude responses. The differences between these filters lie in their phase responses $\phi(f)$.

DEFINITION

5.3: Minimum-phase Filter

A digital filter $H(z)$ is a *minimum-phase* filter if and only if all of its zeros lie inside or on the unit circle. Otherwise it is a *nonminimum-phase* filter.

Every IIR filter $H(z)$ can be converted to a minimum-phase filter with the same magnitude response by replacing the zeros outside the unit circle with their reciprocals. The term "minimum phase" arises from the fact that the net phase change of a minimum-phase filter, over the frequency range $[0, f_s/2]$, is

$$\phi(f_s/2) - \phi(0) = 0 \qquad (5.32)$$

Nonminimum-phase filters have at least one zero outside the unit circle. It can be shown (Proakis and Manolakis, 1992) that if $H(z)$ has p zeros outside the unit circle, then the net phase change is $\phi(f_s/2) - \phi(0) = -p\pi$. Thus, among filters that have the same magnitude response, the minimum-phase filter is the filter that has the smallest amount of phase lag.

Example 5.5 **Minimum-phase Filter**

To illustrate how different filters can have identical magnitude responses, consider the following second-order IIR filter.

$$H_{00}(z) = \frac{2(z + 0.5)(z - 0.5)}{(z - 0.5)^2 + 0.25}$$

This is a stable IIR filter with poles at $z = 0.5 \pm j0.5$. It is also a minimum-phase filter because the zeros at $z = \pm 0.5$ are both inside the unit circle. There are three other IIR filters that have an identical magnitude response, which we obtain by using the reciprocal of the first zero, the second zero, or both and multiplying by the negative of the original zero as in Eq. 5.31. Thus, the transfer functions of other three filters are.

$$H_{10}(z) = \frac{(z + 2)(z - 0.5)}{(z - 0.5)^2 + 0.25}$$

$$H_{01}(z) = \frac{-(z + 0.5)(z - 2)}{(z - 0.5)^2 + 0.25}$$

$$H_{11}(z) = \frac{-0.5(z + 2)(z - 2)}{(z - 0.5)^2 + 0.25}$$

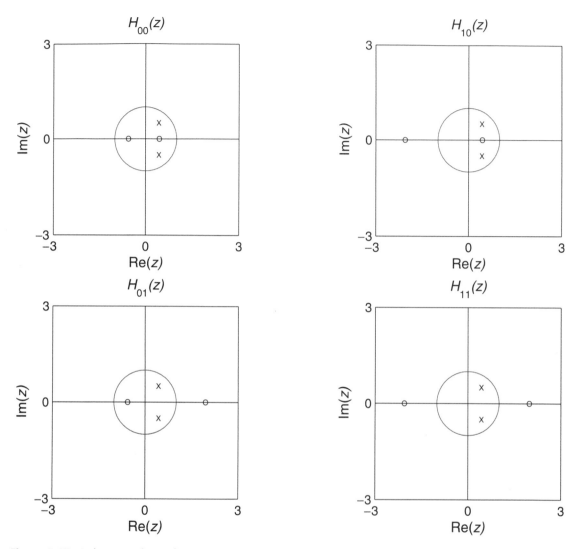

Figure 5.15 Pole-zero Plots of Four Filters Having Same $A(f)$

Pole-zero plots of the four equivalent filters are shown in Figure 5.15. Only filter $H_{00}(z)$ is a minimum-phase filter. Since filter $H_{11}(z)$ has all of its zeros outside the unit circle, it is called a *maximum-phase* filter, while $H_{10}(z)$ and $H_{01}(z)$ are called *mixed-phase* filters.

Plots of the four magnitude responses are shown in Figure 5.16, where it is evident that they are all identical. However, the four phase responses are distinct from one another as can be seen in the plot in Figure 5.17. Note that the net phase change for the minimum-phase filter is zero. For the two mixed-phase filters, $H_{10}(z)$ and $H_{01}(z)$, the net phase change is $-\pi$. Finally, the maximum-phase filter, $H_{11}(z)$, has a net phase change of -2π.

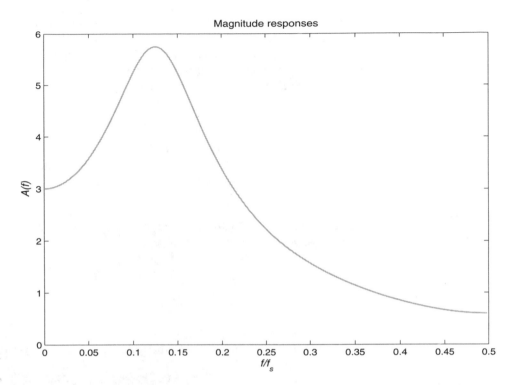

Figure 5.16 Identical Magnitude Responses of the Four Filters in Example 5.5

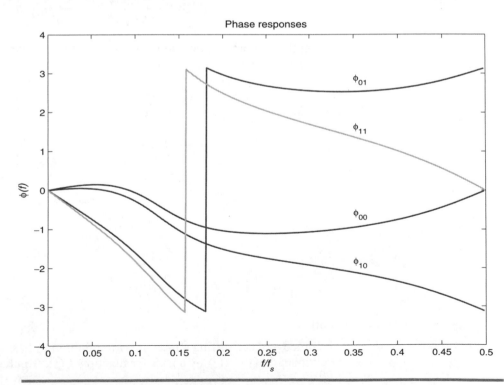

Figure 5.17 The Phase Responses of the Four Filters in Example 5.5

Every rational IIR filter can be converted to minimum-phase form by replacing each zero outside the unit circle by its reciprocal and scaling by the negative of the zero. If the original filter has a pair of complex conjugate zeros at $z = r \exp(\pm j\theta)$ with $r > 1$, then both zeros must be replaced by zeros at $z = r^{-1} \exp(\mp j\theta)$ in order for the coefficients of the new filter to remain real.

5.4.2 Allpass Filters

Another important class of IIR filters that can be used to provide phase compensation is the allpass filter. As the name implies, an *allpass* filter is a filter that passes all spectral components equally because it has a flat magnitude response.

DEFINITION

5.4: Allpass Filter

A digital filter $H(z)$ with real coefficients is an *allpass* filter if and only if it has the following magnitude response.

$$A(f) = 1, \quad 0 \le f \le \frac{f_s}{2}$$

There are no special constraints on the phase response of an allpass filter. Allpass filters have transfer functions with the following *reflective symmetry*.

Allpass structure

$$H_{\text{all}}(z) = \frac{a_n + a_{n-1}z^{-1} + \cdots + z^{-n}}{1 + a_1 z^{-1} + \cdots + a_n z^{-n}} \tag{5.33}$$

Notice that the numerator polynomial is just the denominator polynomial, $a(z)$, but with the coefficients reversed. To see how this gives a flat magnitude response, let $A(f)$ denote the magnitude response of the FIR filter $H(z) = a(z)$ corresponding to the denominator in Eq. 5.33. Then, using the fact that the magnitude response is an even function, the magnitude response of the filter in Eq. 5.33 is

$$A_{\text{all}}(f) = |H_{\text{all}}(z)|_{z=\exp(j2\pi fT)}$$

$$= \left| \frac{z^{-n}a(z^{-1})}{a(z)} \right|_{z=\exp(j2\pi fT)}$$

$$= \frac{|z^{-n}| \cdot |a(z^{-1})|}{|a(z)|} \Bigg|_{z=\exp(j2\pi fT)}$$

$$= \frac{A(-f)}{A(f)}$$

$$= 1 \tag{5.34}$$

The process of converting a filter $H(z)$ to minimum-phase form can be thought of as multiplication of $H(z)$ by a transfer function $F(z)$. To illustrate, suppose $H(z)$

has a single zero at $z = c$ where c lies outside the unit circle. Then replacing this zero with one at $z = c^{-1}$ and multiplying by $-c$ is equivalent to multiplying $H(z)$ by

$$F(z) = \frac{-c(z - c^{-1})}{z - c} \tag{5.35}$$

If $z = c$ is the only zero of $H(z)$ outside the unit circle, then the minimum-phase version of $H(z)$ can be expressed as

$$H_{\min}(z) = F(z)H(z) \tag{5.36}$$

Next, consider the characteristics of the transfer function $F(z)$ used to convert $H(z)$ to minimum-phase form. Note from Eq. 5.35 that

$$F(z) = \frac{-c(z - c^{-1})}{z - c}$$

$$= \frac{-cz + 1}{z - c}$$

$$= \frac{-c + z^{-1}}{1 - cz^{-1}} \tag{5.37}$$

Comparing Eq. 5.37 with Eq. 5.33, we see that $F(z)$ is an allpass filter with $a = [1, -c]^T$. Although $F(z)$ was developed using only one zero outside the unit circle, the process can be repeated for any number of zeros with the resulting allpass filter having factors similar to Eq. 5.35.

The magnitude response of the inverse of a filter is just the inverse of the magnitude response of the filter. Hence, if $H_{\text{all}}(z) = F^{-1}(z)$, then $H_{\text{all}}(z)$ is an allpass filter. Furthermore, multiplying both sides of Eq. 5.36 on the left by $H_{\text{all}}(z)$, we conclude that the every rational IIR transfer function $H(z)$ can be decomposed into the product of an allpass filter $H_{\text{all}}(z)$ times a minimum-phase filter $H_{\min}(z)$.

Minimum-phase
decomposition

$$H(z) = H_{\text{all}}(z)H_{\min}(z) \tag{5.38}$$

A block diagram of the decomposition into allpass and minimum-phase parts is shown in Figure 5.18. The minimum-phase part is the minimum-phase form of $H(z)$. Therefore, the magnitude response of $H_{\min}(z)$ is identical to the magnitude response of $H(z)$. The allpass part is the inverse of the system that transforms $H(z)$ into its minimum-phase form. Consequently, the allpass part $H_{\text{all}}(z)$ is always stable.

Figure 5.18 Decomposition of IIR Filter into Allpass and Minimum-phase Parts

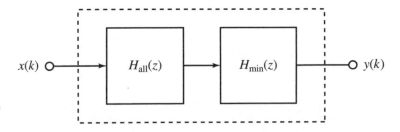

Another way to characterize an allpass filter is in terms of its poles and zeros. For each pole of $H_{all}(z)$ at $z = c$, there is a matching zero at its reciprocal, $z = c^{-1}$. Thus, allpass filters always have the same number of poles and zeros, with the poles and zeros forming reciprocal pairs. The following algorithm summarizes the steps needed to decompose a general IIR filter into its minimum-phase and allpass parts.

ALGORITHM 5.1:
Minimum-phase
Decomposition

1. Set $H_{min}(z) = H(z)$, and $H_{all}(z) = 1$. Factor the numerator polynomial of $H(z)$ as follows.

$$b(z) = b_0(z - z_1)(z - z_2)\cdots(z - z_m)$$

2. For $i = 1$ to m do
{

If $|z_i| > 1$ then compute

$$F(z) = \frac{-z_i z + 1}{z - z_i}$$

$$H_{min}(z) = F(z)H_{min}(z)$$

$$H_{all}(z) = F^{-1}(z)H_{all}(z)$$

}

Example 5.6   **Minimum-phase Decomposition**

As an illustration of the decomposition of an IIR transfer function into allpass and minimum-phase parts, consider the following digital filter.

$$H(z) = \frac{0.2[(z + 0.5)^2 + 1.5^2]}{z^2 - 0.64}$$

This is a stable IIR filter with real poles at $p_{1,2} = \pm 0.8$ and complex-conjugate zeros at $z_{1,2} = -0.5 \pm j1.5$, as shown in Figure 5.19. Since the zeros are both outside the unit circle, this is a maximum-phase filter. The minimum-phase form is obtained by replacing the zeros by their reciprocals and multiplying by the negatives of the zeros. The new zeros are

$$z_{3,4} = \frac{1}{z_{1,2}}$$

$$= \frac{1}{-0.5 \pm j1.5}$$

$$= \frac{-0.5 \mp j1.5}{.25 + 2.25}$$

$$= -0.2 \mp j0.6$$

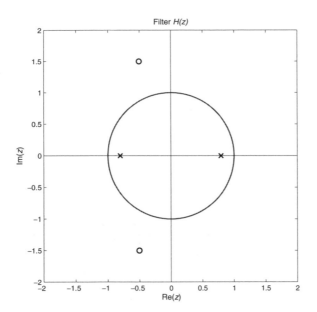

Figure 5.19   Pole-zero Plot of System in Example 5.6

The product is $z_1 z_2 = |z_1|^2$. Thus, the minimum-phase form of $H(z)$ is

$$H_{\min}(z) = \frac{|z_1|^2 0.2(z - z_3)(z - z_4)}{z^2 - 0.64}$$

$$= \frac{(.25 + 2.25)0.2[(z + 0.2)^2 + 0.6^2]}{z^2 - 0.64}$$

$$= \frac{0.5[(z + 0.2)^2 + 0.6^2]}{z^2 - 0.64}$$

Since both zeros were replaced, the allpass part is just the original numerator divided by the numerator of $H_{\min}(z)$.

$$H_{\text{all}}(z) = \frac{0.2[(z + 0.5)^2 + 1.5^2]}{0.5[(z + 0.2)^2 + 0.6^2]}$$

$$= \frac{0.4[(z + 0.5)^2 + 1.5^2]}{(z + 0.2)^2 + 0.6^2}$$

A plot of the original and decomposed magnitude responses and the original and decomposed phase responses is shown in Figure 5.20. Note that $A_{\min}(f) = A(f)$, and $A_{\text{all}}(f) = 1$ as expected. It is also evident that $\phi_{\min}(f)$ has less phase lag than $\phi(f)$.

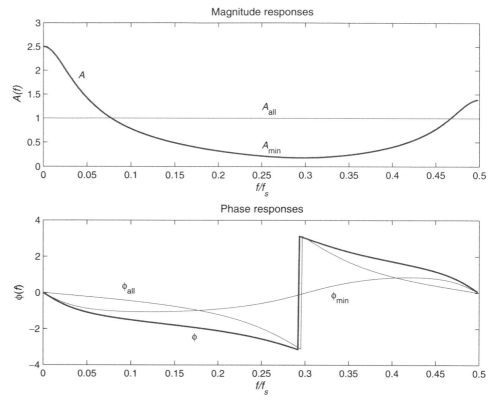

Figure 5.20 Magnitude and Phase Plots of $H(z)$, $H_{\text{all}}(z)$, and $H_{\text{min}}(z)$

MATLAB Toolbox

The FDSP toolbox contains the following function for decomposing a transfer function into minimum-phase and allpass parts.

```
[b_min,a_min,b_all,a_all] = f_minall (b,a);        % minimum-phase decomposition
```

On entry to *f_minall*, b is a vector of length $m + 1$ containing the numerator coefficients, and a is a vector of length $n + 1$ containing the denominator coefficients with $a(1) = 1$. Thus, the transfer function to be decomposed is

$$H(z) = \frac{b_0 + b_1 z^{-1} + \cdots + b_m z^{-m}}{1 + a_1 z^{-1} + \cdots a_n z^{-n}} \tag{5.39}$$

On exit from *f_minall*, *b_min* and *a_min* are vectors containing the numerator and denominator coefficients, respectively, of the minimum-phase part, $H_{min}(z)$. Similarly, *b_all* and *a_all* are vectors containing the numerator and denominator coefficients, respectively, of the allpass part, $H_{all}(z)$. The decomposition of $H(z)$ into minimum-phase and allpass parts is a cascade-form decomposition.

$$H(z) = H_{all}(z)H_{min}(z) \tag{5.40}$$

5.5 FIR Filter Realization Structures

In this section, we investigate alternative configurations that can be used to realize FIR filters with signal flow graphs. These filter realization structures differ from one another with respect to storage requirements, computational time, and sensitivity to finite word length effects. An mth-order FIR filter has the following transfer function.

$$H(z) = b_0 + b_1 z^{-1} + \cdots + b_m z^{-m} \tag{5.41}$$

5.5.1 Direct Forms

Tapped Delay Line

Taking the inverse Z-transform of $Y(z) = H(z)X(z)$ using Eq. 5.41 and the delay property, we get the following time-domain representation of an mth order FIR filter.

$$y(k) = \sum_{i=0}^{m} b_i x(k - i) \tag{5.42}$$

The representation in Eq. 5.42 is a *direct* representation because the coefficients of the difference equations are obtained directly from inspection of the transfer function. A signal flow graph of a direct-form realization, for the case $m = 3$, is shown in Figure 5.21. This structure is called a *tapped delay line* or a *transversal filter*.

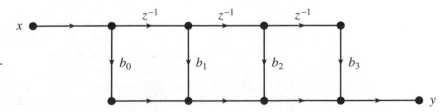

Figure 5.21 Tapped-delay-line Realization of an FIR Filter, $m = 3$

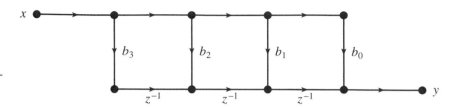

Figure 5.22 Transposed Tapped-delay-line Realization of an FIR Filter, $m = 3$

Transposed Tapped Delay Line

Every signal flow graph satisfies something called the flow graph reversal theorem. This theorem states that if each arc of a signal flow graph is reversed and the input and output labels are interchanged, then the resulting signal flow graph is equivalent. Applying the flow-graph reversal theorem to the direct-form graph in Figure 5.21, and drawing the final result with the input on the left rather than the right, we arrive at the FIR filter realization shown in Figure 5.22, which is called a *transposed tapped delay line*.

Linear-phase Form

The tapped-delay-line realizations in Figures 5.21 and 5.22 are general realizations that are valid for any FIR filter. However, many FIR filters are designed to be linear-phase FIR filters. Recall that $b_i = h(i)$ for an FIR filter. If the linear-phase symmetry constraint in Eq. 5.23 is recast in terms of the FIR filter coefficients, this yields

$$b_k = \pm b_{m-k}, \quad 0 \le k \le m \tag{5.43}$$

By exploiting this symmetry constraint, we can develop a filter realization that requires only about half as many floating-point multiplications (FLOPs). To illustrate this method, consider the most general linear-phase filter, the type 1 filter, with even symmetry and even order. Letting $r = m/2$, it is possible to rewrite Eq. 5.42 as

Linear-phase
realization

$$y(k) = b_r x(k - r) + \sum_{i=0}^{r-1} b_i [x(k - i) + x(k - m + i)] \tag{5.44}$$

Thus, the number of multiplications has been reduced from m to $m/2 + 1$. Similar expressions for $y(k)$ can be developed for the other three types of linear-phase FIR filters (see Problem P5.18). A signal flow graph realization of a type 1 linear-phase FIR filter is shown in Figure 5.23 for the case $m = 6$.

Observe from Figure 5.23 that there are $m/2 + 1$ floating-point multiplications, one for each distinct coefficient. However, the number of delay elements is m, so there is no reduction in the storage requirement. A comparison of the direct form FIR filter realizations is summarized in Table 5.2, where it can be seen that the three realizations are identical in terms of storage requirements. In each case, the number of floating-point operations or FLOPs grows linearly with the order of the filter. For large values of m, the linear-phase form has approximately half the multiplications, but about 50 percent more additions.

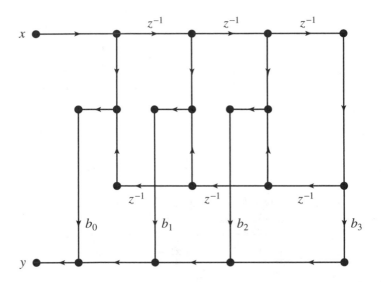

Figure 5.23 Direct-form Realization of a Type 1 Linear-phase FIR Filter, $m = 6$

Table 5.2: ▶
Comparison of Direct-form Realizations of FIR Filter of Order m

Direct Form	Storage Elements	Additions	Multiplications
Tapped delay line	m	$m + 1$	$m + 1$
Transposed tapped delay line	m	$m + 1$	$m + 1$
Linear phase	m	$3m/2 + 1$	$m/2 + 1$

5.5.2 Cascade Form

The direct forms have the virtue that they can be implemented easily from direct inspection of the transfer function. However, the direct forms also suffer from a practical drawback. As the order of the filter, m, increases, the direct-form filters become increasingly sensitive to finite word length effects. For example, the roots of a polynomial can be very sensitive to small changes in the coefficients of the polynomial, particularly as the degree of the polynomial increases. To develop a realization that will be less sensitive to the effects of finite word length, it is helpful to first recast the transfer function in Eq. 5.41 in terms of positive powers of z. If the numerator is then factored, this yields the factored form

$$H(z) = \frac{b_0(z - z_1)(z - z_2) \cdots (z - z_m)}{z^m} \quad (5.45)$$

Since $H(z)$ has m poles at $z = 0$, an FIR filter is always stable. The coefficients of $H(z)$ are assumed to be real, so complex zeros occur in conjugate pairs. The representation in Eq. 5.45 can be recast as a product of M second-order subsystems as follows where $M = \text{floor}[(m + 1)/2]$.

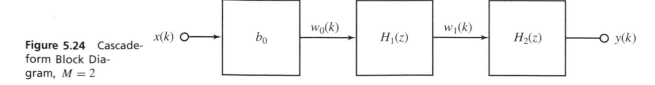

Figure 5.24 Cascade-form Block Diagram, $M = 2$

Cascade form

$$H(z) = b_0 H_1(z) \cdots H_M(z) \qquad (5.46)$$

This is called a *cascade-form* realization, and a block diagram for the case $M = 2$ is shown in Figure 5.24. Second-order block $H_i(z)$ is constructed from two poles at $z = 0$ and either two real zeros or a complex conjugate pair of zeros. This way, the coefficients of $H_i(z)$ are guaranteed to be real.

$$H_i(z) = 1 + b_{i1}z^{-1} + b_{i2}z^{-2}, \quad 1 \le i \le M \qquad (5.47)$$

If $H_i(z)$ is constructed from zeros z_i and z_j, then the two coefficients can be computed using sums and products as follows for $1 \le i \le M$.

$$b_{i1} = -(z_i + z_j) \qquad (5.48a)$$

$$b_{i2} = z_i z_j \qquad (5.48b)$$

Let w_i denote the output of the ith second-order block. Then from Eq. 5.46 and Eq. 5.47, a cascade-form realization is characterized by the following time domain equations.

$$w_0(k) = b_0 x(k) \qquad (5.49a)$$

$$w_i(k) = w_{i-1}(k) + b_{i1}w_{i-1}(k - 1) + b_{i2}w_{i-1}(k - 2), \quad 1 \le i \le M \quad (5.49b)$$

$$y(k) = w_M(k) \qquad (5.49c)$$

If m is even, there will be M subsystems, each of order two, and if m is odd, there will be $M - 1$ second-order subsystems plus one first-order subsystem. The coefficients of a first-order subsystem are obtained from Eq. 5.48 by setting $z_j = 0$.

Either of the tapped-delay-line forms can be used to realize the second-order blocks in Eq. 5.47. A block diagram of the overall structure of a cascade-form realization, for the case $M = 3$, is shown in Figure 5.25. Since the cascade-form coefficients must be computed using Eq. 5.48, rather than obtained directly from inspection of $H(z)$, the cascade-form realization is an example of an *indirect* form.

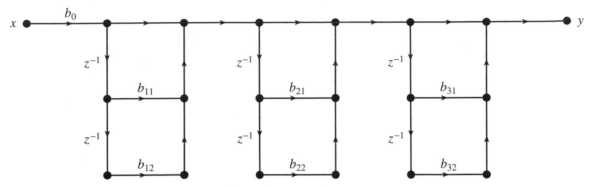

Figure 5.25 Cascade-form Realization of FIR Filter, $M = 3$

<table>
<tr><td>Example 5.7</td><td>**FIR Cascade Form**</td></tr>
</table>

As an illustration of a cascade-form realization of an FIR filter, consider the following fifth-order transfer function.

$$H(z) = \frac{3[(z - 0.4)^2 + 0.25](z + 0.3)(z - 0.6)(z + 0.9)}{z^5}$$

Inspection of $H(z)$ reveals that the zeros are

$$z_{1,2} = 0.4 \pm j0.5$$

$$z_3 = -0.3$$

$$z_4 = 0.6$$

$$z_5 = -0.9$$

Suppose $H_1(z)$ is a block associated with the complex-conjugate pair of zeros, $H_2(z)$ is associated with the real zeros, $z = -0.3$ and $z = 0.6$, and $H_3(z)$ is a first-order block associated with the zero, $z = -0.9$. Running script *exam5_7* on the distribution CD, we get the following subsystems.

$$b_0 = 3$$

$$H_1(z) = 1 - 0.8z^{-1} + 0.41z^{-2}$$

$$H_2(z) = 1 - 0.3z^{-1} - 0.18z^{-2}$$

$$H_3(z) = 1 + 0.9z^{-1}$$

The signal flow graph of the resulting cascade-form realization of the fifth-order FIR filter is as shown in Figure 5.26.

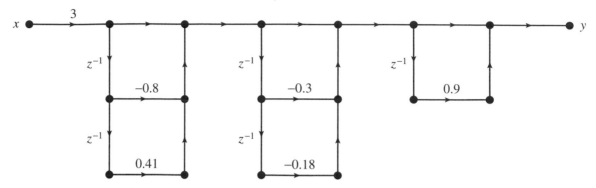

Figure 5.26 Cascade-form Realization of Filter in Example 5.7

5.5.3 Lattice Form

Another indirect form that finds applications in speech processing and adaptive systems is the lattice-form realization shown in Figure 5.27 for the case $m = 2$. The time-domain equations for a lattice-form realization of order m are expressed in terms of the intermediate variables u_i and v_i. From Figure 5.27, we have

$$u_0(k) = b_0 x(k) \tag{5.50a}$$

$$v_0(k) = u_0(k) \tag{5.50b}$$

$$u_i(k) = u_{i-1}(k) + K_i v_{i-1}(k-1), \quad 1 \le i \le m \tag{5.50c}$$

$$v_i(k) = K_i u_{i-1}(k) + v_{i-1}(k-1), \quad 1 \le i \le m \tag{5.50d}$$

$$y(k) = u_m(k) \tag{5.50e}$$

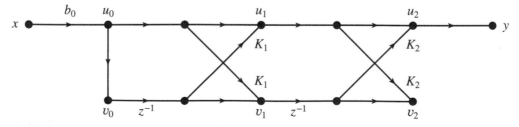

Figure 5.27 Lattice-form Realization of FIR Filter, $m = 2$

Thus, an mth order lattice-form realization has m stages. Coefficient K_i of the ith stage is called a *reflection coefficient*. To determine the vector of m reflection coefficients, it is useful to introduce the following operation which is applicable to an mth order FIR filter.

$$z^{-m} H(z^{-1}) = \sum_{i=0}^{m} b_{m-i} z^{-i} \qquad (5.51)$$

Comparing Eq. 5.51 with Eq. 5.41, we see that $z^{-m} H(z^{-1})$ is simply the polynomial obtained by *reversing* the coefficients of $H(z)$. The following algorithm can be used to compute the reflection coefficients.

ALGORITHM 5.2:
Lattice-form
Realization

1. Factor $H(z)$ into $H(z) = b_0 A_m(z)$ and compute

$$B_m(z) = z^{-m} A_m(z^{-1})$$

$$K_m = \lim_{z \to \infty} B_m(z)$$

2. For $i = m$ down to 2 compute
   {

$$A_{i-1}(z) = \frac{A_i(z) - K_i B_i(z)}{1 - K_i^2}$$

$$B_{i-1}(z) = z^{-(i-1)} A_{i-1}(z^{-1})$$

$$K_{i-1} = \lim_{z \to \infty} B_{i-1}(z)$$

   }

Algorithm 5.2 produces b_0 and an $m \times 1$ vector of reflection coefficients, K, as long as $|K_i| \neq 1$ for $1 \leq i \leq m$. If $|K_i| = 1$, then $A_{i-i}(z)$ has a zero on the unit circle. In this case, this zero can be factored out and the algorithm applied to the reduced-order polynomial.

Example 5.8   **FIR Lattice Form**

As an illustration of a lattice-form realization of an FIR filter, consider the following second-order transfer function.

$$H(z) = 2 + 6z^{-1} - 4z^{-2}$$

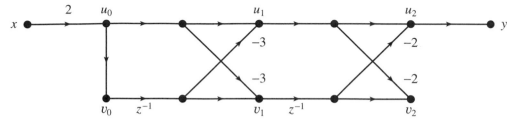

Figure 5.28   Lattice-form Realization of FIR Filter in Example 5.8

Applying step 1 of Algorithm 5.2, we have $b_0 = 2$ and

$$A_2(z) = 1 + 3z^{-1} - 2z^{-2}$$

$$B_2(z) = -2 + 3z^{-1} + z^{-2}$$

$$K_2 = -2$$

Next, applying step 2 with $i = 2$, we have

$$A_1(z) = \frac{1 + 3z^{-1} - 2z^{-2} + 2(-2 + 3z^{-1} + z^{-2})}{1 - 4}$$

$$= \frac{-3 + 9z^{-1}}{-3}$$

$$= 1 - 3z^{-1}$$

$$B_1(z) = -3 + z^{-1}$$

$$K_1 = -3$$

Thus, $b_0 = 2$, and the reflection coefficient vector is $K = [-3, -2]^T$. A signal flow graph of the lattice-form realization is shown in Figure 5.28.

There are additional indirect forms that have been proposed and are used. Included among these are the frequency-sampling realization and a parallel-form realization with complex coefficients (Proakis and Manolakis, 1992).

MATLAB Toolbox

The FDSP toolbox contains the following functions for computing and evaluating indirect form realizations of an FIR transfer function.

```
[B,A,b_0] = f_cascade (b);          % compute cascade-form realization
[K,b_0]   = f_lattice (b);          % compute lattice-form realization
y = f_filtcas (B,A,b_0,x);          % evaluate cascade-form  filter
y = f_filtlat (K,b_0,x);            % evaluate lattice-form FIR filter
```

Function *f_cascade* computes the cascade-form realization of an FIR filter. On entry to *f_cascade*, b is a vector of length $m+1$ containing the FIR filter coefficients. The transfer function of the FIR system to be realized is

$$H(z) = b_0 + b_1 z^{-1} + \cdots + b_m z^{-m} \tag{5.52}$$

On exit from *f_cascade*, B is an $M \times 3$ matrix containing the coefficients of the numerator polynomials, A is an $M \times 3$ matrix containing the coefficients of the denominator polynomials, and b_0 is a scalar containing the filter gain. The number of second-order blocks is $M = \text{floor}[(m + 1)/2]$, and the ith block is specified by the ith row of A and B. The cascade-form realization is

$$H(z) = b_0 H_1(z) \cdots H_M(z) \tag{5.53}$$

The function *f_lattice* computes the lattice-form realization of the transfer function in Eq. 5.52. On entry to *f_lattice*, b is a vector of length $m + 1$ containing the coefficients of the numerator polynomial of $H(z)$. On exit from *f_lattice*, K is a vector of length m containing the reflection coefficients, and b_0 is a scalar containing the filter gain.

Functions *f_filtcas* and *f_filtlat* are analogous to the built-in MATLAB function *filter*. Function *f_filtcas* uses the cascade-form realization constructed by *f_cascade* to compute the output corresponding to the input vector x. Similarly, function *f_filtlat* uses the lattice-form realization constructed by *f_lattice* to compute the output corresponding to the input vector x. In both cases, output y is a vector of the same length as x containing the zero-state response.

• • • • • • • • • • • • • • •

5.6 IIR Filter Realization Structures

In this section, we investigate alternative configurations that can be used to realize IIR filters with signal flow graphs. These filter realizations differ from one another with respect to storage requirements, computational time, and sensitivity to finite word length effects.

5.6.1 Direct Forms

An nth-order IIR filter has a transfer function $H(z)$ that can be expressed as a ratio of two polynomials.

$$H(z) = \frac{b_0 + b_1 z^{-1} + \cdots + b_n z^{-n}}{1 + a_1 z^{-1} + \cdots + a_n z^{-n}} \tag{5.54}$$

For convenience, we have assumed that the degree of the numerator is equal to the degree of the denominator because this simplifies and streamlines the treatment. This is not a serious restriction because one can always pad the numerator coefficient vector with zeros, as needed, to make it the same length as the denominator coefficient vector.

Direct Form I

The simplest realization of $H(z)$ is based on factoring the transfer function into its autoregressive and moving average parts as follows.

$$H(z) = \underbrace{\left(\frac{1}{1 + a_1 z^{-1} + \cdots + a_n z^{-n}} \right)}_{H_{ar}(z)} \underbrace{\left(\frac{b_0 + b_1 z^{-1} + \cdots + b_n z^{-n}}{1} \right)}_{H_{ma}(z)} \tag{5.55}$$

If $V(z)$ denotes the output of the moving average subsystem, $H_{ma}(z)$, then the input-output description of the IIR filter can be written in terms of the intermediate variable $V(z)$ as follows.

$$V(z) = H_{ma}(z) X(z) \tag{5.56a}$$

$$Y(z) = H_{ar}(z) V(z) \tag{5.56b}$$

Taking the inverse Z-transform of Eq. 5.56 using Eq. 5.55 and the delay property, we find that an IIR filter can be represented in the time domain by the following pair of difference equations.

$$v(k) = \sum_{i=0}^{n} b_i x(k - i) \tag{5.57a}$$

$$y(k) = v(k) - \sum_{i=1}^{n} a_i y(k - i) \tag{5.57b}$$

The representation in Eq. 5.57 is called a *direct form I* realization. It is a *direct* representation because the coefficients of the difference equations can be obtained directly from inspection of the transfer function. A signal flow graph of a direct form I realization, for the case $n = 3$, is shown in Figure 5.29. Note how the left side of the signal flow graph implements the moving average part in Eq. 5.57a, and the right side implements the autoregressive part in Eq. 5.57b.

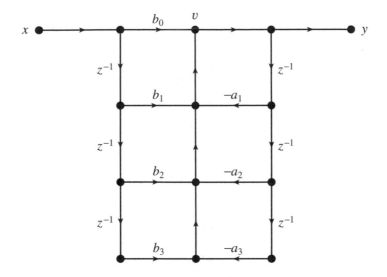

Figure 5.29 Direct Form I Realization, $n = 3$

Direct Form II

There is a very simple change that can be made to Eq. 5.55 to generate an alternative direct-form realization that has some advantages over the direct form I structure. Suppose the order of the autoregressive and moving average subsystems is interchanged. Clearly, this does not affect the overall transfer function.

$$H(z) = \underbrace{\left(\frac{b_0 + b_1 z^{-1} + \cdots + b_n z^{-n}}{1}\right)}_{H_{\mathrm{ma}}(z)} \underbrace{\left(\frac{1}{1 + a_1 z^{-1} + \cdots + a_n z^{-n}}\right)}_{H_{\mathrm{ar}}(z)} \tag{5.58}$$

Next, let $V(z)$ denote the output of the autoregressive subsystem, $H_{\mathrm{ar}}(z)$. Then the input-output description of the overall filter in terms of the intermediate variable $V(z)$ is as follows.

$$V(z) = H_{\mathrm{ar}}(z)X(z) \tag{5.59a}$$

$$Y(z) = H_{\mathrm{ma}}(z)V(z) \tag{5.59b}$$

Taking the inverse Z-transform of Eq. 5.59, using Eq. 5.58 and the delay property, we find that an IIR filter can be represented in the time domain by the following difference equations.

$$v(k) = x(k) - \sum_{i=1}^{n} a_i x(k - i) \tag{5.60a}$$

$$y(k) = \sum_{i=0}^{n} b_i v(k - i) \tag{5.60b}$$

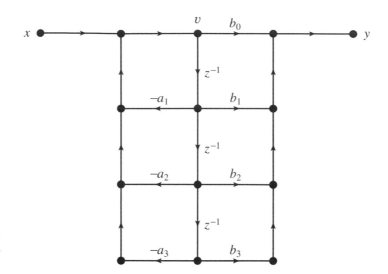

Figure 5.30 Direct Form II Realization, $n = 3$

This alternative representation of an IIR filter is called a *direct form II* realization. A signal flow graph of a direct form II realization, for the case $n = 3$, is shown in Figure 5.30. In this case, the left side of the signal flow graph implements the autoregressive part in Eq. 5.60a, and the right side implements the moving average part in Eq. 5.60b.

It is of interest to compare the signal flow graphs in Figures 5.29 and 5.30. Each arc associated with a delay element requires one memory or storage element to implement. Consequently, direct form I requires a total of $2n$ storage elements, whereas direct form II requires only n storage elements. Since the minimum number of storage elements required for an nth order filter is n, direct form II is an example of a *canonic* representation.

Transposed Direct Form II

Just as was the case with FIR filters, we can apply the signal flow graph reversal theorem to generate a transposed direct-form realization by reversing the directions of all arcs and interchanging the labels of the input and output. After redrawing the signal flow graph so the input is on the left, this yields the *transposed direct form II* realization shown in Figure 5.31 for the case $m = 3$.

The difference equations describing the transposed direct form II realization can be obtained directly from inspection of Figure 5.31. Here a vector of intermediate variables $v = [v_1, v_2, \ldots, v_n]^T$ is used. The intermediate variables are defined recursively starting at the bottom of the signal flow graph and moving up the center. The last equation is then the output equation.

$$v_1(k) = b_n x(k) - a_n y(k) \tag{5.61a}$$

$$v_i(k) = b_i x(k) - a_i y(k) + v_{i-1}(k), \quad 2 \le i \le n \tag{5.61b}$$

$$y(k) = b_0 x(k) + v_n(k-1) \tag{5.61c}$$

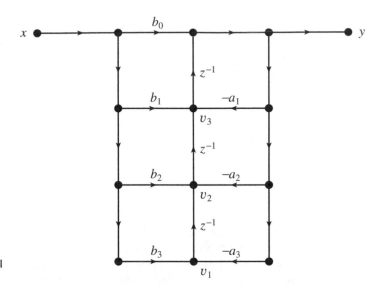

Figure 5.31 Transposed Direct Form II Realization, $n = 3$

The direct-form realizations also can be compared in terms of the required computational effort. Each summing junction node with m inputs requires $m - 1$ floating-point additions, and each arc with a constant nonunity gain requires one floating-point multiplication. The results of the comparison are summarized in Table 5.3, where it can be seen that the three direct-form realizations are identical in terms of the computational time, measured in floating-point operations or FLOPs. In each case, the number of FLOPs grows linearly with the order of the filter. However, the direct form II realizations have the advantage that they require only half as much memory.

5.6.2   Parallel Form

Just as with FIR filters, there are a number of IIR indirect forms whose coefficients are derived from the original coefficient vectors, a and b. To develop the indirect-form realization structures, we begin by factoring the denominator polynomial of the transfer function.

$$H(z) = \frac{b(z)}{(z - p_1)(z - p_2) \cdots (z - p_n)} \tag{5.62}$$

Here p_k is the kth pole of the filter. If $H(z)$ has poles at $z = 0$, then these poles represent pure delays that can be removed and treated separately. Consequently,

Table 5.3: ▶
Comparison of Direct-form Realization of IIR Filter

Direct Form	Storage Elements	Additions	Multiplications
I	$2n$	$2n$	$2n - 1$
II	n	$2n$	$2n - 1$
Transposed II	n	$2n$	$2n - 1$

we assume that $p_k \neq 0$ for $1 \leq k \leq n$. For most practical filters, the nonzero poles are distinct from one another. Suppose $H(z)/z$ is expressed in partial fraction form. Multiplying both sides by z then results in the following representation of $H(z)$, assuming the poles are distinct.

$$H(z) = R_0 + \sum_{i=1}^{n} \frac{R_i z}{z - p_i} \tag{5.63}$$

Here R_i is the residue of $H(z)/z$ at the ith pole with $p_0 = 0$. From Eq. 5.63, we have

$$R_0 = H(0) \tag{5.64a}$$

$$R_i = \left. \frac{(z - p_i)H(z)}{z} \right|_{z=p_i}, \quad 1 \leq i \leq n \tag{5.64b}$$

The problem with using Eq. 5.63 directly for a signal flow graph realization is that the poles and residues are often complex. If $H(z)$ has real coefficients, then the complex poles and residues will appear in conjugate pairs. Consequently, we can rewrite $H(z)$ as a sum of N second-order subsystems with real coefficients as follows, where $N = \text{floor}[(n + 1)/2]$.

Parallel-form realization

$$H(z) = R_0 + \sum_{i=1}^{N} H_i(z) \tag{5.65}$$

This is called a *parallel-form* realization. The ith second-order subsystem is constructed by combining pairs of terms in Eq. 5.63 associated with either real poles or complex conjugate pairs of poles. This way, the second-order coefficients will be real. Combining the terms associated with poles p_i and p_j, simplifying, and expressing the final result in terms of negative powers of z yields:

$$H_i(z) = \frac{b_{i0} + b_{i1}z^{-1}}{1 + a_{i1}z^{-1} + a_{i2}z^{-2}}, \quad 1 \leq i \leq m \tag{5.66}$$

Note that $b_{i2} = 0$. The real coefficients of the second-order block can be expressed in terms of the poles and residues as follows for $1 \leq i \leq N$.

$$b_{i0} = R_i + R_j \tag{5.67a}$$

$$b_{i1} = -(R_i p_j + R_j p_i) \tag{5.67b}$$

$$a_{i1} = -(p_i + p_j) \tag{5.67c}$$

$$a_{i2} = p_i p_j \tag{5.67d}$$

Let w_i denote the output of the ith second-order block. Then from Eqs. 5.65 and 5.66, a parallel-form realization is characterized in the time domain by the following difference equations.

$$w_i(k) = b_{i0}x(k) + b_{i1}x(k - 1) - a_{i1}w_i(k - 1) - a_{i2}w_i(k - 2), \quad 1 \leq i \leq N \tag{5.68a}$$

$$y(k) = R_0 x(k) + \sum_{i=1}^{N} w_i(k) \tag{5.68b}$$

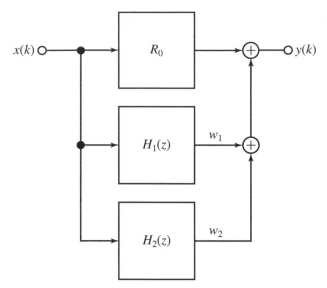

Figure 5.32 Parallel-form Block Diagram, $N = 2$

Note that if n is even, then there will be N subsystems, each of order two, whereas if n is odd, there will be $N - 1$ second-order subsystems and one first-order subsystem. The coefficients of the first-order subsystem are obtained from Eq. 5.67 by setting $R_j = 0$ and $p_j = 0$.

Any of the direct forms can be used to realize the second-order blocks in Eq. 5.66. A block diagram showing the overall structure of a parallel-form realization, for the case $N = 2$, is shown in Figure 5.32. Since the parallel-form coefficients must be computed using Eq. 5.67, rather than obtained directly from inspection of $H(z)$, the parallel-form realization is an example of an *indirect* form.

Example 5.9   **IIR Parallel Form**

As an illustration of a parallel-form realization of an IIR filter, consider the following fourth-order transfer function.

$$H(z) = \frac{2z(z^3 + 1)}{[(z + 0.3)^2 + 0.16](z - 0.8)(z + 0.7)}$$

Inspection of $H(z)$ reveals that the poles are

$$p_{1,2} = -0.3 \pm j0.4$$

$$p_3 = 0.8$$

$$p_4 = -0.7$$

Suppose $H_1(z)$ is a block associated with the complex conjugate pair of poles, and $H_2(z)$ is associated with the real poles. Running script *exam5_9* on the distribution CD, we get the following three subsystems.

$$R_0 = 0$$

$$H_1(z) = \frac{3.266 - 2.134z^{-1}}{1 + 0.6z^{-1} + 0.25z^{-2}}$$

$$H_2(z) = \frac{-1.266 + 3.220z^{-1}}{1 - 0.1z^{-1} - 0.56z^{-2}}$$

Suppose a direct form II realization is used for the second-order blocks. The resulting signal flow graph of a parallel-form realization of the fourth-order filter is as shown in Figure 5.33.

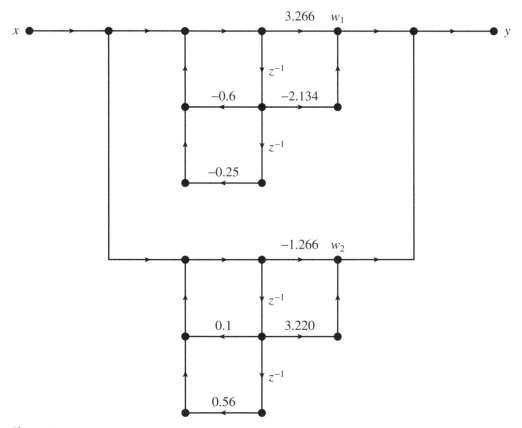

Figure 5.33 Parallel-form Realization from Example 5.9

5.6.3 Cascade Form

An even simpler way to decompose $H(z)$ into lower-order subsystems is to factor both the numerator and the denominator of $H(z)$ as follows.

$$H(z) = \frac{b_0(z - z_1)(z - z_2) \cdots (z - z_n)}{(z - p_1)(z - p_2) \cdots (z - p_n)} \tag{5.69}$$

Note that if the degree of the numerator in Eq. 5.54 is $m < n$, then the factored representation in Eq. 5.69 will have $n - m$ zeros at $z_k = 0$. Since the coefficients of $H(z)$ are assumed to be real, complex poles and zeros will occur in conjugate pairs. The representation in Eq. 5.69 can be recast as a product of N second-order subsystems as follows where $N = \text{floor}[(n + 1)/2]$.

Cascade-form realization

$$H(z) = b_0 H_1(z) \cdots H_N(z) \tag{5.70}$$

This is called a *cascade-form* realization. Second-order block $H_i(z)$ is constructed from either two real zeros or a complex-conjugate pair of zeros (and similarly for the poles). This way, the coefficients of $H_i(z)$ are guaranteed to be real.

$$H_i(z) = \frac{1 + b_{i1}z^{-1} + b_{i2}z^{-2}}{1 + a_{i1}z^{-1} + a_{i2}z^{-2}}, \quad 1 \le i \le N \tag{5.71}$$

If $H_i(z)$ is constructed from zeros z_i and z_j and poles p_q and p_r, then the four coefficients can be computed using sums and products as follows for $1 \le i \le N$.

$$b_{i1} = -(z_i + z_j) \tag{5.72a}$$

$$b_{i2} = z_i z_j \tag{5.72b}$$

$$a_{i1} = -(p_q + p_r) \tag{5.72c}$$

$$a_{i2} = p_q p_r \tag{5.72d}$$

Let w_i denote the output of the ith second-order block. Then from Eq. 5.70 and Eq. 5.71, a cascade-form realization is characterized by the following time domain equations.

$$w_0(k) = b_0 x(k) \tag{5.73a}$$

$$w_i(k) = w_{i-1}(k) + b_{i1}w_{i-1}(k - 1) + b_{i2}w_{i-1}(k - 2) -$$
$$a_{i1}w_i(k - 1) - a_{i2}w_i(k - 2), \quad 1 \le i \le N \tag{5.73b}$$

$$y(k) = w_N(k) \tag{5.73c}$$

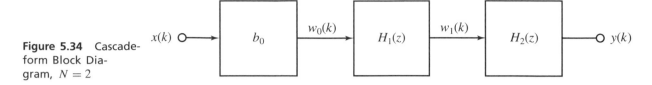

Figure 5.34 Cascade-form Block Diagram, $N = 2$

Just as with the parallel form, if n is even, there will be N subsystems, each of order two, while if n is odd, there will be $N - 1$ second-order subsystems plus one first-order subsystem. The coefficients of a first-order subsystem are obtained from Eq. 5.72 by setting $z_j = 0$ and $p_r = 0$.

Any of the direct forms can be used to realize the second-order blocks in Eq. 5.71. A block diagram of the overall structure of a cascade-form realization, for the case $N = 2$, is shown in Figure 5.34. Since the cascade-form coefficients must be computed using Eq. 5.72, rather than obtained directly from inspection of $H(z)$, the cascade-form realization is another example of an *indirect* form.

Example 5.10   **IIR Cascade Form**

To compare the cascade-form realization with the parallel-form realization, consider the fourth-order transfer function introduced earlier in Example 5.9.

$$H(z) = \frac{2z(z^3 + 1)}{[(z + 0.3)^2 + 0.16](z - 0.8)(z + 0.7)}$$

The poles are listed in Example 5.9. There is a single zero at $z = 0$, and the remaining zeros are the three roots of -1, which are equally spaced around the unit circle.

$$z_1 = -1$$

$$z_{2,3} = \cos(\pi/3) \pm j \sin(\pi/3)$$

$$z_4 = 0$$

Suppose $H_1(z)$ is a block associated with the complex-conjugate pairs of zeros $\{z_2, z_3\}$ and poles $\{p_1, p_2\}$, and $H_2(z)$ is associated with the real zeros and poles. Running script *exam5_10* on the distribution CD, we get the following three subsystems.

$$b_0 = 2$$

$$H_1(z) = \frac{1 - z^{-1} + z^{-2}}{1 + 0.6z^{-1} + 0.25z^{-2}}$$

$$H_2(z) = \frac{1 + z^{-1}}{1 - 0.1z^{-1} - 0.56z^{-2}}$$

If a direct form II realization is used for the second-order blocks, then the resulting signal flow graph of a cascade-form realization of the fourth-order filter is as shown in Figure 5.35.

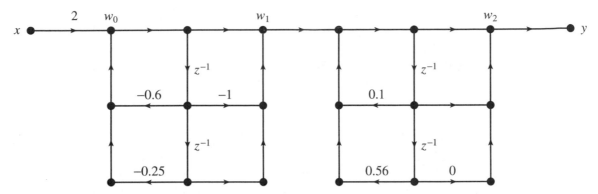

Figure 5.35 Cascade-form Realization from Example 5.10

One advantage that the cascade form has over the parallel form is that there is considerable flexibility in the way the zeros and poles can be grouped together to form the second-order blocks. Let $b_i(z)$ be the numerator associated with the ith pair of zeros, and let $a_j(z)$ be the denominator associated with the jth pair of poles. Since i and j range from 1 to N, there are a total of $N!$ possible orderings of the numerators (and similarly for the denominators). Hence, the total number of ways the numerators and denominators can be combined to form second-order blocks is $p = (N!)^2$. All of the orderings are equivalent if infinite-precision arithmetic is used. For finite-precision filters, it is recommended that pairs of zeros and poles that are close to one another be grouped together in order to reduce the occurrence of block outputs that are very large or very small. First, the pole closest to the unit circle is paired with the nearest zero. This process is repeated until all of the poles and zeros are paired. Finally, it is recommended that the blocks be ordered either in terms of increasing pole distance from the unit circle or in terms of decreasing pole distance (Jackson, 1996).

MATLAB Toolbox

The FDSP toolbox contains the following functions for computing and evaluating the indirect-form realizations of an IIR transfer function.

```
[B,A,R_0] = f_parallel (b,a);        % compute parallel-form realization
[B,A,b_0] = f_cascade (b,a);         % compute cascade-form realization
y = f_filtpar (B,A,R_0,x);           % evaluate parallel-form filter
y = f_filtcas (B,A,b_0,x);           % evaluate cascade-form  filter
```

Function $f\_parallel$ computes the parallel-form realization of an IIR filter. On entry to $f\_parallel$, b is a vector of length $m + 1$ containing the numerator coefficients, and a is a vector of length $n + 1$ containing the denominator coefficients with $a(1) = 1$. The transfer function of the IIR system to be realized is

$$H(z) = \frac{b_0 + b_1 z^{-1} + \cdots + b_m z^{-m}}{1 + a_1 z^{-1} + \cdots + a_n z^{-n}} \tag{5.74}$$

On exit from $f\_parallel$, B is an $N \times 3$ matrix containing the coefficients of the numerator polynomials, A is an $N \times 3$ matrix containing the coefficients of the denominator polynomials, and R_0 is a scalar containing the constant term of the parallel form. The number of second-order blocks is $N = \text{floor}[(n + 1)/2]$, and the ith block is specified by the ith row of A and B. The parallel-form realization is

$$H(z) = R_0 + \sum_{i=1}^{N} H_i(z) \tag{5.75}$$

The function $f\_cascade$ computes the cascade-form realization of the transfer function in Eq. 5.74. The input and output arguments for $f\_cascade$ are identical to those of function $f\_parallel$, except that R_0 is replaced by the scalar gain b_0. The cascade-form realization is

$$H(z) = b_0 H_1(z) \cdots H_N(z) \tag{5.76}$$

Functions $f\_filtpar$ and $f\_filtcas$ are analogous to the built-in MATLAB function $filter$. Function $f\_filtpar$ uses the parallel-form realization constructed by $f\_parallel$ to compute the output corresponding to the input vector x. Similarly, function $f\_filtcas$ uses the cascade-form realization constructed by $f\_cascade$ to compute the output corresponding to the input vector x. In both cases, output y is a vector of the same length as x containing the zero-state response.

● ● ● ● ● ● ● ● ● ● ● ● ● ● ● ●

*5.7 FIR Finite Word Length Effects

When a filter is implemented, finite precision must be used to represent the values of the signals and the filter coefficients. The resulting reductions in filter performance caused by going from infinite precision to finite precision are called finite word length effects. If a filter is implemented in software using MATLAB, then double-precision floating-point arithmetic is used. On a PC, this corresponds to 64 bits of precision or about 16 significant decimal digits. In most instances, this is a sufficiently good approximation to infinite-precision arithmetic that no significant finite word length effects are apparent. However, if a filter is implemented on specialized DSP hardware (Kuo and Gan, 2005), or storage or speed requirements dictate the need to use single-precision floating-point arithmetic or fixed-point arithmetic, then finite word length effects can begin to manifest themselves.

5.7.1 Binary Number Representation

Finite word length effects depend on the method used to represent numbers. If a filter is implemented in software on a PC, then typically a binary *floating-point representation* of numbers is used, where some of the bits are used to represent the mantissa, or fractional part, and the remaining bits are reserved to represent the exponent. For example, MATLAB running on a PC uses $N = 64$ bits with 53 bits reserved for the mantissa and 11 bits for the exponent. Floating-point representations have the advantage that they can represent very large and very small numbers; hence, issues of overflow and scaling typically do not come into play. However, the spacing between adjacent floating-point values is not constant, instead it is proportional to the magnitude of the number.

An alternative to the floating-point representation is the *fixed-point representation* which does not have a field of bits reserved to represent the exponent. Both floating-point and fixed-point representations can be found on specialized DSP hardware. The fixed-point representation is faster and more efficient than the floating-point representation. Furthermore, the precision, or spacing between adjacent values, is uniform throughout the range of numerical values represented. Unfortunately, this range typically is much smaller than for floating-point numbers, so one may have to introduce scaling to prevent overflow. Although finite word length effects appear in both floating-point and fixed-point arithmetic, they are less severe for floating-point representations, particularly double-precision representations such as those used in MATLAB.

In this section, we will focus our attention on the use of fixed-point numerical representations. An N-bit binary fixed-point number b consisting of N binary dig*its*, or bits, can be expressed as follows.

$$b = b_0 b_1 \cdots b_{N-1}, \quad 0 \le b_i \le 1 \tag{5.77}$$

Typically, the number b is normalized with the binary point appearing between bits b_0 and b_1. In this case, the decimal equivalent of a positive number is

$$x = \sum_{i=1}^{N-1} b_i 2^{-i} \tag{5.78}$$

Bit b_0 is reserved to hold the sign of x, with $b_0 = 0$ indicating a positive number and $b_0 = 1$ a negative number. There are several schemes for encoding negative numbers including sign magnitude, one's complement, offset binary, and two's complement (Gerald and Wheatley, 1989). The most popular method is the two's complement representation. The two's complement representation of a negative number is obtained by complementing each bit, and then adding one to the least significant bit. Any carry past the sign bit is ignored. One of the virtues of two's complement arithmetic occurs when we add several numbers. If the sum fits within N bits, then the total will be correct even if the intermediate results or partial sums overflow!

The range of values that can be represented in Eq. 5.78 is $-1 \le x < 1$. Of course, many values of interest may fall outside this normalized range. Larger values can be accommodated by using a fixed scale factor c. Notice from Eq. 5.78 that scaling

Figure 5.36 Fixed-point Representation of N-bit Number using Scale Factor $c = 2^M$

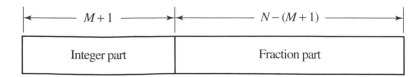

by $c = 2^M$ effectively moves the binary point M places to the right. This increases the range to $-c \leq x < c$, but causes a loss of precision. The precision, or spacing between adjacent values, is constant and is called the *quantization level*.

Quantization level

$$q = \frac{c}{2^{N-1}} \tag{5.79}$$

When a scale factor of $c = 2^M$ is used, this corresponds to reserving $M + 1$ bits for the integer part including the sign, and the remaining $N - (M + 1)$ bits for the fraction part as shown in Figure 5.36.

5.7.2 Input Quantization Error

Recall from Chapter 1 that the value of the input signal $x(k)$ is *quantized* to a finite number of bits as a result of analog-to-digital (ADC) conversion. To model quantization, it is helpful to introduce the following operator.

DEFINITION

5.5: Quantization Operator

Let N be the number of bits used to represent a real value x. The *quantized* version of x is denoted $Q_N(x)$ and defined

$$Q_N(x) \triangleq q \text{ floor} \left(\frac{x + q/2}{q} \right)$$

The form of quantization in Definition 5.5 uses *rounding*, which can be seen from the $q/2$ term in the numerator. If this term is removed, then the resulting quantization operation uses *truncation*. For convenience, we will assume that quantization by rounding to N bits is used. A graph of the nonlinear input-output characteristic of the quantization operator for the case $N = 4$ is shown in Figure 5.37.

If an ADC has a precision of N bits, then the quantized output from the ADC, denoted $x_q(k)$, is computed as follows.

$$x_q(k) = Q_N[x(k)] \tag{5.80}$$

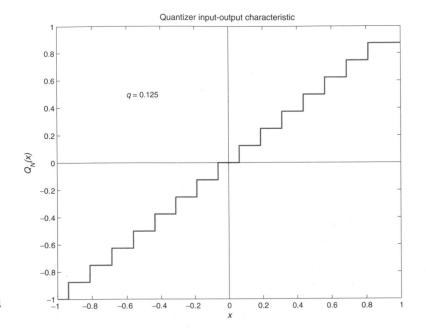

Quantizer input-output characteristic

$q = 0.125$

Figure 5.37 Input-output Characteristic of Quantization Operator for $N = 4$ using Rounding

Although the nonlinear deterministic model in Eq. 5.80 is accurate, it is more useful for analysis purposes to replace it by an equivalent linear statistical model. The quantized signal can be regarded as an infinite-precision signal, $x(k)$, plus a *quantization error* term, $\Delta x(k)$, as shown in Figure 5.38. Thus, the quantized signal is

$$x_q(k) = x(k) + \Delta x(k) \qquad (5.81)$$

Recall from Chapter 1 that if $|x_a(t)| \le c$, then $|\Delta x(k)| \le q/2$, where q in Eq. 5.79 is the ADC quantization level. Consequently, when rounding is used, the quantization error can be modeled as *white noise* that is uniformly distributed over the interval $[-q/2, q/2]$. The probability density of ADC quantization noise is as shown in Figure 5.39.

A convenient measure of the size of the quantization noise is the average power. For zero-mean noise, the average power is the variance $\sigma_x^2 = E[\Delta x^2]$. Using Definition 3.2 and the probability density in Figure 5.39, we find that the average power

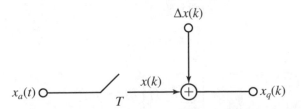

$\Delta x(k)$

$x(k)$

$x_a(t)$ T $x_q(k)$

Figure 5.38 Linear Statistical Model of Input Quantization

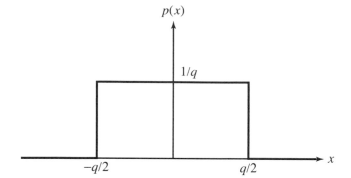

Figure 5.39 Probability Density of ADC Quantization Noise using Rounding

of the ADC quantization noise is

$$\sigma_x^2 = \int_{-q/2}^{q/2} x \, dx$$

$$= \frac{q^2}{12} \tag{5.82}$$

The quantization noise associated with the input signal is filtered by the system $H(z)$ and appears in the system output as

$$y_q(k) = y(k) + \Delta y(k) \tag{5.83}$$

To determine the average power of the output noise, first note that for a linear system the response to the input noise $\Delta x(k)$ is the output noise $\Delta y(k)$. That is, if $h(k)$ is the impulse response of the system, then subtracting the noise-free output from the complete output yields:

$$\Delta y(k) = \sum_{i=0}^{m} h(i) \Delta x(k - i) \tag{5.84}$$

The input quantization noise $\Delta x(k)$ is zero-mean uncorrelated white noise. Consequently, the average power of the output noise is

$$E[\Delta y^2(k)] = E\left[\sum_{i=0}^{m} h(i) \Delta x(k - i) \sum_{p=0}^{m} h(p) \Delta x(k - p) \right]$$

$$= \sum_{i=0}^{m} \sum_{p=0}^{m} h(i)h(p)E[\Delta x(k - i)\Delta x(k - p)]$$

$$= \sum_{i=0}^{m} h^2(i)E[\Delta x^2(k - i)]$$

$$= \left[\sum_{i=0}^{m} h^2(i) \right] E[\Delta x^2(k)] \tag{5.85}$$

Thus, the average power of the output noise is proportional to the average power of the input quantization noise as follows.

$$\sigma_y^2 = \Gamma \sigma_x^2 \tag{5.86}$$

The constant of proportionality Γ is called the *system power gain*. From Eq. 5.85, the system power gain is computed from the impulse response $h(k)$ as follows.

System power gain

$$\Gamma \triangleq \sum_{k=0}^{\infty} h^2(k) \tag{5.87}$$

Notice that we have replaced the upper limit of the sum by infinity. This can be done for an FIR filter because all of the terms beyond the mth term are zero. For an IIR filter, the power gain expression in Eq. 5.87 continues to be valid if the filter is stable and the impulse response is square integrable.

Example 5.11   **Input Quantization Noise**

As an illustration of ADC quantization effects, suppose an ADC with a precision of $N = 8$ bits is used to sample an input signal in the range $|x_a(t)| \le 10$. From Eq. 5.79, the ADC quantization level is

$$q = \frac{10}{2^7}$$

$$= 0.0781$$

It then follows from Eq. 5.82 that the average power of the ADC quantization noise at the input is

$$\sigma_x^2 = \frac{0.0781^2}{12}$$

$$= 5.0863 \times 10^{-4}$$

Next, suppose the quantization noise is passed through the following first-order FIR filter.

$$H(z) = 10 \sum_{i=0}^{30} (0.9)^i z^{-i}$$

The impulse response of this FIR filter is

$$h(k) = 10(0.9)^k [u(k) - u(k - 31)]$$

Using the geometric series and Eq. 5.87, the power gain of the filter is

$$\Gamma = \sum_{k=0}^{30}[10(0.9)^k]^2$$

$$= 100\left[\sum_{k=0}^{\infty}(0.81)^k - \sum_{k=31}^{\infty}(0.81)^k\right]$$

$$= 100\left[\frac{1}{1-0.81} - \frac{(0.81)^{31}}{1-0.81}\right]$$

$$= 525.37$$

Finally, using Eq. 5.86, the average power of the ADC quantization noise appearing at the filter output is

$$\sigma_y^2 = \Gamma\sigma_x^2$$

$$= 526.37(5.0863 \times 10^{-4})$$

$$= 0.2672$$

Observe that even though the noise power at the input is relatively small, the noise power at the output can be large.

5.7.3   Coefficient Quantization Error

FIR filter coefficients are also quantized when they are stored in fixed-length memory locations. Let the unquantized or infinite-precision version of the transfer function be

$$H(z) = \sum_{i=0}^{m} b_i z^{-i} \tag{5.88}$$

Suppose the elements of the coefficient vector are quantized to N bits to yield $b_q = Q_N(b)$, where Q_N is the N-bit quantization operator introduced in Definition 5.5. The quantized coefficient vector b_q can be expressed as follows.

$$b_q = b + \Delta b \tag{5.89}$$

Here Δb is the *coefficient quantization error*. If $|b_i| \leq c$, then the elements of the vector Δb can be modeled as random numbers uniformly distributed over $[-q/2, q/2]$ where q is the quantization level given in Eq. 5.79. For an FIR filter, the effects of coefficient quantization are relatively easy to model. Let $H_q(z)$ denote the transfer function using quantized coefficients. From Eq. 5.88 and Eq. 5.89, we have

$$H_q(z) = \sum_{i=0}^{m}(b_i + \Delta b_i)z^{-i}$$

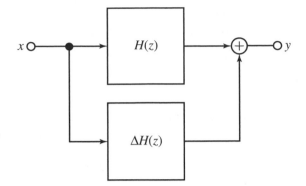

Figure 5.40 The Effects of Coefficient Quantization on an FIR Filter $H(z)$

$$= \sum_{i=0}^{m} b_i z^{-i} + \sum_{i=0}^{m} \Delta b_i z^{-i}$$

$$= H(z) + \Delta H(z) \tag{5.90}$$

Here the subsystem $\Delta H(z)$ is the coefficient quantization error transfer function

$$\Delta H(z) = \sum_{i=0}^{m} \Delta b_i z^{-i} \tag{5.91}$$

From Eq. 5.90, we see that quantization of the FIR coefficients is equivalent to introducing the error system $\Delta H(z)$ in parallel with the unquantized system as shown in Figure 5.40.

One can place a simple upper bound on the effects of coefficient quantization error on the frequency response $H(f)$. Using $|\Delta b_i| \le q/2$, the error in the system magnitude response is bounded as follows.

$$\Delta A(f) = |\sum_{i=0}^{m} \Delta b_i \exp(-ji2\pi fT)|$$

$$\le \sum_{i=0}^{m} |\Delta b_i \exp(-ji2\pi fT)|$$

$$= \sum_{i=0}^{m} |\Delta b_i|$$

$$\le \frac{(m+1)q}{2} \tag{5.92}$$

It then follows from Eq. 5.79 that for an mth order FIR filter with coefficients $|b_i| \le c$ that are quantized to N bits, the error in the magnitude of the frequency response satisfies

FIR magnitude response error

$$\Delta A(f) \le \frac{(m+1)c}{2^N} \tag{5.93}$$

The upper bound in Eq. 5.93 is a conservative one that assumes a worst case in which each of the individual coefficient quantization errors has the same sign and is the maximum possible value.

Example 5.12 | **FIR Coefficient Quantization Error**

As an illustration of FIR coefficient quantization noise, consider a bandpass filter of order $m = 64$ designed using the least-squares method, a technique that is covered in Chapter 6. Suppose the coefficients are quantized using $N = 8$ bits. Evaluation of the $(m + 1)$ coefficients reveals that they lie in the range $|b_i| \leq c$, where $c = 1$. Using Eq. 5.93, the error in the magnitude response is bounded as follows.

$$\Delta A(f) \leq \frac{64 + 1}{2^8}$$

$$= 0.2539$$

When script *exam5_12* on the distribution CD is executed, it produces the two magnitude response plots shown in Figure 5.41. The first plot approximates the unquantized case using double precision floating-point arithmetic. The second plot uses a tapped-delay-line direct-form realization with quantized coefficients. It is

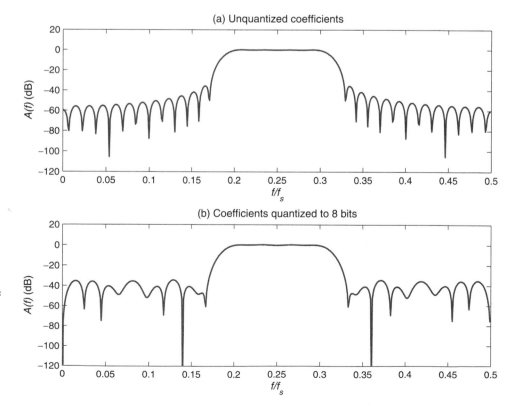

Figure 5.41 Magnitude Responses of Least-squares Bandpass Filter of Order $m = 64$ for (a) Unquantized Coefficients and (b) Coefficients Quantized to $N = 8$ Bits

apparent from inspection that quantizing to $N = 8$ bits introduces error, particularly in the stopband. The signal attenuation for the quantized case is not nearly as good as it is for the unquantized case.

Unit Circle Zeros

Another way to evaluate the effects of coefficient quantization is to look at the locations of the zeros. The roots of a polynomial can be very sensitive to small changes in the coefficients of the polynomial, particularly for higher degree polynomials (Gerald and Wheatley, 1989). As a consequence, higher order direct-form realizations of $H(z)$ can be sensitive to coefficient quantization error. For FIR filters, the most important case corresponds to zeros on the unit circle. Zeros on the unit circle are important because they produce complete attenuation of the input signal at specific frequencies. For example, suppose it is desired to remove the frequency F_0. This is achieved by placing zeros at $z = \pm \exp(j\theta_0)$, where $\theta_0 = 2\pi F_0 T$. A second-order FIR filter that completely attenuates frequency F_0 is then

$$
\begin{aligned}
H(z) &= \frac{[z - \exp(j\theta_0)][z - \exp(-j\theta_0)]}{z^2} \\
&= \frac{z^2 - [\exp(j\theta_0) + \exp(-j\theta_0)]z + 1}{z^2} \\
&= \frac{z^2 - 2\cos(\theta_0)z + 1}{z^2} \\
&= 1 - 2\cos(\theta_0)z^{-1} + z^{-2}
\end{aligned}
\tag{5.94}
$$

Note that $H(z)$ has a coefficient vector of $b = [1, -2\cos(\theta_0), 1]^T$. If b is quantized, this will result in a small change in θ_0, but no change b_0 or b_2 because they can be represented exactly. Consequently, the quantized zero will *remain* on the unit circle, although the angle (i.e., the frequency) may change. It follows that, if $H(z)$ is realized using a cascade of quantized second-order blocks, then zeros on the unit circle will be *preserved*, although their frequencies will change slightly.

Linear-phase Block

Most FIR filters are linear-phase filters, so it is useful to examine what effect coefficient quantization has on this important property. Recall that the most general type 1 linear-phase FIR filter of order m can be realized with the direct form shown in Figure 5.23. From Eq. 5.44, the difference equation of this direct-form linear-phase realization is as follows, where $r = m/2$.

$$
y(k) = b_r x(k - r) + \sum_{i=0}^{r-1} b_i [x(k - i) + x(k - m + i)]
\tag{5.95}
$$

To preserve the linear-phase response of a type 1 filter, it is necessary that $b_i = b_{m-i}$ for $0 \leq i \leq m$. But from the structure in Eq. 5.95, it is apparent that b_i and b_{m-i}

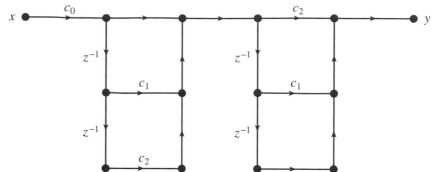

Figure 5.42 Cascade-form Realization of a Fourth-order Linear-phase Block

are implemented with the *same* coefficient. Consequently, if the $r+1$ coefficients in Eq. 5.95 are quantized, the resulting quantized filter will still be a linear-phase filter, only its magnitude response will be affected. That is, the linear-phase feature of the direct-form realization in Eq. 5.95 is *unaffected* by coefficient quantization.

As it turns out, a linear-phase response can be preserved, even when a cascade-form realization is used. Recall that linear-phase filters have the property that if $z = r \exp(j\phi)$ is a zero with $r \neq 1$, then so is its reciprocal, $z = r^{-1} \exp(-j\phi)$. For a filter with real coefficients, complex zeros appear in conjugate pairs. Consequently, complex zeros that are not on the unit circle appear in groups of four. To preserve this grouping, we can use the following fourth-order block in a cascade-form realization.

$$
\begin{aligned}
H(z) &= \frac{[z - r\exp(j\phi)][z - r\exp(-j\phi)][z - r^{-1}\exp(-j\phi)][z - r^{-1}\exp(j\phi)]}{z^4} \\
&= \frac{[z^2 - 2r\cos(\phi)z + r^2][z^2 - 2r^{-1}\cos(\phi)z + r^{-2}]}{z^4} \\
&= \frac{[z^2 - 2r\cos(\phi)z + r^2][r^2 z^2 - 2r\cos(\phi)z + 1]}{r^2 z^4} \\
&= c_0(1 + c_1 z^{-1} + c_2)(c_2 + c_1 z^{-1} + 1)
\end{aligned}
\tag{5.96}
$$

Note that this factored fourth-order linear-phase block can be realized as a cascade of two second-order blocks where the distinct coefficients are $c = [r^{-2}, -2r\cos(\phi), r^2]^T$. If c is quantized, the reciprocal zero relationship is preserved. Hence the quantized fourth-order block is still a linear-phase system. A block diagram of a cascade-form realization of a fourth-order linear-phase block is shown in Figure 5.42.

5.7.4 Roundoff Error, Overflow, and Scaling

The arithmetic used to compute the filter output must also be performed using finite precision. The time-domain representation of an FIR filter using a tapped-delay-line direct-form realization is

$$
y(k) = \sum_{i=0}^{m} b_i x(k - i)
\tag{5.97}
$$

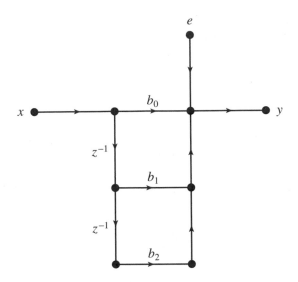

Figure 5.43 Linear Model of Product Roundoff Error in an FIR Filter, $m = 2$

If the coefficients are quantized to N bits and the signals are quantized to N bits, then the product terms in Eq. 5.97 will each be of length $2N$ bits. When the products are then rounded to N bits, the resulting error is called *roundoff error*. It can be modeled as white noise, with one white noise source for each product. Assuming the roundoff noise sources are statistically independent of one another, the noise sources associated with the products can be combined into a single error term, as follows where Q_N denotes the N-bit quantization operator.

$$e(k) = \sum_{i=0}^{m} Q_N[b_i x(k-i)] - b_i x(k-i) \qquad (5.98)$$

For a tapped-delay-line direct-form realization, this results in the equivalent linear model of roundoff error shown in Figure 5.43 for the case $m = 2$.

Suppose both the coefficient b_i and the input $x(k-i)$ lie in the range $[-c, c]$. Then the quantization level is q, as given in Eq. 5.79. Each roundoff error noise is uniformly distributed over $[-q/2, q/2]$ and has average power $\sigma_x^2 = q^2/12$. Since the $m+1$ sources are assumed to be statistically independent, their contributions can be added, which means that the average power of the noise appearing at the filter output is

FIR roundoff noise power

$$\sigma_y^2 = \frac{(m+1)q^2}{12} \qquad (5.99)$$

The roundoff noise can be reduced further if a special hardware architecture is used. Some DSP processors use a $2N$-bit double-length accumulator to store the results of the multiplications in Eq. 5.97. When this hardware configuration is used, it is only the final sum, $y(k)$, that is quantized to N bits. As a consequence, the noise term in Eq. 5.98 simplifies to

$$e(k) = Q_N \left[\sum_{i=0}^{n} b_i x(k-i) \right] - y(k) \tag{5.100}$$

In this case, there is only one roundoff noise source instead of $m + 1$ as in Eq. 5.98. The end result is that the average power of the roundoff error output noise in Eq. 5.99 is reduced by a factor of $m + 1$. This makes the use of a double-length accumulator an attractive hardware option for implementing a direct form FIR filter.

Overflow

Another source of error occurs as a result of the summing operation in Eq. 5.97. The sum of several N-bit numbers will not necessarily fit within N bits. When the sum is too large to fit, this results in *overflow error*. Overflow errors can cause a significant change in the filter output. This type of error can be eliminated, or significantly reduced, by scaling either the input or the filter coefficients. If $|x(k)| \leq c$, the FIR filter output in Eq. 5.97 is also bounded as follows.

$$|y(k)| = | \sum_{i=0}^{m} b_i x(k-i)|$$

$$\leq \sum_{i=0}^{m} |b_i x(k-i)|$$

$$= \sum_{i=0}^{m} |b_i| \cdot |x(k-i)|$$

$$\leq c \sum_{i=0}^{m} |b_i| \tag{5.101}$$

Thus, $|y(k)| \leq c\|b\|_1$, where $\|b\|_1$ is the L_1 *norm* of the coefficient vector b. That is,

$$\|b\|_1 \triangleq \sum_{i=0}^{m} |b_i| \tag{5.102}$$

From Eq. 5.101, we see that addition overflow at the output is eliminated (i.e., $|y(k)| \leq c$) when the input signal $x(k)$ is scaled by s_1, where scale factor $s_1 = 1/\|b\|_1$. A signal flow graph of a third-order direct form FIR filter realization that uses scaling to prevent overflow is shown in Figure 5.44.

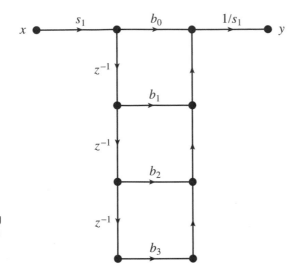

Figure 5.44 Scaling to Prevent Addition Overflow in an FIR Filter, $m = 3$

Scaling using the L_1 norm is effective in preventing overflow, but it does suffer from a practical drawback. Roundoff noise and ADC quantization noise are not affected significantly by scaling. As a result, when the input is scaled s_1, the resulting reduction in signal strength can cause a corresponding reduction in the signal-to-noise ratio. Less severe forms of scaling can be used that eliminate most, but not all, overflow. For example, if the input signal is a pure sinusoid, then overflow from this type of periodic input can be eliminated by using scaling that is based on the filter magnitude response.

$$\|b\|_\infty \stackrel{\Delta}{=} \max_{0 \le f \le f_s/2} \{A(f)\} \tag{5.103}$$

Overflow from a pure sinusoidal input is prevented if the input signal $x(k)$ is scaled by $s_\infty = 1/\|b\|_\infty$. Another commonly used form of scaling uses the L_2 or Euclidean norm.

$$\|b\|_2 \stackrel{\Delta}{=} \left(\sum_{i=0}^{m} |b_i|^2 \right)^{1/2} \tag{5.104}$$

Again, the scale factor is $s_2 = 1/\|b\|_2$. One advantage of the L_2 norm is that, like the L_1 norm, it is easy to compute. The three norms can be shown to satisfy the following relationship.

$$\|b\|_2 \le \|b\|_\infty \le \|b\|_1 \tag{5.105}$$

Example 5.13 **FIR Overflow and Scaling**

As an illustration of the prevention of overflow by scaling, suppose $|x(k)| \le 5$, and consider the following FIR filter.

$$H(z) = \sum_{i=0}^{20} \frac{z^{-i}}{1+i}$$

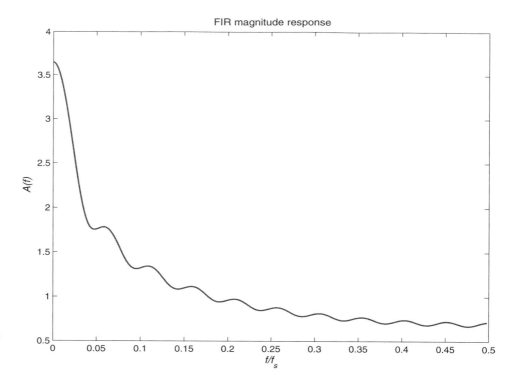

Figure 5.45 Magnitude Responses of FIR Filter in Example 5.13

Here $c = 5$. Running script *exam5_13* on the distribution CD, we get the following values for the scale factors for this filter of order $m = 20$.

$$s_1 = 3.6454$$

$$s_2 = 1.2643$$

$$s_\infty = 3.6454$$

In this instance, it turns out that the L_1 and L_∞ scale factors are identical. For this FIR system, $s_\infty = A(0)$ because the magnitude response achieves its peak at $f = 0$, as can be seen from the plot shown in Figure 5.45.

*5.8 IIR Finite Word Length Effects

IIR filters are more general than FIR filters and, because of this, there are certain additional finite word length effects that are peculiar to IIR filters. We begin by examining some finite word length effects that the two types of filters have in common. For example, input quantization error for an IIR filter is the same as that for an FIR filter. The only difference is that when the power gain Γ in Eq. 5.87 is used to compute the power of the quantization noise at the output, there is an infinite number of terms to sum for an IIR filter instead of the finite number for an FIR filter.

5.8.1 Coefficient Quantization Error

The effects of coefficient quantization error are somewhat more involved for an IIR filter because one has to consider both the poles and the zeros, not just the zeros as with an FIR filter. Recall that if the coefficients range over the interval $[-c, c]$ and N bits are used to represent the coefficients, then the quantization level is

$$q = \frac{c}{2^{N-1}} \tag{5.106}$$

If $c = 2^M$, then for a fixed-point representation, $M + 1$ bits are used for the integer part (including the sign), and $N - (M+1)$ bits are used for the fraction part. Consider an IIR filter with the following transfer function.

$$H(z) = \frac{b_0 + b_1 z^{-1} + \cdots + b_m z^{-m}}{1 + a_1 z^{-1} + \cdots + a_n z^{-n}} \tag{5.107}$$

The filter parameters, a and b, must be quantized because they are stored in fixed-length memory locations. Assuming the coefficients of $H(z)$ are quantized to N bits, this results in the following quantized transfer function.

$$H_q(z) = \frac{Q_N(b_0) + Q_N(b_1)z^{-1} + \cdots + Q_N(b_m)z^{-m}}{1 + Q_N(a_1)z^{-1} + \cdots + Q_N(a_n)z^{-n}} \tag{5.108}$$

Pole Locations

One way to evaluate the effects of coefficient quantization is to look at the locations of the poles and zeros. Recall that the roots of a polynomial can be very sensitive to small changes in the coefficients of the polynomial, particularly for higher degree polynomials. As a consequence, high-order direct-form realizations of $H(z)$ can be very sensitive to coefficient quantization error. For example, if $H(z)$ is a narrowband lowpass or highpass filter with poles clustered just inside the unit circle, then some of those poles may migrate across the unit circle and render a direct-form realization unstable. Even if the poles do not cross the unit circle, movement of a pole or a zero near the unit circle can cause a significant change in the magnitude response.

Given the sensitivity of the poles and zeros to coefficient quantization, the preferred realizations are the indirect parallel and cascade forms based on second-order blocks. For both of these realizations, the poles are decoupled from one another with each pair of poles associated with its own second-degree polynomial. For the cascade realization, this is also true for the zeros. However, for the parallel-form realization, the zeros are more sensitive to coefficient quantization because the residues in Eq. 5.64 depend on all of the coefficients of $H(z)$.

It is of interest to examine a typical second-order block in more detail. Suppose a complex-conjugate pair of poles is located at $p = r \exp(\pm j\phi)$. Then the transfer function of this block can be written as follows.

$$H(z) = \frac{b(z)}{[z - r\exp(j\phi)][z - r\exp(-j\phi)]}$$

$$= \frac{b(z)}{z^2 - r[\exp(j\phi) + \exp(-j\phi)]z + r^2}$$

$$= \frac{b(z)}{z^2 - 2r\cos(\phi)z + r^2} \tag{5.109}$$

The coefficient vector of the denominator of a second-order block is $a = [1, -2r\cos(\phi), r^2]^T$. We can recast the denominator coefficient vector in terms of the pole p as follows.

$$a = [1, -2\mathrm{Re}(p), |p|^2]^T \tag{5.110}$$

Note that the real part of the pole is proportional to a_1, but the radius of the pole is proportional to $\sqrt{a_2}$. This nonlinear dependence of the pole radius on coefficient a_2 means that achievable pole locations using quantized versions of a will not be equally spaced. For stable poles, coefficient a_1 is in the range $(-c, c)$ where $c = 2$. The distribution of possible pole locations of stable poles for the case $N = 5$ is shown in Figure 5.46. Note that not only is the grid of possible pole locations not uniform, it is also very sparse in the vicinity of the real axis.

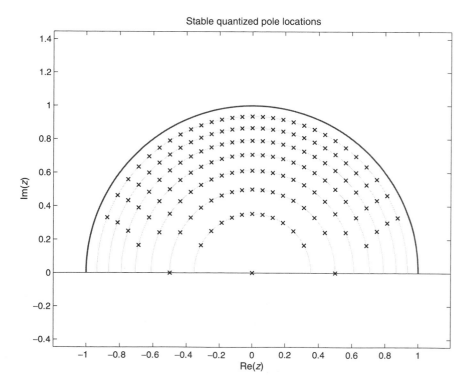

Figure 5.46 Realizable Stable Pole Locations for a Quantizing Second-order Block with $c = 2$ and $N = 5$

The problem of nonuniform placement of poles can be circumvented by using a coupled-form realization of the denominators of the second-order blocks (Rabiner *et al.*, 1970). This realization features parameters that correspond directly to the real and imaginary parts of the poles. As a result, coefficient quantization produces a uniform grid of realizable pole locations. A disadvantage of the coupled-form realization is that the number of multiplications is increased in comparison with a direct-form realization.

Zero Placement

The placement of zeros of a second-order block is also constrained as in Figure 5.46. For some filters of interest, the zeros are on the unit circle. In these cases, $r = 1$ and the second-order block transfer function simplifies to

$$H(z) = \frac{b_0[z^2 - 2\cos(\phi)z + 1]}{a(z)} \tag{5.111}$$

Since $b_2 = b_0$, the zeros of the quantized transfer function remain on the unit circle, only their angles (frequencies) change. Thus, zeros of cascade-form filters that are on the unit circle are relatively insensitive to coefficient quantization. Examples of filters with zeros on the unit circle include notch filters and inverse comb filters which are discussed in Chapter 8.

Example 5.14   **IIR Coefficient Quantization**

As an illustration of the detrimental effects of coefficient quantization error, consider a comb filter which is designed to extract a finite number of isolated equally spaced frequencies. The design of comb filters is discussed in Chapter 8. For a filter of order $n = 9$ with a pole radius of $r = 0.98$, the comb-filter transfer function is

$$H(z) = \frac{0.1663}{1 - 0.8337z^{-9}}$$

This is a relatively benign example because all but two of the coefficients can be represented exactly: only b_0 and a_9 have quantization errors. Suppose $c = 2$ and $N = 4$ are used to quantize the coefficients. Thus, the coefficient values are in the range $[-2, 2]$, and the quantization level is $q = 0.25$. A comparison of the magnitude responses for the double-precision floating-point case (64 bits) and the N-bit fixed-point case is shown in Figure 5.47, where it is evident that there is a difference in the magnitude of responses as would be expected given the low precision. The attenuation between the frequencies to be extracted is not as good for the quantized filter. When $N \geq 12$, the two magnitude responses are more or less indistinguishable. For $N < 4$, the quantized system $H_q(z)$ becomes unstable.

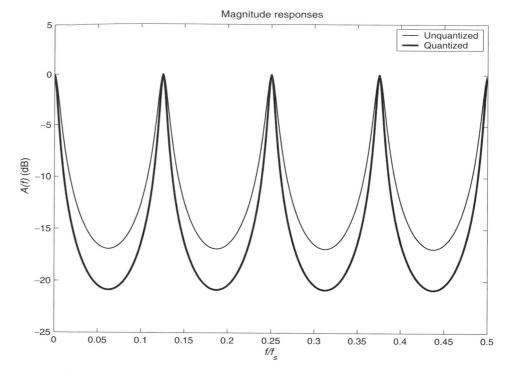

Figure 5.47 Magnitude Responses of Comb Filter using a Double-precision Floating-point Implementation and a Fixed-point Representation with $c = 2$ and $N = 4$

5.8.2 Roundoff Error, Overflow, and Scaling

As with FIR filters, the arithmetic used to compute an IIR filter output must be performed with finite precision. For example, the output of a general IIR filter can be computed as follows using a direct form II realization.

$$v(k) = \sum_{i=1}^{n} a_i v(k - i) + x(k) \tag{5.112a}$$

$$y(k) = \sum_{i=0}^{m} b_i v(k - i) \tag{5.112b}$$

If the coefficients are quantized to N bits and the signals are quantized to N bits, then the product terms will each be of length $2N$ bits. When the products are then rounded to N bits, the resulting *roundoff error* can be modeled as uniformly distributed white noise. In this instance, there are several sources of noise. Assuming the roundoff noise sources are statistically independent of one another, the noise sources

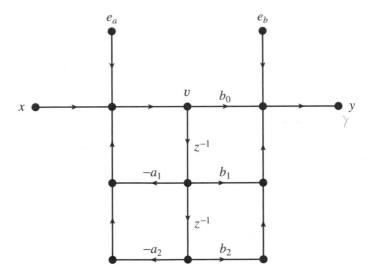

Figure 5.48   Linear Model of Product Roundoff Error in a Direct Form II Realization

associated with the input terms can be combined (similarly for the noise sources associated with the output terms).

$$e_a(k) \overset{\Delta}{=} \sum_{i=1}^{n} Q_N[a_i v(k-i)] - a_i v(k-i) \tag{5.113a}$$

$$e_b(k) \overset{\Delta}{=} \sum_{i=0}^{n} Q_N[b_i v(k-i)] - b_i v(k-i) \tag{5.113b}$$

Using a second-order direct form II realization, this results in the equivalent linear model of roundoff error displayed in Figure 5.48 for the case $m = n = 2$.

For an IIR filter, it can be shown (Oppenheim *et al.*, 1999) that the average power of the roundoff noise appearing at the filter output is as follows, where q is the signal quantization level in Eq. 5.106, and Γ is the system power gain in Eq. 5.87.

IIR roundoff noise power

$$\sigma_y^2 = \frac{(\Gamma n + m + 1)q^2}{12} \tag{5.114}$$

Overflow

Another source of error occurs as a result of the summing operations in Eq. 5.112. The sum of several N-bit numbers will not always fit within N bits. When the sum is too large to fit, this results in *overflow error*. A single overflow error can cause a significant change in filter performance. This is because when a two's-complement representation overflows, even by a small amount, it goes from a large positive number to a large negative number or conversely. This can be seen from the overflow characteristic shown in Figure 5.49, which assumes that the numbers are fractions in the interval $-1 \leq x < 1$.

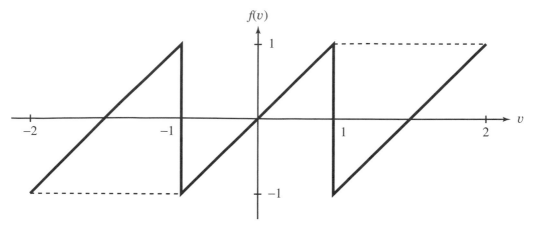

Figure 5.49   Overflow Characteristic of Two's Complement Addition

There are a number of approaches to compensating for overflow. One way is to detect overflow and then *clip* the summing junction output to the maximum value that can be represented. This clipping or saturation characteristic is shown with the dashed line in Figure 5.49. Clipping tends to reduce, but not eliminate, the detrimental effects of overflow error.

Scaling

Another approach is to eliminate overflow from occurring at all by the use of *scaling*. Let $h_i(k)$ be the impulse response measured at the output of the ith summing node. For the second-order block shown in Figure 5.48, there are two summing nodes with

$$h_1(k) = Z^{-1}\left\{\frac{1}{a(z)}\right\} \tag{5.115a}$$

$$h_2(k) = Z^{-1}\left\{\frac{b(z)}{a(z)}\right\} \tag{5.115b}$$

Notice that $h_1(k)$ is the impulse response of the auto-regressive or all-pole part, and $h_2(k)$ is the complete impulse response of $H(z)$. If $y_i(k)$ is the output of the ith summing node, and $|x(k)| \leq c$, then

$$|y_i(k)| = |\sum_{p=0}^{\infty} h_i(p)x(k-p)|$$

$$\leq \sum_{p=0}^{\infty} |h_i(p)x(k-p)|$$

$$= \sum_{p=0}^{\infty} |h_i(p)| \cdot |x(k-p)|$$

$$\leq c \sum_{p=0}^{\infty} |h_i(p)| \tag{5.116}$$

Thus, $|y_i(k)| \leq c\|h_i\|_1$, where $\|h_i\|$ is the L_1 norm of h_i. That is,

$$\|h_i\|_1 \overset{\Delta}{=} \sum_{k=0}^{\infty} |h_i(k)| \tag{5.117}$$

Notice that the L_1 norm of h_i in Eq. 5.117 is an infinite series version of the L_1 norm of the coefficient vector b in Eq. 5.102. Recall from Chapter 2 that if the system $H_i(z)$ is BIBO stable, then the impulse response $h_i(k)$ is absolutely summable. Consequently, for stable filters, the infinite series in Eq. 5.117 will converge.

Suppose there are a total of r summing nodes in the signal flow graph. Then addition overflow can be prevented if the input signal $x(k)$ is scaled by s_1, where scale factor s_p is defined as follows.

Scaling to avoid overflow

$$s_p = \frac{1}{\max_{i=1}^{r}\{\|h_i\|_p\}}, \quad 1 \leq p \leq \infty \tag{5.118}$$

A signal flow graph of a second-order direct form II realization that uses scaling to prevent overflow is shown in Figure 5.50.

Although scaling using the L_1 norm is effective in preventing overflow, it does suffer from a practical drawback. Roundoff noise and input quantization noise are not significantly affected by scaling. As a consequence, when the input is scaled by s_1, the resulting reduction in signal strength can cause a corresponding reduction in the signal-to-noise ratio. A less severe form of scaling can be used that eliminates most, but not all, overflow. In those instances where overflow does occur, it can be compensated for by using clipping. If the input signal is a pure sinusoid, then overflow from this type of periodic input can be eliminated by using scaling that is based on the filter magnitude response. Here the L_∞ norm of h_i is used, where

$$\|h_i\|_\infty \overset{\Delta}{=} \max_{0 \leq f \leq f_s/2} \{A_i(f)\} \tag{5.119}$$

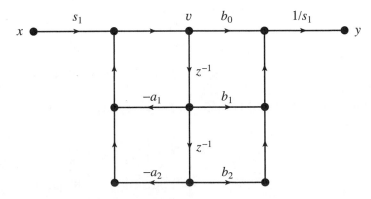

Figure 5.50 Scaling to Prevent Addition Overflow in a Second-order Direct Form II Block

Addition overflow from a pure sinusoidal input is prevented if the input signal $x(k)$ is scaled by s_∞, where s_∞ is computed as in Eq. 5.118. The most common form of scaling uses the L_2 or energy norm, which is defined as follows.

$$||h_i||_2 \overset{\Delta}{=} \left(\sum_{k=0}^{\infty} |h_i(k)|^2 \right)^{1/2} \tag{5.120}$$

Again, the scale factor s_2 is computed using Eq. 5.118. One advantage of the L_2 norm is that it is relatively easy to compute. In fact, for certain realizations, closed-form expressions for s_2 can be computed in terms of the filter coefficients (Ifeachor and Jervis, 2002). Note that the L_2 norm of h_i in Eq. 5.120 is a generalization of the Euclidean norm of the coefficient vector b in Eq. 5.104. Like their finite-dimensional counterparts, the L_1, L_2, and L_∞ norms can be shown to satisfy the following relationship.

$$||h||_2 \leq ||h||_\infty \leq ||h||_1 \tag{5.121}$$

Example 5.15   **IIR Overflow and Scaling**

As an illustration of the prevention of overflow by scaling, suppose $|x(k)| \leq 5$, and consider the following IIR filter.

$$H(z) = \frac{4z^{-1}}{1 - 0.64z^{-2}}$$

Here $c = 5$, $b = [0, 4, 0]^T$, and $a = [1, 0, -0.64]^T$. Expressing $H(z)$ in terms of positive powers of z and factoring the denominator yields:

$$H(z) = \frac{4z}{(z - 0.8)(z + 0.8)}$$

From Eq. 5.115a, the impulse response of the auto-regressive part of $H(z)$ is

$$h_1(k) = Z^{-1} \left\{ \frac{1}{(z - 0.8)(z + 0.8)} \right\}$$
$$= 0.625[(0.8)^{k-1} - (-0.8)^{k-1}]u(k - 1)$$
$$= 0.78125[(0.8)^k - (-0.8)^k]u(k)$$

Similarly, from Eq. 5.115b, the impulse response of $H(z)$ is

$$h_2(k) = Z^{-1} \left\{ \frac{4z}{(z - 0.8)(z + 0.8)} \right\}$$
$$= 2.5[(0.8)^k - (-0.8)^k]u(k)$$

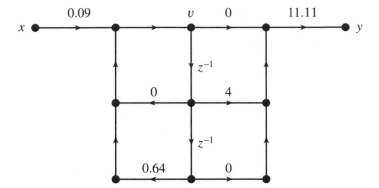

Figure 5.51 Scaling to Prevent Addition Overflow in a Second-order Direct Form II Block

Notice that $h_2(k) \geq h_1(k)$ for $k \geq 0$. Therefore, it is sufficient to compute the norm of $h(k) = h_2(k)$. From Eq. 5.117, the L_1 norm is

$$\|h\|_1 = 2.5 \sum_{k=0}^{\infty} |(0.8)^k - (-0.8)^k|$$

$$= 5 \sum_{k=0}^{\infty} (0.8)^{2k+1}$$

$$= 5(0.8) \sum_{k=0}^{\infty} (0.64)^k$$

$$= \frac{4}{1 - 0.64}$$

$$= 11.11$$

Thus, a scale factor which will eliminate fixed-point overflow is $s_1 = 1/11.11$ or

$$s_1 = 0.09$$

A signal flow graph of a direct form II realization with scaling to avoid overflow is shown in Figure 5.51.

5.8.3   Limit Cycles

For IIR filters, there is an unusual finite word length effect that is sometimes observed when the input goes to zero. Recall that if a filter is stable and the input goes to zero, then the output should approach zero as the natural-mode terms die out. However, for finite-precision IIR filters the output sometimes approaches a nonzero constant or it oscillates. These zero-input oscillations, which are called *limit cycles*, are a nonlinear phenomena. There are two types of limit cycles. The first is a limit cycle that is caused by overflow error. Overflow limit cycles can be quite large in amplitude, but they can be eliminated if the outputs of the summing junctions are clipped. Of course, scaling by s_1 will also eliminate this type of limit cycle. The second type of limit cycle is a small limit cycle of amplitude q that can occur as a result of product roundoff error.

<table>
<tr><td>**Example 5.16**</td><td>**Limit Cycle**</td></tr>
</table>

As an illustration of a limit cycle caused by product roundoff error, consider the following quantized first-order IIR system.

$$y(k) = Q_N[-0.7y(k-1)] + 3x(k)$$

Suppose $N = 4$ bits are used with a scale factor of $c = 4$. In this case, the quantization level is

$$q = \frac{4}{2^3} = 0.5$$

Next, consider the impulse response of the quantized system, $h_q(k)$. The result, computed using script *exam5_16* on the distribution CD, is shown in Figure 5.52. Note that even though $H(z)$ clearly is stable with a pole at $p = -0.7$, the steady-state response does not go to zero.   Instead, it oscillates with period two and amplitude q because of the product roundoff error. For comparison, the unquantized impulse response, $h(k)$, is also displayed using a solid line.

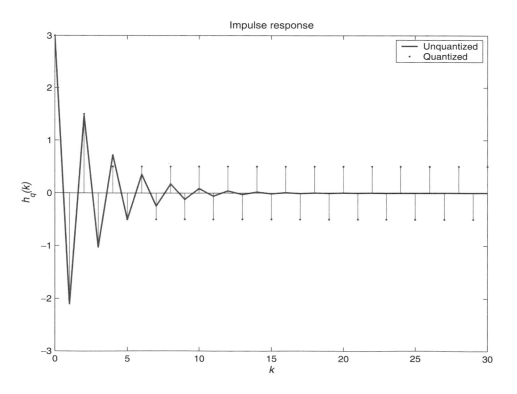

Figure 5.52 Limit Cycle in Impulse Response Caused by Product Roundoff Error using $N = 4$ Bits and a Scale Factor of $c = 4$

In IIR filters, limit-cycle solutions can be small limit cycles associated with round-off error, as in Figure 5.52, or large amplitude limit cycles associated with overflow. For FIR filters, limit cycles are *not* possible because there are no feedback paths to sustain an oscillation. Indeed, in comparison with IIR filters, FIR filters are relatively less sensitive to finite word length effects in general. Along with their guaranteed stability, this is one of the advantages of FIR filters that accounts for their popularity.

MATLAB Toolbox

The FDSP toolbox contains the following functions for evaluating finite word length effects using different filter realization structures.

```
y = f_filter (b,a,x,bits,realize);       % Quantized filter output
h = f_impulse (b,a,M,bits,realize);      % Quantized filter impulse response
[H,f] = f_freqz (b,a,N,fs,bits,realize); % Quantized filter frequency response
```

The function *f_filter* is a generalization of the built-in MATLAB function *filter* that is used to compute the zero-state response of a system with the following transfer function.

$$H(z) = \frac{b_0 + b_1 z^{-1} + \cdots + b_m z^{-m}}{1 + a_1 z^{-1} + \cdots + a_n z^{-n}} \qquad (5.122)$$

On entry to *f_filter*, b is a vector of length $m + 1$ containing the numerator coefficients, a is a vector of length $n + 1$ containing the denominator coefficients with $a(1) = 1$, x is a vector of length P containing the samples of the input, *bits* is an integer specifying the number of bits to be used for coefficient quantization, and *realize* is an integer specifying the realization structure to use as summarized in Table 5.4.

Inputs *bits* and *realize* are optional. If they are not present, then a direct-form realization with no coefficient quantization is used. This is the same as the MATLAB function *filter* which uses a double-precision floating-point implementation. Quantization can be removed by setting *bits* = [ ] or setting *bits* to a high value such as *bits* = 64. On exit from *f_filter*, y is a $P \times 1$ vector containing the zero-state response to the input x using coefficient quantization and the selected filter realization structure.

Table 5.4: ▶
Selection of the
Filter Realization
Structure

realize	FIR	IIR
0	Direct	Direct
1	Cascade	Cascade
2	Lattice	Parallel

The function *f_impulse* computes the quantized impulse response of $H(z)$ using the specified filter realization structure. The inputs to *f_impulse* are identical to those of *f_filter* except that input x is replaced by the desired number of samples M. On exit from *f_impulse*, h is an $M \times 1$ vector containing the quantized impulse response using the selected filter realization structure.

The function *f_freqz* computes the quantized frequency response of $H(z)$ using the specified filter realization structure. The inputs to *f_freqz* are identical to those of *f_filter* except that input x is replaced by the number of evaluation points N and the sampling frequency fs. On exit from *f_freqz*, H is an $N \times 1$ complex vector containing the quantized frequency response using the selected filter realization structure, and f is an $N \times 1$ vector of evaluation frequencies with $f(i) = (i - 1)fs/(2N)$ for $1 \leq i \leq N$.

5.9 Software Applications

5.9.1 GUI Module: *g_filters*

FDSP Toolbox

The FDSP toolbox includes a graphical user interface module called *g_filters* that allows the user to construct filters from design specifications and evaluate finite word length effects, all without any need for programming. GUI module *g_filters* features a display screen with tiled windows as shown in Figure 5.53.

The *Block diagram* window in the upper-left corner contains a block diagram of the filter under investigation which can be an FIR filter, an IIR filter, or a user-defined filter. The FIR filters are designed with the windowing method to be discussed in Chapter 6, while the IIR filters are Butterworth filters created with the bilinear transformation to be discussed in Chapter 8. The following transfer function is used where $a = 1$ for the FIR filters.

$$H(z) = \frac{b_0 + b_1 z^{-1} + \cdots b_m z^{-m}}{1 + a_1 z^{-1} + \cdots + a_n z^{-n}} \tag{5.123}$$

The *Parameters* window below the block diagram displays edit boxes containing the filter parameters. The contents of each edit box can be directly modified by the user with the changes activated with the Enter key. Parameters F_0, F_1, B, and fs are the lower cutoff frequency, upper cutoff frequency, transition bandwidth, and sampling frequency, respectively. The lowpass filter uses cutoff frequency F_0, the highpass filter uses cutoff frequency F_1, and the bandpass and bandstop filters use both F_0 and F_1. The parameters *delta_p* and *delta_s* specify the passband ripple and stopband attenuation, respectively.

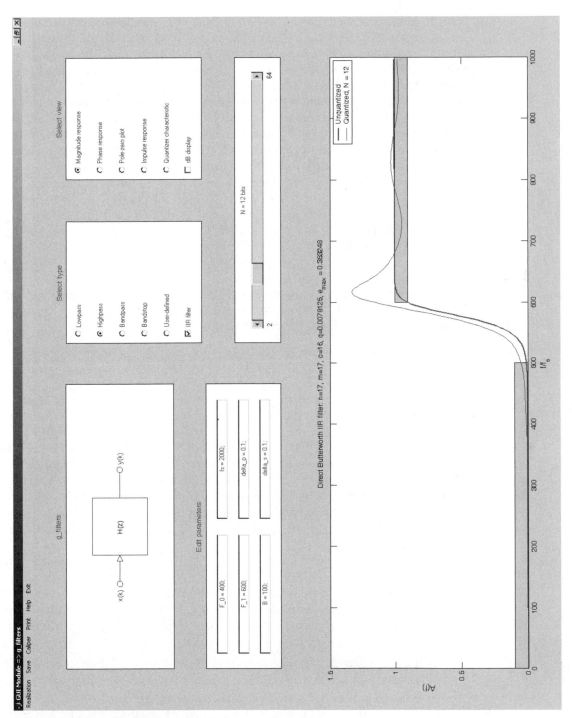

Figure 5.53 Display Screen of Chapter GUI Module *g_filters*

The *Type* and *View* windows in the upper-right corner of the screen allow the user to select both the type of filter and the viewing mode. There are two categories of filter types. The user can select either an FIR filter or an IIR filter using the IIR checkbox. Within each of these types, the user can then select the basic filter type: lowpass, highpass, bandpass, or bandstop. There is also a user-defined filter whose parameters are defined in a user-supplied MAT-file that contains a, b, and fs. The *View* options include magnitude responses, phase responses, pole-zero plots, impulse responses, and the quantizer input-output characteristic. The dB checkbox toggles between logarithmic (dB) and linear displays of the magnitude response. When dB is checked, the passband ripple and stopband attenuation factors in the *Parameter* window also change to their logarithmic equivalents, A_p and A_s.

Just below the *Type* and *View* windows is a horizontal slider bar, which controls N: the number of bits of precision used for coefficient quantization. To determine the quantization level, first $c_{\max}$ is computed such that $|b_i| \leq c_{\max}$ for $0 \leq i \leq m$ and $|a_i| \leq c_{\max}$ for $0 \leq i \leq n$. Scale factor c is then set to the next higher power of two as follows.

$$c = 2^{\text{ceil}[\log_2(c_{\max})]} \tag{5.124}$$

This way, a fixed-point representation of the filter coefficients uses $M + 1$ bits for the integer part and $N - M - 1$ bits for the fraction part where $M = \log_2(c)$. This corresponds to the following coefficient quantization level.

$$q = \frac{c}{2^{N-1}} \tag{5.125}$$

All of the view options, except the quantizer characteristic, display two cases for comparison. The first is a double-precision floating-point filter which approximates the *unquantized* case, and the second is a fixed-point N-bit *quantized* filter using coefficient quantization. Higher order IIR filters can become unstable for relatively small values of N. When this happens, the migration of the quantized poles outside the unit circle can be viewed directly using the pole-zero plot option. The *Plot* window along the bottom half of the screen shows the selected view.

The *Menu* bar at the top of the screen includes several menu options. The *Realization* option allows the user to choose between direct- and cascade-form realization structures. The *Save* option is used to save the current a and b and fs in a user-specified MAT-file for future use. Files created in this manner can be loaded with the User-defined input option. The *Caliper* option allows the user to measure any point on the current plot by moving the mouse crosshairs to that point and clicking. The *Print* option prints the contents of the plot window. Finally, the *Help* option provides the user with some helpful suggestions on how to use module *g_filters* effectively.

5.9.2   Highpass Elliptic Filter

To illustrate the use of filter design specifications and the effects of finite word length, suppose $f_s = 200$ Hz and consider the problem of constructing a highpass digital filter to meet the following design specifications.

$$F_s = 40 \text{ Hz} \qquad (5.126\text{a})$$

$$F_p = 42 \text{ Hz} \qquad (5.126\text{b})$$

$$\delta_p = 0.05 \qquad (5.126\text{c})$$

$$\delta_s = 0.05 \qquad (5.126\text{d})$$

There are many filters that can meet or exceed these specifications. Design techniques for FIR filters are discussed in Chapter 6 and for IIR filters are discussed in Chapter 8. Notice that the width of the transition band is relatively small at

$$B = |F_p - F_s|$$

$$= 2 \text{ Hz} \qquad (5.127)$$

As B, δ_p, and δ_s are decreased, the required filter order increases. As we shall see, the filter with the smallest order that still meets the specifications is an elliptic filter. An elliptic filter is an IIR filter that is optimal in the sense that the magnitude response contains ripples of equal amplitude in both the passband and the stopband. The coefficients of an elliptic filter can be obtained by using the FDSP toolbox function $f\_ellipticz$, as shown in the following script, labeled $exam5\_17$ on the distribution CD. The design of elliptic filters is discussed in detail in Chapter 8, where the theory behind $f\_ellipticz$ is introduced.

```
function exam5_17

% Example 5.17: Highpass elliptic filter

clear
clc
fprintf('Example 5.17: Highpass elliptic filter\n')

% Initialize

fs = 200;
F_s = f_prompt ('Enter stopband cutoff frequency',0,fs/2,40);
F_p = f_prompt ('Enter passband cutoff frequency',F_s,fs/2,42);
delta_p = f_prompt ('Enter passband ripple factor',0,0.5,0.05);
delta_s = f_prompt ('Enter stopband attenuation factor',0,0.5,0.05);
N = f_prompt ('Enter number of bits of precision',2,64,10);

% Compute filter coefficients

f_type = 2;
[b,a] = f_ellipticz (F_p,F_s,delta_p,delta_s,f_type,fs);
n = length(a) - 1
```

```
% Compute quantized filter coefficients

c = max(abs([a(:) ; b(:)]));
c = 2^ceil(log(c)/log(2))
q = c/2^(N-1)
a_q = f_quant (a,q,c,0);
b_q = f_quant (b,q,c,0);

% Compute and plot magnitude responses

p = 250;
[H,f] = f_freqz (b,a,p,fs);
A = abs(H);
[H_q,f] = f_freqz (b_q,a_q,p,fs);
A_q = abs(H_q);
figure
h = plot (f,A,f,A_q);
set (h(1),'LineWidth',1.5)
q_str = sprintf ('Quantized, N = %d',N);
legend ('Unquantized',q_str)
hold on
fill ([0 F_s F_s 0],[0 0 delta_s delta_s],'c')
fill ([F_p fs/2 fs/2 F_p],[1-delta_p 1-delta_p 1 1],'c')
h = plot (f,A,f,A_q);
set (h(1),'LineWidth',1.5)
f_wait

% Pole-zeros plots

figure
subplot (2,2,1)
f_pzplot(b,a,'Unquantized Filter');
axis ([-1.5 1.5 -1.5 1.5])
for N = [10 8 6]
    q = c/(2^(N-1));
    a_q = f_quant (a,q,c,0);
    b_q = f_quant (b,q,c,0);
```

```
        switch N
            case 10, subplot (2,2,2);
            case 8, subplot (2,2,3);
            case 6, subplot (2,2,4)
        end
        caption = sprintf ('Quantized Filter, N = %d',N);
        f_pzplot (b_q,a_q,caption);
        axis ([-1.5 1.5 -1.5 1.5])
end
f_wait
```

When script *exam5_17* is run, it produces an elliptic filter of order $n = 6$. It then computes and plots two magnitude responses as shown in Figure 5.54. The second magnitude response is that of a quantized elliptic highpass filter using coefficient quantization with a scale factor of $c = 4$ and $N = 10$ bits. Note that, whereas the unquantized (double-precision floating-point) filter meets the design specifications, the quantized filter clearly does not.

The loss in the fidelity of the magnitude response can be attributed to the fact that the poles and zeros of the quantized filter tend to move from their optimal positions as the quantization level increases. Script *exam5_17* generates four plots of the poles and zeros corresponding to different levels of quantization as shown in

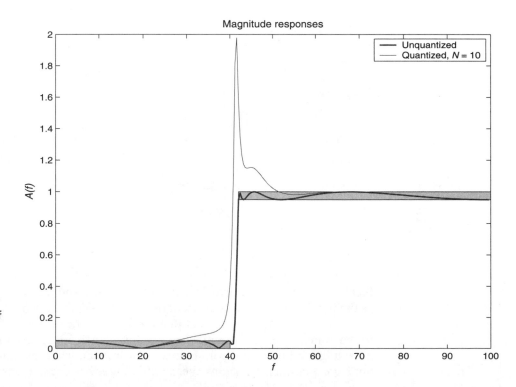

Figure 5.54 Magnitude Responses of Unquantized and Quantized Elliptic Highpass Filters of Order $n = 6$

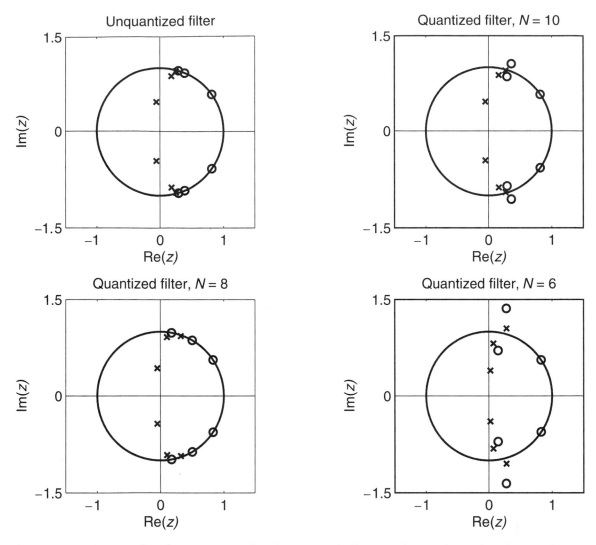

Figure 5.55 Pole-zero Plots for Unquantized and Quantized Elliptic Highpass Filters of Order $n = 6$

Figure 5.55. Inspection reveals that both the poles and the zeros move significantly as the number of bits for coefficient quantization decreases from $N = 10$ to $N = 6$. This can be attributed to the fact that the roots of the polynomial are very sensitive to the values of the polynomial coefficients. For the final case of $N = 6$, the poles have migrated outside the unit circle, which means that this implementation does not have a frequency response because the filter has become unstable.

• • • • • • • • • • • • • • • •

5.10 Chapter Summary

This chapter focused on filter design specifications, realization structures, and finite word length effects. Both FIR and IIR filters were examined. The following IIR transfer function reduces to an FIR transfer function when $a_i = 0$ for $1 \leq i \leq n$.

$$H(z) = \frac{b_0 + b_1 z^{-1} + \cdots + b_m z^{-m}}{1 + a_1 z^{-1} + \cdots + a_n z^{-n}} \qquad (5.128)$$

Magnitude Response Specifications

Filter design specifications typically involve the desired magnitude response, $A(f)$. The basic types of frequency-selective filters are lowpass, highpass, bandpass, and bandstop filters. For a lowpass filter, the design specifications are as follows.

$$1 - \delta_p \leq A(f) \leq 1, \quad 0 \leq f \leq F_p \qquad (5.129a)$$

$$0 \leq A(f) \leq \delta_s, \quad F_s \leq f \leq f_s/2 \qquad (5.129b)$$

Here, Eq. 5.129a is the passband specification, and Eq. 5.129b is the stopband specification. For the passband, $0 < F_p < f_s/2$ is the passband cutoff frequency and $\delta_p > 0$ is the passband ripple factor. For the stopband, $F_p < F_s < f_s/2$ is the stopband cutoff frequency, and $\delta_s > 0$ is the stopband attenuation factor. Left unspecified is the band of frequencies between the passband and the stopband, which is called the transition band. The width of the transition band is

$$B = |F_s - F_p| \qquad (5.130)$$

As the transition bandwidth B, passband rippled δ_p, and stopband attenuation δ_s all approach zero, the order of the filter required to meet the specifications approaches infinity. The limiting special case of $B = 0$, $\delta_p = 0$, and $\delta_s = 0$ is an ideal lowpass filter. Bandpass filters have two stopbands bracketing a passband, and bandstop filters have two passbands bracketing a stopband. The filter magnitude response is often represented using a logarithmic scale of decibels (dB) as follows.

$$A(f) = 20 \log_{10}\{|H(f)|\} \text{ dB} \qquad (5.131)$$

The ripple and attenuation factors, δ_p and δ_s, have logarithmic equivalents, A_p and A_s, that are expressed in units of dB. The logarithmic scale can be useful for showing the degree of attenuation in the stopband.

Phase Response Specifications

Often the desired phase response of a filter is left unspecified. One noteworthy exception is the design of linear-phase filters. Linear-phase filters delay all spectral components of the input by the same amount and therefore do not distort the spectral components that pass through the filter. Although a linear phase characteristic can be approximated in the passband with an IIR Bessel filter, the simplest way to design an exact linear-phase filter is to use an FIR filter with an impulse response that satisfies the following linear-phase *symmetry constraint.*

$$h(k) = \pm h(m - k), \quad 0 \le k \le m \tag{5.132}$$

The symmetry constraint in Eq. 5.132 is a direct constraint on the FIR coefficients, because for an FIR filter, the nonzero part of the impulse response is specified by the numerator coefficients.

$$h(k) = \begin{cases} b(k), & 0 \le k \le m \\ 0, & m < k < \infty \end{cases} \tag{5.133}$$

If the plus sign is used in Eq. 5.132, the impulse response $h(k)$ is a palindrome that exhibits even symmetry about the midpoint $k = m/2$, otherwise it exhibits odd symmetry. There are four types of linear-phase FIR filters depending on whether the symmetry is even or odd and the filter order m is even or odd. The most general linear-phase filter is a type 1 filter with even symmetry and even order. The other three filter types have zeros at one or both ends of the frequency range, and are sometimes used for specialized applications. The zeros of a linear-phase FIR filter that are not on the unit circle occur in groups of four, because for every zero at $z = r \exp(j\phi)$, there is a reciprocal zero at $z = r^{-1} \exp(j\phi)$.

Every IIR transfer function $H(z)$ has a minimum-phase form, $H_{\min}(z)$, whose magnitude response is the same as that of $H(z)$, but whose phase response has the least amount of phase lag possible. The minimum-phase form of $H(z)$ can be obtained as follows.

$$H_{\min}(z) = H_{\text{all}}^{-1}(z) H(z) \tag{5.134}$$

Here $H_{\text{all}}(z)$ is an allpass filter that is constructed from the zeros of $H(z)$ that lie outside the unit circle. The allpass filter is stable and passes all spectral components equally with a magnitude response of one.

Filter Realization Structures

There are a number of alternative signal flow graph realizations of FIR and IIR filters. Direct-form realizations have the property that the gains in the signal flow graphs and the difference equations are obtained directly from inspection of the transfer function. For FIR filters, these include the tapped delay line, the transposed tapped delay line, and a direct-form realization for linear-phase filters that requires only about half as many floating-point multiplications or FLOPs. For IIR filters, direct-form realizations include the direct form I, direct form II, and transposed direct form II structures. The direct form II realizations are canonic in the sense that they require the minimum number of memory locations to store past signal samples.

There are also a number of indirect realizations whose parameters must be computed from the original transfer function. The indirect forms decompose the original transfer function into lower-order blocks by combining complex conjugate pairs of poles and zeros. For example, both FIR and IIR filters can be realized with the following cascade-form realization which is based on factoring $H(z)$.

$$H(z) = b_0 H_1(z) \cdots H_M(z) \tag{5.135}$$

Here $M = \text{floor}[(m+1)/2]$ and $H_i(z)$ are second-order blocks with real coefficients, except for $H_M(z)$ which is a first-order block when the filter order m is odd. Another FIR filter realization is the lattice-form realization that consists of m blocks and a signal flow graph that resembles a lattice ladder structure. For IIR filters, an additional indirect filter structure is the parallel-form realization that is based on a partial fraction expansion of $H(z)$.

$$H(z) = R_0 + \sum_{i=1}^{N} H_i(z) \tag{5.136}$$

Again, $N = \text{floor}[(n+1)/2]$, and each $H_i(z)$ is a second-order block with real coefficients, except $H_N(z)$ which is a first order block when n is odd. All of the filter realizations are equivalent to one another in terms of their overall input-output behavior if infinite-precision arithmetic is used.

Finite Word Length Effects

Finite word length effects arise when a filter is implemented in either hardware or software. They are caused by the fact that both the filter parameters and the filter signals must be represented using a finite number of bits of precision. Both floating-point and fixed-point numerical representations can be used. MATLAB uses a double-precision floating-point representation. When an N-bit fixed-point representation is used for values in the range $[-c, c]$, the quantization level, or spacing between adjacent values, is

$$q = \frac{c}{2^{N-1}} \tag{5.137}$$

Typically, the scale factor is $c = 2^M$ for some integer $M \geq 0$. This way, $M+1$ bits are used to represent the integer part including the sign, and the remaining $N - (M+1)$ bits are used for the fraction part.

Quantization error, which can arise from input or ADC quantization, coefficient quantization, and product roundoff quantization, typically is modeled using additive white noise uniformly distributed over $[-q/2, q/2]$. Another source of error is overflow error which can occur when several finite-precision numbers are added. Overflow error can be eliminated by proper scaling of the input. The roots of a polynomial are very sensitive to the changes in the polynomial coefficients, particularly for high-degree polynomials. It is for this reason that IIR filter implementations can become unstable when their poles migrate across the unit circle as a result of quantization error. In addition, overflow error and roundoff error can cause steady-state oscillations in an IIR filter output after the input goes to zero. These oscillations are called limit cycles. Quantized FIR filters do not become unstable or exhibit limit-cycle oscillations, and they generally are less sensitive to finite word length effects than IIR filters. For both FIR and IIR filters, the indirect-form realizations tend to be less sensitive to finite word length effects because the block transfer functions are only of second order.

The FDSP toolbox includes a GUI module called *g_filters* that allows the user to construct filters using design specifications and examine finite word length effects, in each case without any need for programming. The filters include FIR and IIR filters, lowpass, highpass, bandpass, and bandstop filters, plus user-defined filters. The effects of coefficient quantization on filter performance can be investigated as a function of the number of bits of precision.

● ● ● ● ● ● ● ● ● ● ● ● ● ●

5.11 Problems

The problems are divided into Analysis problems that can be solved by hand or with a calculator, GUI Simulation problems that are solved using GUI module *g_filters*, and Computation problems. The Computation problems require the student to write a MATLAB script using the FDSP toolbox functions summarized in Appendix B. Solutions to selected problems are available on the distribution CD. Students are encouraged to use these problems, which are identified with a ☑, as a check on their understanding of the material.

5.11.1 Analysis

Section 5.2

P5.1. Consider the following first-order IIR filter.

$$H(z) = \frac{0.4(1 - z^{-1})}{1 + 0.2z^{-1}}$$

(a) Compute and sketch the magnitude response $A(f)$.
(b) What type of filter is this (lowpass, highpass, bandpass, bandstop)?
(c) Suppose $F_p = 0.4f_s$. Find the passband ripple δ_p.
(d) Suppose $F_s = 0.2f_s$. Find the stopband attenuation δ_s.

☑ **P5.2.** A bandpass filter has a sampling frequency of $f_s = 2000$ Hz and satisfies the following design specifications.

$$[F_{s1}, F_{p1}, F_{p2}, F_{s2}, \delta_p, \delta_s] = [200, 300, 600, 700, 0.15, 0.05]$$

(a) Find the logarithmic passband ripple, A_p.
(b) Find the logarithmic stopband attenuation, A_s.
(c) Using a logarithmic scale, sketch the shaded passband and stopband regions that $A(f)$ must lie within.

P5.3. A bandstop filter has a sampling frequency of $f_s = 200$ Hz and satisfies the following design specifications.

$$[F_{p1}, F_{s1}, F_{s2}, F_{p2}, A_p, A_s] = [30, 40, 60, 80, 2, 30]$$

(a) Find the linear passband ripple, δ_p.
(b) Find the linear stopband attenuation, δ_s.
(c) Using a linear scale, sketch the shaded passband and stopband regions that $A(f)$ must lie within.

Section 5.3

P5.4. Consider the following FIR filter of order m known as a *running average* filter.

$$H(z) = \frac{1 + z^{-1} + \cdots + z^{-m}}{m + 1}$$

(a) Find the impulse response of this filter.

(b) Is this a linear-phase filter? If so, what type?

(c) Find the group delay of this filter.

P5.5. A linear-phase FIR filter $H(z)$ of order $m = 8$ has zeros at $z = \pm j0.5$ and $z = \pm 0.8$.

(a) Find the remaining zeros of $H(z)$ and sketch the poles and zeros in the complex plane.

(b) The DC gain of the filter is 2. Find the filter transfer function $H(z)$.

(c) Suppose the input signal gets delayed by 20 msec as it passes through this filter. What is the sampling frequency, f_s?

P5.6. Construct a type 1 linear-phase filter of order $m = 2$ with coefficients satisfying $|b_k| = 1$ for $0 \leq k \leq m$.

(a) Find the transfer function, $H(z)$.

(b) Find the amplitude response, $A_r(f)$.

(c) Find the zeros of $H(z)$.

P5.7. Construct a type 2 linear-phase filter of order $m = 1$ with coefficients satisfying $|b_k| = 1$ for $0 \leq k \leq m$.

(a) Find the transfer function, $H(z)$.

(b) Find the amplitude response, $A_r(f)$.

(c) Find the zeros of $H(z)$.

P5.8. Construct a type 3 linear-phase filter of order $m = 2$ with coefficients satisfying $|b_k| = 1$ for $0 \leq k \leq m$.

(a) Find the transfer function, $H(z)$.

(b) Find the amplitude response, $A_r(f)$.

(c) Find the zeros of $H(z)$.

P5.9. Construct a type 4 linear-phase filter of order $m = 1$ with coefficients satisfying $|b_k| = 1$ for $0 \leq k \leq m$.

(a) Find the transfer function, $H(z)$.

(b) Find the amplitude response, $A_r(f)$.

(c) Find the zeros of $H(z)$.

P5.10. Consider the following FIR filter.

$$H(z) = 3 + z^{-1} - 4z^{-2} + 5z^{-3} - 7z^{-4}$$

(a) Is this a linear-phase filter? If so, what is the type?

(b) Sketch a signal flow graph showing a tapped-delay-line direct-form realization of $H(z)$.

(c) Sketch a signal flow graph showing a transposed tapped-delay-line direct-form realization of $H(z)$.

P5.11. Let $H(z)$ be an arbitrary FIR transfer function of order m. Show that $H(z)$ can be written as a sum of two linear-phase transfer functions $H_e(z)$ and $H_o(z)$ where $h_e(k)$ exhibits even symmetry about $k = m/2$ and $h_o(k)$ exhibits odd symmetry about $k = m/2$.

$$H(z) = H_e(z) + H_o(z)$$

Section 5.4

P5.12. Consider the following IIR filter.

$$H(z) = \frac{2(z + 1.25)(z^2 + 0.25)}{z(z^2 - 0.81)}$$

(a) Find the minimum-phase version of this system, and sketch its poles and zeros.

(b) Find the maximum-phase version of this system, and sketch its poles and zeros.

(c) How many transfer functions with real coefficients have the same magnitude response as $H(z)$? ∞ or 2

P5.13. The following IIR filter has two parameters α and β. For what values of these parameters is this an allpass filter?

$$H(z) = \frac{1 + 3z^{-1} + (\alpha + \beta)z^{-2} + 2z^{-3}}{2 + (\alpha - \beta)z^{-1} + 3z^{-2} + z^{-3}}$$

☑ **P5.14.** Consider the following IIR filter.

$$H(z) = \frac{10(z^2 - 4)(z^2 + 0.25)}{(z^2 + 0.64)(z^2 - 0.16)}$$

(a) Find $H_{min}(z)$, the minimum-phase version of $H(z)$.

(b) Sketch the poles and zeros of $H_{min}(z)$.

(c) Find an allpass filter $H_{all}(z)$ such that $H(z) = H_{all}(z)H_{min}(z)$.

(d) Sketch the poles and zeros of $H_{all}(z)$.

P5.15. Let $H(z)$ be a nonzero linear-phase FIR filter of order $m = 2$.

(a) Is it possible for $H(z)$ to be a minimum-phase filter? If so, construct an example. If not, why not?

(b) Is it possible for $H(z)$ to be an allpass filter? If so, construct an example. If not, why not?

Section 5.5

P5.16. Consider the following FIR filter. Find a cascade-form realization of this filter and sketch the signal flow graph.

$$H(z) = \frac{10(z^2 - 0.6z - 0.16)[(z - 0.4)^2 + .25]}{z^4}$$

P5.17. Consider the following FIR filter. Find a lattice-form realization of this filter and sketch the signal flow graph.

$$H(z) = 1 + 2z^{-1} + 3z^{-2} + 4z^{-3}$$

P5.18. Find an efficient direct-form realization for a linear-phase filter of order $m = 2r$ similar to Eq. 5.44, but applicable to a type 3 filter. Sketch the signal flow graph for the case $m = 4$.

Section 5.6

P5.19. Sketch a direct form I signal flow graph realization of the following IIR transfer function.

$$H(z) = \frac{0.8 - 1.2z^{-1} + 0.4z^{-3}}{1 - 0.9z^{-1} + 0.6z^{-2} + 0.3z^{-3}}$$

P5.20. Sketch a direct form II signal flow graph realization of the following difference equation.

$$y(k) = 10x(k) + 2x(k - 1) - 4x(k - 2) + 5x(k - 3) - 0.7y(k - 2) + 0.4y(k - 3)$$

P5.21. Sketch a transposed direct form II signal flow graph realization of the following transfer function.

$$H(z) = \frac{1 - 2z^{-1} + 3z^{-2} - 4z^{-3}}{1 + 0.8z^{-1} + 0.6z^{-2} + 0.4z^{-3}}$$

P5.22. Consider the following IIR system.

$$H(z) = \frac{z^3}{(z - 0.8)(z^2 - z + 0.24)}$$

(a) Expand $H(z)$ into partial fractions.

(b) Sketch a parallel-form signal flow graph realization by combining the two poles that are closest to the unit circle into a second-order block.

☑ **P5.23.** Consider the following IIR system.

$$H(z) = \frac{2(z^2 + 0.64)(z^2 - z + 0.24)}{(z^2 + 1.2z + 0.27)(z^2 + 0.81)}$$

(a) Sketch the poles and zeros of $H(z)$.

(b) Sketch a cascade-form signal flow graph realization by grouping the complex zeros with the complex poles. Use a direct form II realization for each block.

Section 5.7

P5.24. Suppose a 12-bit fixed-point representation is used to represent values in the range $-10 \leq x < 10$.

(a) How many distinct values of x can be represented?

(b) What is the quantization level, or spacing between adjacent values?

P5.25. Consider the system shown in Figure P5.25. The ADC has a precision of 10 bits and an input range of $|x_a(t)| \leq 10$. The transfer function of the digital filter is

$$H(z) = \frac{3z^2 - 2z}{z^2 - 1.2z + 0.32}$$

(a) Find the quantization level of the ADC.

(b) Find the average power of the quantization noise at the input x.

(c) Find the power gain of $H(z)$.

(d) Find the average power of the quantization noise at the output y.

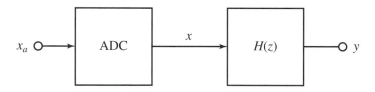

P5.25 ADC Quanti-
zation Noise

P5.26. Suppose a 16-bit fixed-point representation is used for values in the range $|x| \leq 8$.

(a) How many distinct values of x can be represented?

(b) What is the quantization level, or spacing, between adjacent values?

(c) How many bits are used to represent the integer part (including the sign)?

(d) How many bits are used to represent the fraction part?

P5.27. Suppose the coefficients of an FIR filter of order $m = 30$ all lie within the range $|b_i| \leq 4$. Assuming they are quantized to $N = 12$ bits, find an upper bound on the error in the magnitude of the frequency response caused by coefficient quantization.

P5.28. A high-order FIR filter is realized as a cascade of second-order blocks.

(a) Suppose the filter has a sampling rate of $f_s = 300$ Hz and a zero at $z_0 = \exp(j\pi/3)$. Find a nonzero periodic signal $x(k)$ that gets attenuated completely by the filter.

(b) If a zero of a second-order block starts out on the unit circle, will the radius of the zero change as a result of the coefficient quantization? That is, will the zero still be on the unit circle?

(c) If a zero of a second-order block starts out on the unit circle, will the angle of the zero change as a result of the coefficient quantization? That is, will the frequency of the zero change?

P5.29. Consider the following FIR filter.

$$H(z) = \frac{(z^2 + 25)(z^2 + 0.04)}{z^4}$$

(a) Show that this is a type 1 linear-phase filter.

(b) Sketch a signal flow graph realization of $H(z)$ that is still a linear-phase system even when the coefficients are quantized.

P5.30. Consider the following FIR filter.

$$H(z) = 3 + 4z^{-1} + 6z^{-2} + 4z^{-3} + 3z^{-4}$$

Suppose the input signal lies in the range $|x(k)| \leq 10$. Find the scale factor for the input which ensures that the filter output will not overflow the range $|y(k)| \leq 10$.

Section 5.8

P5.31. Consider the following IIR filter. Suppose 8-bit fixed-point arithmetic is used to implement this filter using a scale factor of $c = 4$.

$$H(z) = \frac{2z}{z + 0.7}$$

(a) Find the quantization level q.

(b) Find the power gain of this filter.

(c) Find the average power of the product roundoff error.

P5.32. Consider the following IIR filter.

$$H(z) = \frac{0.5}{z + 0.9}$$

(a) Sketch a direct form II signal flow graph of $H(z)$.

(b) Suppose all filter variables are represented as fixed-point fractions, and the input is constrained to $|x(k)| \leq 2$. Find a scale factor, s_1, that eliminates summing junction overflow error.

(c) Sketch a modified direct form II signal flow graph of $H(z)$ that implements scaling to eliminate summing junction overflow.

P5.33. For the system in Problem P5.32, find a scale factor s_∞ that will eliminate summing junction overflow when the input is a pure sinusoid of amplitude $\alpha \leq 5$.

P5.34. Let $f_{\text{clip}}(x)$ be the following unit clipping nonlinearity.

$$f_{\text{clip}}(x) \triangleq \begin{cases} -1, & -\infty < x < -1 \\ x, & -1 \leq x \leq 1 \\ 1, & 1 < x < \infty \end{cases}$$

Show how f_{clip} can be used to eliminate limit cycles due to overflow error by sketching a modified direct form II signal flow graph of a second-order IIR block. You can assume all values are represented as fractions.

5.11.2   GUI Simulation

Section 5.9

P5.35. Use the GUI module *g_filters* to analyze an IIR bandpass filter. Use the default parameter values, but adjust the number of bits of precision N to the lowest value possible without the quantized filter going unstable.

(a) Plot the magnitude response.

(b) Plot the pole-zero pattern.

☑ **P5.36.** Use the GUI module *g_filters* and select an IIR highpass filter. Use the default parameter values, but adjust the number of bits of precision N to highest value that still makes the quantized filter go unstable.

(a) Plot the unstable pole-zero plot.

(b) Increase N by one so the quantized filter becomes stable. Then plot the impulse response.

P5.37. Use the GUI module *g_filters* and select the User-defined filter option. Load the filter in MAT-file *u_filters1*. Use the default parameter values, but set the number of bits for coefficient quantization to $N = 8$.

(a) Plot the magnitude response using a direct-form realization.

(b) Plot the phase response using a direct-form realization.

(c) Plot the magnitude response using a cascade-form realization.

(d) Plot the phase response using a cascade-form realization.

P5.38. A notch filter is a filter that is designed to remove a single frequency (see Chapter 8). Consider the following transfer function for a notch filter.

$$H(z) = \frac{0.9766(1 + z^{-1} + z^{-2})}{1 + 0.9764z^{-1} + 0.9534z^{-2}}$$

Create a MAT-file that contains $fs = 1000$ and the a and b for this filter. Then use the User-defined option of GUI module *g_filters* to load this filter. Choose the smallest value of N that gives a maximum error of less than 0.001.

(a) Plot the magnitude response. Use the *Caliper* option to estimate the notch frequency.

(b) Plot the phase response.

(c) Plot the pole-zero pattern.

P5.39. Use the GUI module *g_filters* and select an FIR bandstop filter. Use the default parameter values, but adjust the number of bits of precision N until the quantization level q is larger than 0.05.

(a) Plot the magnitude response.

(b) Plot the impulse response.

(c) Plot the quantizer characteristic.

P5.40. Use the GUI module *g_filters* and select an FIR lowpass filter. Adjust the parameter values to $f_s = 100$ Hz, $F_0 = 30$ Hz, and $B = 10$ Hz. Use the default value for N.

(a) Plot the magnitude response using the dB scale.

(b) Plot the phase response.

(c) Plot the impulse response. Is this a linear-phase filter? If so, what type?

P5.41. Consider the following *running average filter*. Create a MAT-file that contains $fs = 300$, a, and b for this filter.

$$y(k) = \frac{1}{10} \sum_{i=0}^{9} x(k - i)$$

Use the GUI module *g_filters* with the User-defined option to load this filter, and use the default value for N.

(a) Plot the magnitude response.

(b) Plot the phase response.

(c) Plot the pole-zero plot.

(d) Plot the impulse response. Is this a linear-phase filter? If so, what type?

P5.42. The derivative of an analog signal $x_a(t)$ can be approximated numerically by taking differences between the samples of the signal using the following first-order *backwards Euler differentiator*.

$$y(k) = \frac{x(k) - x(k-1)}{T}$$

Create a MAT-file that contains $fs = 10$, a, and b for this filter. Then use GUI module *g_filters* with the User-defined option to load this filter. Use the default value for N.

(a) Plot the magnitude response.

(b) Plot the phase response.

(c) Plot the impulse response. Is this a linear-phase filter? If so, what type?

5.11.3 Computation

Section 5.4

P5.43. Consider the following IIR filter.

$$H(z) = \frac{1 + 1.75z^{-2} - 0.5z^{-4}}{1 + 0.4096z^{-4}}$$

(a) Write a MATLAB script that uses *f_minall* to compute and print the coefficients of the minimum-phase and allpass parts of $H(z)$.

(b) Use the MATLAB *subplot* command to plot the magnitude responses $A(f)$, $A_{min}(f)$, and $A_{all}(f)$ on a single screen using three separate plots.

(c) Repeat part (b), but for the phase responses.

(d) Use *f_pzplot* to plot the poles and zeros of $H(z)$, $H_{min}(z)$ and $H_{all}(z)$ on one screen using three separate square plots.

Sections 5.5 and 5.7

P5.44. Consider the following FIR transfer function.

$$H(z) = \sum_{i=0}^{20} \frac{z^{-i}}{1+i}$$

(a) Write a MATLAB script that uses *f_lattice* to compute a lattice-form realization of this filter. Print the gain and the reflection coefficients of the blocks.

(b) Suppose the sampling frequency is $f_s = 600$ Hz. Use *f_freqz* to compute the frequency response using a lattice-form realization. Compute both the unquantized frequency response (e.g., 64 bits), and the frequency response with coefficient quantization using $N = 8$ bits. Plot both magnitude responses on a single plot using the dB scale and a legend.

P5.45. Consider the following FIR impulse response. Suppose the filter order is $m = 30$.

$$h(k) = \frac{k+1}{m}, \quad 0 \le k \le m$$

(a) Write a MATLAB script that uses $f\_cascade$ to compute a cascade-form realization of this filter. Print the gain b_0 and the block coefficients, B and A.

(b) Suppose the sampling frequency is $f_s = 400$ Hz. Use f_freqz to compute the frequency response using a cascade-form realization. Compute both the unquantized frequency response (set bits = 64), and the frequency response with coefficient quantization using 8 bits. Plot both magnitude responses on a single plot using the dB scale and a legend.

Sections 5.6 and 5.8

☑ **P5.46.** A *comb filter* (see Chapter 8) is a filter that extracts a set of isolated equally spaced frequencies from a signal. Consider the following comb filter that has n *teeth*.

$$H(z) = \frac{b_0}{1 - r^n z^{-n}}$$

Here the filter gain is $b_0 = 1 - r^n$. Suppose $n = 10$, $r = 0.98$, and $f_s = 300$ Hz. Write a MATLAB script that uses f_freqz to compute the frequency response using a direct-form realization. Compute both the unquantized frequency response (set bits = 64), and the frequency response with coefficient quantization using $N = 4$ bits. Plot both magnitude responses on a single plot using the linear scale and a legend.

P5.47. An *inverse comb filter* (see Chapter 8) is a filter that removes a set of isolated equally spaced frequencies from a signal. Consider the following inverse comb filter that has n *teeth*.

$$H(z) = \frac{b_0(1 - z^{-n})}{1 - r^n z^{-n}}$$

Here the filter gain is $b_0 = (1 + r^n)/2$. Suppose $n = 8$ and $r = 0.96$.

(a) Write a MATLAB script that uses $f\_parallel$ to compute a parallel-form realization of this filter. Print the parameter R_0 and the block coefficients, B and A.

(b) Suppose the sampling frequency is $f_s = 200$ Hz. Use f_freqz to compute the frequency response using a parallel-form realization. Compute both the unquantized frequency response (set bits = 64), and the frequency response with coefficient quantization using $N = 8$ bits. Plot both magnitude responses on a single plot using the linear scale and a legend.

Section 5.8

P5.48. Write a MATLAB function called *f_filtnorm* that returns the L_p norm, $\|h\|_p$, of a digital filter. The function *f_filtnorm* should use the following calling sequence.

```
d = f_filtnorm (b,a,p);          % Return L_p norm of filter
```

On entry to *f_filtnorm*, b is a coefficient vector of length $m+1$ specifying the numerator polynomial, a is a coefficient vector of length $n+1$ specifying the denominator polynomial with $a(1) = 1$, and p is an integer specifying the norm type. To get the L_∞ norm, the user should set $p = $ Inf. On exit from *f_filtnorm*, d is the L_p norm $\|h\|_p$. Test *f_filtnorm* by writing a MATLAB script that computes and prints the L_1, L_2, and L_∞ norms of the comb filter in Problem P5.46. Verify that Eq. 5.121 holds in this case.

CHAPTER 6

FIR Filter Design

Learning Objectives

- Understand the relative advantages and disadvantages of FIR filters in comparison with IIR filters (Section 6.1)

- Know how to measure the signal-to-noise ratio (Section 6.1)

- Be able to design a linear-phase FIR filter with a prescribed magnitude response using the windowing method (Section 6.2)

- Be able to design a linear-phase FIR filter with a prescribed magnitude response using the frequency-sampling method (Section 6.3)

- Be able to design a linear-phase FIR filter with a prescribed magnitude response using the least-squares method (Section 6.4)

- Be able to design an optimal linear-phase equiripple frequency-selective filter using the Parks–McClellan algorithm (Section 6.5)

- Know how to design FIR differentiators and Hilbert transformers (Section 6.6)

- Know how to use the GUI module g_fir to design and analyze digital FIR filters without any programming (Section 6.7)

● ● ● ● ● ● ● ● ● ● ● ● ● ● ● ⋯ ⋯

6.1 **Motivation**

A digital filter is a discrete-time system that is designed to reshape the spectrum of the input signal in order to produce desired spectral characteristics in the output signal. In this chapter, we focus on a specific type of digital filter, the finite impulse response or FIR filter. An mth order FIR filter is a discrete-time system with the following generic transfer function.

FIR filter

$$H(z) = b_0 + b_1 z^{-1} + \cdots + b_m z^{-m}$$

FIR filters offer a number of important advantages in comparison with IIR filters. The nonzero part of the impulse response of an FIR filter is simply $h(k) = b_k$ for $0 \le k \le m$. Consequently, an FIR impulse response can be obtained directly from inspection of the transfer function or the difference equation. Since the poles of an FIR filter are all located at the origin, FIR filters are always stable. FIR filters tend to be less sensitive to finite word length effects. Unlike IIR filters, quantized FIR filters cannot become unstable or exhibit limit-cycle oscillations. FIR filters can be designed to closely approximate arbitrary magnitude responses if the order of the filter is allowed to get sufficiently large. Furthermore, the phase response of an FIR filter can be made to be linear. This is an important characteristic, because it means that different spectral components of the input signal are delayed by the same amount as they are processed by the filter. A linear-phase filter does not distort a signal within the passband; it only delays it.

Although impressive control of the frequency response can be achieved with FIR filters, these filters do suffer from some limitations in comparison with IIR filters. One fundamental drawback occurs when we attempt to design a sharp frequency-selective filter with a narrow transition band, similar to an IIR elliptic filter. To meet the same design specifications with an FIR filter, a much higher-order filter is required. The increased order means larger storage requirements and a longer computational time. Computational time is particularly important in real-time applications where intersample signal processing must be completed before each new ADC sample arrives. High-order FIR filters are more suitable for offline batch processing where the entire input signal is available ahead of time.

We begin this chapter by introducing an example of an application of FIR filters. Next, several methods for FIR filter design are presented. They all produce linear-phase FIR filters with a propagation delay of $\tau = mT/2$, where m is the filter order, and T is the sampling interval. The first design method is the windowing method, a simple and effective technique that is based on soft or gradual truncation of the impulse response. The second design method is the frequency-sampling method, which uses the inverse DFT to compute the filter coefficients. The frequency-sampling method can be optimized with the inclusion of transition-band samples. The third design method is the least-squares method. This is an optimization method that uses an arbitrary set of discrete frequencies and a user-selectable weighting function. All three of these methods can be used to produce a filter with a prescribed magnitude response. The last method is also an optimization method that uses the

Parks–McClellan algorithm to design an equiripple frequency-selective filter that is analogous to the IIR elliptic filter. Specialized linear-phase FIR filters including differentiators and Hilbert transformers are then presented. Finally, a GUI module called *g fir* is introduced that allows the user to design and evaluate a variety of FIR filters without any need for programming. The chapter concludes with a presentation of an application example, and a summary of FIR filter design techniques. The selected application example compares four design methods applied to a bandstop filter.

6.1.1 First-order Differentiator

In engineering applications, there are instances where it is useful to obtain the derivative of an analog signal from samples of the signal. For example, an estimate of velocity might be obtained from position measurements, or an acceleration estimate might be obtained from velocity measurements. Numerical differentiation is a difficult practical problem because the process is highly sensitive to the effects of noise. As an introduction to the topic, we examine the use of simple low-order FIR filters to numerically approximate the differentiation process. The objective is to design a digital equivalent of the following analog system.

$$H_a(s) = s \qquad (6.1)$$

Suppose the analog signal to be differentiated, $x_a(t)$, is approximated with a linear polynomial in the neighborhood of $t = kT$.

$$x_a(t) \approx x(k) + c(t - kT) \qquad (6.2)$$

Recall that $x(k) = x_a(kT)$. Let $\dot{x}_a(t) = dx_a(t)/dt$ denote the derivative of $x_a(t)$. Then from Eq. 6.2, we have $\dot{x}_a(kT) = c$. Evaluating Eq. 6.2 at $t = (k - 1)T$ and solving for c, this yields the following first-order approximation for the derivative.

First-order
differentiator

$$y_1(k) = \frac{x(k) - x(k - 1)}{T} \qquad (6.3)$$

The formulation in Eq. 6.3 is called a *backward Euler* numerical approximation to the derivative. Note that it is an FIR filter of order $m = 1$ with coefficient vector $b = [1/T, -1/T]$. A block diagram of the first-order backward differentiator is shown in Figure 6.1.

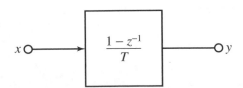

Figure 6.1 Block Diagram of a First-order Backward Euler Differentiator

6.1.2   Second-order Differentiator

The FIR filter model of the differentiation process can be improved if we approximate $x_a(t)$ with a quadratic polynomial as follows.

$$x_a(t) \approx x(k) + c(t - kT) + d(t - kT)^2 \tag{6.4}$$

Again, $\dot{x}_a(kT) = c$. To determine the parameter c, we evaluate Eq. 6.4 at $t = (k-1)T$ and at $t = (k-2)T$ which yields

$$x(k-1) = x(k) - Tc + T^2 d \tag{6.5a}$$

$$x(k-2) = x(k) - 2Tc + 4T^2 d \tag{6.5b}$$

If Eq. 6.5b is subtracted from four times Eq. 6.5a, the dependence on d drops out. The resulting equation can then be solved for c which yields the following second-order differentiator.

Second-order
differentiator

$$y_2(k) = \frac{3x(k) - 4x(k-1) + x(k-2)}{2T} \tag{6.6}$$

This FIR filter is called a second-order backward differentiator. It is backward, because it makes use of current and past samples, not future samples. This is important if the differentiator is to be used in real-time applications such as feedback control. A block diagram of the second-order backward differentiator is shown in Figure 6.2.

It would be tempting to continue this process using higher-order polynomial approximations of $x_a(t)$. Unfortunately, one quickly reaches a point of diminishing return in terms of accuracy. This is because higher-degree polynomials are not effective in modeling the underling trend of signals. Even though they can be made to go through all of the samples, they tend to do so by oscillating wildly between the samples as the polynomial degree increases. Later, we will examine a different higher-order design approach that can be used to approximate a delayed version of a differentiator. To examine the effectiveness of the differentiator in Figure 6.2, consider the following noise-free input signal.

$$x_a(t) = \sin(\pi t) \tag{6.7}$$

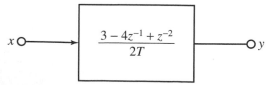

Figure 6.2 Block Diagram of a Second-order Backwards Differentiator

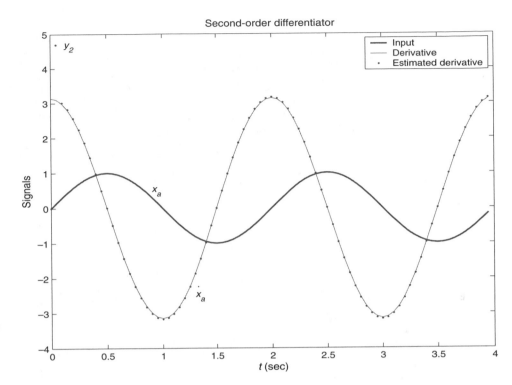

Figure 6.3 Numerical Approximation to the Derivative of a Sinusoidal Input using a Second-order FIR Filter

Plots of $x_a(t)$, $\dot{x}_a(t)$, and $y_2(k)$ are shown in Figure 6.3. Here a sampling frequency of $f_s = 20$ Hz is used. After a short start-up transient, the approximation is effective in this case.

6.1.3 Signal-to-noise Ratio

Although the approximation to the derivative in Figure 6.3 is effective for the noise-free signal in Eq. 6.7, when noise is added to $x_a(t)$ the numerical approximations to the derivative rapidly deteriorate. To illustrate, consider the noise-corrupted signal

$$y(k) = x(k) + v(k) \tag{6.8}$$

Here $v(k)$ is white noise uniformly distributed over the interval $[-c, c]$. To specify the size of the noise relative to the size of the signal, the following concept be used.

DEFINITION

6.1: Signal-to-noise Ratio

Let $y(k) = x(k) + v(k)$ represent a signal $x(k)$ that is corrupted with noise $v(k)$. If P_x is the average power of the signal, and P_v is the average power of the noise, then the *signal-to-noise ratio* is defined as

$$\text{SNR}(y) \triangleq 10 \log_{10} \left(\frac{P_x}{P_v} \right) \ \text{dB}$$

The average power is the mean or expected value of the square of the signal, $P_x = E[x^2(k)]$. A signal that is all noise has a signal-to-noise ratio of minus infinity, while a noise-free signal has a signal-to-noise ratio of plus infinity. When the signal-to-noise ratio is zero dB, the power of the signal is equal to the power of the noise.

Using the trigonometric identities from Appendix D, one can show that the average power of a sinusoid of amplitude A is $A^2/2$. From Eq. 3.85, the average power of white noise uniformly distributed over $[-c, c]$ is $P_v = c^2/3$. Suppose $c = 0.01$, which represents only a modest amount of noise. Then from Definition 6.1, the signal-to-noise ratio of the noise corrupted sinusoid in Eq. 6.8 is

$$\text{SNR}(y) = 10 \log_{10}\left[\frac{1/2}{(0.01)^2/3}\right]$$

$$= 10 \log_{10}\left(\frac{10^4}{6}\right)$$

$$= 32.22 \text{ dB} \qquad (6.9)$$

When the noise-corrupted sinusoid in Eq. 6.8 is processed with the second-order differentiator in Figure 6.2, the results are as shown in Figure 6.4. Note that $x_a(t)$ itself is not significantly distorted because the signal-to-noise ratio is relatively large. However, the numerical approximation to $\dot{x}_a(t)$ has deteriorated significantly in comparison with the noise-free case in Figure 6.3. This is because the differentiation process

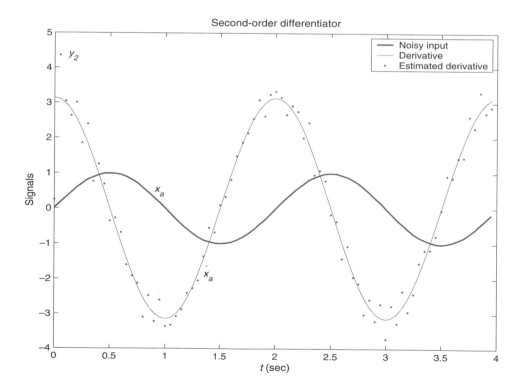

Figure 6.4 Numerical Approximations to the Derivative of a Noisy Sinusoidal Input using a Second-order FIR Filter

amplifies the high-frequency part of the noise. In this particular instance, most of the noise could have been removed by first preprocessing $y(k)$ with a narrowband resonator filter with resonant frequency $F_0 = 0.5$ Hz. Resonator filters are discussed in Chapter 8. However, for a more general broadband signal, $x_a(t)$, this is not a viable option because the filtering would remove important spectral components of $x_a(t)$ itself.

Typically, additive white noise $v(k)$ is uncorrelated with the signal $x(k)$ which means that $E[x(k)v(k)] = E[x(k)]E[v(k)]$. For uncorrelated noise, consider the relationship between the average power of the noise-corrupted signal and the average power of the noise-free signal. Here,

$$
\begin{aligned}
P_y &= E[y^2(k)] \\
&= E[\{x(k) + v(k)\}^2] \\
&= E[x^2(k) + 2x(k)v(k) + v^2(k)] \\
&= E[x^2(k)] + E[2x(k)v(k)] + E[v^2(k)] \\
&= P_x + 2E[x(k)]E[v(k)] + P_v
\end{aligned}
\tag{6.10}
$$

The relationship is particularly simple for zero-mean noise where $E[v(k)] = 0$. In this case, Eq. 6.10 reduces to

Uncorrelated zero-mean noise

$$
P_y = P_x + P_v
\tag{6.11}
$$

Thus, for a signal $x(k)$ that is corrupted with uncorrelated zero-mean white noise $v(k)$, the average power of the noise-corrupted signal $y(k)$ is just the sum of the average power of the signal plus the average power of the noise. Typically, $y(k)$ is known or can be measured. If the average power of either the signal $x(k)$ or the noise $v(k)$ are known, or can be computed, then the average power of the other can be determined using Eq. 6.11. The signal-to-noise ratio of $y(k)$ can then be determined using Definition 6.1.

● ● ● ● ● ● ● ● ● ● ● ● ● ● ● ●

6.2 Windowing Method

In this section, we examine a simple technique for designing a linear-phase FIR filter with a prescribed magnitude response. First, we briefly revisit the filter design specifications introduced in Chapter 5. For linear-phase FIR filters, the design specifications are often formulated in terms of the desired *amplitude response*, $A_r(f)$, introduced in Eq. 5.20. Recall that the amplitude response is real but can be positive or negative. An FIR lowpass design specification based on the desired amplitude response is shown in Figure 6.5. Note that the passband ripple parameter, δ_p, now represents the radius of a region centered about $A_r(f) = 1$ within which the amplitude response must lie. The same is true for the stopband attenuation. These specifications based on the amplitude response also carry over to highpass, bandpass, and bandstop filters.

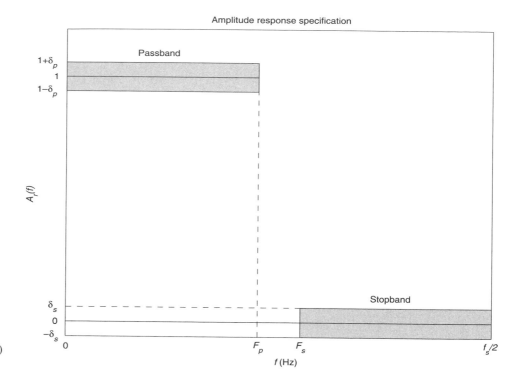

Amplitude response specification

Figure 6.5 Amplitude Response Specification for an FIR Lowpass Filter with $|A_r(f)| = A(f)$

6.2.1 Truncated Impulse Response

To develop a linear-phase filter with a prescribed magnitude response, recall that the frequency response is defined in terms of the transfer function as follows.

$$H(f) = H(z)|_{z=\exp(j2\pi fT)} \tag{6.12}$$

The filters that we have considered thus far all have been causal filters with $h(k) = 0$ for $k < 0$. In order to develop a general design technique that can be used in a wide variety of applications, we shall consider noncausal filters as well. If $h(k)$ is noncausal, then the Z-transform used in Eq. 6.12 should be a two-sided Z-transform that includes both positive and negative powers of z. This yields the following more general expression for the desired frequency response.

$$H(f) = \sum_{k=-\infty}^{\infty} h(k) \exp(-j2\pi kfT) \tag{6.13}$$

Using Euler's identity in Eq. 6.13, we see that $H(f)$ is periodic with period f_s. Consequently, we can interpret Eq. 6.13 as a complex Fourier series of the periodic function $H(f)$. The kth coefficient of the Fourier series is

$$h(k) = \frac{1}{f_s} \int_{-f_s/2}^{f_s/2} H(f) \exp(j2\pi kfT) df, \quad -\infty < k < \infty \tag{6.14}$$

Using Eq. 6.14, one can recover the impulse response $h(k)$ from a desired frequency response $H(f)$. Since we are interested in designing a linear-phase filter, the frequency response will take on one of two forms.

Type 1 and Type 2 Filters

Let m be the filter order, and suppose $h(k)$ exhibits even symmetry about $k = m/2$. Then from Eq. 5.20 and Table 5.1, the desired frequency response for a type 1 or type 2 causal linear-phase filter with group delay $\tau = mT/2$ is

Even amplitude response

$$H(f) = A_r(f) \exp(-j\pi mfT) \; \Big\} \text{ type 1 and 2} \qquad (6.15)$$

For a type 1 or type 2 linear-phase filter, the amplitude response, $A_r(f)$, is a real even function that is specified by the filter designer. The desired magnitude response is $|A_r(f)| = A(f)$. Using Eq. 6.15, Euler's identity, and the fact that $A_r(f)$ is even, the expression for $h(k)$ in Eq. 6.14 becomes

$$
\begin{aligned}
h(k) &= \frac{1}{f_s} \int_{-f_s/2}^{f_s/2} A_r(f) \exp(-j\pi mfT) \exp[j2\pi kfT] df \\
&= T \int_{-f_s/2}^{f_s/2} A_r(f) \exp[j2\pi(k - 0.5m)fT] df \\
&= T \int_{-f_s/2}^{f_s/2} A_r(f) \{\cos[2\pi(k - 0.5m)fT] + j\sin[2\pi(k - 0.5m)fT]\} df \\
&= T \int_{-f_s/2}^{f_s/2} A_r(f) \cos[2\pi(k - 0.5m)fT] df \qquad (6.16)
\end{aligned}
$$

Since the integrand in Eq. 6.16 is even, the integral can be performed over the positive frequencies and doubled. This results in the following impulse response for a type 1 or type 2 linear-phase filter order of m.

Truncated impulse response, even

$$h(k) = 2T \int_{0}^{f_s/2} A_r(f) \cos[2\pi(k - 0.5m)fT] df, \quad 0 \le k \le m \qquad (6.17)$$

Type 3 and Type 4 Filters

Next, suppose the impulse response exhibits odd symmetry about $k = m/2$. Then from Eq. 5.20 and Table 5.1, the desired frequency response for a causal type 3 or type 4 filter with group delay $\tau = mT/2$ is

Odd amplitude response

$$H(f) = jA_r(f) \exp(-j\pi mfT) \; \Big\} \text{ type 3 and 4} \qquad (6.18)$$

Note that we have used the fact that $\exp(j\pi/2) = j$. For a type 3 or type 4 linear-phase filter, the amplitude response, $A_r(f)$, is a real, odd function that is specified by the designer to get the desired magnitude response, $|A_r(f)| = A(f)$. Using Eq. 6.18, Euler's identity, and the fact that $A_r(f)$ is odd, the expression for the impulse response in Eq. 6.14 becomes

$$
\begin{aligned}
h(k) &= \frac{j}{f_s} \int_{-f_s/2}^{f_s/2} A_r(f) \exp[j2\pi(k-0.5m)fT]df \\
&= jT \int_{-f_s/2}^{f_s/2} A_r(f)\{\cos[2\pi(k-0.5m)fT] + j\sin[2\pi(k-0.5m)fT]\}df \\
&= -T \int_{-f_s/2}^{f_s/2} A_r(f)\sin[2\pi(k-0.5m)fT]df \quad (6.19)
\end{aligned}
$$

Here we have used the fact that $\exp(j\pi/2) = j$. Again, the integrand in Eq. 6.19 is even, so the integral can be performed over the positive frequencies and doubled. This yields the following impulse response for a type 3 or type 4 linear-phase filter order of m.

Truncated impulse response, odd

$$
h(k) = -2T \int_{0}^{f_s/2} A_r(f)\sin[2\pi(k-0.5m)fT]df, \quad 0 \le k \le m \quad (6.20)
$$

Recall that the type 1 linear-phase filter (even symmetry, even order) is the most general in the sense that there are no zeros at $f = 0$ or $f = f_s/2$. Using Eq. 6.17 with order $m = 2p$, the impulse responses for the four ideal frequency-selective filters are summarized in Table 6.1.

Table 6.1: ▶
Impulse Responses
of Ideal
Frequency-selective
Linear-phase Type 1
Filters of Order
$m = 2p$

Filter	$h(p)$	$h(k)$, $0 \le k \le m$, $k \ne p$
Lowpass	$2F_0T$	$\dfrac{\sin[2\pi(k-p)F_0T]}{\pi(k-p)}$
Highpass	$1 - 2F_0T$	$\dfrac{-\sin[2\pi(k-p)F_0T]}{\pi(k-p)}$
Bandpass	$2[F_1 - F_0]T$	$\dfrac{\sin[2\pi(k-p)F_1T] - \sin[2\pi(k-p)F_0T]}{\pi(k-p)}$
Bandstop	$1 - 2[F_1 - F_0]T$	$\dfrac{\sin[2\pi(k-p)F_0T] - \sin[2\pi(k-p)F_1T]}{\pi(k-p)}$

| **Example 6.1** | **Truncated Impulse Response Filter** |

As an illustration of an FIR filter designed by the truncated impulse response method, suppose $f_s = 100$ Hz, and consider the problem of designing a lowpass filter with cutoff frequency $F_0 = f_s/4$. Suppose $m = 40$ is used to approximate $H(f)$. Then $p = 20$, and from Table 6.1, the filter coefficients are $h(p) = 0.5$, and

$$h(k) = \frac{\sin[0.5\pi(k-p)]}{\pi(k-p)}, \quad 0 \le k \le m, \ k \ne p$$

The delay in this case is

$$\tau = pT$$

$$= 0.2 \text{ sec}$$

The impulse response of this filter can be obtained by running script *exam6_1* on the distribution CD. From the resulting plot, shown in Figure 6.6, we see that this is indeed a type 1 linear-phase FIR filter with a palindrome-like impulse response, $h(k)$, centered about $k = 20$.

The magnitude response and phase response generated by script *exam6_1* are shown in Figure 6.7. Although the magnitude response is an effective approximation to the ideal lowpass characteristic, it is evident that there is *ringing* or oscillation,

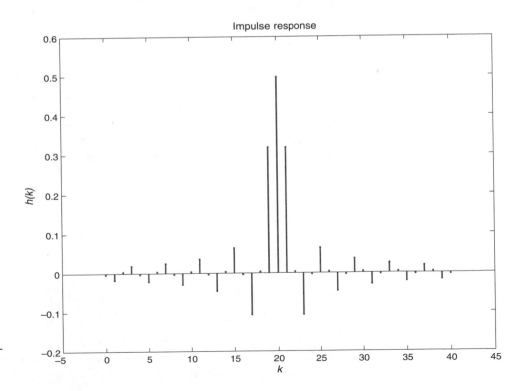

Figure 6.6 Palindrome Impulse Response of a Type 1 Linear-phase FIR Filter with $m = 40$

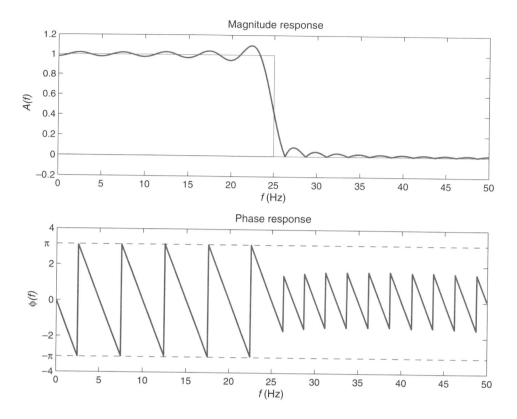

Figure 6.7 Frequency Response of a Truncated Impulse Response FIR Filter with $m = 40$

particularly in the neighborhood of the cutoff frequency, $F_0 = 25$ Hz. The approximation to an ideal lowpass characteristic can be improved by increasing the filter order. However, even for large values of p, significant ringing persists near $f = F_0$. This is an inherent characteristic of the truncated impulse response that is present whenever there is a jump discontinuity in the function being approximated. The oscillation near a jump discontinuity is referred to as *Gibb's phenomenon*.

It is also of interest to observe the phase response carefully. The jump discontinuities of amplitude 2π in the passband occurring at $\phi = -\pi$ are an artifact of the fact that the phase is computed modulo 2π, so at $-\pi$ it wraps around to π. However, the jump discontinuities of amplitude π in the stopband are discontinuities that occur because of a sign change in the amplitude response, $A_r(f)$. Notice that all of these jumps occur at points in the set F_z where $A(f) = 0$.

6.2.2 Windowing

The beauty of the truncated impulse response method is that it allows us to design a magnitude response of prescribed shape. However, if the desired magnitude response contains one or more jump discontinuities, then the oscillations caused by

Gibb's phenomenon effectively prohibit the design of filters having a very small passband ripple or stopband attenuation. Fortunately, there is a tradeoff we can make that reduces the ripples at the expense of increasing the width of the transition band. To see how this can be done, first notice that the filter transfer function can be rewritten as follows.

$$H(z) = \sum_{i=-\infty}^{\infty} w_R(i)h(i)z^{-i} \tag{6.21}$$

This corresponds to a causal filter order of m when $w_R(i)$ is the following *rectangular window* order of m.

$$w_R(i) \triangleq \begin{cases} 1, & 0 \leq i \leq m \\ 0, & \text{otherwise} \end{cases} \tag{6.22}$$

Thus, truncation of the impulse response is equivalent to multiplication of the impulse response by a rectangular window. It is the abrupt truncation of the impulse response that causes the oscillations associated with Gibb's phenomenon. The amplitude of the oscillations can be decreased by tapering the impulse response to zero more gradually. Recall that for an FIR filter the numerator coefficients are $b_i = h(i)$. Therefore, if $w(i)$ represents a window order of m, then the tapered or windowed filter coefficients are

Windowed filter coefficients

$$\boxed{b_i = w(i)h(i), \quad 0 \leq i \leq m} \tag{6.23}$$

There are many windows that have been proposed. Some of the more popular ones are summarized in Table 6.2. They include the rectangular window (also called the boxcar window), the Hanning window, the Hamming window, and the Blackman window. Recall from Chapter 3 that these are the same data windows that were used with Welch's method to estimate the power density spectrum of a signal.

Plots of the windows are shown in Figure 6.8 for the case $m = 60$. Notice that they all are symmetric about $i = m/2$ and attain a peak value of $w(m/2) = 1$

Table 6.2: ▶
Windows of Order
m

Type	Name	$w(i)$, $0 \leq i \leq m$
0	Rectangular	1
1	Hanning	$0.5 - 0.5\cos\left(\dfrac{\pi i}{0.5m}\right)$
2	Hamming	$0.54 - 0.46\cos\left(\dfrac{\pi i}{0.5m}\right)$
3	Blackman	$0.42 - 0.5\cos\left(\dfrac{\pi i}{0.5m}\right) + 0.08\cos\left(\dfrac{2\pi i}{0.5m}\right)$

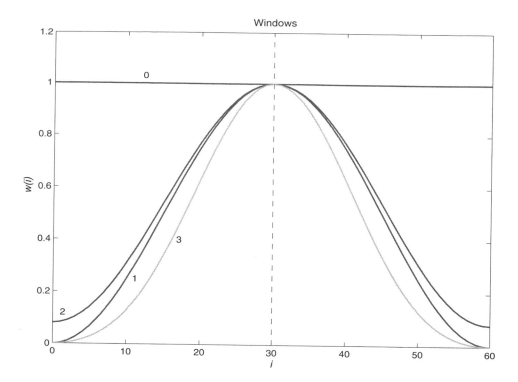

Figure 6.8 Windows used to Taper Truncated Impulse Response: 0 = rectangular, 1 = Hanning, 2 = Hamming, and 3 = Blackman

at the midpoint. With the exception of the rectangular window, they all gradually taper to zero at the end points $i = 0$ and $i = m$, except for the Hamming window which is near zero. Multiplication by a tapered window can be thought of as a form of *soft* truncation as opposed to the hard or abrupt truncation of the rectangular window. Since the windows are all palindromes with $w(i) = w(m - i)$, filters using the windowed coefficients in Eq. 6.23 continue to be linear-phase filters.

Example 6.2 | **Windowed Lowpass Filter**

To illustrate the effects of the different windows, consider the design of a lowpass filter with cutoff frequency $F_0 = f_s/4$. Suppose $m = 40$ and $p = m/2$. Then from Tables 6.1 and 6.2 and from Eq. 6.23, the filter coefficients using the rectangular window are $b_p = 0.5$ and

$$b_i = \frac{\sin[0.5\pi(i - p)]}{\pi(i - p)}, \quad 0 \le i \le m, \ i \ne p$$

A plot of the rectangular magnitude response, using normalized frequency $\hat{f} = f/f_s$, is shown in Figure 6.9. Here the logarithmic dB scale is used for the magnitude response because it better illustrates the amount of attenuation in the stopband which is 21 dB or approximately a factor of 10 in this case.

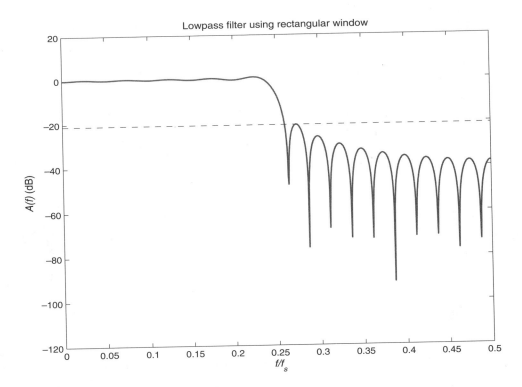

Figure 6.9 Lowpass Magnitude Response using a Rectangular Window, $m = 40$

Next consider the same filter, but using the Hanning window. From Table 6.2, the filter coefficients are $b_p = 0.5$ and

$$b_i = \frac{[0.5 + 0.5\cos(\pi i / p)]\sin[0.5\pi(i - p)]}{\pi(i - p)}, \quad 0 \le i \le m, \ i \ne p$$

A plot of the Hanning magnitude response is shown in Figure 6.10. Notice that the attenuation in the stopband is now 44 dB, and furthermore the response is also more flat in the passband in comparison with the rectangular window in Figure 6.9. However, this improvement in passband ripple and stopband attenuation is achieved at the expense of a what is clearly a wider transition band.

Still better stopband attenuation can be obtained using the Hamming window. From Table 6.2, the filter coefficients are $b_p = 0.5$ and

$$b_i = \frac{[0.54 + 0.46\cos(\pi i / p)]\sin[0.5\pi(i - p)]}{\pi(i - p)}, \quad 0 \le i \le m, \ i \ne p$$

A plot of the Hamming magnitude response is shown in Figure 6.11. In this case, the stopband attenuation is 53 dB and is constant throughout most of the stopband.

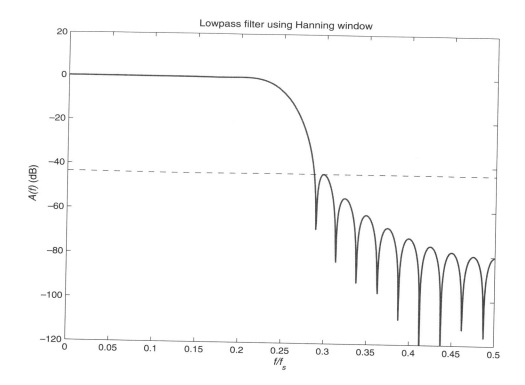

Figure 6.10 Lowpass Magnitude Response using a Hanning Window, $m = 40$

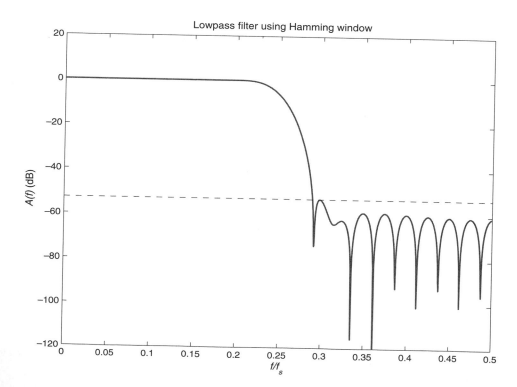

Figure 6.11 Lowpass Magnitude Response using a Hamming Window, $m = 40$

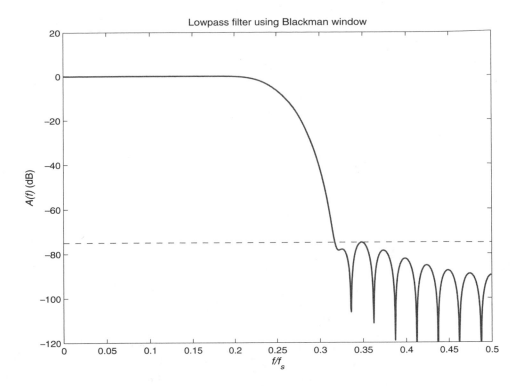

Figure 6.12 Low-pass Magnitude Response using a Blackman Window, $m = 40$

Finally, the maximum stopband attenuation can be obtained using the Blackman window. From Table 6.2, the filter coefficients are $b_p = 0.5$ and

$$b_i = \frac{[0.42 + 0.5\cos(\pi i/p) + 0.08\cos(2\pi i/p)]\sin[0.5\pi(i-p)]}{\pi(i-p)}, \quad 0 \le i \le m, \ i \ne p$$

A plot of the Blackman magnitude response is shown in Figure 6.12. Here the stopband attenuation is 75 dB, which is 54 dB better than with the rectangular window. However, the choice of the Blackman window is not clear-cut because the transition band is largest in this case.

The improvements in passband ripple and stopband attenuation achieved by using windowing come at the cost of a wider transition band. However, the width of the transition band for each window can be controlled by the filter order m. A summary of filter design characteristics using the different windows can be found in Table 6.3.

Table 6.3: ▶
Design Characteristics of Windows

Window Type	Transition Bandwidth $\hat{B} = |F_s - F_p|/f_s$	Passband Ripple δ_p	Stopband Attenuation δ_s	Passband Ripple A_p (dB)	Stopband Attenuation A_s (dB)
Rectangular	$\dfrac{0.9}{m}$	0.0819	0.0819	0.742	21
Hanning	$\dfrac{3.1}{m}$	0.0063	0.0063	0.055	44
Hamming	$\dfrac{3.3}{m}$	0.0022	0.0022	0.019	53
Blackman	$\dfrac{5.5}{m}$	0.00017	0.00017	0.0015	75.4

Kaiser Windows

Additional windows have been proposed including the Bartlett, Lanczos, Tukey, and Kaiser windows. The Kaiser window is a near-optimal window based in the use of zeroth-order modified Bessel functions of the first kind (Kaiser, 1966). The Kaiser window includes a shape parameter β which can be used to achieve a trade-off between the width of the transition band (main-lobe width) and the stopband attenuation (side-lobe amplitude). The family of Kaiser windows includes the rectangular window as a limiting special case (Kaiser, 1974). We summarize the steps of the windowed FIR filter design procedure with the following algorithm.

ALGORITHM 6.1:
Windowed FIR
Filter Design

1. Pick $m > 0$ and a window w.
2. For $i = 0$ to m compute
   {
$$b_i = w(i)2T \int_0^{f_s/2} A_r(f) \cos[2\pi(i - 0.5m)fT]df$$
   }
3. Set
$$H(z) = \sum_{i=0}^{m} b_i z^{-i}$$

The FIR filter produced by Algorithm 6.1 is a type 1 or type 2 linear-phase filter with a delay of $\tau = mT/2$ and an even amplitude response, $A_r(f)$. To design a type 3 or type 4 filter, the desired amplitude response should be odd and the expression for b_i in step 2 should be based on Eq. 6.20 instead of Eq. 6.17.

Example 6.3 **Windowed Bandpass Filter**

Consider the problem of designing a bandpass filter with cutoff frequencies $F_0 = f_s/8$ and $F_1 = 3f_s/8$. Suppose the Blackman window is used. Using Tables 6.1 and 6.2 and Algorithm 6.1, the filter coefficients are $b_p = 0.5$ and

$$b_i = \frac{[0.42 + 0.5\cos(\pi i/p) + 0.08\cos(2\pi i/p)]\{\sin[0.75\pi(i-p)] - \sin[0.25\pi(i-p)]\}}{\pi(i-p)},$$

$$0 \le i \le m, \; i \ne p$$

Suppose $m = 80$. A plot of the magnitude response, obtained by running script *exam6_3* on the distribution CD, is shown in Figure 6.13. It is clear that the passband and stopband ripples have been effectively reduced in comparison with Figure 6.7. From Table 6.3, the normalized width of the transition band in this case is

$$\frac{\Delta F}{f_s} = \frac{5.5}{80}$$

$$= 0.069$$

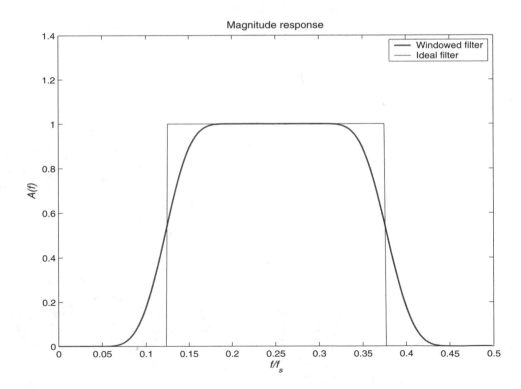

Figure 6.13 Magnitude Response of Bandpass Filter using a Hamming Window, $m = 80$

MATLAB Toolbox

The FDSP toolbox contains the following functions for designing a linear-phase FIR filter using the windowing method.

```
b = f_firideal (f_type,F,m,fs,win)   % Ideal linear-phase FIR windowed filter
b = f_firwin (fun,fs,win,sym,p);     % General linear-phase FIR windowed filter
```

Function *f_firideal* designs a linear-phase windowed FIR filter using one of the four ideal frequency-selective filter specifications. On entry to *f_firideal*, input *f_type* is an integer specifying the filter type: 0 for lowpass, 1 for highpass, 2 for bandpass, and 3 for bandstop. Input F specifies the desired cutoff frequency or frequencies in Hz. For lowpass and highpass filters, F is a scalar, and for bandpass and bandstop filters F is a vector of length two. The remaining inputs to *f_firideal* are the filter order m, which must be even, the sampling frequency fs, and the window type, win, as summarized in Table 6.2. On exit from *f_firideal*, b is a $1 \times (m + 1)$ vector containing the filter coefficients.

$$H(z) = \sum_{i=0}^{m} b_i z^{-i} \qquad (6.24)$$

The function *f_firwin* provides an alternative way to design a windowed FIR filter. Function *f_firwin* is more general than function *f_firideal* in that the desired amplitude response $A_r(f)$ can be arbitrary. On entry to *f_firwin*, *fun* is a string containing the name of a user-supplied M-file function that specifies the desired amplitude response, $A_r(f)$. The calling sequence for the function *fun* is as follows.

```
A = fun (f,fs,p);                    % User-defined amplitude response
```

On entry to *fun*, f is the evaluation frequency which may be a vector, fs is the sampling frequency, and p is an optional parameter vector that can contain such things as cutoff frequencies and gains. On exit from *fun*, A is the desired amplitude response evaluated at f. The remaining calling arguments of the function *f_firwin* include the filter order m, the sampling frequency fs, the window type win as defined in Table 6.2, the impulse-response symmetry type sym, and the optional parameter vector p used in the function *fun*. If $sym = 0$, then $h(k)$ exhibits even symmetry about $k = m/2$, and if $sym \neq 0$, it exhibits odd symmetry. When $sym = 0$, the $A_r(f)$ returned by the function *fun* should be an even function, and when $sym \neq 0$ it should be an odd function. On exit from *f_firwin*, b is a $1 \times (m + 1)$ vector containing the filter coefficients. See *exam6_2* and *exam6_3* on the distribution CD for examples that use *f_firideal* and *f_firwin*, respectively.

6.3 **Frequency-sampling Method**

An alternative technique for designing a linear-phase FIR filter with a prescribed magnitude response is the frequency-sampling method. As the name implies, this method is based on using samples of the desired frequency response.

6.3.1 Frequency Sampling

Suppose there are N frequency samples equally spaced over the range $0 \leq f < f_s$ with the ith discrete frequency being

$$f_i = \frac{if_s}{N}, \quad 0 \leq i < N \tag{6.25}$$

Recall from Eq. 3.95 that the samples of the frequency response can be obtained directly from the DFT of the impulse response. In particular, for an FIR filter order of $m = N - 1$ we have $H(f_i) = H(i)$ for $0 \leq i < N$ where $H(i) = \text{DFT}\{h(k)\}$. Taking the inverse DFT, we then arrive at the following expression for the impulse response of the desired FIR filter.

Frequency-sampled
impulse response

$$h(k) = \text{IDFT}\{H(f_i)\}, \quad 0 \leq k < N \tag{6.26}$$

The filter in Eq. 6.26 has a frequency response $H(f)$ that interpolates, or passes through, the N samples. Next consider the problem of placing constraints on $H(f)$ that ensure that the filter is a linear-phase filter. Suppose $h(k)$ is a linear-phase filter exhibiting even symmetry about the midpoint $k = m/2$. Then from Eq. 6.15, the frequency response of this type 1 or type 2 filter is as follows

$$H(f) = A_r(f) \exp(-j\pi mfT) \tag{6.27}$$

Here the amplitude response, $A_r(f)$, is a real, even function that specifies the desired magnitude response, $|A_r(f)| = A(f)$. Recalling the expression for the IDFT in Eq. 3.34, and using Eq. 6.25 through Eq. 6.27, we can write the kth sample of the impulse response as

$$h(k) = \frac{1}{N} \sum_{i=0}^{N-1} H(f_i) \exp(j2\pi ik/N)$$

$$= \frac{1}{N} \sum_{i=0}^{N-1} A_r(f_i) \exp(-j\pi mf_iT) \exp(j2\pi ik/N)$$

$$= \frac{1}{N} \sum_{i=0}^{N-1} A_r(f_i) \exp(-j\pi mi/N) \exp(j2\pi ik/N)$$

$$= \frac{1}{N} \sum_{i=0}^{N-1} A_r(f_i) \exp[j2\pi i(k - 0.5m)/N]$$

$$= \frac{1}{N} \sum_{i=0}^{N-1} A_r(f_i)\{\cos[2\pi i(k - 0.5m)/N] + j \sin[2\pi i(k - 0.5m)/N]\} \quad (6.28)$$

For a real $h(k)$, the sine terms in Eq. 6.28 cancel one another. The $i = 0$ term can be treated separately, in which case the expression for $h(k)$ in Eq. 6.28 reduces to

$$h(k) = \frac{A_r(f_0)}{N} + \frac{1}{N} \sum_{i=1}^{N-1} A_r(f_i) \cos[2\pi i(k - 0.5m)/N] \quad (6.29)$$

Using the symmetry property of the DFT in Eq. 3.50, one can show that the contributions of the i term and the $N - i$ term are identical. Thus, we can sum half of the terms and double the result. Recalling that $b_k = h(k)$ for an FIR filter, we arrive at the following expression for the coefficients of a linear-phase frequency-sampled filter order of m, where $m = N - 1$.

Frequency-sampled filter, even

$$b_k = \frac{A_r(0)}{m+1} + \frac{2}{m+1} \sum_{i=1}^{\text{floor}(m/2)} A_r(f_i) \cos\left[\frac{2\pi i(k - 0.5m)}{m+1}\right], \quad 0 \le k \le m \quad (6.30)$$

Note that for a type 1 filter the order m is even, in which case floor$(m/2) = m/2$. For a type 2 filter, m is odd.

Example 6.4   Frequency-sampled Lowpass Filter

To illustrate the frequency-sampling method, consider the problem of designing a lowpass filter with cutoff frequency $F_0 = f_s/4$. Suppose a filter order of $m = 20$ is used. In this case, the samples of the desired amplitude response are

$$A_r(f_i) = \begin{cases} 1, & 0 \le i \le 5 \\ 0, & 6 \le i \le 10 \end{cases}$$

Next from Eq. 6.30, the filter coefficients are

$$b_k = \frac{1}{21} + \frac{2}{21} \sum_{i=1}^{5} \cos\left[\frac{2\pi i(k - 10)}{21}\right], \quad 0 \le k \le 20$$

A plot of the magnitude response, obtained by running script *exam6_4* on the distribution CD, is shown in Figure 6.14. Notice that the magnitude response does go through the frequency samples as required. However, there are significant ripples in the magnitude response between the samples. The stopband attenuation is more easily seen in the logarithmic plot which reveals that $A_s = 15.6$ dB.

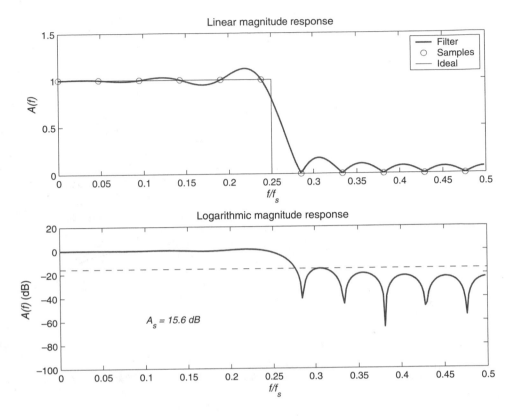

Figure 6.14 Magnitude Response of a Frequency-sampled Lowpass Filter with $m = 20$

6.3.2 Transition-band Optimization

The ringing or ripple in the magnitude response evident in Figure 6.14 is caused by the abrupt transition from passband to stopband in the desired magnitude response. One way to reduce these oscillations is to taper the filter coefficients using a data window. Recall that the tradeoff involved in using a window to reduce ripple is a wider transition band. With the frequency-sampling method we can forego the use of a window and instead explicitly specify $A(f)$ in the transition band by including one or more transition-band samples. This increase in the width of the transition band has the effect of improving the passband ripple and stopband attenuation as can be seen from the following example.

Example 6.5

Filter with Transition-band Sample

Again consider the problem of designing a lowpass filter with cutoff frequency $F_0 = f_s/4$. Suppose a filter order of $m = 20$ is used as in Example 6.4. However, in this case we insert a single transition-band sample as follows.

$$A_r(f_i) = \begin{cases} 1, & 0 \le i \le 5 \\ 0.5, & i = 6 \\ 0, & 7 \le i \le 10 \end{cases}$$

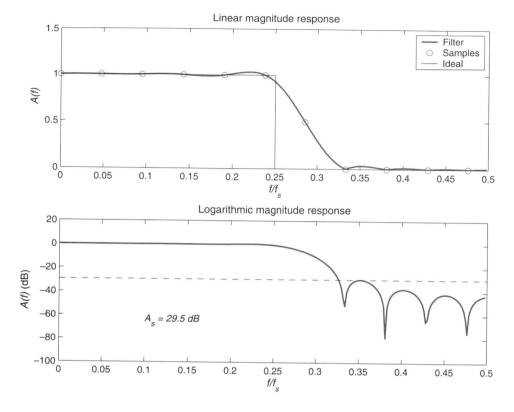

Figure 6.15 Magnitude Response of a Lowpass Filter with $m = 20$ and One Transition-band Sample, $A_r(f_6) = 0.5$

Using Eq. 6.30, the filter coefficients are

$$b_k = \frac{1}{21} + \frac{2}{21} \left\{ \sum_{i=1}^{5} \cos\left[\frac{2\pi i(k-10)}{21}\right] + 0.5\cos\left[\frac{2\pi 6(k-10)}{21}\right] \right\}, \quad 0 \le k \le 20$$

A plot of the logarithmic magnitude response, obtained by running script *exam6_5* on the distribution CD, is shown in Figure 6.15. Comparing Figure 6.15 with Figure 6.14, we see that by inserting a transition-band sample we have increased the stopband attenuation from 15.6 dB to 29.5 dB. The passband ripple has also been reduced.

The transition-band sample that was inserted in the desired magnitude response in Example 6.5 was $A_r(f_6) = 0.5$ which corresponds to a straight line interpolation between the end of the ideal passband and the start of the ideal stopband. Clearly, other choices for the value of the transition-band sample are also possible. We can use this extra degree of freedom to shape the magnitude response in the transition band in order to maximize the stopband attenuation as can be seen in the following example.

| **Example 6.6** | **Filter with Optimal Transition-band Sample** |

Again consider the problem of designing a lowpass filter order of $m = 20$ with a cutoff frequency of $F_0 = f_s/4$. In this case, we insert a general transition-band sample as follows.

$$A_r(f_i) = \begin{cases} 1, & 0 \le i \le 5 \\ x, & i = 6 \\ 0, & 7 \le i \le 10 \end{cases}$$

Using Eq. 6.30, the filter coefficients are

$$b_k(x) = \frac{1}{21} + \frac{2}{21} \left\{ \sum_{i=1}^{5} \cos \left[\frac{2\pi i(k-10)}{21} \right] + x \cos \left[\frac{2\pi 6(k-10)}{21} \right] \right\}, \quad 0 \le k \le 20$$

The problem is to find a value for x which maximizes the stopband attenuation A_s. A simple way to achieve this is to compute the stopband attenuation, $A_s(x)$, for three distinct values of x in the range $0 < x < 1$. For example, suppose the values $x = [0.25, 0.5, 0.75]^T$ are used. We can then pass a quadratic polynomial through these three data points.

$$A_s(x) = c_1 + c_2 x + c_3 x^2$$

The coefficient vector, $c = [c_1, c_2, c_3]^T$, of the polynomial which interpolates the data points must satisfy the following system of three linear algebraic equations.

$$\begin{bmatrix} 1 & x_1 & x_1^2 \\ 1 & x_2 & x_2^2 \\ 1 & x_3 & x_3^2 \end{bmatrix} \begin{bmatrix} c_1 \\ c_2 \\ c_3 \end{bmatrix} = \begin{bmatrix} A_s(x_1) \\ A_s(x_2) \\ A_s(x_3) \end{bmatrix}$$

Solving this system for c then yields a polynomial model of the stopband attenuation as a function of the transition-band sample. We can then maximize the stopband attenuation by differentiating $A_s(x)$, setting the result to zero, and solving for x. This yields the following optimal transition sample value.

$$x_{max} = \frac{-c_2}{2c_3}$$

This optimization procedure is implemented in script *exam6_6* on the distribution CD. Running script *exam6_6* produces a coefficient vector of $c = [18.35, 62.74, -80.83]^T$.

Thus, the optimal transition-band sample is

$$A_r(f_6) = x_{max}$$

$$= \frac{-62.74}{2(-80.83)}$$

$$= 0.388$$

A plot of the optimum magnitude response is shown in Figure 6.16. By using an optimal value for the transition sample, the stopband attenuation has increased to $A_s = 39.9$ dB.

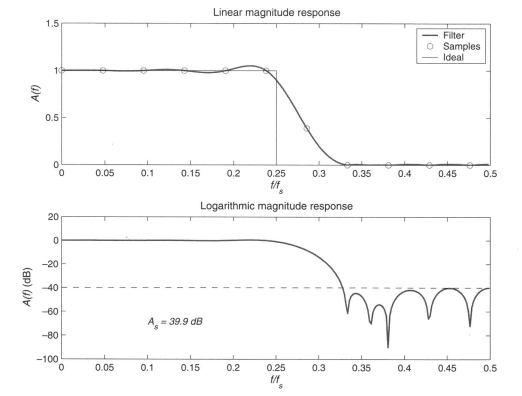

Figure 6.16 Magnitude Response of a Lowpass Filter with $m = 20$ and an Optimal Transition-band Sample, $A_r(f_6) = 0.388$

The notion of using samples in the transition band can be extended to more than one sample with a corresponding improvement in the stopband attenuation. Rabiner, *et al.* (1970) have developed tables of optimal transition-band samples for FIR filters of different lengths and different numbers of transition-band samples. For example, using a filter order of $n = 15$ with five passband samples and two transition-band samples, a stopband attenuation in excess of 100 dB can be achieved.

The frequency-sampling method also can be used to design linear-phase FIR filters whose impulse responses exhibit odd symmetry about $k = m/2$. From Eq. 6.18, this corresponds to a filter with the following type of frequency response.

$$H(f) = j A_r(f) \exp(-j\pi m f T) \tag{6.31}$$

Using Eq. 6.25 and Eq. 6.26, we can write the kth sample of the impulse response as follows where the number of samples is $N = m + 1$.

$$h(k) = \frac{1}{N} \sum_{i=0}^{N-1} H(f_i) \exp(j 2\pi i k/N)$$

$$= \frac{j}{N} \sum_{i=0}^{N-1} A_r(f_i) \exp(-j\pi m f_i T) \exp\{j 2\pi i (k/N)\}$$

$$= \frac{j}{N} \sum_{i=0}^{N-1} A_r(f_i) \exp\{j [2\pi i (k - 0.5m)/N]\}$$

$$= \frac{j}{N} \sum_{i=0}^{N-1} A_r(f_i)\{\cos[2\pi i (k - 0.5m)/N] + j \sin[2\pi i (k - 0.5m)/N]\} \tag{6.32}$$

For a real $h(k)$, the cosine terms in Eq. 6.32 cancel one another and we have

$$h(k) = \frac{-1}{N} \sum_{i=0}^{N-1} A_r(f_i) \sin[2\pi i (k - 0.5m)/N] \tag{6.33}$$

Since $A_r(f)$ is an odd function, the $i = 0$ term drops out because $A_r(0) = 0$. Using the symmetry property of the DFT in Eq. 3.50, one can show that the contributions of the i term and the $N - i$ term are identical. Thus, we can sum half of the terms and double the result. Recalling that $b_k = h(k)$ for an FIR filter, we arrive at the following expression for the coefficients of a linear-phase frequency-sampled filter order of m, where $m = N - 1$.

Frequency-sampled filter, odd

$$b_k = \frac{-2}{m+1} \sum_{i=1}^{\text{floor}(m/2)} A_r(f_i) \sin\left[\frac{2\pi i (k - 0.5m)}{m+1}\right], \quad 0 \le k \le m \tag{6.34}$$

Note that for a type 3 filter, the order m is even, in which case $\text{floor}(m/2) = m/2$. For a type 4 filter, m is odd.

MATLAB Toolbox

The FDSP toolbox contains the following function for designing a linear-phase FIR filter using the frequency-sampling method.

```
b = f_firsamp (A_r,m,fs,sym);    % FIR filter design using frequency sampling
```

On entry to *f_firsamp*, A_r, is a vector of length $p = \text{floor}(m/2) + 1$ containing the samples of the desired amplitude response. That is, if $A(f)$ is the desired magnitude response, then

$$|A_r(i+1)| = A(f_i), \quad 0 \le i < p \tag{6.35}$$

Here $f_i = i f_s/N$ for $0 \le i < p$. Next, m is the order of the filter, fs is the sampling frequency in Hz, and sym specifies the impulse response symmetry type. If $sym = 0$, then $h(k)$ exhibits even symmetry about $k = m/2$, and if $sym \ne 0$, it exhibits odd symmetry. Thus, the four linear-phase filter types can be implemented depending on the choice of sym and m. When $sym \ne 0$, the amplitude response is odd, so $A\_r(1) = 0$. On exit from *f_firsamp*, b is a vector of length $m+1$ containing the filter coefficients.

● ● ● ● ● ● ● ● ● ● ● ● ● ● ● ● ●

6.4 Least-squares Method

The frequency response of a digital filter is periodic with period f_s. Because the windowing method uses a truncated Fourier series expansion of the desired amplitude response, $A_d(f)$, this method produces a filter that is optimal, in the sense that it minimizes the following objective.

$$J = \int_0^{f_s/2} [A_d(f) - A_r(f)]^2 df \tag{6.36}$$

The actual amplitude response $A_r(f)$ is real, and for a type 1 or type 2 linear-phase filter, it is even, while for a type 3 or type 4 linear-phase filter, it is odd. An alternative approach to filter design is to use a discrete version of the objective J. Let $\{F_0, F_1, \ldots, F_p\}$ be a set of $p+1$ distinct frequencies with $F_0 = 0$, $F_p = f_s/2$, and

$$F_0 < F_1 < \cdots < F_p \tag{6.37}$$

The spacing between the discrete frequencies is often uniform, as in $F_i = i f_s/(2p)$, but this is not required. Next, let $w(i) > 0$ be a *weighting function* where $w(i)$ specifies the relative importance of discrete frequency F_i. The special case, $w(i) = 1$ for $0 \le i \le p$ is referred to as *uniform weighting*. A weighted discrete version of the objective function in Eq. 6.36 can be formulated as follows.

$$J_p = \sum_{i=0}^{p} w^2(i)[A_r(F_i) - A_d(F_i)]^2 \tag{6.38}$$

A filter design technique that minimizes J_p is called a *least-squares method*. The filter amplitude response depends on the filter coefficient vector b, but the exact form of this dependence depends on the type of linear-phase filter used. To illustrate the least-squares method, we will use the most general linear-phase filter: a type 1 filter order of m, where $m \leq 2p$.

$$H(z) = \sum_{i=0}^{m} b_i z^{-i} \qquad (6.39)$$

Recall from Table 5.1 that for a type 1 filter, m is even, and the impulse response satisfies the even symmetry condition $h(m - k) = h(k)$. For FIR filters in general, we have $b_i = h(i)$, which means $b_{m-i} = b_i$ for $0 \leq i \leq m$. For convenience, let $\theta = 2\pi f T$ and let $m = 2r$. We can then write the frequency response of $H(z)$ as follows.

$$H(f) = H(z)|_{z=\exp(j\theta)}$$

$$= \sum_{i=0}^{m} b_i \exp(-ji\theta)$$

$$= \exp(-jr\theta) \sum_{i=0}^{m} b_i \exp[-j(i-r)\theta] \qquad (6.40)$$

Since m is even, the middle or rth term can be separated out from the sum. The remaining terms then can be combined in pairs using the symmetry condition, $b_{m-i} = b_i$, and Euler's identity.

$$H(f) = \exp(-jr\theta)\{b_r + \sum_{i=0}^{r-1} b_i \exp[-j(i-r)\theta] + b_{m-i} \exp[-j(m-i-r)\theta]\}$$

$$= \exp(-jr\theta)\{b_r + \sum_{i=0}^{r-1} b_i [\exp[-j(i-r)\theta] + \exp[j(i-r-m+2r)\theta]]\}$$

$$= \exp(-jr\theta)\{b_r + \sum_{i=0}^{r-1} b_i [\exp[-j(i-r)\theta] + \exp[j(i-r)\theta]]\}$$

$$= \exp(-jr\theta)\{b_r + 2\sum_{i=0}^{r-1} b_i \cos[(i-r)\theta]\}$$

$$= \exp(-jr\theta)A_r(f) \qquad (6.41)$$

For convenience, define $c_i = b_i$ for $i \neq r$ and $c_r = b_r/2$. Recalling that $\theta = 2\pi f T$, we then arrive at the following amplitude response for a type 1 linear-phase FIR filter order of $2r$.

Amplitude response

$$A_r(f) = 2\sum_{i=0}^{r} c_i \cos[2\pi(i-r)fT] \qquad (6.42)$$

Now that we have an expression which shows the dependence of $A_r(f)$ on c, we can proceed to find an optimal value for c and therefore, b. From Eq. 6.38, the objective function is

$$J_p(c) = \sum_{i=0}^{p} w^2(i) \left[2 \sum_{k=0}^{r} c_k \cos[2\pi(k-r)F_iT] - A_d(F_i) \right]^2 \tag{6.43}$$

To find an optimal value for the coefficient vector c, we set $\partial J_p(c)/\partial c = 0$ and solve for c. Rather than take explicit partial derivatives, it is helpful to reformulate Eq. 6.43 in terms of vector notation. Let the $(p+1) \times (r+1)$ matrix G and the $(p+1) \times 1$ column vector d be defined as follows.

$$G_{ik} \overset{\Delta}{=} 2w(i)\cos[2\pi(k-r)F_iT], \quad 0 \le i \le p, \; 0 \le k \le r \tag{6.44a}$$

$$d_i \overset{\Delta}{=} w(i)A_d(F_i), \quad 0 \le i \le p \tag{6.44b}$$

If $c = [c_0, c_1, \ldots, c_r]^T$ is the coefficient vector, then the objective function in Eq. 6.43 can be written in compact vector form as

$$J_p(c) = (Gc - d)^T(Gc - d) \tag{6.45}$$

Since $p \ge r$, the linear algebraic system $Gc = d$ is overdetermined with more equations than unknowns, so there is no c such that $Gc = d$. To find the coefficient vector c which minimizes $J_p(c)$, we multiply $Gc = d$ on the left by G^T, which yields the *normal equations*.

Normal equations

$$G^T Gc = G^T d \tag{6.46}$$

The solution to Eq. 6.46 is the coefficient vector of the least-squares filter: the filter that minimizes $J_p(c)$. Note that $G^T G$ is a square $(r+1) \times (r+1)$ matrix of full rank. Consequently, the optimal coefficient vector can be expressed as

$$c = (G^T G)^{-1} G^T d \tag{6.47}$$

The matrix $G^+ = (G^T G)^{-1} G^T$ is the call *pseudo-inverse* of G. Normally, we do not solve for c using the pseudo-inverse. Instead, the normal equations in Eq. 6.46 are solved directly because this takes only about one third as many FLOPs for large r. The normal equations can be ill-conditioned for large values of r. Consequently, when r is large, special techniques such as the Levinson-Durbin algorithm (Levinson, 1947; Durbin, 1959) should be used to solve Eq. 6.46. Once the $(r+1) \times 1$ vector c is determined, the original $(m+1) \times 1$ coefficient vector b is then obtained as follows.

$$b_i = \begin{cases} c_i, & 0 \le i < r \\ 2c_r, & i = r \\ c_{2r-i}, & r < i \le 2r \end{cases} \tag{6.48}$$

| Example 6.7 | **Least-squares Bandpass Filter** |

To illustrate the least-squares method, consider the problem of designing a bandpass filter with a piecewise linear-magnitude response that features a passband of $3f_s/16 \le |f| \le 5f_s/16$ and transition bands of width $f_s/32$. Suppose $p = 40$, and uniformly spaced discrete frequencies are used.

$$F_i = \frac{if_s}{2p}, \quad 0 \le i \le p$$

Two cases are considered. The first one uses uniform weighting, and the second weights the passband samples by $w(i) = 10$. In both cases, the order of the FIR filter is $m = 40$. The results, obtained by running script *exam6_7* on the distribution CD, are shown in Figure 6.17. Note how, by weighting the passband samples more heavily, the passband ripple is reduced, but at the expense of less attenuation in the stopband.

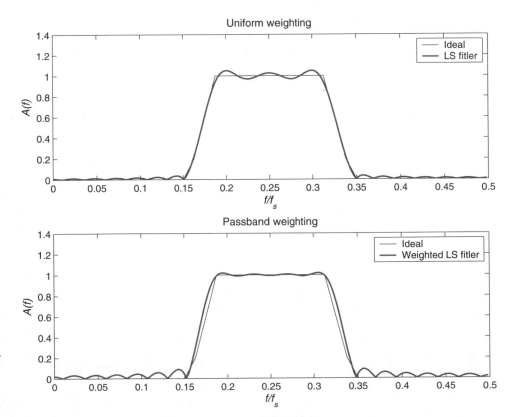

Figure 6.17 Magnitude Responses of a Least-squares Bandpass Filter of Order $m = 64$ using $p + 1 = 129$ Discrete Frequencies and both Uniform Weighting and Increased Passband Weighting

MATLAB Toolbox

The FDSP toolbox contains the following function for designing a type 1 linear-phase FIR filter using the least-squares method.

```
b = f_firls (F,A,m,fs,w);     % FIR filter design using least squares
```

On entry to *f_firls*, F and A are vectors of length $p + 1$ containing the discrete frequencies; the desired amplitude response, $0 \leq m \leq 2p$, is the filter order; fs is the sampling frequency; and w is an optional vector of length $p + 1$ containing the weighting factors. The default value is uniform weighting. On exit from *f_firls*, b is a vector of length $p + 1$ containing the filter coefficients. The function *f_firls* produces a type 1 or type 2 linear-phase FIR filter order of m that minimizes the objective J_p in Eq. 6.38.

• • • • • • • • • • • • • •

*6.5 Optimal Equiripple FIR Filters

The FIR filter design methods discussed thus far all involve optimization. The windowing method (with a rectangular window) minimizes the mean-squared error in Eq. 6.36. The frequency-sampling method uses optimal placement of transition-band samples to maximize the stopband attenuation. The least-squares method minimizes a weighted sum of squares of errors at discrete frequencies as in Eq. 6.38. In this section, we examine an optimization technique that is based on minimizing the maximum of the absolute value of the error within the passband and the stopband.

6.5.1 Minimax Error Criterion

To simplify the development, we focus our attention on the most general type of linear-phase FIR filter, a type 1 filter. Thus, we assume that the filter order is even, with $m = 2p$, and the impulse response $h(k)$ exhibits even symmetry about the midpoint $k = p$. From Eq. 6.15, this means that the frequency response can be represented as follows.

$$H(f) = A_r(f) \exp(-j2\pi pfT) \tag{6.49}$$

Here the filter amplitude response, $A_r(f)$, is a real, even function of f. Like the frequency response, the amplitude response is periodic with period f_s. Therefore, the amplitude response can be approximated with the following truncated trigonometric Fourier series.

$$A_r(f) = \sum_{i=0}^{p} d_i \cos(2\pi i fT) \tag{6.50}$$

Notice that there are no sine terms because $A_r(f)$ is an even function. For a windowed filter with the rectangular window, the coefficients d_i are the Fourier coefficients. In this case, the resulting filter minimizes the mean-square error in Eq. 6.36.

However, there is another way to compute the d_i, which minimizes an alternative objective. To see this, it is helpful to reformulate the truncated cosine series in Eq. 6.50 using Chebyshev polynomials. The first two *Chebyshev polynomials* are $T_0(x) = 1$ and $T_1(x) = x$. The remaining Chebyshev polynomials can be defined recursively as follows.

$$T_k(x) = 2xT_{k-1}(x) - T_{k-2}(x), \quad k \geq 2 \tag{6.51}$$

Thus, $T_k(x)$ is a polynomial of degree k. The Chebyshev polynomials are a classic family of orthogonal polynomials that have many useful properties (Schilling and Lee, 1988). One such property becomes evident when the kth Chebyshev polynomial is evaluated at $x = \cos(\theta)$, which yields

$$T_k[\cos(\theta)] = \cos(k\theta), \quad k \geq 0 \tag{6.52}$$

In view of this harmonic generating property, the truncated cosine series in Eq. 6.50 can be recast in terms of Chebyshev polynomials as

$$A_r(f) = \sum_{k=0}^{p} d_k T_k(x) \Bigg|_{x=\cos(2\pi fT)} \tag{6.53}$$

Thus, $A_r(f)$ can be thought of as a trigonometric polynomial: a polynomial in $x = \cos(2\pi fT)$. To formulate an alternative error criterion, let $A_d(f)$ be the desired amplitude response, and let $W(f) > 0$ be a weighting function. Then the weighted *error* at frequency f is defined as

$$E(f) \triangleq W(f)[A_d(f) - A_r(f)] \tag{6.54}$$

A logical way to select the weighting function is to set $W(f) = 1/\delta_p$ in the passband and $W(f) = 1/\delta_s$ in the stopband. This way, if one of the specifications is more stringent than the other, it will be given a higher weight because it is more difficult to satisfy. The weighting function in Eq. 6.54 can be scaled by a positive constant without changing the nature of the optimization problem. This leads to the following normalized weighting function.

$$W(f) = \begin{cases} \dfrac{\delta_s}{\delta_p}, & f \in \text{passband} \\ 1, & f \in \text{stopband} \end{cases} \tag{6.55}$$

Frequency-selective filter specifications place constraints on the amplitude response in the passband and the stopband, but not in the transition band. Let F denote the set of frequencies over which the amplitude response is specified. Then F is the following compact subset of $[0, f_s/2]$, where $\cup$ denotes the union of the two sets.

$$F \triangleq \text{passband} \cup \text{stopband} \tag{6.56}$$

Filter Type	F
Lowpass	$[0, F_p] \cup [F_s, f_s/2]$
Highpass	$[0, F_s] \cup [F_p, f_s/2]$
Bandpass	$[0, F_{s1}] \cup [F_{p1}, F_{p2}] \cup [F_{s2}, f_s/2]$
Bandstop	$[0, F_{p1}] \cup [F_{s1}, F_{s2}] \cup [F_{p2}, f_s/2]$

The specification frequencies for the four basic frequency-selective filter types are summarized in Table 6.4.

Given the filter-specification frequencies and the weighting function, the design objective is then to find a Chebyshev coefficient vector, $d \in R^{p+1}$, which solves the following optimization problem.

$$\min_{d \in R^{p+1}} \left[\max_{f \in F} \{ |E(f)| \} \right] \tag{6.57}$$

The performance criterion in Eq. 6.57 is called the *minimax* criterion because it minimizes the maximum of the absolute value of the error over the passband and the stopband.

Next consider the problem of determining the filter impulse response once an optimal Chebyshev coefficient vector d has been found. This can be achieved using the discrete-time Fourier transform introduced in Chapter 3. Recall from Eq. 3.23 that if $H(f) = \text{DTFT}\{h(k)\}$ is the frequency response, then the impulse response can be recovered from $H(f)$ using the inverse DTFT as follows.

$$h(k) = \frac{1}{f_s} \int_{-f_s/2}^{f_s/2} H(f) \exp(jk2\pi f T) df \tag{6.58}$$

If the amplitude response is as in Eq. 6.50, then from Eq. 6.49 we have

$$h(k) = \frac{1}{f_s} \int_{-f_s/2}^{f_s/2} A_r(f) \exp(-jp2\pi f T) \exp(jk2\pi f T) df$$

$$= \frac{1}{f_s} \int_{-f_s/2}^{f_s/2} A_r(f) \exp[j(k-p)2\pi f T] df$$

$$= \frac{1}{f_s} \int_{-f_s/2}^{f_s/2} \sum_{i=0}^{p} d_i \cos(i2\pi f T) \exp[j(k-p)2\pi f T] df$$

$$= \frac{1}{f_s} \sum_{i=0}^{p} d_i \int_{-f_s/2}^{f_s/2} \cos(i2\pi f T) \exp[j(k-p)2\pi f T] df \tag{6.59}$$

First, consider the case $k = p$. Here the exponent in Eq. 6.59 is zero, so the exponential factor disappears, and the impulse response simplifies to

$$h(p) = \frac{1}{f_s} \sum_{i=0}^{p} d_i \int_{-f_s/2}^{f_s/2} \cos(i2\pi fT)df$$

$$= d_0 \tag{6.60}$$

Next, consider the case $k \neq p$. When Euler's identity is used, the terms in Eq. 6.59 involving odd functions of f disappear because the range of integration is symmetric about $f = 0$. Using the product of cosines trigonometric identity from Appendix D then yields the following.

$$h(k) = \frac{1}{f_s} \sum_{i=0}^{p} d_i \int_{-f_s/2}^{f_s/2} \cos(i2\pi fT)\{\cos[(k-p)2\pi fT] + j\sin[(k-p)2\pi fT]\}df$$

$$= \frac{1}{f_s} \sum_{i=0}^{p} d_i \int_{-f_s/2}^{f_s/2} \cos(i2\pi fT) \cos[(k-p)2\pi fT]df$$

$$= \frac{1}{2f_s} \sum_{i=0}^{p} d_i \int_{-f_s/2}^{f_s/2} \{\cos[(i+[k-p])2\pi fT) + \cos[(i-[k-p])2\pi fT]\}df$$

$$= \frac{d_{p-k}}{2}, \quad 0 \leq k < p \tag{6.61}$$

Similarly, for the range $p < k \leq 2p$, the $i = k - p$ term survives, and we have $h(k) = d_{k-p}/2$. To summarize, the impulse response of a type 1 linear-phase filter can be recovered from the Chebyshev coefficient vector d as follows.

$$h(k) = \begin{cases} \dfrac{d_{p-k}}{2}, & 0 \leq k < p \\ d_p, & k = p \\ \dfrac{d_{k-p}}{2}, & p < k \leq 2p \end{cases} \tag{6.62}$$

6.5.2 Parks–McClellan Algorithm

The formulation of the amplitude response $A_r(f)$ as a polynomial in $x = \cos(2\pi fT)$ effectively transforms the filter design problem into a polynomial approximation problem over the set F. Let δ denote the optimal value of the minimax performance criterion.

$$\delta = \min_{d \in R^{p+1}} [\max_{f \in F} \{|E(f)|\}] \tag{6.63}$$

Parks and McClellan (1972, Refs. 54 and 55) applied the *alternation theorem* from the theory of polynomial approximation to solve for d. This theorem is due to Remez (1957).

The function $A_r(f)$ in Eq. 6.50 solves the minimax optimization problem in Eq. 6.63 if and only if there exist at least $p+2$ extremal frequencies $F_0 < F_1 < \cdots < F_{p+1}$ in F, such that $E(F_{i+1}) = -E(F_i)$ and

$$|E(F_i)| = \delta, \quad 0 \leq i < p+2$$

Extremal frequencies are frequencies at which the magnitude of the error achieves its extreme or maximum value within the passband and the stopband. Extremal frequencies include local minima and local maxima, and they can also include band edge frequencies. The name, alternation theorem, arises from the fact that the sign of the error alternates as one traverses the extremal frequencies. An example of an optimal amplitude response for a lowpass filter is shown in Figure 6.18. Notice that the local extrema cause ripples in the amplitude response and, within each band, these ripples are of the same size. It is for this reason that an optimal minimax filter is called an *equiripple* filter.

For the equiripple filter shown in Figure 6.18, the passband ripple is $\delta_p = 0.06$, the stopband attenuation is $\delta_s = 0.04$, and the filter order is $m = 12$. Observe that there are four extrema frequencies in the passband and another four in the stopband. Thus, the number of extrema frequencies is $p + 2 = 8$, so the amplitude response in

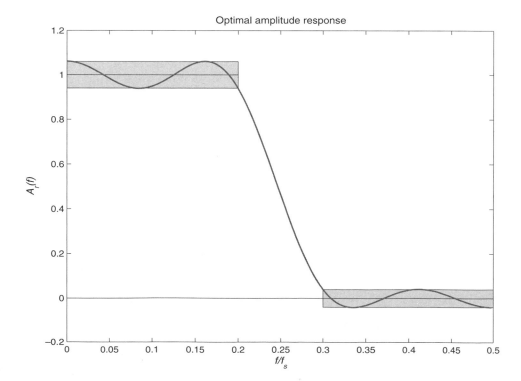

Figure 6.18 Optimal Amplitude Response using the Minimax Optimization Criterion with $m = 12$

Figure 6.18 is optimal. The need for at least $p+2$ extremal frequencies arises from the following observations. Since $A_r(f)$ is a polynomial of degree p, there can be up to $p-1$ local minima and local maxima where the slope of $A_r(f)$ is zero. For a lowpass or highpass filter, the optimal amplitude response also goes through the interior passband and stopband edge points.

$$A_r(F_p) = 1 - \delta_p \qquad (6.64a)$$

$$A_r(F_s) = \delta_s \qquad (6.64b)$$

That makes $p+1$ extremal frequencies. In addition, at least one, and perhaps both, of the endpoint frequencies, $F = 0$ and $F = f_s/2$, are also extremal frequencies. Therefore, for a lowpass or a highpass filter, the number of extremal frequencies in the optimal amplitude response will be either $p+2$ or $p+3$. Bandpass and bandstop filters can have up to $p + 5$ extremal frequencies because there are two additional band edge frequencies. In any event, we know from Theorem 6.1 that an optimal equiripple amplitude response must have at least $p + 2$ extremal frequencies.

In order to determine the optimal Chebyshev coefficient vector $d \in R^{p+1}$, we start with the alternation theorem. Using the definition of $E(f)$ in Eq. 6.54, we have

$$W(F_i)[A_d(F_i) - A_r(F_i)] = (-1)^i \delta, \quad 0 \le i < p+2 \qquad (6.65)$$

These equations can be recast as

$$A_r(F_i) + \frac{(-1)^i \delta}{W(F_i)} = A_d(F_i), \quad 0 \le i < p+2 \qquad (6.66)$$

Let $\theta_i = 2\pi F_i T$ for $0 \le i < p+2$. Then substituting for $A_r(F_i)$ using Eq. 6.50 yields

$$\sum_{k=0}^{p} d_k \cos(k\theta_i) + \frac{(-1)^i \delta}{W(F_i)} = A_d(F_i), \quad 0 \le i < p+2 \qquad (6.67)$$

Next let $c = [d_0, \cdots, d_p, \delta]^T$ represent a vector of unknown parameters. Then the $p + 2$ equations in Eq. 6.67 can be rewritten as the following matrix equation.

$$\begin{bmatrix} 1 & \cos(\theta_0) & \cos(2\theta_0) & \cdots & \cos(p\theta_0) & \frac{1}{W(F_0)} \\ 1 & \cos(\theta_1) & \cos(2\theta_1) & \cdots & \cos(p\theta_1) & \frac{-1}{W(F_1)} \\ & & \vdots & & & \\ 1 & \cos(\theta_p) & \cos(2\theta_p) & \cdots & \cos(p\theta_p) & \frac{(-1)^p}{W(F_p)} \\ 1 & \cos(\theta_{p+1}) & \cos(2\theta_{p+1}) & \cdots & \cos(p\theta_{p+1}) & \frac{(-1)^{p+1}}{W(F_{p+1})} \end{bmatrix} \underbrace{\begin{bmatrix} d_0 \\ d_1 \\ \vdots \\ d_p \\ \delta \end{bmatrix}}_{c} = \begin{bmatrix} A_d(F_0) \\ A_d(F_1) \\ \vdots \\ A_d(F_p) \\ A_d(F_{p+1}) \end{bmatrix} \qquad (6.68)$$

Given a set of extremal frequencies, we can solve Eq. 6.68 for the Chebyshev coefficient vector d and the parameter δ. Unfortunately, we do not know the extremal frequencies, so one has to estimate them using an iterative process known as the Remez exchange algorithm (Remez, 1957). To locate the $p + 2$ extremal frequencies, we start by choosing an initial guess. For example, they might be equally spaced over the set F. Next, Eq. 6.68 is solved for d and δ. Once d is known, the error function $E(f)$ in Eq. 6.54 can be evaluated. Typically, $E(f)$ is evaluated on a dense grid of at least $16m$ points in F. If $|E(f)| < \delta + \epsilon$ for some small tolerance ϵ, then convergence has been achieved. Otherwise, a new set of $m + 2$ extremal frequencies is determined from the $p + 2$ largest peaks of $|E(f)|$, and the process is repeated. This iterative process is summarized as follows.

ALGORITHM 6.2:
Equiripple FIR
Filter

1. Pick a filter order $m = 2p$, $N > 0$, and $\epsilon > 0$. Set $k = 1$, and pick $M \geq 16m$. Compute the initial extrema frequencies F_i for $0 \leq i < p + 2$ to be equally spaced over the set F.

2. Do
   {
   (a) Solve Eq. 6.68 for d and δ.
   (b) Evaluate $E_k = E(F_k)$ for $0 \leq k < M$ where the dense F_k are equally spaced over F. Compute the peak error.

   $$\|E\| = \max_{k=0}^{M-1}\{|E_k|\}$$

   (c) If $\|E\| \geq \delta + \epsilon$ then
       (1) Find the extrema points (local minima and maxima) in the dense set E_k. If there are more than $p + 2$ extrema points, choose the $p + 2$ points whose magnitudes are largest.
       (2) Update the $p + 2$ extremal frequencies F_i using the frequencies of the extrema points.
   (d) Set $k = k + 1$.

   }

3. While $\|E\| \geq \delta + \epsilon$ and $k < N$
4. Compute h using Eq. 6.62, and set

   $$H(z) = \sum_{i=0}^{m} h(i)z^{-i}$$

If Algorithm 6.2 terminates with $k = N$, then either ϵ or N should be increased. The computed minimax value δ may or may not satisfy the stopband specification $\delta \leq \delta_s$, depending on the filter order. If the ripples in the resulting amplitude response exceed the filter specifications, then the filter order m should be increased. Kaiser has proposed the following estimate for the equiripple filter order needed to meet a given design specification (Rabiner *et al.*, 1975).

$$m \approx \text{ceil} \left\{ \frac{-[10 \log_{10}(\delta_p \delta_s) + 13]}{14.6\hat{B}} + 1 \right\} \tag{6.69}$$

Here $\hat{B} = |F_s - F_p|/f_s$ is the normalized width of the transition band. This value can be used as a starting point for choosing m. If the filter specifications are not met, then m can be gradually increased until they are met or exceeded.

Before we examine an example of equiripple filter design, it should be pointed out that step 2a in Algorithm 6.2 can be made more efficient. To see this, let α_i be defined as follows where $\prod$ denotes the product.

$$\alpha_i = \frac{(-1)^i}{\displaystyle\prod_{k=0, k \neq i}^{p+1} [\cos(\theta_i) - \cos(\theta_k)]} \tag{6.70}$$

Parks and McClellan (1972, Refs. 48 and 49) have shown that the parameter δ can then be computed separately as follows.

$$\delta = \frac{\displaystyle\sum_{i=0}^{p+1} \alpha_i A_d(F_i)}{\displaystyle\sum_{i=0}^{p+1} \alpha_i / W(F_i)} \tag{6.71}$$

Given δ, the terms in Eq. 6.67 involving δ can be brought over to the right-hand side. The new augmented right-hand side vector is then

$$h_i = A_d(F_i) - \frac{(-1)^i \delta}{W(F_i)}, \quad 0 \leq i < p+1 \tag{6.72}$$

It then follows that Eq. 6.67 can be rewritten as

$$A_r(F_i) = h_i, \quad 0 \leq i < p+1 \tag{6.73}$$

Thus, the value of $A_r(f)$ is known at $p+1$ of the extremal frequencies. Since $A_r(f)$ is known to be a polynomial of degree p, there is no need to solve Eq. 6.68 for the coefficient vector d at each iteration. This is a potentially time consuming step when p is large. Instead, the points $A_r(F_i)$ can be used to construct a Lagrange interpolating polynomial (Parks and McClellan, 1972, Refs. 48 and 49). Using $\theta_i = 2\pi F_i T$ and $\theta = 2\pi f T$, the kth Lagrange interpolating polynomial is

$$L_i(f) = \frac{\displaystyle\prod_{k=0, k \neq i}^{p} [\cos(\theta) - \cos(\theta_k)]}{\displaystyle\prod_{k=0, k \neq i}^{p} [\cos(\theta_i) - \cos(\theta_k)]} \tag{6.74}$$

Note that $L_k(F_i) = \delta(i - k)$ where $\delta(i)$ is the unit impulse. Using this orthogonality property and Eq. 6.73, the amplitude response can be represented as follows.

$$A_r(f) = \sum_{i=0}^{p+1} h_i L_i(f) \tag{6.75}$$

This removes the need to solve the linear algebraic system in step 2b, a process that takes about $(p + 1)^3/3$ FLOPs. Once Algorithm 6.2 has converged, Eq. 6.68 can be solved once for the final Chebyshev coefficient vector d and then Eq. 6.62 can be used to find h.

Example 6.8   **Equiripple Filter**

As an illustration of the equiripple filter design technique, consider the problem of designing a bandpass filter with $f_s = 200$ Hz and the following design specifications.

$$(F_{s1}, F_{p1}, F_{p2}, F_{s2}) = (36, 40, 60, 64) \text{ Hz}$$

$$(\delta_p, \delta_s) = (0.02, 0.02)$$

The transition band is fairly narrow in this case with a normalized width of

$$\hat{B} = \frac{F_{p1} - F_{s1}}{f_s}$$

$$= 0.04$$

From Eq. 6.69, an initial estimate of the filter order required to meet these specifications is

$$m \approx \text{ceil} \left\{ \frac{-[10 \log_{10}(0.0004) + 13]}{14.6(0.04)} + 1 \right\}$$

$$= 73$$

The coefficients of the equiripple filter can be computed by running script *exam6_8* on the distribution CD. The estimate of $m \approx 73$ is a bit low, and the resulting filter does not quite meet the specifications. The specifications are satisfied for $m \approx 78$ in this case. The magnitude response for the case $m = 82$ is shown in Figure 6.19, where the equiripple nature of the magnitude response is apparent.

It is somewhat difficult to visualize the degree to which the specifications are satisfied while viewing Figure 6.19. Instead, the stopband attenuation is easier to see when the magnitude response is plotted using the logarithmic dB scale. Using Eq. 5.15, the passband ripple and stopband attenuation in dB are

$$A_p = 0.1755 \text{ dB}$$

$$A_s = 33.98 \text{ dB}$$

A plot of the magnitude response in units of dB is shown in Figure 6.20. Here it is clear that, for a filter of order $m = 82$, the specified stopband attenuation of $A_s = 33.98$ dB is exceeded. Therefore, the filter order could be reduced somewhat and still meet the specification.

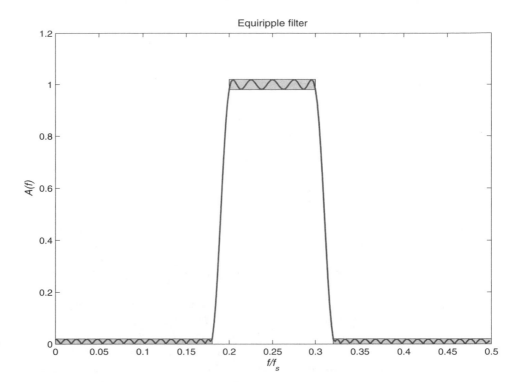

Figure 6.19 Magnitude Response of an Optimal Equiripple Bandpass Filter, $m = 82$

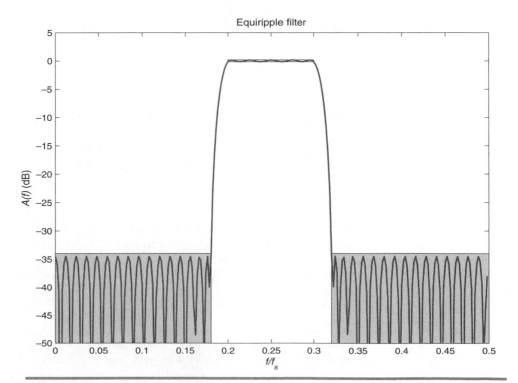

Figure 6.20 Magnitude Response of an Optimal Equiripple Bandpass Filter in dB, $m = 82$

ftype	Filter	F_p	F_s
0	Lowpass	F_p	F_s
1	Highpass	F_p	F_s
2	Bandpass	$[F_{p1}, F_{p2}]$	$[F_{s1}, F_{s2}]$
3	Bandstop	$[F_{p1}, F_{p2}]$	$[F_{s1}, F_{s2}]$

MATLAB Toolbox

The FDSP toolbox contains the following function for designing an optimal equiripple FIR filter using the Parks–McLellan algorithm.

```
b = f_firparks (m,F_p,F_s,delta_p,delta_s,ftype,fs);   % FIR equiripple filter
```

On entry to *f_firparks*, m is the desired filter order, F_p specifies the passband frequency or frequencies, F_s specifies the stopband frequency or frequencies, *delta_p* is the passband ripple, *delta_s* is the stopband attenuation, $0 \leq ftype \leq 3$ is the frequency-selective filter type, and fs is the sampling frequency. The definitions of F_p and F_s depend on the frequency-selective filter type and are summarized in Table 6.5. On exit form *f_firparks*, b is a vector of length $m + 1$ containing the FIR filter coefficients. It is the responsibility of the user to pick a sufficiently large value for the filter order m to meet the design specifications. The estimate of m in Eq. 6.69 can be used as a starting point.

● ● ● ● ● ● ● ● ● ● ● ● ● ● ● ● ● ● ●

*6.6   Differentiators and Hilbert Transformers

In this section, we investigate some specialized linear-phase FIR filters that have impulse responses and amplitude responses that exhibit odd symmetry.

6.6.1   Differentiators

Recall that in Section 6.1 we considered the problem of designing low-order differentiators based on numerical approximations to the slope of the input signal. An alternative approach is to design a filter whose frequency response corresponds to that of a differentiator. Recall that an analog *differentiator* is a system with the following frequency response

$$H_{\text{diff}}(f) = j2\pi f \tag{6.76}$$

Thus, a differentiator has a constant-phase response of $\phi_a(f) = \pi/2$, and a magnitude response that increases linearly with f. If a causal linear-phase FIR filter is used to approximate a differentiator, then we must allow for a group delay of τ. Thus, the problem is to design a digital filter that approximates the following analog system.

$$y_a(t) = \frac{d}{dt} x_a(t - \tau) \tag{6.77}$$

Recall that a delay of τ has a frequency response of $\exp(-j2\pi\tau f)$. Consequently, the frequency response of the system that we want to approximate is

$$H_a(f) = j2\pi f \exp(-j2\pi\tau f) \tag{6.78}$$

Observe that Eq. 6.78 now begins to look like the frequency response of a linear-phase FIR filter with an impulse response that exhibits odd symmetry. In particular, comparing Eq. 6.78 with Eq. 6.18, we see that $H_a(f)$ can be implemented with an mth order linear-phase FIR filter with the frequency response:

$$H(f) = j2\pi f T \exp(-j\pi m f T) \tag{6.79}$$

Note that a T is included in the factor $A_r(f) = 2\pi f T$ because a digital differentiator only processes frequencies in the range $0 \le f \le f_s/2$. The group delay in this case is $\tau = mT/2$.

In order to choose between a type 3 linear-phase FIR filter and a type 4 filter, it is helpful to look at the zeros of $H(f)$. Recall from Table 5.1 that a type 3 filter with even order m has a zero at $f = 0$ and a zero at $f = f_s/2$. This is in contrast to a type 4 filter with an odd order m that only has a zero at $f = 0$. If $A_r(f) = 2\pi f T$, then $A_r(0) = 0$ and $A_r(f_s/2) = \pi$. Thus, the type 4 linear-phase filter appears to be the filter of choice. The effectiveness of the type 4 filter in comparison with the type 3 filter is illustrated with the following example.

Example 6.9   **Differentiator**

Consider a linear-phase FIR filter or order m with an impulse response $h(k)$ that exhibits odd symmetry about the midpoint $k = m/2$. Suppose the windowing method is used to design $H(z)$. Since $H(f)$ does not contain any jump discontinuities, a rectangular window should suffice. From Eq. 6.20, the filter coefficients are

$$b_i = -2T \int_0^{f_s/2} A_r(f) \sin[2\pi(i - 0.5m)fT]df$$

$$= -2T \int_0^{f_s/2} 2\pi f T \sin[2\pi(i - 0.5m)fT]df$$

To simplify the integral, let $\theta = 2\pi f T$. Then the expression for b_i becomes

$$b_i = \frac{-1}{\pi} \int_0^{\pi} \theta \sin[(i - 0.5m)\theta]d\theta$$

$$= \frac{-1}{\pi} \left\{ \frac{\sin[(i - 0.5m)\theta]}{(i - 0.5m)^2} - \frac{\theta \cos[(i - 0.5m)\theta]}{i - 0.5m} \right\} \Bigg|_0^{\pi}$$

Thus, the filter coefficients for the differentiator are

$$b_i = \frac{\cos[(i - 0.5m)\pi]}{i - 0.5m} - \frac{\sin[(i - 0.5m)\pi]}{\pi(i - 0.5m)^2}, \quad 0 \le i \le m$$

Note that when m is odd as in a type 4 filter, the factor $i - 0.5m$ never goes to zero. In this case, $(i - 0.5m)\pi$ is an odd multiple of $\pi/2$ and the cosine term drops out. For a type 3 filter with m even, $(i - 0.5m)\pi$ is a multiple of π and the sine term drops out. Using L'Hospital's rule, the case $i = m/2$ yields $b_{m/2} = 0$. Plots of the magnitude and impulse responses for the type 3 case with $m = 12$ are shown in Figure 6.21. These were generated by running script *exam6_9* on the distribution CD. Notice that the zero at $f = f_s/2$ causes significant ringing in the magnitude response, particularly at higher frequencies. In this case, the impulse response goes to zero relatively slowly, suggesting a poor fit.

When a type 4 filter is used, the results are more effective. Plots of the magnitude and impulse responses for the case $m = 11$ are shown in Figure 6.22. Even though this is a lower-order filter than the type 3 filter, it is clearly a better fit, with some error evident at higher frequencies. Notice how the impulse response goes to zero relatively quickly.

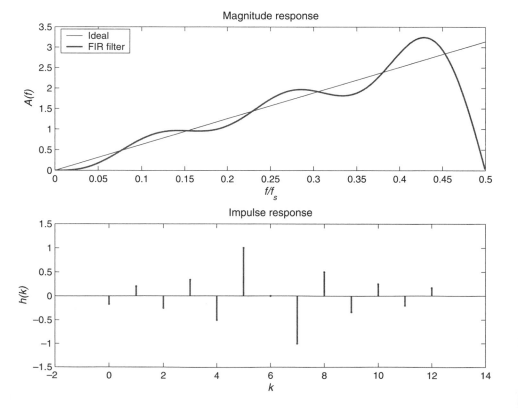

Figure 6.21 A Differentiator using a Type 3 Linear-phase FIR Filter of Order $m = 12$ and a Rectangular Window

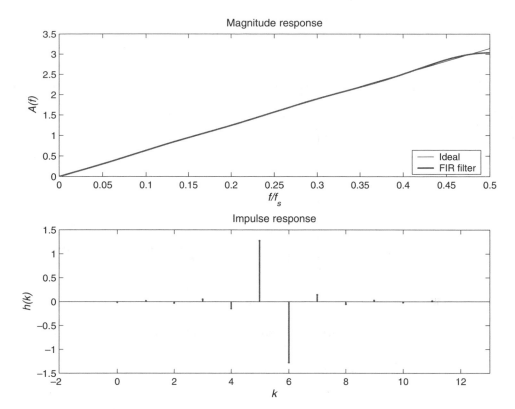

Figure 6.22 A Differentiator using a Type 4 Linear-phase FIR Filter of Order $m = 11$ and a Rectangular Window

It should be pointed out that the phase response of the FIR differentiator is $\phi(f) = \pi/2 - \pi m f T$, which corresponds to $\phi(0) = \pi/2$ and a group delay of $\tau = mT/2$. A constant phase response of $\phi(f) = \pi/2$, corresponding to the pure differentiator in Eq. 6.76, can be achieved if a noncausal implementation is used. A noncausal filter is obtained by multiplying $H(z)$ by z^p, where $p = \text{floor}(m/2)$. Thus, the transfer function of the noncausal filter is

$$H_0(z) = \sum_{i=0}^{m} b_i z^{p-i} \tag{6.80}$$

A noncausal filter can be used if offline batch processing is used with the entire input signal available ahead of time. If the differentiator is part of a real-time implementation, then a causal filter must be used.

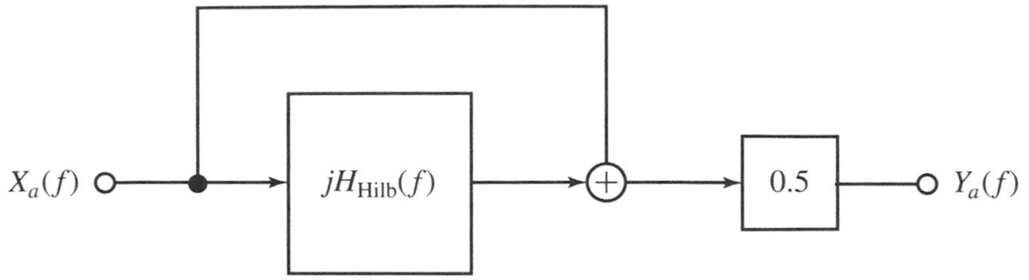

Figure 6.23 Generation of a Half-band Signal using a Hilbert Transformer

6.6.2 Hilbert Transformers

Another example of a specialized filter that can be effectively realized with a type 3 or type 4 linear-phase FIR filter is the Hilbert transformer. An analog version of a *Hilbert transformer* is a system with the following frequency response.

$$H_{\text{Hilb}}(f) = -j\,\text{sgn}(f) \tag{6.81}$$

Here sgn(f) denotes the *sign* or signum function which is defined as

$$\text{sgn}(f) \triangleq \begin{cases} 1, & f > 0 \\ 0, & f = 0 \\ -1, & f < 0 \end{cases} \tag{6.82}$$

Thus, a Hilbert transformer is essentially an allpass filter that imparts a constant phase shift of $-\pi/2$ for $f > 0$ and $\pi/2$ for $f < 0$. Hilbert transformers are used in a number of applications in communications and speech processing. For example, consider the system shown in Figure 6.23. Note from Eq. 6.81 that $j H_{\text{Hilb}}(f) = \text{sgn}(f)$. Consequently, the spectrum of output, $Y_a(f)$, is related to the spectrum of the input $X_a(f)$ as follows.

$$Y_a(f) = \left[\frac{1 + \text{sgn}(f)}{2}\right] X_a(f)$$

$$= \begin{cases} X_a(f), & f > 0 \\ 0.5 X_a(0), & f = 0 \\ 0, & f < 0 \end{cases} \tag{6.83}$$

The complex signal $y_a(t)$ is called an analytic signal because its spectrum is zero for $f < 0$. If the original signal $x_a(t)$ is real and bandlimited, then the signal $y_a(t)$ contains all of the information needed to reconstruct $x_a(t)$, but it occupies only *half* of the bandwidth. As a result, $y_a(t)$ can be modulated, sampled, and transmitted more efficiently than $x_a(t)$. The signal $y_a(t)$ is called a half-band signal.

To implement a Hilbert transformer with a linear-phase FIR filter, we insert a delay of $\tau = mT/2$, where m is the filter order. Using Eq. 6.81, this yields the following frequency response for the digital filter.

$$H(f) = -j\,\text{sgn}(f)\exp(-j\pi m f T) \tag{6.84}$$

From Eq. 6.31, this is the frequency response of a linear-phase FIR filter with odd symmetry about $k = m/2$. Thus, a type 3 or a type 4 filter with $A_r(f) = -\text{sgn}(f)$ should be used.

Example 6.10   **Hilbert Transformer**

Consider a linear-phase FIR filter of order m with an impulse response $h(k)$ that exhibits odd symmetry about $k = m/2$. Suppose the windowing method is used to design $H(z)$. From Eq. 6.20 the filter coefficients, using $b_i = h(i)$, are

$$b_i = -2T \int_0^{f_s/2} A_r(f)\sin[2\pi(i - 0.5m)fT]df$$

$$= 2T \int_0^{f_s/2} \sin[2\pi(i - 0.5m)fT]df$$

$$= \left\{\frac{-2T\cos[2\pi(i - 0.5m)fT]}{2\pi(i - 0.5m)T}\right\}\Bigg|_0^{f_s/2}$$

Thus, the filter coefficients for the Hilbert transformer using a rectangular window are

$$b_i = \frac{1 - \cos[\pi(i - 0.5m)]}{\pi(i - 0.5m)}, \quad 0 \le i \le m$$

When m is odd, as in a type 4 filter, the factor $i - 0.5m$ never goes to zero. In this case, $(i - 0.5m)\pi$ is an odd multiple of $\pi/2$, and the cosine term drops out. For a type 3 filter with m even, the case $i = m/2$ yields $b_{m/2} = 0$. Plots of the magnitude and impulse responses for the type 3 case with $m = 30$ are shown in Figure 6.24, where a rectangular window is used. These were generated by running script *exam6_10* on the distribution CD. Notice that this approximates a Hilbert transformer, but only over a limited range of frequencies. Often this is sufficient depending on the application. The ripples can be reduced by tapering the coefficients with a data window. Plots of the magnitude and impulse responses for a type 3 filter with $m = 30$ and the Hamming window are shown in Figure 6.25. It is evident that the magnitude response is smoother in the passband, but the width of the passband has been reduced somewhat in this case.

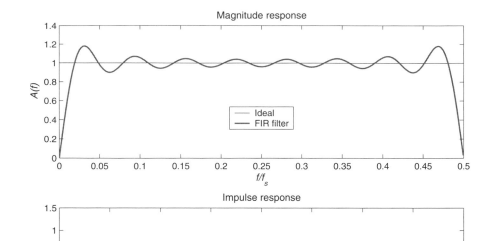

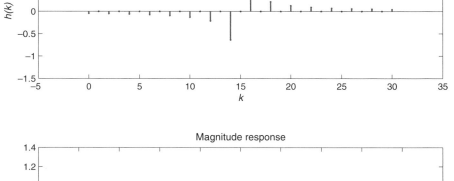

Figure 6.24 A Hilbert Transformer using a Type 3 Linear-phase FIR Filter of Order $m = 30$ and a Rectangular Window

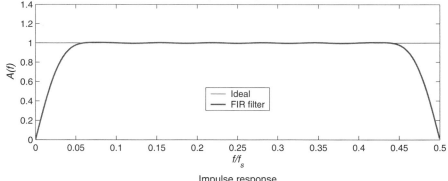

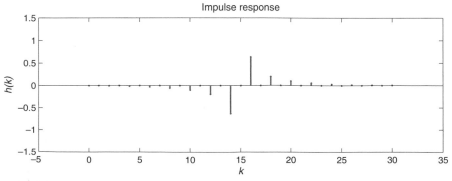

Figure 6.25 A Hilbert Transformer using a Type 3 Linear-phase FIR Filter of Order $m = 30$ and a Hamming Window

6.7 Software Applications

FDSP Toolbox

6.7.1 GUI Module: *g_fir*

The FDSP toolbox includes a graphical user interface module called *g_fir* that allows the user to design and evaluate FIR filters without any need for programming. GUI module *g_fir* features a display screen with tiled windows as shown in Figure 6.26.

The upper-left *Block diagram* window contains a block diagram of the FIR filter under investigation. It is an mth order filter with the following transfer function.

$$H(z) = b_0 + b_1 z^{-1} + \cdots + b_m z^{-m} \tag{6.85}$$

The *Parameters* window below the block diagram displays edit boxes containing the filter parameters. The contents of each edit box can be directly modified by the user with the Enter key used to activate changes. The parameters F_0, F_1, B, and f_s are the lower cutoff frequency, upper cutoff frequency, transition bandwidth, and sampling frequency, respectively. The lowpass filter uses cutoff frequency F_0, the highpass filter uses cutoff frequency F_1, and the bandpass and bandstop filters use both F_0 and F_1. For bandpass and banstop filters, it is assumed that the two transition bands have the same width. The parameters *delta_p* and *delta_s* are the passband ripple factor and the stopband attenuation factor, respectively.

The *Type* and *View* windows in the upper-right corner of the screen allow the user to select both the type of filter and the viewing mode. The filter types include lowpass, highpass, bandpass, and bandstop filters. Also included is a user-defined filter. The desired amplitude response of the user-defined filter is specified in a user-supplied M-file function that has the following calling sequence where *u_amp* is a user-supplied name.

```
A = u_amp (f,fs);
```

When *u_amp* is called with frequency f and sampling frequency fs, it must evaluate the desired amplitude response at the vector f and return the result in the amplitude vector A.

The *View* options include the magnitude response, the phase response, the impulse response, a pole-zero plot, and the window used with the windowed design method. The dB checkbox toggles the magnitude response display between linear and logarithmic scales. When it is checked, the passband ripple and stopband attenuation in the *Parameters* window change to their logarithmic equivalents, A_p and A_s, respectively. The *Plot* window along the bottom half of the screen shows the selected view. Below the *Type* and *View* windows is a horizontal slider bar that allows the user to directly control the filter order m.

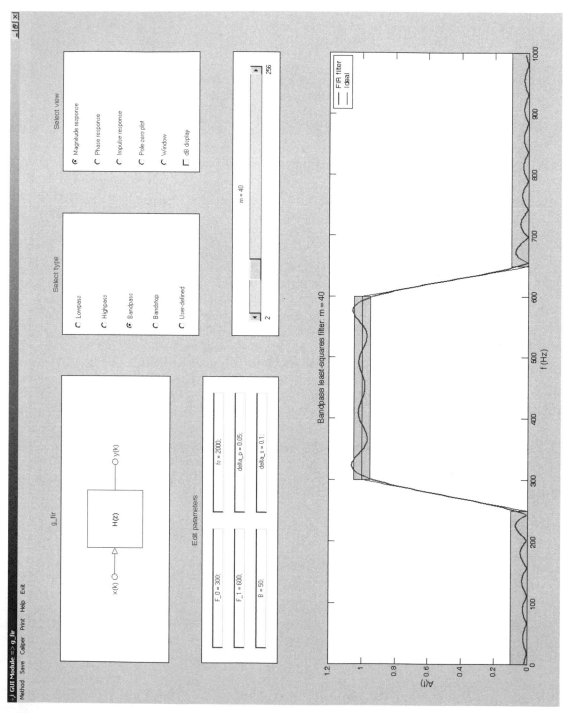

Figure 6.26 Display Screen of Chapter GUI Module *g_fir*

The *Menu* bar at the top of the screen includes several menu options. The *Method* option allows the user to select the filter design method from the windowed (i.e., rectangular, Hanning, Hamming, or Blackman), frequency-sampled, least-squares, and equiripple methods. The *Save* option is used to save the current *a* and *b* and f_s in a user specified MAT-file for future use. Files created in this manner can be loaded into other GUI modules including *g_filters*. The *Caliper* option allows the user to measure any point on the current plot by moving the mouse crosshairsto that point and clicking. The *Print* option prints the contents of the plot window. Finally, the *Help* option provides the user with some helpful suggestions on how to effectively use module *g_fir*.

6.7.2 Bandstop Filter Design: A Comparison

In order to illustrate the different design methods for constructing an FIR filter, suppose $f_s = 2000$ Hz, and consider the problem of designing a bandstop filter to meet the following specifications.

$$(F_{p1}, F_{s1}, F_{s2}, F_{p2}) = (200, 300, 700, 800) \text{ Hz} \tag{6.86a}$$

$$(\delta_p, \delta_s) = (0.04, 0.02) \tag{6.86b}$$

Using Eq. 5.15, the corresponding passband ripple and stopband attenuation in dB are

$$A_p = 0.36 \text{ dB} \tag{6.87a}$$

$$A_s = 33.98 \text{ dB} \tag{6.87b}$$

To facilitate a comparison of the four design methods covered in this chapter, a common filter order of $m = 80$ is used for all cases. Plots of the resulting magnitude responses can be obtained by running the following script, labeled *exam6_11* on the distribution CD.

```
function exam6_11

% Example 6.11 FIR bandstop filter design: A comparison

clear
clc
fprintf ('Example 6.11: FIR bandstop filter design: A comparison\n')
```

```
% Filter specifications

delta_p = 0.04
delta_s = 0.02
A_p = -20*log10(1 - delta_p)
A_s = -20*log10(delta_s)
fs = 2000;
T = 1/fs;
F_p = [200 800]
F_s = [300 700]
a = 1;
N = 250;
sym = 0;
A_min = 80;
A_max = 20;
m = f_prompt ('Enter filter order',0,120,80);

% Compute windowed filter

win = f_prompt ('Enter window type',0,3,3);
p = [F_p(1), F_s(1), F_s(2), F_p(2)];
b = f_firwin (@bandstop,m,fs,win,sym,p);
[H,f] = f_freqz (b,a,N,fs);
AdB = 20*log10(abs(H));
i_stop = (f >= F_s(1)) & (f <= F_s(2));
A_stop = -max(AdB(i_stop))
figure
caption = sprintf ('Windowed Filter, m = %d, A_s = %.1f dB',m,A_stop);
f_labels (caption,'f (Hz)','A(f) (dB)')
axis ([0 fs/2 -A_min A_max])
hold on
box on
fill ([0 F_p(1) F_p(1) 0],[-A_p -A_p A_p A_p],'c')
fill ([F_s(1) F_s(2) F_s(2) F_s(1)],[-A_min -A_min -A_s -A_s],'c')
fill ([F_p(2) fs/2 fs/2 F_p(2)],[-A_p -A_p A_p A_p],'c')
plot (f,AdB)
f_wait
```

```
% Compute frequency-sampled filter

M = floor(m/2) + 1;
F = linspace (0,fs/2,M);
A = bandstop (F,fs,p);
b = f_firsamp (A,m,fs,sym);
[H,f] = f_freqz (b,a,N,fs);
AdB = 20*log10(abs(H));
figure
A_stop = -max(AdB(i_stop))
caption = sprintf ('Frequency-Sampled Filter, m = %d, A_s = %.1f dB',m,A_stop);
f_labels (caption,'f (Hz)','A(f) (dB)')
axis ([0 fs/2 -A_min A_max])
hold on
box on
fill ([0 F_p(1) F_p(1) 0],[-A_p -A_p A_p A_p],'c')
fill ([F_s(1) F_s(2) F_s(2) F_s(1)],[-A_min -A_min -A_s -A_s],'c')
fill ([F_p(2) fs/2 fs/2 F_p(2)],[-A_p -A_p A_p A_p],'c')
plot (f,AdB)
f_wait

% Compute least-squares filter

F = linspace (0,fs/2,m);
A = bandstop (F,fs,p);
w = delta_s/delta_p;
W = w*ones(size(A));
W(i_stop) = 1;
b = f_firls (F,A,m,fs,W);
[H,f] = f_freqz (b,a,N,fs);
AdB = 20*log10(abs(H));
figure
A_stop = -max(AdB(i_stop))
caption = sprintf ('Least-Squares Filter, m = %d, A_s = %.1f dB',m,A_stop);
f_labels (caption,'f (Hz)','A(f) (dB)')
axis ([0 fs/2 -A_min A_max])
hold on
box on
fill ([0 F_p(1) F_p(1) 0],[-A_p -A_p A_p A_p],'c')
fill ([F_s(1) F_s(2) F_s(2) F_s(1)],[-A_min -A_min -A_s -A_s],'c')
fill ([F_p(2) fs/2 fs/2 F_p(2)],[-A_p -A_p A_p A_p],'c')
plot (f,AdB)
f_wait
```

```
% Compute equiripple filter

ftype = 3;
b = f_firparks (m,F_p,F_s,delta_p,delta_s,ftype,fs);
[H,f] = f_freqz (b,a,N,fs);
AdB = 20*log10(abs(H));
A_stop = -max(AdB(i_stop))
figure
caption = sprintf ('Equiripple Filter, m = %d, A_s = %.1f dB',m,A_stop);
f_labels (caption,'f (Hz)','A(f) (dB)')
axis ([0 fs/2 -A_min A_max])
hold on
box on
fill ([0 F_p(1) F_p(1) 0],[-A_p -A_p A_p A_p],'c')
fill ([F_s(1) F_s(2) F_s(2) F_s(1)],[-A_min -A_min -A_s -A_s],'c')
fill ([F_p(2) fs/2 fs/2 F_p(2)],[-A_p -A_p A_p A_p],'c')
plot (f,AdB)
f_wait

function A = bandstop (f,fs,p)
% Description: Piecewise-Linear Amplitude Response of Bandstop Filter
%
% p(1) = F_p1
% p(2) = F_s1
% p(3) = F_s2
% p(4) = F_p2

A = zeros(size(f));
for i = 1 : length(f)
    if (f(i) <= p(1) | f(i) >= p(4))
        A(i) = 1;
    elseif (f(i) > p(1) & f(i) < p(2))
        A(i) = 1 - (f(i) - p(1))/(p(2)-p(1));
    elseif (f(i) > p(3) & f(i) < p(4))
        A(i) = (f(i) - p(3))/(p(4) - p(3));
    end
end
```

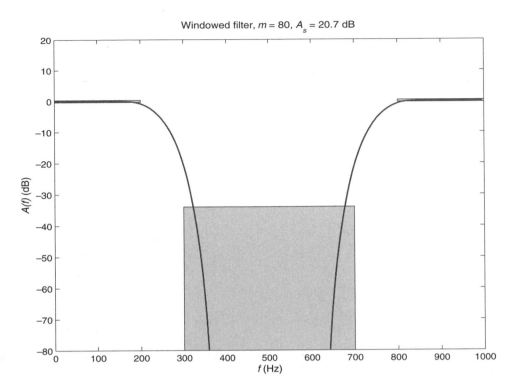

Figure 6.27 Magnitude Response of a Windowed Bandstop Filter using a Blackman Window, $m = 80$

The first plot generated by *exam6_11* corresponds to a windowed filter and is shown in Figure 6.27. In this case, a Blackman window was used. It is apparent that the windowed filter does not meet the stopband specification because the transition band is too wide. The stopband attenuation in this case is $A_s = 20.7$ dB.

The second plot generated by *exam6_11*, which corresponds to the frequency-sampled method, is shown in Figure 6.28. Here the spacing between the frequency samples is

$$\Delta F = \frac{f_s}{m}$$

$$= 25 \text{ Hz} \tag{6.88}$$

From Eq. 6.86, the width of the transition band is $B = 100$ Hz. Consequently, there are three samples in the interior of each transition band. On the surface it appears that the frequency-sampled magnitude response may have met the stopband specification. However a close inspection of Figure 6.28 reveals that near $f = F_{s2}$, the magnitude response is larger than the A_s specification, so the stopband attenuation ends up being only $A_s = 24.5$ dB in this case.

The third plot generated by *exam6_11*, corresponding to the least-squares method, is as shown in Figure 6.29.   To facilitate comparison with the equiripple

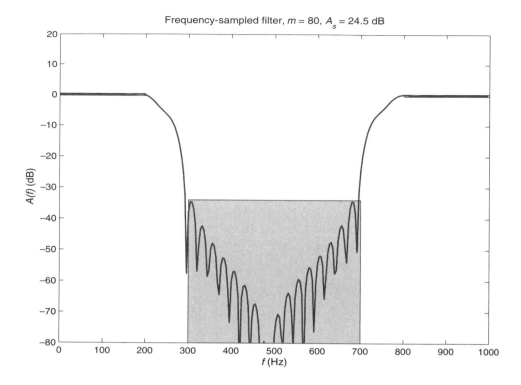

Figure 6.28 Magnitude Response of a Frequency-sampled Bandstop Filter, $m = 80$

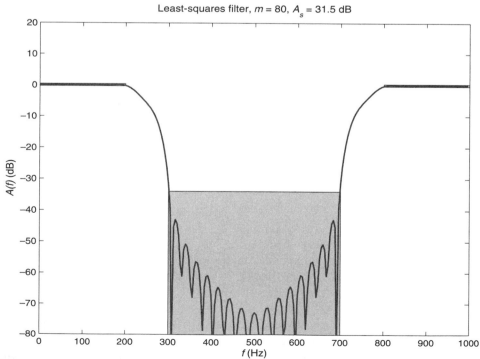

Figure 6.29 Magnitude Response of a Least-squares Bandstop Filter, $m = 80$

method, the weighting vector for the least-squares method was constructed using Eq. 6.55. From Eq. 6.86 we have $\delta_s/\delta_p = 0.5$, so this resulted in the following weighting vector.

$$w(i) = \begin{cases} 0.5, & f_i \in \text{passband} \\ 1, & f_i \in \text{stopband} \end{cases} \tag{6.89}$$

The least-squares method also requires that the weights be specified in the transition band, and in this case, they were set to the passband value. The resulting magnitude response in Figure 6.29 appears to meet the design specifications using a filter order of $m = 80$, but an actual calculation of the stopband attenuation reveals that it does not quite, because the stopband attenuation is $A_s = 31.5$ dB in comparison with the specification of 33.98 dB.

The last plot generated by *exam6_11* corresponds to the optimal equiripple design method and is as shown in Figure 6.30. It is apparent from Figure 6.30 that the equiripple filter easily meets, and in fact exceeds, the stopband specification. The stopband attenuation achieved, in this case, is $A_s = 72.8$ dB. Recall that each 20 dB corresponds to a reduction in gain by a factor of 10. Therefore, the stopband gain is somewhere between 10^{-4} and 10^{-3} for this filter.

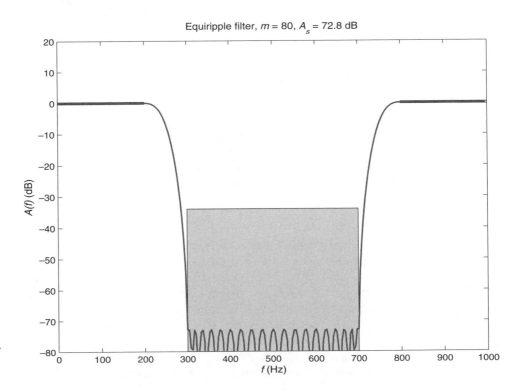

Figure 6.30 Magnitude Response of an Equiripple Bandstop Filter, $m = 80$

• • • • • • • • • • • • • • • • • •

6.8 Chapter Summary

Finite Impule Response Filters

This chapter focused on the design of finite impulse response or FIR digital filters having the following transfer function.

$$H(z) = b_0 + b_1 z^{-1} + \cdots + b_m z^{-m} \tag{6.90}$$

FIR filters offer a number of important advantages in comparison with IIR filters. FIR filters are always stable, regardless of the values of the filter coefficients. FIR filters are also less sensitive to finite word length effects. The impulse response of an FIR filter can be obtained directly from inspection of the transfer function or the difference equation as follows.

$$h(k) = \begin{cases} b_k, & 0 \le k \le m \\ 0, & m < k < \infty \end{cases} \tag{6.91}$$

Techniques are available for designing FIR filters that closely approximate arbitrary magnitude responses if the order of the filter is allowed to be sufficiently large. One drawback of FIR filters is that they require higher order filters than IIR filters that satisfy the same design specifications. This implies larger storage requirements and longer computational times, a consideration that may be important for real-time signal processing applications.

Linear-phase Filters

The phase response of FIR filters can be made to be linear. A linear phase response is an important characteristic because it means that different spectral components of the input signal are delayed by the same amount as they are processed by the filter. A linear-phase filter does not distort a signal within the passband, it only delays it by $\tau = mT/2$. The symmetry constraint on the impulse response that ensures that an FIR filter has linear phase is

$$h(k) = \pm h(m - k), \quad 0 \le k \le m \tag{6.92}$$

If the plus sign is used, the impulse response $h(k)$ is a palindrome that exhibits even symmetry about the midpoint $k = m/2$, otherwise it exhibits odd symmetry. There are four types of linear-phase FIR filters, depending on whether the symmetry is even or odd and the filter order is even or odd. The most general linear-phase filter is a type 1 filter with even symmetry and even order. The other three filter types have zeros at one or both ends of the frequency range, and they are sometimes used for specialized applications such as the design of differentiators and Hilbert transformers. A differentiator has a linear amplitude response, while a Hilbert transformer is an allpass filter with a piecewise-constant amplitude response.

Filter Design Methods

Four techniques were presented for designing an mth order linear-phase FIR filter. They all introduce a delay of $m/2$ samples to make the impulse response causal. The windowing method is a truncated impulse response technique that tapers the coefficients with a data window to reduce ringing in the amplitude response caused by the Gibb's phenomenon. There is a tradeoff between the reduction in ringing and the width of the transition band. Popular windows include the rectangular, Hanning, Hamming, and Blackman windows. When the rectangular window is used, the resulting filter minimizes the following mean-square error, where $A_d(f)$ is the desired amplitude response and $A_r(f)$ is the actual amplitude response.

$$J = \int_0^{f_s/2} |A_d(f) - A_r(f)|^2 df \tag{6.93}$$

The frequency-sampling method uses m equally spaced frequencies and the IDFT to compute the filter coefficients. Oscillations in the resulting magnitude response can be reduced by including one or more frequency samples in a transition band. The values of the transition-band samples can be optimized to maximize the stopband attenuation.

The least-squares method is a direct optimization method that uses an arbitrary set of distinct discrete frequencies. It minimizes a weighted sum of squares of the error between the desired and the actual amplitude responses. Finding the coefficients requires solving a linear algebraic system order of $r + 1$, where $r = m/2$. If significant error exists at certain frequencies, these frequencies can be given additional weight to redistribute the error. The windowed, frequency-sampled, and least-squares methods can be used to design general linear-phase FIR filters with prescribed amplitude responses. The magnitude of the amplitude response is the filter magnitude response.

The final filter design technique was the optimal equiripple method, a technique which minimizes the maximum of the absolute value of the error in the passband and the stopband. Equiripple filters have amplitude responses that have ripples of equal magnitude in the passband and in the stopband. For a given set of frequency-selective filter design specifications, equiripple filters tend to be of lower order than filters designed with the windowed, frequency-sampled, and least-squares methods. To minimize the maximum of the absolute value of the error in the passband and the stopband, the amplitude response of an optimal equiripple filter must satisfy the following equations.

$$A_r(F_i) + \frac{(-1)^i \delta}{W(F_i)} = A_d(F_i), \quad 0 \le i < p + 2 \tag{6.94}$$

Here $W(f) > 0$ is a weighting function, and the F_i are extremal frequencies in the passband and the stopband where the magnitude of the error achieves its maximum value of δ. The FIR amplitude response $A_r(f)$ can be shown to be a polynomial in $x = \cos(2\pi f T)$ of degree $p = \text{floor}(m/2)$. The following normalized weighting function is used to design an equiripple frequency-selective filter.

$$W(f) = \begin{cases} \frac{\delta_s}{\delta_p}, & f \in \text{passband} \\ 1, & f \in \text{stopband} \end{cases} \tag{6.95}$$

The FDSP toolbox includes a GUI module called g_fir that allows the user to design and evaluate FIR filters without any need for programming. The filters include lowpass, highpass, bandpass, and bandstop filters plus a user-defined filter whose amplitude response is specified in an M-file function. Different design methods can be compared with each other and against the design specifications as the user varies the filter order m. The design methods include the windowed method, the frequency-sampled method, the least-squares method, and the optimal equiripple method.

● ● ● ● ● ● ● ● ● ● ● ● ● ● ● ●

6.9 Problems

The problems are divided into Analysis problems that can be solved by hand or with a calculator, GUI Simulation problems that are solved using GUI module g_fir, and Computation problems. The Computation problems require the student to write a MATLAB script using the FDSP toolbox functions summarized in Appendix B. Solutions to selected problems are available on the distribution CD. Students are encouraged to use these problems, which are identified with a ☑, as a check on their understanding of the material.

6.9.1   Analysis

Section 6.1

P6.1. Consider the following noise-corrupted periodic signal. Here $v(k)$ is white noise uniformly distributed over $[-0.5, 0.5]$.

$$x(k) = 3 + 2\cos(0.2\pi k)$$
$$y(k) = x(k) + v(k)$$

(a) Find the average power of the noise-free signal, $x(k)$.

(b) Find the signal-to-noise ratio of $y(k)$.

(c) Suppose $y(k)$ is sent through an ideal lowpass filter with cutoff frequency $F_0 = 0.15 f_s$ to produce $z(k)$. Is the signal $x(k)$ affected by this filter? Find the signal-to-noise ratio of $z(k)$.

Section 6.2

P6.2. Consider the problem of designing an mth-order type 3 linear-phase FIR filter having the following amplitude response.

$$A_r(f) = \sin(2\pi f T), \quad 0 \le f \le f_s/2$$

(a) Assuming $m = 2p$ for some integer p, find the coefficients using the windowing method with the rectangular window.

(b) Find the filter coefficients using the windowing method with the Hamming window.

P6.3. Suppose a lowpass filter order of $m = 10$ is designed using the windowing method with the Hanning window and $f_s = 2000$ Hz.

(a) Estimate the width of the transition band.

(b) Estimate the linear passband ripple and stopband attenuation.

(c) Estimate the logarithmic passband ripple and stopband attenuation.

☑ **P6.4.** Consider the problem of using the windowing method to design a lowpass filter to meet the following specifications.

$$(f_s, F_p, F_s) = (100, 30, 50) \text{ Hz}$$

$$(A_p, A_s) = (0.02, 50) \text{ dB}$$

(a) Which types of windows can be used to satisfy these design specifications?

(b) For each of the windows in part (a), find the minimum filter order of m that will satisfy the design specifications.

(c) Assuming an ideal piecewise-constant amplitude response is used, find an appropriate value for the cutoff frequency F_c.

P6.5. Suppose the windowing method is used to design an mth-order lowpass FIR filter. The candidate windows include rectangular, Hanning, Hamming, and Blackman.

(a) Which window has the smallest transition band?

(b) Which window has the smallest passband ripple, A_p?

(c) Which window has the largest stopband attenuation, A_s?

P6.6. A linear-phase FIR filter is designed with the windowing method using the Hanning window. The filter meets its transition bandwidth specification of 200 Hz exactly with a filter order of $m = 30$.

(a) What is the sampling rate, f_s?

(b) Find the filter order needed to achieve the same transition bandwidth using the Hamming window.

(c) Find the filter order needed to achieve the same transition bandwidth using the Blackman window.

P6.7. Consider the problem of designing a linear-phase ideal bandstop FIR filter with the windowing method using the Blackman window. Find the coefficients of a filter order of $m = 40$ using the following cutoff frequencies.

$$(f_s, F_{s1}, F_{s2}) = (10, 2, 4) \text{ kHz}$$

P6.8. Consider the problem of designing a type 1 linear-phase windowed FIR filter with the following desired amplitude response.

$$A_r(f) = \cos(\pi f T), \quad 0 \le |f| \le f_s/2$$

Suppose the filter order is even, with $m = 2p$. Find the impulse response $h(k)$ using a rectangular window. Simplify the expression for $h(k)$ as much as possible.

Section 6.3

P6.9. Consider the problem of designing a type 1 linear-phase bandpass FIR filter using the frequency-sampling method. Suppose the filter order is $m = 60$. Find a simplified expression for the filter coefficients using the following ideal design specifications.

$$(f_s, F_{p1}, F_{p2}) = (1000, 100, 300) \text{ Hz}$$

Section 6.4

☑ **P6.10.** Consider a type 3 linear-phase FIR filter order of $m = 2r$. Find a simplified expression for the amplitude response $A_r(f)$ similar to Eq. 6.42, but for a type 3 linear-phase FIR filter.

P6.11. Use the results of Problem P6.10 to derive the normal equations for the coefficients of a least-squares type 3 linear-phase filter. Specifically, find expressions for the matrix G and the right-hand side vector d, and show how to obtain the filter coefficients from the solution to the normal equations.

Section 6.5

P6.12. Suppose the equiripple design method is used to construct a highpass filter to meet the following specifications. Estimate the required filter order.

$$(f_s, F_s, F_p) = (100, 20, 30) \text{ kHz}$$

$$(A_p, A_s) = (0.2, 32) \text{ dB}$$

P6.13. Consider the problem of constructing an equiripple bandstop filter of order $m = 40$. Suppose the design specifications are as follows.

$$(f_s, F_{p1}, F_{s1}, F_{s2}, F_{p2}) = (200, 20, 30, 50, 60) \text{ Hz}$$

$$(\delta_p, \delta_s) = (0.05, 0.03)$$

(a) Let r be the number of extremal frequencies in the optimal amplitude response. Find a range for r.

(b) Find the set of specification frequencies, F.

(c) Find the weighting function $W(f)$.

(d) Find the desired amplitude response $A_d(f)$.

(e) The amplitude response $A_r(f)$ is a polynomial in x. Find x in terms of f, and find the polynomial degree.

P6.14. Consider the problem of constructing an equiripple lowpass filter of order $m = 4$ satisfying the following design specifications.

$$(f_s, F_p, F_s) = (10, 2, 3) \text{ Hz}$$

$$(\delta_p, \delta_s) = (0.05, 0.1)$$

Suppose the initial guess for the extremal frequencies is as follows.

$$(F_0, F_1, F_2, F_3) = (0, F_p, F_s, f_s/2)$$

(a) Find the weights $W(F_i)$ for $0 \le i \le 3$.
(b) Find the desired amplitude response values $A_d(F_i)$ for $0 \le i \le 3$.
(c) Find the extremal angles $\theta_i = 2\pi F_i T$ for $0 \le i \le 3$.
(d) Write down the vector equation that must be solved to find the Chebyshev coefficient vector d and the parameter δ. You do not have to solve the equation, just formulate it.

Section 6.6

P6.15. Consider the problem of designing a filter to approximate a differentiator. Use the frequency-sampling method to design a type 3 linear-phase filter order of $m = 40$ that approximates a differentiator, but with a delay of $m/2$ samples. That is, find simplified expressions for the coefficients of a filter with the following desired amplitude response.

$$A_r(f) = 2\pi f T$$

6.9.2   GUI Simulation

Section 6.7

P6.16. Use the GUI module *g_fir* to design a windowed lowpass filter. Use the default parameter values except for the width of the transition band which should be set to $B = 150$ Hz. For each of the following cases, find the lowest value for the filter order m which meets the specifications. Plot the linear magnitude response in each case.

(a) Rectangular window.
(b) Hanning window.
(c) Hamming window.
(d) Blackman window.

P6.17. Use the GUI module *g_fir* to construct a windowed highpass filter using the Hamming window and the default parameter values.

(a) Plot the linear magnitude response and use the Caliper option to measure the actual width of the transition band.
(b) Plot the phase response.
(c) Plot the impulse response.

☑ **P6.18.** Use the GUI module *g_fir* to design a least-squares bandpass filter to meet the following specifications. Adjust the filter order to the lowest value that meets the design specifications.

$$(f_s, F_{s1}, F_{p1}, F_{p2}, F_{s2}) = (2000,300,400,600,700) \text{ Hz}$$

$$(A_p, A_s) = (0.4, 30) \text{ dB}$$

(a) Plot the magnitude response using the dB scale.

(b) Save filter parameters a, b, and f_s. Then use GUI module *g_filters* to load these as a user-defined filter. Adjust the number of bits used for coefficient quantization to a level that shows a clear difference between the quantized and unquantized linear magnitude responses using a direct-form realization. Plot the linear magnitude responses.

P6.19. Use the GUI module *g_fir* to design an optimal equiripple bandpass filter to meet the following specifications. Adjust the filter order to the lowest value that meets the design specifications.

$$(f_s, F_{s1}, F_{p1}, F_{p2}, F_{s2}) = (2000,300,400,600,700) \text{ Hz}$$

$$(A_p, A_s) = (0.4, 30) \text{ dB}$$

(a) Plot the magnitude response using the dB scale.

(b) Save filter parameters a, b, and f_s. Then use GUI module *g_filters* to load these as a user-defined filter. Adjust the number of bits used for coefficient quantization to a level that shows a clear difference between the quantized and unquantized linear magnitude responses using a direct-form realization. Plot the linear magnitude responses.

P6.20. Use the GUI module *g_fir* to design a windowed bandstop filter with the Hanning window to meet the following specifications. Adjust the filter order to the lowest value that meets the design specifications.

$$(f_s, F_{p1}, F_{s1}, F_{s2}, F_{p2}) = (100, 20, 25, 35, 40) \text{ Hz}$$

$$(\delta_p, \delta_s) = (0.05, 0.05)$$

(a) Plot the magnitude response using the linear scale.

(b) Save filter parameters a, b, and f_s. Then use GUI module *g_filters* to load these as a user-defined filter. Adjust the number of bits used for coefficient quantization to a level that shows a clear difference between the quantized and unquantized linear magnitude responses using a direct-form realization. Plot the linear magnitude responses.

P6.21. Use the GUI module *g_fir* to design a frequency-sampled bandstop filter to meet the following specifications. Adjust the filter order to the lowest value that meets the design specifications.

$$(f_s, F_{p1}, F_{s1}, F_{s2}, F_{p2}) = (100, 20, 25, 35, 40) \text{ Hz}$$

$$(\delta_p, \delta_s) = (0.05, 0.05)$$

(a) Plot the magnitude response using the linear scale.

(b) Save filter parameters a, b, and f_s. Then use GUI module *g_filters* to load these as a user-defined filter. Adjust the number of bits used for coefficient quantization to a level that shows a clear difference between the quantized and unquantized linear magnitude responses using a direct-form realization. Plot the linear magnitude responses.

P6.22. Use the GUI module *g_fir* and the User-defined option to load the filter in file *u_fir1*. Adjust the filter order to $m = 90$. Plot the linear magnitude response for each of the following cases.

(a) Windowed filter with Blackman window.

(b) Least-squares filter.

P6.23. Write an amplitude response function for the following user-defined filter (see *u_fir1* for an example).

$$A_r(f) = 2 \left| \cos\left(\frac{2\pi f}{f_s} \right) \right|$$

Then use the User-defined option of GUI module *g_fir* to load this filter. Select a least-squares filter. Plot the linear magnitude response for the following three cases.

(a) $m = 10$.

(b) $m = 20$.

(c) $m = 40$.

P6.24. Write an amplitude response function for the following user-defined filter (see *u_fir1* for an example).

$$A_r(f) = \frac{\cos(\pi f^2/100)}{1 + f^2}, \quad 0 \le f \le 10 \text{ Hz}$$

Then use the User-defined option of GUI module *g_fir* to load this filter. Set $f_s = 20$ Hz, and select a frequency-sampled filter. Plot the following cases.

(a) Magnitude response, $m = 10$.

(b) Magnitude response, $m = 20$.

(c) Magnitude response, $m = 40$.

(d) Impulse response, $m = 40$.

6.9.3 Computation

Section 6.1

P6.25. Write a MATLAB script that constructs the following signal where $f_s = 200$ Hz. Here $v(k)$ is white noise uniformly distributed over $[-1, 1]$, $F_1 = 10$ Hz, $F_2 = 30$ Hz, and $N = 4096$. Use a random number generator seed of 100 to produce $v(k)$.

$$x(k) = 4 \sin(2\pi F_1 kT) \cos(2\pi F_2 kT), \quad 0 \le k < N$$

$$y(k) = x(k) + v(k), \quad 0 \le k < N$$

(a) Compute P_x and P_v directly from the samples. Use Definition 6.1 to compute and print the signal-to-noise ratio of $y(k)$.

(b) Compute P_y directly from the samples. Use P_v, Eq. 6.1, and Definition 6.1 to compute and print the signal-to-noise ratio of $y(k)$.

(c) Compute and print the percent error of the estimate of the SNR found in part (b) relative to the SNR found in part (a).

(d) Plot the magnitude spectrum of $y(k)$ showing the signal and the noise.

Section 6.2

☑ **P6.26.** Write a MATLAB script that uses *f_firideal* to design a linear-phase lowpass FIR filter of order $m = 40$ with passband cutoff frequency $F_p = f_s/5$ and stopband cutoff frequency $F_s = f_s/4$, where the sampling frequency is $f_s = 100$ Hz. Use a rectangular window, and set the ideal cutoff frequency to the middle of the transition band. Use *f_freqz* to compute and plot the magnitude response using the linear scale. Then use Table 6.3, the *hold on* command, and the *fill* function to add the following items to your magnitude response plot.

(a) A shaded area showing the passband ripple, δ_p.

(b) A shaded area showing the stopband attenuation, δ_s.

P6.27. Write a MATLAB script that uses *f_firideal* to design a linear-phase highpass FIR filter of order $m = 30$ with stopband cutoff frequency $F_s = 20$ Hz, passband cutoff frequency $F_p = 30$, and sampling frequency $f_s = 100$ Hz. Use a Hanning window, and set the ideal cutoff frequency to the middle of the transition band.

(a) Use *f_freqz* to compute and plot the magnitude response using the dB scale.

(b) Use Table 6.3, the *hold on* command, and the *fill* function to add a shaded area showing the predicted stopband attenuation, A_s.

P6.28. Write a MATLAB script that uses *f_firideal* to design a linear-phase highpass FIR filter of order $m = 40$ with stopband cutoff frequency $F_s = 20$ Hz, passband cutoff frequency $F_p = 30$ and sampling frequency $f_s = 100$ Hz. Use a Hamming window, and set the ideal cutoff frequency to the middle of the transition band.

(a) Use *f_freqz* to compute and plot the magnitude response using the dB scale.

(b) Use Table 6.3, the *hold on* command, and the *fill* function to add a shaded area showing the predicted stopband attenuation, A_s.

P6.29. Write a MATLAB script that uses *f_firwin* to design a linear-phase highpass FIR filter of order $m = 60$ with stopband cutoff frequency $F_s = 20$ Hz, passband cutoff frequency $F_p = 30$, and sampling frequency $f_s = 100$ Hz. Use a Blackman window, and make the desired amplitude response piecewise-constant with cutoff $F_c = (F_s + F_p)/2$.

(a) Use *f_freqz* to compute and plot the magnitude response using the dB scale.

(b) Use Table 6.3, the *hold on* command, and the *fill* function to add a shaded area showing the predicted stopband attenuation, A_s.

P6.30. Write a MATLAB script that uses *f_firwin* to design a type 1 linear-phase FIR filter of order $m = 80$ using $f_s = 1000$ Hz and the Hamming window to approximate the following amplitude response. Use *f_freqz* to compute the magnitude response.

$$A_r(f) = \begin{cases} \left(\dfrac{f}{250}\right)^2, & 0 \le |f| < 250 \\ 0.5\cos\left[\dfrac{\pi(f - 250)}{500}\right], & 250 \le |f| < 500 \end{cases}$$

(a) Plot the linear magnitude response.

(b) On the same graph, add the desired magnitude response and a legend.

Section 6.3

✅ **P6.31.** Write a MATLAB script that uses function *f_firsamp* to design a linear-phase bandpass FIR filter of order $m = 40$ using the frequency sampling method. Use a sampling frequency of $f_s = 200$ Hz, and a passband of $F_p = [20, 60]$ Hz. Use *f_freqz* to compute and plot the linear magnitude response. Add the frequency samples using a separate plot symbol and a legend. Do the following cases.

(a) No transition band samples (ideal amplitude response).

(b) One transition band sample of amplitude 0.5 on each side of the passband.

P6.32. Write a MATLAB script that uses function *f_firsamp* to design a linear-phase bandstop FIR filter of order $m = 60$ using the frequency sampling method. Use a sampling frequency of $f_s = 20$ kHz, and a stopband of $F_s = [3, 8]$ kHz. Use *f_freqz* to compute and plot the linear magnitude response. Add the frequency samples using a separate plot symbol and a legend. Do the following cases.

(a) No transition band samples (ideal amplitude response).

(b) One transition band sample of amplitude 0.5 on each side of the stopband.

Section 6.4

P6.33. Write a MATLAB script that uses function *f_firls* to design a least-squares linear-phase FIR filter of order $m = 30$ with sampling frequency $f_s = 400$ and the following amplitude response.

$$A_r(f) = \begin{cases} \dfrac{f}{100}, & 0 \le |f| < 100 \\[2mm] \dfrac{200 - f}{100}, & 100 \le |f| \le 200 \end{cases}$$

Select $2m$ equally spaced discrete frequencies, and use uniform weighting. Use *f_freqz* to compute and plot both magnitude responses (ideal and actual) on the same graph.

Section 6.5

P6.34. The Chebyshev polynomials have several interesting properties. Write a MATLAB script that uses the FDSP toolbox function *f_chebpoly* and the *subplot* command to construct a 2×2 array of plots of the Chebyshev polynomials, $T_k(x)$ for $1 \le k \le 4$. Use the plot range, $-1 \le x \le 1$. Using induction and your observations of the plots, list as many general properties of $T_k(x)$ as you can. Use the help command for instructions on how to use *f_chebpoly*.

P6.35. Write a MATLAB function called *u_firorder* which estimates the order of an equiripple filter required to meet given design specifications using Eq. 6.69. The calling sequence for *u_firorder* should be as follows.

```
m = u_firorder (delta_p,delta_s,B_hat);
```

On entry to *u_firorder*, *delta_p* is the passband ripple, *delta_s* is the stopband attenuation, and *B_hat* is the normalized width of the transition band. On exit from *u_firorder*, the integer m is the estimate of the required equiripple filter order based on Eq. 6.69. Test your function by plotting a family of curves on one graph. For the kth curve use $delta\_p = delta\_s = \delta$, where $\delta = 0.03k$ for $1 \le k \le 3$. Plot m versus *B_hat* for $0.01 \le B\_hat \le 0.1$ and include a legend.

P6.36. Write a MATLAB script that uses the function *f_firparks* to design an equiripple lowpass filter to meet the following design specifications where $f_s = 4000$ Hz. Find the lowest order filter that meets the specifications.

$$(F_p, F_s) = (1200, 1400) \text{ Hz}$$

$$(\delta_p, \delta_s) = (0.03, 0.04)$$

(a) Print the minimum filter order and the estimated order based on Eq. 6.69.

(b) Plot the linear magnitude response.

(c) Use *fill* to add shaded areas to the plot showing the design specifications.

P6.37. Write a MATLAB script that uses the function *f_firparks* to design an equiripple highpass filter to meet the following design specifications where $f_s = 300$ Hz. Find the lowest order filter that meets the specifications.

$$(F_s, F_p) = (90, 110) \ Hz$$

$$(\delta_p, \delta_s) = (0.02, 0.03)$$

(a) Print the minimum filter order and the estimated order based on Eq. 6.69.

(b) Plot the linear magnitude response.

(c) Use *fill* to add shaded areas to the plot showing the design specifications.

Multirate Signal Processing

Learning Objectives

- Know how to design and implement integer decimators and interpolators (Section 7.2)

- Know how to design and implement rational sampling rate converters, both single stage and multistage (Section 7.3)

- Be able to efficiently implement sampling rate converters using polyphase filter realizations (Section 7.4)

- Be able to apply multirate techniques to design narrowband filters and filter banks (Section 7.5)

- Understand how to design an oversampling ADC and what benefits are achieved (Section 7.6)

- Understand how to design an oversampling DAC and what benefits are achieved (Section 7.7)

- Know how to use the GUI module *g_multirate* to design and evaluate multirate DSP systems (Section 7.8)

● ● ● ● ● ● ● ● ● ● ● ● ● ● ● ● ●

7.1 **Motivation**

All of the discrete-time systems that we have encountered thus far have signals that are sampled at a single fixed sampling rate f_s. When we relax this assumption and allow some of the signals to be sampled at one rate while others are sampled at a higher or a lower rate, this leads to a *multirate* system. As we shall see, multirate systems can offer important advantages over fixed-rate systems in terms of overall performance. One of the simplest examples of a multirate system is a sampling rate *decimator* which decreases the sampling rate of a discrete-time signal by an integer factor, M.

Integer decimator

$$y(k) = \sum_{i=0}^{m} b_i x(Mk - i)$$

Here output $y(k)$ is a filtered version of the input $x(k)$, but the input is evaluated only at every Mth sample. Extracting every Mth sample effectively reduces the sampling rate by M. The filtering operation is needed in order to preserve the spectral characteristics of the down-sampled signal. It is also possible to increase the sampling rate by an integer factor L using an *interpolator*. More generally, sampling rate converters can be designed where the ratio of the output sampling frequency to the input sampling frequency is an arbitrary rational number, L/M. Modern high-performance DSP systems exploit the benefits of multirate systems. For example, multirate techniques can be used to design analog-to-digital converters (ADCs) and digital-to-analog converters (DACs) with improved noise immunity. Another important class of applications is the design of banks of narrowband filters, such as those used in frequency multiplexing and demultiplexing applications.

   We begin this chapter by introducing some examples of applications of multirate systems. Next, the design of integer sampling rate decimators and interpolators is presented. These rate-converter building blocks are then used to construct rational sampling rate converters, both single stage and multistage. This is followed by a discussion of efficient realization structures for rate converters based on time-varying polyphase filters. Next, the discussion turns to multirate system applications starting with the design of narrowband filters and filter banks. The improved performance characteristics of oversampling ADCs are then presented. This is followed by an analogous presentation applied to oversampling DACs. Finally, a GUI module called *g_multirate* is introduced that allows the user to design and evaluate multirate DSP systems without any need for programming. The chapter concludes with an application example and a summary of multirate signal processing techniques. The selected example features the design of a multirate system that converts music from compact disc (CD) format to digital audio tape (DAT) format.

7.1.1 Narrowband Filter Banks

Several signals can be transmitted simultaneously over a single communication channel by allocating a separate band of frequencies for each signal. This technique,

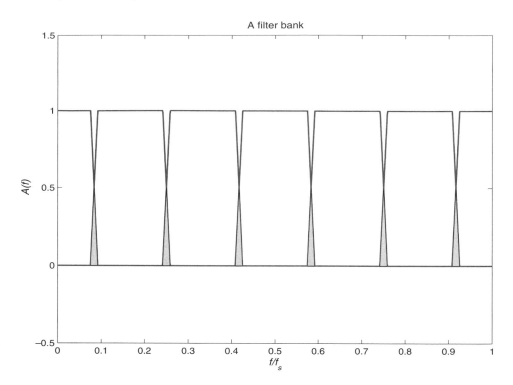

Figure 7.1 Magnitude Responses of a Bank of Six Narrowband Filters

known as frequency division multiplexing or subband processing, requires the use of a bank of narrowband filters. This way each filter can be used to extract a different signal. The magnitude responses of a bank of six narrowband filters are shown in Figure 7.1. Notice that to maximize the number of filters in the bank, their transition bands overlap with one another, as shown in the shaded regions.

A filter is referred to as a *narrowband* filter when the width of its passband (or its stopband) is small in comparison with the sampling frequency f_s. For example, let $B_{\text{pass}} = F_{p2} - F_{p1}$ denote the width of the passband of a bandpass filter. This filter is a narrowband filter if

$$B_{\text{pass}} \ll f_s \tag{7.1}$$

Narrowband lowpass and highpass filters can be defined in an analogous way. The challenge in designing a bank of narrowband filters arises when one considers the required width of the transition band. Suppose we design a bank of N narrowband filters. Since the discrete-time frequency response is periodic with period f_s, we can take the frequency range to be $[0, f_s]$ rather than $-f_s/2$ to $f_s/2$. Then the ith filter will be centered at $F_i = if_s/N$ for $0 \le i \le N$ and will have passband width of

$$B_{\text{pass}} < \frac{f_s}{N} \tag{7.2}$$

In order to maximize the use of the spectrum, the width of the transition band should be small in comparison with the width of the passband. Consequently, for a narrowband filter, the width of the transition band is very small in comparison with f_s. As an illustration, suppose a filter bank of $N = 20$ filters is to be designed, and suppose the width of the transition band is set to $B_{\text{trans}} = f_s/(40N)$. Then the normalized width of the transition band is

$$B = \frac{B_{\text{tran}}}{f_s}$$

$$= \frac{1}{40N}$$

$$= 0.00125 \tag{7.3}$$

This design requirement is quite severe. To see what it implies, suppose the passband ripple and stopband attenuation are as follows for each filter in the bank.

$$(\delta_p, \delta_s) = (0.01, 0.02) \tag{7.4}$$

If the equiripple FIR filter design method is used to design the filters, then from Eq. 6.69, the estimated order of the filters required to meet the design specification is

$$m \approx \text{ceil} \left\{ \frac{-[10 \log_{10}(\delta_p \delta_s) + 13]}{14.6B} + 1 \right\}$$

$$= \text{ceil} \left\{ \frac{-[10 \log_{10}(0.0002) + 13]}{14.6(0.00125)} + 1 \right\}$$

$$= 1316 \tag{7.5}$$

Clearly, a very high-order filter is needed to meet the narrowband design specification in this case. Recall that if an alternative FIR design method is used, such as a windowed, frequency-sampled, or least-squares filter, the required filter order will be even higher. Implementing a filter of such a high order brings with it a host of practical problems, including significant storage requirements, lengthy processing time, and potentially debilitating finite word length effects. Fortunately, by using a multirate design with a multistage polyphase realization, these difficulties can be reduced significantly and the performance of the narrowband filter can be improved.

7.1.2 Delay Systems

A design task that occurs repeatedly in different applications is the problem of delaying a discrete-time signal without otherwise distorting it. If the desired delay is an integer multiple of the sampling interval T, then this is achieved easily. One can allocate a memory buffer in the form of a shift register of length M, as shown in Figure 7.2. Here the signal shifted out the other end will be a delayed version of the input with a delay of $\tau = MT$.

Figure 7.2 Delay of Discrete-time Signal using an M-sample Shift Register

More challenging is the problem of designing a system where the delay is not an integer multiple of the sampling interval, but instead involves an intersample delay. In effect, what is required is an allpass filter with a phase response of

$$\phi(f) = -2\pi f \tau \tag{7.6}$$

Recall from Eq. 5.33 that allpass IIR filters can be designed easily by enforcing a reflective symmetry constraint on the coefficients. However, the design of fixed-rate allpass IIR filters with an arbitrary group delay is a challenging task. Fortunately, by using multirate techniques, this design problem becomes more manageable. The basic idea is to first increase the sampling rate by a factor, L. We then delay the up-sampled signal by $0 < M < L$ samples using a shift register. This is followed by decreasing the sampling rate by a factor L to restore the original sampling frequency. The processing steps are summarized in Figure 7.3.

The factor L interpolator in Figure 7.3 increases the sampling rate by L so that the intermediate signal $r(k)$ is sampled at the rate $f_r = L f_s$. A brute-force analog approach to changing the sampling rate is to convert $x(k)$ from digital to analog with a DAC, and then sample the result at the new rate with an ADC. A drawback of this analog approach is that it introduces additional quantization and aliasing errors. As we shall see, sampling rate converters that avoid these types of error can be designed by working strictly in the discrete-time domain. Once $x(k)$ has been up-sampled to produce $r(k)$, this intermediate signal is then delayed by an integer number of samples M using the shift register in Figure 7.2. If the length of the shift register is in the range $0 < M < L$, this produces a *fractional* or intersample delay when viewed in terms of the original sampling rate, f_s. In particular, the delay introduced by the shift register block in Figure 7.3 is

$$\tau = \left(\frac{M}{L}\right) T \tag{7.7}$$

Figure 7.3 Intersample Delay of Discrete-time Signal using a Multirate System

Finally, the factor L decimator down-samples the delayed signal $v(k - M)$ by L, thereby restoring the original sampling frequency. It should be pointed out that the interpolator and decimator blocks in Figure 7.3 include linear-phase FIR lowpass filters. So while these processing steps will also introduce delays, these delays are integer multiples of the original sampling interval.

● ● ● ● ● ● ● ● ● ● ● ● ● ● ● ● ● ● ●

7.2   **Integer Sampling Rate Converters**

Modern high-performance DSP systems often make use of *multirate* systems: systems where some of the signals are sampled at one frequency and others are sampled at another frequency. For example, the need for a sharp high-order analog anti-aliasing prefilter for an ADC can be avoided if oversampling is used and the sampling rate is later reduced to the desired value.

A conceptually simple way to change the sampling rate is to convert a discrete-time signal from digital to analog with a DAC, and then resample the analog signal with an ADC using the desired sampling frequency. This brute force approach to sampling rate conversion has the advantage that the new sampling rate can be any value achievable by the ADC. However, a drawback is that the DAC and ADC introduce additional quantization noise and aliasing errors. In this section, we introduce techniques that avoid these drawbacks by implementing sampling rate converters entirely in the discrete-time domain.

7.2.1   Sampling Rate Decimator

Let us begin with a relatively simple problem, namely, a reduction in the sampling rate by an integer factor M. A sampling rate converter that reduces the sampling rate is called a *decimator* because one is removing samples. Let $x(k)$ be the discrete-time signal obtained by sampling an analog signal $x_a(t)$ at the rate f_s. If $T = 1/f_s$ is the sampling interval, then

$$x(k) = x_a(kT), \quad k \geq 0 \tag{7.8}$$

The objective is to start with $x(k)$ and synthesize a new discrete-time signal $y(k)$ that corresponds to sampling $x_a(t)$ at the reduced rate $f_M = f_s/M$, where M is a positive integer. Since M is an integer, it would appear that this can be accomplished by simply extracting every Mth sample of $x(k)$ as follows.

$$x_M(k) = x(Mk), \quad k \geq 0 \tag{7.9}$$

The problem with this basic approach is that it does not take into account the frequency content of the two signals. If the original signal $x(k)$ is sampled in a manner that avoids aliasing, then from the sampling theorem, the analog signal $x_a(t)$ must be bandlimited to less than $f_s/2$ Hz. However, to avoid aliasing with the reduced-rate signal, $x_M(k)$, the analog signal must be bandlimited to less than $f_s/(2M)$ Hz. Thus,

Figure 7.4 Sampling Rate Decimation by an Integer Factor, M

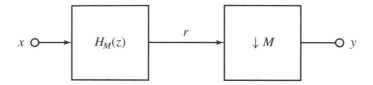

to eliminate aliasing in $x_M(k)$, we must first pass $x(k)$ through a lowpass filter with a cutoff frequency of $F_M = f_s/(2M)$.

$$H_M(f) \triangleq \begin{cases} 1, & 0 \le |f| < F_M \\ 0, & F_M \le |f| \le f_s/2 \end{cases} \quad (7.10)$$

Unlike an analog anti-aliasing filter associated with an ADC, the filter in Eq. 7.10 is a *digital* anti-aliasing filter. Sampling rate decimation by an integer factor, M, is summarized in the block diagram shown in Figure 7.4. Note that it is standard practice to denote sampling rate reduction, also called *down-sampling*, with the down-arrow notation $\downarrow$.

   Since the anti-aliasing filter in Figure 7.4 is a digital filter, any of the linear-phase FIR filter design techniques discussed in Chapter 6 can be applied to design this lowpass filter. If $H_M(z)$ is implemented as an FIR filter of order m, then the output of the sampling rate decimator can be expressed in the time domain as follows.

Integer decimator

$$y(k) = \sum_{i=0}^{m} b_i x(Mk - i) \quad (7.11)$$

Example 7.1 **Integer Decimator**

As an illustration of sampling rate decimation, consider the following analog input signal.

$$x_a(t) = \sin(2\pi t) - 0.5\cos(4\pi t)$$

Let the sampling frequency be $f_s = 20$ Hz. Suppose the objective is to decimate the samples $x(k)$ by a factor of $M = 2$. From Eq. 7.10, the required lowpass filter has a gain of $H_M(0) = 1$ and a cutoff frequency of $F_M = 5$ Hz. Suppose a windowed linear-phase filter of order $m = 20$ with a Hanning window is used. Since the original signal is already bandlimited to 2 Hz, the FIR filter does not have any appreciable effect in this instance. The decimator output is obtained by running script *exam7_1* on the distribution CD. Plots of the original samples and the decimated samples are shown in Figure 7.5. It is apparent from inspection that, after an initial start-up transient, the decimated samples faithfully reproduce the original signal. Note that there is a delay of $m/2 = 10$ of the original samples caused by the linear-phase filter.

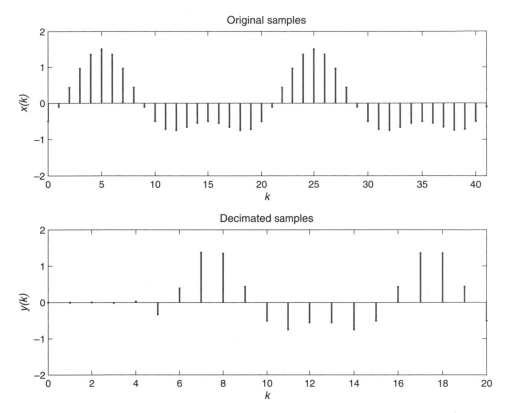

Figure 7.5 Sampling Rate Decimation by an Integer Factor, $M = 2$, using an FIR Filter of Order $m = 20$ with a Hamming Window

7.2.2   Sampling Rate Interpolator

Next, consider the dual problem of designing a converter that increases the sampling rate by an integer factor, L. A sampling rate converter that increases the sampling rate is called an *interpolator* because one is inserting new samples that interpolate between the original samples. Here the objective is to synthesize a discrete-time signal $y(k)$ that corresponds to sampling $x_a(t)$ at the increased rate of $f_L = Lf_s$, where L is a positive integer. Since L is an integer, every Lth sample of the new signal $x_L(k)$ will correspond to a sample of the original signal $x(k)$. There are potentially many ways to interpolate between the original samples. The easiest is to simply insert $L - 1$ zero samples between each of the original samples as follows.

$$x_L(k) = \begin{cases} x(k/L), & k = 0, L, 2L, \cdots \\ 0, & \text{otherwise} \end{cases} \tag{7.12}$$

A helpful way to view $x_L(k)$ is in terms of the following periodic pulse train with period L.

$$\delta_L(k) = \sum_{i=-\infty}^{\infty} \delta(k - Li) \tag{7.13}$$

The signal $x_L(k)$ is the signal $x(k/L)$ amplitude modulated by the periodic pulse train, $\delta_L(k)$. That is,

$$x_L(k) = x(k/L)\delta_L(k) \tag{7.14}$$

Note that $x(k/L)$ is not defined except when k is an integer multiple of L. However, the product in Eq. 7.14 is well defined for all k because $\delta_L(k) = 0$ when k is not a multiple of L. One can always replace k/L in Eq. 7.14 by floor(k/L) without changing the result.

The effect of using zero samples for interpolation can be seen when we look at the Z-transform of the interpolated signal. Using the change of variable $i = k/L$, we have

$$\begin{aligned}
X_L(z) &= \sum_{k=0}^{\infty} x(k/L)\delta_L(k)z^{-k} \\
&= \sum_{i=0}^{\infty} x(i)z^{-Li} \\
&= \sum_{i=0}^{\infty} x(i)(z^L)^{-i} \\
&= X(z^L) \tag{7.15}
\end{aligned}$$

Recall that the spectrum of a discrete-time signal can be obtained from the Z-transform by evaluating the Z-transform along the unit circle. Replacing z in Eq. 7.15 by $\exp(j2\pi fT)$, we find that the spectrum of the interpolated signal is as follows.

$$X_L(f) = X(Lf), \quad 0 \le f \le f_s \tag{7.16}$$

Thus, the spectrum of the interpolated signal, $x_L(k)$, is an L-fold replication of the spectrum of the original signal, $x(k)$, with each replication centered at a multiple of f_s/L. These $L-1$ images of the original spectrum must be removed by passing $x_L(k)$ through a lowpass anti-imaging filter with a cutoff frequency of $F_L = f_s/(2L)$.

$$H_L(f) \triangleq \begin{cases} L, & 0 \le |f| < F_L \\ 0, & F_L \le |f| \le f_s/2 \end{cases} \tag{7.17}$$

Note that the passband gain of the anti-imaging filter has been set to $H_L(0) = L$. This is done to compensate for the fact that the average value of $x_L(k)$ is $1/L$ times

Figure 7.6 Sampling Rate Interpolation by an Integer Factor, L

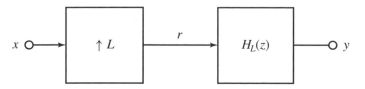

the average value of $x(k)$ due to the presence of the zero samples. Unlike an analog anti-imaging filter associated with a DAC, the filter in Eq. 7.17 is a *digital* anti-imaging filter. Sampling rate interpolation by an integer factor, L, is summarized in the block diagram shown in Figure 7.6. Again, it is standard practice to denote a sampling rate increase, also called *up-sampling*, with the up-arrow notation $\uparrow L$.

Since the anti-imaging filter in Figure 7.6 is a digital filter, any of the linear-phase FIR-filter design techniques introduced in Chapter 6 can be used to design this lowpass filter. If $H_L(z)$ is implemented as an FIR filter of order m, then the output of the sampling rate interpolator can be expressed in the time domain as follows.

Integer interpolator

$$y(k) = \sum_{i=0}^{m} b_i \delta_L(k-i)x\left(\frac{k-i}{L}\right) \tag{7.18}$$

Example 7.2 Integer Interpolator

As an illustration of sampling rate interpolation, consider the same analog input signal used in Example 7.1.

$$x_a(t) = \sin(2\pi t) - 0.5\cos(4\pi t)$$

Again, suppose the sampling frequency is $f_s = 20$ Hz. Consider the problem of interpolating the samples $x(k)$ by a factor of $L = 3$. Using Eq. 7.17, the required lowpass filter has a gain of $H_L(0) = 3$ and a cutoff frequency of $F_L = 10/3$ Hz. Suppose a windowed linear-phase filter of order $m = 20$, with a Hanning window, is used. The insertion of two zero samples between each of the original samples causes high-frequency images of the original spectrum to appear that must be removed by the anti-imaging filter. The interpolator output is obtained by running script *exam7_2* on the distribution CD. Plots of the original samples and the interpolated samples are shown in Figure 7.7. It is apparent from inspection that the interpolated samples have filled in between the original samples and preserved the wave shape in this case. It may seem counter-intuitive that inserting a run of zero samples can interpolate between existing samples. It is the inclusion of the lowpass filter that effectively recovers the unique underlying bandlimited analog signal.

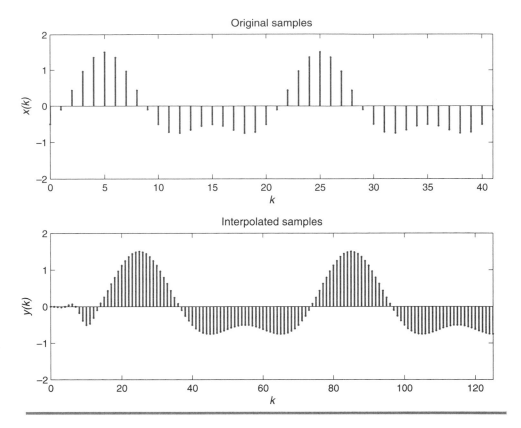

Figure 7.7 Sampling Rate Interpolation by an Integer Factor, $L = 3$, using an FIR Filter of Order $m = 20$ with a Hanning Window

MATLAB Toolbox

The FDSP toolbox contains the following functions for performing integer sampling rate conversions.

```
[y,b] = f_decimate (x,fs,M,m,ftype);    % Sampling rate decimator
[y,b] = f_interpol (x,fs,L,m,ftype);    % Sampling rate interpolator
```

Function *f_decimate* is a sampling rate decimator. On entry to *f_decimate*, x is a vector of length P containing the samples of the signal to be converted, fs is the sampling frequency of x, M is the rate conversion factor, m is the FIR filter order, and $ftype$ is the FIR filter type for the lowpass filter with the selections summarized in Table 7.1.

On exit from *f_decimate*, y is vector of length N containing the rate-converted samples where $N = floor(P/M)$, and b is an optional vector of length $m + 1$ containing the coefficients of the anti-aliasing FIR filter. If relatively large values of M are to be used, then it is the responsibility of the user to factor M into lower-order factors and perform the rate conversion in stages with multiple calls to *f_decimate*.

ftype	FIR Filter
0	Windowed (rectangular)
1	Windowed (Hanning)
2	Windowed (Hamming)
3	Windowed (Blackman)
4	Frequency-sampled
5	Least-squares
6	Equiripple

Table 7.1: ▶
Lowpass FIR Filter
Types

Function *f_interpol* is a sampling rate interpolator. The calling sequence and outputs for *f_interpol* are identical to those for *f_decimate* except that the rate conversion factor is L rather than M. In this case, output y is a vector of length N, where $N = LP$, and the lowpass FIR filter is an anti-imaging filter.

● ● ● ● ● ● ● ● ● ● ● ● ● ● ● ● ●

7.3 Rational Sampling Rate Converters

7.3.1 Single-stage Converters

Sampling rate conversion by integer factors is useful, but can be too restrictive in some practical applications. For example, digital audio tape (DAT) used in sound recording studios has a sampling rate of $f_s = 48$ kHz, while a compact disc (CD) is recorded at a sampling rate of $f_s = 44.1$ kHz. In order to convert music from one format to the other, a noninteger change in the sampling rate is required. Fortunately, we have all the tools in place to design a much larger set of sampling rate converters. The basic approach is to first interpolate the signal by a factor of L and then decimate the result by a factor of M. The net effect of this cascade configuration of an interpolator followed by a decimator is to change the sampling rate by a rational factor, L/M. That is,

$$f_S = \left(\frac{L}{M} \right) f_s \tag{7.19}$$

A block diagram of a rational sampling rate converter is shown in Figure 7.8. If $L/M < 1$, then the system in Figure 7.8 is a rational decimator, and if $L/M > 1$, it is a rational interpolator.

The interpolation in Figure 7.8 is done first in order to work at the higher sampling rate, thereby preserving the original spectral characteristics of $x(k)$. Moreover, this ordering has an added benefit because the cascade configuration of the two lowpass filters can be combined into a signal equivalent lowpass filter with a frequency response of $H_0(f) = H_L(f)H_M(f)$. The simplified configuration is shown in

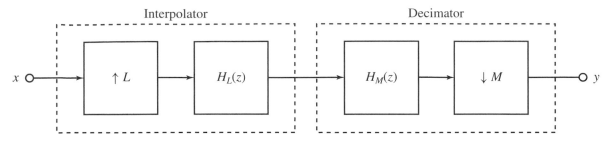

Figure 7.8 Rational Sampling Rate Converter with a Conversion Factor, L/M

Figure 7.9 Simplified Rational Sampling Rate Converter with a composite Anti-aliasing and Anti-imaging Filter and a Conversion Factor, L/M

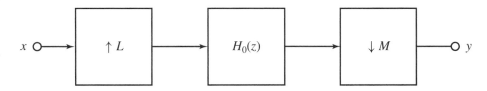

Figure 7.9. The passband gain of this composite anti-aliasing and anti-imaging filter is $H_0(0) = L$, and the cutoff frequency is F_0, where

$$F_0 = \min\left\{\frac{f_s}{2L}, \frac{f_s}{2M}\right\} \tag{7.20}$$

Thus, the frequency response of the composite digital anti-aliasing and anti-imaging filter is

$$H_0(f) = \begin{cases} L, & 0 \le |f| < F_0 \\ 0, & F_0 < |f| \le f_s/2 \end{cases} \tag{7.21}$$

If the composite filter is a linear-phase FIR filter of order m, then the output of a rational sampling rate converter can be expressed as follows in the time domain.

Rational rate converter

$$y(k) = \sum_{i=0}^{m} b_i \delta_L(Mk - i) x\left(\frac{Mk - i}{L}\right) \tag{7.22}$$

As a partial check, observe that Eq. 7.22 reduces to the decimator special case in Eq. 7.11 when $L = 1$ because $\delta_1(k) = 1$. Similarly, Eq. 7.22 reduces to the interpolator special case in Eq. 7.18 when $M = 1$.

Example 7.3 **Rational Sampling Rate Converter**

As an illustration of sampling rate conversion by a rational factor, consider the following analog input signal.

$$x_a(t) = \cos(2\pi t) + 0.8\sin(4\pi t)$$

Suppose the sampling frequency is $f_s = 20$ Hz, and consider the problem of changing the sampling rate of $x(k)$ by a factor of $L/M = 3/2$. In this case, the required lowpass filter has a gain of $H_0(0) = 3$ and a cutoff frequency of

$$F_0 = \min\left\{\frac{20}{6}, \frac{20}{4}\right\}$$

$$= \frac{10}{3} \text{ Hz}$$

Suppose a windowed linear-phase filter of order $m = 20$ with a Hamming window is used. The converter output is obtained by running script *exam7_3* on the distribution CD. Plots of the original samples and the rate-converted samples are shown in Figure 7.10. It is apparent from inspection that the interpolated samples have filled in between the original samples with three new samples for each pair of original samples.

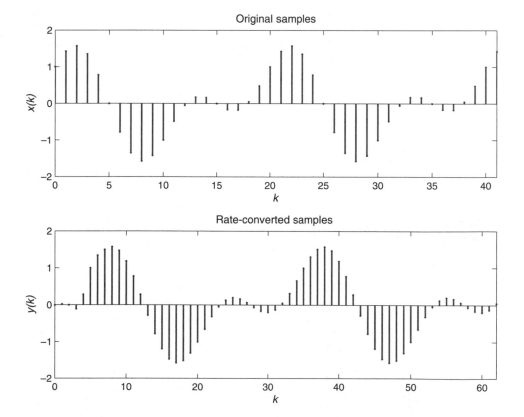

Figure 7.10 Sampling Rate Conversion by a Rational Factor, $L/M = 3/2$, using an FIR Filter of Order $m = 20$, with a Hamming Window.

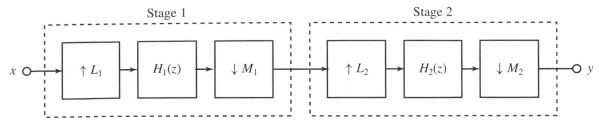

Figure 7.11 A Multistage Sampling Rate Converter with $r = 2$ Stages

7.3.2 Multistage Converters

In some practical applications, the values for L or M can be relatively large. This presents some special challenges when it comes to implementation. If either L or M is large, the composite anti-aliasing and anti-imaging filter, $H_0(z)$, will be a narrowband lowpass filter with a cutoff frequency of $F_0 \ll f_s$. Narrowband linear-phase filter specifications are difficult to meet and can require very high-order FIR filters. This in turn can mean increased storage space, a large computational time, and detrimental finite word length effects. The latter drawback can be mitigated by using a *multistage* sampling rate converter. The basic idea is to factor the desired conversion ratio into a product of ratios, each of which uses smaller values for L and M.

$$\frac{L}{M} = \left(\frac{L_1}{M_1}\right)\left(\frac{L_2}{M_2}\right) \cdots \left(\frac{L_r}{M_r}\right) \tag{7.23}$$

One can then implement r lower-order *stages* separately and configure them in a cascade as shown in Figure 7.11 for the case $r = 2$. The optimal number of stages and the optimal factoring of L/M can be determined based on minimizing the computational time and the storage requirements (Crochiere and Rabiner; 1975,1976).

Example 7.4 **DAT-to-CD Converter**

Consider the problem of designing a sampling rate converter that will convert a signal $x(k)$ that was sampled using the standard digital audio tape (DAT) format to a signal $y(k)$ that is suitable for recording on a compact disc (CD). Since the CD sampling rate of 44.1 kHz is smaller than the DAT sampling rate of 48 kHz, this requires a rational decimator. The required frequency conversion ratio is

$$\frac{L}{M} = \frac{44.1}{48}$$

$$= \frac{441}{480}$$

$$= \frac{147}{160}$$

Consequently, this application requires a rational decimator with $L = 147$, and $M = 160$. From Eqs. 7.20 and 7.21, a single-stage composite anti-aliasing and anti-imaging filter $H_0(f)$ must have a passband gain of $H_0(0) = 147$ and a cutoff frequency of

$$F_0 = \min \left\{ \frac{24}{147}, \frac{24}{160} \right\} \text{ kHz}$$

$$= 150 \text{ Hz}$$

Thus, the ideal frequency response for the composite anti-aliasing and anti-imaging filter is

$$H_0(f) = \begin{cases} 147, & 0 \leq |f| < 150 \\ 0, & 150 \leq |f| < 24000 \end{cases}$$

This is clearly a narrowband lowpass filter with $F_0 = 0.003125 f_s$. A direct single-stage implementation would require a very high-order linear-phase FIR filter. This can be avoided if a multistage implementation is used. For example, the following three conversion ratios all have single-digit integer factors.

$$\frac{147}{160} = \left(\frac{7}{8} \right) \left(\frac{7}{5} \right) \left(\frac{3}{4} \right)$$

Using this multistage approach, a DAT-to-CD converter can be implemented using two decimators and one interpolator. To convert from CD to DAT format, the reciprocals can be used, which yields two interpolators and a decimator. A detailed design of a CD-to-DAT sampling rate converter is presented later.

MATLAB Toolbox

The FDSP toolbox contains the following function for performing rational sampling rate conversion. It includes integer decimators and interpolators as special cases.

```
[y,b] = f_rateconv (x,fs,L,M,m,ftype);    % Rational sampling rate converter
```

On entry to *f_rateconv*, x is a vector of length P containing the samples of the signal to be converted, fs is the sampling frequency of x, L is the numerator of the rate conversion factor, M is the denominator of the rate conversion factor, m is the FIR filter order, and *ftype* is the FIR filter type for the lowpass filter with the selections previously summarized in Table 7.1.

On exit from $f\_rateconv$, y is a vector of length N containing the rate-converted samples, where $N = floor(LP/M)$ and b is an optional vector of length $m + 1$ containing the coefficients of the combined anti-aliasing and anti-imaging FIR filter. If relatively large values of L or M are to be used, then it is the responsibility of the user to factor L and M into lower-order factors and perform the rate conversion in stages with multiple calls to $f\_rateconv$.

7.4 Multirate Filter Realization Structures

Sampling rate converters have a considerable amount of built-in redundancy in terms of the required computational effort. In the case of an interpolator with $L \gg 1$, most of the samples that are being processed by the lowpass filter are zero samples inserted between the original samples. Consequently, many of the floating-point operations are multiplications by zero. An analogous observation holds for a decimator with $M \gg 1$. Here, all of the input samples are processed by the lowpass filter, but then only every Mth sample of the filter output is used.

7.4.1   Polyphase Interpolator

To develop more efficient implementations of rate converters, we begin with an interpolator. For the lowpass FIR anti-imaging filter $h(i) = b_i, 0 \le i \le m$. Thus, from Eq. 7.18, the time-domain representation of an interpolator with a rate conversion factor L can be expressed as follows.

$$y(k) = \sum_{r=0}^{m} h(r)\delta_L(k - r)x\left(\frac{k - r}{L}\right) \tag{7.24}$$

Suppose the filter order m is selected such that $m + 1$ is an integer multiple of the conversion factor L. That is, pad $h(k)$ with zeros, if needed, such that $m + 1 = pL$ for some integer p where

$$p = \frac{m + 1}{L} \tag{7.25}$$

The interpolator computation in Eq. 7.24 can be restructured by using L subfilters, each of order $p - 1$. To see how this is done, we decompose the computation into L separate cases depending on the value of k.

Case 0

$$k = Lq$$

Suppose $k = Lq$ for some integer q. Recall that $\delta_L(k)$ is a periodic train of impulses starting at $k = 0$ with period L. Therefore, $\delta_L(k - r)$ is zero except when $k - r$ is a multiple of L. Since $k = Lq$, this means that the terms in Eq. 7.24 will be zero except when r is a multiple of L. Setting $r = Li$ in Eq. 7.24 extracts every Lth sample of $h(r)$ starting with $r = 0$, which yields

$$
\begin{aligned}
y(Lq) &= \sum_{i=0}^{p-1} h(Li) x \left(\frac{Lq - Li}{L} \right) \\
&= \sum_{i=0}^{p-1} h(Li) x(q - i)
\end{aligned}
\tag{7.26}
$$

Case 1

$$k = Lq + 1$$

Next, consider the case when $k = Lq + 1$ for some integer q. Here the factor $\delta_L(k - r)$ will be zero except when r is one plus a multiple of L. Setting $r = Li + 1$ extracts every Lth sample of $h(r)$, but this time the samples start with $h(1)$ which yields

$$
\begin{aligned}
y(Lq + 1) &= \sum_{i=0}^{p-1} h(Li + 1) x \left[\frac{Lq + 1 - (Li + 1)}{L} \right] \\
&= \sum_{i=0}^{p-1} h(Li + 1) x(q - i)
\end{aligned}
\tag{7.27}
$$

Case L − 1

$$k = Lq + L - 1$$

This process can be continued until we get to the last case where $k = Lq + L - 1$ for some integer q. Here the factor $\delta_L(k - r)$ will be zero except when r is $L - 1$ plus a multiple of L. As before, setting $r = Li + L - 1$ extracts every Lth sample of $h(r)$, this time starting with sample $L - 1$, which yields

$$y(Lq + L - 1) = \sum_{i=0}^{p-1} h(Li + L - 1)x\left[\frac{Lq + L - 1 - (Li + L - 1)}{L}\right]$$

$$= \sum_{i=0}^{p-1} h(Li + L - 1)x(q - i) \tag{7.28}$$

Notice that for each of the L separate cases, the number of floating-point multiplications required is p, rather than $m + 1$, because we are only using every Lth sample of the original impulse response. Thus, the computational effort has been reduced by a factor of L. The decomposition of an interpolator into L separate cases is illustrated with the following example.

Example 7.5   **Polyphase Interpolator**

Consider the problem of implementing an interpolator with a rate conversion factor of $L = 3$. Suppose the lowpass filter order is $m = 8$. Then the order of each subfilter is

$$p = \frac{m + 1}{L}$$

$$= 3$$

For this interpolator, there are three cases, and for each case, a subfilter is constructed by using every third sample of $h(k)$. A signal flow graph of the resulting discrete-time system is shown in Figure 7.12. The *commutator* switch selects which subfilter output to use as it rotates at a rate of $3f_s$ steps per second in the counterclockwise direction. This way, there are three samples of the output $y(i)$ generated for each sample of the input $x(k)$.

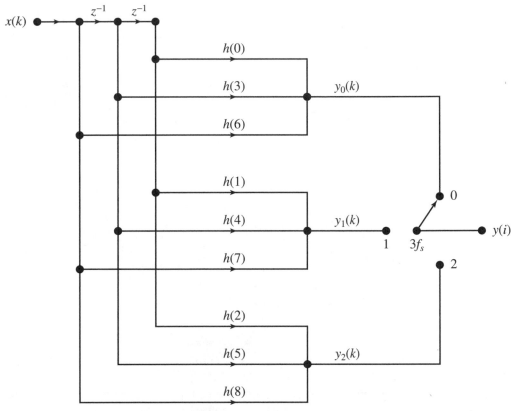

Figure 7.12   Polyphase Filter Realization of an Integer Interpolator with $L = 3$ and $m = 8$. The Commutator Switch Rotates Counterclockwise at a Rate of $3f_s$ Steps per Second

The subfilters in Figure 7.12 and in Eq. 7.26 through Eq. 7.28 are called *polyphase* filters. The impulse response of the ith polyphase filter is denoted as $g_i(k)$ and is defined as the following decimated version of the original impulse response $h(k)$.

**Polyphase
interpolator filters**

$$g_i(k) \triangleq h(Lk + i), \quad 0 \leq i < L, \, 0 \leq k < p \tag{7.29}$$

Thus, $g_i(k)$ is offset by i samples, but otherwise extracts every Lth sample of the original impulse response $h(k)$. For the interpolator in Figure 7.12 where $L = 3$, the three polyphase subfilters have the following impulse responses.

$$g_0 = [h(0), h(3), h(6)]^T \tag{7.30a}$$

$$g_1 = [h(1), h(4), h(7)]^T \tag{7.30b}$$

$$g_2 = [h(2), h(5), h(8)]^T \tag{7.30c}$$

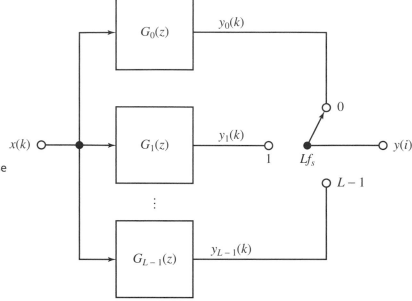

Figure 7.13 Polyphase Filter Realization of an Interpolator with a Conversion Factor, L. The Commutator Switch Rotates Counterclockwise at a Rate of Lf_s Steps per Second

A block diagram of a polyphase filter implementation of an interpolator with a conversion factor, L, is shown in Figure 7.13. The presence of the commutator switch, which selects which subfilter output to route to the output, indicates that this rate converter is a *time-varying* discrete-time system. The commutator switch rotates counterclockwise starting from the position shown, which corresponds to $i = 0$. It has a rate of rotation of Lf_s steps per second, so there are L output samples generated for each input sample.

7.4.2 Polyphase Decimator

Just as an interpolator has a time-varying polyphase filter realization, so does a decimator. Recall from Eq. 7.11 that the time-domain representation of a decimator with a rate conversion factor M can be expressed as follows.

$$y(k) = \sum_{r=0}^{m} h(r)x(Mk - r) \tag{7.31}$$

Again, suppose the filter order m is selected such that $m + 1$ is an integer multiple of the conversion factor M. That is, pad $h(k)$ with zeros, as needed, such that $m + 1 = pM$ for some integer p, where

$$p = \frac{m + 1}{M} \tag{7.32}$$

The decimator computation in Eq. 7.31 can be restructured by using M subfilters, each of order $p - 1$. The impulse responses of the polyphase filters for the decimator

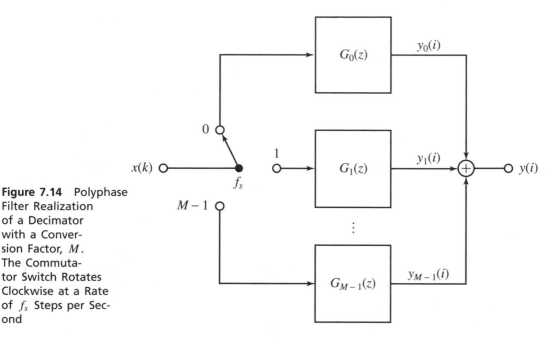

Figure 7.14 Polyphase Filter Realization of a Decimator with a Conversion Factor, M. The Commutator Switch Rotates Clockwise at a Rate of f_s Steps per Second

are very similar to those of the interpolator in Eq. 7.29. The only difference is that the conversion rate factor is M.

Polyphase decimator filters

$$g_i(k) \triangleq h(Mk + i), \quad 0 \le i < M, \ 0 \le k < p \qquad (7.33)$$

The polyphase filter realization of a decimator is the transpose of the polyphase filter realization of an interpolator. Recall from the signal flow graph reversal theorem in Chapter 5 that the transposed structure is obtained by reversing the direction of each arc and interchanging the labels of the inputs and the outputs. If we then redraw the block diagram with the input on the left, this results in the polyphase decimator structure shown in Figure 7.14. Here the commutator switch rotates clockwise at a rate of f_s steps per second. Since it takes M steps for each subfilter to be serviced with one input sample, this means the output is updated at a rate of f_s/M Hz.

• • • • • • • • • • • • • • •

7.5 Subband Processing

Now that we have a means of changing the sampling rate of a discrete-time signal, we can put this technique to work in a number of practical ways.

7.5.1 Narrowband Filters

A narrowband filter is a sharp filter whose passband or stopband is small in comparison with the sampling frequency. Implementations of linear-phase narrowband

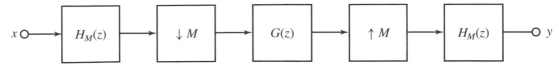

Figure 7.15   A Multirate Narrowband Filter using a Rate Conversion Factor, M

filters typically require very high-order FIR filters. This implies increased storage requirements, longer computational times, and significant finite word length effects. The latter problem can be reduced by using a multirate design of a narrowband filter. Suppose the ideal filter specification is to pass frequencies in the range $0 \leq |f| \leq F_0$, where $F_0 \ll f_s$.

$$H(f) = \begin{cases} 1, & 0 \leq |f| \leq F_0 \\ 0, & F_0 < |f| \leq f_s/2 \end{cases} \tag{7.34}$$

The first step of the multirate method is to reduce the sampling rate by a factor, M. This has the effect of increasing the relative width of the passband by a factor of M. Setting $M F_0 \leq f_s/4$ yields the following upper bound on the decimation factor M.

$$M \leq \frac{f_s}{4 F_0} \tag{7.35}$$

For the maximum value of M the new cutoff frequency is $M F_0 = 0.25 f_s$. Thus, by reducing the sampling rate, we transform the narrowband filter, $H(z)$, into a wideband filter, $G(z)$, with a cutoff frequency that is up to one fourth of the sampling rate.

$$G(f) = \begin{cases} 1, & 0 \leq |f| \leq M F_0 \\ 0, & M F_0 < |f| \leq f_s/2 \end{cases} \tag{7.36}$$

The wideband filter $G(z)$ is easier to implement than a narrowband filter. To complete the process, the original sampling frequency must be restored using sampling rate interpolation by a factor of M. The resulting overall implementation of a multirate narrowband filter is shown in the block diagram in Figure 7.15. The following example compares a multirate narrowband design with a conventional fixed-rate design.

Example 7.6   **Multirate Narrowband Filter**

To illustrate the multirate filter design technique, consider the problem of designing an ideal lowpass filter with a cutoff frequency of $F_0 = f_s/32$. Using Eq. 7.35, the decimation factor must satisfy

$$M \leq \frac{f_s}{4 F_0}$$

$$= 8$$

Suppose $M = 8$, and the windowing method is used to design both $H(z)$ in Eq. 7.34 and $G(z)$ in Eq. 7.36. Script *exam7_6* on the distribution CD generates the two magnitude responses shown in Figure 7.16. To clarify the display, only the first quarter of the frequency range, $0 \leq f \leq f_s/8$, is shown. The fixed-rate magnitude response, $H(f)$, corresponds to a windowed FIR filter of order $m = 240$ using the

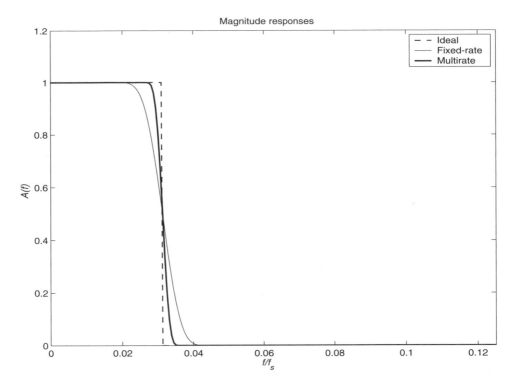

Figure 7.16 Magnitude Responses of Narrowband Lowpass Filters using a Fixed-rate Design with $m = 240$ and a Multirate Design with $m = 80$ and Rate Conversion Factor, $M = 8$

Blackman window. For comparison, the multirate magnitude response based on $G(f)$ uses a windowed FIR filter of order $m = 80$ with the Blackman window. The anti-aliasing and anti-imaging filters $H_M(z)$ are also FIR filters of order $m = 80$. Thus, the two approaches are roughly comparable in terms of storage requirements and computational time. However, the multirate design is less sensitive to finite word length effects because it is a cascade of three filters of order $m = 80$, instead of one filter of order $m = 240$. It is evident from inspection of Figure 7.16 that the multirate design is superior to the fixed-rate design in terms of the width of the transition band. The passband ripple of the fixed-rate design can be reduced by decreasing m, but this is achieved at the expense of further increases in the width of the transition band.

7.5.2 Filter Banks

The narrowband lowpass filter designed in Example 7.6 could instead have been a narrowband, bandpass, or highpass filter. By using a combination of lowpass, bandpass, and highpass filters, the entire spectrum, $[-f_s/2, f_s/2]$, can be covered with a bank of N *subband filters*. The magnitude responses for a filter bank consisting of $N = 3$ subband filters is shown in Figure 7.17. Since the discrete-time frequency response is periodic with period f_s, the spectrum in Figure 7.17 is plotted over the positive frequencies $[0, f_s]$ rather than $[-f_s/2, f_s/2]$. Notice that the transition bands of the subband filters have nonzero widths and that adjacent transition bands overlap. This way the entire spectrum is used, and the overall filter bank is effectively

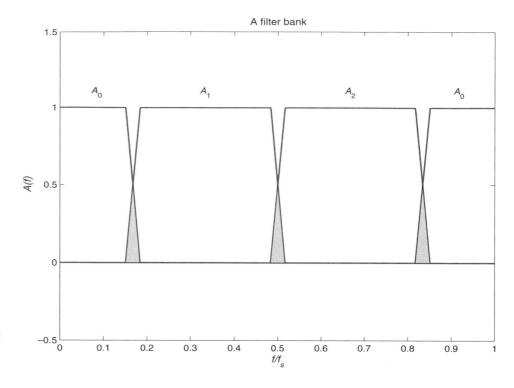

Figure 7.17 Magnitude Responses of a Bank of $N = 3$ Subband Filters

an allpass filter. The ith frequency band is called the ith *channel*, and breaking the entire spectrum into N channels is called *frequency division multiplexing*.

A filter bank is implemented as a parallel configuration of filters, as shown in Figure 7.18. The parallel configuration on the left side of Figure 7.18 is called an

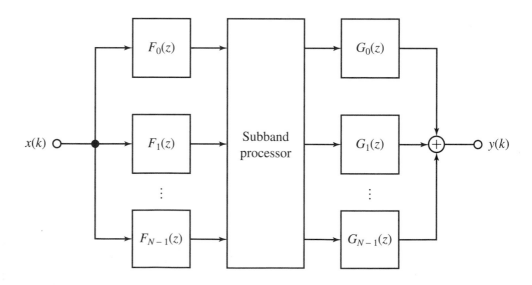

Figure 7.18 Analysis and Synthesis Filter Banks

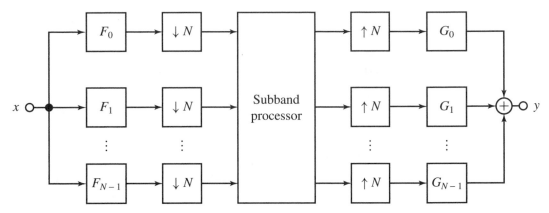

Figure 7.19 Decimated and Interpolated Filter Banks

analysis bank because it decomposes the overall spectrum into N subbands. Each subband can then be processed separately. Depending on the application, there may also be a second parallel configuration of N filters, as shown on the right side of Figure 7.18. This is called a *synthesis bank* because it recombines the subsignals into a single composite signal $y(k)$.

For the bank of N filters shown in Figure 7.18, the width of each subband is $1/N$ times the width of the overall spectrum, $[0, f_s]$. Since each subband is of width f_s/N, the bandlimited subsignals can be down-sampled or decimated by a factor of N. This makes processing of the separate channels more efficient. Following the subband processing, the subsignals are then up-sampled or interpolated by a factor of N to restore the original sampling rate. The up-sampled signals are then recombined into one signal in the synthesis filter bank on the right. The resulting configurations, called decimated and interpolated filter banks, are shown in Figure 7.19.

Normally, the time signals that we work with are real-valued. If we relax this assumption and consider the possibility of complex-valued time signals, then there is a simple way to synthesize a high-bandwidth composite signal $x(k)$ that contains several low-bandwidth subsignals. The essential step is to shift the spectrum of each subsignal so that it occupies a particular band in the overall spectrum. This can be achieved by using the following *frequency shift* property of the discrete-time Fourier transform (DTFT). This property can be verified by direct substitution using Definition 3.1.

Frequency shift property

$$\text{DTFT}\{\exp(jk2\pi F_i T)x(k)\} = X(f - F_i) \qquad (7.37)$$

Thus, if we scale the kth sample of a signal $x(k)$ by the complex quantity, $\exp(jk2\pi F_i T)$, this shifts the spectrum of $x(k)$ to the right by F_i Hz. For example, suppose $x_i(k)$ is a bandlimited *subsignal* with bandwidth $B < f_s/N$ for $0 \le i < N$.

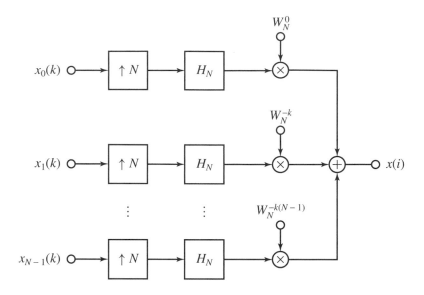

Figure 7.20 Signal Synthesis using a Uniform DFT Filter Bank

Then the spectrum of $x_i(k)$ can be shifted to the right and centered at the frequency $F_i = i f_s/N$ by creating the following complex-valued signal.

$$y_i(k) = \exp(jk2\pi F_i T)x_i(k)$$

$$= \exp\left(\frac{jki2\pi}{N}\right)x_i(k)$$

$$= W_N^{-ki}x_i(k) \tag{7.38}$$

Here we have made use of the notation $W_N = \exp(-j2\pi/N)$ used previously in Chapter 3 with the DFT. Because $x_i(k)$ occupies only $1/N$ times the total bandwidth, we first up-sample $x_i(k)$ by a factor of N before modulating it as in Eq. 7.37. When resulting subsignals are then combined, this results in the synthesis filter bank shown in Figure 7.20. This is called a *uniform DFT* filter bank. Observe that the same prototype lowpass filter, $H_N(z)$, can be used to remove the images generated by the up-sampling. Modulation by W_N^{-ki} then causes the spectrum of $x_i(k)$ to be shifted to the ith subband of $[0, f_s]$. Whereas the subsignal $x_i(k)$ may be real, the *composite* high bandwidth signal $x(i)$ will be complex.

The signal synthesized by the filter bank in Figure 7.20 can be decomposed with an analysis filter bank. First, the signal $x(i)$ is modulated by W_N^{ik}. This has the effect of shifting the kth subband of $x(i)$ back to the origin. This subsignal is then down-sampled to cancel the effects of the up-sampling that was used in the synthesis bank. The end result is the uniform DFT analysis filter bank shown in Figure 7.21.

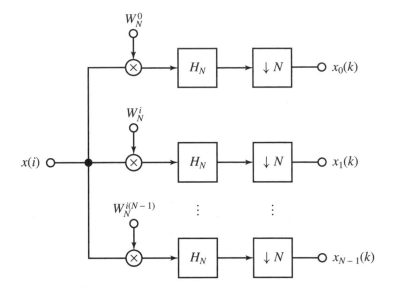

Figure 7.21 Signal Analysis using a Uniform DFT Filter Bank

Example 7.7   **Signal Synthesis**

To illustrate the process of signal synthesis using a uniform DFT filter bank, let the number of filters in the bank be $N = 4$, and let the sampling frequency be $f_s = 10$ Hz. Suppose the subsignals are of length $p = 64$, and suppose all subsignals are bandlimited to $|f| \leq F_0$, where $F_0 = f_s/4$. We define the subsignals in terms of their spectra as follows, where $f_i = if_s/p$ for $0 \leq i < p$.

$$X_0(i) = \cos\left(\frac{\pi f_i}{2F_0}\right)$$

$$X_1(i) = 1 - |f_i|/F_0$$

$$X_2(i) = \left|\sin\left(\frac{\pi f_i}{F_0}\right)\right|$$

$$X_3(i) = 1 - (|f_i|/F_0)^2$$

In each case, the phase spectra are zero. Plots of the magnitude spectra are shown in Figure 7.22. After interchanging the first and second halves of the DFT spectra using the MATLAB function *fftshift*, the time signals are then recovered from the spectra as follows using the inverse DFT.

$$x_q(k) = \text{IDFT}\{X_q(i)\}, \quad 0 \leq q < 4$$

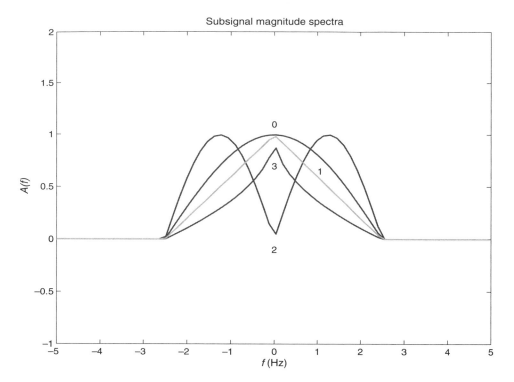

Figure 7.22 Spectra of Four Bandlimited Subsignals with a Bandwidth of $B = f_s/4$

Next, $x_q(k)$ is up-sampled by a factor of $N = 4$ as in Figure 7.20. The lowpass anti-imaging filter used was a windowed filter of order $m = 120$ using a Blackman window. For $N = 4$, the modulation factor is

$$W_4 = \exp(-j2\pi/4)$$

$$= \cos(\pi/2) - j\sin(\pi/2)$$

$$= j$$

Thus, from Figure 7.20, the complex composite signal $x(k)$ is

$$x(k) = x_0(k) + W_4 x_1(k) + W_4^2 x_2(k) + W_4^3 x_3(k)$$

$$= x_0(k) + j x_1(k) - x_2(k) - j x_3(k)$$

$$= x_0(k) - x_2(k) + j[x_1(k) - x_3(k)]$$

Plots of the real and imaginary parts of the composite signal $x(k)$, obtained by running script *exam7_7* on the distribution CD, are shown in Figure 7.23. Note that due to

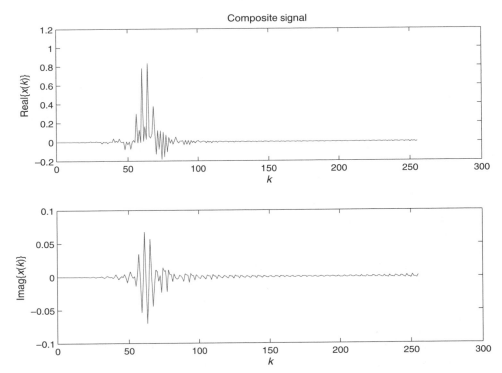

Figure 7.23 Real and Imaginary Parts of the Composite High-bandwidth Signal $x(k)$ Containing Four Subsignals using Frequency-division Multiplexing

the up-sampling, $x(k)$ is now of length $Np = 256$. Next, the magnitude spectrum of $x(k)$ is computed using

$$A(i) = |\text{FFT}\{x(k)\}|, \quad 0 \le i < Np$$

After interchanging the first and second halves of $A(i)$ using the MATLAB function *fftshift*, the resulting magnitude spectrum of $x(k)$ is shown in Figure 7.24. It is clear that the spectra of $x_i(k)$ have been shifted and centered at $F_i = if_s/N$ for $0 \le i < N$. There is some overlap between subspectra, perhaps caused by the nonideal nature of the anti-imaging filter. Notice from Figure 7.22, that each of the subsignals occupies only half of the bandwidth $[-f_s/2, f_s/2]$. Since this is the case, it should be possible to reduce the cutoff frequency of the anti-imaging filter $H_N(z)$ by a factor of $\alpha = 0.5$, so that

$$F_c = \frac{\alpha f_s}{2N}$$

The resulting magnitude spectrum of $x(k)$ using this lower cutoff frequency is shown in Figure 7.25. Notice that this has effectively eliminated the spectral overlap between the subbands. It should now be possible to use an analysis filter bank such as the uniform DFT filter bank in Figure 7.21 to extract the individual subsignals from $x(k)$.

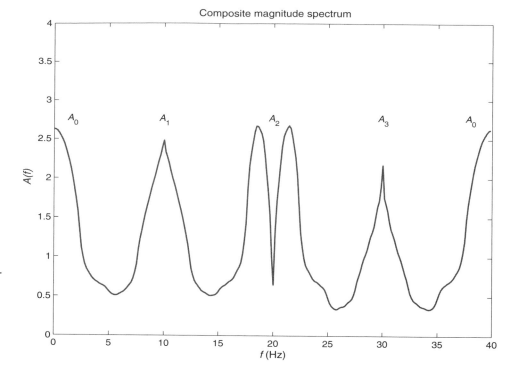

Figure 7.24 Magnitude Spectrum of the Composite High-bandwidth Signal Showing the Spectra of Subsignals in Each of Four Bands using a Windowed Blackman Anti-imaging Filter

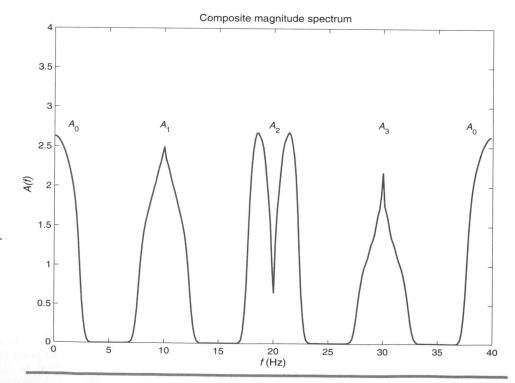

Figure 7.25 Magnitude Spectrum of the Composite High-bandwidth Signal Showing the Spectra of Subsignals in Each of Four Bands using a Windowed Blackman Anti-imaging Filter with the Cutoff Frequency Reduced by Factor $\alpha = 0.5$

*7.6 Oversampling ADC

One of the practical difficulties associated with analog-to-digital conversion is the need for a lowpass analog anti-aliasing prefilter to bandlimit the signal to less than half of the sampling rate. High-order analog filters are expensive, and they are also difficult to keep in calibration. Using multirate techniques, we can transfer some of the anti-aliasing function into the digital domain, and thereby use a simpler low-order analog filter.

Suppose the range of frequencies of interest for an analog signal $x_a(t)$ is $0 \leq |F| \leq F_a$. Normally, to avoid aliasing, we would bandlimit $x_a(t)$ to F_a Hz with a sharp analog lowpass filter, and then sample at a rate f_s that is greater than twice the bandwidth. Suppose that we instead *oversample* $x_a(t)$ with a sampling rate of $f_s = 2MF_a$, where M is an integer greater than one. Recall that this corresponds to oversampling by a factor of $f_s/(2F_a) = M$. Oversampling by a factor of M significantly reduces the requirements for the anti-aliasing filter. The analog anti-aliasing filter now has to satisfy the following frequency response specification.

$$H_a(f) = \begin{cases} 1, & 0 \leq |f| \leq F_a \\ 0, & MF_a \leq |f| < \infty \end{cases} \tag{7.39}$$

Even though $H_a(f)$ is still an ideal filter with no passband ripple and complete stopband attenuation, the width of the transition band is no longer zero but is instead $\Delta f = (M-1)F_a$. Given a large transition band, $H_a(s)$ can be approximated with a simple inexpensive low-order filter such as a first- or second-order Butterworth filter. Recall from Eq. 1.50 that the magnitude response of an nth-order analog Butterworth lowpass filter with a cutoff frequency of F_a is

$$A_n(f) = \frac{1}{\sqrt{1 + (f/F_a)^{2n}}} \tag{7.40}$$

The tradeoff of an increased sampling rate for a simpler analog anti-aliasing filter does leave us with a discrete-time signal that is sampled at a rate that is higher than twice the maximum frequency of interest. Following the sampling operation, we can reduce this sampling rate to the minimum value using a decimator. The resulting structure, called an *oversampling ADC*, is shown in the block diagram in Figure 7.26. Note that it has *two* anti-aliasing filters, a low-order analog filter $H_a(s)$ with cutoff frequency F_a, and a high-order digital filter $H_M(z)$, also with cutoff frequency F_a.

The use of a simpler analog anti-aliasing filter is the main benefit of the oversampling ADC, but it is not the only one. Another advantage becomes apparent when

Figure 7.26 Oversampling ADC using Sampling Rate Decimation by an Integer Factor, M

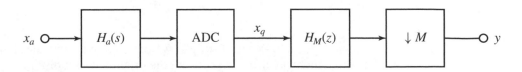

we look at the ADC quantization noise. Recall from Eq. 5.79 that if $|x_a(t)| \leq c$, then the quantization level or precision of an N-bit ADC is

$$q = \frac{c}{2^{N-1}} \tag{7.41}$$

If rounding is used, then the quantized output of the ADC is $x_q(k) = x(k) + \Delta x(k)$, where the quantization error, $\Delta x(k)$, can be modeled as white noise uniformly distributed over the interval $[-q/2, q/2]$. It was shown previously in Eq. 5.82 that the average power of the ADC quantization noise, $\sigma_x^2 = E[\Delta x(k)^2]$, can be expressed in terms of q as

$$\sigma_x^2 = \frac{q^2}{12} \tag{7.42}$$

The expression in Eq. 7.42 is the average power of the noise-corrupted signal, $x_q(k)$, in Figure 7.26. This signal is subsequently processed by the digital anti-aliasing filter, $H_M(z)$, and then down-sampled. The down-sampling process does not affect the average power. This is because, even though we are extracting only every Mth sample and thereby lowering the total energy, the resulting signal is also shorter by a factor of M. To examine the effects of the filter $H_M(z)$ on the noise, recall from Eq. 5.86 that the average power of the noise at the filter output is

$$\sigma_y^2 = \Gamma \sigma_x^2 \tag{7.43}$$

Here Γ is the power gain of $H_M(z)$. The digital anti-aliasing filter $H_M(z)$ is an FIR filter of order m. Thus, from Eq. 5.87, the power gain can be expressed in terms of the impulse response $h_M(k)$ as

$$\Gamma = \sum_{k=0}^{m} h_M^2(k) \tag{7.44}$$

The expression for the power gain can be simplified further. Using Parseval's theorem from Table 3.4, the power gain can be recast in terms of $H_M(i) = \text{DFT}\{h_M(k)\}$ as

$$\Gamma = \frac{1}{m+1} \sum_{i=0}^{m} |H_M(i)|^2 \tag{7.45}$$

Here $H_M(i)$ is the ith sample of the frequency response of $H_M(z)$. From Eq. 3.95, we have $H_M(i) = H_M(f)$ with $f = if_s/(m+1)$. Since $H_M(z)$ is a lowpass filter with a cutoff frequency of $F_0 = f_s/(2M)$, it follows that the sum in Eq. 7.45 is just $1/M$ times $(m+1)$. Consequently, $\Gamma = 1/M$, and from Eqs. 5.86 and 7.43, the average power of the quantization noise appearing at the output of the oversampling ADC is

Oversampling ADC noise power

$$\sigma_y^2 = \frac{q^2}{12M} \tag{7.46}$$

Comparing Eq. 7.46 with Eq. 7.42, we observe that oversampling by a factor of M has the beneficial effect of reducing the quantization noise power by a factor of M.

This is achieved because oversampling by M spreads the noise power out over the frequency range $[0, MF_0]$. The digital anti-aliasing filter $H_M(z)$ with cutoff frequency F_0 then removes most of the quantization noise.

The reduction in the quantization noise power in Eq. 7.46 can be interpreted as an increase in the number of effective bits of precision. For example, let P be the number of bits of precision using oversampling by a factor of M, and let N be the number of bits of precision without oversampling. Using Eq. 7.41 and Eq. 7.42, and equating the quantization noise power for the two cases yields

$$\frac{c^2}{12M[2^{2(P-1)}]} = \frac{c^2}{12[2^{2(N-1)}]} \tag{7.47}$$

Canceling the common terms, taking reciprocals, and then taking the base-2 logarithm of each side, we arrive at the following expression for the required precision when oversampling is used.

Oversampling ADC precision

$$P = N - \frac{\log_2(M)}{2} \tag{7.48}$$

Thus, we see that oversampling by a factor of $M = 4$ decreases the required precision of the ADC by 1 bit. That is, the quantization noise power of an N-bit ADC without oversampling is the same as the quantization noise power of an $(N-1)$-bit ADC with oversampling by a factor of $M = 4$. More generally, oversampling by $M = 4^r$ reduces the required ADC precision by r bits.

Using direct quantization, the power spectrum of the quantization noise is flat over the frequency range $0 \le |f| \le f_s/2$. By using a different quantization scheme, called *sigma-delta modulation*, the spectrum of the quantization noise can be reshaped such that most of the power lies outside the frequency range, $0 \le |f| \le F_a$ (Candy and Temes, 1992). When a sigma-delta ADC is used, oversampling can generate an even greater savings in the precision of approximately 3 bits when oversampling by a factor of $M = 4$. In the limit as the oversampling ratio is made very high, the required precision can be reduced to a single bit, and this results in a 1-bit sigma-delta ADC (Mitra, 2001).

Example 7.8   **Oversampling ADC**

To illustrate the effectiveness of oversampling in the analog-to-digital conversion process, let the analog anti-aliasing filter be an nth-order Butterworth filter with the magnitude response in Eq. 7.40. The objective of this example is to use oversampling to ensure that the maximum aliasing error is sufficiently small. The maximum of the magnitude of the aliasing error occurs at the folding frequency, $f_d = f_s/2$. At this frequency, the *aliasing error scale factor* is

$$\alpha_n \overset{\Delta}{=} A_n(f_d)$$

If oversampling by a factor of M is used, then $f_s = 2MF_a$, which means $f_d = MF_a$. Using Eq. 7.40 to achieve an aliasing error scale factor of ϵ, we need

$$\frac{1}{\sqrt{1 + M^{2n}}} \le \epsilon$$

Taking reciprocals and squaring both sides then yields:

$$1 + M^{2n} \geq \epsilon^{-2}$$

Solving for M, we then have

$$M \geq (\epsilon^{-2} - 1)^{0.5/n}$$

For example, if a second-order Butterworth filter is used, then $n = 2$. Suppose the desired aliasing error scale factor is $\epsilon = 0.01$. In this case, the required sampling rate conversion factor is

$$M = \text{ceil}\{(10^4 - 1)^{1/4}\}$$

$$= \text{ceil}(9.9997)$$

$$= 10$$

The calculation of M can be verified graphically using Figure 7.27, which was generated by running script *exam7_8* on the distribution CD. The curves display the aliasing-error scale factor, α_n, versus the oversampling factor, M, for the first four Butterworth filters. For clarity of display, the ordinate is the base-10 logarithm of the aliasing-error scale factor. It is evident that a very low aliasing error can be achieved by using a combination of oversampling and a sufficiently high-order Butterworth anti-aliasing filter.

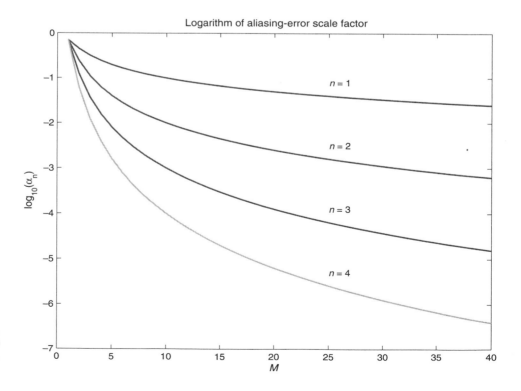

Figure 7.27 Logarithm of Aliasing-error Scale Factor α_n versus the Oversampling Factor, M, using a Butterworth Anti-aliasing Filter of Order n

● ● ● ● ● ● ● ● ● ● ● ● ● ● ● ●

*7.7 Oversampling DAC

Just as oversampling can be used to ease the requirements on the analog anti-aliasing prefilter of an ADC, it can also be used to ease the requirements on the analog anti-imaging postfilter of a DAC. Recall that a DAC can be modeled as a zero-order hold with a transfer function of

$$H_0(s) = \frac{1 - \exp(-Ts)}{2} \tag{7.49}$$

The DAC output is a piecewise-constant signal containing spectral images centered at multiples of the sampling frequency. Suppose the range of frequencies of interest for the analog output signal $y_a(t)$ is $0 \le |f| \le F_a$, where $f_s = 2F_a$. Normally, we would pass the piecewise-constant DAC output through a sharp, analog lowpass filter with a cutoff frequency of $f_s/2$ to remove the spectral images that are centered at multiples of f_s. Suppose that we instead increase the sampling rate to $f_s = 2LF_a$, where L is an integer greater than one, and then perform the digital-to-analog conversion. This oversampling by a factor of L spreads the spectral images in the DAC output out so that now they are centered at multiples of Lf_s, thus making them easier to filter out. In particular, the analog anti-imaging filter now has to satisfy the following frequency response specification.

$$H_a(f) = \begin{cases} 1, & 0 \le |f| \le F_a \\ 0, & LF_a \le |f| < \infty \end{cases} \tag{7.50}$$

Even though $H_a(f)$ is still an ideal filter with no passband ripple and complete stopband attenuation, the width of the transition band is no longer zero but is instead $\Delta f = (L - 1)F_a$. Given a large transition band, $H_a(s)$ can be approximated with a simple inexpensive low-order filter such as a first- or second-order analog Butterworth filter.

The two-step process of sampling rate interpolation followed by a DAC is called *oversampling digital-to-analog conversion*. The overall structure of an oversampling DAC is shown in the block diagram in Figure 7.28. Note that it has *two* anti-imaging filters, a high-order digital filter $H_L(z)$ with cutoff frequency F_a, and a low-order analog filter $H_a(s)$, also with cutoff frequency F_a.

Just as with an ADC, the digital anti-imaging filter, $H_L(z)$, effectively reduces the quantization noise power because it removes frequencies in the expanded stopband,

Figure 7.28 Oversampling DAC using Sampling Rate Interpolation and Passband Compensation

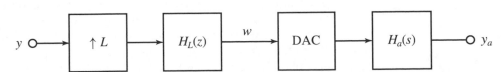

$F_a \leq |f| \leq Lf_s/2$. As a result, the average power of the quantization noise at the DAC output is as in Eq. 7.46, but with M replaced by L. That is,

Oversampling DAC
noise power

$$\sigma_y^2 = \frac{q^2}{12L} \tag{7.51}$$

The use of a sampling rate interpolator provides us with some additional opportunities when it comes to improving the system performance in the passband, $0 \leq |f| \leq F_a$. From Eq. 7.49 and Euler's identity, the magnitude response of the zero-order hold model of the DAC using a sampling frequency of Lf_s is

$$A_0(f) = |H_0(s)|_{s=j2\pi f}$$

$$= \left| \frac{1 - \exp(-j2\pi fT/L)}{j2\pi f} \right|$$

$$= \left| \frac{\exp(-j\pi fT/L)[\exp(j\pi fT/L) - \exp(-j\pi fT/L)]}{j2\pi f} \right|$$

$$= \left| \frac{\exp(-j\pi fT/L)\sin(\pi fT/L)}{\pi f} \right|$$

$$= \left| \frac{\sin(\pi fT/L)}{\pi f} \right|$$

$$= \frac{T}{L} \left| \text{sinc}\left(\frac{\pi fT}{L} \right) \right| \tag{7.52}$$

A plot of the magnitude response of the DAC is shown in Figure 7.29. The side lobes are what cause the images of the baseband spectrum to appear at the output. These images must be removed with the analog anti-imaging filter, $H_a(s)$.

Within the passband, the DAC shapes the magnitude spectrum of the signal using the scale factor $A_0(f)$. The effects of the DAC within the passband can be compensated for by using a more general version of the anti-imaging filter $H_L(z)$. In particular, we can *equalize* the effects of the DAC, within the frequency band $0 \leq |f| \leq F_a$, by using a digital anti-imaging filter with the following frequency response.

$$H_L(f) = \begin{cases} \dfrac{L}{T\,|\text{sinc}(\pi fT/L)|}, & 0 \leq |f| \leq F_a \\ 0, & F_a < |f| < f_s/2 \end{cases} \tag{7.53}$$

When $L \gg 1$, the spectral distortion in the passband due to the DAC is small because $\text{sinc}(\pi fT/L) \approx 1$. However, if a low-order analog anti-imaging filter is used, then there may be a fairly significant ripple within the passband. This ripple can be compensated for as well with the digital filter $H_L(z)$. Suppose an nth-order Butterworth filter is used for the analog anti-imaging filter. Then from Eq. 7.40, the effects of both the DAC and the analog postfilter can be equalized, within the

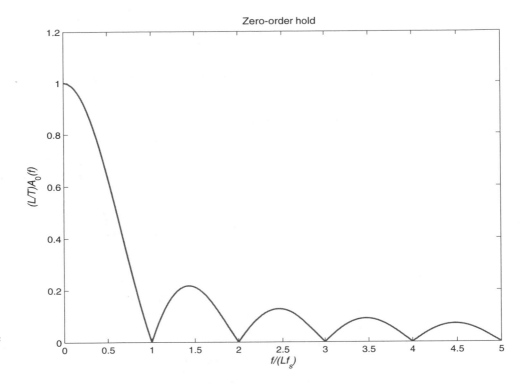

Figure 7.29 Magnitude Response of a Zero-order Hold Model of a DAC

passband, by using a digital anti-imaging filter with the following frequency response. This way, the overall magnitude response is flat within the passband.

DAC passband equalization

$$H_L(f) = \begin{cases} \dfrac{L\sqrt{1+(f/F_a)^{2n}}}{T\,|\mathrm{sinc}(\pi f T/L)|}, & 0 \le |f| \le F_a \\ 0, & F_a < |f| < f_s/2 \end{cases} \qquad (7.54)$$

Example 7.9   **Oversampling DAC**

To illustrate the effectiveness of oversampling in the digital-to-analog conversion process, let the analog anti-imaging filter be an nth-order Butterworth filter with the magnitude response in Eq. 7.40. The objective of this example is to use oversampling to ensure that the magnitude spectra of the images are sufficiently small. Since the Butterworth magnitude response decreases monotonically, the spectral images are all scaled by a factor of at least $A_n(f_d)$, where $f_d = f_s/2$ is the folding frequency. Thus, the scaling factor for the spectral images is

$$\beta_n \triangleq A_n(f_d)$$

If oversampling by a factor of L is used, then $f_s = 2LF_a$, which means $f_d = LF_a$.

Using Eq. 7.40 to achieve a spectral image scaling factor of ϵ, we need

$$\frac{1}{\sqrt{1 + L^{2n}}} \leq \epsilon$$

Taking reciprocals and squaring both sides then yields

$$1 + L^{2n} \geq \epsilon^{-2}$$

Solving for L, we then have

$$L \geq (\epsilon^{-2} - 1)^{0.5/n}$$

For example, if a first-order Butterworth filter is used, then $n = 1$. Suppose the spectral image scaling factor is $\epsilon = 0.05$. In this case, the sampling rate conversion factor required is

$$L = \text{ceil}\{(400 - 1)^{1/2}\}$$

$$= \text{ceil}(19.975)$$

$$= 20$$

The magnitude response of the equalized digital anti-imaging filter $H_L(z)$ is shown in Figure 7.30. It was generated by running script *exam7_9* on the distribution CD. For clarity of display, only the range $0 \leq |f| \leq f_s/L$ is shown. Since $L = 20$ in this case, almost all of the passband compensation is included to cancel the effects of the first-order Butterworth filter. The DAC (zero-order hold) magnitude response is essentially flat in the passband because $L \gg 1$.

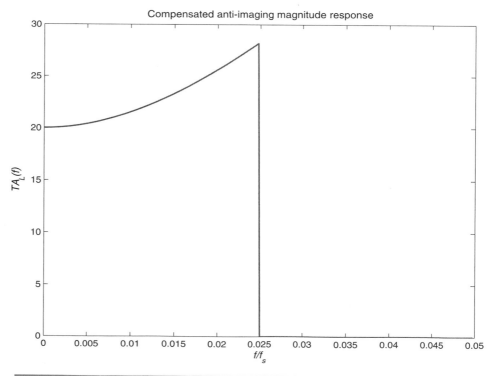

Figure 7.30 Magnitude Response of Equalized Digital Anti-imaging Filter $H_L(z)$, Compensating for the Effects of the DAC and the Analog Anti-imaging Filter when $n = 1$ and $L = 20$

FDSP Toolbox

7.8 **Software Applications**

7.8.1   **GUI Module:** *g_multirate*

The FDSP toolbox includes a graphical user interface module called *g_multirate* that allows the user to design and evaluate multirate discrete-time systems without any need for programming. GUI module *g_multirate* features a display screen with tiled windows as shown in Figure 7.31.

The upper-left *Block diagram* window contains a block diagram of the multirate system under investigation. It is a single-stage rational sampling rate converter with a rate conversion factor, L/M, described by the following time domain equation.

$$y(k) = \sum_{i=0}^{m} b_i \delta_L(Mk - i)x\left(\frac{Mk - i}{L}\right) \tag{7.55}$$

The *Parameters* window below the block diagram displays edit boxes containing simulation parameters. The contents of each edit box can be directly modified by the user with the Enter key used to activate the changes. Parameter fs is the sampling frequency, L is the interpolation factor, M is the decimation factor, and c is a damping factor used with the damped-cosine input. There are also two pushbutton controls that play the input signal $x(k)$ and the output signal $y(i)$, respectively, on the PC speaker.

The *Type* and *View* windows in the upper-right corner of the screen allow the user to select both the type of input and the viewing mode. The input types include uniformly distributed white noise, a damped cosine with a frequency of $f_s/10$ and a damping factor c^k, an amplitude-modulated sine wave, and a frequency-modulated sine wave. The Record x option prompts the user to record one second of audio data using the PC microphone. Finally, the User-defined option prompts the user for the name of a MAT-file containing a vector of samples x and the sampling frequency fs.

The *View* options include the Time signals $x(k)$ and $y(i)$ and their Magnitude spectra. Additional viewing options include the Magnitude response, Phase response, and Impulse response of the combined anti-aliasing and anti-imaging filter.

$$H_0(z) = \sum_{i=0}^{m} b_i z^{-i} \tag{7.56}$$

The impulse response plot also plots the poles and zeros of the filter. There is also a dB checkbox control that toggles the magnitude spectra and magnitude response plots between linear and logarithmic scales. The filter order, m, is directly controllable with the horizontal slider bar below the *Type* and *View* windows. The *Plot* window along the bottom half of the screen shows the selected view.

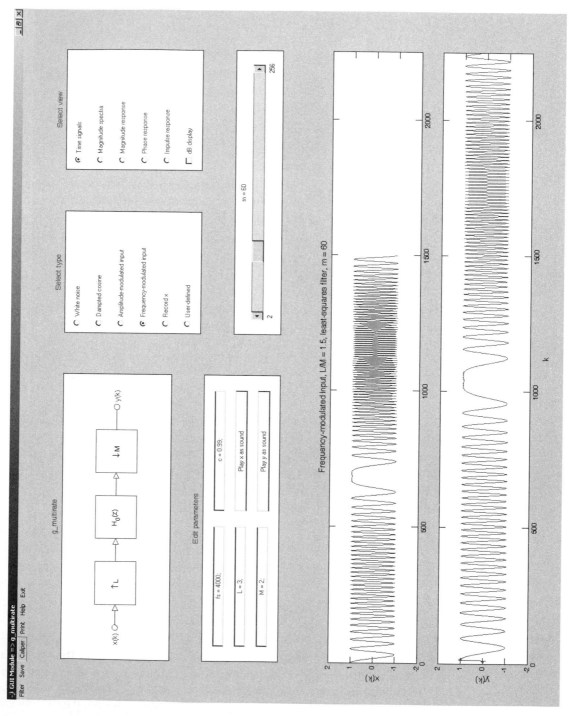

Figure 7.31 Display Screen of Chapter GUI Module *g_multirate*

The *Menu* bar along the top of the screen includes several menu options. The *Filter* option allows the user to select the filter type for the combined anti-aliasing and anti-imaging filter, $H_0(z)$. The choices include windowed (rectangular, Hanning, Hamming, or Blackman), frequency-sampled, least-squares, and equiripple filters. The *Save* option is used to save the current x and fs in a user-specified MAT-file for future use. Files created in this manner subsequently can be loaded with the User-defined input option. The *Caliper* option allows the user to measure any point on the current plot by moving the mouse crosshairs to that point and clicking. The *Print* option prints the contents of the *Plot* window. Finally, the *Help* option provides the user with some helpful suggestions on how to effectively use module *g_multirate*.

7.8.2 Sampling Rate Converter (CD to DAT)

A multistage sampling rate converter is a good vehicle for illustrating the topics covered in this chapter. Consider the problem of converting music from compact disc (CD) format to digital audio tape (DAT) format. A CD is sampled at a rate of 44.1 kHz, while a DAT is sampled at a somewhat higher rate of 48 kHz. Thus, the required frequency conversion ratio to go from CD format to DAT format is

$$\frac{L}{M} = \frac{48}{44.1}$$
$$= \frac{480}{441}$$
$$= \frac{160}{147} \tag{7.57}$$

First, suppose a single-stage sampling rate converter is to be used. From Eq. 7.20, the combined anti-aliasing and anti-imaging filter, $H_0(f)$, must have a cutoff frequency of

$$F_0 = \min\left\{\frac{44100}{2(147)}, \frac{44100}{2(160)}\right\}$$
$$= 137.8 \text{ Hz} \tag{7.58}$$

The required passband gain is $H_0(0) = 160$. Hence, from Eq. 7.21, the desired frequency response for the ideal lowpass filter is

$$H_0(f) = \begin{cases} 160, & 0 \le |f| < 137.8 \\ 0, & 137.8 \le |f| < 22500 \end{cases} \tag{7.59}$$

Filter $H_0(z)$ is a *narrowband* filter with $F_0 = 0.003025 f_s$. Since a direct single-stage implementation would require a very high-order linear-phase FIR filter, we will instead use a multi-stage implementation. Example 7.4 examined the case of

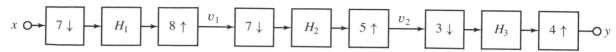

Figure 7.32 A Three-stage CD-to-DAT Sampling Rate Converter

converting from DAT to CD. Taking the reciprocals of those conversion factors, we arrive at the following three-stage implementation.

$$\frac{160}{147} = \left(\frac{8}{7}\right)\left(\frac{5}{7}\right)\left(\frac{4}{3}\right) \tag{7.60}$$

Thus, a CD-to-DAT converter can be implemented using two rational interpolators and one rational decimator. A block diagram of the three-stage sampling rate converter is shown in Figure 7.32. Here $v_1(k)$ and $v_2(k)$ are intermediate output signals from stages one and two, respectively. The sampling rates of $v_1(k)$ and $v_2(k)$ are

$$f_1 = \left(\frac{8}{7}\right)44.1 = 50.4 \text{ kHz} \tag{7.61a}$$

$$f_2 = \left(\frac{5}{7}\right)50.4 = 36.0 \text{ kHz} \tag{7.61b}$$

Using Eqs. 7.20 and 7.60, the three cutoff frequencies of the combined anti-aliasing and anti-imaging filters are as follows.

$$F_1 = \min\left\{\frac{44100}{2(8)}, \frac{44100}{2(7)}\right\} = 2756.3 \text{ Hz} \tag{7.62a}$$

$$F_2 = \min\left\{\frac{50400}{2(5)}, \frac{50400}{2(7)}\right\} = 3600 \text{ Hz} \tag{7.62b}$$

$$F_3 = \min\left\{\frac{436000}{2(4)}, \frac{36000}{2(3)}\right\} = 4500 \text{ Hz} \tag{7.62c}$$

It then follows from Eq. 7.21 that the ideal frequency responses of the three stage filters are

$$H_1(f) = \begin{cases} 8, & 0 \leq |f| < 2756.3 \\ 0, & 2756.3 \leq |f| < 22500 \end{cases} \tag{7.63a}$$

$$H_2(f) = \begin{cases} 5, & 0 \leq |f| < 3600 \\ 0, & 3600 \leq |f| < 25200 \end{cases} \tag{7.63b}$$

$$H_3(f) = \begin{cases} 4, & 0 \leq |f| < 4500 \\ 0, & 4500 \leq |f| < 18000 \end{cases} \tag{7.63c}$$

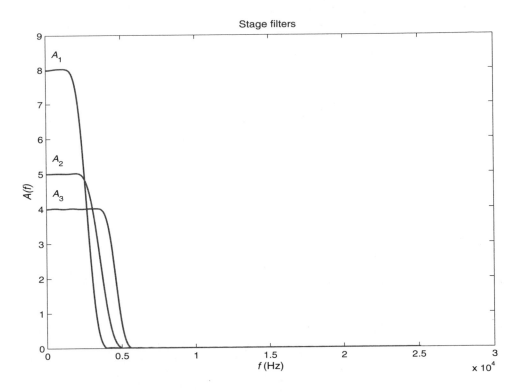

Figure 7.33 Magnitude Responses of the Three Stage Filters

Suppose the stage filters are each implemented using a windowed linear-phase FIR filter of order $m = 60$ with the Hamming window. Plots of the magnitude responses of the three filters are shown in Figure 7.33.

Recall that the FDSP toolbox contains a function called *f_rateconv* for performing a single-stage rational sampling rate conversion by L/M. The three-stage conversion is implemented with multiple calls to *f_rateconv* by running the following MATLAB script, labeled *exam7_10* on the distribution CD.

```
function exam7_10

% Example 7.10: CD-to-DAT sampling rate converter

clear
clc
fprintf('Example 7.10: CD-to-DAT sampling rate converter\n\n')
f_CD = 44100;
f_DAT = 48000;
```

```
L = [8 5 4];
M = [7 7 3];
m = 60;
win = 2;
sym = 0;
fs1 = L(1)*f_CD/M(1)
fs2 = L(2)*fs1/M(2)
fs = [f_CD,fs1,fs2];

% Compute stage filter magnitude responses

stg = f_prompt ('\nCompute stage filters separately (0=no,1=yes)',0,1,0);
if stg
   r = 250;
   for i = 1 : 3
      F(i) = (fs(i)/2)*min(1/L(i),1/M(i));
      p = [0 F(i) F(i) 0];
      b = L(i)*f_firwin ('f_firamp',m,fs(i),win,sym,p);
      [H,f(i,:)] = f_freqz (b,1,r,fs(i));
      A(i,:) = abs(H);
   end
   figure
   plot (f',A','LineWidth',1.5)
   f_labels ('Stage filters','f (Hz)','A(f)')
   x = 400;
   text (x,L(1)+.4,'A_1')
   text (x,L(2)+.4,'A_2')
   text (x,L(3)+.4,'A_3')
   f_wait
end

% Sample an input signal at CD rate

d = 2;
x = f_getsound (d,f_CD);
f_wait ('Press any key to play back sound at 44.1 kHz ...')
soundsc (x,f_CD)
f_wait ('Press any key to play back sound at 48 kHz ...')
soundsc (x,f_DAT)
```

```
figure
plot(x);
f_labels ('Use the mouse to select the start of a short speech segment ...','k','x(k)')
[k1,x1] = f_caliper(1);

% Convert segment of it to DAT rate

f_wait ('Press any key to rate convert the selected segment...')
p = floor(k1) : floor(k1)+400;
v1 = f_rateconv (x(p),fs(1),L(1),M(1),m,win);
v2 = f_rateconv (v1,fs(2),L(2),M(2),m,win);
y = f_rateconv (v2,fs(3),L(3),M(3),m,win);
% Plot segments

figure
subplot(4,1,1)
k = p - p(1);
plot(k,x(p),'LineWidth',1.5)
f_labels ('','','x(k)')
subplot(4,1,2)
plot(0:length(v1)-1,v1,'LineWidth',1.5)
f_labels ('','','v_1(k)')
subplot(4,1,3)
plot(0:length(v2)-1,v2,'LineWidth',1.5)
f_labels ('','','v_2(k)')
subplot(4,1,4)
plot(0:length(y)-1,y,'LineWidth',1.5)
f_labels ('','k','y(k)')
f_wait
```

When script *exam7_10* is run, it first computes the three-stage filter magnitude responses previously shown in Figure 7.33. It then prompts the user to speak into the microphone and records the response at the CD rate. The recorded speech is then played back at both the CD rate and the higher DAT rate for comparison. Next, the recorded speech is displayed in a plot and crosshairs appear. The user should use the mouse to select the start of a speech segment. The selected segment is then converted from the CD rate to the DAT rate in three stages with the results as shown in Figure 7.34. Note that although all the segments appear to be the same shape, a close inspection of the scales along the horizontal axes shows that each is at a different sampling rate.

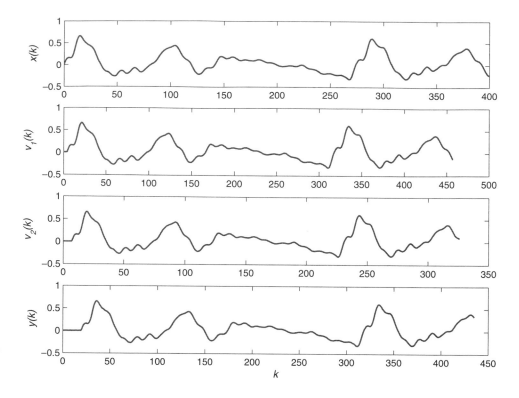

Figure 7.34 Rate-converted Segments of Recorded Speech

● ● ● ● ● ● ● ● ● ● ● ● ● ●

7.9 Chapter Summary

Rate Converters

This chapter focused on multirate signal processing techniques. Every multirate system contains a sampling rate converter, a system that changes the sampling rate of a discrete-time signal without converting the signal back to analog form. Instead, the resampling is done in the digital domain. The simplest sampling rate converter decreases the sampling frequency by an integer factor M. This is called a *decimator*, and it can be implemented with the following time-domain equation.

$$y(k) = \sum_{i=0}^{m} b_i x(Mk - i) \tag{7.64}$$

The FIR filter with impulse response $h_M(k) = b_k$ for $0 \le k \le m$ is a lowpass filter with a passband gain of one and a cutoff frequency of $F_M = f_s/(2M)$. This filter is included to preserve the spectral characteristics of $x(k)$. Decreasing the sampling rate by a factor of M is called *down-sampling*, and it is represented in block diagrams with the symbol $\downarrow M$. The effect of down-sampling is to spread the spectrum out by a factor of M. Hence to avoid aliasing $H_M(z) = Z\{h_M(k)\}$ is inserted as a digital anti-aliasing filter.

It is also possible to increase the sampling rate by an integer factor L. This is called an *interpolator*, and it can be implemented with the following time-domain equation.

$$y(k) = \sum_{i=0}^{m} b_i \delta_L(k-i)x\left(\frac{k-i}{L}\right) \tag{7.65}$$

Here $\delta_L(k)$ is a periodic train of impulses of period L starting at $k = 0$. The FIR filter with impulse response $h_L(k) = b_k$ for $0 \le k \le m$ is a lowpass filter with a passband gain of L and a cutoff frequency of $F_L = f_s/(2L)$. Again, this filter is included to preserve the spectral characteristics of $x(k)$. Increasing the sampling rate by a factor of L is called *up-sampling*, and it is represented in block diagrams with the symbol $\uparrow L$. The effect of up-sampling is to compress the spectrum by a factor of L. Because the spectrum is periodic, this generates images of the original spectrum which must be removed by $H_L(z) = Z\{h_L(k)\}$, which is a digital anti-imaging filter.

More general rate conversion by a rational factor L/M can be achieved by using a cascade configuration of an interpolator with rate conversion factor L followed by a decimator with rate conversion factor M. This is called a rational rate converter, and it can be implemented using the following time-domain equation.

$$y(k) = \sum_{i=0}^{m} b_i \delta_L(Mk-i)x\left(\frac{Mk-i}{L}\right) \tag{7.66}$$

Because the interpolator is followed by the decimator, the anti-aliasing pre-filter of the decimator can be combined with the anti-imaging post-filter of the interpolator. The cutoff frequency of the cascade combination of the two filters is $F_0 = \min\{F_L, F_M\}$ and the passband gain is L. Thus, the desired frequency response of the ideal combined anti-aliasing and anti-imaging filter is

$$H_0(f) = \begin{cases} L, & 0 \le |f| \le F_0 \\ 0, & F_0 < |f| < f_s/2 \end{cases} \tag{7.67}$$

Typically, anti-aliasing and anti-imaging filters are implemented using FIR filters because this way linear-phase filters can be employed which delay the signal in the passband, but do not otherwise distort it. It is also possible to use IIR filters which are discussed in Chapter 8. To simplify the filter design, rational sampling rate converters with large values for L or M are typically implemented as a cascade configuration of lower-order rate converters. This is called a *multistage* rate converter, and it is based on the following factorization of the rate conversion factor.

$$\frac{L}{M} = \left(\frac{L_1}{M_1}\right)\left(\frac{L_2}{M_2}\right)\cdots\left(\frac{L_p}{M_p}\right) \tag{7.68}$$

Both interpolators and decimators can be implemented efficiently using *polyphase filters*. If the rate conversion factor is N, then a parallel combination of N polyphase filters is used where the impulse response of the ith polyphase filter is obtained by extracting every Nth sample of the original impulse response of the anti-aliasing or anti-imaging filter. The output samples are then obtained from each of the polyphase filters in turn using a commutator switch. A polyphase rate converter realization reduces the number of floating-point operations by a factor of N. It also makes explicit the observation that a rate converter is a time-varying linear discrete-time system.

Applications

There are many applications of multirate systems in modern DSP systems. For example, if it is necessary to delay a discrete-time signal by a fraction of a sample, the signal can be up-sampled by L, delayed by $0 < M < L$ using a shift register, and then down-sampled by L to restore the sampling rate. This achieves an intersample delay of $\tau = (m + L/M)T$. Another example is the design of a narrowband lowpass filter, a filter with a cutoff frequency satisfying $F_0 \ll f_s$. Here the signal is down-sampled to spread out the spectrum, filtered with an easier to implement wideband filter, and then up-sampled to restore the original sampling rate. Narrowband filters centered at submultiples of f_s can be configured in parallel to form filter banks. This technique, called *subband processing*, can be used to transmit several low-bandwidth signals over a single high-bandwidth channel using frequency division multiplexing.

Another class of applications of multirate systems arises in the implementation of high-performance analog-to-digital and digital-to-analog converters. An oversampling ADC uses oversampling by a factor of M followed by a sampling rate decimator. This technique reduces the requirements on the analog anti-aliasing filter by inserting a transition band of width $B = (M - 1)F_a$, where F_a is the bandwidth of the analog signal. This means that a high-order analog anti-aliasing filter with a sharp cutoff can be replaced with a less expensive low-order analog anti-aliasing filter. The oversampling ADC also has the added benefit that the power of the quantization noise appearing at the output is reduced by a factor of M.

Oversampling can also be used to implement a high-performance DAC. An oversampling DAC consists of a sampling rate interpolator with a conversion factor of L followed by a DAC. The effect of the up-sampling is to reduce the requirements on the analog anti-imaging filter by inserting a transition band of width $B = (L-1)F_a$, where F_a is the bandwidth of the signal. Again this means that a high-order analog anti-imaging filter with a sharp cutoff can be replaced with a less expensive, low-order analog anti-imaging filter. Like the oversampling ADC, the oversampling DAC also has the added benefit that the power of the quantization noise appearing at the output is reduced by a factor of L. For an oversampling DAC, the effects of the zero-order hold and the analog anti-imaging filter $H_a(s)$ can be *equalized*, within the passband, by using a more general digital anti-imaging filter for the interpolator. For example, suppose the analog anti-imaging filter is an nth-order Butterworth filter. Then the overall magnitude response of the oversampling DAC is flat within the passband if the following digital anti-imaging filter is used.

$$
H_L(f) = \begin{cases} \dfrac{L\sqrt{1 + (f/F_a)^{2n}}}{T \, |\text{sinc}(\pi f T/L)|}, & 0 \le |f| \le F_a \\[2mm] 0, & F_a < |f| < f_s/2 \end{cases}
\tag{7.69}
$$

The FDSP toolbox includes a GUI module called *g_multirate* that allows the user to design and evaluate rational sampling rate converters. Rate conversion can be applied to a variety of inputs, including recorded sound and user-defined inputs stored in MAT-files. The choices for the anti-aliasing anti-imaging filters include windowed, frequency-sampled, least-squares, and equiripple linear-phase FIR filters.

7.10 Problems

The problems are divided into Analysis problems that can be solved by hand or with a calculator, GUI Simulation problems that are solved using GUI module *g_multirate*, and Computation problems. The Computation problems require the student to write a MATLAB script using the FDSP toolbox functions summarized in Appendix B. Solutions to selected problems are available on the distribution CD. Students are encouraged to use these problems, which are identified with a ☑, as a check on their understanding of the material.

7.10.1 Analysis

Section 7.1

P7.1. Consider the variable delay system shown in Figure P7.1, where the sampling rate of $x(k)$ is f_s. Suppose $H_L(z)$ is a linear-phase FIR filters of order m. Find an expression for the total delay that takes into account the interpolator delay, the shift register delay, and the decimator delay.

P7.1   A Variable Delay System

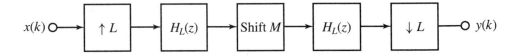

$$x(k) \ \bigcirc\!\!\longrightarrow\!\!\boxed{\uparrow L}\longrightarrow\boxed{H_L(z)}\longrightarrow\boxed{\text{Shift } M}\longrightarrow\boxed{H_L(z)}\longrightarrow\boxed{\downarrow L}\longrightarrow\!\!\bigcirc\ y(k)$$

P7.2. Suppose a signal is sampled at a rate of $f_s = 10$ Hz. Consider the problem of using the variable delay system in Figure P7.1 to implement an overall delay of $\tau = 2.38$ sec.

(a) Find the smallest interpolation factor L that is needed.

(b) Suppose the linear-phase FIR filters are of order $m = 50$. How much delay is introduced by the two lowpass filters?

(c) What length of shift register, M, is needed to achieve the overall delay?

Section 7.2

☑ **P7.3.** Consider the problem of designing a sampling rate decimator with a decimation factor of $M = 8$.

(a) Sketch a block diagram of the sampling rate decimator.

(b) Find the required frequency response of the ideal anti-aliasing digital filter. Here f_s is the sampling rate of $x(k)$.

(c) Using Tables 6.1 and 6.2, design an anti-aliasing filter of order $m = 40$, using the windowing method with a Hanning window.

(d) Find the difference equation for the sampling rate decimator.

P7.4. Consider the problem of designing a sampling rate interpolator with an interpolation factor of $L = 10$.

(a) Sketch a block diagram of the sampling rate interpolator.

(b) Find the required frequency response of the ideal anti-imaging digital filter. Here f_s is the sampling rate of $x(k)$.

(c) Using Tables 6.1 and 6.2, design an anti-imaging filter of order $m = 30$, using the windowing method with a Hamming window.

(d) Find the difference equation for the sampling rate interpolator.

Section 7.3

P7.5. Consider the problem of designing a rational sampling rate converter with a frequency conversion factor of $L/M = 0.8$.

(a) Sketch a block diagram of the sampling rate converter.

(b) Find the required frequency response of the ideal anti-aliasing and anti-imaging digital filter. Here f_s is the sampling rate of $x(k)$.

(c) Design an anti-aliasing and anti-imaging filter of order $m = 50$ using the windowing method with the Blackman window using Tables 6.1 and 6.2.

(d) Find the difference equation for the sampling rate converter.

P7.6. Suppose a multirate signal processing application requires a sampling rate conversion factor of $L/M = 0.525$.

(a) Find the required frequency response of the ideal anti-aliasing and anti-imaging digital filter assuming a single-stage converter is used.

(b) Factor L/M into a product of two or more rational numbers whose numerators and denominators are less than or equal to 10.

(c) Sketch a block diagram of a multi-stage sampling rate converter based on your factoring of L/M from part (b).

(d) Find the required frequency responses of the ideal combined anti-aliasing and anti-imaging digital filters for each of the stages in part (c).

Section 7.4

P7.7. Consider an integer decimator with rate conversion factor of M and a linear-phase FIR anti-aliasing filter of order m. Suppose m is selected such that $m + 1 = pM$ for some integer p.

(a) Find n_M, the number of floating-point multiplications (FLOPs) needed to compute each sample of the output. Express your answer in terms of p.

(b) Suppose a polyphase filter realization is used to implement the decimator. Find N_M, the number of FLOPs needed to compute each sample of the output.

(c) Express N_M as a percentage of n_M.

P7.8. Consider the problem of designing a decimator with $f_s = 60$ Hz and a rate conversion factor of $M = 3$.

(a) What is the sampling rate of the output signal?

(b) Sketch the desired magnitude response of the ideal anti-aliasing filter, $H_M(z)$.

(c) Suppose the anti-aliasing filter is a windowed filter of order $m = 32$ using the Hamming window. Use Tables 6.1 and 6.2 to find the impulse response, $h_M(k)$.

(d) Suppose a polyphase realization is used. How many polyphase filters are needed, and what is the order of each polyphase filter?

(e) Sketch a polyphase filter realization of the decimator.

P7.9. Consider an integer interpolator with rate conversion factor of L and a linear-phase FIR anti-imagining filter of order m. Suppose m is selected such that $m + 1 = pL$ for some integer p.

(a) Find n_L, the number of floating-point multiplications (FLOPs) needed to compute each sample of the output. Express your answer in terms of p.

(b) Suppose a polyphase filter realization is used to implement the interpolator. Find N_L, the number of FLOPs needed to compute each sample of the output.

(c) Express N_L as a percentage of n_L.

P7.10. Consider the problem of designing an interpolator with $f_s = 12$ Hz and a rate conversion factor $L = 3$.

(a) What is the sampling rate of the output signal?

(b) Sketch the desired magnitude response of the ideal anti-imaging filter $H_L(z)$.

(c) Suppose the anti-imaging filter is a windowed filter of order $m = 20$ using the Hanning window. Use Tables 6.1 and 6.2 to find the impulse response, $h_L(k)$.

(d) Suppose a polyphase realization is used. How many polyphase filters are needed, and what is the order of the each polyphase filter?

(e) Sketch a polyphase filter realization of the interpolator.

P7.11. Sketch a polyphase filter realization of a rational rate converter with a rate conversion factor of $L/M = 3/2$. Indicate how fast each commutator switch rotates and the direction of rotation.

Section 7.5

P7.12. Consider the problem of designing a multirate narrowband lowpass FIR filter as shown in Figure P7.12. Suppose the sampling frequency is $f_s = 8000$ Hz and the cutoff frequency is $F_0 = 200$ Hz.

   (a) Find the largest integer frequency conversion factor M that can be used.

   (b) Using Eq. 6.30, design an anti-aliasing filter $H_M(z)$ of order $m = 32$ using the frequency-sampled method. Do not use any transition band samples.

P7.12 A Multirate Narrowband FIR Filter

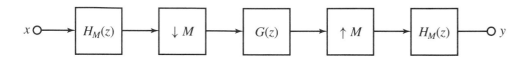

P7.13. Consider the problem of designing a passband filter with the following ideal magnitude response.

$$A(f) = \begin{cases} 0, & 0 \le f < F_0 \\ 1, & F_0 \le f \le F_1 \\ 0, & F_1 < f < f_s \end{cases}$$

   (a) Let $B = F_1 - F_0$ be the width of the passband, and consider the problem of designing a lowpass filter $G(z)$ with cutoff frequency of $F_c = B/2$. Using Tables 6.1 and 6.2, find the filter impulse response $g(k)$ for a filter of order $m = 60$ using the windowing method with the Hamming window.

   (b) Using the frequency shift property in Eq. 7.37 and $g(k)$, find the impulse response $h(k)$ of the complex passband filter with cutoff frequencies F_0 and F_1.

   (c) Is the magnitude response of $H(z)$ an even function of f? Why, or why not?

   (d) Is the magnitude response of $H(z)$ a periodic function of f? If so, what is the period?

P7.14. Consider the problem of designing a highpass filter with the following ideal magnitude response.

$$A(f) = \begin{cases} 0, & 0 \le f < F_0 \\ 1, & F_0 \le f < f_s \end{cases}$$

   (a) Let $B = f_s - F_0$ be the width of the passband, and consider the problem of designing a lowpass filter $G(z)$ with a cutoff frequency of $F_c = B/2$. Using Tables 6.1 and 6.2, find the filter impulse response $g(k)$ for a filter of order $m = 50$ using the windowing method with the Blackman window.

   (b) Using the frequency shift property in Eq. 7.37 and $g(k)$, find the impulse response $h(k)$ of the complex highpass filter with a cutoff frequency of F_0.

   (c) Is the magnitude response of $H(z)$ an even function of f? Why, or why not?

   (d) Is the magnitude response of $H(z)$ a periodic function of f? If so, what is the period?

Section 7.6

P7.15. Consider a 10-bit oversampling ADC with analog inputs in the range $|x_a(t)| \leq 5$, as shown in Figure P7.15.

(a) Find the average power of the quantization noise of the quantized input, $x_q(k)$.

(b) Suppose a second-order Butterworth filter is used for the analog anti-aliasing prefilter. The objective is to reduce the aliasing error by a factor of at least $\alpha = 0.005$. Find the minimum required oversampling factor M.

(c) Find the average power of the quantization noise at the output, $y(k)$, of the oversampling ADC.

(d) Suppose $f_s = 1000$ Hz. Sketch the ideal magnitude response of the digital anti-aliasing filter $H_M(f)$.

(e) Using Tables 6.1 and 6.2, design a linear-phase FIR filter of order $m = 80$ whose frequency response approximates $H_M(f)$, using the windowing method with a Hanning window.

P7.15 An Oversampling ADC with an Oversampling Factor, M

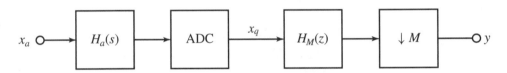

P7.16. A 12-bit oversampling ADC oversamples by a factor of $M = 64$. To achieve the same average power of the quantization noise at the output, but without using oversampling, how many bits are required?

P7.17. Suppose an analog signal in the range $|x_a(t)| \leq 5$ is sampled with a 10-bit oversampling ADC with an oversampling factor of $M = 16$. The output of the ADC is passed through an FIR filter $H(z)$, as shown in Figure P7.17, where

$$H(z) = 1 - 2z^{-1} + 3z^{-2} - 2z^{-3} + z^{-4}$$

(a) Find the quantization level q.

(b) Find the power gain of the filter $H(z)$.

(c) Find the average power of the quantization noise at the system output, $y(k)$.

(d) To get the same quantization noise power, but without using oversampling, how many bits are required?

P7.17 A Discrete-time Multirate System

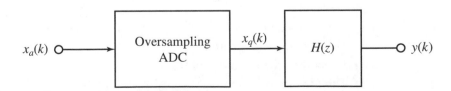

Section 7.7

P7.18. Consider a 10-bit oversampling DAC with analog outputs in the range $|y_a(t)| \leq 10$, as shown in Figure P7.18.

(a) Suppose a first-order Butterworth filter is used for the analog anti-imaging post-filter. The objective is to reduce the imaging error by a factor of at least $\beta = 0.05$. Find the minimum required oversampling factor L.

(b) Find the average power of the quantization noise at the output of the DAC.

(c) Suppose $f_s = 2000$ Hz. Find the ideal frequency response of the digital anti-imaging filter $H_L(f)$. Include equalizer compensation for both the analog anti-imaging filter and the zero-order hold.

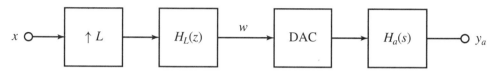

P7.18 An Oversampling DAC with an Oversampling Factor, L

7.10.2 GUI Simulation

Section 7.8

P7.19. Using the GUI module *g_multirate*, select the damped cosine input. Set the damping factor to $c = 0.995$, the interpolation factor to $L = 2$, and the decimation factor to $M = 3$. Keep all other parameters at the default values. Plot the following.

(a) The time signals.

(b) The magnitude spectra.

P7.20. Using the GUI module *g_multirate*, select the amplitude-modulated input. Reduce the sampling rate of the input using an integer decimator with a rate conversion factor of $M = 2$. Use a windowed filter with the Hanning window, and plot the following.

(a) The time signals.

(b) Their magnitude spectra.

(c) The filter magnitude response.

(d) The filter impulse response.

P7.21. Using the GUI module *g_multirate*, select the frequency-modulated input. Increase the sampling frequency of the input using an interpolator with a rate conversion factor of $L = 3$. Use a windowed filter with the Hamming window, and plot the following.

(a) The time signals.

(b) Their magnitude spectra.

(c) The filter magnitude response.

(d) The filter phase response.

P7.22. Use the GUI module *g_multirate* with the default filter order. Select the damped cosine input. Print the magnitude responses of the following anti-aliasing and anti-imaging filters using the linear scale.

(a) Windowed filter with the Blackman window.

(b) Frequency-sampled filter.

(c) Least-squares filter.

P7.23. Use the GUI module *g_multirate*, and adjust the filter order to $m = 80$. Select the white noise input. Print the magnitude responses of the following anti-aliasing and anti-imaging filters using the dB scale.

(a) Windowed filter with the Hanning window.

(b) Windowed filter with a Hamming window.

(c) Equiripple filter.

P7.24. Using the GUI module *g_multirate*, record the word *hello* in x. Play it back to make sure it is a good recording. Save the recording in a MAT-file named *my_sound* using the Save option. Then reload it using the User-defined option. Convert the sampling rate using $L = 3$ and $M = 4$, and a least-squares filter of order $m = 60$. Plot the following.

(a) The time signals.

(b) Their magnitude spectra.

(c) The filter magnitude response.

(d) The filter impulse response.

☑ **P7.25.** Use the GUI module *g_multirate* and the User-defined input option to load the MAT-file *u_multirate1*. Convert the sampling rate using $L = 4$ and $M = 3$ using a frequency-sampled filter of order $m = 60$. Plot the following.

(a) The time signals with the peak of $x(k)$ marked with the *Caliper* option.

(b) Their magnitude spectra.

(c) The filter magnitude response.

7.10.3 Computation

Section 7.2

P7.26. Consider the following periodic analog signal with three harmonics.

$$x_a(t) = \cos(2\pi t) - 0.8\sin(4\pi t) + 0.6\cos(6\pi t)$$

Suppose this signal is sampled at $f_s = 64$ Hz, using $N = 120$ samples to produce a discrete-time signal $x(k) = x_a(kT)$ for $0 \le k < N$. Write a MATLAB script that uses *f_decimate* to decimate this signal by converting it to a sampling rate of $f_S = 32$ Hz. For the anti-aliasing filter, use a windowed filter of order $m = 40$ with the Hamming window. Use the *subplot* command and the *stem* function to plot the following discrete-time signals on one screen.

(a) The original signal $x(k)$.

(b) The resampled signal $y(k)$ below it using a different color.

☑ **P7.27.** Consider the following periodic analog signal with three harmonics.

$$x_a(t) = \sin(2\pi t) - 3\cos(4\pi t) + 2\sin(6\pi t)$$

Suppose this signal is sampled at $f_s = 24$ Hz using $N = 50$ samples to produce a discrete-time signal $x(k) = x_a(kT)$ for $0 \le k < N$. Write a MATLAB script that uses *f_interpol* to interpolate this signal by converting it to a sampling rate of $f_s = 72$ Hz. For the anti-imaging filter, use a least-squares filter of order $m = 50$. Use the *subplot* command and the *stem* function to plot the following discrete-time signals on the same screen.

(a) The original signal $x(k)$.

(b) The resampled signal $y(k)$ below it using a different color.

Section 7.3

P7.28. Consider the following periodic analog signal with three harmonics.

$$x_a(t) = 2\cos(2\pi t) + 3\sin(4\pi t) - 3\sin(6\pi t)$$

Suppose this signal is sampled at $f_s = 30$ Hz using $N = 50$ samples to produce a discrete-time signal $x(k) = x_a(kT)$ for $0 \le k < N$. Write a MATLAB script that uses *f_rateconv* to convert it to a sampling rate of $f_s = 50$ Hz. For the anti-aliasing and anti-imaging filter, use a frequency-sampled filter of order $m = 60$. Use the *subplot* command and the *stem* function to plot the following discrete-time signals on the same screen.

(a) The original signal $x(k)$.

(b) The resampled signal $y(k)$ below it using a different color.

Section 7.5

P7.29. Write a MATLAB function called *u_narrowband* that uses the FDSP toolbox functions *f_firideal* and *f_multirate* to compute the zero-state response of the multirate narrowband lowpass filter shown in Figure P7.1. The calling sequence for *u_narrowband* is as follows.

```
[y,M] = u_narrowband (x,F_0,win,fs,m);
```

On entry to *u_narrowband*, x is a vector of length N containing the input samples, F_0 is the lowpass cutoff frequency where $F\_0 \le fs/4$, *win* is an integer specifying the window type as in Table 6.2, *fs* is the sampling frequency, and m is the filter order which can be assumed to be even. Use a windowed filter of order m and window type *win* for the anti-aliasing and anti-imaging filters and the wideband lowpass filter $G(z)$. On exit from *u_narrowband*, y is a vector of length N containing the filter output, and M is the integer frequency conversion factor. Use the maximum frequency conversion factor possible.

Test function *u_narrowband* by writing a script that uses it to design a lowpass filter with a cutoff frequency of $F_0 = 10$ Hz, a sampling frequency of $f_s = 400$ Hz, and a filter order of $m = 50$. Plot the following.

(a) The narrowband filter impulse response.

(b) The narrowband filter magnitude response and the ideal magnitude response on the same graph with a legend.

☑ **P7.30.** Write a function called *u_synbank* that synthesizes a composite signal $x(i)$ from N low-bandwidth subsignals $x_i(k)$ using a uniform DFT synthesis filter bank. The calling sequence for *u_synbank* is as follows.

```
x = u_synbank (X,m,alpha,win,fs);
```

On entry to *u_synbank*, X is a p by N matrix with the ith subsignal in column i, m is the anti-imaging filter order, *alpha* is a factor between 0 and 1 which controls the filter cutoff frequency, *win* is the window type as in Table 6.2, and fs is the sampling frequency. Use a lowpass cutoff frequency of $F_0 = alpha(fs)/(2N)$. On exit from *u_synbank*, x is the complex composite signal of bandwidth $N(fs)/2$ containing the frequency translated subsignals in its subbands.

Test function *u_synbank* by writing a script that uses the FDSP toolbox function *f_subsignals* to construct a 32 by 4 matrix X with the samples of the kth subsignal in column k. The function *f_subsignals* produces signals whose spectra are given in Example 7.7. Use *alpha* = 0.5, $f_s = 200$ Hz, and a windowed filter of order $m = 90$ with a Hamming window. Save x and fs in a MAT-file named *prob7_30* and plot the following

(a) The real and imaginary parts of the complex composite signal $x(i)$. Use *subplot* to construct a 2 × 1 array of plots on one screen.

(b) The magnitude spectrum $A(f) = |X(f)|$ for $0 \le f \le f_s$.

P7.31. Write a function called *u_analbank* that analyzes a composite signal $x(i)$ and decomposes it into N low-bandwidth subsignals $x_i(k)$ using a uniform DFT analysis filter bank. The calling sequence for *u_analbank* is as follows.

```
X = u_analbank (x,N,m,alpha,win,fs);
```

On entry to *u_analbank*, x is the complex composite signal of bandwidth, $N(fs)/2$, containing the frequency translated subsignals in its subbands, N is the number of subsignals, m is the anti-aliasing filter order, *alpha* is a factor between 0 and 1 which controls the filter cutoff frequency, *win* is the window type as in Table 6.2, and f is the sampling frequency. Use a lowpass cutoff frequency of $F_0 = alpha(fs)/(2N)$. On exit from *u_analbank*, X is a p by N matrix with the ith subsignal in column i.

Test function *u_analbank* by writing a script that analyzes the composite signal $x(i)$ obtained from the solution to Problem P7.30. That is, load MAT-file *prob7_30*. Use *alpha* = 0.5, and a windowed filter of order $m = 90$ with a Hamming window. Plot the following.

(a) The magnitude spectrum $A(f) = |X(f)|$ for $0 \le f \le f_s$.

(b) The magnitude spectra of the subsignals extracted from X. Use *subplot* to construct a 2 × 2 array of plots on one screen.

IIR Filter Design

Learning Objectives

- Know how to design a tunable plucked-string filter for music synthesis (Section 8.1)
- Be able to design resonators, notch filters, and comb filters using pole-zero placement with gain matching (Section 8.2)
- Understand the characteristics and relative advantages of the classical lowpass analog filters: Butterworth, Chebyshev I and II, and elliptic (Section 8.3)
- Know how to convert an analog filter into an equivalent digital filter using the bilinear transformation method (Section 8.4)
- Be able to convert a lowpass filter into a prescribed lowpass, highpass, bandpass, or bandstop filter using a frequency transformation (Section 8.5)
- Know how to add special effects to music and speech using reverb filters (Section 8.7)
- Know how to use the GUI module g_iir to design and analyze digital IIR filters without any need for programming (Section 8.7)

● ● ● ● ● ● ● ● ● ● ● ● ● ● ● ● ●

8.1 Motivation

Just as mechanical filters are used in pipes to block the flow of certain particles, discrete-time systems can be used as digital filters to block the flow of certain signals. A digital filter is a discrete-time system that is designed to reshape the spectrum of the input signal in order to produce desired spectral characteristics in the output signal. Thus, a digital filter is a frequency-selective filter that modifies the magnitude spectrum and the phase spectrum by selecting or enhancing certain spectral components and inhibiting others. In this chapter, we investigate the design of digital infinite impulse response (IIR) filters having the following generic transfer function.

IIR filter

$$H(z) = \frac{b_0 + b_1 z^{-1} + \cdots + b_m z^{-m}}{1 + a_1 z^{-1} + \cdots + a_n z^{-n}}$$

Recall from Chapter 2 that $H(z)$ is an IIR filter if $a_i \neq 0$ for some $1 \leq i \leq n$. Otherwise, $H(z)$ is an FIR filter. Specialized IIR filters can be designed directly using gain matching and a judicious placement of poles and zeros. A more general technique is to start with a normalized lowpass analog filter and put it through a series of transformations to convert it to a desired digital filter. First, a frequency transformation is used to map the normalized lowpass analog filter into a specified frequency-selective filter. This is followed by a bilinear transformation which converts the analog filter into an equivalent digital filter.

We begin this chapter by introducing some practical examples of applications of IIR filters. Next, simple techniques based on pole-zero placement and gain matching are introduced for direct design of specialized IIR filters such as resonators, notch filters, and comb filters. This is followed by a presentation of the classical analog lowpass filters including Butterworth, Chebyshev, and elliptic filters. An analog-to-digital filter transformation technique, called the bilinear transformation method, is then presented. Next, we examine frequency transformations which map normalized lowpass filters into lowpass, highpass, bandpass, and bandstop filters. Finally, a GUI module called *g_iir* is introduced that allows the user to design and evaluate a variety of digital IIR filters without any need for programming. The chapter concludes with an application example and a summary of IIR filter design techniques. The selected application example features the design of a reverb filter, a high-order IIR filter that adds special sound effects to music and speech.

8.1.1   Tunable Plucked-string Filter

Computer-generated music is a natural application area for IIR filter design (Steiglitz, 1996). A simple, yet highly effective, building block for the synthesis of musical sounds is the tunable plucked-string filter shown in Figure 8.1. The output from this type of filter can be used, for example, to synthesize the sound from a stringed instrument, such as a guitar.

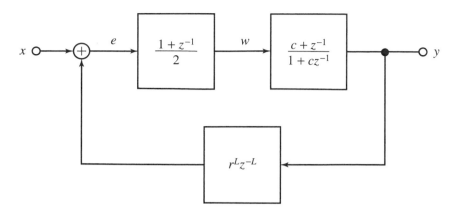

Figure 8.1 A Tunable Plucked-string Filter

The design parameters of the plucked-string filter in Figure 8.1 are the sampling frequency f_s, the pitch parameter $0 < c < 1$, the feedback delay L, and the feedback attenuation factor $0 < r < 1$. Later, we will examine the components of this type of filter in detail. For now, consider the question of developing an expression for the overall transfer function $H(z) = Y(z)/X(z)$ of the plucked-string filter. The block with input $e(k)$ and output $w(k)$ is a first-order lowpass filter with transfer function

$$F(z) \triangleq \frac{W(z)}{E(z)}$$

$$= \frac{1 + z^{-1}}{2} \tag{8.1}$$

Recall from Eq. 5.33 that the block with input $w(k)$ and output $y(k)$ is a first-order allpass filter. Allpass filters pass all frequencies equally because they have flat magnitude responses. The purpose of an allpass filter is to change the phase of the input and thereby introduce some delay. The transfer function of the allpass filter in Figure 8.1 is as follows, where $0 < c < 1$ is the pitch parameter.

$$G(z) \triangleq \frac{Y(z)}{W(z)}$$

$$= \frac{c + z^{-1}}{1 + cz^{-1}} \tag{8.2}$$

The key to finding the overall transfer function is to compute the Z-transform of the summing junction output, $E(z)$. From Eqs. 8.1 and 8.2 and Figure 8.1, we have

$$E(z) = X(z) + r^L z^{-L} Y(z)$$

$$= X(z) + r^L z^{-L} G(z) W(z)$$

$$= X(z) + r^L z^{-L} G(z) F(z) E(z) \tag{8.3}$$

Solving Eq. 8.3 for $E(z)$ then yields

$$E(z) = \frac{X(z)}{1 - r^L z^{-L} G(z) F(z)} \tag{8.4}$$

From Eqs. 8.1 and 8.2, the Z-transform of the output can then be expressed as

$$Y(z) = G(z)W(z)$$
$$= G(z)F(z)E(z)$$
$$= \frac{G(z)F(z)X(z)}{1 - r^L z^{-L} G(z)F(z)} \tag{8.5}$$

It follows that the overall transfer function of the tunable plucked-string filter is

$$H(z) = \frac{Y(z)}{X(z)}$$
$$= \frac{G(z)F(z)}{1 - r^L z^{-L} G(z)F(z)} \tag{8.6}$$

To verify that this is an IIR filter, we substitute the expressions for $F(z)$ and $G(z)$, from Eqs. 8.1 and 8.2, respectively, into Eq. 8.6 which yields

$$H(z) = \frac{\left(\dfrac{c + z^{-1}}{1 + cz^{-1}}\right)\left(\dfrac{1 + z^{-1}}{2}\right)}{1 - r^L z^{-L}\left(\dfrac{c + z^{-1}}{1 + cz^{-1}}\right)\left(\dfrac{1 + z^{-1}}{2}\right)} \tag{8.7}$$

Multiplying the top and bottom of Eq. 8.7 by $(1 + cz^{-1})$ and combining terms then leads to the following simplified transfer function for the tunable plucked-string filter.

Tunable
plucked-string filter

$$H(z) = \frac{0.5[c + (1 + c)z^{-1} + z^{-2}]}{1 + cz^{-1} - 0.5r^L[cz^{-L} + (1 + c)z^{-(L+1)} + z^{-(L+2)}]} \tag{8.8}$$

Since the denominator polynomial satisfies $a(z) \neq 1$, this is indeed an IIR filter. The plucked-string sound is generated by the filter output when the input is an impulse or a short burst of white noise.

The frequency response of the plucked-string filter consists of a series of peaks or *resonances* that decay gradually depending on the value of the feedback attenuation parameter r. To tune the first resonant frequency, we use the parameters L and c. Suppose the sampling frequency is f_s and the desired value for the first resonance, or *pitch*, is F_0. Then L and c can be computed as follows (Jaffe and Smith, 1983).

$$L = \text{floor}\left(\frac{f_s - 0.5F_0}{F_0}\right) \tag{8.9a}$$

$$\delta = \frac{f_s - (L + 0.5)F_0}{F_0} \tag{8.9b}$$

$$c = \frac{1 - \delta}{1 + \delta} \tag{8.9c}$$

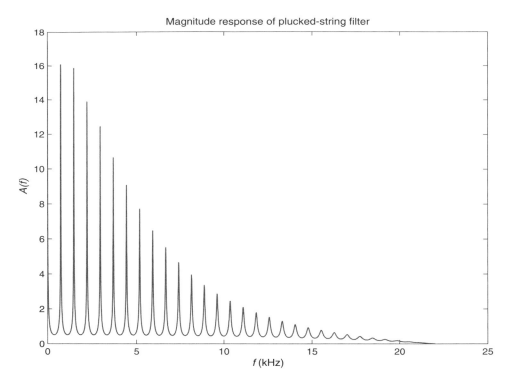

Figure 8.2 Magnitude Response of Plucked-string Filter: $L = 59$, $c = 0.8272$, and $r = 0.999$

To make the discussion specific, suppose the sampling frequency is $f_s = 44.1$ kHz which is a value commonly used in digital recording. Next, suppose the desired location of the first resonance is $F_0 = 740$ Hz. Applying Eq. 8.9 then yields $L = 59$ and $c = 0.8272$. If the feedback attenuation factor is set to $r = 0.999$, then this results in the plucked-string filter magnitude response shown in Figure 8.2. Interestingly enough, the resonant frequencies are almost, but not quite, harmonically related (Steiglitz, 1996). When Figure 8.2 is generated by running script *fig8_2* on the distribution CD, the sound made by the filter output is also played on the PC speakers. Give it a try, and let your ears be the judge!

8.1.2 Colored Noise

As a second example of an application of IIR filters, consider the problem of creating a test signal with desired spectral characteristics. One of the most popular test signals is white noise because it contains power at all frequencies. In particular, an N-point white noise signal $v(k)$ with average power P_v has the following power-density spectrum.

$$S_N(f) \approx P_v, \quad 0 \le f \le \frac{f_s}{2} \tag{8.10}$$

The term *white* arises from the fact $x(k)$ contains power at all frequencies just as white light is composed of all colors. If the natural frequencies of a linear system

Figure 8.3 Generation of Colored Noise $x(k)$ from White Noise $v(k)$

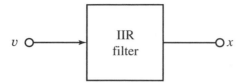

are confined to an interval $[F_0, F_1]$, then it is more efficient to excite these natural modes with a signal that has its power restricted to the frequency range $[F_0, F_1]$. Since only a subset of the entire range of frequencies is represented, this type of signal is sometimes referred to as *colored* noise. For example, low-frequency noise might be thought of as *red* and high-frequency noise as *blue*. The desired power density spectrum for colored noise in the interval $[F_0, F_1]$ is

$$S_N(f) = \begin{cases} 0, & 0 \le f < F_0 \\ P_x, & F_0 \le f \le F_1 \\ 0, & F_1 \le f < f_s/2 \end{cases} \tag{8.11}$$

A simple scheme for generating colored noise with a desired power-density spectrum is shown in Figure 8.3. The basic approach is to start with white noise, which is easily constructed, and then pass it through a digital filter that removes the undesired spectral components. The digital filter can be either an IIR filter or an FIR filter. However, because a linear phase response is not crucial, the colored noise can be generated more efficiently using an IIR filter.

The ideal filter is a bandpass filter with a low-frequency cutoff of F_0, a high-frequency cutoff of F_1, and a passband gain of one as shown in Figure 8.4. The removal of signal power below F_0 Hz and above F_1 Hz means than the average power of the colored noise $x(k)$ will be the following fraction of the average power of the white noise $v(k)$.

$$P_x = \frac{2(F_1 - F_0)P_v}{f_s} \tag{8.12}$$

To make the example specific, suppose the sampling frequency is $f_s = 800$ Hz, and the desired frequency band for the colored noise is $[F_0, F_1] = [150,300]$ Hz. Let $v(k)$ consist of $N = 2048$ samples of white noise uniformly distributed over $[-1, 1]$. Using a design technique covered later in the chapter, a tenth-order elliptic bandpass filter can be constructed. When the white noise is passed through this IIR filter, this results in colored noise with the power-density spectrum shown in Figure 8.5. It is evident that there is still a small amount of power outside the desired range $[150,300]$ Hz. This is due to the fact that practical filters, unlike ideal filters, have a transition band between the passband and the stopband. The width of the transition band can be reduced by going to a higher-order filter, but it can not be eliminated completely.

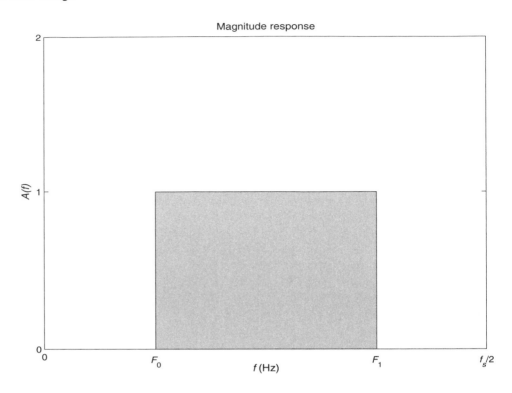

Figure 8.4 Magnitude Response of an Ideal Bandpass Filter

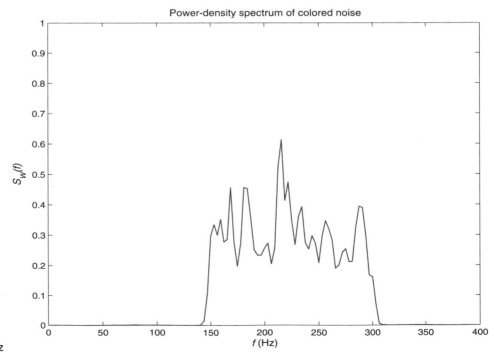

Figure 8.5 Power-density Spectrum of Colored Noise Created by Sending White Noise through a Bandpass Filter with $[F_0, F_1] = [150, 300]$ Hz

• • • • • • • • • • • • • • • •

8.2 Filter Design by Pole-zero Placement

In addition to the basic frequency-selective filters, there are a number of specialized filters that arise in applications. A direct design procedure is available for these filters based on gain matching and the judicious placement of poles and zeros.

8.2.1   Resonator

Recall that a bandpass filter is a filter that passes signals whose frequencies lie within an interval $[F_0, F_1]$. When the width of the passband is small in comparison with f_s, we say that the filter is a *narrowband* filter. An important limiting special case of a narrowband filter is a filter designed to pass a single frequency, $0 < F_0 < f_s/2$. Such a filter is called a *resonator* with a *resonant frequency* of F_0. Thus, the frequency response of an ideal resonator is

$$H_{\text{res}}(f) = \delta_0(f - F_0), \quad 0 \le f \le f_s/2 \tag{8.13}$$

Here $\delta_0(f)$ denotes the unit pulse $\delta(k)$ but with the integer argument k replaced by a real argument f. That is,

$$\delta_0(f) \triangleq \begin{cases} 1, & f = 0 \\ 0, & f \neq 0 \end{cases} \tag{8.14}$$

A resonator can be used to extract a single frequency component, or a very narrow range of frequencies, from a signal. A simple way to design a resonator is to place a pole near the point on the unit circle that corresponds to the resonant frequency, F_0. Recall that as the angles of the points along the top half of the unit circle go from 0 to π, the frequency f ranges from 0 to $f_s/2$. Thus, the angle corresponding to frequency F_0 is

Critical angle

$$\theta_0 = \frac{2\pi F_0}{f_s} \tag{8.15}$$

The radius of the pole must be less than one for the filter transfer function, $H_{\text{res}}(z)$, to be stable. Furthermore, if the coefficients of the denominator of $H_{\text{res}}(z)$ are to be real, then complex poles must occur in conjugate pairs. We can ensure that the resonator completely attenuates the two end frequencies, $f = 0$ and $f = f_s/2$, by placing zeros at $z = 1$ and $z = -1$, respectively. These constraints yield a resonator transfer function with the following factored form.

$$H_{\text{res}}(z) = \frac{b_0(z - 1)(z + 1)}{[z - r\exp(j\theta_0)][z - r\exp(-j\theta_0)]} \tag{8.16}$$

Using Euler's identity, the resonator transfer function can be simplified and expressed as a ratio of two polynomials.

$$H_{res}(z) = \frac{b_0(z^2 - 1)}{z^2 - r[\exp(j\theta_0) + \exp(-j\theta_0)]z + r^2}$$

$$= \frac{b_0(z^2 - 1)}{z^2 - r2\text{Re}\{\exp(j\theta_0)\}z + r^2}$$

$$= \frac{b_0(z^2 - 1)}{z^2 - 2r\cos(\theta_0)z + r^2} \tag{8.17}$$

There are two design parameters that remain to be determined, the pole radius r, and the gain factor b_0. To achieve a sharp filter that is highly selective, we need $r \approx 1$, but for stability, it is essential that $r < 1$. Suppose ΔF denotes the radius of the 3-dB passband of the filter. Thus, $|H_{res}(f)| \geq 1/\sqrt{2}$ for f in the range $[F_0 - \Delta F, F_0 + \Delta F]$. For a narrowband filter, $\Delta F \ll f_s$. In this case, the following approximation can be used to estimate the pole radius (Ifeachor and Jervis, 2002).

Pole radius
$$r \approx 1 - \frac{\Delta F \pi}{f_s} \tag{8.18}$$

The gain factor b_0 in the numerator of $H_{res}(z)$ is inserted to ensure that the passband gain is one. The value of z corresponding to the nominal center of the passband is $z_0 = \exp(j2\pi F_0 T)$. Setting $|H(z_0)| = 1$ in Eq. 8.17 and solving for b_0 yields the following expression for the gain factor.

Resonator filter gain
$$b_0 = \frac{|\exp(j2\theta_0) - 2r\cos(\theta_0)\exp(j\theta_0) + r^2|}{|\exp(j2\theta_0) - 1|} \tag{8.19}$$

The resonator transfer function, in terms of negative powers of z, is then

Resonator filter
$$H_{res}(z) = \frac{b_0(1 - z^{-2})}{1 - 2r\cos(\theta_0)z^{-1} + r^2z^{-2}} \tag{8.20}$$

Example 8.1 **Resonator Filter**

Suppose the sampling frequency is $f_s = 1200$ Hz, and it is desired to design a resonator to meet the following specifications.

$$F_0 = 200 \text{ Hz}$$

$$\Delta F = 6 \text{ Hz}$$

From Eq. 8.15, the pole angle is

$$\theta_0 = \frac{2\pi(200)}{1200}$$

$$= \frac{\pi}{3}$$

Clearly, $\Delta F \ll f_s$ in this case. Thus, from Eq. 8.18, the pole radius is

$$r = 1 - \frac{6\pi}{1200}$$

$$= 0.9843$$

Next, from Eq. 8.19, the scalar multiplier b_0 required for a passband gain of one is

$$b_0 = \frac{|\exp(j2\pi/3) - 2(0.9843)\cos(\pi/3)\exp(j\pi/3) + (0.9843)^2|}{|\exp(j2\pi/3) - 1|}$$

$$= 0.0156$$

Finally, from Eq. 8.20, the transfer function of the resonator is

$$H_{\text{res}}(z) = \frac{0.0156(1 - z^{-2})}{1 - 2(0.9843)\cos(\pi/3)z^{-1} + (0.9843)^2 z^{-2}}$$

$$= \frac{0.0156(1 - z^{-2})}{1 - 0.9843z^{-1} + 0.9688z^{-2}}$$

A plot of the poles and zeros of the resonator, obtained by running script *exam8_1* on the distribution CD, is shown in Figure 8.6. The complex conjugate pair of poles just inside the unit circle cause the magnitude response of the resonator to peak near $f = F_0$. This is evident from the plot of the resonator magnitude response shown Figure 8.7.

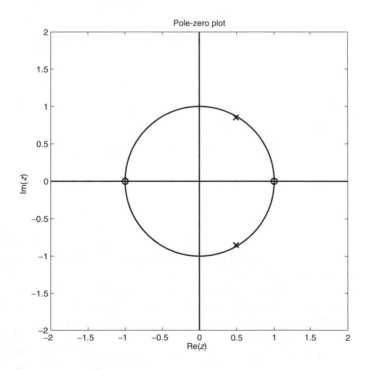

Figure 8.6   Poles and Zeros of a Resonator

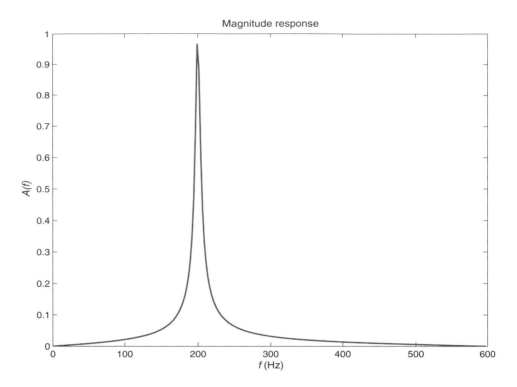

Figure 8.7 Magnitude Response of a Resonator with $F_0 = 200$ Hz

8.2.2 Notch Filter

Another specialized filter that occurs in applications is the *notch filter*. A notch filter can be thought of as a limiting special case of a bandstop filter where the width of the stopband goes to zero. That is, a notch filter is a filter designed to remove a single frequency, F_0, called the *notch frequency*. The frequency response of an ideal notch filter is

$$H_{\text{notch}}(f) = 1 - \delta_0(f - F_0), \quad 0 \le f \le f_s/2 \tag{8.21}$$

To design a notch filter, we start by placing a zero at the point on the unit circle that corresponds to the notch frequency, F_0. The angle, θ_0, associated with frequency F_0 was given previously in Eq. 8.15. Placing a zero at $z_0 = \exp(j2\pi F_0 T)$ ensures that $H_{\text{notch}}(F_0) = 0$, as desired. However, it does not leave us with a design parameter to control the 3-dB bandwidth of the stopband. This is achieved by placing a pole at the same angle θ_0, but just inside the unit circle with $r < 1$. Since the poles and zeros must occur in complex-conjugate pairs (for real coefficients) this yields a transfer function for a notch filter with the following factored form.

$$H_{\text{notch}}(z) = \frac{b_0[z - \exp(j\theta_0)][z - \exp(-j\theta_0)]}{[z - r\exp(j\theta_0)][z - r\exp(-j\theta_0)]} \tag{8.22}$$

Using Euler's identity, the transfer function in Eq. 8.22 again can be simplified and expressed as a ratio of two polynomials with the final result being

$$H_{\text{notch}}(z) = \frac{b_0(z^2 - 2\cos(\theta_0)z + 1)}{z^2 - 2r\cos(\theta_0)z + r^2} \tag{8.23}$$

There are two design parameters to be specified, the pole radius r, and the gain factor b_0. Just as with the resonator, to achieve a sharp filter that is highly selective we need $r \approx 1$, but for stability and to avoid canceling the zero it essential that $r < 1$. If ΔF denotes the radius of the 3-dB stopband of the filter, and if $\Delta F \ll f_s$, then the approximation for r in Eq. 8.18 can be used.

The gain factor b_0 in the numerator of $H_{\text{notch}}(z)$ is inserted to ensure that the passband gain is one. The passband includes both $f = 0$ and $f = f_s/2$. To set the DC gain to one we set $|H_{\text{notch}}(1)| = 1$ in Eq. 8.23 and solve for b_0 which yields

Notch filter gain

$$b_0 = \frac{|1 - 2r\cos(\theta_0) + r^2|}{2|1 - \cos(\theta_0)|} \tag{8.24}$$

Alternatively, the high-frequency gain can be set to one using $|H_{\text{notch}}(-1)| = 1$. The notch filter transfer function, in terms of negative powers of z, is then

Notch filter

$$H_{\text{notch}}(z) = \frac{b_0[1 - 2\cos(\theta_0)z^{-1} + z^{-2}]}{1 - 2r\cos(\theta_0)z^{-1} + r^2z^{-2}} \tag{8.25}$$

Example 8.2   **Notch Filter**

Suppose the sampling frequency is $f_s = 2400$ Hz, and it is desired to design a notch filter to meet the following specifications.

$$F_0 = 800 \text{ Hz}$$

$$\Delta F = 18 \text{ Hz}$$

From Eq. 8.15, the angle of the zero is

$$\theta_0 = \frac{2\pi(800)}{2400}$$

$$= \frac{2\pi}{3}$$

In this case, $\Delta F \ll f_s$. Thus, from Eq. 8.18, the pole radius is

$$r = 1 - \frac{18\pi}{2400}$$

$$= 0.9764$$

Next, from Eq. 8.24, the scalar multiplier b_0 required for a DC passband gain of one is

$$b_0 = \frac{|1 - 2(0.9764)\cos(2\pi/3) + (0.9764)^2|}{2|1 - \cos(2\pi/3)|}$$

$$= 0.9766$$

Finally, from Eq. 8.25, the transfer function of the notch filter is

$$H_{\text{notch}}(z) = \frac{0.9766[1 - 2\cos(2\pi/3)z^{-1} + z^{-2}]}{1 - 2(0.9764)\cos(2\pi/3)z^{-1} + (0.9764)^2 z^{-2}}$$

$$= \frac{0.9766(1 + z^{-1} + z^{-2})}{1 + 0.9764z^{-1} + 0.9534z^{-2}}$$

A plot of the poles and zeros of the notch filter, obtained by running script *exam8_2* on the distribution CD, is shown in Figure 8.8. Note how the poles "almost" cancel the zeros. The complex conjugate pair of zeros on the unit circle cause the magnitude response of the notch filter to go to zero at $f = F_0$. This is apparent from the plot of the notch filter magnitude response shown in Figure 8.9.

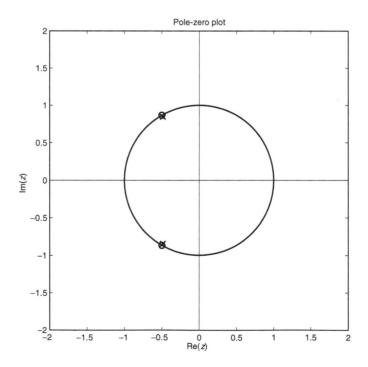

Figure 8.8 Poles and Zeros of a Notch Filter

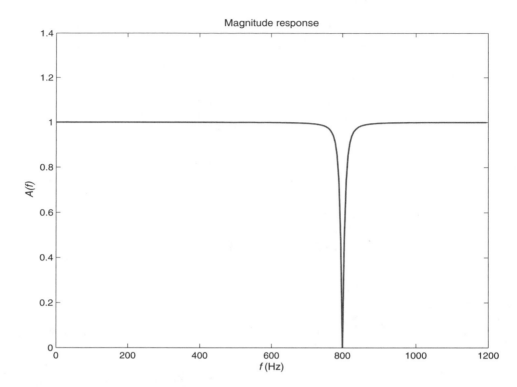

Figure 8.9 Magnitude Response of a Notch Filter with $F_0 = 800$ Hz

8.2.3   Comb Filters

Another specialized filter, one that includes a resonator as a special case, is the *comb filter*. A comb filter is a narrowband filter that has several equally spaced passbands starting at $f = 0$. In the limit, as the widths of the passbands go to zero, the comb filter is a filter that passes DC, a fundamental frequency F_0, and several of its harmonics. Thus, an ideal comb filter of order n has the following frequency response where $F_0 = f_s/n$.

$$H_{\text{comb}}(f) = \sum_{i=0}^{\text{floor}(n/2)} \delta_0(f - iF_0), \quad 0 \le f \le f_s/2 \tag{8.26}$$

Note that if n is even, then *floor*$(n/2) = n/2$. Consequently, for a comb filter of even order, there are $n/2 + 1$ resonant frequencies in the range $[0, f_s/2]$, and for a comb filter of odd order, there are only $(n - 1)/2 + 1$ resonant frequencies. Odd-order comb filters do not have a resonant frequency at $f = f_s/2$. Since the resonant frequencies are equally spaced, a comb filter of order n has a very simple transfer function, namely,

Comb filter

$$H_{\text{comb}}(z) = \frac{b_0}{1 - r^n z^{-n}} \tag{8.27}$$

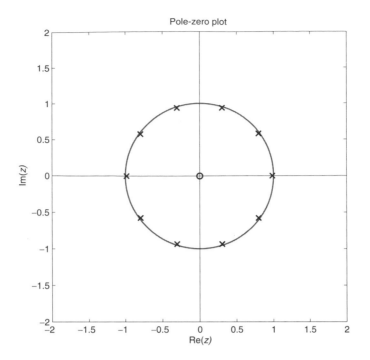

Figure 8.10 Poles and Zeros of a Comb Filter of Order $n = 10$

Thus, $H_{\text{comb}}(z)$ has n zeros at the origin, and the poles of $H_{\text{comb}}(z)$ correspond to the n roots of r^n. That is, the poles are equally spaced around a circle of radius $r < 1$. The distribution of poles for the even case $n = 10$ and $r = 0.9843$ is shown in Figure 8.10.

To achieve a highly selective comb filter, we require $r \approx 1$ using Eq. 8.18, but for stability, it is necessary that $r < 1$. The gain constant can be selected such that the passband gain at DC is one. Setting $|H_{\text{comb}}(1)| = 1$ in Eq. 8.27 and solving for b_0 yields

Comb filter gain

$$b_0 = 1 - r^n \tag{8.28}$$

A plot of the magnitude response of the comb filter in Figure 8.10, corresponding to the case $n = 10$, is shown in Figure 8.11. Here the sampling frequency was selected to be $f_s = 200$ Hz, and the 3-dB radius for slecting r was $\Delta F = 1$ Hz.

Just as the comb filter is a generalization of the resonator, there is a generalization of the notch filter called an *inverse comb filter* that eliminates DC, the fundamental notch frequency F_0, and several of its harmonics. Thus, an ideal inverse comb filter of order n has the following frequency response, where $F_0 = f_s/n$.

$$H_{\text{inv}}(f) = 1 - \sum_{i=0}^{\text{floor}(n/2)} \delta_0(f - iF_0), \quad 0 \le f \le f_s/2 \tag{8.29}$$

In addition to having zeros equally spaced around the unit circle, the inverse comb filter also has equally spaced poles just inside the unit circle. Thus, the transfer

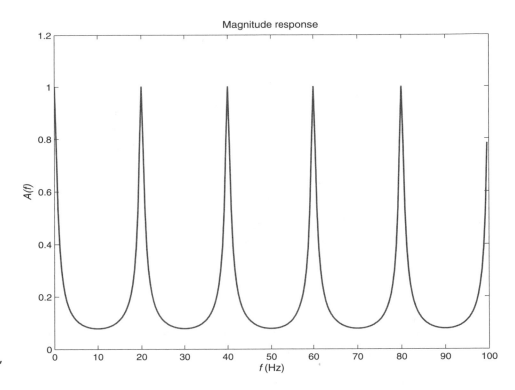

Figure 8.11 Magnitude Response of a Comb Filter with $n = 10$, $f_s = 200$ Hz, and $\Delta F = 1$ Hz

function of an inverse comb filter of order n has the following form.

Inverse comb filter

$$H_{\text{inv}}(z) = \frac{b_0(1 - z^{-n})}{1 - r^n z^{-n}} \qquad (8.30)$$

The distribution of poles and zeros for the odd-order case, $n = 11$ and $r = 0.9857$, is shown in Figure 8.12. Note how there are no poles or zeros at $z = -1$, because n is odd.

To pass the frequencies between the harmonics of F_0, we need $r \approx 1$ using Eq. 8.18, but for stability and to avoid pole-zero cancellation, it is necessary that $r < 1$. The gain constant can be selected such that the passband gain at $f = F_0/2$ is one, where $F_0 = f_s/n$. The point on the unit circle corresponding to the middle of the first passband is $z_1 = \exp(j\pi/n)$. Setting $|H_{\text{inv}}(z_1)| = 1$ in Eq. 8.30 and solving for b_0 yields

Inverse comb filter gain

$$b_0 = \frac{1 + r^n}{2} \qquad (8.31)$$

A plot of the magnitude response of the inverse comb filter in Figure 8.12, corresponding to the case $n = 11$, is shown in Figure 8.13. Here the sampling frequency was selected to be $f_s = 2200$, and the 3-dB radius for selecting r was $\Delta F = 10$ Hz.

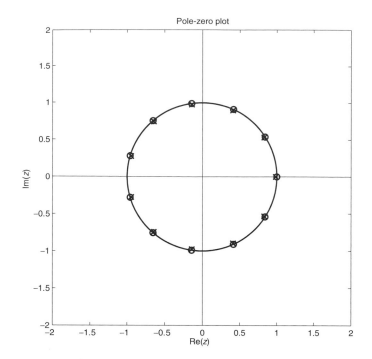

Figure 8.12 Poles and Zeros of an Inverse Comb Filter of Order $n = 11$

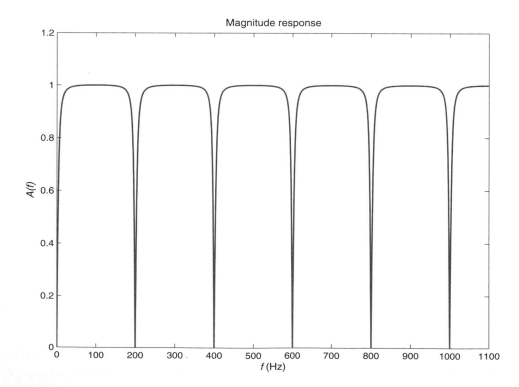

Figure 8.13 Magnitude Response of an Inverse Comb Filter with $n = 11$, $f_s = 2200$ Hz, and $\Delta F = 10$ Hz

There are a number of applications of comb filters and inverse comb filters. For example, suppose the input signal is a noise-corrupted periodic signal with a known fundamental frequency of F_0. Let $H_{\text{comb}}(z)$ be a comb filter of order n, and suppose we pick the sampling frequency such that

$$f_s = nF_0 \tag{8.32}$$

The first resonant frequency of the comb filter is then F_0. Consequently, the comb filter can be used to extract the first $n/2$ harmonics of the periodic component of the input signal in this case.

Another application arises when an input signal includes a periodic noise component that needs to be removed. For example, a sensitive acoustic or a biomedical measurement may be corrupted by the 60 Hz "hum" of overhead fluorescent lights. In this case, an inverse comb filter can be used to remove this periodic noise.

MATLAB Toolbox

The FDSP toolbox contains the following functions for designing IIR filters by gain matching and pole-zero placement.

```
[b,a] = f_iirres    (F_0,Delta_F,fs);    % digital resonator filter
[b,a] = f_iirnotch  (F_0,Delta_F,fs);    % digital notch filter
[b,a] = f_iircomb   (n,Delta_F,fs);      % digital comb filter
[b,a] = f_iirinv    (n,Delta_F,fs);      % digital inverse comb filter
```

Functions *f_iirres* and *f_iirnotch* design resonators and notch filters, respectively. The more general functions *f_iircomb* and *f_iirinv* are used to design comb filters and inverse comb filters, respectively. On entry to *f_iirres* and *f_iirnotch*, F_0 is the resonant or the notch frequency, *Delta_F* is the 3-dB radius of the peak or the notch, and *fs* is the sampling frequency. It is assumed that $Delta\_F \ll f_s/2$. On exit from *f_iirres* and *f_iirnotch*, *b* and *a* are vectors of length 3 containing the filter numerator and denominator coefficients, respectively.

The calling arguments for *f_iircomb* and *f_iirinv* are the same as those for *f_iirres* and *f_iirnotch*, except that F_0 is replaced by the filter order, n. Here it is assumed that $Delta\_F \ll f_s/(2n)$. On exit from *f_iircomb* and *f_iirinv*, *b* and *a* are vectors of length $n+1$ containing the filter numerator and denominator coefficients, respectively.

• • • • • • • • • • • • • • •

8.3 Filter Design Parameters

The most widely used design procedure for digital IIR filters starts with a normalized lowpass analog filter called a *prototype* filter. One then transforms the prototype filter into a desired frequency-selective digital filter. There are four classical families of analog lowpass filters that are typically used as prototype filters, and each family

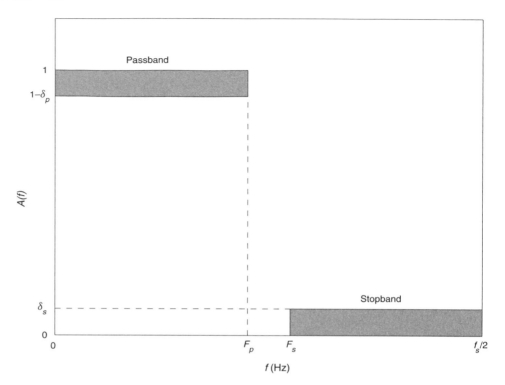

Figure 8.14 Design Specifications of an IIR Lowpass Filter

is optimal in some sense. Before we examine them, it is helpful to introduce two filter design parameters that are derived from the filter design specifications. For convenience, the lowpass filter design specifications first introduced in Chapter 5 are displayed in Figure 8.14. Recall that F_p is the passband cutoff frequency, F_s is the stopband cutoff frequency, δ_p is the passband ripple, and δ_s is the stopband attenuation. Left unspecified is the magnitude response in the transition band which is of width $B = F_s - F_p$. Let $A_a(f)$ denote the desired analog magnitude response. Then the passband and stopband specifications can be represented by separate inequalities as follows.

$$1 - \delta_p \leq A_a(f) \leq 1, \quad 0 \leq |f| \leq F_p \tag{8.33a}$$

$$0 \leq A_a(f) \leq \delta_s, \quad F_s \leq |f| < \infty \tag{8.33b}$$

The design specifications in Eq. 8.33 are linear design specifications. In order to better reveal the amount of attenuation in the stopband, the magnitude response is sometimes plotted in units of dB using the logarithmic scale, $A(f) = 20 \log_{10}\{|H(f)|\}$. The logarithmic equivalents of the passband ripple and stopband attenuation are

$$A_p = -20 \log_{10}(1 - \delta_p) \text{ dB} \tag{8.34a}$$

$$A_s = -20 \log_{10}(\delta_s) \text{ dB} \tag{8.34b}$$

For example, a stop attenuation of $\delta_s = 0.01$ corresponds to $A_s = 40$ dB, and each reduction by a factor of ten generates an increase of 20 dB. The design procedures for the classical analog filters are based on linear specifications. However, if the user instead starts out with logarithmic specifications, then they can be converted to equivalent linear specifications as follows.

$$\delta_p = 1 - 10^{-A_p/20} \tag{8.35a}$$

$$\delta_s = 10^{-A_s/20} \tag{8.35b}$$

The development of design formulas for the classical analog filters can be streamlined by introducing the following two filter design parameters that are obtained from the filter design specifications (Porat, 1997).

Selectivity factor

$$r \triangleq \frac{F_p}{F_s} \tag{8.36a}$$

Discrimination factor

$$d \triangleq \left[\frac{(1 - \delta_p)^{-2} - 1}{\delta_s^{-2} - 1} \right]^{1/2} \tag{8.36b}$$

The first parameter $0 < r < 1$ is called the *selectivity factor*. Note that for an ideal filter there is no transition band so $F_s = F_p$. Consequently, for an ideal filter, the selectivity factor is $r = 1$ whereas for a practical filter, $r < 1$. The second parameter, $d > 0$, is called the *discrimination factor*. Observe that when $\delta_p = 0$, the numerator in Eq. 8.36b goes to zero, so $d = 0$. Similarly when $\delta_s = 0$, the denominator in Eq. 8.36b goes to infinity so again $d = 0$. Hence, for an ideal filter, the discrimination factor is $d = 0$, whereas as for a practical filter $d > 0$.

Example 8.3   **Filter Design Parameters**

As a simple illustration of filter design parameters, consider the problem of designing a lowpass analog filter to meet the following logarithmic design specifications.

$$(F_p, F_s) = (400, 500) \text{ Hz}$$

$$(A_p, A_s) = (0.5, 35) \text{ dB}$$

First, we convert from logarithmic to linear specifications. From Eq. 8.35a, the required passband ripple is

$$\delta_p = 1 - 10^{-0.5/20}$$

$$= 0.0559$$

Similarly, from Eq. 8.35b, the required stopband attenuation is

$$\delta_s = 10^{-35/20}$$

$$= 0.0178$$

For this filter, the width of the transition band is $B = 100$ Hz. Thus, the selectivity factor r is less than one, and from Eq. 8.36a we have

$$r = 0.8$$

Finally, both the passband ripple and the stopband attenuation are positive. Thus, the discrimination factor d will also be positive. From Eq. 8.36b, we have

$$d = \left[\frac{(1 - 0.0559)^{-2} - 1}{(0.0178)^{-2} - 1} \right]^{1/2}$$

$$= 0.006213$$

Classical analog filters can be designed by starting with a desired magnitude response and working backwards to determine the poles, zeros, and gain. To apply this reverse procedure, it is necessary to first develop a relationship between an analog transfer function, $H_a(s)$, and the square of its magnitude response. Recall that the frequency response of an analog filter is defined as follows.

$$H_a(f) = H_a(s)|_{s=j2\pi f} \tag{8.37}$$

The frequency response, $H_a(f)$, can be expressed in polar form as $H_a(f) = A_a(f) \exp[j\phi_a(f)]$, where $A_a(f)$ is the magnitude response, and $\phi_a(f)$ is the phase response. Since the coefficients of $H_a(s)$ are real, the square of the magnitude response can be expressed as follows where $H_a^*(s)$ denotes the complex conjugate of $H_a(s)$.

$$A_a^2(f) = |H_a(f)|^2$$

$$= |H_a(s)|^2_{s=j2\pi f}$$

$$= \{H_a(s)H_a^*(s)\}|_{s=j2\pi f}$$

$$= \{H_a(s)H_a(-s)\}|_{s=j2\pi f} \tag{8.38}$$

Instead of replacing s by $j2\pi f$ on the right-hand side of Eq. 8.38, we can replace f by $s/(j2\pi)$ on the left-hand side. This yields the following fundamental relationship between the transfer function and its squared magnitude response.

Squared magnitude response

$$H_a(s)H_a(-s) = A_a^2(f)|_{f=s/(j2\pi)} \tag{8.39}$$

Each of the classical analog filters can be characterized by its squared magnitude response. The relationship in Eq. 8.39 is then employed to synthesize the filter transfer function.

● ● ● ● ● ● ● ● ● ● ● ● ● ● ●●●●

8.4 Classical Analog Filters

The most popular design procedure for digital IIR filters is to start with a normalized lowpass analog filter and then transform it into an equivalent frequency-selective digital filter.

8.4.1 Butterworth Filters

The first family of analog filters that we consider are the Butterworth filters. A Butterworth filter of order n is a lowpass analog filter with the following squared magnitude response.

Butterworth squared magnitude response

$$A_a^2(f) = \frac{1}{1 + (f/F_c)^{2n}} \qquad (8.40)$$

Notice from Eq. 8.40 that $A_a^2(F_c) = 0.5$. Frequency F_c is called the *3-dB cutoff frequency* because

$$20 \log_{10}\{A_a(F_c)\} \approx -3 \text{ dB} \qquad (8.41)$$

A plot of the squared magnitude response for a Butterworth filter of order $n = 4$ with a 3-dB cutoff frequency of $F_c = 1$ Hz is shown in Figure 8.15.

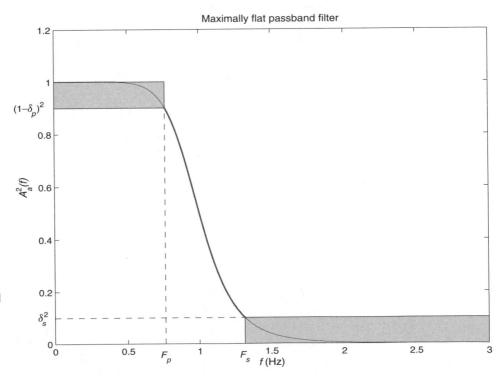

Figure 8.15 Squared Magnitude Response of a Lowpass Butterworth Filter of Order $n = 4$ with $F_c = 1$ Hz

The poles of $H_a(s)$ can be recovered from the squared magnitude response. Using Eq. 8.40 and the relationship in Eq. 8.39, we have

$$
\begin{aligned}
H_a(s)H_a(-s) &= A_a^2(f)|_{f=s/(j2\pi)} \\
&= \frac{1}{1 + [s/(j2\pi F_c)]^{2n}} \\
&= \frac{(j2\pi F_c)^{2n}}{s^{2n} + (j2\pi F_c)^{2n}} \\
&= \frac{(-1)^n (2\pi F_c)^{2n}}{s^{2n} + (-1)^n (2\pi F_c)^{2n}}
\end{aligned}
\tag{8.42}
$$

Thus the poles, p_k, of $H_a(s)H_a(-s)$ lie on a circle of radius $2\pi F_c$ at angles θ_k where

$$
\theta_k = \frac{(2k + 1 + n)\pi}{2n}, \qquad 0 \le k < 2n
\tag{8.43a}
$$

$$
p_k = 2\pi F_c \exp(j\theta_k), \qquad 0 \le k < 2n
\tag{8.43b}
$$

A *normalized* lowpass filter is a lowpass filter whose cutoff frequency is $F_c = 1/(2\pi)$ Hz, which corresponds to a radian cutoff frequency of $\Omega_c = 1$ rad/sec. For a normalized lowpass Butterworth filter, the poles are equally spaced around the unit circle with a separation of π/n radians. Two cases are illustrated in Figure 8.16, corresponding to an odd order ($n = 5$) and an even order ($n = 6$). Note that in either case, the first n poles all lie in the left half of the complex plane. We will associate the left-half plane poles $\{p_0, p_1, \ldots, p_{n-1}\}$ with $H_a(s)$ in Eq. 8.42, and the right-half

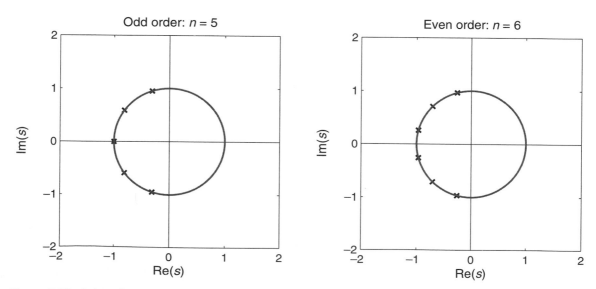

Figure 8.16   Poles of Normalized Lowpass Butterworth Filters

plane poles $\{p_n, p_{n+1}, \ldots, p_{2n-1}\}$ with $H_a(-s)$. This way, filter $H_a(s)$ is guaranteed to be stable. The transfer function of an nth order lowpass Butterworth filter with cutoff frequency F_c is then

Butterworth filter

$$H_a(s) = \frac{(2\pi F_c)^n}{(s - p_0)(s - p_1) \cdots (s - p_{n-1})} \tag{8.44}$$

Butterworth filters have a number of useful qualitative properties. One is that the magnitude response decreases *monotonically* starting from $A_a(0) = 1$. For high frequencies, the asymptotic attenuation of an nth order filter is $20n$ dB per decade. That is,

$$20 \log_{10}\{A_a(10f)\} \approx 20 \log_{10}\{A_a(f)\} - 20n \ \text{dB}, \quad f \gg F_c \tag{8.45}$$

Perhaps the most noteworthy property of Butterworth filters is that the first $2n - 1$ derivatives of the squared magnitude response are zero at $f = 0$. Consequently, among filters of order n, the Butterworth filter magnitude response is as flat as possible at $f = 0$. For this reason, Butterworth filters are called *maximally flat filters*.

The two design parameters available with Butterworth filters are the filter order n and the cutoff frequency F_c. Suppose it is desired to develop a lowpass Butterworth filter satisfying the linear design specification in Figure 8.14. Then from Eq. 8.40, the passband and stopband specification constraints are

$$\frac{1}{1 + (F_p/F_c)^{2n}} = (1 - \delta_p)^2 \tag{8.46a}$$

$$\frac{1}{1 + (F_s/F_c)^{2n}} = \delta_s^2 \tag{8.46b}$$

The passband constraint in Eq. 8.46a and the stopband constraint in Eq. 8.46b can each be solved for F_c^{2n}. By setting these two expressions for F_e^{2n} equal to one another, we eliminate the cutoff frequency parameter, F_c. Solving the resulting equation for the filter order n, and using the design parameters in Eq. 8.36, then yields

Butterworth filter order

$$n = \text{ceil} \left[\frac{\ln(d)}{\ln(r)} \right] \tag{8.47}$$

Thus, the required filter order can be expressed directly in terms of the discrimination factor d and the selectivity factor r defined in Eq. 8.36. The function *ceil* in Eq. 8.47 is used because the expression in the square brackets may not be an integer. The *ceil* function rounds up to the next integer value. Since n is rounded up, typically this means that the passband and the stopband specifications will be exceeded (more than met). To meet the passband constraint exactly, we solve Eq. 8.46a for F_c, which yields

$$F_{cp} = \frac{F_p}{[(1 - \delta_p)^{-2} - 1]^{1/(2n)}} \tag{8.48}$$

In this case, the stopband constraint is exceeded. Similarly, to meet the stopband constraint exactly, we solve Eq. 8.46b for F_c, which yields the slightly simpler expression

$$F_{cs} = \frac{F_s}{(\delta_s^{-2} - 1)^{1/(2n)}} \tag{8.49}$$

In this case, the passband constraint is exceeded. Finally, we can exceed both constraints (assuming the expression for n is not already an integer) if the cutoff frequency is set to the average.

$$F_c = \frac{F_{cp} + F_{cs}}{2} \qquad \text{use } F_c = \sqrt{F_{cp} F_{cs}} \tag{8.50}$$

The design formulas in Eq. 8.47 through Eq. 8.50 are all based on the linear design specifications in Figure 8.14. If the logarithmic design specifications are used instead, then Eq. 8.35 should be applied first to convert A_p and A_s into δ_p and δ_s, respectively.

Example 8.4 **Butterworth Filter**

As an illustration of the use of the design formulas, consider the problem of designing a lowpass Butterworth filter to meet the following linear design specifications.

$$F_p = 1000 \text{ Hz}$$

$$F_s = 2000 \text{ Hz}$$

$$\delta_p = 0.05$$

$$\delta_s = 0.05$$

From Eq. 8.36, the selectivity and discrimination factors are

$$r = 0.5$$

$$d = \left(\frac{0.95^{-2} - 1}{0.05^{-2} - 1} \right)^{1/2}$$

$$= 0.0165$$

Thus, from Eq. 8.47 the minimum filter order is

$$n = \text{ceil}\left[\frac{\ln(0.0165)}{\ln(0.5)} \right]$$

$$= \text{ceil}(5.9253)$$

$$= 6$$

Next, from Eq. 8.48, the cutoff frequency for which the passband specification is met exactly is

$$F_{cp} = \frac{1000}{(0.95^{-2} - 1)^{1/12}}$$

$$= 1203.8 \text{ Hz}$$

Similarly, from Eq. 8.49, the cutoff frequency for which the stopband specification is met exactly is

$$F_{cs} = \frac{2000}{(0.05^{-2} - 1)^{1/12}}$$

$$= 1209.0 \text{ Hz}$$

Any cutoff frequency in the range $F_{cp} \leq F_c \leq F_{cs}$ will meet or exceed both specifications. For example, $F_c = 1206$ Hz will suffice. Using Eq. 8.34, the equivalent logarithmic passband and stopband specifications are $A_p = 0.4455$ dB, and $A_s = 26.02$ dB.

Butterworth filter transfer functions can be designed directly using Eqs. 8.43 and 8.44. There is also an alternative table-based approach that works well for low-order filters. It starts with a normalized lowpass Butterworth filter, and then makes use of a simple *frequency transformation*. Let $H_n(s)$ denote the transfer function of a normalized nth-order Butterworth lowpass filter, a filter with a 3-dB radian cutoff frequency of $\Omega_c = 1$ rad/sec.

$$H_n(s) = \frac{a_n}{s^n + a_1 s^{n-1} + \cdots + a_n} \tag{8.51}$$

The coefficients of the denominator polynomials for the first few normalized Butterworth lowpass filters are summarized in Table 8.1.

Next, let F_c denote the desired 3-dB cutoff frequency in Hz. The transfer function $H_a(s)$ can be obtained by replacing s in Eq. 8.51 with s/Ω_c where $\Omega_c = 2\pi F_c$. Thus, if $a(s)$ is as given in Table 8.1, then an nth-order Butterworth lowpass filter with radian cutoff frequency Ω_c has the following transfer function.

Butterworth filter using Table 8.1

$$H_a(s) = \frac{\Omega_c^n a_n}{s^n + \Omega_c a_1 s^{n-1} + \cdots + \Omega_c^n a_n} \tag{8.52}$$

Table 8.1: ▶
Denominator Coefficients of Normalized Lowpass Butterworth Filters

n	a_0	a_1	a_2	a_3	a_4	a_5	a_6	a_7	a_8
1	1	1	0	0	0	0	0	0	0
2	1	1.414214	1	0	0	0	0	0	0
3	1	2	2	1	0	0	0	0	0
4	1	2.613126	3.414214	2.613126	1	0	0	0	0
5	1	3.236068	5.236068	5.236068	3.236068	1	0	0	0
6	1	3.863703	7.464102	9.14162	7.464102	3.863703	1	0	0
7	1	4.493959	10.09783	14.59179	14.59179	10.09783	4.493959	1	0
8	1	5.125831	13.13707	21.84615	25.68836	21.84615	13.13707	5.125831	1

The replacement of s with s/Ω_c is an example of a frequency transformation which maps a normalized lowpass filter into a general lowpass filter. Later, we will examine other examples of frequency transformations which convert normalized lowpass filters into highpass, bandpass, and bandstop filters.

Example 8.5	**Butterworth Transfer Function**

As an illustration of the frequency-transformation method using Table 8.1, consider the problem of designing a transfer function for a third-order lowpass Butterworth filter with a cutoff frequency of $F_c = 10$ Hz. In this case, $\Omega_c = 20\pi$, and from Table 8.1 we have

$$H_a(s) = \frac{2.481 \times 10^5}{s^3 + 125.7s^2 + 7896s + 2.481 \times 10^5}$$

8.4.2   Chebyshev-I Filters

The magnitude responses of Butterworth filters are smooth and flat because of the maximally flat property. However, a drawback of the maximally flat property is that the transition band of a Butterworth filter is not as narrow as it could be. An effective way to decrease the width of the transition band is to allow ripples or oscillations in the passband or the stopband. The following Chebyshev-I filter of order n is designed to allow n ripples within the passband.

Chebyshev-I squared
magnitude response

$$A_a^2(f) = \frac{1}{1 + \epsilon^2 T_n^2(f/F_p)} \tag{8.53}$$

Here n is the filter order, F_p is the passband frequency, $\epsilon > 0$ is a *ripple factor* parameter, and $T_n(x)$ is a polynomial of degree n called a *Chebyshev polynomial* of the first kind. Recall from Chapter 6 that the Chebyshev polynomials can be generated recursively. The first two Chebyshev polynomials are $T_0(x) = 1$ and $T_1(x) = x$. The remaining polynomials are then computed from the previous two according to the recurrence relation

$$T_{k+1}(x) = 2x T_k(x) + T_{k-1}(x), \quad k \geq 1 \tag{8.54}$$

Therefore, $T_2(x) = 2x^2 - 1$ and so on. The first few Chebyshev polynomials are summarized in Table 8.2.

The Chebyshev polynomials have many interesting properties. For example, $T_n(x)$ is an odd function when n is odd, and an even function when n is even. Furthermore, $T_n(1) = 1$ for all n. For the purpose of filter design, the most important property is that $T_n(x)$ oscillates in the interval $[-1, 1]$ when $|x| \leq 1$ and $T_n(x)$ is monotonic when $|x| > 1$. This oscillation causes the square of the magnitude response of a Chebyshev-I filter to have ripples of equal size in the passband and be monotonically decreasing outside of the passband. A plot of the squared magnitude

n	$T_n(x)$
0	1
1	x
2	$2x^2 - 1$
3	$4x^3 - 3x$
4	$8x^4 - 8x^2 + 1$
5	$16x^5 - 20x^3 + 5x$
6	$32x^6 - 48x^4 + 18x^2 - 1$
7	$64x^7 - 112x^5 + 56x^3 - 7x$
8	$128x^8 - 256x^6 + 160x^4 - 32x^2 + 1$

response is shown in Figure 8.17 for the case $n = 4$ with $F_p = 1$ Hz. Note that because $T_n(1) = 1$, it follows from Eq. 8.53 that, at the edge of the passband,

$$A_a^2(F_p) = \frac{1}{1 + \epsilon^2} \tag{8.55}$$

Therefore, the ripple factor parameter, ϵ, specifies the size of the passband ripple of the filter. Setting $1/(1 + \epsilon^2) = (1 - \delta_p)^2$ and solving for ϵ, we find that a desired passband rippled δ_p can be achieved by setting the ripple factor parameter as follows.

Chebyshev-I ripple
factor

$$\epsilon = \left[(1 - \delta_p)^{-2} - 1 \right]^{1/2} \tag{8.56}$$

Notice from Figure 8.17 that not only are the n ripples in $A_a^2(f)$ confined to the passband, but they are all of the same amplitude, δ_p. Because of this characteristic, Chebyshev filters are called *equiripple* filters. More specifically, Chebyshev-I filters are optimal in the sense that they are equiripple in the passband. At the start of the passband, the squared magnitude response is either one or $1/(1 + \epsilon^2)$ depending on whether n is odd or even, respectively.

$$A_a^2(0) = \begin{cases} 1, & n \text{ odd} \\ \dfrac{1}{1 + \epsilon^2}, & n \text{ even} \end{cases} \tag{8.57}$$

Unlike the poles of a Butterworth filter that are on a circle, the poles of a Chebyshev-I filter are on an ellipse. The minor and major axes of the ellipse are

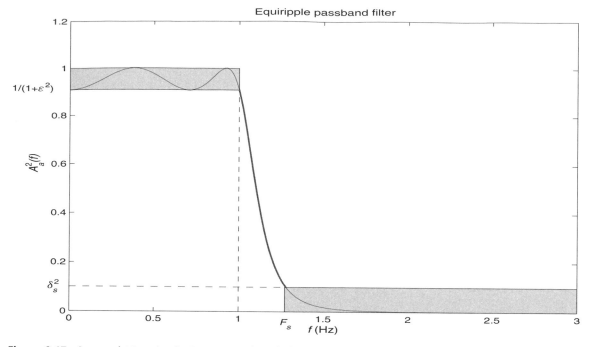

Figure 8.17 Squared Magnitude Response of a Chebyshev-I Lowpass Filter of Order $n = 4$ with $F_p = 1$ Hz

computed as follows where $F_0 = F_p$.

$$\alpha = \epsilon^{-1} + \sqrt{\epsilon^{-2} + 1} \qquad (8.58a)$$

$$r_1 = \pi F_0(\alpha^{1/n} - \alpha^{-1/n}) \qquad (8.58b)$$

$$r_2 = \pi F_0(\alpha^{1/n} + \alpha^{-1/n}) \qquad (8.58c)$$

The angles at which the poles are located are the same as those for a Butterworth filter, namely

$$\theta_k = \frac{(2k + 1 + n)\pi}{2n}, \quad 0 \le k < n \qquad (8.59)$$

If the poles are expressed in rectangular form as $p_k = \sigma_k + j\omega_k$, then the real and imaginary parts of the poles are

$$\sigma_k = r_1 \cos(\theta_k), \quad 0 \le k < n \qquad (8.60a)$$

$$\omega_k = r_2 \sin(\theta_k), \quad 0 \le k < n \qquad (8.60b)$$

The DC gain of the Chebyshev-I filter is $A_a(0)$ as given in Eq. 8.57. Let $\beta = (-1)^n p_0 p_1 \cdots p_{n-1}$. Then the transfer function of an nth-order Chebyshev-I filter is as follows.

Chebyshev-I filter

$$H_a(s) = \frac{\beta A_a(0)}{(s - p_0)(s - p_1) \cdots (s - p_{n-1})} \tag{8.61}$$

The only design parameter that remains to be determined for a Chebyshev-I filter is the filter order n. The minimal filter order depends on the filter design specifications. Using the selectivity and discrimination factors in Eq. 8.36, we have

Chebyshev filter order

$$n = \text{ceil} \left[\frac{\ln(d^{-1} + \sqrt{d^{-2} - 1})}{\ln(r^{-1} + \sqrt{r^{-2} - 1})} \right] \tag{8.62}$$

Unlike a Butterworth filter, a Chebyshev-I filter always meets the passband specification exactly as long as the ripple factor ϵ is chosen as in Eq. 8.56. The stopband specification will be exceeded when the expression inside the square brackets in Eq. 8.62 is less than the integer filter order n.

Example 8.6   **Chebyshev-I Filter**

As an illustration of the use of the Chebyshev design formulas, consider the problem of designing a lowpass Chebyshev-I filter to meet the same design specifications as in Example 8.4. From Example 8.4, the selectivity and discrimination factors are

$$r = 0.5$$

$$d = 0.0165$$

Then from Eq. 8.62 the minimum filter order is

$$n = \text{ceil} \left[\frac{\log[(0.0165)^{-1} + \sqrt{(0.0165)^{-2} - 1}]}{\log[(0.5)^{-1} + \sqrt{(0.5)^{-2} - 1}]} \right]$$

$$= \text{ceil}(3.6449)$$

$$= 4$$

Thus, we find that, by permitting ripples in the passband, the design specification can be met with a Chebyshev-I filter of order $n = 4$. This is in contrast to the maximally flat Butterworth filter response in Example 8.4 which required a filter of order $n = 6$ for the same specifications.

8.4.3 Chebyshev-II Filters

The use of the term Chebyshev-I filter suggests that there must also be a Chebyshev-II filter, and this is indeed the case. A Chebyshev-II filter is an equiripple filter that has the ripples in the stopband rather than the passband. This is achieved by having the following squared magnitude response.

Chebyshev-II squared
magnitude response

$$A_a^2(f) = \frac{\epsilon^2 T_n^2(F_s/f)}{1 + \epsilon^2 T_n^2(F_s/f)} \tag{8.63}$$

The design parameters for the Chebyshev-II filter are the same as those for the Chebyshev-I filter. However, in this case the magnitude response oscillates in the stopband and is monotonically decreasing outside the stopband. A plot of the squared magnitude response is shown in Figure 8.18 for the case $n = 4$ with $F_s = 1$ Hz. Recalling that $T_n(1) = 1$, it follows from Eq. 8.63 that at the edge of the stopband

$$A_a^2(F_s) = \frac{\epsilon^2}{1 + \epsilon^2} \tag{8.64}$$

In this case, the ripple factor parameter, ϵ, specifies the size of the stopband attenuation of the filter. Setting $\epsilon^2/(1 + \epsilon^2) = \delta_s^2$ and solving for ϵ, we find that a

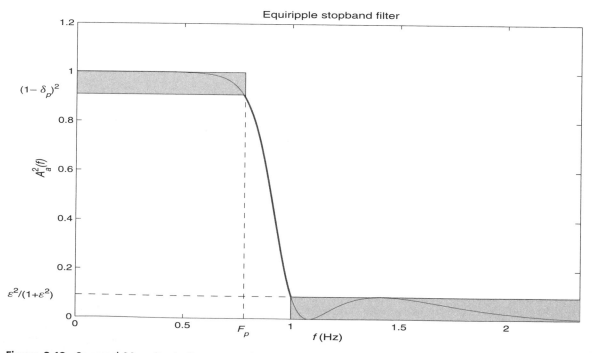

Figure 8.18 Squared Magnitude Response of a Chebyshev-II Lowpass Filter of Order $n = 4$ with $F_s = 1$ Hz

desired stopband attenuation δ_s can be achieved by setting the ripple factor parameter as follows.

Chebyshev-II ripple factor

$$\epsilon = \delta_s (1 - \delta_s^2)^{-1/2} \tag{8.65}$$

Again notice from Figure 8.18 that there are n ripples in $A_a^2(f)$ confined to the stopband, and they are all of the same amplitude, δ_s. Consequently, we say that Chebyshev-II filters are optimal in the sense that they are equiripple in the stopband. At the end of the stopband, the squared magnitude response is either zero or $\epsilon^2/(1 + \epsilon^2)$, depending on whether n is odd or even, respectively. That is,

$$\lim_{f \to \infty} A_a^2(f) = \begin{cases} 0, & n \text{ odd} \\ \dfrac{\epsilon^2}{1 + \epsilon^2}, & n \text{ even} \end{cases} \tag{8.66}$$

Because f/F_p in the Chebyshev-I magnitude response has been replaced by F_s/f in the Chebyshev-II magnitude response, the poles of the Chebyshev-II filter are located at the reciprocals of the poles of the Chebyshev-I filter. That is, if $p_k = \sigma_k + j\omega_k$ are the poles defined in Eq. 8.60, but with $F_0 = F_s$, then the Chebyshev-II poles are

$$q_k = \frac{(2\pi F_s)^2}{p_k}, \quad 0 \le k < n \tag{8.67}$$

Note from Eq. 8.63 that the numerator of $A_a^2(f)$ is not constant. This means that a Chebyshev-II filter also has either n or $n-1$ finite zeros. They are located along the imaginary axis at

$$r_k = \frac{j2\pi F_s}{\sin(\theta_k)}, \quad 0 \le k < n \tag{8.68}$$

When n is even, there are n finite zeros, as indicated in Eq. 8.68. However, when n is odd there are only $n-1$ finite zeros. This is because when n is odd we observe from Eq. 8.59 that $\theta_{(n-1)/2} = \pi$. Thus, $r_{(n-1)/2}$ is an infinite zero in this case.

For every Chebyshev-II filter, the DC gain is $A_a(0) = 1$. Let $\beta = q_0 q_1 \cdots q_{n-1}/(r_0 r_1 \cdots r_{n-1})$, where $r_{(n-1)/2}$ is left out if n is odd. Then the transfer function of an nth-order Chebyshev-II filter is

Chebyshev-II filter

$$H_a(s) = \frac{\beta(s - r_0)(s - r_1) \cdots (s - r_{n-1})}{(s - q_0)(s - q_1) \cdots (s - q_{n-1})} \tag{8.69}$$

Again, when n is odd, the numerator factor $(s - r_{(n-1)/2})$ is left out of Eq. 8.69. The minimum order for a Chebyshev-II filter is the same as the minimum order for a Chebyshev-I filter and is given in Eq. 8.62. Thus, the Chebyshev-II filter has a smaller transition band than the Butterworth filter, but like the Butterworth filter it is monotonic in the passband. A Chebyshev-II filter will always meet the stopband specification exactly as long as the ripple factor ϵ is chosen, as in Eq. 8.65. The passband specification will be exceeded when the expression inside the square brackets in Eq. 8.62 is less than the integer filter order n.

8.4.4   Elliptic Filters

The last classical lowpass analog filter that we consider is the elliptic or Cauer filter. Elliptic filters are filters that are equiripple in both the passband and the stopband. Therefore, elliptic IIR filters are similar to the optimal equiripple FIR filters constructed in Chapter 6 using the Parks-McLellan algorithm. The squared magnitude response of an nth-order elliptic filter is as follows.

Elliptic squared magnitude response

$$A_a^2(f) = \frac{1}{1 + \epsilon^2 U_n^2(f/F_p)} \tag{8.70}$$

Here U_n is an nth-order Jacobian elliptic function, also called a Chebyshev rational function (Porat, 1997). By permitting ripples in both the passband and the stopband, elliptic filters achieve very narrow transition bands. A plot of the squared magnitude response is shown in Figure 8.19 for the case $n = 4$ with $F_p = 1$ Hz.

The design parameters for an elliptic filter are similar to those of the Chebyshev filters. Since $U_n(1) = 1$ for all n, it follows from Eq. 8.70 that Eq. 8.55 holds. This in turn means that the passband specification can be met exactly if the ripple factor ϵ is set to satisfy Eq. 8.56. Elliptic filters are considerably more complex to analyze and design than Butterworth and Chebyshev filters. Finding the zeros and poles of an elliptic filter involves the iterative solution of nonlinear algebraic equations, equations whose terms include integrals (Parks and Burrus, 1987). Let

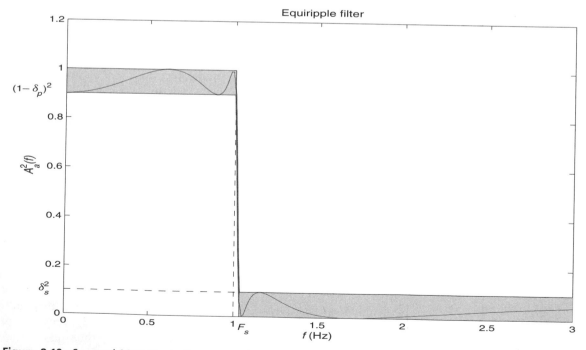

Figure 8.19   Squared Magnitude Response of an Elliptic Lowpass Filter of Order $n = 4$ with $F_p = 1$ Hz

us focus, instead, on the remaining design parameter: the minimal filter order. Let $g(x)$ denote the following function, which is called a complete elliptic integral of the first kind.

$$g(x) = \int_0^{\pi/2} \frac{d\theta}{\sqrt{1 - x^2 \sin^2(\theta)}} \tag{8.71}$$

Recalling the definitions of the selectivity and discrimination factors in Eq. 8.36, the required order for an elliptic filter to meet the design specifications is

Elliptic filter order

$$n = \text{ceil}\left[\frac{g(r^2)g(\sqrt{1 - d^2})}{g(\sqrt{1 - r^2})g(d^2)}\right] \tag{8.72}$$

Elliptic filters always meet the passband specification exactly as long as the ripple factor ϵ is chosen as in Eq. 8.56. The stopband specification will be exceeded when the expression in the square brackets in Eq. 8.72 is smaller than the filter order n.

Example 8.7 **Elliptic Filter**

For comparison, suppose we design an elliptic lowpass filter that meets the same specifications that were used in Examples 8.4 and 8.6. From Example 8.4, the selectivity and discrimination factors are

$$r = 0.5$$

$$d = 0.0165$$

The elliptic integral function in Eq. 8.71 can be evaluated numerically using the MATLAB function *ellipke*. Running script *exam8_7* on the distribution CD, the filter order is

$$n = \text{ceil}\left[\frac{g(0.25)g(\sqrt{1 - (0.0165)^2})}{g(0.75)g[(0.0165)^2]}\right]$$

$$= \text{ceil}(2.9061)$$

$$= 3$$

In this case, we find that by permitting ripples in both the passband and the stopband, the design specification can be met with an elliptic filter of order $n = 3$. This is in contrast to the Chebyshev filters which required $n = 4$ and the Butterworth filter which required $n = 6$.

Although the elliptic filter is the filter of choice if the only criteria is to minimize the order, the other classical analog filters are often used as well because they tend to have better (more linear) phase response characteristics. A summary of the essential characteristics of the classical analog filters is shown in Table 8.3.

Table 8.3: ▶
Summary of Classical
Analog Filters

Analog Filter	Passband	Stopband	Transition Band	Exact Specification
Butterworth	Monotonic	Monotonic	Broad	Either
Chebyshev-I	Equiripple	Monotonic	Narrow	Passband
Chebyshev-II	Monotonic	Equiripple	Narrow	Stopband
Elliptic	Equiripple	Equiripple	Very narrow	Passband

MATLAB Toolbox

The FDSP toolbox contains four functions for computing the coefficients of the classical lowpass analog filters, and a function for computing the analog frequency response.

```
[b,a] = f_butters  (F_p,F_s,delta_p,delta_s,n);   % analog Butterworth filter
[b,a] = f_chebyls  (F_p,F_s,delta_p,delta_s,n);   % analog Chebyshev-I filter
[b,a] = f_cheby2s  (F_p,F_s,delta_p,delta_s,n);   % analog Chebyshev-II filter
[b,a] = f_elliptics (F_p,F_s,delta_p,delta_s,n);  % analog elliptic filter
[H,f] = f_freqs    (b,a,N,fmax);                  % analog frequency response
```

The first four functions compute the coefficients of the Butterworth, Chebyshev-I, Chebyshev-II, and elliptic filters, respectively. On entry to each of these functions, F_p is the passband frequency, F_s is the stopband frequency, $delta\_p$ is the passband ripple, $delta\_s$ is the stopband attenuation, and n is an optional argument specifying the filter order. If n is not present, then the order is determined automatically using the lowest-order filter that will meet the specifications. When n is present, the filter may or may not meet all of the specifications. On exit from the four filter design functions, b is a vector of length $m + 1$ containing the numerator coefficients of the filter, and a is a vector of length $n + 1$ containing the denomonator coefficients of the filter with $a(1) = 1$.

$$H_a(s) = \frac{b_0 s^m + b_1 s^{m-1} + \cdots + b_m}{s^n + a_1 s^{n-1} + \cdots a_n} \tag{8.73}$$

The function f_freqs is analogous to f_freqz but used to compute the frequency response of a continuous-time transfer function, $H_a(s)$, where $m \leq n$. On entry to f_freqs, b is a vector of length $m + 1$ containing the numerator coefficients, a is a vector of length $n + 1$ containing the denominator coefficients with $a(1) = 1$, N is the number of points at which to evaluate the frequency response, and $fmax$ is the maximum frequency. The frequency response $H_a(f)$ in Eq. 8.37 is computed at N frequencies equally spaced over the interval $0 \leq f \leq fmax$. On exit from f_freqs, H is a vector of length N containing the samples of $H_a(f)$, and f is a vector of length N containing the frequencies at which H is evaluated. That is, $H(i) = H_a[f(i)]$ for $1 \leq i \leq N$.

8.5 **Bilinear-transformation Method**

Now that we have a collection of analog prototype filters to select from, the next problem is to transform an analog filter into an equivalent digital filter. Although a number of approaches are available, they all must satisfy the fundamental qualitative constraint that a stable analog filter, $H_a(s)$, transform into a stable digital filter, $H(z)$. The classical analog nth-order filters discussed in Section 8.4 each have n distinct poles p_k. Therefore, $H_a(s)$ can be written in partially factored form as

$$H_a(s) = \frac{b(s)}{(s - p_0)(s - p_1) \cdots (s - p_{n-1})} \tag{8.74}$$

A design technique that is highly effective is based on replacing integration with a discrete-time numerical approximation. Recall that an integrator has the continuous-time transfer function $H_0(s) = 1/s$. The time-domain input-output representation of an integrator is

$$y_a(t) = \int_0^t x_a(\tau) d\tau \tag{8.75}$$

Suppose we approximate the area under the curve $x_a(t)$ numerically using the samples $x(k) = x_a(kT)$, where T is the sampling interval. Consider the trapezoids formed by connecting the samples with straight lines, as shown in Figure 8.20. This is equivalent to using a piecewise-linear approximation to $x_a(t)$. Let $y(k)$ denote the approximation to the integral at time $t = kT$. The approximation at time kT is the approximation at time $(k - 1)T$ plus the area of the kth trapezoid. From Figure 8.20, the kth trapezoid has a width T and an average height $[x(k - 1) + x(k)]/2$. Thus,

$$y(k) = y(k - 1) + T \left[\frac{x(k - 1) + x(k)}{2} \right] \tag{8.76}$$

The approximation in Eq. 8.76 is called a trapezoid rule integrator. Taking the Z-transform of both sides of Eq. 8.76 and using the delay property, we get $(1 - z^{-1})Y(z) = (T/2)(1 + z^{-1})X(z)$. Thus, the transfer function of a trapezoid rule integrator is

$$H_0(z) = \frac{T}{2} \left(\frac{1 + z^{-1}}{1 - z^{-1}} \right) \tag{8.77}$$

If we replace $1/s$ by $H_0(z)$, this is equivalent to replacing s by $1/H_0(z)$. Therefore, we can approximate the integration process with a trapezoid rule integrator by making the following substitution for s in the analog filter transfer function $H_a(s)$.

$$H(z) = H_a(s)|_{s=g(z)} \tag{8.78}$$

Here $g(z) = 1/H_0(z)$. That is, the substitution $s = g(z)$ in Eq. 8.78 uses

**Bilinear
transformation**

$$g(z) = \frac{2}{T} \left(\frac{z - 1}{z + 1} \right) \tag{8.79}$$

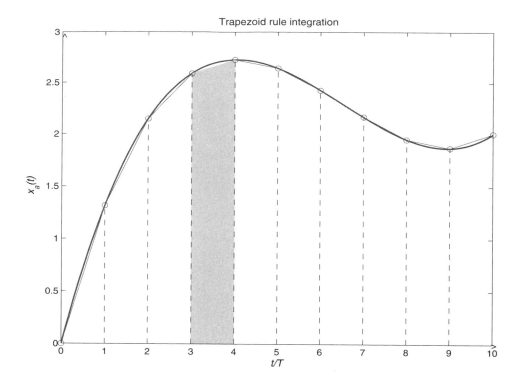

Figure 8.20 Piece-wise-linear Approximation of Integration using Trapezoids

The transformation from $H_a(s)$ to $H(z)$ in Eq. 8.78 is called a bilinear mapping, and filter designs based on it use the *bilinear-transformation* method. Before we design a filter using the bilinear transformation, it is helpful to examine the relationship between z and $s = g(z)$ in more detail. First, note that the transformation can be inverted. That is, we can solve Eq. 8.79 for z, which yields

$$z = \frac{2 + sT}{2 - sT} \tag{8.80}$$

Next, suppose s is expressed in rectangular form as $s = \sigma + j\omega$. Substituting this into Eq. 8.80 and taking the magnitude of both sides then yields

$$|z| = \frac{\sqrt{(2 + \sigma T)^2 + (\omega T)^2}}{\sqrt{(2 - \sigma T)^2 + (\omega T)^2}} \tag{8.81}$$

Notice that if $\sigma = 0$, then $|z| = 1$. Thus, the bilinear transformation maps the imaginary axis of the s plane into the unit circle of the z plane. Furthermore, if $\sigma < 0$, then $|z| < 1$, which means that the left half of the s plane is mapped into the interior of the unit circle in the z plane. Similarly, when $\sigma > 0$, the right half of the s plane is mapped into the exterior of the unit circle in the z plane. Therefore, the bilinear transformation satisfies the fundamental property that it is guaranteed to transform a stable analog filter $H_a(s)$ into a stable digital filter $H(z)$. A graphical summary of the bilinear transformation is shown in Figure 8.21.

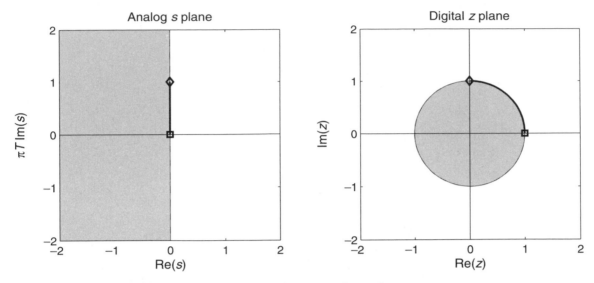

Figure 8.21   Bilinear Transformation from the s Plane onto the z Plane

The bilinear transformation in Figure 8.21 maps the entire imaginary axis of the s plane onto the unit circle of the z plane, as in Eq. 8.80. In so doing, the *analog frequencies*, $0 \leq F < \infty$, get compressed or warped into the *digital frequency* range, $0 \leq f < f_s/2$. To develop the relationship between F and f, let $s = j2\pi F$ denote a point on the imaginary axis of the s plane, and let $z = \exp(j2\pi fT)$ denote the corresponding point on the unit circle of the z plane. Setting $s = g(z)$ in Eq. 8.79 and using Euler's identity, we get

$$
\begin{aligned}
j2\pi F &= \frac{2}{T}\left[\frac{\exp(j2\pi fT) - 1}{\exp(j2\pi fT) + 1}\right] \\
&= \frac{2}{T}\left\{\frac{\exp(j\pi fT)[\exp(j\pi fT) - \exp(-j\pi fT)]}{\exp(j\pi fT)[\exp(j\pi fT) + \exp(-j\pi fT)]}\right\} \\
&= \frac{2}{T}\left[\frac{j2\sin(\pi fT)}{2\cos(\pi fT)}\right] \\
&= \frac{j2\tan(\pi fT)}{T} \tag{8.82}
\end{aligned}
$$

Solving Eq. 8.82 for F then yields the following relationship between the digital filter frequency, f, and the analog filter frequency, F.

Frequency warping

$$
F = \frac{\tan(\pi fT)}{\pi T} \tag{8.83}
$$

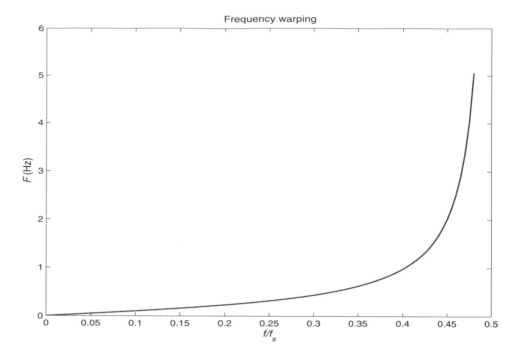

Figure 8.22 Frequency Warping Caused by the Bilinear Transformation

The transformation from f to F in Eq. 8.83 is called *frequency warping* because it represents an expansion of the finite digital frequency range, $0 \le f < f_s/2$, into the infinite analog frequency range, $0 \le F < \infty$. The nonlinear frequency-warping curve is shown in Figure 8.22. Note that there is an asymptote at the folding frequency, $f_d = f_s/2$.

The mapping from digital frequency, f, to analog frequency, F, in Eq. 8.83 can be inverted by solving Eq. 8.83 for f. Multiplying Eq. 8.83 by πT, taking the arctangent of both sides, and dividing by πT yields

$$f = \frac{\tan^{-1}(\pi F T)}{\pi T} \tag{8.84}$$

When the bilinear transformation from s to z in Eq. 8.80 is performed, the analog frequencies in the range $0 \le F < \infty$ get compressed into digital frequencies in the range $0 \le f < f_s/2$, as indicated in Eq. 8.84. We can take this nonlinear compression into account in filter design by first *prewarping* each desired digital cutoff frequency, f_c, into a corresponding analog cutoff frequency, F_c, using Eq. 8.83. When the bilinear transformation is then performed, these prewarped cutoff frequencies get warped back into the original desired digital cutoff frequencies as in Eq. 8.84. The overall design procedure for the bilinear-transformation method is summarized in the following algorithm where it is assumed that $m \le n$.

ALGORITHM 8.1:
Bilinear-
transformation
Method

1. Prewarp all digital cutoff frequencies, f_i, into corresponding analog cutoff frequencies, F_i, using Eq. 8.83.

2. Construct an analog prototype filter, $H_a(s)$, using the prewarped cutoff frequencies.

3. If $H_a(s)$ is low order, compute $H(z) = H_a[g(z)]$ using Eq. 8.79. For a higher-order $H_a(s)$, the following steps can be used.

   (a) Factor the numerator and denominator of $H_a(s)$ as follows.

   $$H_a(s) = \frac{\beta_0(z - u_1) \cdots (z - u_m)}{(z - v_1) \cdots (z - v_n)}$$

   (b) Compute the digital zeros and poles as follows using Eq. 8.80.

   $$z_i = \frac{2 + u_i T}{2 - u_i T}, \quad 1 \le i \le m$$

   $$p_i = \frac{2 + v_i T}{2 - v_i T}, \quad 1 \le i \le n$$

   (c) Compute the digital filter gain using

   $$b_0 = \frac{\beta_0 T^{n-m}(2 - u_1 T) \cdots (2 - u_m T)}{(2 - v_1 T) \cdots (2 - v_n T)}$$

   (d) Construct the factored form of the digital filter as follows.

   $$H(z) = \frac{b_0(z + 1)^{n-m}(z - z_1) \cdots (z - z_m)}{(z - p_1) \cdots (z - p_n)}$$

4. Express $H(z)$ as a ratio of two polynomials in z^{-1}.

Algorithm 8.1 assumes that $H_a(s)$ is a proper rational polynomial, which means that $m \le n$. If $m < n$, then $H_a(s)$ will have $n - m$ zeros at $s = \infty$. Note from step 3(d) that these $n - m$ high-frequency zeros get mapped into zeros at $z = -1$, the highest digital frequency that $H(z)$ can process.

Example 8.8   **Bilinear-transformation Method**

As an illustration of using Algorithm 8.1 to design a digital lowpass filter, suppose the sampling frequency is $f_s = 20$ Hz, and consider the following lowpass design specifications.

$$(f_0, f_1) = (2.5, 7.5) \text{ Hz}$$

$$(\delta_p, \delta_s) = (0.1, 0.1)$$

Here f_0 and f_1 denote the desired passband and stopband frequencies, respectively. From step 1 of Algorithm 8.1, the prewarped passband and stopband frequencies are

$$F_0 = \frac{\tan(2.5\pi/20)}{\pi/20}$$

$$= 2.637 \text{ Hz}$$

$$F_1 = \frac{\tan(7.5\pi/20)}{\pi/20}$$

$$= 15.37 \text{ Hz}$$

Suppose the analog prototype filter used is a lowpass Butterworth filter. From Eq. 8.47, the minimum order for the filter is

$$n = \text{ceil} \left\{ \frac{\log \left[\dfrac{0.9^{-2} - 1}{0.1^{-2} - 1} \right]}{2 \log \left(\dfrac{2.637}{15.37} \right)} \right\}$$

$$= \text{ceil}(1.715)$$

$$= 2$$

Next, suppose the cutoff frequency, F_c, is selected to meet the passband specification exactly. Then from Eq. 8.48, the required cutoff frequency is

$$F_c = \frac{2.637}{(0.9^{-2} - 1)^{1/4}}$$

$$= 3.789 \text{ Hz}$$

Thus, the radian cutoff frequency is $\Omega_c = 2\pi F_c = 23.81$ rad/sec. From Table 8.1, and Eq. 8.52, the transfer function of a prewarped Butterworth filter of order $n = 2$ is

$$H_a(s) = \frac{\Omega_c^2}{s^2 + \sqrt{2}\Omega_c s + \Omega_c^2}$$

Since $H_a(s)$ is a low-order filter, we apply step 3 of Algorithm 8.1 using direct substitution. Thus, from Eq. 8.79, the discrete-equivalent transfer function, $H(z)$, is

$$H(z) = H_a[g(z)]$$

$$= \frac{\Omega_c^2}{g^2(z) + \sqrt{2}\Omega_c g(z) + \Omega_c^2}$$

$$= \frac{\Omega_c^2}{\left[\dfrac{2(z-1)}{T(z+1)}\right]^2 + \sqrt{2}\Omega_c\left[\dfrac{2(z-1)}{T(z+1)}\right] + \Omega_c^2}$$

$$= \frac{(T\Omega_c)^2(z+1)^2}{4(z-1)^2 + 2\sqrt{2}T\Omega_c(z-1)(z+1) + (T\Omega_c)^2(z+1)^2}$$

$$= \frac{(T\Omega_c)^2(z+1)^2}{4(z^2 - 2z + 1) + 2\sqrt{2}T\Omega_c(z^2 - 1) + (T\Omega_c)^2(z^2 + 2z + 1)}$$

Next, the terms in the denominator are combined, the denominator is normalized, and the numerator and denominator are multiplied by z^{-2}. The final result after substitution of $T = 1/20$ and $\Omega_c = 23.81$ is then

$$H(z) = \frac{0.1613(1 + 2z^{-1} + z^{-2})}{1 - 0.5881z^{-1} + 0.2334z^{-2}}$$

A plot of the digital filter magnitude response is shown in Figure 8.23. Note how the passband specification, $1 - \delta_p \le A(f) \le 1$ for $0 \le f \le 2.5$ Hz, is met exactly, and the stopband specification, $0 \le A(f) \le \delta_s$ for $7.5 \le f \le 10$ Hz, is exceeded.

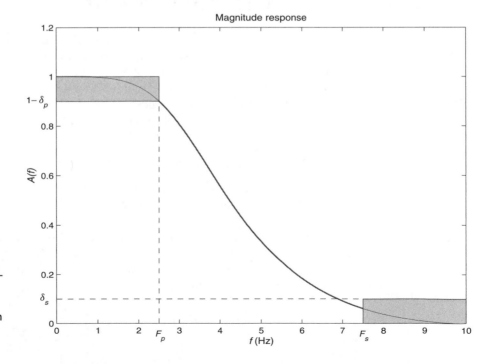

Figure 8.23 Magnitude Response of a Digital IIR Filter Obtained by a Bilinear Transformation of the Prewarped Analog Butterworth Filter: $n = 2$ and $F_c = 3.789$ Hz

MATLAB Toolbox

The FDSP toolbox contains the following function for performing a digital-to-analog filter transformation using the bilinear-transformation method.

```
[B,A] = f_bilin  (b,a,fs);          % bilinear transformation method
```

On entry to f_bilin, b is a vector of length $m+1$ containing the analog numerator coefficients, a is a vector of length $n+1$ containing the analog denominator coefficients, and fs is the sampling frequency. Here $n \geq m$ and $a(1) = 1$.

$$H_a(s) = \frac{b_0 s^m + b_1 s^{m-1} + \cdots + b_m}{s^n + a_1 s^{n-1} + \cdots a_n} \tag{8.85}$$

On exit from f_bilin, B and A are vectors containing the coefficients of the numerator polynomial and denominator polynomial, respectively, of the equivalent digital filter using the bilinear-transformation method. Note that all critical frequencies of $H_a(s)$ must be prewarped as in step 1 of Algorithm 8.1 before f_bilin is called. There can be more than two critical frequencies, and $H_a(s)$ can be any stable analog prototype filter.

8.6 Frequency Transformations

At this point, the tools are in place to design digital lowpass filters. We start with a normalized classical analog lowpass filter, $H_a(s)$, prewarp the cutoff frequency using Eq. 8.83, and then apply the bilinear transformation, $s = g(z)$, in Eq. 8.79 to produce a digital equivalent filter, $H(z)$. In this section, we use frequency transformations to extend this technique so it is also applicable to the other frequency-selective filters such as highpass, bandpass, and bandstop filters.

8.6.1 Analog Frequency Transformations

Recall that one way to design a lowpass Butterworth filter with a radian cutoff frequency of Ω_0 is to start with a normalized lowpass Butterworth filter and replace s with s/Ω_0. This is an example of a frequency transformation. Using this same basic approach, we can transform a normalized lowpass filter into other types of frequency-selective filters such as highpass, bandpass, and bandstop filters. To illustrate the procedure, consider the following normalized lowpass Butterworth filter of order $n = 1$, taken from Table 8.1.

$$H_{\text{norm}}(s) = \frac{1}{s+1} \tag{8.86}$$

Suppose s is replaced, not by s/Ω_0, but by Ω_0/s. The resulting transfer function is then

$$H_a(s) = H_{\text{norm}}(\Omega_0/s)$$

$$= \frac{1}{\Omega_0/s + 1}$$

$$= \frac{s}{s + \Omega_0} \tag{8.87}$$

Notice that $A_a(0) = 0$ and $A_a(\infty) = 1$, which means that $H_a(s)$ is a highpass filter. Furthermore, $A_a(\Omega_0) = 1/\sqrt{2}$. Thus, the frequency transformation, $D(s) = \Omega_0/s$, maps a normalized lowpass filter into a highpass filter with a 3-dB cutoff frequency of Ω_0 rad/sec.

It is also possible to transform a normalized lowpass filter into a bandpass filter. Recall that a bandpass filter has a low-frequency cutoff, Ω_0, and a high-frequency cutoff, Ω_1. Since a bandpass filter has two cutoff frequencies, the complex frequency variable s must be replaced by a quadratic polynomial of s in order to double the order of the transfer function. In particular, if s is replaced by $D(s) = (s^2 + \Omega_0\Omega_1)/[(\Omega_1 - \Omega_0)s]$, then the resulting filter is a bandpass filter with the desired cutoff frequencies.

Just as the highpass transformation is the reciprocal of the lowpass transformation, the bandstop transformation is the reciprocal of the bandpass transformation. A summary of the four basic frequency transformations can be found in Table 8.4. Using these transformations, a normalized lowpass transfer function $H_{\text{norm}}(s)$ with a cutoff frequency of $\Omega_c = 1$ rad/sec can be converted into an arbitrary lowpass, highpass, bandpass, or bandstop transfer function, $H_a(s)$.

Table 8.4: ▶
Analog Frequency
Transformations,
$H_a(s) = H_{\text{norm}}[D(s)]$

$H_a(s)$	$D(s)$
Lowpass with cutoff Ω_0	$\dfrac{s}{\Omega_0}$
Highpass with cutoff Ω_0	$\dfrac{\Omega_0}{s}$
Bandpass with cutoffs Ω_0, Ω_1	$\dfrac{s^2 + \Omega_0\Omega_1}{(\Omega_1 - \Omega_0)s}$
Bandstop with cutoffs Ω_0, Ω_1	$\dfrac{(\Omega_1 - \Omega_0)s}{s^2 + \Omega_0\Omega_1}$

<table>
<tr><td>**Example 8.9**</td><td>**Lowpass to Bandpass**</td></tr>
</table>

As an illustration of the frequency-transformation method, consider the problem of designing an analog bandpass filter. Suppose the desired cutoff frequencies are $F_0 = 5$ Hz and $F_1 = 15$ Hz. Thus, the corresponding radian cutoff frequencies are $\Omega_0 = 10\pi$ rad/sec and $\Omega_1 = 30\pi$ rad/sec. Suppose we start with the first-order lowpass Butterworth filter in Eq. 8.86. Using the third entry of Table 8.4, the bandpass filter transfer function is

$$H_a(s) = H_{\text{norm}}[D(s)]$$

$$= \frac{1}{\dfrac{s^2 + \Omega_0\Omega_1}{(\Omega_1 - \Omega_0)s} + 1}$$

$$= \frac{(\Omega_1 - \Omega_0)s}{s^2 + (\Omega_1 - \Omega_0)s + \Omega_0\Omega_1}$$

$$= \frac{20\pi s}{s^2 + 20\pi s + 300\pi^2}$$

Given the classical analog lowpass filters in Section 8.4 and the frequency transformations in Table 8.4, it is possible to design a variety of analog frequency-selective filters. The bilinear analog-to-digital filter transformation in Section 8.5 then can be applied to convert these analog filters to equivalent digital filters. Before the bilinear transformation is applied, all cutoff frequencies must be prewarped using Eq. 8.83. The following example illustrates the use of the bilinear-transformation method to construct a bandpass filter.

<table>
<tr><td>**Example 8.10**</td><td>**Digital Bandpass Filter**</td></tr>
</table>

Consider the second-order analog bandpass filter from Example 8.9. That is, suppose the desired cutoff frequencies are $F_0 = 5$ Hz and $F_1 = 15$ Hz, and the sampling rate is $f_s = 50$ Hz. Applying Eq. 8.83, the prewarped radian cutoff frequencies are

$$\Omega_0 = \frac{2\pi \tan(\pi F_0 T)}{\pi T}$$

$$= 100\tan(0.2\pi)$$

$$= 32.49$$

$$\Omega_1 = \frac{2\pi \tan(\pi F_1 T)}{\pi T}$$

$$= 100\tan(0.6\pi)$$

$$= 137.64$$

From Example 8.9, the transfer function of an analog second-order Butterworth bandpass filter with cutoff frequencies of Ω_0 and Ω_1 is

$$H_a(s) = \frac{(\Omega_1 - \Omega_0)s}{s^2 + (\Omega_1 - \Omega_0)s + \Omega_0\Omega_1}$$

$$= \frac{105.15s}{s^2 + 105.15s + 4472.1}$$

Next, we apply the bilinear transformation to convert $H_a(s)$ into an equivalent digital filter. Using Eq. 8.79, this yields

$$H(z) = H_a[g(z)]$$

$$= \frac{105.15g(z)}{g^2(z) + 105.15g(z) + 4472.1}$$

$$= \frac{105.15\left[\dfrac{2(z-1)}{T(z+1)}\right]}{\left[\dfrac{2(z-1)}{T(z+1)}\right]^2 + 105.15\left[\dfrac{2(z-1)}{T(z+1)}\right] + 4472.1}$$

$$= \frac{1.0515(z-1)(z+1)}{(z-1)^2 + 1.0515(z-1)(z+1) + 0.44721(z+1)^2}$$

$$= \frac{1.0515(z^2-1)}{(z^2 - 2z + 1) + 1.0515(z^2 - 1) + 0.44721(z^2 + 2z + 1)}$$

Combining like terms, normalizing the denominator, and multiplying top and bottom by z^{-2} then results in the following digital bandpass filter using the bilinear-transformation method

$$H(z) = \frac{0.4208(1 - z^{-2})}{1 - 0.4425z^{-1} + 0.1584z^{-2}}$$

A plot of the magnitude response of $H(z)$, obtained by running script *exam8_10* on the distribution CD, is shown in Figure 8.24.

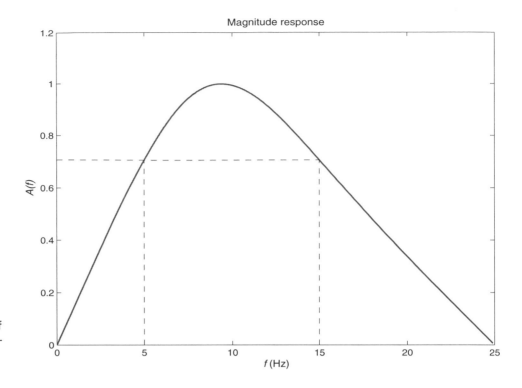

Figure 8.24 Frequency Response of a Second-order Digital Bandpass Filter from Example 8.10

The design technique based on an analog frequency transformation of a lowpass prototype filter is summarized in Figure 8.25.

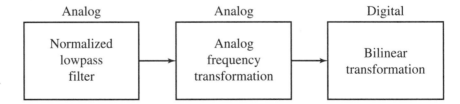

Figure 8.25 Digital Filter Design using an Analog Frequency Transformation

8.6.2 Digital Frequency Transformations

Frequency transformations from lowpass filters to other frequency-selective filters can also be done in the digital domain. In this case, the bilinear analog-to-digital filter transformation is applied to a normalized lowpass prototype filter. The resulting lowpass digital filter is then transformed to a lowpass, highpass, bandpass, or bandstop filter using a digital frequency transformation.

Let $H_{\text{low}}(z)$ be a digital lowpass filter with a cutoff frequency of F_c. This is converted to another frequency-selective filter by replacing z with a frequency transformation, $D(z)$, as follows.

$$H(z) = H_{\text{low}}[D(z)] \tag{8.88}$$

The transformation $D(z)$ must satisfy certain qualitative properties. First, it must map a rational polynomial, $H_{\text{low}}(z)$, into a rational polynomial $H(z)$. This means that $D(z)$ itself must be a ratio of polynomials. Because $D(z)$ is a frequency-response transformation, it should map the unit circle into the unit circle. Evaluating $D(z)$ along the unit circle yields the frequency response $D(f)$. Thus, the magnitude response must satisfy

$$|D(f)| = 1, \quad 0 \leq f < f_s \tag{8.89}$$

The magnitude-response constraint in Eq. 8.89 is an allpass characteristic. Hence, $D(z)$ must be an allpass filter as in Eq. 5.33. To maintain stability, the transformation $D(z)$ must also map the interior of the unit circle into the interior of the unit circle. Constantinides (1970) has developed four basic digital frequency transformations that are summarized in Table 8.5. The source filter is a lowpass filter with a cutoff frequency of F_c, and the destination filter is the filter listed in column one. The design technique based on a digital frequency transformation of a lowpass filter is summarized in Figure 8.26.

Table 8.5: ▶
Digital Frequency
Transformations,
$H(z) = H_{\text{low}}[D(z)]$

$H(z)$	$D(z)$	**Coefficients**
Lowpass with cutoff F_0	$\dfrac{-(z - a_0)}{a_0 z - 1}$	$a_0 = \dfrac{\sin[\pi(F_c - F_0)]}{\sin[\pi(F_c + F_0)]}$
Highpass with cutoff F_0	$\dfrac{z - a_0}{a_0 z - 1}$	$a_0 = \dfrac{\cos[\pi(F_c + F_0)]}{\cos[\pi(F_c - F_0)]}$
Bandpass with cutoffs F_0, F_1	$\dfrac{-(z^2 + a_0 z + a_1)}{a_1 z^2 + a_0 z + 1}$	$\alpha = \dfrac{\cos[\pi(F_1 + F_0)]}{\cos[\pi(F_1 - F_0)]}$ $\beta = \tan(\pi F_c) \cot[\pi(F_1 - F_0)]$ $a_0 = \dfrac{-2\alpha\beta}{\beta + 1}$ $a_1 = \dfrac{\beta - 1}{\beta + 1}$
Bandstop with cutoffs F_0, F_1	$\dfrac{z^2 + a_0 z + a_1}{a_1 z^2 + a_0 z + 1}$	$\alpha = \dfrac{\cos[\pi(F_1 + F_0)]}{\cos[\pi(F_1 - F_0)]}$ $\beta = \tan(\pi F_c) \tan[\pi(F_1 - F_0)]$ $a_0 = \dfrac{-2\alpha}{\beta + 1}$ $a_1 = \dfrac{1 - \beta}{1 + \beta}$

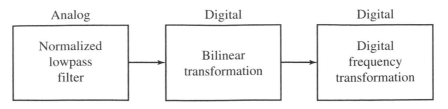

Figure 8.26 Digital Filter Design using a Digital Frequency Transformation

Table 8.6: ▶ Interpretation of Cutoff Frequency calling Arguments Based on Frequency-selective Filter Type

ftype	Filter	F_p	F_s
0	Lowpass	F_p	F_s
1	Highpass	F_p	F_s
2	Bandpass	$[F_{p1}, F_{p2}]$	$[F_{s1}, F_{s2}]$
3	Bandstop	$[F_{p1}, F_{p2}]$	$[F_{s1}, F_{s2}]$

MATLAB Toolbox

The FDSP toolbox contains the following functions for designing classical frequency-selective IIR digital filters using the method summarized in Figure 8.25. Also listed is a function for computing the digital frequency response.

```
[b,a] = f_butterz  (F_p,F_s,delta_p,delta_s,ftype,fs,n); % digital Butterworth
[b,a] = f_cheby1z  (F_p,F_s,delta_p,delta_s,ftype,fs,n); % digital Chebyshev-I
[b,a] = f_cheby2z  (F_p,F_s,delta_p,delta_s,ftype,fs,n); % digital Chebyshev-II
[b,a] = f_ellipticz (F_p,F_s,delta_p,delta_s,ftype,fs,n); % digital elliptic
[H,f] = f_freqz    (b,a,N,fs);                           % digital frequency response
```

The first four functions compute the coefficients of the Butterworth, Chebyshev-I, Chebyshev-II, and elliptic filters, respectively. On entry to each of these functions, F_p is the passband cutoff frequency, F_s is the stopband cutoff frequency, *delta_p* is the passband ripple, *delta_s* is the stopband attenuation, *ftype* specifies the frequency-selective filter type in Table 8.6, *fs* is the sampling frequency, and *n* is an optional argument specifying the filter order. If *n* is not present, then the order is determined automatically using an estimate of the order needed to meet the specifications. When *n* is present, the filter may or may not meet all of the specifications. Arguments of F_p and F_s are either scalars or vectors of length 2, depending on *ftype* as summarized in Table 8.6. On exit from the four filter design functions, *b* is a vector of length $m + 1$ containing the numerator coefficients, and *a* is a vector of length $n + 1$ containing the denomonator coefficients of the digital filter with $a(1) = 1$.

$$H(z) = \frac{b_0 + b_1 z^{-1} + \cdots + b_m z^{-m}}{1 + a_1 z^{-1} + \cdots a_n z^{-n}} \tag{8.90}$$

The function *f_freqz*, previously described in Chapter 3, is analogous to *f_freqs* but used to compute the frequency response of a discrete-time system. On entry to *f_freqz*, b is a vector of length $m+1$ containing the numerator coefficients, a is a vector of length $n+1$ containing the denominator coefficients with $a(1) = 1$, N is the number of evaluation frequencies, and fs is the sampling frequency. The frequency response, $H(f)$, is computed at N frequencies equally spaced over the interval $0 \le f < f_s/2$. On exit from *f_freqz*, H is a vector of length N containing the samples of $H(f)$, and f is a vector of length N containing the frequencies at which H is evaluated.

8.7 Software Applications

8.7.1 GUI Module: *g_iir*

FDSP Toolbox

The FDSP toolbox includes a graphical user interface module called *g_iir* that allows the user to design and evaluate IIR filters without any need for programming. GUI module *g_iir* features a display screen with titled windows as shown in Figure 8.27.

The upper-left *Block diagram* window contains a block diagram of the IIR filter under investigation. It is an nth-order filter with the following transfer function.

$$H(z) = \frac{b_0 + b_1 z^{-1} + \cdots + b_n z^{-m}}{1 + a_1 z^{-1} + \cdots + a_n z^{-n}} \tag{8.91}$$

The *Parameters* window below the block diagram displays edit boxes containing the filter parameters. The contents of each edit box can be directly modified by the user with the changes activated with the Enter key. Parameters $F0$, $F1$, B, and fs are the lower cutoff frequency, upper cutoff frequency, transition bandwidth, and sampling frequency, respectively. The lowpass filter uses cutoff frequency $F0$, the highpass filter uses cutoff frequency $F1$, and the bandpass and bandstop filters use both $F0$ and $F1$. For resonators and notch filters, the average of $F0$ amd $F1$ is used. The parameters *delta_p* and *delta_s* specify the passband ripple and stopband attenuation, respectively.

The *Type* and *View* windows in the upper-right corner of the screen allow the user to select both the type of frequency-selective filter and the viewing mode. The filter types include a resonator filter, a notch filter, lowpass, highpass, bandpass, and bandstop filters, and a user-defined filter whose coefficients are defined in a user-supplied MAT-file that contains a, b, and fs. The *View* options include the magnitude response, the phase response, a pole-zero plot, and the impulse response. The *Plot* window along the bottom half of the screen shows the selected view.

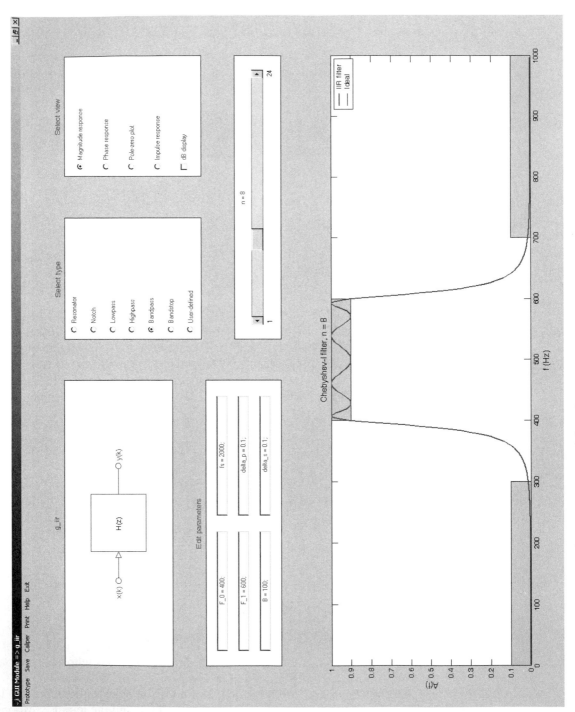

Figure 8.27 Display Screen of Chapter GUI Module *g_iir*

Just below the view options is a checkbox control that toggles the magnitude-response display between linear and logarithmic scales. When it is checked, the passband ripple and stopband attenuation in the *Parameters* window change to their logarithmic equivalents, A_p and A_s, respectively. Below the *Type* and *View* windows is a horizontal slider bar that allows the user to directly control the filter order, n. Note that the filter order may or may not meet all of the design specifications depending on the value of n. Furthermore, for some filters, when n is set too high the filter implementation can become unstable due to finite word length effects.

The *Menu* bar along the top of the screen includes several menu options. The *Prototype* option allows the user to select the type of analog prototype filter (Butterworth, Chebyshev-I, Chebyshev-II, or elliptic) for use with the lowpass, highpass, bandpass, and bandstop filter types. The *Save* option is used to save the filter parameters a, b, and fs in a user-specified MAT-file for future use. Files created in this manner subsequently can be loaded with the User-defined filter option. The *Caliper* option allows the user to measure any point on the current plot by moving the mouse crosshairs to that point and clicking. The *Print* option prints the contents of the *Plot* window. Finally, the *Help* option provides the user with some helpful suggestions on how to effectively use module *g_iir*.

8.7.2 Reverb Filter

Music generated in a concert hall sounds rich and full because it arrives at the listener along multiple paths, both direct and through a series of reflections. This reverberation effect can be simulated by processing the sound signal with a *reverb* filter (Steiglitz, 1996). The speed of sound in air at room temperature is about $v = 345$ m/sec. Suppose the distance from the music source to the listener is d m. If f_s denotes the sampling rate, then the distance in samples is

$$L = \text{floor}\left(\frac{f_s d}{v}\right) \tag{8.92}$$

Sound is attenuated as it travels through air and becomes dispersed. Let $0 < r < 1$ be the factor by which sound is attenuated as it propagates from the source to the listener. To roughly approximate a reverberation effect, suppose the echoes from one or more reflections arrive at multiples of L samples and are attenuated by powers of r. If there are n echoes, or paths of increasing length, then this effect can be modeled by the following difference equation.

$$y(k) = \sum_{i=1}^{n} r^i x(k - Li) \tag{8.93}$$

In the limit as the number of echoes, n, approaches infinity, we arrive at the following geometric-series transfer function for multiple echoes.

$$F(z) = \sum_{i=1}^{\infty} r^i z^{-Li}$$

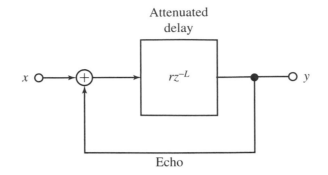

Figure 8.28 Basic Comb Filter used to Model Multiple Echoes

$$= \frac{rz^{-L}}{1 - rz^{-L}} \tag{8.94}$$

Observe that $F(z)$ is essentially a comb filter with poles equally spaced around a circle of radius r. A block diagram of the comb filter, that explicitly shows the feedback path taken by the echoes, is shown in Figure 8.28.

To develop a more refined model of the reverberation effect, we should take into consideration the observation that high-frequency sounds tend to get absorbed more than low-frequency sounds. This effect can be included by inserting a first-order lowpass filter, $D(z)$, into the feedback path of the basic comb filter (Moorer, 1979).

$$D(z) = \frac{1 - g}{1 - gz^{-1}} \tag{8.95}$$

Here the real pole, $0 < g < 1$, controls the cutoff frequency of the lowpass characteristic. A block diagram showing this more refined lowpass comb filter is shown in Figure 8.29. The transfer function of the lowpass comb filter can be obtained from Figure 8.29 by solving for the summing junction output signal $V(z)$. Working

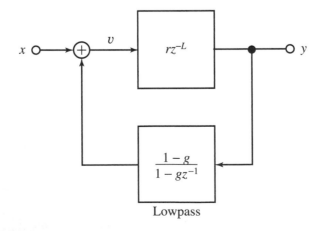

Figure 8.29 Lowpass Comb Filter that Includes Effects of Frequency-dependent Sound Absorption

backwards around the loop we have

$$V(z) = X(z) + D(z)Y(z)$$

$$= X(z) + D(z)rz^{-L}V(z) \tag{8.96}$$

Solving Eq. 8.96 for $V(z)$ then yields

$$V(z) = \frac{X(z)}{1 - rz^{-L}D(z)} \tag{8.97}$$

Finally, from Figure 8.29, Eq. 8.97, and Eq. 8.95, the output of the lowpass comb filter is

$$Y(z) = rz^{-L}V(z)$$

$$= \frac{rz^{-L}X(z)}{1 - rz^{-L}D(z)}$$

$$= \frac{rz^{-L}X(z)}{1 - rz^{-L}(1 - g)/(1 - gz^{-1})}$$

$$= \frac{rz^{-L}(1 - gz^{-1})X(z)}{1 - gz^{-1} - r(1 - g)z^{-L}} \tag{8.98}$$

Multiplying both sides of Eq. 8.98 by z^{L+1}, the overall transfer function of the lowpass comb filter, in terms of positive powers of z, is

Lowpass comb filter

$$C(z) = \frac{r(z - g)}{z[z^L - gz^{L-1} - r(1 - g)]} \tag{8.99}$$

The reverberation effect can be made much fuller and richer when multiple lowpass comb filters with different delays and cutoff frequencies are used. Moorer (1979) has proposed using multiple lowpass comb filters in parallel, followed by an allpass filter. The allpass filter inserts a frequency-dependent phase shift, or delay, but does not change the magnitude response. Recall from Eq. 5.33 that an allpass filter is a filter whose numerator and denominator exhibit reflective symmetry. For example, the following Mth-order allpass filter can delay signals up to M samples.

Allpass filter

$$G(z) = \frac{c + z^{-M}}{1 + cz^{-M}} \tag{8.100}$$

The overall configuration for a reverb filter, featuring P lowpass comb sections, is shown in Figure 8.30. The composite transfer function of the reverb filter is as follows.

$$H(z) = G(z) \sum_{i=1}^{P} C_i(z) \tag{8.101}$$

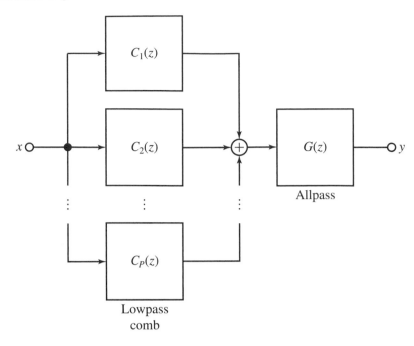

Figure 8.30 A Reverb Filter

To implement the reverb filter in Figure 8.30, the following parameters can be used (Moorer, 1979). For the allpass filter, set

$$c = 0.7 \tag{8.102a}$$

$$M = \text{floor}(0.006 f_s) \tag{8.102b}$$

This corresponds to a delay of 0.006 sec. Next, use $P = 6$ lowpass comb filters with the following attenuation factor, delays, and poles.

$$r = 0.83 \tag{8.103a}$$

$$L = \text{floor}\{[0.050, 0.056, 0.061, 0.068, 0.072, 0.078] f_s\} \tag{8.103b}$$

$$g = [0.24, 0.26, 0.28, 0.29, 0.30, 0.32] \tag{8.103c}$$

The FDSP toolbox contains a function called *f_reverb* that computes the reverb filter output using the parameters in Eq. 8.102 and Eq. 8.103. The reverb filter can be tested by running the following MATLAB script, labeled *exam8_11* on the distribution CD.

```
function exam8_11

% Example 8.11: Reverb filter

clear
clc
fprintf('Example 8.11: Reverb filter\n\n')
```

```
% Plot impulse response

fs = 8000;
N = 8192;
x = [1,zeros(1,N-1)];
y = x;
[h,n] = f_reverb(x,fs);
n
figure
plot ([1:N-1],h(2:N))
f_labels ('Impulse response','\it{k}','\it{h(k)}')
f_wait

% Plot magnitude response

H = fft(h);
A = abs(H);
f = linspace (0,(N-1)*fs/N,N);
figure
plot (f(1:N/2),A(1:N/2))
f_labels ('Magnitude response','{\itf} (Hz)','\it{A(f)}')
f_wait

% Get sound and put it through reverb filter

tau = 3.0;
choice = 0;
p = floor(8000/tau);
z = zeros(1,p);
while choice ~= 4
   choice = menu ('Please select one','record sound','play back (normal)',...
             'play back (reverb)','exit');
   switch (choice)
   case 1,
      z = f_getsound (tau,fs);
      y = f_reverb (z,fs);
   case 2,
      soundsc (z,fs)
   case 3,
      soundsc (y,fs);
   end
end
```

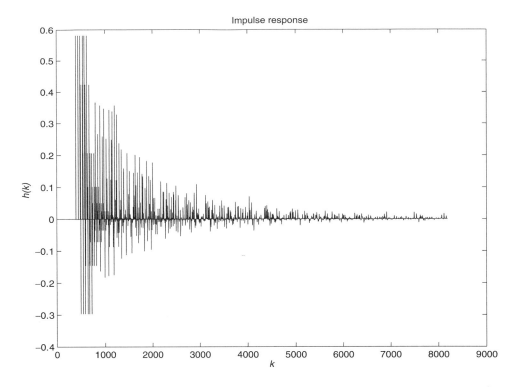

Figure 8.31 Impulse Response of a Reverb Filter

When script *exam8_11* is run, it first computes the impulse response shown in Figure 8.31. Unlike many of the discrete-time systems we have investigated, the impulse response of this system takes a very long time to die out because of the significant delays representing echoes in the system.

The second segment of script *exam8_11* computes the magnitude response shown in Figure 8.32. The interactions of the multiple lowpass comb filters provide a magnitude response that is broadband, yet exhibits detailed variation. This is due, in part, to the very high order of the reverb filter. From Eq. 8.99, Eq. 8.100, and Figure 8.30, the total filter order is as follows.

$$n = M + P + L_1 + \cdots + L_P \tag{8.104}$$

Recall from Eq. 8.102 and Eq. 8.103 that the allpass order, M, and the delays, L_i, are proportional to the sampling frequency which is set to $f_s = 8000$ Hz in *exam8_11*. Using Eqs. 8.102 through 8.104, this results in an IIR reverb filter order n, where

$$n = \text{floor}(0.006 f_s) + 6 + \text{floor}\{[0.050 + 0.056 + 0.061 + 0.068 + 0.072 + 0.078] f_s\}$$

$$= 3134$$

Since the reverb filter is stable, this means that all 3134 poles must be inside the unit circle! The final segment of script *exam8_11* displays a menu that allows the user to record up to four seconds of sound from the PC microphone and then play it back

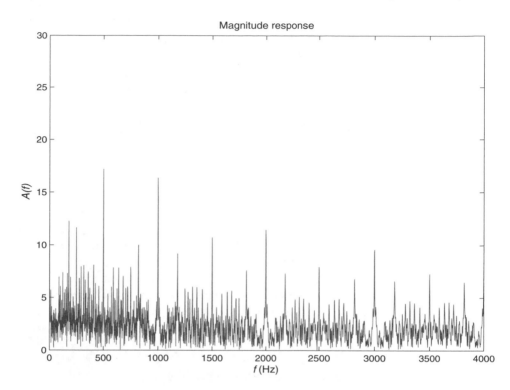

Figure 8.32　Magnitude Response of a Reverb Filter

on the PC speaker both with and without reverb filtering. Even with a small PC speaker, the differences in the sound are distinct, and quite striking. Give it a try!

● ● ● ● ● ● ● ● ● ● ● ● ● ● ● ●

8.8　Chapter Summary

Infinite Impulse Response Filters

This chapter focused on the design of infinite impulse response digital filters having the following transfer function.

$$H(z) = \frac{b_0 + b_1 z^{-1} + \cdots + b_m z^{-m}}{1 + a_1 z^{-1} + \cdots + a_n z^{-n}} \qquad (8.105)$$

The four basic frequency-selective filter types are lowpass, highpass, bandpass, and bandstop filters. For an ideal filter, the passband gain is $A(f) = 1$, the stopband is $A(f) = 0$, and there is no transition band. However, for a practical filter of finite order, there must be a transition band separating the passband and stopband. Furthermore, for a physically realizable filter, the passband gain is not constant but lies instead within an interval $[1 - \delta_p, 1]$, where $\delta_p > 0$ is the passband ripple. Similarly, the stopband gain lies within an interval $[0, \delta_s]$, where $\delta_s > 0$ is the stopband attenuation. The filter magnitude response is often represented using a logarithmic scale of decibels (dB) as follows.

$$A(f) = 20 \log_{10}\{|H(f)|\} \text{ dB} \tag{8.106}$$

The ripple and attenuation specifications, δ_p and δ_s, have logarithmic equivalents, A_p and A_s, that are expressed in units of dB. The logarithmic scale is useful for showing the amount of attenuation in the stopband.

Pole-zero Placement with Gain Matching

There are a number of specialized IIR filters that can be designed using pole-zero placement and gain matching. These include a resonator which is designed to pass a single frequency, and a notch filter which is designed to reject a single frequency. Generalizations of these two basic filters include the comb filter which is designed to pass several harmonically related frequencies, and the inverse comb filter which is designed to reject several harmonically related frequencies. Comb filters and inverse comb filters can be used to pass or reject a periodic input that is corrupted with noise if the first resonant frequency is set to match the fundamental frequency of the periodic input.

Classical Analog Prototype Filters

A highly effective and widely used approach for designing frequency-selective IIR filters starts with an analog prototype filter and then transforms it to an equivalent digital filter. There are four classical families of analog filters that are used for prototype filters, and each is optimal in some sense. Butterworth filters have the property that their magnitude responses are as flat as possible in the passband. Butterworth filters are easy to design, but they have a relatively wide transition band. The transition band can be made more narrow by allowing ripples in the magnitude response. Chebyshev-I filters have ripples of equal amplitude in the passband and meet the passband specifications exactly, while Chebyshev-II filters have ripples of equal amplitude in the stopband and meet the stopband specification exactly. Elliptic filters have a very narrow transition band that is achieved by allowing ripples of equal size in both the passband and the stopband. Elliptic IIR filters have magnitude responses that are similar to the optimal equiripple FIR filters designed with the Parks-McLellan algorithm.

For the classical filters, the filter order required to meet a given design specification can be computed using two design parameters called the *selectivity factor*, r, and the *discrimination factor*, d. For an ideal filter, $r = 1$ and $d = 0$, whereas for a practical filter, $r < 1$ and $d > 0$. The classical analog lowpass filters, $H_a(s)$, can be designed by starting with the magnitude response, $A_a(f)$, and working backwards to determine the poles, zeros, and gain using the following fundamental relationship.

$$H_a(s)H_a(-s) = A_a^2(f)|_{f=s/(j2\pi)} \tag{8.107}$$

Analog-to-digital Filter Transformation

Analog frequency-selective prototype filters of various types can be obtained by applying frequency transformations to a normalized lowpass filter, a filter whose radian cutoff frequency is $\Omega_0 = 1$ rad/sec. For bandpass and bandstop filters, these frequency transformations double the order of the filter. It is also possible to perform frequency transformations on digital lowpass filters. The most commonly used analog-to-digital filter transformation technique is the bilinear-transformation method. This technique maps a stable analog filter, $H_a(s)$, into a stable digital filter, $H(z)$, using the following change of variables in $H_a(s)$.

$$s = \frac{2}{T}\left(\frac{z-1}{z+1}\right) \tag{8.108}$$

The bilinear-transformation method maps the imaginary axis of the s plane onto the unit circle of the z plane. The resulting compression of frequencies is called frequency warping, and it must be taken into account when specifying the critical frequencies of the filter. That is, in constructing the analog prototype filter, each of the desired cutoff frequencies should first be prewarped using

$$F = \frac{\tan(\pi f T)}{\pi T} \tag{8.109}$$

IIR filters can be used for music synthesis and to create special sound effects. For example, the impulse response of the tunable plucked-string filter can be adjusted to emulate the sound produced by a variety of stringed instruments. The rich full sound of concert halls can be reproduced by introducing special sound effects using a reverb filter. A reverb filter is a high-order IIR filter that is implemented as a parallel configuration of comb filters, each having a lowpass filter in their feedback path, followed by an allpass filter.

The FDSP toolbox includes a GUI module called g_iir that allows the user to design, evaluate, and compare IIR filters without any need for programming. The featured filters include resonators, notch filters, lowpass, highpass, bandpass, and bandstop filters, plus user-defined filters specified in a MAT-file. The filter order can be adjusted directly using a slider bar.

• • • • • • • • • • • • • • • •

8.9 Problems

The problems are divided into Analysis problems that can be solved by hand or with a calculator, GUI Simulation problems that are solved using GUI module g_iir, and Computation problems. The Computation problems require the student to write a MATLAB script using the FDSP toolbox functions summarized in Appendix B. Solutions to selected problems are available on the distribution CD. Students are encouraged to use these problems, which are identified with a ☑, as a check on their understanding of the material.

8.9.1 Analysis

Section 8.1

P8.1. Consider the problem of designing a filter whose impulse response emulates the sound from a stringed musical instrument. Suppose the sampling frequency is $f_s = 44.1$ kHz, and the desired resonant frequency or pitch is $F_0 = 480$ Hz.

(a) Find the feedback parameter, L, and the pitch parameter, c, in Figure 8.1.

(b) Suppose the attenuation factor is $r = 0.998$. Find the tunable plucked-string filter transfer function $H(z)$.

Section 8.2

P8.2. Consider the problem of designing a resonator that passes the frequency $F_0 = 100$ Hz.

(a) Find a sampling frequency, f_s, that places the resonator pole at an angle of $\theta_0 = \pi/2$.

(b) Design a resonator, $H_{res}(z)$, that has a 3-dB passband radius of $\Delta F = 2$ Hz.

(c) Sketch a signal flow graph using a direct form II realization.

P8.3. Consider the problem of designing a resonator that has two resonant frequencies. Suppose the sampling frequency is $f_s = 360$ Hz.

(a) Design a resonator, $H_0(z)$, that has a resonant frequency at $F_0 = 90$ Hz and a 3-dB passband radius of 3 Hz.

(b) Design a resonator, $H_1(z)$, that has a resonant frequency of $F_1 = 120$ Hz and a 3-dB passband radius of 4 Hz.

(c) Combine $H_0(z)$ and $H_1(z)$ to produce a resonator, $H(z)$, that has resonant frequencies at $F_0 = 80$ Hz and $F_1 = 120$ Hz. *Hint*: Use one of the indirect forms.

(d) Sketch the signal flow graph of $H(z)$ using direct form II realizations for the blocks $H_0(z)$ and $H_1(z)$.

P8.4. Consider the problem of designing a notch filter that rejects the frequency $F_0 = 60$ Hz.

(a) Suppose the notch-filter pole is at the angle $\theta_0 = \pi/3$. Find the sampling frequency f_s.

(b) Design a notch filter, $H_{notch}(z)$, that has a 3-dB stopband radius of $\Delta F = 1$ Hz.

(c) Sketch the signal flow graph using a transposed direct form II realization.

P8.5. Consider the problem of designing a notch filter that has two notch frequencies. Suppose the sampling frequency is $f_s = 360$ Hz.

   (a) Design a notch filter, $H_0(z)$, that has a resonant frequency at $F_0 = 60$ Hz and a 3-dB stopband radius of 2 Hz.

   (b) Design a notch filter, $H_1(z)$, that has a resonant frequency at $F_0 = 90$ Hz and a 3-dB stopband radius of 2 Hz.

   (c) Combine $H_0(z)$ and $H_1(z)$ to produce a notch filter $H(z)$ that has notches at $F_0 = 60$ Hz and $F_1 = 90$ Hz. *Hint*: Use one of the indirect forms.

   (d) Sketch the signal flow graph of $H(z)$ using direct form II realizations for the blocks $H_0(z)$ and $H_1(z)$.

P8.6. Consider an input signal, $y(k)$, that consists of a periodic component, $x(k)$, plus a random white noise component, $v(k)$.

$$y(k) = x(k) + v(k), \quad 0 \le k < 256$$

Suppose the sampling rate is f_s, and this results in a signal $x(k)$ that is periodic with a period of $L = 16$. Design a comb filter, $H_{comb}(z)$, that passes harmonics zero through $L/2$ of $x(k)$. Use a 3-dB passband radius of $\Delta F = f_s/100$.

P8.7. Consider an input, $y(k)$, that consists of a signal of interest, $x(k)$, plus a disturbance, $d(k)$.

$$y(k) = x(k) + d(k), \quad 0 \le k < N$$

Suppose that when the sampling rate is f_s, the disturbance $d(k)$ is periodic with a period of $L = 12$. Design an inverse comb filter, $H_{inv}(z)$, that removes harmonics zero through $L/2$ of $d(k)$ from $y(k)$. Use a 3-dB passband radius of $\Delta F = f_s/200$.

Section 8.3

P8.8. Consider the problem of designing a lowpass analog filter, $H_a(s)$, to meet the following specifications.

$$[F_p, F_s, \delta_p, \delta_s] = [1000, 1200, 0.05, 0.02]$$

   (a) Find the passband ripple and stopband attenuation in units of dB.

   (b) Find the selectivity factor, r.

   (c) Find the discrimination factor, d.

P8.9. Consider the following design specifications for a lowpass analog filter.

$$[F_p, F_s, \delta_p, \delta_s] = [50, 60, 0.05, 0.02]$$

Find the minimum-order filter needed to meet these specifications using the following classical analog filters.

   (a) Butterworth filter.

   (b) Chebyshev-I filter.

   (c) Chebyshev-II filter.

Section 8.4

P8.10. Consider the problem of designing a lowpass analog Butterworth filter to meet the following specifications.

$$[F_p, F_s, \delta_p, \delta_s] = [300, 500, 0.1, 0.05]$$

(a) Find the minimum filter order n.

(b) For what cutoff frequency, F_c, is the passband specification exactly met?

(c) For what cutoff frequency, F_c, is the stopband specification exactly met?

(d) Find a cutoff frequency, F_c, for which $H_a(s)$ exceeds both the passband and the stopband specification.

P8.11. Find the transfer function, $H(s)$, of a third-order analog lowpass Butterworth filter that has a 3-dB cutoff frequency of $F_c = 4$ Hz.

P8.12. Sketch the poles and zeros of an analog lowpass Butterworth filter of order $n = 8$ that has a 3-dB cutoff frequency of $F_c = 1/\pi$ Hz.

P8.13. Consider the problem of designing an analog lowpass Chebyshev-I filter to meet the following design specifications. Find the minimum order of the filter.

$$[F_p, F_s, \delta_p, \delta_s] = [100, 200, .03, .05]$$

P8.14. Design a second-order analog lowpass Chebyshev-I filter, $H_a(s)$, using $F_p = 10$ Hz and $\delta_p = 0.1$.

P8.15. Find the minimum order n of an analog elliptic filter that will meet the following design specifications. You can use the MATLAB function *ellipke* to evaluate an elliptic integral of the first kind.

$$[F_p, F_s, \delta_p, \delta_s] = [100, 200, .03, .05]$$

Section 8.5

P8.16. Consider the following first-order analog filter.

$$H_a(s) = \frac{s}{s + 4\pi}$$

(a) What type of frequency-selective filter is this (lowpass, highpass, bandpass, or bandstop)?

(b) What is the 3-dB cutoff frequency f_0 of this filter?

(c) Suppose $f_s = 10$ Hz. Find the prewarped cutoff frequency F_0.

(d) Design a digital-equivalent filter, $H(z)$, using the bilinear-transformation method.

P8.17. The simplest digital equivalent filter is one that preserves the impulse response of $H_a(s)$. Let $h_a(t)$ denote the desired impulse response.

$$h_a(t) = L^{-1}\{H_a(s)\}$$

Next, let T be the sampling interval. The objective is to design a digital filter, $H(z)$, whose impulse response, $h(k)$, satisfies

$$h(k) = h_a(kT), \quad k \geq 0$$

Thus, the impulse response of $H(z)$ consists of samples of the impulse response of $H_a(s)$. This design technique, which preserves the impulse response, is called the *impulse-invariant* method. Suppose $H_a(s)$ is a stable, strictly proper, rational polynomial with n distinct poles $\{p_1, p_2, \cdots, p_n\}$.

(a) Expand $H_a(s)/s$ into partial fractions.
(b) Find the impulse response $h_a(t)$.
(c) Sample $h_a(t)$ to find the impulse response $h(k)$.
(d) Find the transfer function $H(z)$.

P8.18. Consider the following analog prototype filter of order $n = 2$.

$$H_a(s) = \frac{6}{s^2 + 5s + 6}$$

(a) Find the poles of $H_a(s)/s$.
(b) Find the residues of $H_a(s)/s$ at each pole.
(c) Find a digital-equivalent transfer function using the impulse-invariant method in Problem P8.17. You can assume the sampling interval is $T = 0.5$ sec.

P8.19. Consider the following analog filter that has n poles and m zeros with $m \leq n$.

$$H_a(s) = \frac{\beta(s - z_1)(s - z_2) \cdots (s - z_m)}{(s - p_1)(s - p_2) \cdots (s - p_n)}$$

An alternative way to convert an analog filter into a digital filter is to map each pole and zero of $H_a(s)$ into a corresponding pole and zero of $H(z)$ using $z = \exp(sT)$. This yields:

$$H(z) = \frac{b_0(z + 1)^{n-m}[z - \exp(z_1 T)][z - \exp(z_2 T)] \cdots [z - \exp(z_m T)]}{[z - \exp(p_1 T)][z - \exp(p_2 T)] \cdots [z - \exp(p_n T)]}$$

Note that if $n > m$, then $H_a(s)$ has $n - m$ zeros at $s = \infty$. These zeros are mapped into the highest digital frequency, $z = -1$. The gain factor b_0 is selected such that the two filters have the same passband gain. For example, if $H_a(s)$ is a lowpass filter, then $H_a(0) = H(1)$. This method is called the *matched Z-transform* method. Use the matched Z-transform method to find a digital equivalent of the following analog filter. You can assume $T = 0.2$. Match the gains at DC.

$$H_a(s) = \frac{10s + 1}{s^2 + 3s + 2}$$

Section 8.6

P8.20. Find the transfer function, $H(s)$, of a second-order highpass Butterworth filter that has a 3-dB cutoff frequency of $F_c = 5$ Hz.

P8.21. Find the transfer function, $H(s)$, of a fourth-order bandpass Butterworth filter that has 3-dB cutoff frequencies of $F_0 = 2$ Hz and $F_1 = 4$ Hz.

8.9.2   GUI Simulation

Section 8.7

P8.22. Use the GUI module *g_iir* to design a resonator filter with a resonant frequency of $F_0 = 100$ Hz, and a sampling frequency of $f_s = 300$ Hz.

(a) Plot the magnitude response. Use the *Caliper* option to mark the peak.

(b) Plot the phase response. Is this a linear-phase filter?

(c) Plot the pole-zero plot.

P8.23. Use the GUI module *g_iir* to design a notch filter with a notch frequency of $F_0 = 200$ Hz, and a sampling frequency of $f_s = 900$ Hz.

(a) Plot the magnitude response.

(b) Plot the phase response. Is this a linear-phase filter?

(c) Plot the impulse response.

P8.24. Use the GUI module *g_iir* to construct a Chebyshev-I lowpass filter using the default parameter values. Plot the magnitude response for the following cases.

(a) The filter order is adjusted manually to the highest value that does not meet the specifications.

(b) The filter order is adjusted manually to the lowest value that meets or exceeds the specifications.

P8.25. Use the GUI module *g_iir* to design a lowpass Butterworth filter using the default parameter values. Adjust the filter order to the lowest value that meets or exceeds the specifications. Plot the following.

(a) The magnitude response.

(b) The phase response. Is this a linear-phase filter?

(c) The pole-zero plot.

☑ **P8.26.** Use the GUI module *g_iir* to design a highpass Chebyshev-I filter using the default parameter values. Adjust the filter order to the lowest value that meets or exceeds the specifications. Plot the following.

(a) The magnitude response.

(b) The phase response. Is this a linear-phase filter?

(c) The pole-zero plot.

P8.27. Use the GUI module *g_iir* to design a bandpass Chebyshev-II filter using the default parameter values. Adjust the filter order to the lowest value that meets or exceeds the specifications. Plot the following.

(a) The magnitude response.

(b) The phase response. Is this a linear-phase filter?

(c) The pole-zero plot.

P8.28. Use the GUI module *g_iir* to design a bandstop elliptic filter using the default parameter values. Adjust the filter order to the lowest value that meets or exceeds the specifications. Plot the following.

(a) The magnitude response.

(b) The phase response. Is this a linear-phase filter?

(c) The pole-zero plot.

P8.29. Use the GUI module *g_iir* to design a Butterworth bandpass filter. Find the smallest order filter that meets or exceeds the following design specifications.

$$[f_s, F_{s1}, F_{p1}, F_{p2}, F_{s2}] = [2000, 300, 400, 600, 700] \text{ Hz}$$

$$[A_p, A_s] = [0.6, 30] \text{ dB}$$

(a) Plot the magnitude response using the dB scale.

(b) Plot the pole-zero pattern.

(c) Save a, b, and fs in a MAT-file named *prob8_29*. Then use GUI module *g_filters* to load this as a user-defined filter. Adjust the number of bits used for coefficient quantization to a level that shows a significant difference between the quantized and unquantized linear magnitude responses. Plot the magnitude responses.

P8.30. Use the GUI module *g_iir* to design a Chebyshev-I bandpass filter. Find the smallest order filter that meets or exceeds the following design specifications.

$$[f_s, F_{s1}, F_{p1}, F_{p2}, F_{s2}] = [2000, 300, 400, 600, 700] \text{ Hz}$$

$$[\delta_p, \delta_s] = [0.05, 0.03]$$

(a) Plot the magnitude response.

(b) Plot the pole-zero pattern.

(c) Save a, b, and fs in a MAT-file named *prob8_30*. Then use GUI module *g_filters* to load this as a user-defined filter. Adjust the number of bits used for coefficient quantization to a level that shows a significant difference between the quantized and unquantized linear magnitude responses. Plot the magnitude responses.

P8.31. Use the GUI module *g_iir* to design an elliptic bandpass filter. Find the smallest order filter that meets or exceeds the following design specifications.

$$[f_s, F_{s1}, F_{p1}, F_{p2}, F_{s2}] = [2000, 350, 400, 600, 650] \text{ Hz}$$

$$[\delta_p, \delta_s] = [0.04, 0.02]$$

(a) Plot the magnitude response.

(b) Plot the pole-zero pattern.

(c) Save a, b, and fs in a MAT-file named *prob8_31*. Then use GUI module *g_filters* to load this as a user-defined filter. Adjust the number of bits used for coefficient quantization to a level that shows a significant difference between the quantized and unquantized linear magnitude responses. Plot the magnitude responses.

P8.32. Use the GUI module *g_iir* to design a Butterworth bandpass filter. Find the smallest order filter that meets or exceeds the following design specifications.

$$[f_s, F_{p1}, F_{s1}, F_{s2}, F_{p2}] = [100, 20, 25, 35, 40] \text{ Hz}$$

$$[\delta_p, \delta_s] = [0.05, 0.02]$$

(a) Plot the magnitude response using the dB scale.

(b) Plot the pole-zero pattern.

(c) Save a, b, and fs in a MAT-file named *prob8_32*. Then use GUI module *g_filters* to load this as a user-defined filter. Adjust the number of bits used for coefficient quantization to a level that shows a significant difference between the quantized and unquantized linear magnitude responses. Plot the magnitude responses.

☑ **P8.33.** Use the GUI module *g_iir* to design a Chebyshev-II bandstop filter. Find the smallest order filter that meets or exceeds the following design specifications.

$$[f_s, F_{p1}, F_{s1}, F_{s2}, F_{p2}] = [20000, 2500, 3000, 4000, 4500] \text{ Hz}$$

$$[\delta_p, \delta_s] = [0.04, 0.03]$$

(a) Plot the magnitude response.

(b) Plot the pole-zero pattern.

(c) Save a, b, and fs in a MAT-file named *prob8_33*. Then use GUI module *g_filters* to load this as a user-defined filter. Adjust the number of bits used for coefficient quantization to a level that shows a significant difference between the quantized and unquantized linear magnitude responses. Plot the magnitude responses.

P8.34. Use the GUI module *g_iir* to design an elliptic bandstop filter. Find the smallest order filter that meets or exceeds the following design specifications.

$$[f_s, F_{p1}, F_{s1}, F_{s2}, F_{p2}] = [20, 6.5, 7, 8, 8.5] \text{ Hz}$$

$$[\delta_p, \delta_s] = [0.02, 0.015]$$

(a) Plot the magnitude response.

(b) Plot the pole-zero pattern.

(c) Save a, b, and fs in a MAT-file named *prob8_34*. Then use GUI module *g_filters* to load this as a user-defined filter. Adjust the number of bits used for coefficient quantization to a level that shows a significant difference between the quantized and unquantized linear magnitude responses. Plot the magnitude responses.

P8.35. Use the GUI module *g_iir* and the User-defined option to load the filter in MAT-file *u_iir1*.

(a) Plot the magnitude response. What type of filter is this?

(b) Plot the phase response

(c) Plot the impulse response.

P8.36. Create a MAT-file called *my_iir* that contains b, a, and fs for an inverse comb filter of order $n = 12$, using $fs = 1000$ Hz and a 3-dB radius of $\Delta F = 2$ Hz. Then use the GUI module *g_iir* and the User-defined option to load this filter.

(a) Plot the magnitude response.

(b) Plot the phase response.

(c) Plot the pole-zero pattern.

8.9.3   Computation

Section 8.4

P8.37. Write a MATLAB script that uses *f_butters* to design an analog Butterworth lowpass filter to meet the following design specifications.

$$[F_p, F_s, \delta_p, \delta_s] = [10, 20, 0.04, 0.02]$$

(a) Print the filter order.

(b) Use *f_freqs* to compute and plot the magnitude response for $0 \le f \le 2F_s$.

(c) Use *fill* to add shaded areas showing the design specifications on the magnitude response plot.

P8.38. Write a MATLAB script that uses $f\_cheby1s$ to design an analog Chebyshev-I lowpass filter to meet the following design specifications.

$$[F_p, F_s, \delta_p, \delta_s] = [10, 20, 0.04, 0.02]$$

(a) Print the filter order.

(b) Use f_freqs to compute and plot the magnitude response for $0 \leq f \leq 2F_s$.

(c) Use $fill$ to add shaded areas showing the design specifications on the magnitude response plot.

P8.39. Write a MATLAB script that uses $f\_cheby2s$ to design an analog Chebyshev-II lowpass filter to meet the following design specifications.

$$[F_p, F_s, \delta_p, \delta_s] = [10, 20, 0.04, 0.02]$$

(a) Print the filter order.

(b) Use f_freqs to compute and plot the magnitude response for $0 \leq f \leq 2F_s$.

(c) Use $fill$ to add shaded areas showing the design specifications on the magnitude response plot.

☑ **P8.40.** Write a MATLAB script that uses $f\_elliptics$ to design an analog elliptic lowpass filter to meet the following design specifications.

$$[F_p, F_s, \delta_p, \delta_s] = [10, 20, 0.04, 0.02]$$

(a) Print the filter order.

(b) Use f_freqs to compute and plot the magnitude response for $0 \leq f \leq 2F_s$.

(c) Use $fill$ to add shaded areas showing the design specifications on the magnitude response plot.

Section 8.6

P8.41. Write a MATLAB script that uses $f\_butters$ and $f\_low2highs$ to design an analog Butterworth highpass filter to meet the following design specifications.

$$[F_s, F_p, A_p, A_s] = [4, 6, 0.5, 24]$$

(a) Print the filter order, δ_p, and δ_s.

(b) Use f_freqs to compute and plot the magnitude response for $0 \leq f \leq 2F_p$ using the linear scale.

(c) Use $fill$ to add shaded areas showing the design specifications on the magnitude response plot.

P8.42. Write a MATLAB script that uses $f\_cheby1s$ and $f\_low2bps$ to design an analog Chebyshev-I bandpass filter to meet the following design specifications.

$$[F_{s1}, F_{p1}, F_{p2}, F_{s2}, A_p, A_s] = [35, 45, 60, 70, 0.4, 28]$$

(a) Print the filter order, δ_p, and δ_s.

(b) Use f_freqs to compute and plot the magnitude response for $0 \leq f \leq 2F_{s2}$ using the linear scale.

(c) Use $fill$ to add shaded areas showing the design specifications on the magnitude response plot.

Section 8.7

☑ **P8.43.** Write a MATLAB script that uses *f_butters* and *f_bilin* to find the digital equivalent, $H(z)$, of a sixth-order lowpass Butterworth filter using the bilinear transformation method. Suppose the sampling frequency is $f_s = 10$ Hz. Prewarp the analog cutoff frequency so that the digital cutoff frequency comes out to be $F_c = 1$ Hz.

(a) Plot the impulse response, $h(k)$.

(b) Use *f_pzplot* to plot the poles and zeros of $H(z)$.

(c) Use *f_freqz* to compute and plot the magnitude response, $A(f)$. Add the ideal magnitude response and a plot legend.

P8.44. Write a MATLAB script that uses *f_butterz* to design a digital Butterworth bandstop filter that meets the following design specifications.

$$[f_s, F_{p1}, F_{s1}, F_{s2}, F_{p2}, \delta_p, \delta_s] = [2000, 200, 300, 600, 700, 0.05, 0.03]$$

(a) Find the smallest filter order that meets the specifications. Print the order.

(b) Use *f_freqz* to compute and plot the magnitude response.

(c) Use *fill* to add shaded areas showing the design specifications.

P8.45. Write a MATLAB script that uses *f_cheby1z* to design a digital Chebyshev-I bandstop filter that meets the following design specifications.

$$[f_s, F_{p1}, F_{s1}, F_{s2}, F_{p2}, \delta_p, \delta_s] = [2000, 200, 300, 600, 700, 0.05, 0.03]$$

(a) Find the smallest filter order that meets the specifications. Print the order.

(b) Use *f_freqz* to compute and plot the magnitude response.

(c) Use *fill* to add shaded areas showing the design specifications.

P8.46. Write a MATLAB script that uses *f_cheby2z* to design a digital Chebyshev-II bandpass filter that meets the following design specifications.

$$[f_s, F_{s1}, F_{p1}, F_{p2}, F_{s2}, \delta_p, \delta_s] = [1600, 250, 350, 550, 650, 0.06, 0.04]$$

(a) Find the smallest filter order that meets the specifications. Print the order.

(b) Use *f_freqz* to compute and plot the magnitude response.

(c) Use *fill* to add shaded areas showing the design specifications.

P8.47. Write a MATLAB script that uses *f_ellipticz* to design a digital elliptic bandpass filter that meets the following design specifications.

$$[f_s, F_{s1}, F_{p1}, F_{p2}, F_{s2}, \delta_p, \delta_s] = [1600, 250, 350, 550, 650, 0.06, 0.04]$$

(a) Find the smallest filter order that meets the specifications. Print the order.

(b) Use *f_freqz* to compute and plot the magnitude response.

(c) Use *fill* to add shaded areas showing the design specifications.

Adaptive Signal Processing

Learning Objectives

- Understand how to use adaptive filters to perform system identification, channel equalization, signal prediction, and noise cancellation (Section 9.1)

- Know how to compute the mean square error and how to find an optimal weight vector that minimizes the mean square error (Section 9.2)

- Understand how to implement the least mean square (LMS) method for updating the weight vector (Section 9.3)

- Know how to find bounds on the step size and estimate the rate of convergence and the steady-state error of the LMS method (Section 9.4)

- Understand how to modify the basic LMS method to enhance performance using the normalized method, the correlation method, and the leaky method (Section 9.5)

- Be able to design FIR filters using pseudo-filter input-output specifications (Section 9.6)

- Know how to apply the recursive least squares (RLS) method (Section 9.7)

- Know how to apply the filtered-x LMS and signal-synthesis methods to achieve active control of acoustic noise (Section 9.8)

- Be able to identify nonlinear discrete-time systems using an adaptive radial basis function or RBF network (Section 9.9)

- Know how to use the GUI module g_adapt to perform system identification (Section 9.10)

● ● ● ● ● ● ● ● ● ● ● ● ● ● ●●●●

9.1 Motivation

The digital filters investigated in previous chapters have one fundamental character-istic in common. The coefficients of these filters are fixed; they do not evolve with time. When the filter parameters are allowed to vary with time, this leads to a power-ful new family of digital filters called *adaptive* filters. In this chapter, we investigate an adaptive FIR type of filter called a *transversal* filter that has the following generic form.

Adaptive transversal filter

$$y(k) = \sum_{i=0}^{m} w_i(k)x(k-i)$$

Notice that the constant FIR coefficient vector b has been replaced by a time-varying vector, $w(k)$, of length $m + 1$, called the *weight* vector. The adaptive filter design problem consists of developing an algorithm for updating the weight vector, $w(k)$, to ensure that the filter satisfies some design criterion. For example, the objective might be to get the filter output, $y(k)$, to track a desired output, $d(k)$, as time increases. The transversal filter structure has an important qualitative advantage over an IIR filter structure. Once the weight vector has converged, the resulting filter is guaranteed to be stable.

We begin this chapter by introducing some examples of applications of adaptive filters. Four broad classes of applications are examined, including system identifi-cation, channel equalization, signal prediction, and noise cancellation. Next, the mean-square-error design criterion is formulated and a closed-form filter solution called the Weiner filter is developed. A simple and elegant method for updating the filter weights called the least mean square or LMS method is then developed. The performance characteristics of the LMS method are investigated, including bounds on the step size that ensure convergence, estimates of the convergence rate, and estimates of the steady-state error. A number of modifications to the basic LMS method are then presented, including the normalized LMS method, the correlation LMS method, and the leaky LMS method. A simple design technique for FIR filters using pseudo-filter input-output specifications is then introduced. Next, an efficient recursive adaptive filter implementation called the recursive least squares or RLS method is presented. A more general LMS technique called the filtered-x LMS method is then developed, along with a signal-synthesis method. Both techniques are then applied to the problem of active control of acoustic noise. Next, attention turns to the problem of identifying nonlinear discrete-time systems using radial ba-sis function or RBF networks. Finally, a GUI module called g_adapt is introduced that allows the user to perform system identification without any need for program-ming. The chapter concludes with an application example and a summary of adaptive signal-processing techniques. The selected application example features the identi-fication of a differential-difference system commonly used in the chemical-process control industry.

There are a variety of applications of adaptive filters that arise in fields ranging from equalization of telephone channels to geophysical exploration for oil and gas deposits. In this section, we outline some general classes of applications of adaptive signal processing. A brief history of adaptive signal processing and its applications can be found in Haykin (2002).

9.1.1   System Identification

The success enjoyed by engineers in applying analysis and design techniques to practical problems often can be traced to the effective use of mathematical models of physical phenomena. In many instances, a mathematical model can be developed using underlying physical principles and an understanding of the components of the system and how they are interconnected. However, there are other instances where this bottom-up approach is less effective, because the physical system or phenomenon is too complex and is not well understood. In these cases, it is often useful to think of the unknown system as a *black box*, where measurements can be taken of the input and output, but little is known about the details of what is inside the box (hence the term black). Typically, we assume that the unknown system can be modeled as a linear discrete-time system. The problem of obtaining a model of the system from input and output measurements is called the *system identification* problem. Adaptive filters are highly effective for performing system identification using the configuration shown in Figure 9.1.

It is standard practice to represent an adaptive filter in a block diagram using a diagonal arrow through the block. The arrow can be thought of as a needle on a dial that is adjusted as the parameters of the adaptive filter are changed. The system identification configuration in Figure 9.1 shows the adaptive filter in parallel with the unknown black-box system. Both systems are driven with the same test input, $x(k)$. The objective is to adjust the parameters or coefficients of the adaptive filter so that its output mimics the response of the unknown system. Thus, the *desired output*, $d(k)$, is the output of the unknown system, and the difference between the desired output and the adaptive filter output, $y(k)$, is the *error* signal $e(k)$.

$$e(k) \overset{\Delta}{=} d(k) - y(k) \tag{9.1}$$

The algorithm for updating the parameters of the adaptive filter uses the error $e(k)$ and the input $x(k)$ to adjust the weights so as to reduce the square of the error.

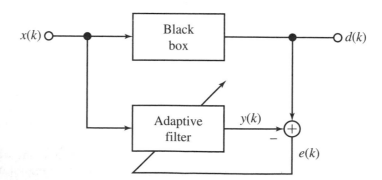

Figure 9.1   System Identification

Later, we will explore the adaptive-filter block in more detail. For now, notice that if the error signal can be made to go to zero, then the adaptive-filter output is an exact reproduction of the unknown system output. In this case, the adaptive filter becomes an exact model of the unknown black-box system. This model can be used in simulation studies, and it also can be used to predict the response of the unknown system to new inputs.

9.1.2 Channel Equalization

Another important class of applications of adaptive filters can be found in the communication industry. Consider the problem of transmitting information over a communication channel. At the receiving end, the signal will be distorted due to the effects of the channel itself. For example, the channel invariably will exhibit some type of frequency-response characteristic with some spectral components of the input attenuated more than others. In addition, there will be phase distortion and delay, and the signal may be corrupted with additive noise. To remove, or at least minimize, the detrimental effects of the communication channel, we should pass the received signal through a filter that approximates the inverse of the channel so that the cascade or series connection of the two systems restores the original signal. The technique of inserting an inverse system in series with an original unknown system is called *equalization*, because it results in an overall system with a transfer function of one. Equalization, or inverse modeling, can be achieved with an adaptive filter using the configuration shown in Figure 9.2.

Here the black-box system, which represents the unknown communication channel, is in series with the adaptive filter. This series combination is in parallel with a delay element corresponding to a delay of M samples. Thus, the desired output in this case is simply a delayed version of the transmitted signal.

$$d(k) = x(k - M) \tag{9.2}$$

The reason for inserting a delay is that the black-box system typically imparts some delay to the signal $x(k)$ as it is processed by the system. Therefore, an exact inverse system would have to include a corresponding time advance, something that is not feasible for a causal filter. Furthermore, if the unknown black-box system does

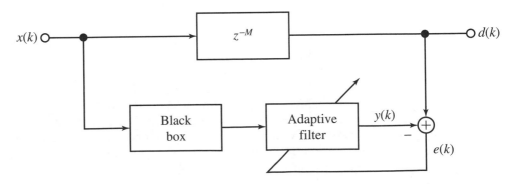

Figure 9.2 Channel Equalization

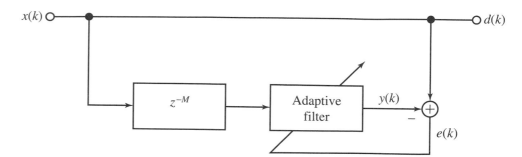

$x(k)$ ○

z^{-M}

Adaptive filter

$y(k)$

$d(k)$ ○

$e(k)$

Figure 9.3 Signal Prediction

represent a communication channel, then delaying the signal by M samples will not distort the information that arrives at the receiver. Recall that a constant group delay can be achieved by using a linear-phase FIR filter.

9.1.3 Signal Prediction

As an illustration of another class of applications, consider the problem of encoding speech for transmission or storage. The direct technique is to encode the speech samples themselves. An effective alternative is to use the past samples of the speech to predict the values of future samples. Typically, the error in the prediction has a smaller variance than the original speech signal itself. Consequently, the prediction error can be encoded using a smaller number of bits than a direct encoding of the speech. In this way, an efficient encoding system can be implemented. An adaptive filter can be used to predict future samples of speech, or other signals, by using the configuration shown in Figure 9.3.

In this case, the desired output is the input itself. Since the adaptive filter processes a delayed version of the input, the only way the error can be made to go to zero is if the adaptive filter successfully predicts the value of the input M samples into the future. Of course, an exact prediction of a completely random input is not possible with a causal system. Typically, the input consists of an underlying deterministic component plus additive noise. In these cases, information from the past samples can be used to minimize the square of the prediction error.

9.1.4 Noise cancellation

Still another broad class of applications of adaptive filters focuses on the problem of interference or noise cancellation. As an illustration, suppose the driver of a car places a call using a cell phone. The cell phone microphone will pick up both the driver's voice plus ambient road noise that varies with the car speed and the driving conditions. To make the speaker's voice more intelligible at the receiving end, a second reference microphone can be placed in the car to measure the ambient road noise. An adaptive filter can then be used to process this reference signal and subtract the result from the signal detected by the primary microphone using the configuration shown in Figure 9.4.

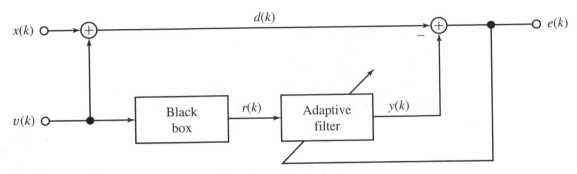

Figure 9.4 Noise cancellation

Note that the desired output, $d(k) = x(k) + v(k)$, consists of speech plus road noise. The reference signal, $r(k)$, is a filtered version of the noise. The presence of an unknown black-box system takes into account the fact that the primary microphone and the reference microphone are placed at different locations, and therefore, the reference signal $r(k)$ is different from, but correlated to, the noise $v(k)$ appearing at the primary microphone. The error in this case is

$$e(k) = x(k) + v(k) - y(k) \tag{9.3}$$

If the speech, $x(k)$, and the additive road noise, $v(k)$, are uncorrelated with one another, then the minimum possible value for $e^2(k)$ occurs when $y(k) = v(k)$, which corresponds to the road noise being removed completely from the transmitted speech signal, $e(k)$.

● ● ● ● ● ● ● ● ● ● ● ● ● ● ●

9.2 Mean Square Error

9.2.1 Adaptive Transversal Filters

An mth-order adaptive transversal filter is a linear time-varying discrete-time system that can be represented by the following difference equation.

$$y(k) = \sum_{i=0}^{m} w_i(k)x(k - i) \tag{9.4}$$

Note that the filter output is a time-varying linear combination of the past inputs. A signal flow graph of an adaptive transversal filter is shown in Figure 9.5 for the case $m = 4$. Given the structure shown in Figure 9.5, a transversal filter is sometimes referred to as a tapped delay line.

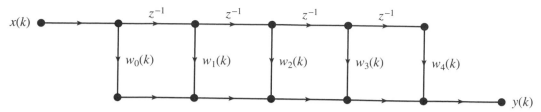

Figure 9.5 Signal Flow Graph of an Adaptive Transversal Filter

A compact formulation of a transversal filter can be obtained by introducing the following pair of $(m + 1) \times 1$ column vectors.

$$\theta(k) \triangleq [x(k), x(k-1), \ldots, x(k-m)]^T \tag{9.5a}$$

$$w(k) \triangleq [w_0(k), w_1(k), \ldots, w_m(k)]^T \tag{9.5b}$$

Here $\theta(k)$ is a vector of past inputs called the *state* vector, and $w(k)$ is the current value of the *weight* vector. Combining Eqs. 9.4 and 9.5, we find that the adaptive filter output can be expressed as a dot product of the two vectors.

Adaptive filter output

$$y(k) = w^T(k)\theta(k), \quad k \geq 0 \tag{9.6}$$

In view of Eq. 9.6, an adaptive filter can be thought of as having two inputs: the time-varying weight vector $w(k)$ and the vector of past inputs $\theta(k)$. The vector $w(k)$ is itself the output of a weight-update algorithm, as shown in Figure 9.6. Recall that $d(k)$ in Figure 9.6 represents the *desired output*, and the difference between $d(k)$ and the filter output $y(k)$ is the *error*, $e(k)$. That is,

$$e(k) = d(k) - y(k) \tag{9.7}$$

The details of how the desired output $d(k)$ and the filter input $x(k)$ are generated depend on the type of adaptive filter application. Examples of different classes of applications were presented in Section 9.1.

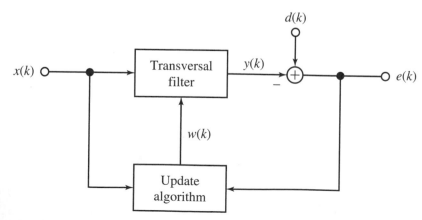

Figure 9.6 Adaptive Filter Block Showing the Weight-update Algorithm

9.2.2 Cross-correlation Revisited

Typically, the filter input $x(k)$ and the desired output $d(k)$ are modeled as random signals or, more formally, random processes. This makes $y(k)$ and $e(k)$ random as well. For the purpose of this analysis, let us assume that $x(k)$ and $d(k)$ are *stationary* random signals, meaning that their statistical properties do not change with time. The notion of cross-correlation, first introduced in Chapter 4, can be extended to random signals using the expected value operator.

DEFINITION

9.1: Random
Cross-correlation

Let $x(k)$ be an L-point random signal and let $y(k)$ be an M-point random signal, where $M \leq L$. Then the *cross-correlation* of $x(k)$ with $y(k)$ is denoted as $r_{xy}(i)$ and defined as

$$r_{xy}(i) \triangleq E[x(k)y(k-i)], \quad 0 \leq i < M$$

Recall from Eq. 4.67, that if a random signal has the property that it is *ergodic*, then the expected value operation can be computed using a time average. Consequently, for a signal $x(k)$ the expected, or mean, value can be approximated as

$$E[x(k)] \approx \frac{1}{N} \sum_{n=0}^{N-1} x(k-n), \quad N \gg 1 \tag{9.8}$$

If the signal $x(k)$ is periodic, then the expected value can be determined exactly by computing the average value over one period.

When the approximation in Eq. 9.8 is used to evaluate the cross-correlation of two random signals, Definition 9.1 reduces to the deterministic definition of linear cross-correlation introduced previously in Definition 4.3. It is in this sense that Definition 9.1 is a generalization of the notion of cross-correlation to random signals.

There are two properties of cross-correlation that we will make use of in our analysis of adaptive filters. Since $x(k)$ is assumed to be *stationary*, the statistical properties of $x(k)$ do not change with time. Consequently, if $x(k)$ is translated in time, its expected or mean value does not change.

$$E[x(k-i)] = E[x(k)], \quad i \geq 0 \tag{9.9}$$

Another fundamental property concerns the expected value of the product of two signals. If two random signals $x(k)$ and $y(k)$ are *statistically independent* of one another or *uncorrelated*, then the expected value of their product is equal to the product of their expected values.

$$E[x(k)y(k)] = E[x(k)]E[y(k)] \tag{9.10}$$

Note that if $x(k)$ and $y(k)$ are uncorrelated and either $x(k)$ or $y(k)$ has zero mean, then the expected value of their product is zero.

Suppose $v(k)$ denotes white noise uniformly distributed over $[a, b]$. This particular broadband signal turns out to be an excellent input signal for system identification purposes due to its flat power density spectrum. White noise was examined in detail in Chapter 3. For convenient reference, the following expression denotes the average power of uniform white noise as developed in Chapter 3.

$$P_v = \frac{b^3 - a^3}{3(b - a)} \tag{9.11}$$

9.2.3   Mean Square Error

The average of the square of the error of the system in Figure 9.6 is referred to as the *mean square error* of the system. The mean square error, $\epsilon(w)$, can be expressed in terms of the expected value operator as follows.

$$\epsilon(w) \stackrel{\Delta}{=} E[e^2(k)] \tag{9.12}$$

To see how the mean square error is affected by the filter weights, consider the case when the weight vector w is held constant. Using Eqs. 9.6 and 9.7, we find that the square of the error can be expressed as follows.

$$
\begin{aligned}
e^2(k) &= [d(k) - w^T \theta(k)]^2 \\
&= d^2(k) - 2d(k)w^T\theta(k) + [w^T\theta(k)]^2 \\
&= d^2(k) - 2w^T d(k)\theta(k) + w^T\theta(k)\theta^T(k)w
\end{aligned}
\tag{9.13}
$$

Since the expected value operation is linear, the expected value of the sum is the sum of the expected values, and scaling a variable simply scales its expected value. Taking the expected value of both sides of Eq. 9.13 then yields the following expression for the mean square error.

$$\epsilon(w) = E[d^2(k)] - 2w^T E[d(k)\theta(k)] + w^T E[\theta(k)\theta^T(k)]w \tag{9.14}$$

To develop a compact formulation of $\epsilon(w)$, we introduce an $(m + 1) \times 1$ column vector p and an $(m + 1) \times (m + 1)$ matrix R as follows.

$$p \stackrel{\Delta}{=} E[d(k)\theta(k)] \tag{9.15a}$$

$$R \stackrel{\Delta}{=} E[\theta(k)\theta^T(k)] \tag{9.15b}$$

The vector p is referred to as the *cross-correlation* vector of the desired output, $d(k)$, with the vector of past inputs, $\theta(k)$. From Eq. 9.5a, we have $p_i = E[d(k)x(k-i)]$. It then follows from Definition 9.1 that

Cross-correlation
vector

$$p_i = r_{dx}(i), \quad 0 \leq i \leq m \tag{9.16}$$

The square matrix R, obtained by taking the expected value of the outer product of $\theta(k)$ with itself, is referred to as the *auto-correlation* matrix of the past inputs. Again, note from Eq. 9.5a that

$$R_{ij} = E[x(k - i)x(k - j)], \quad 0 \leq i, j \leq m \tag{9.17}$$

Since $x(k)$ is assumed to be stationary, the signal $x(k - i)x(k - j)$ can be translated in time without changing its expected value. Replacing k with $k + i$ yields $R_{ij} = E[x(k)x(k + i - j)]$. Thus, from Definition 9.1 we have

Auto-correlation matrix

$$R_{ij} = r_{xx}(j - i), \quad 0 \leq i, j \leq m \tag{9.18}$$

The auto-correlation matrix has a number of interesting and useful properties. First, notice from Eq. 9.17 that since $x(k - i)x(k - j) = x(k - j)x(k - i)$, it follows that R is symmetric. Next, observe from Eq. 9.18 that $j = i$ yields $R_{ii} = r_{xx}(0)$. But the auto-correlation evaluated at a lag of zero is just the average power. That is,

$$R_{ii} = r_{xx}(0)$$
$$= E[x^2(k)]$$
$$= P_x, \quad 0 \leq i \leq m \tag{9.19}$$

Consequently, when $x(k)$ is stationary, the diagonal elements of R are all identical and equal to the average power of the input. More generally, it is clear from Eq. 9.18 that the symmetric autocorrelation matrix R has diagonal *bands* of equal elements, above and below the diagonal, as can be seen from the case $m = 4$ shown in Eq. 9.20. A matrix with this diagonal striped structure is referred to as a *Toeplitz* matrix.

$$R = \begin{bmatrix} r_{xx}(0) & r_{xx}(1) & r_{xx}(2) & r_{xx}(3) & r_{xx}(4) \\ r_{xx}(1) & r_{xx}(0) & r_{xx}(1) & r_{xx}(2) & r_{xx}(3) \\ r_{xx}(2) & r_{xx}(1) & r_{xx}(0) & r_{xx}(1) & r_{xx}(1) \\ r_{xx}(3) & r_{xx}(2) & r_{xx}(1) & r_{xx}(0) & r_{xx}(1) \\ r_{xx}(4) & r_{xx}(3) & r_{xx}(2) & r_{xx}(1) & r_{xx}(1) \end{bmatrix} \tag{9.20}$$

Using the definitions of the cross-correlation vector p and the input-correlation matrix R in Eq. 9.15, the expression for the mean-square-error performance function in Eq. 9.14 simplifies to

Mean square error

$$\epsilon(w) = P_d - 2w^T p + w^T R w \tag{9.21}$$

Here $P_d = E[d^2(k)]$ is the average power of the desired output. It is clear from Eq. 9.21 that the mean square error is a *quadratic* function of the weight vector w. Note that when $m = 1$, the mean square error can be thought of as a surface over the w plane. Our objective is to locate the lowest point on this error surface.

To find an optimal value for w, one that minimizes the mean square error, we start by computing the gradient vector, $\nabla\epsilon(w)$, of partial derivatives of $\epsilon(w)$ with respect to the elements of w. Taking the derivatives of $\epsilon(w)$ in Eq. 9.21 with respect to w_i and combining the results for $0 \leq i \leq m$, it is possible to show that

$$\nabla\epsilon(w) = 2(Rw - p) \tag{9.22}$$

Consider the case when the input-correlation matrix R is invertible. Setting $\nabla\epsilon(w) = 0$ in Eq. 9.22 and solving for w, we then arrive at the following optimal value for the weight vector.

Optimal weight vector

$$w^* = R^{-1}p \tag{9.23}$$

The optimal weight vector w^* in Eq. 9.23 is referred to as the *Wiener* solution (Levinson, 1947).

Example 9.1 **Optimal Weight Vector**

As an illustration of the mean square error and the optimal weight vector, consider the following example adapted from Widrow and Sterns (1985). Suppose $m = 1$ and the input and desired output are the following periodic functions of period N where $N > 2$.

$$x(k) = 2\cos\left(\frac{2\pi k}{N}\right)$$

$$d(k) = \sin\left(\frac{2\pi k}{N}\right)$$

Thus, the adaptive filter must perform a scaling and phase-shifting operation on the input. First, consider the cross-correlation vector p. Since $d(k)$ and $x(k)$ are periodic, we can compute the expected values by averaging over one period. For convenience, let $\psi = 2\pi/N$. Using Eq. 9.16 and some trigonometric identities from Appendix D yields

$$
\begin{aligned}
p_i &= E[d(k)x(k-i)] \\
&= E[2\sin(k\psi)\cos\{(k-i)\psi\}] \\
&= 2E[\sin(k\psi)\{\cos(k\psi)\cos(i\psi) + \sin(k\psi)\sin(i\psi)\}] \\
&= 2\cos(i\psi)E[\sin(k\psi)\cos(k\psi)] + 2\sin(i\psi)E[\sin^2(k\psi)] \\
&= \cos(i\psi)E[\sin(2k\psi)] + \sin(i\psi)E[1 - \cos(2k\psi)] \\
&= \sin(i\psi), \quad 0 \leq i \leq 1
\end{aligned}
$$

Thus, the cross-correlation between the desired output and the vector of past inputs is

$$p = [0, \sin(\psi)]^T$$

Next, consider the auto-correlation matrix of the past inputs. Here

$$
\begin{aligned}
E[x(k)x(k-i)] &= E[4\cos(k\psi)\cos\{(k-i)\psi\}] \\
&= 4E[\cos(k\psi)\{\cos(k\psi)\cos(i\psi) + \sin(k\psi)\sin(i\psi)\}] \\
&= 4\cos(i\psi)E[\cos^2(k\psi)] + 4\sin(i\psi)E[\cos(k\psi)\sin(k\psi)] \\
&= 2\cos(i\psi)E[1 + \cos(2k\psi)] + 2\sin(i\psi)E[\sin(2k\psi)] \\
&= 2\cos(i\psi)
\end{aligned}
$$

Thus, from Eq. 9.20, the auto-correlation matrix is

$$
R = 2 \begin{bmatrix} 1 & \cos(\psi) \\ \cos(\psi) & 1 \end{bmatrix}
$$

Since $N > 2$, it is evident that R is symmetric, banded, and nonsingular with

$$
\begin{aligned}
\det(R) &= 4[1 - \cos^2(\psi)] \\
&= 4\sin^2(\psi)
\end{aligned}
$$

From Eq. 9.23, the optimal value for the weight vector in this case is

$$
\begin{aligned}
w^* &= \begin{bmatrix} 2 & 2\cos(\psi) \\ 2\cos(\psi) & 2 \end{bmatrix}^{-1} \begin{bmatrix} 0 \\ \sin(\psi) \end{bmatrix} \\
&= \frac{1}{4\sin^2(\psi)} \begin{bmatrix} 2 & -2\cos(\psi) \\ -2\cos(\psi) & 2 \end{bmatrix} \begin{bmatrix} 0 \\ \sin(\psi) \end{bmatrix} \\
&= \frac{1}{2\sin^2(\psi)} \begin{bmatrix} -\cos(\psi)\sin(\psi) \\ \sin(\psi) \end{bmatrix} \\
&= 0.5[-\cot(\psi), \quad \csc(\psi)]^T
\end{aligned}
$$

To make the problem more specific, suppose $N = 4$, in which case $\psi = \pi/2$. A plot of the resulting mean-square-error surface is shown in Figure 9.7. In this case, the optimal weight vector that minimizes the mean square error is

$$w^* = [0, 0.5]^T$$

Mean square error

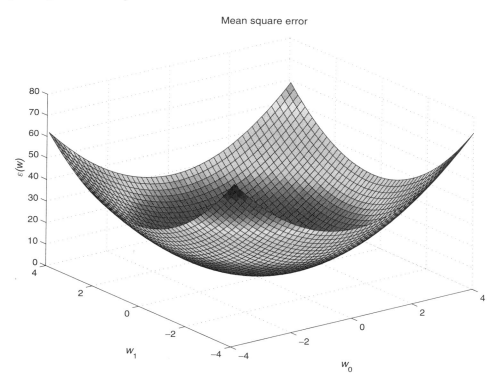

Figure 9.7 Mean-square-error Surface for Example 9.1

When adaptive filters are used for system identification, the user often has direct control of the input signal $x(k)$. To get reliable results, the spectral content of the input should be sufficiently rich that it excites all of the natural modes of the system being identified. One input that is particularly rich in frequency content is a random white noise input, a signal whose power density spectrum is flat.

Example 9.2 | **White Noise Input**

Suppose $x(k)$ is zero-mean white noise with an average power of

$$P_x = E[x^2(k)]$$

To determine the input-correlation matrix, first note that for white noise, the signals $x(k)$ and $x(k-i)$ are uncorrelated for $i \neq 0$. For uncorrelated signals, the expected value of the product is equal to the product of the expected values. Since $x(k)$ is zero-mean white noise, we have $E[x(k)] = 0$. Thus, for $i \neq 0$,

$$E[x(k)x(k-i)] = E[x(k)]E[x(k-i)]$$

$$= 0$$

It then follows from Eq. 9.19 that for zero-mean white noise with average power P_x, the input-correlation matrix is simply

$$R = P_x I$$

Consequently, zero-mean white noise produces a nonsingular, diagonal input-correlation matrix with the average power of $x(k)$ along the diagonal. It follows from Eq. 9.23 that the optimal weight in this case is simply

$$w^* = \frac{p}{P_x}$$

• • • • • • • • • • • • • • •

9.3   **The Least Mean Square (LMS) Method**

The optimal weight vector found in Section 9.2 was based on the assumption that the input $x(k)$ and the desired output $d(k)$ are stationary random signals. This assumption is useful for analysis purposes because it allows us to determine, for example, the characteristics of the input that are required to ensure that an optimal weight vector exists and is unique. However, in actual applications of adaptive filters, the input and desired output are often not stationary, instead their statistical properties evolve with time. When the input and desired output are not stationary, it is useful to iteratively update an estimate of the optimal weight vector using a numerical search technique.

Suppose $w(0)$ is an initial guess for the optimal weight vector. For example, in the absence of any specialized knowledge about the application, one might simply take $w(0) = 0$. At subsequent time steps, the new weight is set to the old weight plus a correction term as follows.

$$w(k + 1) = w(k) + \Delta w(k), \quad k \geq 0 \tag{9.24}$$

A simple way to compute a correction term, $\Delta w(k)$, is to use the gradient vector of partial derivatives of the mean square error, $\epsilon(w)$, with respect to the elements of w.

$$\nabla \epsilon_i(w) \triangleq \frac{\partial \epsilon(w)}{\partial w_i}, \quad 0 \leq i \leq m \tag{9.25}$$

The gradient vector, $\nabla \epsilon(w)$, points in the direction of maximum increase of $\epsilon(w)$. For example, when $m = 1$, the mean square error is a surface, and $\nabla \epsilon(w)$ is a 2×1 vector that points in the steepest uphill direction: the direction of steepest ascent. Since our objective is to find the minimum point on this surface, we take a step of length $\mu > 0$ in the opposite direction of the gradient. That is, we set $\Delta w(k) = -\mu \nabla \epsilon[w(k)]$, in which case the weight-update algorithm becomes

$$w(k + 1) = w(k) - \mu \nabla \epsilon[w(k)], \quad k \geq 0 \tag{9.26}$$

The weight-update formula in Eq. 9.26 is called the method of *steepest descent*. Note that the *step size*, μ, must be kept small, because as one departs from the point $w(k)$, the direction of steepest descent changes. The main computational difficulty of the steepest-descent method is the need to compute the gradient vector, $\nabla \epsilon(w)$, at each discrete time. Since the ith element of the gradient vector represents the slope of $\epsilon(w)$ along the ith dimension, the gradient vector can be approximated numerically using differences. For example, let i^j denote the jth column of the $(m+1) \times (m+1)$ identity matrix $I = [i^1, i^2, \ldots, i^{m+1}]$. If $\delta > 0$ is small, then the jth element of the gradient vector can be approximated with the following forward difference.

$$\nabla \epsilon_j(w) \approx \frac{\epsilon(w + \delta i^{j+1}) - \epsilon(w)}{\delta}, \quad 0 \le j \le m \tag{9.27}$$

Observe that in the limit as δ approaches zero, the expression in Eq. 9.27 is the partial derivative of $\epsilon(w)$ with respect to w_j. The approximation of the gradient in Eq. 9.27 requires $m+2$ evaluations of the mean square error. Suppose the mean square error is itself approximated by using a time average of N samples of the square of the error. From Eq. 9.6, each sample of $e^2(k)$ requires $m+1$ floating-point multiplications or FLOPs. Thus, the total number of FLOPs required to numerically estimate the gradient of the mean square error is

$$r = N(m+1)(m+2) \quad \text{FLOPs} \tag{9.28}$$

For a large filter, $m \gg 1$, and for an accurate estimate, $N \gg 1$. Consequently, implementing the steepest-descent method using a numerical estimate of the gradient vector can be computationally expensive.

There is an alternative approach to estimating the gradient vector that is much more cost effective (Widrow and Sterns, 1985). Suppose that, for the purpose of computing the gradient, the mean square error is approximated using the *instantaneous* value of the square of the error. That is, for the purpose of computing $\nabla \epsilon(w)$, the following approximation is used for the mean square error.

$$\epsilon(w) \approx e^2(k) \tag{9.29}$$

This is clearly a rough approximation because it is equivalent to using a single sample to estimate the mean. However, this technique yields a dramatic simplification in the expression for the gradient. Let $\hat{\nabla} \epsilon(w)$ be the estimate of the gradient using Eq. 9.29. Then from Eqs. 9.6 and 9.7, we have

$$\hat{\nabla} \epsilon(w) = 2e(k) \frac{\partial e(k)}{\partial w}$$

$$= -2e(k)\theta(k) \tag{9.30}$$

Using this estimate of the gradient, the steepest-descent method in Eq. 9.26 then reduces to the following simplified weight-update algorithm.

LMS method

$$w(k+1) = w(k) + 2\mu e(k)\theta(k), \quad k \ge 0 \tag{9.31}$$

The weight-update formula in Eq. 9.31 is called the *least mean square* or LMS method (Widrow and Hoff, 1960). Note that, unlike the traditional steepest-descent method, the LMS method requires only $m + 1$ FLOPs to estimate the gradient at each time step. The LMS is a highly efficient way to update the weight vector. For example, when $N = 10$ and $m = 10$, the LMS method is more than two orders of magnitude faster than the numerical steepest-descent method.

Although the approximation used to estimate the gradient vector in Eq. 9.29 may appear to be rather crude, experience has shown that the LMS algorithm for updating the weights is quite robust. Indeed, Hassibi *et al.*, (1996) have shown that the LMS algorithm is optimal when a minimax error criterion is used.

The estimate of the gradient in Eq. 9.30 is itself a random signal. It is instructive to examine the mean or expected value of this random signal. Suppose the weight w has converged to its steady-state value and is constant. Starting from Eq. 9.30, and using the definitions of p and R in Eq. 9.15, we have

$$
\begin{aligned}
E[\hat{\nabla}\epsilon(w)] &= -2E[e(k)\theta(k)] \\
&= -2E[d(k)\theta(k) - y(k)\theta(k)] \\
&= -2E[d(k)\theta(k) - \theta(k)\{\theta^T(k)w\}] \\
&= -2\{E[d(k)\theta(k)] - E[\theta(k)\theta^T(k)]w\} \\
&= 2(Rw - p)
\end{aligned}
\tag{9.32}
$$

But from Eq. 9.22, the exact value of the gradient of the mean square error is $\nabla\epsilon(w) = 2(Rw - p)$. Hence,

$$
E[\hat{\nabla}\epsilon(w)] = \nabla\epsilon(w)
\tag{9.33}
$$

That is, the expected value of the estimate of the gradient of the mean square error is equal to the gradient of the mean square. Consequently, we say that $\hat{\nabla}\epsilon(w)$ is an *unbiased* estimate of $\nabla\epsilon(w)$.

Example 9.3   System Identification

To illustrate the LMS method, consider the system identification problem shown in Figure 9.8. To make the example specific, suppose the system to be identified has the following transfer function.

$$
H(z) = \frac{2 - 3z^{-1} - z^{-2} + 4z^{-4} + 5z^{-5} - 8z^{-6}}{1 - 1.6z^{-1} + 1.75z^{-2} - 1.436z^{-3} + 0.6814z^{-4} - 0.1134z^{-5} - 0.0648z^{-6}}
$$

Next, suppose the input $x(k)$ consists of $N = 1000$ samples of white noise uniformly distributed over $[-1, 1]$. Let the order of the adaptive transversal filter be $m = 50$, and suppose the step size is $\mu = 0.01$. A plot of the first 500 samples of

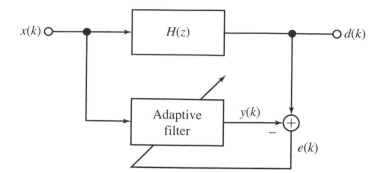

Figure 9.8 System Identification

the square of the error, obtained by running script *exam9_3* on the distribution CD, is shown in Figure 9.9. It is clear that the square of the error converges close to zero after approximately 400 samples.

One way to assess the effectiveness of the adaptive filter is to compare the magnitude response of the system, $H(z)$, with the magnitude response of the adaptive filter, $W(z)$, using the final steady-state weight, $w(N-1)$. The two magnitude responses are plotted in Figure 9.10, where it is evident that they are nearly identical. Note that this is true in spite of the fact that $H(z)$ is a an IIR filter with six poles and six zeros, while the steady-state adaptive filter is an FIR filter of order $m = 50$. By

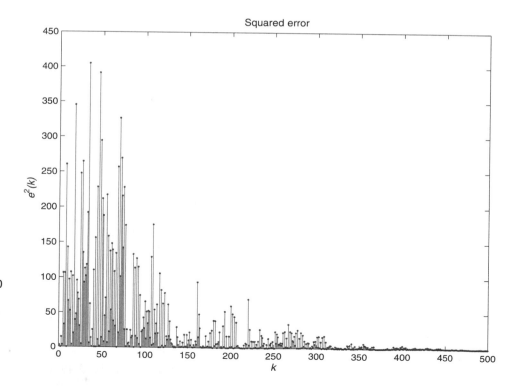

Figure 9.9 First 500 Samples of Squared Error during System Identification using the LMS Method with $m = 50$ and $\mu = 0.01$

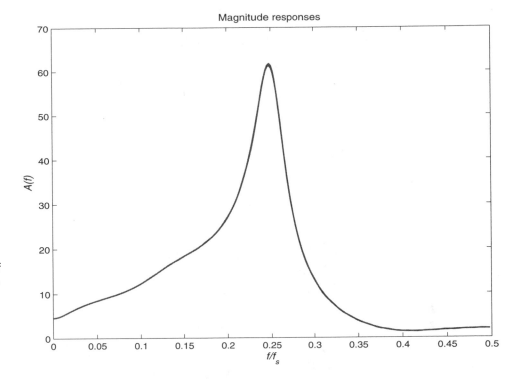

Figure 9.10 Magnitude Responses of the Original IIR System and the Identified System using the LMS Method with $m = 50$, $\mu = 0.01$, and $N = 1000$ Samples

making the order of the adaptive filter sufficiently large, we see that it can model an IIR filter as well. Of course, if the system to be identified is an FIR filter of order p, then an exact fit will be obtained using any adaptive filter of order $m \geq p$, assuming infinite precision arithmetic is used.

MATLAB Toolbox

The FDSP toolbox contains the following function which implements the LMS method.

```
[w,e] = f_lms (x,d,m,mu,w);              % LMS method
```

On entry to *f_lms*, x is a vector of length N containing the input, d is a vector of length N containing the desired output, $m \geq 0$ specifies the order of the adaptive filter, $mu > 0$ is the step size, and w is an optional vector of length $m+1$ containing an initial guess for the weights. The default value is $w = 0$. On exit from *f_lms*, w is a vector of length $m + 1$ containing the estimated value of the optimal weights, and e is an optional vector of length N containing the samples of the error. Thus, $e(k)$ versus k is an approximation to the learning curve.

9.4 Performance Analysis of LMS Method

Although the LMS method is very simple to implement, there remains the question of determining an effective value for the step size, μ. The step size should be small enough to guarantee convergence to an acceptable steady-state error, yet large enough to ensure that the convergence is rapid.

9.4.1 Step Size

Recall that the LMS method uses the error $e(k)$ and the vector of past inputs $\theta(k)$ to update the weight estimate as follows.

$$w(k + 1) = w(k) + 2\mu e(k)\theta(k), \quad k \geq 0 \tag{9.34}$$

Since $w(k)$ is a random signal, let us begin by examining what happens to its mean or expected value as k increases. Taking the expected value of both sides of Eq. 9.34, and noting that $e(k) = d(k) - \theta^T(k)w(k)$, this yields:

$$
\begin{aligned}
E[w(k + 1)] &= E[w(k)] + 2\mu E[e(k)\theta(k)] \\
&= E[w(k)] + 2\mu E[\{d(k) - \theta^T(k)w(k)\}\theta(k)] \\
&= E[w(k)] + 2\mu E[d(k)\theta(k) - \theta(k)\{\theta^T(k)w(k)\}] \\
&= E[w(k)] + 2\mu E[d(k)\theta(k)] - 2\mu E[\theta(k)\theta^T(k)w(k)] \tag{9.35}
\end{aligned}
$$

Recall that if two random signals are statistically independent of one another, the expected value of the product is the product of the expected values. For moderate convergence rates, the past inputs $\theta(k)$ and the weights $w(k)$ can be assumed to be statistically independent. Using the definitions of R and p from Eq. 9.15, we can then rewrite Eq. 9.35 as

$$
\begin{aligned}
E[w(k + 1)] &= E[w(k)] + 2\mu E[d(k)\theta(k)] - 2\mu E[\theta(k)\theta^T(k)]E[w(k)] \\
&= E[w(k)] + 2\mu p - 2\mu R E[w(k)] \\
&= (I - 2\mu R)E[w(k)] + 2\mu p \tag{9.36}
\end{aligned}
$$

To further simplify the expression for the expected value of the weight vector, it is helpful to introduce a signal that represents the variation of the weight from its optimal value in Eq. 9.23. The *weight variation* vector is denoted as $\delta w(k)$ and defined as

$$\delta w(k) \triangleq w(k) - w^* \tag{9.37}$$

We can reformulate Eq. 9.36 in terms of the weight variation by substituting $w(k) = \delta w(k) + w^*$ and using $E[w^*] = w^*$ and $Rw^* = p$. This yields:

$$
\begin{aligned}
E[\delta w(k + 1)] + w^* &= (I - 2\mu R)\{E[\delta w(k)] + w^*\} + 2\mu p \\
&= (I - 2\mu R)E[\delta w(k)] + (I - 2\mu R)w^* + 2\mu p \\
&= (I - 2\mu R)E[\delta w(k)] + w^* \tag{9.38}
\end{aligned}
$$

Thus, the expected value of the weight variation at iteration $k + 1$ is simply

$$E[\delta w(k + 1)] = (I - 2\mu R) E[\delta w(k)] \tag{9.39}$$

The virtue of the formulation in Eq. 9.39 is that it can be solved directly for $E[\delta w(k)]$ using induction.   Note that $E[\delta w(0)] = \delta w(0)$, in which case $E[\delta w(1)] = (I - 2\mu R)\delta w(0)$. More generally,

$$E[\delta w(k)] = (I - 2\mu R)^k \delta w(0), \quad k \geq 0 \tag{9.40}$$

The solution in Eq. 9.40 can be expressed in terms of the original weight vector, $w(k)$, by simply replacing $\delta w(k)$ with $w(k) - w^*$, which yields the following closed-form solution for the expected value of the weight vector at step k.

Weight-vector mean

$$E[w(k)] = w^* + (I - 2\mu R)^k [w(0) - w^*], \quad k \geq 0 \tag{9.41}$$

It is clear from Eq. 9.41 that $E[w(k)]$ will converge to the optimal weight w^* starting from an arbitrary initial guess if and only if $(I - 2\mu R)^k$ converges to the zero matrix as k approaches infinity. At this point, it is helpful to make use of a result from linear algebra. If A is a square matrix, then

$$A^k \to 0 \quad \text{as} \quad k \to \infty \tag{9.42}$$

if and only if the eigenvalues of A all lie strictly inside the unit circle of the complex plane (Noble, 1969). It can be shown by direct substitution that the ith eigenvalue of $A = I - 2\mu R$ is $r_i = 1 - 2\mu\lambda_i$, where λ_i is the ith eigenvalue of R. Thus, $E[w(k)]$ in Eq. 9.41 converges to w^* as $k \to \infty$ if and only if

$$|1 - 2\mu\lambda_i| < 1 \quad \text{for} \quad 1 \leq i \leq m + 1 \tag{9.43}$$

Since R is symmetric and positive-definite, its eigenvalues λ_i are real and positive. Consequently, Eq. 9.43 can be rewritten as $-1 < 1 - 2\mu\lambda_i < 1$. Subtracting one from each term, and dividing each term by $-2\lambda_i$ then yields the inequality $1/\lambda_i > \mu > 0$. This must hold for all $m + 1$ eigenvalues of R. Let

$$\lambda_{\max} \stackrel{\Delta}{=} \max\{\lambda_1, \lambda_2, \ldots, \lambda_{m+1}\} \tag{9.44}$$

It then follows that the range of step sizes over which the LMS method converges, starting from an arbitrary $w(0)$, is

Step-size range

$$0 < \mu < \frac{1}{\lambda_{\max}} \tag{9.45}$$

The convergence of the LMS method for step sizes satisfying Eq. 9.45 is a statistical form of convergence. That is, if μ satisfies Eq. 9.45, then for any initial guess $w(0)$,

$$E[w(k)] \to w^* \quad \text{as} \quad k \to \infty \tag{9.46}$$

Example 9.4 **Step Size**

As a simple illustration of how to find a range of step sizes for the LMS method, consider an adaptive filter of order $m = 1$. Suppose $N \geq 4$, and the input and desired output are as follows.

$$x(k) = 2 \cos\left(\frac{2\pi k}{N}\right)$$

$$d(k) = \sin\left(\frac{2\pi k}{N}\right)$$

For convenience, let $\psi = 2\pi/N$. This two-dimensional adaptive filter was considered previously in Example 9.1, where it was determined that

$$R = 2 \begin{bmatrix} 1 & \cos(\psi) \\ \cos(\psi) & 1 \end{bmatrix}$$

Thus, the characteristic polynomial of the auto-correlation matrix is

$$\Delta(\lambda) = \det\{\lambda I - R\}$$

$$= \det\left\{ \begin{bmatrix} \lambda - 2 & -2\cos(\psi) \\ -2\cos(\psi) & \lambda - 2 \end{bmatrix} \right\}$$

$$= (\lambda - 2)^2 - 4\cos^2(\psi)$$

$$= \lambda^2 - 4\lambda + 4 - 4\cos^2(\psi)$$

$$= \lambda^2 - 4\lambda + 4\sin^2(\psi)$$

Using the quadratic formula, the eigenvalues of R are

$$\lambda_{1,2} = \frac{4 \pm \sqrt{16 - 16\sin^2(\psi)}}{2}$$

$$= 2 \pm 2\sqrt{1 - \sin^2(\psi)}$$

$$= 2 \pm 2\cos(\psi)$$

$$= 2[1 \pm \cos(\psi)]$$

Note that the eigenvalues of R are real and positive. Since $N \geq 4$, we have $0 < \psi \leq \pi/2$, and the largest eigenvalue is $\lambda_{\max} = 2[1 + \cos(\psi)]$. Thus, from Eq. 9.45, the range of step sizes over which the LMS method converges is

$$0 < \mu < \frac{0.5}{1 + \cos(\psi)}$$

A key advantage of the LMS method in Eq. 9.34 is that it is very simple to implement. For example, there is no need to compute the auto-correlation matrix R or the cross-correlation vector p. Unfortunately, the upper bound on the step size in Eq. 9.45 is not nearly as easy to determine. Not only must the auto-correlation matrix R be constructed, but then its eigenvalues must be computed as well. By making use of another result from linear algebra, an alternative step-size bound can be developed, one that is much easier to compute. Recall that the *trace* of a square matrix is just the sum of the diagonal elements. The trace is also equal to the sum of the eigenvalues. That is,

$$\text{trace}(R) = \sum_{i=1}^{m+1} \lambda_i \qquad (9.47)$$

Since R is symmetric and positive-definite, $\lambda_i > 0$. It then follows from Eq. 9.47 that

$$\lambda_{\max} < \text{trace}(R) \qquad (9.48)$$

Substituting Eq. 9.48 into Eq. 9.45, we find that a more conservative range for the step size is $0 < \mu < 1/\text{trace}(R)$. This effectively eliminates the need to compute eigenvalues. The requirement to find R itself also can be eliminated by exploiting the special banded structure of R. Recall from Eq. 9.19 that the diagonal elements of R are all equal to one another with $R_{ii} = P_x$, where $P_x = E[x^2(k)]$ is the average power of the input. Since the trace is just the sum of the $m + 1$ diagonal elements, we have $\text{trace}(R) = (m + 1)P_x$. This leads to the following smaller, but much simpler, range of values for the step size, over which convergence of the LMS method is assured.

Step-size bound

$$0 < \mu < \frac{1}{(m+1)P_x} \qquad (9.49)$$

   Note how the upper bound on μ decreases with the order of the filter and the power of the input signal. In practical applications, it is not uncommon to choose a value for the step size in the range $0.01 < (m + 1)P_x\mu < 0.1$, well below the upper limit (Kuo and Morgan, 1996).

Example 9.5   **Revised Step Size**

Suppose the input for an mth-order adaptive filter consists of zero-mean white noise uniformly distributed over an interval $[-c, c]$. From Eq. 9.11, the average power of $x(k)$ is

$$P_x = \frac{c^2}{3}$$

Applying Eq. 9.49, this yields the following range of step sizes for the LMS method.

$$0 < \mu < \frac{3}{(m+1)c^2}$$

For example, for the system identification application in Example 9.3, the magnitude of the white noise input was $c = 1$, and the filter order was $m = 50$. Thus, the range of step sizes for Example 9.3 is

$$0 < \mu < 0.0588$$

The step size used in Example 9.3, $\mu = 0.01$, was well within this range.

9.4.2 Convergence Rate

The selection of the step size determines not only whether or not the LMS method will converge; it also determines how fast it converges. One way to view convergence is to examine what happens to $w(k)$, as in Eq. 9.41. An equivalent way to view convergence is to examine the *learning curve* which is a plot of the mean square error as a function of the iteration number. Recall from Eq. 9.21 that the mean square error can be expressed as follows

$$\epsilon[w(k)] = P_d - 2w^T(k)p + w^T(k)Rw(k) \tag{9.50}$$

To see how fast the mean square error converges, it is helpful to develop an alternative representation for the auto-correlation matrix R. Let λ_i denote the ith eigenvalue of R, and let q^i be its associated eigenvector. That is,

$$Rq^i = \lambda_i q^i, \quad 1 \leq i \leq m+1 \tag{9.51}$$

Suppose the $m + 1$ eigenvectors are arranged as columns of an $(m + 1) \times (m + 1)$ matrix $Q = [q^1, q^2, \ldots, q^{m+1}]$. Next, let Λ be the diagonal matrix of order $m + 1$ with the eigenvalues along the diagonal.

$$\Lambda \triangleq \begin{bmatrix} \lambda_1 & 0 & \cdots & 0 \\ 0 & \lambda_2 & \cdots & 0 \\ \vdots & \vdots & \ddots & \vdots \\ 0 & 0 & \cdots & \lambda_{m+1} \end{bmatrix} \tag{9.52}$$

Using Q and Λ, the $m + 1$ eigenvector equations in Eq. 9.51 can then be rewritten as a single matrix equation as follows.

$$RQ = \Lambda Q \tag{9.53}$$

Note that the right-hand side of Eq. 9.53 can be replaced with $Q\Lambda$ because Λ is diagonal. Since the auto-correlation matrix R is symmetric, the eigenvectors of R form a linearly independent set which makes the eigenvector matrix Q invertible. If we commute the matrices on the right-hand side of Eq. 9.53 and then post multiply both sides of Eq. 9.53 by Q^{-1}, this yields the following alternative representation of the auto-correlation matrix in terms of its eigenvectors and eigenvalues.

Auto-correlation matrix

$$R = Q\Lambda Q^{-1} \tag{9.54}$$

The representation of R in Eq. 9.54 has a number of useful properties. For example, a direct calculation reveals that $R^2 = Q\Lambda^2 Q^{-1}$. More generally, it is not difficult to show that

$$R^k = Q\Lambda^k Q^{-1}, \quad k \geq 0 \tag{9.55}$$

This property can be used to evaluate the rate of convergence. Substituting the expression for R from Eq. 9.54 into Eq. 9.41 and using Eq. 9.55 we have

$$E[w(k)] = w^* + (I - 2\mu Q \Lambda Q^{-1})^k [w(0) - w^*]$$
$$= w^* + (QIQ^{-1} - 2\mu Q \Lambda Q^{-1})^k [w(0) - w^*]$$
$$= w^* + \{Q(I - 2\mu \Lambda)Q^{-1}\}^k [w(0) - w^*]$$
$$= w^* + Q(I - 2\mu \Lambda)^k Q^{-1} [w(0) - w^*] \tag{9.56}$$

The factor $(I - 2\mu \Lambda)^k$ in Eq. 9.56 consists of the following diagonal matrix, where $r_i = 1 - 2\mu \lambda_i$.

$$(I - 2\mu \Lambda)^k = \begin{bmatrix} r_1^k & 0 & \cdots & 0 \\ 0 & r_2^k & \cdots & 0 \\ \vdots & \vdots & \ddots & \vdots \\ 0 & 0 & \cdots & r_{m+1}^k \end{bmatrix} \tag{9.57}$$

Note that this confirms that the LMS method converges if and only if $|1 - 2\mu \lambda_i| < 1$ for $1 \le i \le m + 1$. The speed of convergence is dominated by the factor, r_i, whose magnitude is largest because this corresponds to the slowest mode. Let

$$\lambda_{\min} \overset{\Delta}{=} \min\{\lambda_1, \lambda_2, \cdots, \lambda_{m+1}\} \tag{9.58}$$

Suppose the step size is constrained to be, at most, half of its maximum value, $\mu \le 0.5/\lambda_{\max}$. Then $0 \le r_i < 1$ and the radius of the slowest or dominant mode is

$$r_{\max} = 1 - 2\mu \lambda_{\min} \tag{9.59}$$

The rate of convergence of the LMS method can be characterized by an exponential time constant, τ_{mse}. Since the mean square error is a quadratic function of the weights, the mean square error converges at a rate of $r_{\max}^{2k}$. Suppose the input and desired output are obtained by sampling with a sampling interval of T. Then the exponential rate of convergence is $\exp(-kT/\tau_{\text{mse}})$. Using Eq. 9.59, this results in the following equation for the mean square error time constant.

$$\exp(-kT/\tau_{\text{mse}}) = (1 - 2\mu \lambda_{\min})^{2k} \tag{9.60}$$

Taking the log of both sides of Eq. 9.60 and solving the resulting equation for τ_{mse} then yields

$$\tau_{\text{mse}} = \frac{-T}{2 \ln(1 - 2\mu \lambda_{\min})} \tag{9.61}$$

If μ is sufficiently small, we can use the approximation $\ln(1 + x) \approx x$. This results in the following simplified approximation for the mean square error *time constant* of the LMS method.

Mean square error time constant

$$\tau_{\text{mse}} \approx \frac{T}{4\mu \lambda_{\min}} \quad \text{sec} \tag{9.62}$$

Note that the time constant can be expressed in units of iterations, rather than seconds, by setting $T = 1$. Furthermore, observe that in order to speed up convergence, we must increase the step size. However, if the step size is made too large, then the LMS will not converge at all.

Example 9.6 **Time Constant**

Again consider the system identification example presented in Example 9.3. There the filter order was $m = 50$, and the input consisted of $N = 1000$ samples of white noise uniformly distributed over $[-1, 1]$. From Eq. 9.11, the average power of the input is $P_x = 1/3$. Recall that for a zero-mean white noise input the auto-correlation matrix is very easy to compute. In particular, from Example 9.2, we have

$$R \approx P_x I$$

$$= \frac{1}{3} I$$

Since R is diagonal, it has a single eigenvalue, $\lambda = 1/3$, repeated $m + 1 = 51$ times. Thus, the minimum eigenvalue is

$$\lambda_{min} = \frac{1}{3}$$

The step size used in Example 9.3 was $\mu = 0.01$. Applying Eq. 9.62, this results in the following time-constant estimate for the system identification example.

$$\tau_{mse} \approx \frac{1}{4\mu\lambda_{min}}$$

$$= \frac{3}{.04}$$

$$= 75$$

Note that we have set $T = 1$, which yields the time constant in iterations. Since $\exp(-5) = 0.007$, the mean square error should be reduced to less than one percent of its peak value after five time constants or $M = 375$ iterations. Inspection of the plot of $e^2(k)$ versus k in Figure 9.9 confirms that this is the case, at least approximately. It should be pointed out that the plot of the squared error in Figure 9.9 is a rough approximation to the learning curve. Recall that the learning curve is a plot of the mean square error, $\epsilon[w(k)]$, and to obtain a better approximation to it, we would have to perform the system identification many times with different white noise inputs and then average the squares of the errors for each run (see Problem P9.34).

9.4.3 Excess Mean Square Error

Our previous analysis of the LMS method appears to suggest that we should choose a step size that is as large as possible, consistent with convergence, in order to reduce the mean square error time constant. As it turns out, there is one more factor, called the excess mean square error, that mitigates against making the step size too large. Once the excess mean square error is taken into account, we find that in selecting μ there is a tradeoff between convergence speed and steady-state accuracy.

Recall that the essential assumption of the LMS method is the approximation of the mean square error with the squared error for the purpose of estimating the gradient vector. This leads to the gradient approximation, $\hat{\nabla}\epsilon(w) = -2e(k)\theta(k)$, found in Eq. 9.30. This estimate differs from the exact value, $\nabla\epsilon(w) = 2(Rw - p)$, in Eq. 9.22. The error in the estimate of the gradient of the mean square error can be modeled as an additive noise term as follows.

$$\nabla\epsilon[w(k)] = \hat{\nabla}\epsilon[w(k)] + v(k) \tag{9.63}$$

The noise term, $v(k)$, causes the steady-state value of the mean square error to be larger than the theoretical minimum value, and the difference is referred to as the *excess mean square error*.

$$\epsilon_{excess}(k) \overset{\Delta}{=} \epsilon[w(k)] - \epsilon_{min} \tag{9.64}$$

To determine an expression for the minimum mean square error, it is useful to reformulate the mean square error in terms of the weight variation, $\delta w(k) = w(k) - w^*$. Using Eq. 9.50 and substituting $w(k) = \delta w(k) + w^*$, we have

$$
\begin{aligned}
\epsilon(w) &= P_d - 2p^T(w^* + \delta w) + (w^* + \delta w)^T R(w^* + \delta w) \\
&= P_d - 2p^T w^* - 2p^T \delta w + (w^*)^T Rw^* + (w^*)^T R\delta w + \delta w^T Rw^* + \delta w^T R\delta w \\
&= P_d - 2p^T w^* - 2p^T \delta w + (w^*)^T p + (R^{-1}p)^T R\delta w + \delta w^T p + \delta w^T R\delta w \\
&= P_d - 2p^T w^* - 2p^T \delta w + p^T w^* + p^T(R^{-1})^T R\delta w + p^T \delta w + \delta w^T R\delta w \\
&= P_d - p^T w^* + \delta w^T R\delta w
\end{aligned}
\tag{9.65}
$$

Here we have made use of the observations that: $R^T = R$, the transpose of the inverse is the inverse of the transpose, the transpose of the product is the product of the transposes in reverse order, and the transpose of a scalar is the scalar. From Eq. 9.65, it is clear that the mean square error achieves its minimum value at $\delta w = 0$. Using $w^* = R^{-1}p$ in Eq. 9.65, we arrive at the following expression for the minimum mean square error.

$$\epsilon_{min} = P_d - p^T R^{-1} p \tag{9.66}$$

Combining Eqs. 9.64 through 9.66 results in the following formulation of the excess mean square error in terms of the weight variation.

$$\epsilon_{excess}(k) = \delta w^T(k) R\delta w(k) \tag{9.67}$$

By examining the statistical properties of the gradient noise term in Eq. 9.63, it is possible to develop the following approximation for the excess mean square error (Widrow and Sterns, 1985).

Excess mean square error

$$\epsilon_{\text{excess}}(k) \approx \mu \epsilon_{\min}(m+1) P_x \tag{9.68}$$

It is apparent from Eq. 9.66 that substantial computational effort is required to determine the minimum mean squared error. For this reason, a normalized version of the excess mean square error is often used. The *misadjustment factor* of the LMS method is denoted M and defined as

$$M \triangleq \frac{\epsilon_{\text{excess}}}{\epsilon_{\min}} \tag{9.69}$$

From Eq. 9.68, the misadjustment factor of the LMS method is

$$M \approx \mu(m+1) P_x \tag{9.70}$$

Note that the normalized excess mean square error increases with the step size, the filter order, and the average power of the input. The dependence on the step size means that in order to reduce M we must decrease μ, but to reduce τ_{mse}, we must increase μ. Thus, there is a trade-off between the convergence speed and the steady-state accuracy of the LMS method.

Example 9.7 **Excess Mean Square Error**

To illustrate the relationship between excess mean square error and step size, consider an adaptive filter or order $m = 1$ with the following input and desired output.

$$x(k) = 2\cos(0.5\pi k) + v(k)$$

$$d(k) = \sin(0.5\pi k)$$

Here $v(k)$ represents white noise uniformly distributed over the interval $[-0.5, 0.5]$ that is statistically independent of $2\cos(0.5\pi k)$. Thus, the average power of the input is

$$
\begin{aligned}
P_x &= E[x^2(k)] \\
&= E[4\cos^2(0.5\pi k) + 4v(k)\cos(0.5\pi k) + v^2(k)] \\
&= 4E[\cos^2(0.5\pi k)] + 4E[v(k)\cos(0.5\pi k)] + E[v^2(k)] \\
&= 2E[\cos^2(\pi k) + 1] + 4E[v(k)]E[\cos(0.5\pi k)] + P_v \\
&= 2 + P_v
\end{aligned}
$$

From Eq. 9.11, the average power of the white noise uniformly distributed over $[-0.5, 0.5]$ is $P_v = (0.5)^2/3$. Thus, the average power of the input is

$$P_x = \frac{25}{12}$$

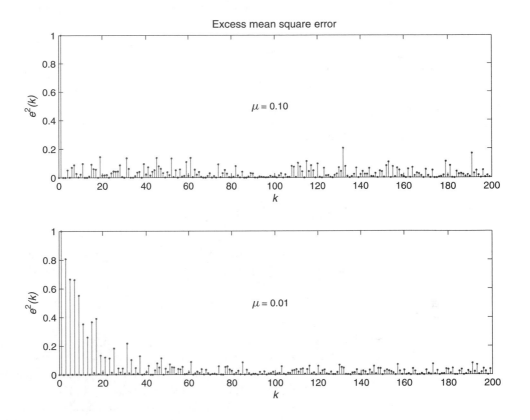

Figure 9.11 Excess Mean Square Error of an Adaptive Filter of Order $m = 1$ using Two Step Sizes

Using Eq. 9.49 with $m = 1$, the range of step sizes needed for convergence of the LMS method is

$$0 < \mu < 0.24$$

Next, from Eq. 9.70, the normalized excess mean square error or misadjustment factor, M, is

$$M \approx \frac{25\mu}{6}$$

To see the effects of μ, suppose $w(0) = 0$ and consider two special cases corresponding to $\mu = 0.1$ and $\mu = 0.01$. Plots of the squared error can be obtained by running script *exam9_7* on the distribution CD. Observe from Figure 9.11 that the plot labeled $\mu = 0.1$ converges very rapidly due to the relatively large step size. However, it is apparent that the steady-state error is also relatively large. The plot labeled $\mu = 0.01$ takes longer to converge, but one is rewarded with a steady-state excess mean square error that is clearly smaller. This illustrates the trade off between convergence speed and steady-state accuracy.

The effects of step size, filter order, and input power on the performance characteristics of the LMS method are summarized in Table 9.1.

Table 9.1: ▶
Performance
Characteristics of an
Adaptive Transversal
Filter of Order m
with Step Size, μ,
and Input Power,
$P_x = E[x^2(k)]$

Property	Value
Convergence range	$0 < \mu < \dfrac{1}{(m+1)P_x}$
Learning-curve time constant	$\tau_{\text{mse}} \approx \dfrac{1}{4\mu\lambda_{\min}}$
Excess mean square error	$\epsilon_{\text{excess}} \approx \mu(m+1)\epsilon_{\min}P_x$

● ● ● ● ● ● ● ● ● ● ● ● ● ●

9.5 Modified LMS Methods

There are a number of interesting modifications that can be made to the LMS method to enhance performance. In this section, we examine three variants of the basic LMS method (Kuo and Morgan, 1996).

9.5.1 Normalized LMS Method

Recall from Table 9.1 that the upper bound on μ needed to ensure convergence depends on the filter order, m, and the input power, P_x. A similar observation holds for the step size needed to achieve a given excess mean square error. To develop a version of the LMS method that has a step size, α, that does not depend on the input power or the filter order, the following *normalized LMS* method has been proposed.

Normalized LMS
method

$$w(k+1) = w(k) + 2\mu(k)e(k)\theta(k), \quad k \geq 0 \tag{9.71}$$

Note how the normalized LMS method differs from the basic LMS method in that the step size, $\mu(k)$, is no longer constant. Instead, it varies with time, as follows.

$$\mu(k) = \frac{\alpha}{(m+1)\hat{P}_x(k)} \tag{9.72}$$

Here $\hat{P}_x(k)$ is a running estimate of the average power of the input. Observe that if $\hat{P}_x(k)$ is replaced with the exact average power, P_x, then from Table 9.1, the range of *constant* step sizes needed to ensure convergence is simply

$$0 < \alpha < 1 \tag{9.73}$$

It is in this sense that the step size has been normalized. The beauty of the normalized approach is that a single value can be used for α, independent of the filter size and the input power.

The simplest way to estimate the average power of the input is to use a rectangular window or running-average filter. The following is an Nth-order running-average filter with input $x^2(k)$.

$$\hat{P}_x(k) = \frac{1}{N+1}\sum_{i=0}^{N} x^2(k-i) \tag{9.74}$$

One of the key features of the LMS method is its highly efficient implementation. In order to preserve this feature, care must be taken to minimize the number of floating-point operations required at each iteration. The number of multiplications and divisions needed to compute $\hat{P}_x$ in Eq. 9.74 is $N + 2$. This can be reduced with a recursive formulation of the running-average filter. Using the change of variable, $j = i - 1$, we can rewrite Eq. 9.74 as

$$
\begin{aligned}
\hat{P}_x(k) &= \frac{1}{N+1} \sum_{j=-1}^{N-1} x^2(k-1-j) \\
&= \frac{1}{N+1} \left[\sum_{j=0}^{N} x^2(k-1-j) \right] + \frac{x^2(k) - x^2(k-N)}{N+1} \\
&= P_x(k-1) + \frac{x^2(k) - x^2(k-N)}{N+1}
\end{aligned}
\tag{9.75}
$$

The recursive formulation in Eq. 9.75 reduces the number of FLOPs per iteration from $N + 2$ to three. Although the implementation in Eq. 9.75 is faster than the one in Eq. 9.74, it is not more efficient in terms of memory requirements because $N + 1$ samples of the input still have to be stored. Recall from Eq. 9.71 that we are already storing $m + 1$ samples of the input in the form of the vector of past inputs θ. If we choose $N = m$, then the following dot product can be used instead to estimate the average power of the input.

$$
\hat{P}_x(k) = \frac{\theta^T(k)\theta(k)}{m+1}
\tag{9.76}
$$

This approach has the advantage that we are not storing any additional values of $x(k)$. Substitution of Eq. 9.76 into Eq. 9.72 results in a further simplification because the $m + 1$ factors cancel. This yields the following simplified expression for the time-varying step size (Slock, 1993).

$$
\mu(k) = \frac{\alpha}{\theta^T(k)\theta(k)}
\tag{9.77}
$$

There is one additional practical difficulty that can arise when a nonstationary input is used. If the input $x(k)$ is zero for $m + 1$ consecutive samples, then $\theta(k) = 0$ and the step size in Eq. 9.77 becomes unbounded. This problem also occurs when the algorithm starts up if $\theta(0) = 0$. To avoid this numerical difficulty, let δ be a small positive value. Then the step size will never be larger than α/δ if the following modified step size is used for the normalized LMS method.

Normalized step size

$$
\mu(k) = \frac{\alpha}{\delta + \theta^T(k)\theta(k)}
\tag{9.78}
$$

Example 9.8   **Normalized LMS Method**

To illustrate how the step size changes with the normalized LMS method, suppose the input is the following amplitude-modulated sine wave.

$$x(k) = \cos\left(\frac{\pi k}{100}\right) \sin\left(\frac{\pi k}{5}\right)$$

Next, suppose the desired output $d(k)$ is produced by the following second-order IIR resonator filter.

$$H(z) = \frac{1 - z^{-2}}{1 + z^{-1} + 0.9z^{-2}}$$

Let the order of the adaptive filter be $m = 5$, and suppose the normalized step size is $\alpha = 0.5$. If we pick a step size bound of $\delta = 0.05$, then the maximum step size is $\alpha/\delta = 10$. Plots of the input $x(k)$, the squared error $e^2(k)$, and the variable step size $\mu(k)$, obtained by running script *exam9_8* on the distribution CD, are shown in Figure 9.12. Notice that the amplitude-modulated input signal reaches its minimum value at $k = 50$, where the cosine factor is zero. The step size hits its peak value somewhat later due to the delay in the running-average estimate of the input power. At $k = 0$, the step size saturates at $\mu(0) = \alpha/\delta$. Observe that, in spite of the increase in step size due to a loss of input signal power, the squared error converges and remains small even when the step size increases.

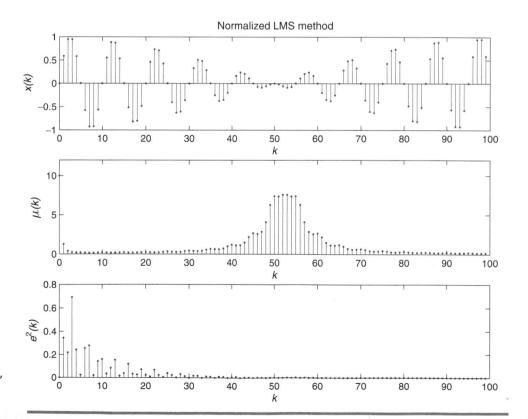

Figure 9.12 Input, Squared Error, and Step Size using the Normalized LMS Method with $m = 5$, $\alpha = 0.5$, and $\delta = 0.05$

9.5.2   Correlation LMS Method

Recall from Table 9.1 that the learning-curve time constant is inversely proportional to the step size. Consequently, the step size μ should be relatively large to ensure rapid convergence. However, the excess mean square error following convergence is directly proportional to μ, which means that μ should be relatively small to improve steady-state accuracy. To avoid this tradeoff, one might use a large step size while convergence is occurring, and then a small step size once convergence has been achieved. The essential task, then, is to detect when convergence has taken place. It is not realistic to use w^* or $\epsilon_{\min}$ to detect convergence because of the computational burden involved. Instead, a less direct means must be employed. Suppose $w(k)$ has converged to w^*, and consider the product of the error $e(k)$ with the vector of past inputs $\theta(k)$. Recalling that $\theta^T(k)w$ is a scalar, we have

$$
\begin{aligned}
e(k)\theta(k) &= [d(k) - y(k)]\theta(k) \\
&= [d(k) - \theta^T(k)w]\theta(k) \\
&= d(k)\theta(k) - \theta(k)\theta^T(k)w
\end{aligned}
\tag{9.79}
$$

Taking the expected value of both sides of Eq. 9.79, and evaluating the result at the optimal weight, $w^* = R^{-1}p$, then yields

$$
\begin{aligned}
E[e(k)\theta(k)] &= E[d(k)\theta(k)] - E[\theta(k)\theta^T(k)]w^* \\
&= p - Rw^* \\
&= 0
\end{aligned}
\tag{9.80}
$$

Given the definition of $\theta(k)$ in Eq. 9.5, the expected value of $e(k)\theta(k)$ can be interpreted as a cross-correlation of the error with the input. In particular, using Definition 9.1 and Eq. 9.5, we can rewrite Eq. 9.80 as

$$
r_{ex}(i) = 0, \quad 0 \le i \le m
\tag{9.81}
$$

From Eq. 9.81, we see that, when the weight vector is optimal, the error is *uncorrelated* with the input. This is an instance of a more general principle that says: When an optimal solution is found, the error in the solution is orthogonal to the data on which the solution is based. From the special case $i = 0$ in Eq. 9.81, we obtain the following scalar relationship which holds when the LMS method has converged.

$$
E[e(k)x(k)] = 0
\tag{9.82}
$$

The basic idea behind the correlation LMS method (Shan and Kailath, 1988) is to choose a step size that is directly proportional to the magnitude of $E[e(k)x(k)]$. This way, the step size will become small when the LMS method has converged, but will be larger during the convergence process. One way to estimate the expected value in Eq. 9.82 is to use a running-average filter with input $e(k)x(k)$. However, this would mean increased storage requirements because the samples of $e(k)$ are not already stored. A less expensive alternative approach to approximating $E[e(k)x(k)]$ is to use a first-order lowpass IIR filter with the following transfer function.

$$H(z) = \frac{(1 - \beta)z}{z - \beta} \qquad (9.83)$$

The scalar $0 < \beta < 1$ is called a *smoothing* parameter, and typically $\beta \approx 1$. The filter output will be an estimate of $E[e(k)x(k)]$ using an exponentially weighted average. The equivalent width of the exponential window is $N = 1/(1 - \beta)$ samples. If $r(k)$ is the filter output, and $e(k)x(k)$ is the filter input, then

Correlation estimate

$$r(k + 1) = \beta r(k) + (1 - \beta)e(k)x(k) \qquad (9.84)$$

Since $r(k)$ becomes small once convergence has taken place, we make the step size proportional to $|r(k)|$ using a proportionality constant or *relative* step size of $\alpha > 0$. This results in the following time-varying step size for the *correlation* LMS method.

Correlation step size

$$\mu(k) = \alpha|r(k)| \qquad (9.85)$$

The correlation LMS method can be interpreted as having two modes of operation. When convergence has been achieved, the step size becomes small, and the algorithm is in the *sleep* mode. Because $\mu(k)$ is small, the excess mean square error is also small. Furthermore, if $y(k)$ contains measurement noise that is uncorrelated with $x(k)$, this noise will not cause an increase in $\mu(k)$. However, if $d(k)$ or $x(k)$ change significantly, this causes $|r(k)|$ to increase and the algorithm then enters the *active* or tracking mode characterized by an increased step size. Once the algorithm converges to the new optimal weight, the step size decreases again, and the algorithm reenters the sleep mode.

| **Example 9.9** | **Correlation LMS Method** |

To illustrate how the correlation method detects convergence and changes modes of operation, let the input be N samples of white noise uniformly distributed over the interval $[-1, 1]$. Suppose this input drives the time-varying feedback system shown in Figure 9.13. Here the open-loop transfer function is

$$G(z) = \frac{1.28}{z^2 - 0.64}$$

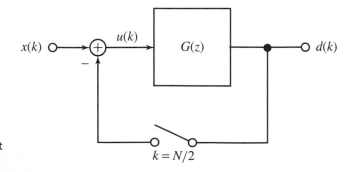

Figure 9.13 A Time-varying Feedback System where the Switch Opens at $k = N/2$

The system in Figure 9.13 is a time-varying linear system because the switch in the feedback path starts out closed but opens starting at sample $k = N/2$. This might correspond, for example, to a feedback sensor malfunctioning. When the switch is closed, the Z-transform of the intermediate signal $u(k)$ is

$$U(z) = X(z) - D(z)$$
$$= X(z) - G(z)U(z)$$

Solving for $U(z)$ yields

$$U(z) = \frac{X(z)}{1 + G(z)}$$

It then follows that, when the switch is closed, the output is

$$D(z)) = G(z)U(z)$$
$$= \frac{G(z)X(z)}{1 + G(z)}$$

Finally, the closed-loop transfer function of the system in Figure 9.13, with the switch closed, is

$$H_{\text{closed}}(z) = \frac{D(z)}{X(z)}$$
$$= \frac{G(z)}{1 + G(z)}$$
$$= \frac{1.28/(z^2 - 0.64)}{1 + 1.28/(z^2 - 0.64)}$$
$$= \frac{1.28}{z^2 + 0.64}$$

Consequently, when the switch is closed for samples $0 \le k < N/2$, the system has an imaginary pair of poles at $z = \pm j0.8$. At time $k = N/2$, the switch opens and the transfer function reduces to $G(z)$ with real poles $z = \pm 0.8$. Suppose this time-varying system is identified with an adaptive filter of order $m = 25$, using the correlation LMS method. Let the relative step size be $\alpha = 0.5$ and the smoothing factor be $\beta = 0.95$. A plot of the squared error and the step size, obtained by running script *exam9_9* on the distribution CD, is shown in Figure 9.14. Observe how the algorithm converges in about 200 samples and enters the sleep mode around 500 samples with a very small step size. At sample $k = 600$, the desired output abruptly changes, and the step size increases, indicating that the active or tracking mode has been entered. The algorithm reconverges around 900 samples and then reenters the sleep mode, this time with a somewhat larger resting step size.

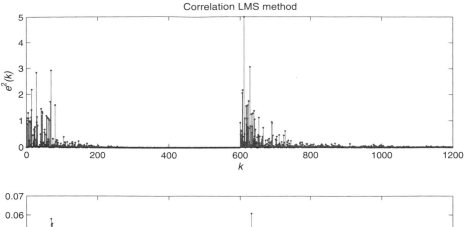

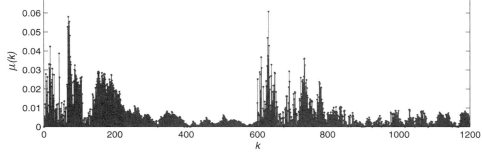

Figure 9.14 Identification of a Time-varying Feedback System using the Correlation LMS Method with $m = 25$, $\alpha = 0.5$, and $\beta = 0.95$

9.5.3 Leaky LMS Method

White noise is a highly effective input for system identification because it has a flat power density spectrum and therefore excites all of the natural modes of the system being identified. When an input with poor spectral content is used, the LMS method can diverge with one or more elements of the weight vector growing without bound. An elegant way to guard against this possibility is to introduce a second term in the objective function. Let $\gamma > 0$, and consider the following *augmented* mean square error.

$$\epsilon_\gamma[w(k)] \overset{\Delta}{=} E[e^2(k)] + \gamma w^T(k)w(k) \tag{9.86}$$

The last term in Eq. 9.86 is called a *penalty function* term because the minimization process tends to penalize any selection of w for which $w^T w$ is large. In this way, the search for a minimum automatically avoids solutions for which $\|w\|$ is large. The parameter $\gamma > 0$ controls how severe the penalty is, and when $\gamma = 0$, the objective function in Eq. 9.86 reduces to the original mean square error, $\epsilon[w(k)]$.

To see what effect the penalty term has on the LMS algorithm, consider the gradient vector of partial derivatives of ϵ_γ with respect to the elements of w. Using the assumption $E[e^2(k)] \approx e^2(k)$ from Eq. 9.29 to compute an estimate of the gradient, we have

$$\hat{\nabla}\epsilon_\gamma(w) = 2e\frac{\partial e}{\partial w} + 2\gamma w$$

$$= -2e\frac{\partial y}{\partial w} + 2\gamma w$$

$$= -2e\theta + 2\gamma w \tag{9.87}$$

Substituting this estimate for the gradient into the steepest-descent method in Eq. 9.26 then yields the following weight-update formula

$$w(k+1) = w(k) - \mu[2\gamma w(k) - 2e(k)\theta(k)]$$

$$= (1 - 2\mu\gamma)w(k) + 2\mu e(k)\theta(k) \tag{9.88}$$

Finally, define $v \triangleq 1 - 2\mu\gamma$. Substituting v into Eq. 9.88 then results in the following simplified formulation called the *leaky* LMS method.

Leaky LMS method

$$w(k+1) = vw(k) + 2\mu e(k)\theta(k), \quad i \geq 0 \tag{9.89}$$

The composite parameter v is called the *leakage* factor. Note that when $v = 1$, the leaky LMS method reduces to the basic LMS method. If $\theta(k) = 0$, then $w(k)$ "leaks" to zero at the rate $w(k) = v^k w(0)$. Typically, the leakage factor is in the range $0 < v < 1$ with $v \approx 1$. It can be shown that including a leakage factor has the same effect as adding low-level white noise to the input (Gitlin, *et al.*, 1982). This makes the algorithm more stable for a variety of inputs. However, it also means that there is a corresponding increase in the excess mean square error due to the presence of the penalty term. Bellanger (1987) has shown that the excess mean square error is proportional to $(1 - v)^2/\mu^2$, which means that this ratio must be kept small. Since $v = 1 - 2\gamma\mu$, this is equivalent to $4\gamma^2 \ll 1$ or

$$v = 1 - 2\mu\gamma \tag{9.90a}$$

$$\gamma \ll 0.5 \tag{9.90b}$$

It is of interest to note that, because $y(k) = w^T(k)\theta(k)$, limiting $w^T(k)w(k)$ has the effect of limiting the magnitude of the output. This can be useful in applications such as active noise control where a large $y(k)$ can overdrive a speaker and distort the sound (Elliott, *et al.*, 1987).

Example 9.10   **Leaky LMS Method**

To illustrate the use of the leaky LMS method, consider the problem of using an adaptive filter to predict the value of an input signal, as shown in Figure 9.15. Suppose the input consists of two sinusoids plus noise.

$$x(k) = 2\sin\left(\frac{\pi k}{12}\right) + 3\cos\left(\frac{\pi k}{4}\right) + v(k)$$

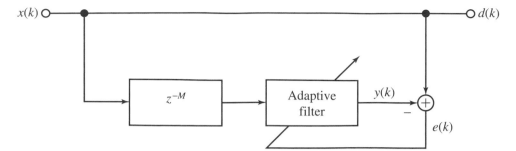

Figure 9.15 Prediction of the Input Signal

Here $v(k)$ is white noise uniformly distributed over the interval $[-0.2, 0.2]$. Next, suppose it is desired to predict the value of the input $M = 10$ samples into the future. Let the order of the adaptive filter be $m = 20$ and the step size be $\mu = 0.002$. To keep the excess mean square error associated with leakage small, we need a leakage factor of $v = 1 - 2\mu\gamma$ where $\gamma \ll 0.5$. Setting $\gamma = 0.05$ yields $v = 0.9998$. Script *exam9_10* on the distribution CD produces the plots shown in Figure 9.16. Observe that, after about 40 samples, the algorithm has converged and the filter output $y(k)$ is an effective approximation of $x(k)$ advanced by $M = 10$ samples.

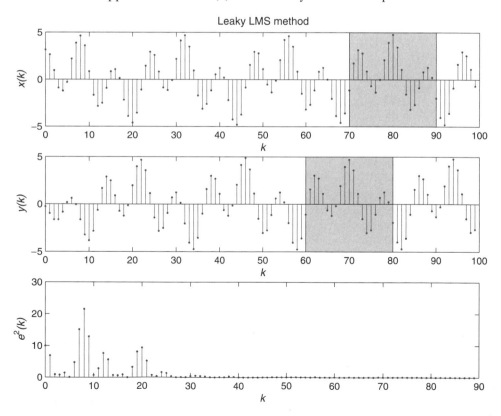

Figure 9.16 Prediction $M = 10$ Samples Ahead using the Leaky LMS Method with $m = 20$, $\mu = 0.002$, and $v = 0.9998$

The normalized, correlation, and leaky LMS methods are three popular variants of the basic LMS method. Other modifications have also been proposed (Kuo and Morgan, 1996). For example, the step size μ can be replaced by a diagonal time-varying step-size matrix, $M(k)$, where the ith diagonal element is $\mu_i(k)$. This modification, with each dimension having its own step size, is called the *variable step size* LMS method. Other modifications, intended to increase speed at the expense of accuracy, include the *signed* LMS methods which replace $e(k)$ with sgn$[e(k)]$ or $\theta(k)$ with sgn$[\theta(k)]$ where *sgn* is the signum or sign function.

MATLAB Toolbox

The FDSP toolbox contains the following functions which correspond to modified versions of the basic LMS method.

```
[w,e,mu] = f_normlms (x,d,m,alpha,delta,w);        % Normalized LMS method
[w,e,mu] = f_corrlms (x,d,m,alpha,beta,w);         % Correlation LMS method
[w,e]    = f_leaklms (x,d,m,mu,nu,w);              % Leaky LMS method
```

Function *f_normlms* implements the normalized LMS method. On entry to *f_normlms*, x is a vector of length N containing the input, d is a vector of length N containing the desired output, $m \geq 0$ is the filter order, *alpha* is the normalized step size, *delta* is the step-size bound, and w is an optional vector of length $m+1$ containing an initial guess for the weights. The default value is $w = 0$. Typically, *alpha* $\ll 1$. The maximum step size μ is *alpha/delta*. Argument *delta* is optional with a default value of *alpha*/100. On exit from *f_normlms*, w is a vector of length m containing the estimated value of the optimal weights, e is an optional vector of length N containing the samples of the error, and *mu* is an optional vector of length N containing the step sizes used.

Function *f_corrlms* implements the correlation LMS method. The input and output arguments for *f_corrlms* are the same as those for *f_normlms* described above, except for the positive scalars *alpha* and *beta*. The scalar *alpha* is the relative step size which must be determined experimentally, and *beta* is the smoothing parameter. The smoothing parameter should be in the range $0 < beta < 1$ with *beta* ≈ 1. The equivalent number of samples used to estimate $E[e(k)x(k)]$ is $M = 1/(1 - beta)$. Parameter *beta* is optional with a default value of *beta* = $1 - 0.5/(m + 1)$.

Function *f_leaklms* implements the leaky LMS method. The input and output arguments for *f_leaklms* are the same as corresponding arguments for *f_normlms* and *f_corrlms*, except for the positive scalars *mu* and *nu*. Scalars *mu* and *nu* specify the step size and the leakage factor, respectively. The step size should be selected to satisfy Eq. 9.49, whereas the leakage factor should be in the range $0 < nu \leq 1$ with *nu* ≈ 1. More specifically, to keep the excess mean square error small, set *nu* as in Eq. 9.90. Parameter *nu* is optional with a default value of $nu = 1 - 0.1 * mu$.

● ● ● ● ● ● ● ● ● ● ● ● ● ●

9.6 Adaptive FIR Filter Design

9.6.1 Pseudo-filters

Recall from Chapter 6 that FIR filters are typically designed to have prescribed magnitude responses and linear phase responses. Filters with a combination of magnitude and phase characteristics can be designed by using the LMS method to estimate the coefficients. The basic idea behind this approach is to use a synthetic *pseudo-filter* to generate the desired output as shown in Figure 9.17.

As the name implies, a pseudo-filter is a fictional linear discrete-time system that may or may not have a physical realization. A pseudo-filter is characterized implicitly by a desired relationship between a periodic input and a steady-state output. Let T be the sampling interval, and suppose the input consists of a sum of N sinusoids as follows.

$$x(k) = \sum_{i=0}^{N-1} C_i \cos(2\pi f_i kT) \tag{9.91}$$

The only constraints on the discrete frequencies $\{f_0, f_1, \cdots, f_{N-1}\}$ are that they be distinct from one another and that they lie within the Nyquist range, $0 \le f_i < f_s/2$. For example, the discrete frequencies are often taken to be uniformly spaced with

$$f_i = \frac{i f_s}{2N}, \quad 0 \le i < N \tag{9.92}$$

The amplitudes or relative weights, $C_i > 0$, are selected based on the importance of each frequency in the overall design specification. For example, if $C_k > C_i$, then frequency f_k will be given more weight than frequency f_i. As a starting point, we often use *uniform* weighting with

$$C_i = 1, \quad 0 \le i < N \tag{9.93}$$

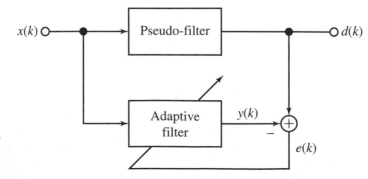

Figure 9.17 Adaptive Filter Design using a Pseudo-Filter

Once the input is selected, the desired steady-state output is then specified. Since the artificial pseudo-filter is assumed to be linear, the steady-state output will be periodic with the same period as the input.

$$d(k) = \sum_{i=0}^{N-1} A_i C_i \cos(2\pi f_i kT + \phi_i) \tag{9.94}$$

Here A_i and ϕ_i denote the desired gain and phase shift, respectively, at frequency f_i. Thus, the design specification consists of a set of four $N \times 1$ vectors, $\{f, C, A, \phi\}$. Here f is the frequency vector, C is the relative weight vector, A is the magnitude vector, and ϕ is the phase vector. In this way, N samples of the desired frequency response, both magnitude and phase, can be specified.

It should be emphasized that the implicit relationship between $x(k)$ and $d(k)$ represents a pseudo-filter because the magnitude $A_i = A(f_i)$ and phase $\phi_i = \phi(f_i)$ of a causal linear discrete-time system are *not* independent of one another. For a causal filter, the real and imaginary parts of the frequency response are interdependent, and so are the magnitude and phase (Proakis and Manolakis, 1992). Consequently, one cannot independently specify both the magnitude and the phase and expect to obtain a causal linear filter realization. Instead, we search for an optimal approximation to the pseudo-filter specifications among causal adaptive transversal filters of order m.

When the order of the adaptive filter is relatively small, the $(m+1) \times 1$ weight vector w can be computed offline by solving $Rw = p$. Given $x(k)$ and $d(k)$, closed-form expressions for the input auto-correlation matrix R and the cross-correlation vector p can be obtained (Problems P9.15 and P9.16). This direct approach can become computationally expensive (and sensitive to roundoff error) as m becomes large. In these instances, it makes more sense to numerically search for the optimal w using the LMS method.

$$w(k + 1) = w(k) + 2\mu e(k)\theta(k) \tag{9.95}$$

Since the objective is to find a fixed FIR filter of order m that best fits the design specifications, the adaptive filter is allowed to run for $M \gg 1$ iterations until the squared error has converged to its steady-state value. The numerator coefficient vector of the fixed FIR filter, $W(z)$, is then set to the final steady-state value of the weights.

$$b = w(M - 1) \tag{9.96}$$

The frequency response of the FIR filter is then computed and compared with the pseudo-filter specifications. If the fit is acceptable, then the process terminates. Otherwise the order m can be increased or the relative weight vector C can be changed at those frequencies where the error is largest. The overall design process is summarized in the flowchart in Figure 9.18.

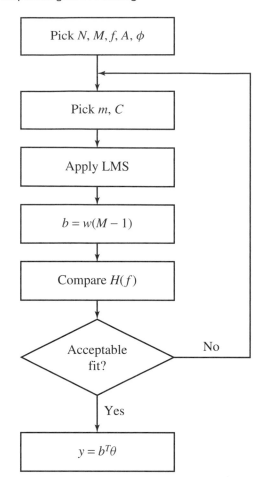

Figure 9.18 Adaptive FIR Filter Design Process

Example 9.11 **Adaptive FIR Filter Design**

To illustrate the use of a pseudo-filter, consider the problem of designing an FIR filter with the following desired magnitude response.

$$
A(f) = \begin{cases}
1.5 - \dfrac{6f}{f_s}, & 0 \le f_s < \dfrac{f_s}{6} \\[2mm]
0.5, & \dfrac{f_s}{6} \le f < \dfrac{f_s}{3} \\[2mm]
0.5 + 0.5 \sin\left(\dfrac{6\pi f}{f_s}\right), & \dfrac{f_s}{3} \le f < \dfrac{f_s}{2}
\end{cases}
$$

Suppose there are $N = 60$ discrete frequencies uniformly distributed, as in Eq. 9.92. Let the relative weights also be uniform, as in Eq. 9.93. Suppose the order of the

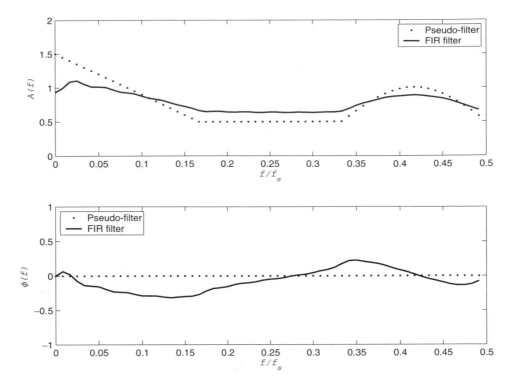

Figure 9.19 Frequency Responses of a Zero-phase Pseudo-filter and an FIR Filter of Order $m = 30$

adaptive FIR filter is $m = 30$, and the step size is $\mu = 0.0001$. Let the LMS method run for $M = 2000$ iterations starting from an initial guess of $w(0) = 0$. Two cases are considered. For the first case, the desired phase response of the pseudo-filter is simply

$$\phi(f) = 0$$

A comparison of the desired and actual frequency responses can be obtained by running script *exam9_11* on the distribution CD, with the results shown in Figure 9.19. Notice that the fit is rather poor because of the unrealistic design specification of zero phase shift. The fit can be improved somewhat by increasing m. Alternatively, a linear-phase pseudo-filter with a constant group delay can be specified. If the group delay is set to $\tau = mT/2$, then this corresponds to a phase response of

$$\phi(f) = -m\pi f$$

This combination of magnitude and phase is much easier to synthesize with a causal FIR filter, as can be seen by the results shown in Figure 9.20. Notice that there is a good fit of both the magnitude response and the phase response in this case.

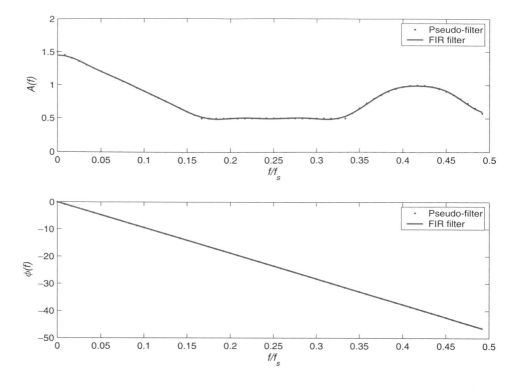

Figure 9.20 Frequency Responses of a Linear-phase Pseudo-filter and an FIR Filter of Order $m = 30$

9.6.2 Linear-phase Pseudo-filters

The special case of linear-phase filters is important because it corresponds to each spectral component of $x(k)$ being delayed by the same amount as it is processed by the filter. Hence, there is no phase distortion of the input, only a delay. Recall from Table 5.1 that an FIR filter of order m with coefficient vector b is a type 1 linear-phase filter if m is even and the filter coefficients satisfy the following symmetry condition.

$$b_i = b_{m-i}, \quad 0 \le i \le m \tag{9.97}$$

A signal flow graph of a transversal filter satisfying this linear-phase symmetry condition is shown in Figure 9.21 for the case $m = 4$. The signal flow graph in Figure 9.21 can be made more efficient by combining branches with identical weights. This results in the equivalent signal flow graph shown in Figure 9.22, which features fewer floating-point multiplications. To develop a concise formulation of the filter output, consider the following pair of $(m/2 + 1) \times 1$ vectors.

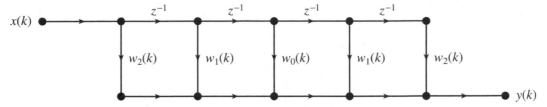

Figure 9.21 A Type 1 Linear-phase Transversal Filter of Order $m = 4$

$$\hat{w}(k) \overset{\Delta}{=} [w_0(k), w_1(k), \cdots, w_{m/2}(k)]^T \tag{9.98a}$$

$$\psi(k) \overset{\Delta}{=} [x(k - m/2), x(k - m/2 - 1) + x(k - m/2 + 1), \cdots, x(k - m) + x(k)]^T \tag{9.98b}$$

Here $\hat{w}(k)$ consists of the first $m/2+1$ weights, while $\psi(k)$ is constructed from the past m inputs. Applying Eq. 9.98 to Figure 9.22, we find that the linear-phase transversal filter has the following compact representation using the dot product.

$$y(k) = \hat{w}(k)^T \psi(k), \quad k \geq 0 \tag{9.99}$$

Note that for large values of m, the linear-phase formulation in Eq. 9.99 requires approximately half as many floating point multiplications or FLOPs as the standard representation in Eq. 9.6. When the LMS method is applied to the linear-phase transversal structure, there is a similar savings in computational effort using

$$\hat{w}(k + 1) = \hat{w}(k) + 2\mu e(k)\psi(k), \quad k \geq 0 \tag{9.100}$$

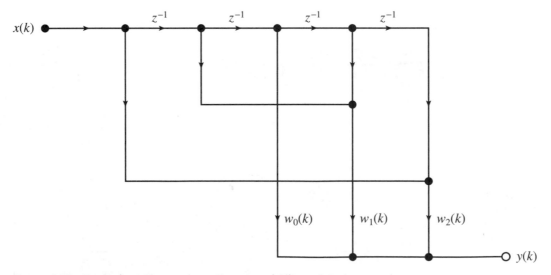

Figure 9.22 Equivalent Linear-phase Transversal Filter of Order $m = 4$

Example 9.12

Adaptive Linear-phase FIR Filter Design

To illustrate the design of a linear-phase FIR filter, consider a pseudo-filter with the following piecewise-continuous magnitude response specification.

$$A(f) = \begin{cases} \left(\dfrac{6f}{f_s}\right)^2, & 0 \le f_s < \dfrac{fs}{6} \\ 0.5, & \dfrac{f_s}{6} \le f < \dfrac{f_s}{3} \\ \left[\dfrac{1 - 2(3f - f_s)}{f_s}\right]^2, & \dfrac{f_s}{3} \le f < \dfrac{f_s}{2} \end{cases}$$

Suppose there are $N = 60$ discrete frequencies uniformly distributed, as in Eq. 9.92. Let the order of the FIR filter be $m = 30$ and the step size be $\mu = 0.0001$. Suppose the LMS method runs for $M = 2000$ iterations starting from an initial guess of $w(0) = 0$. Again, two cases are considered. For the first case, the relative weighting of the discrete frequencies is uniform, as in Eq. 9.93. A comparison of magnitude responses, obtained by running script *exam9_12* on the distribution CD, is shown in Figure 9.23. Observe that the two responses are roughly similar, but that substantial error occurs at multiples of $f_s/6$ where the desired magnitude response is discontinuous. The fit in the vicinity of these frequencies can be improved by using a nonuniform relative weighting. Setting $C_{M/6} = C_{M/3} = 2$ results in the magnitude response plot shown in Figure 9.24, which is somewhat more accurate at the jump discontinuities.

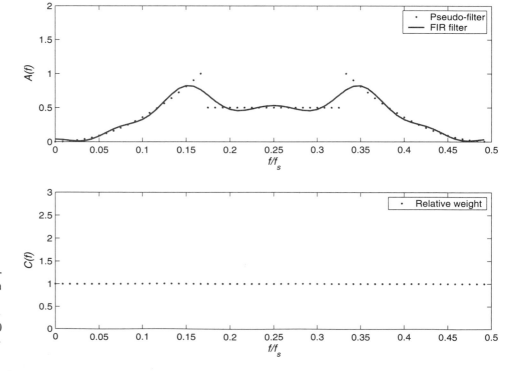

Figure 9.23 Magnitude Responses of a Pseudo-filter and a Linear-phase FIR Filter of Order $m = 30$ using Uniform Relative Weighting

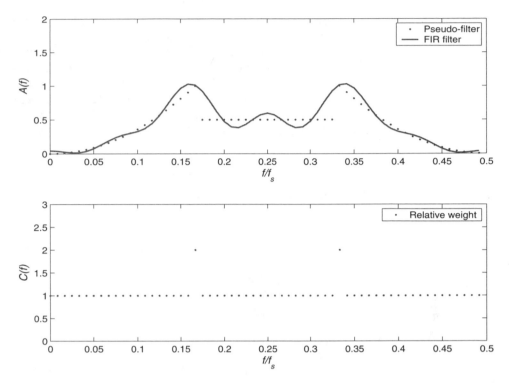

Figure 9.24 Magnitude Responses of a Pseudo-filter and a Linear-phase FIR Filter of Order $m = 30$ using Nonuniform Relative Weighting

9.7 The Recursive Least Squares (RLS) Method

There is an efficient alternative to the LMS method called the recursive least squares or RLS method (Treichler, *et al.*, 2001; Haykin, 2002). The RLS method typically converges much faster than the LMS method, but at a cost of more computational effort per iteration.

9.7.1 Performance Criterion

To formulate the RLS method, we return to the least-squares performance criterion, introduced in Section 9.2, and instead consider the following more general time-varying performance criterion.

$$\epsilon_k(w) = \sum_{i=1}^{k} \gamma^{k-i} e^2(i) + \delta \gamma^k w^T w, \quad k \geq 1 \tag{9.101}$$

Here the exponential weighting factor, $0 < \gamma \le 1$, is also called the *forgetting factor*; because when $\gamma < 1$, it has the effect of reducing the contributions from errors in the remote past. The second term in Eq. 9.101 is a regularization term with $\delta > 0$, called the *regularization parameter*. Note that the second term is similar to the penalty function term in the leaky LMS method in that it tends to prevent solutions for which $w^T w$ grows arbitrarily large. Thus, it has the effect of making the RLS method more stable. When $\gamma = 1$ (no exponential weighting) and $\delta = 0$ (no regularization), the performance criterion in Eq. 9.101 is proportional to the mean square error at time k.

To determine a weight vector w that minimizes $\epsilon_k(w)$, we begin by substituting $e(i) = d(i) - w^T \theta(i)$ into Eq. 9.101, where $d(i)$ is the desired output, and $\theta(i)$ is the vector of past inputs. This results in the following more detailed expression for the performance index.

$$
\begin{aligned}
\epsilon_k(w) &= \sum_{i=1}^{k} \gamma^{k-i} [d(i) - w^T \theta(i)]^2 + \delta \gamma^k w^T w \\
&= \sum_{i=1}^{k} \gamma^{k-i} \{d^2(i) - 2d(i) w^T \theta(i) + [w^T \theta(i)]^2\} + \delta \gamma^k w^T w \\
&= \sum_{i=1}^{k} \gamma^{k-i} [d^2(i) - 2w^T d(i) \theta(i)] + \sum_{i=1}^{k} \gamma^{k-i} w^T \theta(i) [\theta^T(i) w] + \delta \gamma^k w^T w \\
&= \sum_{i=1}^{k} \gamma^{k-i} [d^2(i) - 2w^T d(i) \theta(i)] + w^T \left[\sum_{i=1}^{k} \gamma^{k-i} \theta(i) \theta^T(i) + \delta \gamma^k I \right] w \quad (9.102)
\end{aligned}
$$

The expression for $\epsilon_k(w)$ can be made more concise by introducing the following generalized versions of the auto-correlation matrix and cross-correlation vector, respectively, at time k.

$$
R(k) \triangleq \sum_{i=1}^{k} \gamma^{k-i} \theta(i) \theta^T(i) + \delta \gamma^k I \tag{9.103a}
$$

$$
p(k) \triangleq \sum_{i=1}^{k} \gamma^{k-i} d(i) \theta(i) \tag{9.103b}
$$

Substituting Eq. 9.103 into Eq. 9.102, we find that the exponentially weighted, regularized performance criterion can be expressed as the following quadratic function of the weight vector.

$$
\epsilon_k(w) = \sum_{i=1}^{k} \gamma^{k-i} d^2(i) - 2w^T p(k) + w^T R(k) w \tag{9.104}
$$

Following the same procedure that was used in Section 9.2, the gradient vector of partial derivatives of $\epsilon_k(w)$ with respect to the elements of w can be shown to be $\nabla\epsilon_k(w) = 2[R(k)w - p(k)]$. Setting $\nabla\epsilon_k(w) = 0$ and solving for w, we then arrive at the following expression for the optimal weight at time k.

$$w(k) = R^{-1}(k)p(k) \tag{9.105}$$

Unlike the LMS method, which asymptotically approaches the optimal weight vector using a gradient-based search, the RLS method attempts to find the optimal weight at each iteration.

9.7.2   Recursive Formulation

Although the weight vector in Eq. 9.105 is optimal in terms of minimizing $\epsilon_k(w)$, it is apparent that the computational effort required to find $w(k)$ is large, and it grows more burdensome as k increases. Fortunately, there is a way to reformulate the required computations to make them more economical. The basic idea is to start with the solution at iteration $k - 1$ and add a correction term to obtain the solution at iteration k. In this way, the required quantities can be computed *recursively*. For example, using Eq. 9.103a we can recast the expression for $R(k)$ as follows.

$$\begin{aligned}
R(k) &= \gamma\left[\sum_{i=1}^{k}\gamma^{k-i-1}\theta(i)\theta^T(i) + \delta\gamma^{k-1}I\right] \\
&= \gamma\left[\sum_{i=1}^{k-1}\gamma^{k-i-1}\theta(i)\theta^T(i) + \delta\gamma^{k-1}I\right] + \theta(k)\theta^T(k) \tag{9.106}
\end{aligned}$$

Observe from Eq. 9.103a that the coefficient of γ in Eq. 9.106 is just $R(k - 1)$. Consequently, the exponentially weighted and regularized auto-correlation matrix can be computed recursively as follows.

$$R(k) = \gamma R(k - 1) + \theta(k)\theta^T(k), \quad k \geq 1 \tag{9.107}$$

An initial value for $R(k)$ is required to start the recursion process. Using Eq. 9.103a and assuming that the input $x(k)$ is causal, one can set $R(0) = \delta I$.

A similar procedure can be used to obtain a recursive formulation for the generalized cross-correlation vector in Eq. 9.103b by factoring out a γ and separating the $i = k$ term. This yields the following recursive formulation for $p(k)$ that can be initialized with $p(0) = 0$.

$$p(k) = \gamma p(k - 1) + d(k)\theta(k), \quad k \geq 1 \tag{9.108}$$

Although the recursive computations of $R(k)$ and $p(k)$ greatly simplify the computational effort for these quantities, there remains the problem of inverting $R(k)$ in Eq. 9.105 to find $w(k)$. It is this step that dominates the computational effort, because the number of floating-point operations or FLOPs needed to solve $R(k)w = p(k)$ is proportional to m^3, where m is the order of the transversal filter. As it turns out, it is also possible to compute $R^{-1}(k)$ recursively. To achieve this, we need to make use of a result from linear algebra. Let A and C be square nonsingular matrices, and let B and D be matrices of appropriate dimensions. Then the *matrix inversion lemma* can be stated as follows (Woodbury, 1950; Kailath, 1960).

Matrix inversion lemma

$$(A + BCD)^{-1} = A^{-1} - A^{-1}B(DA^{-1}B + C^{-1})^{-1}DA^{-1} \tag{9.109}$$

This result is precisely what is needed to express $R^{-1}(k)$ in terms of $R^{-1}(k-1)$. Recalling Eq. 9.107, let $A = \gamma R(k-1)$, $B = \theta(k)$, $C = 1$, and $D = \theta^T(k)$. Then applying the matrix inversion lemma in Eq. 9.109, we have

$$R^{-1}(k) = \frac{1}{\gamma}\left[R^{-1}(k-1) - \frac{R^{-1}(k-1)\theta(k)\theta^T(k)R^{-1}(k-1)}{\gamma + \theta^T(k)R^{-1}(k-1)\theta(k)}\right] \tag{9.110}$$

To simplify the final formulation, it is helpful to introduce the following notational quantities.

$$r(k) \triangleq R^{-1}(k-1)\theta(k) \tag{9.111a}$$

$$c(k) \triangleq \gamma + \theta^T(k)r(k) \tag{9.111b}$$

From the expression for $R(k)$ in Eq. 9.103a, it is evident that $R(k)$ is a symmetric matrix. Since the transpose of the inverse equals the inverse of the transpose, this means that $r^T(k) = \theta^T(k)R^{-1}(k-1)$. Substituting Eq. 9.111 into Eq. 9.110, we then find that the inverse of the generalized auto-correlation matrix can be expressed recursively as follows.

$$R^{-1}(k) = \frac{1}{\gamma}\left[R^{-1}(k-1) - \frac{r(k)r^T(k)}{c(k)}\right], \quad k \geq 1 \tag{9.112}$$

The beauty of the formulation in Eq. 9.112 is that no explicit matrix inversions are required. Instead, the expression for the inverse is updated at each step using dot products and scalar multiplications. To start the process, an initial value for the inverse of the auto-correlation matrix is required. Assuming $x(k)$ is causal in Eq. 9.103a, we take

$$R^{-1}(0) = \delta^{-1}I \tag{9.113}$$

Although this initial estimate of the inverse of the auto-correlation matrix is not likely to be accurate (except perhaps for a white noise input), the exponential weighting associated with $\gamma < 1$ tends to minimize the effects of any initial error in the estimate after a sufficient number of iterations. The effective window length associated with the exponential weighting is

$$M = \frac{1}{1-\gamma} \tag{9.114}$$

The steps required to compute the optimal weight at each step using the RLS method are summarized in Algorithm 9.1. To emphasize the fact that no inverses are explicitly computed, the notation Q is used for R^{-1}.

ALGORITHM 9.1:

RLS Method

1. Pick $0 < \gamma \le 1$, $\delta > 0$, $m \ge 0$, $N \ge 1$.

2. Set $w = 0$, $p = 0$, and $Q = I/\delta$. Here w and p are $(m+1) \times 1$ and Q is $(m+1) \times (m+1)$.

3. For $k = 1$ to N compute
   {

$$\theta = [x(k), x(k-1), \ldots, x(k-m)]^T$$

$$r = Q\theta$$

$$c = \gamma + \theta^T r$$

$$p = \gamma p + d(k)\theta$$

$$Q = \frac{1}{\gamma}\left[Q - \frac{rr^T}{c}\right]$$

$$w = Qp$$

   }

The RLS method in Algorithm 9.1 typically converges much faster than the LMS method. However, the computational effort per iteration is larger, even with the efficient recursive formulation. For moderate to large values of the transversal filter order m, the computational effort in Algorithm 9.1 is dominated by the computation of r, w, and Q in step 3. The number of FLOPs required to compute r, θ, and the symmetric Q is approximately $3(m+1)^2$. Thus, the computational effort is proportional to m^2 for large values of m. This makes the RLS method in Algorithm 9.1 an algorithm of order $O(m^2)$. This is in contrast to the much simpler LMS method which is an algorithm of order $O(m)$ with the computational effort proportional to m. There are faster versions of the RLS method that exploit recursive formulations of $r(k)$ and $w(k)$ (Ljung, *et al.*, 1978).

The design parameters associated with the RLS method are the forgetting factor, $0 < \gamma \le 1$, the regularization parameter, $\delta > 0$, and the transversal filter order, $m \ge 0$. The required filter order depends on the application and is often found empirically. Haykin (2002) has shown that the parameter $\mu = 1 - \gamma$ plays a role similar to the step size in the LMS method. Therefore, γ should be close to unity to keep μ small. If a given effective window length for the exponential weighting is desired, then the forgetting factor can be computed using Eq. 9.114. The choice for the regularization parameter depends on the signal-to-noise ratio (SNR) of the input $x(k)$. Moustakides (1997) has shown that when the SNR is high (*e.g.*, 30 dB or higher) the RLS method exhibits fast convergence using the following value for the regularization parameter.

$$\delta = P_x \qquad (9.115)$$

Here $P_x = E[x^2(k)]$ is the average power of $x(k)$, which is assumed to have zero mean, otherwise the variance of $x(k)$ should be used. As the SNR of $x(k)$ decreases, the value of δ should be increased.

Example 9.13   RLS Method

To compare the performance characteristics of the RLS and LMS methods, we revisit the system identification problem first posed in Example 9.3. Here the system to be identified was a sixth-order IIR system with the following transfer function.

$$H(z) = \frac{2 - 3z^{-1} - z^{-2} + 4z^{-4} + 5z^{-5} - 8z^{-6}}{1 - 1.6z^{-1} + 1.75z^{-2} - 1.436z^{-3} + 0.6814z^{-4} - 0.1134z^{-5} - 0.0648z^{-6}}$$

Again, suppose the input $x(k)$ consists of $N = 1000$ samples of white noise uniformly distributed over $[-1, 1]$. Let the order of the adaptive transversal filter be $m = 50$, and suppose a forgetting factor of $\gamma = 0.98$ is used. The regularization parameter δ is set to the average power of the input as in Eq. 9.115. A plot of the first 200 samples of the square of the error, obtained by running script *exam9_13* on the distribution CD, is shown in Figure 9.25. In this case, the square of the error

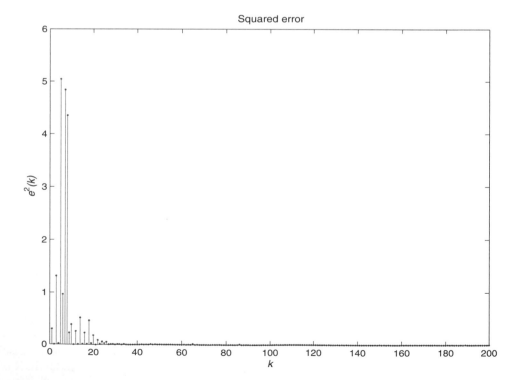

Figure 9.25 First 200 Samples of Squared Error during System Identification using the RLS Method with $m = 50$, $\gamma = 0.99$, and $\delta = P_x$

converges close to zero after approximately 30 samples. This is in contrast to the LMS method, previously shown in Figure 9.9, which took approximately 400 samples to converge. Thus, when measured in iterations, the RLS method is faster than the LMS method by an order of magnitude. However, it should be kept in mind that the LMS method requires about $m = 50$ FLOPs per iteration, whereas the RLS method requires about $3m^2 = 7500$ FLOPs per iteration. In terms of FLOPs, the two methods appear to be roughly equivalent in this case.

Based on Eq. 9.105, one might expect the RLS method to converge even faster, say in one iteration. The reason it took approximately 30 iterations to converge was due to the transients associated with the startup of the algorithm. Since $x(k)$ is a causal signal, the vector of past inputs, θ, continues to be populated with zero samples for the first $m = 50$ iterations. Apparently, the RLS method converges in less than m samples in this instance due to the presence of the forgetting factor, $\gamma = 0.99$, which tends to reduce the influence of samples in the remote past.

An FIR model, $W(z)$, identified with the RLS method, is obtained by using the final steady-state estimate for the weight vector, $w(N-1)$. The magnitude responses of this FIR system and the original system $H(z)$ are plotted in Figure 9.26, where it is evident that they are nearly identical.

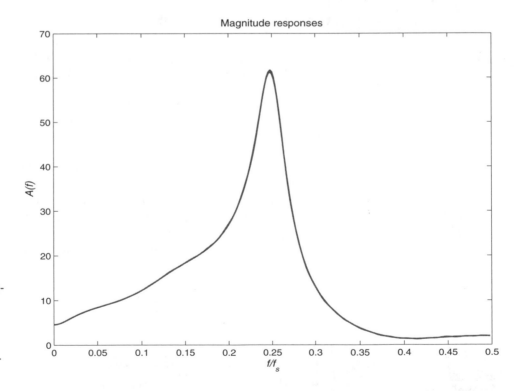

Figure 9.26 Magnitude Responses of the Original IIR System and an Identified System using the RLS Method with $m = 50$, $\gamma = 0.99$, $\delta = P_x$, and $N = 1000$ Samples

MATLAB Toolbox

The FDSP toolbox contains the following function which implements the RLS method in Algorithm 9.1.

```
[w,e] = f_rls (x,d,m,gamma,delta,w);                    % RLS method
```

On entry to f_rls, x is a vector of length N containing the input, d is a vector of length N containing the desired output, $m \geq 0$ is the filter order, $gamma$ is the forgetting factor, $delta$ is the regularization parameter, and w is an optional vector of length $m + 1$ containing an initial guess for the weights. The default value is $w = 0$. The forgetting factor must satisfy $0 < gamma \leq 1$, and typically, $gamma \approx 1$. Recall that $\mu = 1 - gamma$ is the effective step length. The regularization parameter $delta > 0$ is optional with a default value of $delta = P_x$ as in Eq. 9.115. On exit from f_rls, w is a vector of length $m + 1$ containing the estimated value of the optimal weights, and e is an optional vector of length N containing the samples of the error.

9.8 Active Noise Control

One of the practical application areas of adaptive signal processing is active noise control. The basic idea behind the active control of acoustic noise is to inject a secondary sound into an environment so as to cancel the primary sound using destructive interference. Application areas include jet engine noise, road noise in automobiles, blower noise in air ducts, transformer noise, and industrial noise from rotating machines (Kuo and Morgan, 1996). Active noise control requires an adaptive filter configuration as shown in Figure 9.27.

Here the input or *reference signal*, $x(k)$, denotes samples of acoustic noise obtained from a microphone or a non-acoustic sensor. The transfer function, $G(z)$, represents the physical characteristics of the air channel over which the noise travels. The *primary noise*, $d(k)$, is combined with *secondary noise*, $y(k)$, produced by the adaptive filter. The secondary noise, also called *anti-noise*, is designed to destructively interfere with the primary noise so as to produce silence at the error

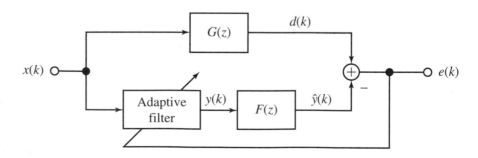

Figure 9.27 Active Control of Acoustic Noise using an Adaptive Filter

microphone, $e(k)$. The feature that makes Figure 9.27 different from a standard system-identification configuration is the appearance of a secondary path system with transfer function $F(z)$. The secondary system represents the hardware used to produce the secondary sound. It includes such things as a power amplifier, a speaker, the secondary air channel, the error microphone, and a preamp. Later we will examine how to model $F(z)$ using offline system-identification techniques.

9.8.1   The Filtered-x LMS Method

The basic LMS method needs to be modified to take into account the presence of the secondary path transfer function, $F(z)$, in Figure 9.27. To this end, suppose the adaptive transversal filter in Figure 9.27 has converged to an FIR system with a constant weight vector, w. The transfer function of the resulting FIR filter is then

$$W(z) = \sum_{i=0}^{m} w_i z^{-i} \tag{9.116}$$

Replacing the adaptive filter in Figure 9.27 with $W(z)$, we then find that, in the steady state, the Z-transform of the error signal is

$$E(z) = D(z) - \hat{Y}(z)$$
$$= G(z)X(z) - F(z)W(z)X(z)$$
$$= [G(z) - F(z)W(z)]X(z) \tag{9.117}$$

For the error to be zero for all inputs $x(k)$, it is required that $G(z) - F(z)W(z) = 0$ or

$$W(z) = F^{-1}(z)G(z) \tag{9.118}$$

On the surface, the expression for $W(z)$ in Eq. 9.118 appears to provide a simple solution to the problem of finding an optimal active-noise-control filter. Unfortunately, this solution is not a practical one. Recall that the secondary-path transfer function, $F(z)$, includes the air channel, over which the sound travels from the speaker to the error microphone. The propagation of sound through air introduces a delay that is proportional to the path length. Given the presence of a delay in $F(z)$, the inverse of $F(z)$ must have a corresponding advance, which means that $F^{-1}(z)$ is not causal and therefore not physically realizable.

To obtain a causal approximation to $W(z)$, we start by examining the error signal in the time domain. Recall that multiplication of Z-transforms in the frequency domain corresponds to convolution of the corresponding signals in the time domain. Consequently, from Figure 9.27 we have

$$e(k) = d(k) - \hat{y}(k)$$
$$= d(k) - f(k) \star y(k)$$
$$= d(k) - f(k) \star [w^T(k)\theta(k)] \tag{9.119}$$

Here $f(k)$ is the impulse response of the secondary system, $F(z)$, and $*$ denotes the linear convolution operation. Recall that $\theta(k)$ is the $(m+1) \times 1$ vector of past inputs at time k, and $w(k)$ is the $(m+1) \times 1$ weight vector at time k. The mean-square-error objective function is

$$\epsilon(w) = E[e^2(k)] \tag{9.120}$$

For the purpose of estimating the gradient of partial derivatives of $\epsilon(w)$ with respect to the elements of w, we use the LMS approximation, $E[e^2(k)] \approx e^2(k)$. Combining this with Eq. 9.119 then yields

$$\nabla \epsilon(w) \approx 2e(k)\nabla e(k)$$
$$= -2e(k)[f(k) \star \theta(k)] \tag{9.121}$$

To simplify the final result, let $\hat{x}(k)$ denote a filtered version of the input using the filter $F(z)$. That is, $\hat{X}(z) = F(z)X(z)$ or

$$\hat{x}(k) = f(k) \star x(k) \tag{9.122}$$

Similarly, let $\hat{\theta}(k)$ denote the $(m+1) \times 1$ vector of filtered past inputs. That is,

$$\hat{\theta}(k) = [\hat{x}(k), \hat{x}(k-1), \cdots, \hat{x}(k-m)]^T \tag{9.123}$$

Combining Eqs. 9.121 through 9.123, it then follows that the gradient of the mean square error can be expressed in terms of the filtered input as

$$\nabla \epsilon(w) \approx -2e(k)\hat{\theta}(k) \tag{9.124}$$

The LMS method uses the steepest-descent method as a starting point. If $\mu > 0$ denotes the step length, then the steepest-descent method for updating the weights is

$$w(k+1) = w(k) - \mu \nabla \epsilon[w(k)] \tag{9.125}$$

If we substitute the approximation for the gradient from Eq. 9.124 into Eq. 9.125, this results in the following weight-update algorithm called the *filtered-x LMS* method or simply the FXLMS method.

FXLMS method

$$w(k+1) = w(k) + 2\mu e(k)\hat{\theta}(k), \quad k \geq 0 \tag{9.126}$$

Note that the only difference between the FXLMS method and the LMS method is that the vector of past inputs is first filtered by the secondary-path transfer function, hence the name filtered-x LMS method. In practice, an approximation to the secondary system, $\hat{F}(z) \approx F(z)$, is used for the prefiltering of $x(k)$, because an exact model of the secondary path is not available. A block diagram of the FXLMS method is shown in Figure 9.28.

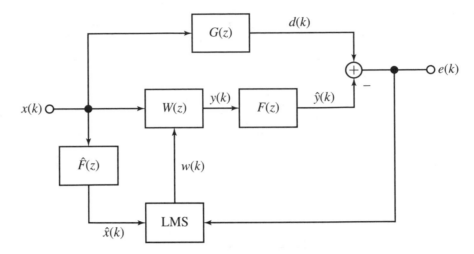

Figure 9.28 The Filtered-x LMS Method

9.8.2 Secondary-path Identification

The key to implementing the FXLMS method in Figure 9.28 is the identification of a model for the secondary path, $F(z)$. To illustrate a secondary path, consider the air duct active noise control system shown in Figure 9.29. The secondary system represents the path traveled by the noise-canceling sound, including the loud speaker and the error microphone.

It should be pointed out that there may be some feedback from the speaker to the reference microphone. For the purpose of this analysis, we will assume that the feedback is negligible. One way to reduce the effects of feedback is to use a directional microphone. Another way to eliminate feedback completely is to use a non-acoustic sensor in place of the reference microphone to produce a signal $x(k)$ that is correlated with the primary noise. Alternatively, the effects of feedback can be taken into account using a feedback-neutralization scheme (Warnaka, *et al.*, 1984).

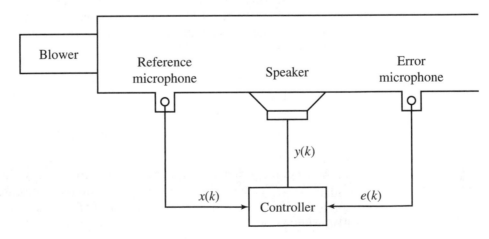

Figure 9.29 Active Noise Control in an Air Duct

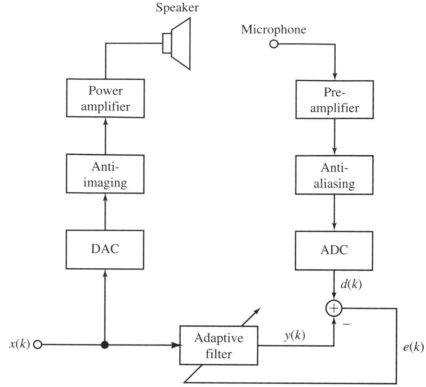

Figure 9.30 Identification of Secondary-path Model, $\hat{F}(z)$

The block diagram in Figure 9.29 is a high-level diagram that leaves many of the details implicit. A more detailed representation, which shows the measurement scheme used to identify a model for the secondary path, is shown in Figure 9.30. An example of a secondary-path model found by this technique can be found in (Schilling, *et al.*, 1998).

Active control of random broadband noise is a challenging problem. Fortunately, the noise that appears in practice often has a significant narrowband or periodic component. For example, rotating machines generate harmonics whose fundamental frequency varies with the speed of rotation. Similarly, electrical transformers and overhead fluorescent lights emit harmonics with a fundamental frequency of $f_0 = 60$ Hz. Let $f_s = 1/T$ denote the sampling frequency. Then the primary noise can be modeled as follows.

$$x(k) = \sum_{i=0}^{r} a_i \cos(2\pi i f_0 kT) + b_i \sin(2\pi i f_0 kT) + v(k), \quad 0 \le k < N \quad (9.127)$$

Here r is the number of harmonics, and f_0 is the frequency of the fundamental harmonic. Since $f_s/2$ is the highest frequency that can be represented without aliasing, the number of harmonics is limited to $r < f_s/(2f_0)$. The broadband term $v(k)$ is additive white noise.

To determine the amount of noise cancellation achieved by active noise control, suppose the controller is not activated for the first $N/4$ samples. The average power over the first $N/4$ samples of $e(k)$ provides a base line relative to which noise cancellation can be measured.

$$P_u = \frac{4}{N} \sum_{k=0}^{N/4-1} e^2(k) \tag{9.128}$$

If the controller is activated at sample $k = N/4$ with the weights updated as in Eq. 9.126, then the system will undergo a transient segment with the weights converging to their optimal values, assuming the step size μ is sufficiently small. If the adaptive filter has reached the steady state by sample $3N/4$, then the average power of the error achieved by the controller is

$$P_c = \frac{4}{N} \sum_{k=3N/4}^{N-1} e^2(k) \tag{9.129}$$

The noise reduction achieved by active noise control can then be expressed in decibels as follows.

Noise reduction

$$E_{\text{anc}} = 10 \log_{10} \left(\frac{P_c}{P_u} \right) \text{ dB} \tag{9.130}$$

Example 9.14   **FXLMS Method**

To illustrate the use of the FXLMS to achieve active noise control, consider the air-duct system shown in Figure 9.29. Suppose the primary path $G(z)$ and secondary path $F(z)$ are each modeled using FIR filters of order $n = 20$. For the purpose of this simulation, let the coefficients of the filters consist of random numbers uniformly distributed over the interval $[-1, 1]$. Suppose the noise-corrupted periodic input in Eq. 9.127 is used, the sampling frequency is $f_s = 2000$ Hz, and the fundamental frequency is $f_0 = 100$ Hz. Let the number of harmonics be $r = 5$, and suppose the additive white noise $v(k)$ is distributed uniformly over $[-0.5, 0.5]$. Finally, suppose the coefficients for each harmonic are generated randomly in the interval $[-1, 1]$, thereby producing random amplitudes and phases for the r harmonics. The FXLMS method in Figure 9.30 can be applied to this system by running script *exam9_14* on the distribution CD. Here $N = 2400$ samples are used with an adaptive filter of order $m = 40$ and a step size of $\mu = 0.0001$. A plot of the resulting squared error is shown in Figure 9.31. Note that the active noise control is activated at sample $k = 600$. Once the transients have decayed to zero, it is evident that a significant amount of noise reduction is achieved. The noise reduction measured using Eq. 9.130 is $E_{\text{anc}} = 40.8$ dB.

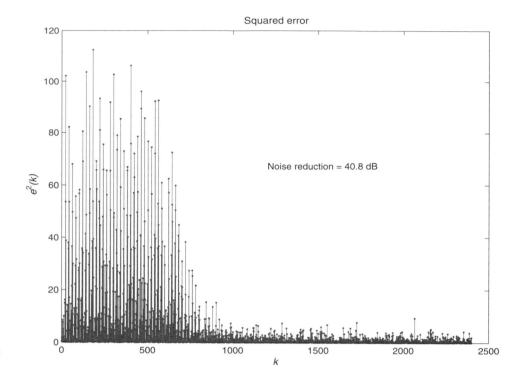

Figure 9.31 Active Noise Control using the FXLMS Method

9.8.3 Signal-synthesis Method

Suppose the input or reference signal is a noise-corrupted periodic signal, as in Eq. 9.127. If the frequency of the fundamental harmonic, f_0, is known or can be measured, then a more direct signal-synthesis approach can be used, as shown in Figure 9.32. For notational convenience, define

$$\psi \overset{\Delta}{=} 2\pi f_0 T \tag{9.131}$$

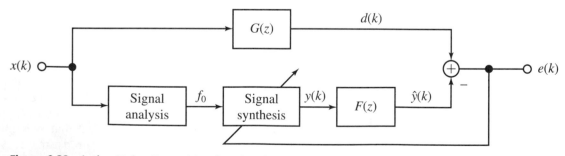

Figure 9.32 Active Noise Control by the Signal-synthesis Method

Since the primary- and secondary-path models are assumed to be linear, the form of the control signal required to cancel the periodic component of the noise is

$$y(k) = \sum_{i=1}^{r} p_i(k) \cos(ik\psi) + q_i(k) \sin(ik\psi) \tag{9.132}$$

Next suppose the secondary path in Figure 9.28 is modeled with an FIR filter of order m with coefficient vector, f. That is,

$$F(z) = \sum_{i=0}^{m} f_i z^{-i} \tag{9.133}$$

To obtain a concise formulation of the secondary-noise signal, let $\phi(k)$ denote the vector of past control signals, $y(k)$.

$$\phi(k) \triangleq [y(k), y(k-1), \cdots, y(k-m)]^T \tag{9.134}$$

It then follows that the secondary-noise signal at time k can be expressed with a dot product as

$$\hat{y}(k) = f^T \phi(k) \tag{9.135}$$

Our objective is to choose values for the coefficients $p(k)$ and $q(k)$ in Eq. 9.132 so as to minimize the mean square error $\epsilon(p,q) = E[e^2(k)]$. As with the LMS method, for the purpose of estimating the gradient, we approximate the mean square error with $E[e^2(k)] \approx e^2(k)$. Then using Eq. 9.132 through Eq. 9.135, the partial derivative of the mean square error with respect to the ith element of the coefficient vector p is

$$\frac{\partial \epsilon(p,q)}{\partial p_i} \approx 2e(k) \frac{\partial e(k)}{\partial p_i}$$

$$= -2e(k) \frac{\partial f^T \phi(k)}{\partial p_i}$$

$$= -2e(k) \sum_{j=1}^{m} f_j \frac{\partial y(k-j)}{\partial p_i}$$

$$= -2e(k) \sum_{j=1}^{m} f_j \cos[i(k-j)\psi] \tag{9.136}$$

Similarly, the partial derivative of the mean square error with respect to the ith element of the coefficient vector q is

$$\frac{\partial \epsilon(p, q)}{\partial q_i} \approx -2e(k) \sum_{j=1}^{m} f_j \sin[i(k-j)\psi] \tag{9.137}$$

To simplify the final result, let $P(k)$ and $Q(k)$ be $r \times 1$ vectors of intermediate variables defined as follows.

$$P_i(k) \overset{\Delta}{=} \sum_{j=0}^{m} f_j \cos[i(k-j)\psi] \tag{9.138a}$$

$$Q_i(k) \overset{\Delta}{=} \sum_{j=0}^{m} f_j \sin[i(k-j)\psi] \tag{9.138b}$$

It then follows that the partial derivatives of the mean square error with respect the elements of p and q are

$$\frac{\partial \epsilon(p, q)}{\partial p_i} \approx -2e(k)P_i(k) \tag{9.139a}$$

$$\frac{\partial \epsilon(p, q)}{\partial q_i} \approx -2e(k)Q_i(k) \tag{9.139b}$$

We can now use the steepest-descent method to update the coefficients of the secondary-sound control signal. Let $\mu > 0$ denote the step size. Using Eq. 9.139, this results in the following *signal-synthesis* active noise control method.

Signal-synthesis
method

$$\begin{bmatrix} p(k+1) \\ q(k+1) \end{bmatrix} = \begin{bmatrix} p(k) \\ q(k) \end{bmatrix} + 2\mu e(k) \begin{bmatrix} P(k) \\ Q(k) \end{bmatrix}, \quad k \geq 0 \tag{9.140}$$

One feature that sets the signal-synthesis method apart from the FXLMS method is that it is typically a lower-dimensional method. The signal-synthesis method has a weight vector of dimension $n = 2r$, where r is the number of harmonics in the periodic component of the primary noise. Since the maximum number of harmonics that can be accommodated with a sampling frequency of f_s is $r < f_s/(2f_0)$, this means that the maximum dimension of the signal-synthesis method is $n < f_s/f_0$, which is often relatively small.

Although the dimension of the signal-synthesis method is small, it is apparent from Eq. 9.138 that there is considerable computational effort required to compute the intermediate variables $P(k)$ and $Q(k)$. At each time step, a total of $2r(m+1)$ floating-point multiplications or FLOPs are required. Note that the trigonometric function evaluations can be precomputed and stored in a look-up table. Since the $2r(m+1)$ FLOPs must be performed in less than T seconds, this could limit the sampling frequency, f_s. Furthermore, for large values of m, the computations in Eq. 9.138 could exhibit significant accumulated round-off error. Fortunately, these limitations can be minimized by developing a recursive formulation for $P(k)$ and $Q(k)$ (Schilling, *et al.*, 1998). Using the cosine of the sum trigonometric identity from Appendix D, we have

$$P_i(k) = \sum_{j=0}^{m} f_j \cos[i(k - j - 1 + 1)\psi]$$

$$= \sum_{j=0}^{m} f_j \cos[i(k - 1 - j)\psi]\cos(i\psi) - \sin[i(k - 1 - j)\psi]\sin(i\psi)$$

$$= \cos(i\psi)P_i(k - 1) - \sin(i\psi)Q_i(k - 1) \tag{9.141}$$

Similarly, using the sine of the sum trigonometric identity from Appendix D, one can show that $Q(k)$ can be expressed recursively as

$$Q_i(k) = \sin(i\psi)P_i(k - 1) + \cos(i\psi)Q_i(k - 1) \tag{9.142}$$

The recursive-update formulas for $P(k)$ and $Q(k)$ can be expressed compactly in vector form as follows.

Recursive update formula

$$\begin{bmatrix} P_i(k) \\ Q_i(k) \end{bmatrix} = \begin{bmatrix} \cos(i\psi) & -\sin(i\psi) \\ \sin(i\psi) & \cos(i\psi) \end{bmatrix} \begin{bmatrix} P_i(k - 1) \\ Q_i(k - 1) \end{bmatrix}, \quad 1 \le i \le r \tag{9.143}$$

Note that the number of FLOPs per iteration has been reduced from $2r(m + 1)$ to $4r$. Thus, the computational effort of the recursive implementation is independent of the size of the filter used to model $F(z)$. To start the recursive computation of $P(k)$ and $Q(k)$, initial values must be used. From Eq. 9.138, the appropriate starting values are

$$P_i(0) = \sum_{j=0}^{m} f_j \cos(ij\psi) \tag{9.144a}$$

$$Q_i(0) = -\sum_{j=0}^{m} f_j \sin(ij\psi) \tag{9.144b}$$

The signal-synthesis method is based on the key assumption that the fundamental frequency of the periodic component of the primary noise is known or can be measured. Examples of applications where f_0 is known include the noise from electrical transformers and from overhead fluorescent lights, where $f_0 = 60$ Hz is a parameter that is regulated carefully by the power company. In those applications where f_0 is not known, it might be measured as shown by the signal-analysis block in Figure 9.31. For example, if the primary noise is produced by a rotating machine or motor, then a tachometer might be used to measure f_0. Another approach is to lock onto the periodic component of the noise using a phase-locked loop (Schilling, *et al.*, 1998). The virtue of measuring the fundamental frequency is that the signal-synthesis method then can track changes in the period of the periodic component of the primary noise.

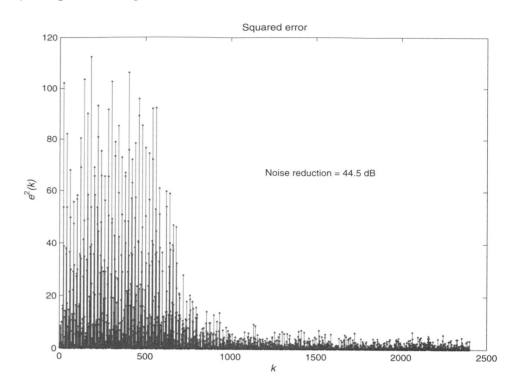

Figure 9.33 Active Noise Control using the Signal-synthesis Method

Example 9.15 **Signal-synthesis Method**

To illustrate the use of the signal-synthesis method to achieve active noise control, consider the air-duct system shown in Figure 9.29. To facilitate comparison with the FXLMS method, suppose the input and system parameters are the same as in Example 9.14. The signal-synthesis method in Figure 9.32 can be applied to this system by running script *exam9_15* on the distribution CD. Again $N = 2400$ samples are used, but this time the selected step size is $\mu = 0.001$. A plot of the resulting squared error is shown in Figure 9.33. Note that the active noise control is activated at sample $k = 600$. Once the transients have decayed to zero, it is apparent a significant amount of noise reduction is achieved. The noise reduction measured using Eq. 9.130 is $E_{anc} = 44.5$ dB. In this case, the final estimates for the control signal coefficients were as follows.

$$p = [-0.0892, -0.1159, 1.0692, -0.9898, -0.1797]^T$$

$$q = [-0.1065, 0.1826, 1.3938, -1.6894, -0.3873]^T$$

Another way to look at the performance of an active-noise-control system is to examine the magnitude spectrum of the error signal with and without the noise control activated. The results are shown in Figure 9.34. The five harmonics are clearly evident in Figure 9.34(a), which corresponds to the first $N/4$ samples before the noise control is activated. The harmonics effectively are eliminated in Figure 9.34(b), which corresponds to the last $N/4$ samples after the noise controller has reached a steady state.

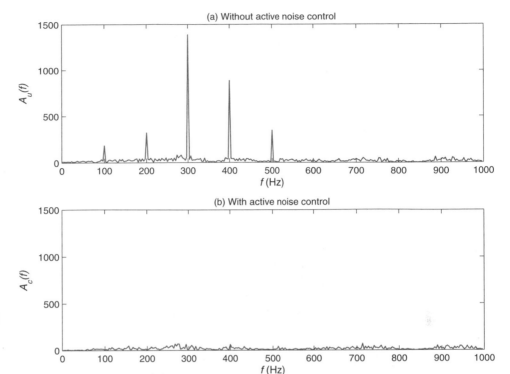

Figure 9.34 Magnitude Spectra of Error (a) without Active Noise Control and (b) with Active Noise Control using the Signal-synthesis Method

MATLAB Toolbox

The FDSP toolbox contains the following functions for implementing active noise control.

```
[w,e]   = f_fxlms (x,g,f,m,mu,w);           % FXLMS method
[p,q,e] = f_sigsyn (x,g,f,f_0,f_s,r,mu);    % Signal synthesis method
```

The function *f_fxlms* implements the filtered-*x* LMS method. On entry to *f_fxlms*, *x* is a vector of length N containing the noise, g and f are vectors of length n containing the FIR coefficients of the primary and secondary paths, respectively, $m \geq 0$ is the filter order, $\mu > 0$ is the step length, and w is an optional vector of length $m + 1$ containing an initial guess for the weights. The default value is $w = 0$. On exit from *f_fxlms*, w contains the computed weights, and e is an optional vector of length N containing the errors.

The function *f_sigsyn* implements the signal-synthesis method of active control of noise that has a periodic component. The calling arguments for *f_sigsyn* are the same as those for *f_fxlms*, except for f_0, f_s, and r. The scalars $f\_0 > 0$ and $f\_s > 2f\_0$ denote the fundamental frequency of the periodic component of the noise and the sampling frequency, respectively. Both are expressed in units of Hz. The integer, $1 \leq r < f\_s/(2f\_0)$, specifies the number of harmonics. On exit from f_sigsyn, p and q are vectors of length r containing the cosine and sine coefficients, respectively, of the synthesized noise control signal. Output e is an optional vector of length N containing the errors.

● ● ● ● ● ● ● ● ● ● ● ● ● ● ●

*9.9 Nonlinear System Identification

9.9.1 Nonlinear Discrete-time Systems

All of the discrete-time systems that we have investigated thus far have been linear systems. One can generalize the notion of a linear discrete-time system to a nonlinear system in the following way.

$$y(k) = f[x(k), \ldots, x(k-m), y(k-1), \ldots, y(k-n)], \quad k \geq 0 \tag{9.145}$$

Here f is a real-valued function which is assumed to be continuous, $x(k)$ is the system input at time k, and $y(k)$ is the system output at time k. Thus, the present output, $y(k)$, depends on the past inputs and the past outputs in some nonlinear fashion. For convenience, we will refer to the nonlinear discrete-time system in Eq. 9.145 as the system S_f. A signal flow graph of the system S_f is shown in Figure 9.35 for the case $m = 2$ and $n = 3$.

A more compact formulation of the system S_f can be obtained by introducing the following generalization of the vector of past inputs called the *state* vector.

$$\theta(k) \triangleq [x(k), \ldots, x(k-m), y(k-1), \ldots, y(k-n)]^T \tag{9.146}$$

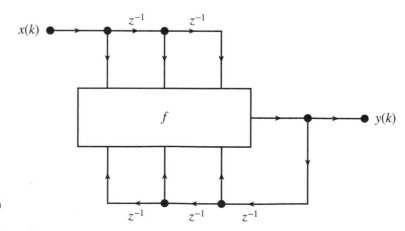

Figure 9.35 Signal Flow Graph of Nonlinear Discrete-time System S_f with $m = 2$ and $n = 3$

Thus, the state vector is a vector containing both the past inputs and the past outputs. The number of elements, p, in the state vector is called the *dimension* of the system.

$$p = m + n + 1 \tag{9.147}$$

Given the $p \times 1$ state vector $\theta(k)$, the output of the nonlinear system S_f at time k is simply

$$y(k) = f[\theta(k)], \quad k \geq 0 \tag{9.148}$$

We will restrict our attention to nonlinear systems that are BIBO stable. Recall from Chapter 2 that a system is BIBO stable if and only if every bounded input, $x(k)$, produces a bounded output, $y(k)$. Let a be a 2×1 vector, and suppose the input is bounded in the following manner.

$$a_1 \leq x(k) \leq a_2 \tag{9.149}$$

If S_f is BIBO stable, this means there exists a 2×1 vector b such that

$$b_1 \leq y(k) \leq b_2 \tag{9.150}$$

Typically, the vector of input bounds a is selected by the user, and then the vector of output bounds b can be estimated experimentally from measurements. If the input is constrained as in Eq. 9.149, then the *domain* of the continuous function f is restricted to the following compact subset of R^p.

$$\Theta = [a_1, a_2]^{m+1} \times [b_1, b_2]^n \tag{9.151}$$

9.9.2   Grid Points

One approach to approximating the function $f : \Theta \rightarrow R$ is to overlay the domain Θ with a grid of elements, and then develop a *local* representation of f valid over each grid element. To that end, suppose there are $d \geq 2$ grid points equally spaced along each of the p dimensions of Θ. Then, along the ith dimension, the jth *grid value* for $0 \leq j < d$ is

$$\Theta_{ij} = \begin{cases} a_1 + j\Delta x, & 1 \leq i \leq m+1 \\ b_1 + j\Delta y, & m+2 \leq i \leq p \end{cases} \tag{9.152}$$

Here Δx and Δy denote the grid-point *spacing* along the x and y directions, respectively. That is,

$$\Delta x \triangleq \frac{a_2 - a_1}{d - 1} \tag{9.153a}$$

$$\Delta y \triangleq \frac{b_2 - b_1}{d - 1} \tag{9.153b}$$

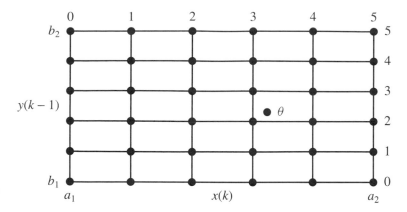

Figure 9.36 Grid Points over the Domain Θ with $m = 0$, $n = 1$, and $d = 6$

As an illustration, suppose $m = 0$ and $n = 1$, which corresponds to a dimension of $p = 2$. If the number of grid points per dimension is $d = 6$, then the grid consists of 36 grid points, or 25 grid elements, as shown in Figure 9.36. Note that in general, the total number of grid points is

$$r = d^p \tag{9.154}$$

It is evident from Eq. 9.154 that as the number of dimensions p and the number of grid points per dimension d grow, the total number of grid points r can be become very large. There are two distinct ways to order this large number of grid points. One is to use something called a vector index. Suppose q is a $p \times 1$ vector with *integer* elements ranging from 0 to $d - 1$. That is, $0 \le q_i < d$ for $1 \le i \le p$. Here q_i selects the coordinate of the qth grid point along the ith dimension. That is, if $\theta(q)$ denotes the qth grid point, then using the grid values in Eq. 9.152 gives

$$\theta_i(q) = \Theta_{iq_i}, \quad 1 \le i \le p \tag{9.155}$$

Here, q can be thought of as *vector index* whose elements select the values of the point $\theta(q)$ along each of the p dimensions. The virtue of this vector-index approach is that, for an arbitrary $\theta \in \Theta$, it is easy to identify the vertices of the grid element that contains θ. For example, the index of the *base* vertex of the grid element containing point θ is

$$v_i(\theta) = \begin{cases} \text{floor}\left(\dfrac{\theta_i - a_1}{\Delta x}\right), & 1 \le i \le m + 1 \\[2mm] \text{floor}\left(\dfrac{\theta_i - b_1}{\Delta y}\right), & m + 2 \le i \le p \end{cases} \tag{9.156a}$$

$$v_i(\theta) = \text{clip}[v_i(\theta), 0, d - 2] \tag{9.156b}$$

The clipping of the computed index to the interval $[0, d - 2]$ in Eq. 9.156b is included in order to account for the possibility that $\theta \notin \Theta$. When clipping occurs, the base vertex of the grid element closest to θ is obtained. Once the index of the base vertex is found, the indices of the other vertices are easily determined by adding combinations of 0 and 1 to the elements of $v(\theta)$. For example, let b^i denote a binary $p \times 1$ vector representing the decimal value i. Then the index of the ith vertex of the grid element containing the point θ is computed as follows.

$$q^i = v(\theta) + b^i, \quad 0 \le i < 2^p \tag{9.157}$$

The following example illustrates how to find the vertices of the local grid element containing an arbitrary state vector θ.

Example 9.16   **Local Grid Element**

Consider the two-dimensional case $m = 0$ and $n = 1$. Suppose $a = [-10, 10]^T$ and $b = [-5, 5]^T$. Let the number of grid elements per dimension be $d = 6$, as shown in Figure 9.36. In this case, the domain of the function f is

$$\Theta = [-10, 10] \times [-5, 5]$$

From Eq. 9.153, the grid element size is $\Delta x \times \Delta y$, where the grid point spacings are

$$\Delta x = \frac{a_2 - a_1}{d - 1} = 4$$

$$\Delta y = \frac{b_2 - b_1}{d - 1} = 2$$

Suppose $\theta = [3.2, -0.4]^T$ represents an arbitrary point in the domain Θ. Then from Eq. 9.156, the index of the base, or lower-left, vertex of the grid element containing θ is

$$v_1(\theta) = \text{floor} \left(\frac{\theta_1 - a_1}{\Delta x} \right) = \text{floor} \left(\frac{13.2}{4} \right) = 3$$

$$v_2(\theta) = \text{floor} \left(\frac{\theta_2 - b_1}{\Delta y} \right) = \text{floor} \left(\frac{4.6}{2} \right) = 2$$

There are a total of $2^p = 4$ vertices per grid element. From Eq. 9.157, the indices of the vertices of the grid element containing θ are

$$q^0 = v(\theta) + [0, 0]^T = [3, 2]^T$$

$$q^1 = v(\theta) + [0, 1]^T = [3, 3]^T$$

$$q^2 = v(\theta) + [1, 0]^T = [4, 2]^T$$

$$q^3 = v(\theta) + [1, 1]^T = [4, 3]^T$$

The point θ is shown in Figure 9.36. Inspection confirms that $\{q^0, q^1, q^2, q^3\}$ do indeed specify the vertices of the local grid element containing θ.

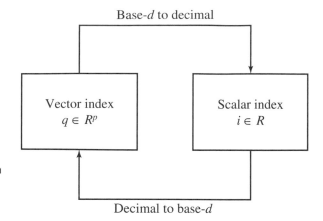

Figure 9.37 Transformations between Vector and Scalar Indices of the Grid Points

9.9.3   Radial Basis Functions

Our objective is to develop a parameterized structure for a nonlinear system that can accurately approximate the function f over the domain Θ. To do this, it is necessary to consider a second way to order the r grid points using a scalar. For grid point $\theta(q)$, the *scalar index* can be computed as follows.

$$i = q_1 + q_2 d + \cdots + q_p d^{p-1} \tag{9.158}$$

As the integer elements of q range from 0 to $d - 1$, the value of i ranges from 0 to $r - 1$, where r is as in Eq. 9.154. The computation in Eq. 9.158 can be seen to be a *base-d to decimal* conversion of q. Consequently, a *decimal to base-d* conversion of i can be used to recover q from i. The relationship between the vector index q and the scalar index i of the grid points is summarized in Figure 9.37.

Using the scalar index i, the following one-dimensional ordering of the r grid points is obtained.

$$\Gamma = \{\theta^0, \theta^1, \cdots, \theta^{r-1}\} \tag{9.159}$$

Given the one-dimensional ordering of the grid point in Eq. 9.159, consider the following structure for approximating the function f on the right-hand side of the nonlinear system S_f.

Zeroth-order approximation

$$f_0(\theta) = w^T g(\theta) \tag{9.160}$$

Here $w = [w_0, \ldots, w_{r-1}]^T$ is an $r \times 1$ *weight* vector whose elements will be determined from input-output measurements of the system S_f. The function $g : R^p \rightarrow R^r$

represents an $r \times 1$ vector of functions called *radial basis* functions with the ith radial basis function centered at grid point θ^i. A radial basis function or RBF is a continuous function, $g_i : R^p \to R$, such that

$$g_i(\theta^i) = 1 \tag{9.161a}$$

$$|g_i(\theta)| \to 0 \quad \text{as} \quad \|\theta - \theta^i\| \to \infty \tag{9.161b}$$

Thus, the value of an RBF goes to zero as the radius from the point about which it is centered goes to infinity. A popular example of a radial basis function is the Gaussian RBF.

$$g_i(\theta) = \exp\left[\frac{(\theta - \theta^i)^T(\theta - \theta^i)}{\sigma^2}\right] \tag{9.162}$$

Here the variance, σ^2, controls the rate at which the RBF goes to zero as the distance from the center point increases. The Gaussian RBF has a number of useful properties, not the least of which is the observation that it has an infinite number of derivatives, all of which are continuous. However, the Gaussian RBF also has some drawbacks. Since $g_i(\theta) > 0$ for all θ, it follows that if σ is too large, there will be many terms in Eq. 9.160 that contribute to $f_0(k)$. This defeats the local nature of the representation which, ideally, has only a few terms contributing to $f_0(k)$. On the other hand, if σ is too small then the value of $g_i(\theta)$ will be near zero midway between the grid points, which reduces the effectiveness of g_i in approximating f.

There are several potential candidates for radial basis functions (Webb and Shannon, 1998). As an alternative to the Gaussian RBF, suppose $p = 1$ and consider the following one-dimensional *raised-cosine* RBF centered about $z = 0$ (Schilling, *et al.*, 2001).

$$G(z) = \begin{cases} \dfrac{1 + \cos(\pi z)}{2}, & |z| \leq 1 \\ 0, & |z| > 1 \end{cases} \tag{9.163}$$

Note that $G(z)$ satisfies the two basic properties in Eq. 9.161. A plot comparing the one-dimensional Gaussian RBF and the raised-cosine RBF is shown in Figure 9.38 for the case $\sigma = 1$. Observe that the raised-cosine RBF is a continuously differentiable function. Furthermore, the set of z over which the raised-cosine RBF is nonzero is contained in the compact (closed and bounded) set $S = [-1, 1]$. In this case, we say that the raised-cosine RBF has *compact support*. This is in contrast to the Gaussian RBF, which is not a function of compact support. The compact support property ensures that a representation based on a sum of RBFs will be a local representation with only a few of the terms contributing to $f_0(k)$.

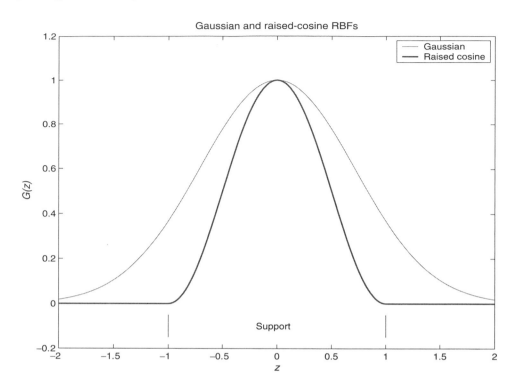

Figure 9.38 Comparison of Gaussian and Raised-cosine RBFs when $p = 1$

The raised-cosine RBF can be generalized from one dimension to p dimensions by forming a product of scalar RBFs, one for each dimension. The scale factors Δx and Δy in Eq. 9.153 can be used to convert the normalized scalar RBF in Eq. 9.163 into a scalar RBF that takes into account the grid-point spacing. This results in the following p-dimensional raised-cosine RBF centered at θ^i. Here the notation $\prod$ denotes the product.

Raised-cosine radial basis functions

$$g_i(\theta) = \prod_{j=1}^{m+1} G\left(\frac{\theta_j - \theta_j^i}{\Delta x}\right) \prod_{j=m+2}^{p} G\left(\frac{\theta_j - \theta_j^i}{\Delta y}\right) \tag{9.164}$$

Example 9.17 **Raised-cosine RBF**

Suppose $m = 0$ and $n = 1$. Let the signal bounds be $a = [-4, 4]$ and $b = [-2, 2]$, and suppose the number of grid points per dimension is $d = 3$. In this case, the grid-point spacing is $\Delta x = 4$ and $\Delta y = 2$, and the domain of f is

$$\Theta = [-4, 4] \times [-2, 2]$$

There are a total of $r = 9$ grid points with an RBF centered about each one. A plot of the RBF centered about the origin can be obtained by running script *exam9_17* on the

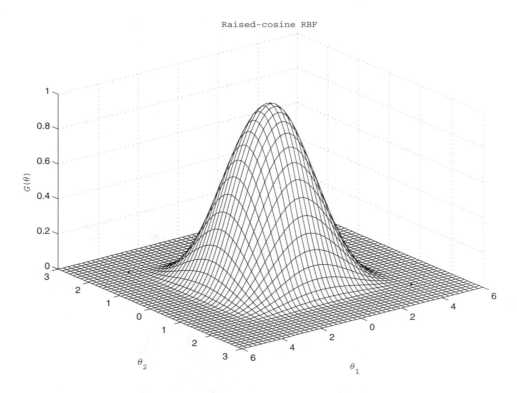

Figure 9.39 A Raised-cosine RBF of Dimension $p = 2$ Centered at $\theta = 0$

distribution CD with the result shown in Figure 9.39. The continuous differentiability and compact support characteristics of this two-dimensional raised-cosine RBF are evident from inspection.

Properties

Raised-cosine RBFs have a number of interesting and useful properties. To start with, they are continuously differentiable functions of θ which means that the approximation to f in Eq. 9.160 is also continuously differentiable. The compact support property, evident in Figures 9.38 and 9.39, means that the RBFs do not interact with one another at the grid points. This leads to the following *orthogonality* property on the grid points.

$$g_j(\theta^i) = \begin{cases} 1, & i = j \\ 0, & i \neq j \end{cases} \tag{9.165}$$

The orthogonality property suggests an easy way to select the $r \times 1$ vector of weights, w. Applying the orthogonality property to Eq. 9.160, we have $f_0(\theta^i) = w_i$.

Thus, the approximation, f_0, to the function f can be made *exact* on the set of r grid points with the following selection of the weights.

RBF network weights

$$w_i = f(\theta^i), \quad 0 \le i < r \tag{9.166}$$

If the number of grid points per dimension, d, is sufficiently large, then the weights in Eq. 9.166 will produce an effective model of the nonlinear discrete-time system S_f over the domain Θ. If the dimension p is too large to permit a relatively large value for d, then the weights in Eq. 9.166 still constitute an excellent initial guess, $w(0)$. A procedure for updating the weights to minimize the mean square error starting from this initial guess is discussed in the next subsection.

The orthogonality property is based on a lack of interaction between RBFs at the grid points. Perhaps the most remarkable property of the raised-cosine RBF occurs between the grid points. Suppose the weights in Eq. 9.160 are all set to unity. It can be shown that the resulting approximation to f has the following property (Schilling, *et al.*, 2001).

$$\sum_{i=0}^{r-1} g_i(\theta) = 1, \quad \theta \in \Theta \tag{9.167}$$

For convenience, we refer to Eq. 9.167 as the *constant interpolation* property. It says that if we weight all of the RBFs equally, then the surface that they produce when their contributions are added is perfectly flat over the domain Θ. This property, which is not shared by Gaussian RBFs, is useful in those instances where the function being approximated is flat over part of its domain. This might occur, for example, if f includes saturation or dead-zone effects.

Example 9.18   Constant Interpolation Property

As an illustration of the constant interpolation property, suppose $m = 0$ and $n = 1$, in which case $p = 2$. Let $a = [-1, 1]$ and $b = [-1, 1]$, and suppose the number of grid points per dimension is $d = 2$. Thus, the grid-point spacing is $\Delta x = 2$ and $\Delta y = 2$, and the domain of the function f is

$$\Theta = [-1, 1] \times [-1, 1]$$

In this case, there are $r = 4$ grid points located at the corners of the square Θ. Suppose the weight vector is $w = [1, 1, 1, 1]^T$. A plot of the resulting interpolated surface, $f_0(\theta)$, obtained by running script *exam9_18* on the distribution CD, is shown in Figure 9.40. It is apparent that on the domain Θ the interpolated surface is flat.

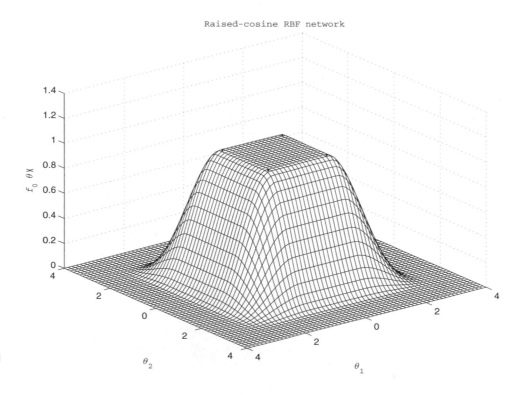

Raised-cosine RBF network

Figure 9.40 Constant Interpolation Produced by Four Equally Weighted Raised-cosine Radial Basis Functions

---

9.9.4   Adaptive RBF Networks

Given the approximation to f in Eq. 9.160, the output of the raised-cosine RBF network at time k can be expressed as follows.

RBF network output

$$y_0(k) = w^T(k)g[\theta(k)], \quad k \geq 0 \tag{9.168}$$

We will refer to Eq. 9.168 as the RBF network S_0. Here the subscript zero is used to denote a zeroth-order RBF network. It is also possible to define a more general first-order RBF network (Al-Ajlouni *et al.*, 2004). Recall that as the dimension p and the number of grid points per dimension d grow, the number of terms r in the dot product in Eq. 9.168 can become very large. Fortunately, for each value of $\theta(k)$, almost all of these terms are zero. Only the local terms in the neighborhood of $\theta(k)$ contribute to $y_0(k)$. Because each raised-cosine RBF has compact support and goes to zero at the adjacent grid points, the only terms contributing to $y_0(k)$ are the terms corresponding to the vertices of the grid element containing $\theta(k)$. Consequently, for each $\theta(k)$, the number of nonzero terms in $y_0(k)$ is at most

$$M = 2^p \tag{9.169}$$

Recall that we can find the vertices of the local grid element containing θ using Eq. 9.156 and Eq. 9.157. This is the basis for the following highly efficient algorithm for evaluating the RBF network output.

ALGORITHM 9.2:

RBF Network Evaluation

1. Set $y_0 = 0$. Compute $p = m = n + 1$ and $M = 2^p$. Compute Δx and Δy using Eq. 9.153.

2. Use Eqs. 9.156 and 9.157 to compute the vector indices of the vertices $\{q^0, \ldots, q^{M-1}\}$ of the grid element containing $\theta(k)$.

3. For $j = 0$ to $M - 1$ do
{

    (a) Convert q^j to the scalar index i using Eq. 9.158.

    (b) Compute grid point θ^i as follows.

$$\theta_s^i = \begin{cases} a_1 + q_s^j \Delta x, & 1 \le s \le m+1 \\ b_1 + q_s^j \Delta y, & m+2 \le s \le p \end{cases}$$

    (c) Using Eq. 9.164 compute

$$y_0(k) = y_0(k) + w_i(k) g_i[\theta(k)]$$

}

The difference in speed between Algorithm 9.2 and a direct brute-force evaluation of Eq. 9.168 can be dramatic. For example, suppose $m = 2$, $n = 3$, and $d = 10$. Then $p = 6$, and there are a total of $r = 10^6$ potential terms to evaluate. However, using Algorithm 9.2 requires the evaluation of only $M = 64$ terms. Consequently, apart from the overhead of step 2, Algorithm 9.2 is faster than direct evaluation in this instance by a factor of $r/M = 15625$; more than four orders of magnitude!

Next consider the problem of updating the weights of the RBF network so as to achieve optimal performance. The *error* between the system S_f and the RBF network S_0 is

$$e(k) = y(k) - y_0(k) \tag{9.170}$$

Our objective is to minimize the mean square error $\epsilon(w) = E[e^2(k)]$. For the purpose of estimating the gradient, we use the fundamental LMS assumption, $E[e^2(k)] \approx e^2(k)$. Then, from Eq. 9.168, the partial derivative of $\epsilon(w)$ with respect to the ith element of w is

$$\frac{\partial \epsilon(w)}{\partial w_i} \approx 2e(k) \frac{\partial e(k)}{\partial w_i}$$

$$= -2e(k)\frac{\partial w^T g[\theta(k)]}{\partial w_i}$$

$$= -2e(k)g_i[\theta(k)], \quad 0 \le i < r \tag{9.171}$$

Thus, the gradient vector is $\nabla \epsilon(w) = -2e(k)g[\theta(k)]$. Using this gradient estimate in the steepest-descent method in Eq. 9.125 then results in the following update formula for the RBF network where $\mu > 0$ is the step size.

RBF network learning

$$w(k+1) = w(k) + 2\mu e(k)g[\theta(k)], \quad k \ge 0 \tag{9.172}$$

Observe that the weight-update algorithm for the nonlinear RBF network S_0 is quite simple and is essentially the same as the LMS method for linear systems, but with θ replaced by $g(\theta)$.

Practical Considerations

Before considering an example of an RBF model of a nonlinear discrete-time system, it is useful to briefly consider some practical issues. The first is the question of how to determine an effective set of output bounds b for the nonlinear system S_f. Suppose x is a $P \times 1$ *test* input of white noise uniformly distributed over $[a_1, a_2]$ where $P \gg 1$. Let $y(k)$ be the resulting output of S_f, and let y_{min} and y_{max} denote the minimum and maximum of $y(k)$, respectively. Then, output bounds can be selected as follows, where $\beta \ge 1$ is a *safety factor*.

$$b_1 = \frac{y_{min} + y_{max}}{2} - \beta \left(\frac{y_{max} - y_{min}}{2} \right) \tag{9.173a}$$

$$b_2 = \frac{y_{min} + y_{max}}{2} + \beta \left(\frac{y_{max} - y_{min}}{2} \right) \tag{9.173b}$$

Note that when $\beta = 1$, the bounds in Eq. 9.173 reduce to $b = [y_{min}, y_{max}]$. Typically, $\beta > 1$ is used because this takes into account the fact that $x(k)$ is of finite length. Furthermore, it is prudent to use $\beta > 1$, because $y_0(k)$ of the RBF network may range outside of the interval $[y_{min}, y_{max}]$ given that $y_0(k)$ is only an approximation to $y(k)$. Of course, making β too large effectively reduces the precision of the RBF model by making the grid-point spacing, Δy, large.

It is also important to take some care in selecting the initial condition, $\theta(0)$, when computing the RBF network output. Recall that the RBF network approximation, $f_0(\theta)$ in Eq. 9.160, has compact support. One can show that $f_0(\theta) = 0$ for $\theta \notin S$ where

$$S = [a_1 - \Delta x, a_2 + \Delta x]^{m+1} \times [b_1 - \Delta y, b_2 + \Delta y]^n \tag{9.174}$$

For example, the support of the RBF network shown in Figure 9.40 is contained in $S = [-3, 3] \times [-3, 3]$. Because of this compact support characteristic, it is important to choose the initial state such that $\theta(0) \in S$. Otherwise, the RBF network output may start out at zero and remain zero for $k \geq 1$.

The final practical issue that needs to be addressed is the question of how to determine whether or not the RBF model represents a good fit to the system S_f. The following *normalized mean square* error will be used.

$$E \triangleq \frac{\displaystyle\sum_{k=0}^{N-1} e^2(k)}{\displaystyle\sum_{k=0}^{N-1} y^2(k)} \tag{9.175}$$

Note that when the RBF network output is $y_0(k) = 0$, the normalized mean square error reduces to $E = 1$. Thus, values of $E \ll 1$ represent a good fit between the system, S_f, and the RBF network model, S_0, with $E = 0$ being a perfect fit.

Example 9.19   Nonlinear System Identification

As practical example of a nonlinear discrete-time system, consider a continuous stirred tank chemical reactor (Al-Ajlouni *et al.*, 2004). Suppose $x(k)$ is the reactant concentration at time kT, and $y(k)$ is the product concentration at time kT. If the input is scaled such that $0 \leq x(k) \leq 1$, then the Van de Vusse reaction can be characterized by the following nonlinear discrete-time system.

$$y(k) = c_1 + c_2 x(k-1) + c_3 y(k-1) + c_4 x^3(k-1) + c_5 y(k-2)x(k-1)x(k-2)$$

The presence of the fourth and fifth terms makes this system nonlinear. If the sampling interval is $T = 0.04$ hours, then the parameters of the system can be taken to be (Hernandez and Arkun, 1996)

$$c = [0.558, 0.538, 0.116, -0.127, -0.034]^T$$

For this system, $m = 2$ and $n = 2$. Thus, the system dimension is $p = 5$. Since the input is non-negative and has been normalized, we have

$$a = [0, 1]$$

To determine the vector of output bounds b, suppose we use a test input consisting of $P = 1000$ points and a safety factor of $\beta = 1.5$, as in Eq. 9.173. This results in $b = [0.4310, 1.1930]$, which means the domain of the function f in this case is

$$\Theta = [0, 1]^3 \times [0.4310, 1.1930]^2$$

Suppose the initial condition for the RBF network is set to $\theta(0) = [0, 0, 0, c, c]^T$ where $c = (b_1 + b_2)/2$. Let the number of grid points per dimension be $d = 5$. From Eq. 9.154, this produces an RBF network with $r = 3125$ terms. However, from Eq. 9.169 only $M = 32$ of these terms are nonzero for each θ. Using Eq. 9.153, the grid-point spacing in the x and y directions is

$$\Delta x = 0.2500$$

$$\Delta y = 0.1905$$

Suppose the elements of the weight vector are computed using Eq. 9.166. Finally, we test the system with $N = 100$ samples of white noise uniformly distributed over $[a_1, a_2]$. A comparison of the two outputs, obtained by running script *exam9_19* on the distribution CD, is shown in Figure 9.41. Note that the solid horizontal lines indicate the boundaries of the domain Θ in the y direction, and the dotted lines show where the grid points are located. Careful inspection reveals that there are only minor differences between the two outputs. Using Eq. 9.175, the normalized mean square error in this case was

$$E = 0.0014$$

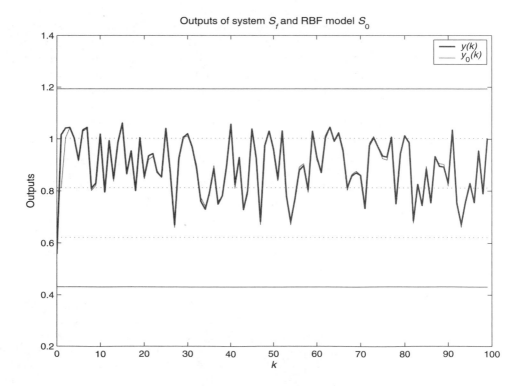

Figure 9.41 Comparison of System S_f and RBF Model with $m = 2$, $n = 2$, and $d = 5$

MATLAB Toolbox

The FDSP toolbox contains the following functions which implement nonlinear system identification using a raised-cosine RBF network.

```
[w,e] = f_rbfw  (f,N,a,b,m,n,d,mu,ic,w);    % RBF network learning algorithm
y     = f_rbf0  (x,w,a,b,m,n,d);            % Raised-cosine RBF network output
theta = f_state (x,y,k,m,n);                % State vector
```

The function *f_rbfw* is the learning algorithm for updating the weights of the raised-cosine RBF network. On entry to *f_rbfw*, f is a string containing the name of a user-supplied MAT-file function that specifies the right-hand side of the nonlinear discrete-time system in Eq. 9.148. The calling sequence for f is

```
y = f (theta,m,n);
```

On entry to f, *theta* is the state vector in Eq. 9.146, $m \geq 0$ is the number of past inputs, and $n \geq 0$ is the number of past outputs. The value returned by f is the system output. The second calling argument of the function *f_rbfw* is $N \geq 0$, which specifies the number of random samples used to train the network. Next, a and b are vectors of length 2 containing the input bounds and output bounds, respectively, $m \geq 0$ is the number of past inputs, $n \geq 0$ is the number of past outputs, $d \geq 2$ is the number of grid points per dimension, $\mu > 0$ is the step size, ic is an initial condition code, and w is an optional vector of length r containing an initial guess for w. If $ic \neq 0$, then the initial value for the weight vector w is computed using Eq. 9.166. Therefore, to compute a w that makes the network output exact at the grid points, use $N = 0$ and $ic = 1$. If $ic = 0$ and w is not present, the default value, $w = 0$, is used. On exit from *f_rbfw*, w is a vector of length r containing the updated weights using N iterations, and e is an optional vector of length N containing the errors during learning. When $ic = 0$, very large values of N may be required to find w. The number of weights is $r = d^p$, where $p = m + n + 1$.

The function *f_rbf0* computes the output of a raised-cosine RBF network. On entry to *f_rbf0*, x is a vector of length N containing the input, w is a vector of length r containing the weights. The remaining calling arguments are the same as defined in *f_rbfw*. On exit from *f_rbf0*, y is a vector of length N containing the RBF network output evaluated at x. A simple way to get a highly effective value for w is to first call *f_rbfw* with $N = 0$ and $ic = 1$ as follows.

```
w = f_rbfw (f,0,a,b,m,n,d,mu,1);    % Computes w satisfying Eq. 9.166
```

The function *f_state* is provided to help the user evaluate the output of a nonlinear discrete-time system S_f using Eq. 9.148. On entry to *f_state*, x is a vector of length N containing the input, y is a vector of length N containing the output, $1 \leq k \leq N$ specifies the discrete time at which the value of the state is to be computed, $m \geq 0$ is the number of past inputs, and $n \geq 0$ is the number of past outputs. When *f_state* is called, the values of $y(i)$ for $i \geq k$ are not used and can be arbitrary. On exit from *f_state*, *theta* is a vector of length $p = m + n + 1$ containing the state at time k.

9.10 **Software Applications**

FDSP Toolbox

9.10.1   **GUI Module:** *g_adapt*

The FDSP toolbox includes a graphical user interface module called *g_adapt* that allows the user to perform system identification using a variety of techniques without any need for programming. GUI module *g_adapt* features the display screen with tiled windows shown in Figure 9.42.

The upper-left *Block diagram* window contains a block diagram of the adaptive system under investigation. It features an mth-order transversal filter characterized by the following time-varying difference equation.

$$y(k) = \sum_{i=0}^{m} w_i(k)x(k-i), \quad 0 \le k < N \tag{9.176}$$

The *Parameters* window below the block diagram displays edit boxes containing the simulation parameters. The contents of each edit box can be directly modified by the user with the Enter key used to activate the changes. The parameters a and b are the coefficient vectors of the black-box system to be identified.

$$\sum_{i=0}^{n} a_i d(k-i) = \sum_{i=0}^{p} b_i x(k-i), \quad 0 \le k < N \tag{9.177}$$

It is important to select a such that the resulting black-box system is BIBO stable. The parameter fs is the sampling frequency, in Hz, and the parameter N is the number of samples of the input. The remaining parameters are all real scalars that control the behavior of the adaptive algorithm. Parameter mu is the step length, and it should be chosen to satisfy the following bound, where P_x is the average power of the input x.

$$0 < \mu < \frac{1}{(m+1)P_x} \tag{9.178}$$

Parameter nu is the leakage factor for the leaky LMS method. Typically, $nu < 1$ with $nu \approx 1$. When $nu = 1$, the leaky LMS method reduces to the LMS method. Parameter $alpha$ is the normalized step size used in the normalized LMS method and the relative step size of the correlation LMS method. The correlation LMS method also uses the smoothing parameter $beta$ where $0 < beta < 1$ with $beta \approx 1$. Inappropriate values for the scalar control parameters can cause some of the methods to diverge.

The *Type* and *View* windows in the upper-right corner of the screen allow the user to select both the type of adaptive algorithm and the viewing mode. The algorithm options include the LMS method, the normalized LMS method, the correlation LMS method, the leaky LMS method, and the RLS method. The *View* options include the input, a comparison of outputs, a comparison of magnitude responses, the learning curve, the step sizes used during learning, and the final weights found. The *Plot* window along the bottom half of the screen shows the selected view.

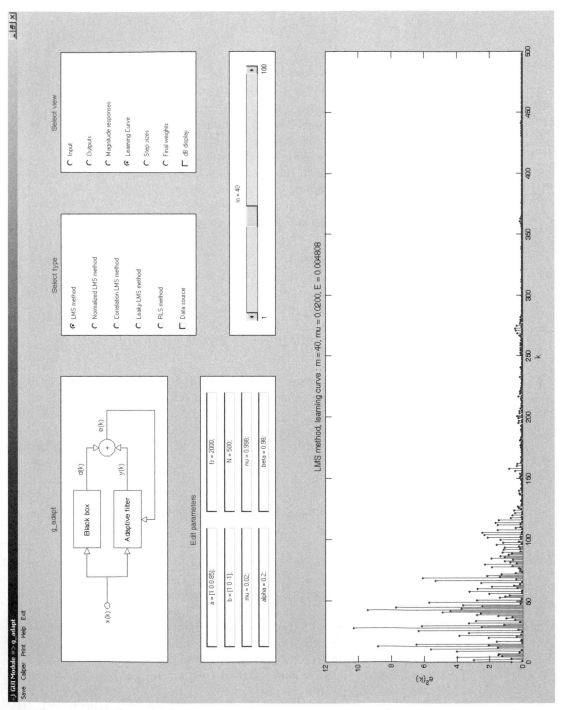

Figure 9.42 Display Screen of Chapter GUI Module *g_adapt*

There are two checkbox controls. The dB checkbox toggles the magnitude response display between linear and logarithmic scales. The Data source checkbox toggles the source of the input and desired output data. When it is checked, the user is prompted for the name of a user-supplied MAT-file that contains the vector of input samples, x, the vector of output samples, y, and the sampling frequency, fs. In this way, systems with input-output data generated offline from another source (*e.g.*, from measurements) can be identified. The black box output in the MAT-file is labeled y, rather than d, in order to preserve compatability with other GUI modules. When the Data source checkbox is not checked, the input consists of white noise uniformly distributed over $[-1, 1]$, and the desired output is computed using the coefficients a and b as in Eq. 9.177. The transversal filter order, m, is controlled with the horizontal slider bar appearing below the *Type* and *View* windows.

The *Menu* bar across the top of the screen includes several menu options. The *Save* option is used to save the current data in a user-specified MAT-file for future use. Files created in this manner can be later loaded using the Data source checkbox control. The *Caliper* option allows the user to measure any point on the current plot by moving the mouse crosshairs to that point and clicking. The *Print* option prints the contents of the *Plot* window. Finally, the *Help* option provides the user with some helpful suggestions on how to effectively use module *g_adapt*.

9.10.2   Identification of a Chemical Process

In the chemical process control industry, it is common to have dynamic systems with time delays due to transportation lags. This leads to process models that are described by both differential and difference equations. As an illustration, consider the following first-order with dead time system that can be used to model, for example, a heated stirred tank (Bequette, 2003).

$$\frac{dy_a(t)}{dt} + py_a(t) = cx_a(t - \tau) \tag{9.179}$$

Here τ is the dead time or delay in the input, and p and c are system parameters. Let T denote the sampling interval. Then the differential-difference system in Eq. 9.179 can be converted to a discrete-time system by approximating the derivative using a backwards difference as follows.

$$\frac{dy_a(t)}{dt} \approx \frac{y(k) - y(k-1)}{T} \tag{9.180}$$

Here it is understood that $y(k) = y_a(kT)$. The input delay can be modeled exactly in discrete time if some care is taken in choosing the sampling interval. Suppose we pick T, such that $T = \tau/M$ for some integer $M \geq 1$. Then $x_a(t - \tau)$ can be replaced by $x(k - M)$, where $x(k) = x_a(kT)$. These substitutions then yield the following equivalent discrete-time model.

$$\frac{y(k) - y(k-1)}{T} + py(k) = cx(k - M) \tag{9.181}$$

Thus, the differential-difference system in Eq. 9.179 can be approximated by an IIR filter if the sampling interval is chosen to be an integer submultiple of the delay τ. To investigate how well the IIR model can be approximated by an adaptive transversal filter, suppose the delay is $\tau = 5$ sec, and the sampling interval is $T = 0.5$ sec. This yields an input delay of $M = 10$ samples. Let the remaining system parameters be $p = 0.2$ and $c = 4$. The normalized LMS method can be used to identify this system using the following script, labeled *exam9_20* on the distribution CD.

```
function exam9_20

% Example 9.20: Identification of a chemical process

clear
clc
fprintf('Example 9.20: Identification of a chemical process\n\n')
tau = 5
T = 0.5
M = tau/T
p = 0.2;
c = 4;
N = 1000;
rand ('seed',1000)
m = f_prompt ('Enter adaptive filter order',0,80,40);
alpha = f_prompt ('Enter normalized step length',0,1,0.1);

% Compute the coefficients of the IIR model

a = [1+p*T, -1];
b = [zeros(1,M) c*T];

% Compute the input and desired output

x = f_randu (N,1,-1,1);
d = filter (b,a,x);

% Identify a model using normalized LMS method

[w,e] = f_normlms (x,d,m,alpha);

% Learning curve

figure
k = 0 : N-1;
stem (k,e.^2,'filled','.')
f_labels ('Learning curve','\it{k}','\it{e^2(k)}')
f_wait
```

```
% Compare responses to new input

P = 200;
x = f_randu (P,1,-1,1);
d = filter (b,a,x);
y = filter (w,1,x);
figure
k = 0 : P-1;
hp = plot (k,d,k,y);
set (hp(1),'LineWidth',1.5)
legend ('\it{d(k)}','\it{y(k)}')
e = d - y;
E = sum(e.^2)/sum(d.^2)
caption = sprintf ('Comparison of outputs, {\\itE} = %.4f',E);
f_labels (caption,'\it{k}','Outputs')
f_wait
```

When script *exam9_20* is run with an adaptive filter of order $m = 40$ using $N = 1000$ samples and a normalized step size of $\alpha = 0.1$, it produces the learning curve shown in Figure 9.43. Observe that the normalized LMS method converges in approximately 700 samples. Next, the final values for the weights are used, and a

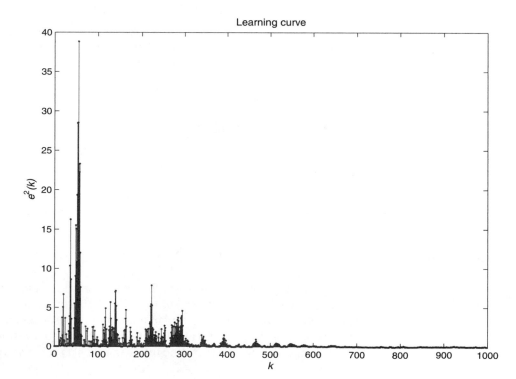

Figure 9.43 Learning Curve using the Normalized LMS Method with $m = 40$ and $\alpha = 0.1$

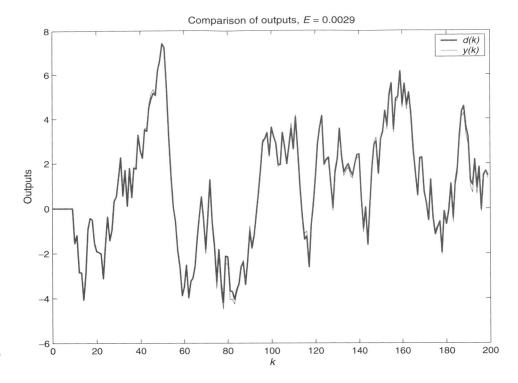

Figure 9.44 Comparison of Desired and Actual Outputs using the Final Values for the Weights

new white noise test input is generated to compare the responses of the two systems. The results, shown in Figure 9.44, confirm that there is a good fit in this case with a normalized mean square error of $E = 0.0029$.

• • • • • • • • • • • • • • • • •

9.11 Chapter Summary

Least Mean Square Techniques

This chapter focused on adaptive signal processing techniques and their applications. The adaptive filter structure used was an mth-order transversal filter of the following form.

$$y(k) = \sum_{i=0}^{m} w_i(k)x(k - i), \quad k \geq 0 \tag{9.182}$$

One important qualitative feature of this structure is that once the weights converge to their steady-state values, the resulting FIR filter is guaranteed to be BIBO stable. Adaptive systems can be configured in a number of ways depending on the application. Examples include system identification, channel equalization, signal prediction, and noise cancellation. Consider the following $(m + 1) \times 1$ *state* vector of past inputs.

$$\theta(k) = [x(k), x(k-1), \ldots, x(k-m)]^T \tag{9.183}$$

A compact expression for the transversal-filter output then can be obtained in terms of $\theta(k)$ using the following dot-product formulation.

$$y(k) = w^T(k)\theta(k), \quad k \geq 0 \tag{9.184}$$

The $(m+1) \times 1$ *weight* vector, $w(k)$, is adjusted so as to minimize the mean square error $\epsilon(w) = E[e^2(k)]$, where E is the expected value operator. Here the system error is $e(k) = d(k) - y(k)$, with $d(k)$ being the desired output. For example, in the system identification application, $d(k)$ is the output of the system to be identified. When the input $x(k)$ is sufficiently rich in frequency content, the mean square error can be shown to be a positive-definite quadratic function of the weight vector w, which means that a unique optimal weight vector exists. The weights can be adjusted by searching the mean-square-error performance function using a steepest-descent search method. For the purpose of computing the gradient of $\epsilon(w)$, one can use the simplifying assumption $E[e^2(k)] \approx e^2(k)$. This leads to the following simple and popular weight-update algorithm called the *least mean square* or LMS method.

$$w(k+1) = w(k) + 2\mu e(k)\theta(k), \quad k \geq 0 \tag{9.185}$$

The scalar parameter μ is the *step size*. If $P_x = E[x^2(k)]$ denotes the average power of the input, then the LMS method will converge for step lengths in the following range.

$$0 < \mu < \frac{1}{(m+1)P_x} \tag{9.186}$$

Since the time constant of the LMS method is inversely proportional to the step size, larger step sizes are preferable for faster convergence. However, once convergence has been achieved, excess mean square error is encountered as a result of the approximation used for the gradient of the mean square error. Because excess mean square error is proportional to the step size, there is a tradeoff in choosing μ between convergence speed and steady-state accuracy. In many applications, it is common to choose values for μ that are about an order of magnitude less than the upper bound in Eq. 9.186.

The basic LMS method in Eq. 9.185 can be modified in a number of ways to enhance performance. These include the normalized LMS method, the correlation LMS method, and the leaky LMS method. The normalized and correlation methods feature step sizes that vary with time. For example, for the normalized LMS method, the step size is as follows, where $0 < \alpha < 1$ is the normalized step size and $\delta > 0$ is included to ensure that the step size never grows beyond α/δ.

$$\mu(k) = \frac{\alpha}{\delta + \theta^T(k)\theta(k)} \tag{9.187}$$

For the correlation LMS method, the step size is relatively large during convergence but becomes small once convergence has taken place. The leaky LMS method tends to make the LMS method more stable by preventing the weights from becoming too large. It is useful for narrowband inputs because it has the effect of adding white noise to the input. However, this tends to increase the excess mean square error once convergence has taken place.

Adaptive Signal Processing Applications

Adaptive signal processing can be used to design FIR filters with prescribed magnitude and phase response characteristics. This is done by applying system identification to a synthetic pseudo-filter. Of course, the magnitude and phase responses of a physical system are not completely independent of one another. Consequently, an optimal transversal filter may or may not produce a close fit to the pseudo-filter specifications.

An important alternative to the LMS family of methods for updating the weights is the *recursive least squares* or RLS method. The RLS method attempts to find the optimal weight at each time step, unlike the LMS method which approaches the optimal weights gradually using a steepest-descent search. As a consequence, the RLS method typically converges much faster than the LMS method, but there is more computational effort per iteration. The computational effort of the RLS method is of order $O(m^2)$, whereas the computational effort for the LMS method is of order $O(m)$, where m is the filter order.

An interesting application area of adaptive signal processing is the active control of acoustic noise. The basic premise is to inject a secondary sound into an environment so as to cancel the primary sound using destructive interference. This application requires adaptive techniques because the nature of the unwanted sound and the characteristics of the environment typically change with time. The secondary sound must be generated by a speaker, transmitted over an air channel, and detected by a microphone. Consequently, active noise control requires a modification of the LMS method called the *filtered-x* or FXLMS method.

$$w(k + 1) = w(k) + 2\mu e(k)\hat{\theta}(k), \quad k \geq 0 \tag{9.188}$$

The FXLMS differs from the LMS method in that the state vector of past inputs first must be filtered by a model of the path traveled by the secondary sound. This model typically is obtained offline using standard system identification techniques. Often, the primary noise consists of a narrowband periodic component plus white noise. When the fundamental frequency of the periodic component of the noise is available, a more direct signal-synthesis approach can be used to eliminate the periodic component of the primary noise.

Nonlinear System Identification

System identification using adaptive signal processing can be extended from linear systems to nonlinear discrete-time systems of the following form.

$$y(k) = f[\theta(k)] \tag{9.189}$$

Here the state vector includes not only the $(m+1)$ past inputs but also n past outputs. Thus f is a real-valued continuous nonlinear function of $p = m + n + 1$ variables. If the nonlinear system in Eq. 9.189 is BIBO stable and the input is bounded, then the domain of the function f is restricted to a closed, bounded region, $\Theta \subset R^p$. This compact domain can be covered with a grid of r points, and over each grid element a simple local representation of $f(\theta)$ can be developed. This leads to the following adaptive nonlinear structure called a *radial basis function* or RBF network.

$$y_0(k) = w(k)^T g[\theta(k)] \qquad (9.190)$$

Here $g:R^r \to R$ is an $r \times 1$ vector of radial basis functions: one centered about each grid point. If a raised-cosine RBF is used, then the resulting network can be shown to have several useful properties. For example, $g(\theta)$ is a continuously differentiable function, and if $w_i = f(\theta^i)$ where θ^i is the ith grid point, then the error between the RBF model and the original nonlinear system is zero at the grid points. When the grid is sufficiently fine, this results in an RBF network that is an effective model of the nonlinear system over the domain Θ.

The FDSP toolbox includes a GUI module called *g_adapt* that allows the user to evaluate and compare several adaptive system identification techniques without any need for programming. The weight-update algorithms featured include the LMS method, the normalized LMS method, the correlation LMS method, the leaky LMS method, and the RLS method. The input and desired output data can be obtained from a user-specified IIR filter, or from a user-supplied MAT-file. The latter option allows for identification based on actual physical measurements.

• • • • • • • • • • • • • • • •

9.12 **Problems**

The problems are divided into Analysis problems that can be solved by hand or with a calculator, GUI Simulation problems that are solved using GUI module *g_adapt*, and Computation problems. The Computation problems require the student to write a MATLAB script using the FDSP toolbox functions summarized in Appendix B. Solutions to selected problems are available on the distribution CD. Students are encouraged to use these problems, which are identified with a ☑, as a check on their understanding of the material.

9.12.1 Analysis

Section 9.1

P9.1. The transversal filter structure used in this chapter is a time-varying FIR filter. One can generalize it by using the following time-varying IIR filter.

$$y(k) = \sum_{i=0}^{m} b_i(k)x(k-i) - \sum_{i=1}^{n} a_i(k)y(k-i)$$

(a) Find suitable definitions for the state vector $\theta(k)$ and the weight vector $w(k)$ such that the output of the time-varying IIR filter can be expressed as a dot product as in Eq. 9.6. That is,

$$y(k) = w^T(k)\theta(k), \quad k \geq 0$$

(b) Suppose the weight vector $w(k)$ converges to a constant. Is the resulting filter guaranteed to be BIBO stable? Why or why not?

P9.2. Suppose a transversal adaptive filter order is $m = 2$ and the input $x(k)$ consists of white noise uniformly distributed over the interval $[a, b]$. Find the input auto-correlation matrix R.

P9.3. Find the constant term of the mean square error when the desired output is the following signal.

$$d(k) = b + \sin\left(\frac{2\pi k}{N}\right) - \cos\left(\frac{2\pi k}{N}\right)$$

P9.4. Consider a transversal filter of order $m = 1$. Suppose the input and desired output are as follows.

$$x(k) = 2 + \sin(\pi k/2)$$
$$d(k) = 1 - 3\cos(\pi k/2)$$

(a) Find the cross-correlation vector p.

(b) Find the input auto-correlation matrix R.

(c) Find the optimal weight vector w^*.

P9.5. Suppose the first row of an auto-correlation matrix R is $r = [9, 7, 5, 3, 1]$.

(a) Find R.

(b) What is the average power of the input?

(c) Suppose $x(k)$ is white noise uniformly distributed over the interval $[0, c]$. Find c.

P9.6. Suppose $v(k)$ is white noise uniformly distributed over $[-c, c]$. Consider the following input.

$$x(k) = 2 + \sin(\pi k/2) + v(k)$$

Find the input auto-correlation matrix R. Does your answer reduce to that of Problem P9.4 when $c = 0$?

P9.7. Suppose an input $x(k)$ and a desired output $d(k)$ have the following auto-correlation matrix and cross-correlation vector. Find the optimal weight vector w^*.

$$R = \begin{bmatrix} 5 & 1 \\ 1 & 5 \end{bmatrix}, \quad p = \begin{bmatrix} 3 \\ -2 \end{bmatrix}$$

Section 9.3

☑ **P9.8.** Suppose the mean square error is approximated using a running-average filter of order $M - 1$ as follows.

$$\epsilon(w) \approx \frac{1}{M} \sum_{i=0}^{M-1} e^2(k - i)$$

(a) Find an expression for the gradient vector $\nabla \epsilon(w)$ using this approximation for the mean square error.

(b) Using the steepest-descent method and the results from part (a), find a weight-update formula.

(c) How many floating-point multiplications (FLOPs) are required per iteration to update the weight vector? You can assume that 2μ is computed ahead of time.

(d) Verify that when $M = 1$ the weight-update formula reduces to the LMS method.

P9.9. There is an offline or batch procedure for computing the optimal weights called the least-squares method (see Problem P9.33). For large values of m, the least-squares method requires approximately $4(m+1)^3/3$ FLOPs to find w. How many iterations are required before the computational effort of the LMS method equals or exceeds the computational effort of the least-squares method?

Section 9.4

P9.10. Suppose an input $x(k)$ has the following auto-correlation matrix.

$$R = \begin{bmatrix} 2 & 1 \\ 1 & 2 \end{bmatrix}$$

(a) Using the eigenvalues of R, find a range of step sizes that ensures convergence of the LMS method.

(b) Using the average power of the input, find a more conservative range of step sizes that ensures convergence of the LMS method.

(c) Suppose the step size is one tenth the maximum in part (b). Find the time constant of the mean square error in units of iterations.

(d) Using the same step size as in part (c), find the excess mean square error. Express your final answer in terms of $\epsilon_{\min}$.

P9.11. Suppose the LMS learning curve converges to within one percent of its final steady-state value in 200 iterations.

(a) Find the learning-curve time constant, τ_{mse}, in units of iterations.

(b) If the minimum eigenvalue of R is $\lambda_{\min} = 0.1$, what is the step size?

(c) If the step size is $\mu = 0.02$, what is the minimum eigenvalue of R?

P9.12. Suppose the normalized excess mean square error using the LMS method is $M = 0.4$ when the input is white noise uniformly distributed over $[-2, 2]$.

(a) Find the average power of the input.

(b) If the step size is $\mu = 0.01$, what is the filter order?

(c) If the filter order is $m = 9$, what is the step size?

Section 9.5

P9.13. Financial considerations dictate that a production system must remain in operation while the system is being identified. During normal operation of the linear system, the input $x(k)$ has relatively poor spectral content.

(a) Which of the modified LMS methods would appear to be an appropriate choice? Why?

(b) How might the input be modified slightly to improve identification without significantly affecting the normal operation of the system?

P9.14. Consider the normalized LMS method.

(a) What is the maximum value of the step size?

(b) Describe an initial condition for the past inputs which will cause the step size to saturate to its maximum value.

Section 9.6

P9.15. Consider the following periodic input that is used as part of the input-output specification for a pseudo-filter. Suppose $f_i = i f_s/(2N)$ for $0 \leq i < N$. Find the auto-correlation matrix R for this input.

$$x(k) = \sum_{i=0}^{N-1} C_i \cos(2\pi f_i kT)$$

☑ **P9.16.** Consider the following periodic input and desired output that form the input-output specification for a pseudo-filter. Suppose $f_i = i f_s/(2N)$ for $0 \leq i < N$. Find the cross-correlation vector p for this input and desired output.

$$x(k) = \sum_{i=0}^{N-1} C_i \cos(2\pi f_i kT)$$

$$d(k) = \sum_{i=0}^{N-1} A_i C_i \cos(2\pi f_i kT + \phi_i)$$

Section 9.7

P9.17. Consider the following expression for the generalized cross-correlation vector used by the RLS method.

$$p(k) = \sum_{i=1}^{k} \gamma^{k-i} d(i)\theta(i)$$

Show that $p(k)$ can be expressed recursively in terms of $p(k-1)$ by deriving the expression for $p(k)$.

Section 9.8

P9.18. Consider the active-noise-control system shown in Figure P9.18. Suppose the secondary path is modeled as a delay with attenuation. That is, for some delay $\tau > 0$ and some attenuation $0 < \alpha < 1$,

$$\hat{y}(t) = \alpha y(t - \tau)$$

(a) Let the sampling interval be $T = \tau/M$. Find the transfer function $F(z)$.

(b) Suppose the primary path $G(z)$ is modeled as follows. Find $W(z)$ using Eq. 9.118.

$$G(z) = \sum_{i=0}^{m} \frac{z^{-i}}{1+i}$$

(c) Is the controller $W(z)$ physically realizable? Why or why not?

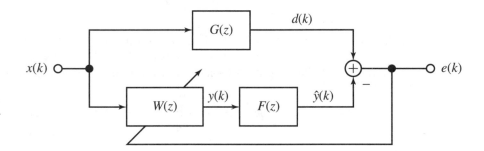

P9.18 Active Control of Acoustic Noise

Section 9.9

P9.19. Consider the problem of identifying the nonlinear discrete-time system in Eq. 9.148 using a raised-cosine RBF network. Let the number of past inputs be $m = 1$ and the number of past outputs be $n = 1$. Suppose the range of values for the inputs is $a = [-2, 2]$, and the range of values for the outputs is $b = [-3, 3]$. Let the number of grid points per dimension be $d = 4$.

(a) Find the domain Θ of the function f.

(b) What is the total number of grid points?

(c) What is the grid-point spacing in the x direction and in the y direction?

(d) For each θ, what is the maximum number of nonzero terms in the RBF network output?

(e) Consider the following state vector. Find the vector indices of the vertices of the grid element containing θ.

$$\theta = [0.3, -1.7, 1.1]^T.$$

(f) Find the scalar indices of the vertices of the grid element containing the θ in part (e).

P9.20. Consider the following candidate for a scalar radial basis function.

$$G_i(z) = \begin{cases} \cos^{2i}\left(\dfrac{\pi z}{2}\right), & |z| \leq 1 \\ 0, & |z| > 1 \end{cases}$$

(a) Show that $G_i(z)$ qualifies as an RBF for $i \geq 1$.

(b) Does $G_i(z)$ have compact support?

(c) Show that $G_i(z)$ reduces to the raised-cosine RBF when $i = 1$.

P9.21. Consider a raised-cosine RBF network with $m = 0$, $n = 0$, $d = 2$, and $a = [0, 1]$. Using the trigonometric identities from Appendix D, show that the constant interpolation property holds in this case, namely,

$$g_0(\theta) + g_1(\theta) = 1, \quad a_1 \leq \theta \leq a_2$$

P9.22. Consider a raised-cosine RBF network with $m = 2$ past inputs and $n = 2$ past outputs. Suppose the range of values for the inputs is $a = [0, 5]$, and the range of values for the outputs is $[-2, 8]$. Let the number of grid points per dimension be $d = 6$.

(a) Find the compact support S of the overall network. That is, find the smallest closed, bounded region, $S \subset R^p$, such that

$$\theta \notin S \Rightarrow f_0(\theta) = 0$$

(b) Show that, in general, $S \rightarrow \Theta$ as $d \rightarrow \infty$ where $\Theta \in R^p$ is the domain of f.

9.12.2   GUI Simulation

Section 9.10

P9.23. Using the GUI module *g_adapt* and the default parameter values, identify the black-box system using the LMS method. Increase the step size in units of 0.01 until the normalized mean square error, E, exceeds 0.1. Then reduce *mu* to half of this value, and plot the following.

(a) The outputs.

(b) The magnitude responses.

(c) The learning curve.

(d) The final weights.

P9.24. Using the GUI module *g_adapt* with the default parameter values, identify the black-box system using the leaky LMS method. Adjust the number of samples to $N = 250$, and plot the following.

(a) The outputs.

(b) The magnitude responses.

(c) The learning curve.

☑ **P9.25.** Use the GUI module *g_adapt* to identify the following black-box system using the normalized LMS method with a filter of order $m = 40$.

$$H(z) = \frac{3}{1 - 0.7z^{-4}}$$

(a) Plot the magnitude responses.

(b) Plot the learning curve.

(c) Plot the step sizes.

P9.26. Use the GUI module *g_adapt* to identify the following black-box system using the correlation LMS method with a filter of order $m = 50$.

$$H(z) = \frac{2}{1 + 0.8z^{-4}}$$

(a) Plot the magnitude responses.

(b) Plot the learning curve.

(c) Plot the step sizes.

P9.27. Using the GUI module *g_adapt* with the default parameter values, identify the black-box system using the leaky LMS method. Plot the learning curve for the following cases corresponding to different values of the leakage factor.

(a) $nu = 0.999$.

(b) $nu = 0.995$.

(c) $nu = 0.990$.

P9.28. Using the GUI module *g_adapt* with the default parameter values, identify the black-box system using the following two methods. Plot the learning curve for each case. Observe the scale of the dependent variable.

(a) The LMS method.

(b) The RLS method.

P9.29. Using the GUI module *g_adapt* and the Data spource option, load the input and desired output from the MAT-file *u_adapt1*. Then identify the system that produced this input-output data using the normalized LMS method. Plot the following.

(a) The learning curve.

(b) The magnitude responses using the dB scale.

(c) The step sizes. Use the *Caliper* option to mark the largest step size.

☑ **P9.30.** Consider the following FIR black-box system. Use the GUI module g_adapt to identify this system using the LMS method.

$$H(z) = 1 - 2z^{-1} + 7z^{-2} + 4z^{-4} - 3z^{-5}$$

Save the data in a MAT-file named my_adapt, and then reload it using the Data source option.

(a) Plot the learning curve when $m = 3$.

(b) Plot the learning curve when $m = 5$.

(c) Plot the learning curve when $m = 7$.

(d) Plot the final weights, $m = 7$.

9.12.3   Computation

Section 9.1

P9.31. Consider the problem of designing an equalizer as shown in Figure P9.31. Suppose the delay is $M = 15$ and $H(z)$ represents a communication channel with the following transfer function.

$$H(z) = \frac{1 + 0.5z^{-1}}{1 + 0.4z^{-1} - 0.32z^{-2}}$$

Write a MATLAB script that uses the FDSP toolbox function f_lms to construct an equalizer of order $m = 30$ for $H(z)$. Suppose $x(k)$ consists of $N = 1000$ samples of white noise uniformly distributed over $[-3, 3]$. Use a step size of $\mu = 0.002$.

(a) Plot the learning curve.

(b) Using the final weights, compute $y(k)$ using input $r(k)$. Then plot $d(k)$ and $y(k)$ for $0 \le k \le N/10$ on the same graph with a legend.

(c) Using the final weights, plot the magnitude responses of $H(z)$, $W(z)$, and $F(z) = H(z)W(z)$ on the same graph using a legend. For the abscissa, use the normalized frequency, f/f_s.

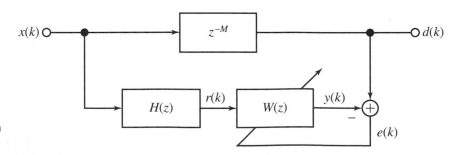

P9.31 Equalization of a Communication Channel, $H(z)$

P9.32. Consider the problem of designing an adaptive noise-cancellation system as shown in Figure P9.32. Suppose the additive noise, $v(k)$, is white noise uniformly distributed over $[-2, 2]$. Let the primary microphone signal be as follows.

$$x(k) = \cos\left(\frac{\pi k}{10}\right) - 0.5 \sin\left(\frac{\pi k}{20}\right) + 0.25 \cos\left(\frac{\pi k}{30}\right)$$

Suppose the path for detecting the noise signal has the following transfer function.

$$H(z) = \frac{0.5}{1 + 0.25 z^{-2}}$$

Write a MATLAB script that uses the FDSP toolbox function f_lms to cancel the noise, $v(k)$, corrupting the signal, $d(k)$. Use an adaptive filter of order $m = 30$, $N = 3000$ samples, and a step size of $\mu = 0.003$.

(a) Plot the learning curve.

(b) Using the final weights, compute $y(k)$ using input $r(k)$. Then plot $x(k)$, $d(k)$, and $e(k)$ for $0 \le k \le N/10$ on the same graph with a legend.

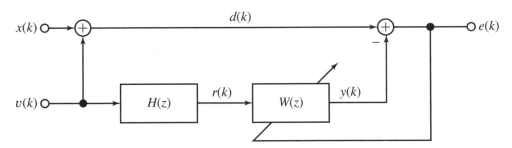

P9.32 Noise cancellation

Section 9.3

P9.33. There is an offline alternative to the LMS method called the *least-squares* method that is available when the entire input signal and desired output signal are available ahead of time. Suppose the weight vector w is constant. Taking the transpose of Eq. 9.6, and replacing the actual output by the desired output, yields

$$\theta^T(k)w = d(k), \quad 0 \le k < N$$

Let $d = [d(0), d(1), \ldots, d(N-1)]^T$ and let X be an $N \times (m+1)$ past-input matrix whose ith row is $\theta^T(i)$ for $0 \le i < N$. Then the N equations can be recast as the following vector equation.

$$Xw = d$$

When $N > (m+1)$, this constitutes an over-determined linear algebraic system of equations. A weight vector that minimizes the squared error $E = (Xw-d)^T(Xw-d)$ is obtained by premultiplying both sides by X^T. This yields the *normal equations*

$$X^T X w = X^T d$$

The coefficient matrix $X^T X$ is $(m+1) \times (m+1)$. If $x(k)$ has adequate spectral content, $X^T X$ will be nonsingular. In this case, the optimal weight vector in a least-squares sense can be obtained by premultiplying by the inverse of $X^T X$, which yields

$$w = (X^T X)^{-1} X^T d$$

Write a MATLAB function called f_lsfit, that computes the optimal least-squares FIR filter weight vector, $b = w$, by solving the normal equations using the MATLAB left-division operator, $\backslash$. The calling sequence should be as follows.

```
b = f_lsfit (x,d,m);
```

On entry to f_lsfit, x and d are vectors of length N containing the input and desired output, respectively, and $m < N$ is the filter order. On exit from f_lsfit, the $(m+1) \times 1$ vector b contains the optimal FIR filter weights. In constructing X, you can assume that $x(k)$ is causal. Test f_lsfit by using $N = 250$ and $m = 30$. Let x be white noise uniformly distributed over $[-1, 1]$, and let d be a filtered version of x using the following IIR filter.

$$H(z) = \frac{1 + z^{-2}}{1 - 0.1z^{-1} - 0.72z^{-2}}$$

(a) Use *stem* to plot the least-squares weight vector b.

(b) Compute $y(k)$ using the weight vector b. Then plot $d(k)$ and $y(k)$ for $0 \le k \le 50$ on the same graph using a legend.

Section 9.4

☑ **P9.34.** A plot of the squared error is only a rough approximation to the learning curve in the sense that $E[e^2(k)] \approx e^2(k)$. Write a MATLAB script that uses the FDSP toolbox function f_lms to identify the following system. For the input, use $N = 500$ samples of white noise uniformly distributed over $[-1, 1]$, and for the filter order use $m = 30$.

$$H(z) = \frac{z}{z^3 + 0.7z^2 - 0.8z - 0.56}$$

(a) Use a step size μ that corresponds to 0.1 of the upper bound in Eq. 9.49. Print the step size used.

(b) Compute and print the mean-square-error time constant in Eq. 9.62, but in units of iterations.

(c) Construct and plot a learning curve by performing the system identification $M = 50$ times with a different white noise input used each time. Plot the average of the M $e^2(k)$ versus k curves and draw vertical lines at integer multiples of the time constant.

Section 9.5

P9.35. Consider the problem of performing system identification as shown in Figure P9.35. Suppose the system to be identified is the following auto-regressive or all-pole filter.

$$H(z) = \frac{1}{z^4 - 0.1z^2 - 0.72}$$

Write a MATLAB script that uses the FDSP toolbox function *f_normlms* to identify a model of order $m = 60$ for this system. Use an input consisting of $N = 1200$ samples of white noise uniformly distributed over $[-1, 1]$, a constant step size of $\alpha = 0.1$, and a maximum step size of $\mu_{max} = 5\alpha$.

(a) Plot the learning curve.

(b) Plot the step sizes.

(c) Plot the magnitude response of $H(z)$ and $W(z)$ on the same graph using a legend where $W(z)$ is the adaptive filter using the final values for the weights.

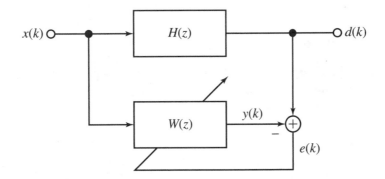

P9.35 Identification of Linear Discrete-time System, $H(z)$

P9.36. Consider the problem of performing system identification, as shown in Figure P9.35. Suppose the system to be identified is the following IIR filter.

$$H(z) = \frac{z^2}{z^3 + 0.8z^2 + 0.25z + 0.2}$$

Write a MATLAB script that uses the FDSP toolbox function *f_corrlms* to identify a model of order $m = 50$ for this system. Use an input consisting of $N = 2000$ samples of white noise uniformly distributed over $[-1, 1]$, a relative step size of $\alpha = 1$, and the default smoothing parameter β.

(a) Plot the learning curve.

(b) Plot the step sizes.

(c) Plot the magnitude response of $H(z)$ and $W(z)$ on the same graph using a legend where $W(z)$ is the adaptive filter using the final values for the weights.

P9.37. Consider the following IIR filter.

$$H(z) = \frac{10(z^2 + z + 1)}{z^4 + 0.2z^2 - 0.48}$$

Write a MATLAB script which used the FDSP toolbox function $f\_leaklms$ to identify a model of order $m = 30$ for this system. Use an input consisting of $N = 120$ samples, a step size of $\mu = 0.005$, and the following periodic input.

$$x(k) = \cos\left(\frac{\pi k}{5}\right) + \sin\left(\frac{\pi k}{10}\right)$$

(a) Plot the learning curve for $\nu = 0.99$.

(b) Plot the learning curve for $\nu = 0.98$.

(c) Plot the learning curve for $\nu = 0.96$.

(d) Using $\nu = 0.995$ and the final value for the weights, plot $d(k)$ and $y(k)$ on the same graph with a legend.

Section 9.6

P9.38. Use the FDSP toolbox to write a MATLAB script that designs an FIR filter to meet the following pseudo-filter design specifications.

$$A(f) = \begin{cases} 2, & 0 \le f < \frac{f_s}{6} \\ 3, & \frac{f_s}{6} \le f < \frac{f_s}{3} \\ 3 - 24\left(f - \frac{f_s}{3}\right), & \frac{f_s}{3} \le f < \frac{5f_s}{12} \\ 1, & \frac{5f_s}{12} \le f \le \frac{f_s}{2} \end{cases}$$

$$\phi(f) = -30\pi f / f_s$$

Suppose there are $N = 80$ discrete frequencies equally spaced over $[0, f_s/2)$, as in Eq. 9.92. Use f_lms with a step size of $\mu = 0.0001$ and $M = 2000$ iterations.

(a) Choose an order for the adaptive filter that best fits the phase specification. Print the order m.

(b) Plot the magnitude response of the filter obtained using the final weights. On the same graph plot the desired magnitude response with isolated plot symbols at each of the N discrete frequencies, and plot a legend.

(c) Plot the phase response of the filter obtained using the final weights. On the same graph plot the desired phase response with isolated plot symbols at each of the N discrete frequencies, and plot a legend.

Section 9.7

☑ **P9.39.** Consider the problem of designing a signal predictor, as shown in Figure P9.39. Suppose the signal whose value is to be predicted is as follows.

$$x(k) = \sin\left(\frac{\pi k}{5}\right)\cos\left(\frac{\pi k}{10}\right) + v(k), \quad 0 \le k < N$$

Here $N = 200$ and $v(k)$ is white noise uniformly distributed over $[-0.05, 0.05]$. Write a MATLAB script that uses the FDSP toolbox function f_rls to predict the value of this signal $M = 20$ samples into the future. Use a filter of order $m = 40$ and a forgetting factor of $\gamma = 0.9$.

(a) Plot the learning curve.

(b) Using the final weights, compute the output, $y(k)$ corresponding to the input, $x(k)$. Then plot $x(k)$ and $y(k)$ on separate graphs above one another using the *subplot* command. Use the *fill* function to shade a section of $x(k)$ of length M starting at $k = 160$. Then shade the corresponding predicted section of $y(k)$ starting at $k = 140$.

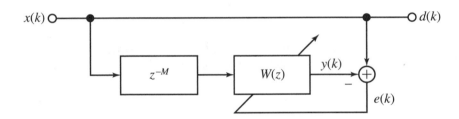

P9.39 Signal Prediction

Section 9.8

P9.40. Consider the active-noise-control system shown in Figure P9.18. Suppose the secondary path is modeled by the following transfer function, that takes into account the delay and attenuation of sound as it travels through air and the characteristics of the microphones, speaker, amplifiers, and DAC.

$$F(z) = \frac{0.2z^{-3}}{1 - 1.4z^{-1} + 0.48z^{-2}}$$

Suppose the sampling frequency is $f_s = 2000$ Hz. Write a MATLAB script that uses the FDSP toolbox f_lms to identify an FIR model of the secondary path $F(z)$ using an adaptive filter of order $m = 25$. Choose an input and a step size which causes the algorithm to converge.

(a) Plot the learning curve to verify convergence.

(b) Plot the magnitude responses of $F(z)$ and the model, $\hat{F}(z)$, on the same graph using a legend.

(c) Plot the phase responses of $F(z)$ and the model, $\hat{F}(z)$, on the same graph using a legend.

P9.41. Consider the active-noise-control system shown in Figure P9.18. Suppose the primary noise, $x(k)$, consists of the following noise-corrupted periodic signal.

$$x(k) = 2 \sum_{i=1}^{5} \frac{\sin(2\pi f_0 i k T)}{1+i} + v(k)$$

Here the fundamental frequency is $f_0 = 100$ Hz and $f_s = 2000$ Hz. The additive noise term, $v(k)$, is white noise uniformly distributed over $[-0.2, 0.2]$. Coefficient vectors for FIR models of the secondary path, $F(z)$, and the primary path, $G(z)$, are contained in MAT-file *prob9_41*. The coefficient vectors are f and g. Write a MATLAB script that loads f and g and uses the FDSP toolbox function *f_fxlms* to apply active noise control with the filtered-x LMS method starting at sample $N/4$ where $N = 2000$. Use a noise controller of order $m = 30$ and a step size of $\mu = 0.002$. Plot the learning curve, including a title that displays the amount of noise cancellation in dB using Eq. 9.130.

P9.42. Consider the active-noise-control system shown in Figure P9.18. Suppose the primary noise, $x(k)$, consists of the following noise-corrupted periodic signal.

$$x(k) = 2 \sum_{i=1}^{5} \frac{\sin(2\pi f_0 i k T)}{1+k} + v(k)$$

Here the fundamental frequency is $f_0 = 100$ Hz and $f_s = 2000$ Hz. The additive noise term, $v(k)$, is white noise uniformly distributed over $[-0.2, 0.2]$. Coefficient vectors for FIR models of the secondary path, $F(z)$, and the primary path, $G(z)$, are contained in MAT-file *prob9_41*. The coefficient vectors are f and g. Write a MATLAB script that loads f and g and uses the FDSP toolbox function *f_sigsyn* to apply active noise control with the signal-synthesis method starting at sample $N/4$ where $N = 2000$. Use a step size of $\mu = 0.04$.

(a) Plot the learning curve. Add a title which displays the amount of noise cancellation in dB using Eq. 9.130.

(b) Plot the magnitude spectra of the noise without cancellation.

(c) Plot the magnitude spectra of the noise with cancellation.

Section 9.9

P9.43. Consider the following nonlinear discrete-time system which has $m = 0$ past inputs and $n = 1$ past outputs.

$$y(k) = 0.8y(k_1) + 0.3[x(k) - y(k-1)]^3$$

Suppose the input $x(k)$ consists of $N = 1000$ samples of white noise uniformly distributed over $[-1, 1]$. Let the number of grid points per dimension be $d = 8$. Write a MATLAB script that performs the following tasks.

(a) Use the FDSP toolbox function *f_state* to compute a set of output-bounds b such that $b_1 \leq y(k) \leq b_2$. Use a safety factor of $\beta = 1.2$, as in Eq. 9.173. Print a, b, Δx, Δy, and the total number of grid points, r.

(b) Plot the output $y(k)$ corresponding to the white noise input $x(k)$. Include dashed lines showing the grid values along the y dimension.

(c) Let $f(\theta)$ denote the right-hand side of the nonlinear difference equation where $\theta(k) = [x(k), y(k-1)]^T$. Plot the surface $f(\theta)$ over the domain $[a_1, a_2] \times [b_1, b_2]$.

P9.44. Consider the following nonlinear discrete-time system which has $m = 0$ past inputs and $n = 1$ past outputs.

$$y(k) = 0.8y(k_1) + 0.3[x(k) - y(k-1)]^3$$

Let the range of inputs be $-1 \leq x(k) \leq 1$ and the number of grid points per dimension be $d = 8$. Write a MATLAB script that does the following.

(a) Use the FDSP toolbox function *f_state* to compute a set of output bounds b such that $b_1 \leq y(k) \leq b_2$. Use $P = 1000$ points of white noise uniformly distributed over $[-1, 1]$ for the test input and a safety factor of $\beta = 1.2$, as in Eq. 9.173. Print a, b, and the total number of grid points, r.

(b) Use FDSP toolbox function *f_rbfw*, with $N = 0$ and $ic = 1$, to compute a weight vector w that satisfies Eq. 9.166. Then use *f_rbf0* to compute the output, $y_0(k)$, to a white noise input with $M = 100$ points uniformly distributed over $[-1, 1]$. Use *f_state* to compute the nonlinear system response, $y(k)$, to the same input. Plot the two outputs on one graph using a legend. Compute the error E using Eq. 9.175, and add this to the graph title.

Elements of MATLAB

This appendix is a brief review of the essentials of MATLAB programming. More comprehensive treatments of the subject can be found, for example, in Chapman (2002) and in Hanselman and Littlefield (2001). The objective of this appendix is to review the elements of MATLAB that are useful for solving the end of chapter Computation problems, and understanding the MATLAB scripts found in the text. It is assumed that students have access to MATLAB Version 6.1 or later.

● ● ● ● ● ● ● ● ● ● ● ● ● ● ● ●

A.1 Workspace

MATLAB is a platform-independent interpretive language that stores variables in a workspace. Commands issued from within the command window are executed immediately, while those stored in an M-file script are executed when the file name is issued as a command. In a script, any text following the % symbol is interpreted by MATLAB as a *comment*. The following commands are useful for program initialization, debugging, and workspace manipulation.

```
clc                 % clear command window
clear               % clear workspace contents
whos                % display workspace contents
what                % display files in current folder
path                % display or modify current MATLAB path
helpwin             % display MATLAB help
helpwin filename    % display help for file filename
which filename      % display location of file filename
lookfor keyword     % find files with keyword in initial comment line
pause               % pause program execution
```

● ● ● ● ● ● ● ● ● ● ● ● ● ● ● ●

A.2 Variables and Initialization

MATLAB variables include arrays (scalars, vectors, and matrices), which can be real or complex, and strings. String constants are enclosed in single quotes, while the elements of arrays are enclosed in square brackets. For scalars, the brackets are optional. Elements within a row are separated by white space or by commas, and rows are separated from one another by the Enter key or by a semicolon. The following are examples of variable initialization using assignment statements.

```
q = pi;                  % area of unit circle
s = 'This is a String.'; % string
a = 4.39;                % real scalar
j = sqrt(-1);            % imaginary scalar
b = 2.1 + j*3.7;         % complex scalar
x = [1 7 5];             % 1 by 3 row vector
y = [3, 6, x];           % 1 by 6 row vector
c = x(3);                % element 3 of x;
z = [3; 8; 2];           % 3 by 1 column vector
q = z';                  % transposed version of z
A = [2 9 7; 1 8 4];      % 2 by 3 matrix
r = A(1,2);              % element in row 1, column 2 of A
B = [1 8 9 7             % 3 by 4 matrix
     3 4 8 2             % row 2
     2 5 1 6];           % row 3
```

Note that the *semicolon* is used in two ways. When used within square brackets, it is a delimiter that separates rows of an array. When used at the end of an assignment statement, it suppresses the printing of the value of the variable. Removing the semicolon causes the variable name and value to be echoed to the screen. Array elements are referenced by using the array name followed by subscripts enclosed in parentheses. Variables can also be initialized using the colon operator and built-in MATLAB functions as follows.

```
x = m:n;                 % x = [m, m+1, ... , n];
y = m:i:n;               % m through n in increments of i
c = x(p:q);              % elements p though q of x
z = linspace (a,b,n);    % n values linearly spaced over [a,b]
A = zeros(m,n);          % m by n matrix of zeros
B = ones(m,n);           % m by n matrix of ones
C = rand(m,n);           % m by n matrix of random numbers in [0,1]
n = length(z);           % number of elements in z
c = A(:,k);              % column k of A
d = B(i,:)               % row i of B
```

The *colon* operator is a powerful operator that is used in two ways. When used with two or three operands, it indicates a range of values starting with the first operand and ending with the last operand. The optional middle operand is the increment which defaults to one. When the colon is used by itself in place of an array subscript, it indicates all values of the subscript.

• • • • • • • • • • • • • •
A.3 **Mathematical Operators**

A.3.1 Scalar Operators

MATLAB is a highly vectorized programming language, which means that, whenever possible, operations apply to entire arrays on an element by element basis. The following are examples of basic mathematical operations where x is an array (scalar, vector, or matrix) and a is a scalar.

```
y = x + a;              % scalar add:      y(i,j) = x(i,j) + a
y = x - a;              % scalar subtract: y(i,j) = x(i,j) - a
y = x * a;              % scalar multiply: y(i,j) = x(i,j) * a
y = x / a;              % scalar divide:   y(i,j) = x(i,j) / a
y = x ^ a;              % scalar power:    y(i,j) = x(i,j) ^ a
```

A.3.2 Array Operators

There is a powerful set of *dot* operators that take two array operands, x and y, and produce an array result, z, by applying the operation element by element as follows.

```
z = x + y;              % array add:      z(i,j) = x(i,j) + y(i,j)
z = x - y;              % array subtract: z(i,j) = x(i,j) - y(i,j)
z = x .* y;             % array multiply: z(i,j) = x(i,j) * y(i,j)
z = x ./ y;             % array divide:   z(i,j) = x(i,j) / y(i,j)
z = x .^ y;             % array power:    z(i,j) = x(i,j) ^ y(i,j)
```

When the dot operators are applied, the two operands must have identical dimensions, or one operand must be scalar.

A.3.3 Matrix Operators

When performing linear algebra, two matrices are compatible for multiplication if the number of columns of the first matrix A equals the number of rows in the second matrix B. Let a and b be two column vectors with m elements, where m is the number of rows of A. The following are examples of matrix operations.

```
C = A * B;              % matrix product
c = a' * b;             % inner product of vectors (dot product)
C = a * b';             % outer product of vectors
x = A \ b;;             % minimum least-squares solution of Ax = b
```

Recall that a' denotes the transpose of a. When the linear algebraic system of equations, $Ax = b$, has a unique solution, the *left-division* operator, \, offers a simple and efficient way to compute x.

● ● ● ● ● ● ● ● ● ● ● ● ● ● ● ●

A.4 Input and Output

A.4.1 Keyboard and Screen

To prompt the user for input from the keyboard and display output on the console screen, the following built-in MATLAB functions can be used.

```
x = input (prompt);        % get keyboard input using prompt string prompt
fprintf (format,list)      % display list of values using format string format
s = sprintf (format,list); % send formated output to string s
```

The arguments *prompt* and *format* are string variables, or string constants, enclosed in single quotes. For the *input* function, arrays can be entered from the keyboard if they are enclosed in square brackets, and strings can be entered if they are enclosed in single quotes. The formatted print function behaves like the corresponding function in the language C. For each variable or expression in the comma-separated *list*, there is a format specifier in the format string indicating where the next value is to be placed and the format used to display it. Examples of format specifiers, or place holders, are $\%d$ for decimal integer, $\%f$ for fixed-point real, $\%e$ for scientific notation real, $\%g$ for general real, $\%s$ for string, and $\%x$ for hexadecimal. In addition, special escape codes can be placed in the format string such as $\backslash n$ for a new line. The *sprintf* function is identical to the *fprintf* function, except that it sends the output to a string s instead of the console screen. Enter *helpwin fprintf* for more details and examples of formatted output.

A.4.2 MAT Files

There is a simple and efficient way to export variables from the workspace and import variables to the workspace using special data files called MAT-files. The following examples illustrate saving data to MAT-files and loading data from MAT-files.

```
save (filename,'x','y')  % Save variables x, y in file filename.MAT
load (filename)          % Load previously saved variables from file filename.MAT
```

Here *filename* is a string variable, or a string constant enclosed in single quotes, containing the name of the MAT-file without the .MAT extension. There is also a version of the *Save* command that saves the variables in a less-efficient format that is readable by non-MATLAB programs. Enter *helpwin save* for more details.

● ● ● ● ● ● ● ● ● ● ● ● ● ● ● ●

A.5 Branching and Loops

A.5.1 Relational and Logical Expressions

MATLAB has a standard collection of relational and logical operators which return values that are true (nonzero) or false (zero). Suppose a and b are numerical arrays of the same size, and *x* and *y* are logical arrays (zeros and ones) of the same size. The following are then examples of the relational and logical expressions.

```
a < b               % a less than b
a <= b              % a less than or equal to b
a > b               % a greater than b
a >= b              % a greater than or equal to b
a == b              % a equal to b
a ~= b              % a not equal to b
x & y               % x and y
x | y               % x or y
~x                  % not x
```

A.5.2   Branching

Relational and logical expressions are used to control the flow of execution within a MATLAB script. The following *if* construct can be used to implement program branching.

```
if expression_1
    ...             % clause 1
elseif expression_2
    ...             % clause 2
elseif expression_3
    ...             % clause 3
else
    ...             % else clause
end
```

If *expression_1* is true (nonzero), then clause 1 is executed. Otherwise, *expression_2* is tested and the process repeats. Consequently, the first clause whose expression is true is executed. If none of the logical expressions are true, then the *else* clause is executed. Clauses consist of a block of one or more statements. The *elseif* and *else* portions of the *if* construct are optional. To implement a menu of options, the following *switch* statement is often used.

```
switch choice
    case value_1,
        ...         % clause 1
    case value_2,
        ...         % clause 2
    otherwise,
        ...         % default clause
end
```

Here *choice* is an integer-valued variable or expression. If *choice* is equal to *value_k*, then program control is transferred to clause k, otherwise it is transferred to the *otherwise* clause. The *otherwise* case is optional.

A.5.3   Loops

Many numerical computations can be done iteratively using loops. The most general loop uses the *while* statement as follows.

```
while expression
    ...                     % loop body
end
```

Here the loop body, a block of statements, is executed repeatedly while the *expression* is true. Typically, the loop body changes the value of a variable in an *expression* so that eventually the *expression* becomes false and the loop terminates. The body of a *while* loop is executed zero or more times, depending on the initial value of the *expression*. When a loop is to be executed a predetermined number of times, say to process each element of an array, the following *for* statement is often used.

```
for k = m:p:n
    ...                     % loop body
end
```

The first time through the loop, the loop body is executed with $k = m$. The second time, the loop body is executed with $k = m + p$, and this process continues while $k \leq n$. Often $k = m : n$ is used instead, in which case the default increment is $p = 1$. Programmers new to MATLAB who have worked with other languages may be tempted to make liberal use of *for* loops. This tends to slow down MATLAB execution speed, because it does not exploit vectorization. Often a *for* loop can be avoided by vectorizing the computation using the dot operators. The percentage of execution time spent on a given statement can be determined by using the *profile* command.

● ● ● ● ● ● ● ● ● ● ● ● ● ● ● ● ● ● ●
A.6   Built-in Functions

There are numerous built-in functions available with MATLAB, and still more functions associated with the MATLAB toolboxes. Some of the more commonly used functions are as follows.

```
x = cos(a);             % cosine of a
x = sin(a);             % sine of a
x = tan(a);             % tangent of a
a = acos(x);            % arc cosine of x
```

```
a = asin(x);            % arc sine of x
a = atan2(y,x);         % arc tangent of y/x
y = exp(x);             % exponential of x
y = log(x);             % natural log of x
y = log10(x);           % common log of x
y = sqrt(x);            % square root of x
y = sum(x);             % sum of elements of x
y = prod(x);            % product of elements of x
z = mod(x,y);           % remainder of x/y
i = floor(x);           % round x down
i = ceil(x);            % round x up
i = round(x);           % round x to nearest integer
s = num2str(x);         % convert number to string
x = str2num(s);         % convert string to number
y = abs(x);             % magnitude of x
y = angle(x);           % angle of x
r = real(x);            % real part of x
i = imag(x);            % imaginary part of x
y = polyval(a,x);       % evaluate polynomial with coefficient vector a at x
r = roots(a);           % roots of polynomial with coefficient vector a
[y,i] = max(x);         % find maximum, y = x(i)
[y,i] = min(x);         % find minimum, y = x(i)
[m,n] = size(A);        % return dimensions of array A
```

Some numerical functions, such as *sqrt*, can produce complex results depending on the input. Complex numbers are handled automatically. Indeed all arrays are assumed to have a real part and an imaginary part. For real arrays, the imaginary part happens to be zero. For the *max* function, there are two outputs enclosed in square brackets. The first is the maximum value of x and the second is the subscript of the maximum value. A similar remark holds for the *min* function. In each case, the second output is optional. When only one output is used, the brackets are optional.

● ● ● ● ● ● ● ● ● ● ● ● ● ● ● ● ⋯

A.7 Graphical Output

MATLAB has a rich and powerful set of functions for creating graphical output. Let x and y be vectors of the sample length, and let Y be a matrix. Then the following commands can be used to create graphical output in a figure window.

```
figure          % create a new figure window
plot(y)         % plot y vs its subscript
plot(x,y)       % plot y vs x
```

```
plot(x,y,s)                 % plot y vs x using string s to specify line type
plot(x1,y1,s1,x2,y2,s2)     % plot yk vs xk using string sk for 1 <= k <= 2
stem (x,y)                  % discrete-time plot
xlabel(s)                   % insert string s for x axis label
ylabel(s)                   % insert string s for y axis label
title(s)                    % insert string s for figure title
axis ([x1 x2 y1 y2])        % set plot limits to [x1,x2] and [y1,y2]
text (x,y,s)                % insert string s at plot coordinates (x,y)
gtext (s)                   % insert string s using mouse
legend (s1,...,sp)          % use strings s1,...,sp for plot legend
subplot(m,n,p)              % create m by n array of plots, draw plot number p
surf(Y)                     % plot matrix Y as a surface
```

The optional argument *s* in the *plot* function is a three-character string that controls the appearance of the curve. The first character specifies the color, the second the type of plot symbol, and the third the type of line connecting the symbols. Enter *helpwin plot* for more details and for examples. The *subplot* command is used to create an array of plots on one screen. The following is an example of a script that creates the family of curves shown in Figure A.1

```
%--------------------------------------------------------------------------------
% Example A.1: A Sample Plotting Program
%--------------------------------------------------------------------------------

% Initialize

clear
clc
n = 250;
T = 5.0;
t = linspace (0,T,n);
y1 = exp(-t) .* sin(2*pi*t);
y2 =  2*(t .^ 3) .* exp(-2*t);

% Plot data

figure
hp = plot (t,y1,t,y2);
set (hp(1),'LineWidth',1.5)
xlabel ('t (sec)')
ylabel ('signals')
title ('A sample plot')
legend ('Damped sine','Damped cubic polynomial')
fprintf ('Press any key to continue...')
pause
%--------------------------------------------------------------------------------
```

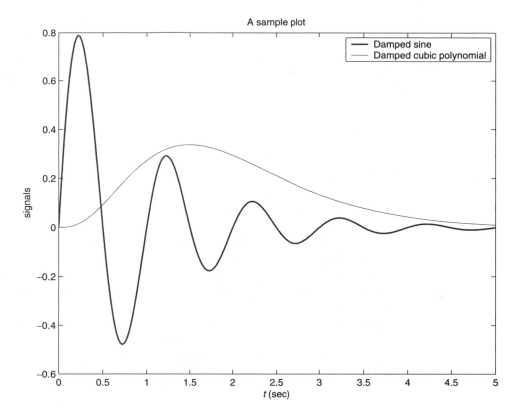

Figure A.1   A Sample Plot

● ● ● ● ● ● ● ● ● ● ● ● ● ●

A.8 User-defined Functions

The real power of MATLAB begins to come into play when the user develops his or her own library of user-defined functions specific to an application area. The template for a user-defined function is as follows.

```
[out1,...,outn] = funcname (in1,...,inm)
% FUNCNAME: This is the lookfor comment
%
% Usage:     [out1,...,outn] = funcname (in1,...,inm);
%
% Inputs:
%            in1 = ...
%            in2 = ...
%
```

```
% Outputs:
%          out1 = ...
%          out2 = ...
%

  ...      % function body goes here
```

The string *funcname* is a user-supplied name for the function. The statements defining the function are saved in an M-file named *funcname.m* in the current working folder or in a folder that is in the MATLAB path. The first comment line, called the *H1* comment, can be searched for keywords using the *lookfor keyword* command. All comments up to the first noncomment line are displayed using the *helpwin funcname* command. The main body of the function uses the *m* input arguments *in1,...,inm* to compute the *n* output arguments, *out1,...,outn*. The MATLAB variables *nargin* and *nargout* can be used within any function to determine the number of actual input and output arguments used when the function is called. If $nargin < m$, then default values can be used for the missing input arguments. Function *funcname* also can be called with fewer than *n* output arguments. If it is called with only one output argument, then the square brackets enclosing the output arguments are optional. The following is an example of a function called *u_sphere* which computes the volume and mass of a sphere given the values for the radius and density.

```
[volume,mass] = u_sphere (radius,density)
% U_SPHERE: Compute volume and mass of a sphere
%
% Usage:    [volume,mass] = u_sphere (radius,density);
%
% Inputs:
%          radius  = radius of sphere
%          density = density of sphere
%
% Outputs:
%          volume = volume of sphere
%          mass   = mass of sphere

volume = (4/3) * pi * (r .^ 3);
mass = density .* volume;
```

Note that the function *u_sphere* has been vectorized by using the dot operators to compute the volume and mass. This way, if *u_sphere* is called with *radius* and *density* as arrays, it will return *volume* and *mass* as arrays of the same size.

● ● ● ● ● ● ● ● ● ● ● ● ● ● ● ● ● ●

A.9 GUIs

Using MATLAB, it is relatively easy to create scripts that use graphical user interface (GUI) objects. A general discussion of handle graphics and graphical user interfaces is beyond the scope of this appendix. However, there are a number of easy-to-use, built-in dialog box GUIs that are available. Some of the more useful ones are listed below. For a more complete list, the reader is referred to Hanselman and Littlefield (2001).

```
menu                % list menu of options and return choice
uigetfile           % open a file for input
uiputfile           % open a file for output
inputdlg            % prompt user for input
helpdlg             % display a help window
printdlg            % print current figure window
waitbar             % show progress of program execution
```

The following is a example of a script that uses some built-in dialog boxes. It uses a menu to create random data, plot it in discrete time, save it in a MAT-file, and load it from a MAT-file.

```
%-------------------------------------------------------------------------------
% Example A.2: A sample GUI program
%-------------------------------------------------------------------------------

% Initialize

clear
clc
n = 30;
k = 0 : n-1;
x = rand(1,n);
choice = 0;

% Implement menu

while choice ~= 5
    choice = menu ('A Sample GUI Program','Generate data','Plot data',...
                   'Save data','Load data','Exit');
```

```
switch choice
    case 1,                        % create data
        x = rand(1,n);
    case 2,                        % plot data
        figure
        stem(k,x)
        xlabel('k')
        ylabel('x(k)')
        title ('A Random Discrete-Time Signal')
    case 3,                        % save data
        fname = uiputfile ('*.MAT','Select MAT-file for output');
        save (fname,'x')
    case 4,                        % load data
        fname = uigetfile ('*.MAT','Select MAT-file for input');
        load (fname)
    end
end
%--------------------------------------------------------------------------------
```

A plot showing the menu produced by the menu function is shown in Figure A.2.

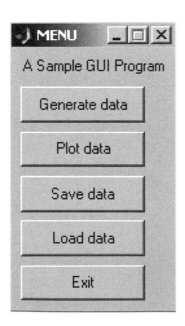

Figure A.2 The Menu Produced by Example A.2

FDSP Toolbox

The distribution CD contains a Fundamentals of Digital Signal Processing (FDSP) toolbox that uses MATLAB to implement the signal processing techniques discussed in the text. This appendix summarizes the main features of the FDSP toolbox. Although MATLAB is platform-independent, the FDSP toolbox was designed primarily for use on a Windows PC with a sound card.

B.1 Installation

The FDSP toolbox was developed to help students solve the GUI Simulation and Computation problems appearing at the end of each chapter. It also provides the instructor and the student with a convenient way to run all of the computational examples and reproduce all of the MATLAB figures that appear in the text. A novel component of the toolbox is a collection of graphical user interface (GUI) modules that allow the user to investigate and explore the signal processing techniques covered in each chapter without any need for programming. The FDSP toolbox can be installed using MATLAB itself (Marwan, 2003). For example, if the FDSP distribution CD is in drive D, then issue the following commands from the MATLAB command prompt.

```
cd d:\
setup
```

The directory structure of the FDSP toolbox is shown in Figure B.1. During installation, *setup* creates and populates the directories in Figure B.1 and then updates the MATLAB file, *startup.m*, so that each of these directories is part of the MATLAB path. It then transfers control to the MATLAB *work* directory. The locations of the FDSP toolbox directories can be seen by executing the MATLAB *path* command.

The main directory, *fdsp*, includes the toolbox installation files. There are four subdirectories below the main FDSP directory. The *fdsp* subdirectory contains the FDSP toolbox functions and the chapter GUI modules. The FDSP functions are named using the convention *f_xxx*, and the chapter GUI modules are named using the convention *g_xxx*. These conventions are adopted in order to ensure compatibility with other MATLAB toolbox boxes, such as the Signal processing and Filter design toolboxes. The software supplied with this text was developed for use with the

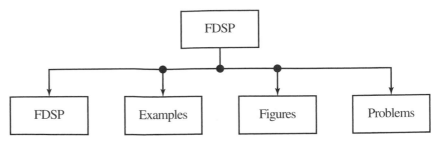

Figure B.1 Directory Structure of the FDSP Toolbox

Student Version of MATLAB, it does not require any auxiliary toolboxes. This was done to keep student expenses to a minimum. However, for those students who do have access to supplementary toolboxes, they can still make use of them without any risk of conflict because of the naming conventions. The *examples* directory contains all of the MATLAB examples that appear in the text, and the *figures* directory contains all of the MATLAB generated figures. The *problems* directory contains solutions to selected end-of-chapter problems in the form of PDF files readable with Adobe Acrobat Reader.

Course Website

Although most students will install the FDSP toolbox directly from the distribution CD using the *setup* command, it is also possible to download the FDSP toolbox from the Internet from the following website.

```
www.clarkson.edu/~schillin/fdsp
```

Students and instructors can download a self-extracting file that contains the latest version of the FDSP toolbox software from this location. Additional supplementary support material for courses which use *Fundamentals of Digital Signal Processing* will be posted on this site as it becomes available.

B.2 Driver Module: *f_dsp*

All of the software on the distribution CD can be accessed conveniently through a driver module called *f_dsp*. The FDSP driver module is launched by entering the following command from the MATLAB command prompt:

```
f_dsp
```

A startup screen for *f_dsp* is shown in Figure B.2. Most of options on the menu toolbar at the top of the screen produce submenus of selections. The *GUI Modules* option is used to run the graphical user interface modules. Using the *Examples* option, all MATLAB examples appearing in the text can be executed. Similarly, the *Figures* option is used to recreate and display all of the MATLAB figures in the text. The *Problems* option is used to display PDF file solutions to selected end-of-chapter problems. The *Help* option provides online help for the

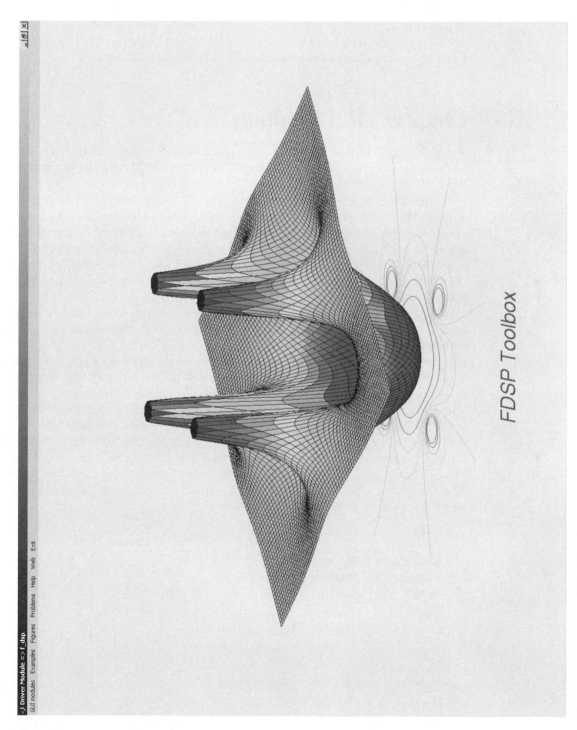

Figure B.2   Driver Module f_dsp for FDSP Toolbox

GUI modules and the FDSP toolbox functions. The *Web* option provides a number of useful links including the course website. At the course website, the user can download a self-extracting file that contains the latest version of the FDSP software. The *Exit* option returns control to the MATLAB command window.

● ● ● ● ● ● ● ● ● ● ● ● ● ● ● ● ●

B.3 Chapter GUI Modules

FDSP Toolbox

When the *GUI Modules* option is selected from the toolbar of the FDSP driver module, the user is provided with the list of chapter GUI modules summarized in Table B.1.

Each of the GUI modules is described in detail near the end of the corresponding chapter. The GUI modules are designed to provide the student with a convenient means of investigating the signal processing concepts covered in that chapter without any need for programming. There is also a set of GUI Simulation problems at the end of each chapter that are designed to be solved using the chapter GUI module. Users who are familiar with MATLAB programming can provide optional MAT-files and M-file functions that interact with the GUI modules.

The chapter GUI modules feature a common user interface that is simple to learn and easy to use. The start-up screen consists of a set of tiled windows using the template shown in Figure B.3.

The upper-left *Block diagram* window shows the system or signal processing operation under investigation. Below the block diagram is a *Parameters* window that contains edit boxes which show the current values of simulation parameters. These values can be edited directly by the user with changes activated with the Enter key. The *Type* window to the right of the block diagram includes radio button controls that allow the user to select the signal or the system type. In addition to predefined choices, there is also a User-defined selection that loads data from a user-supplied MAT-file or M-file. For signal selection, there is an option to record data from the PC microphone. The quality of the recording can be verified by playing the signals on the PC speaker using a pushbutton control.

Table B.1: ▶
Chapter GUI
Modules

Module	Description	Chapter
g_sample	Signal sampling	1
g_reconstruct	Signal reconstruction	1
g_system	Discrete-time systems	2
g_spectra	Signal spectral analysis	3
g_correlate	Signal correlation and convolution	4
g_filters	Filter specifications and structures	5
g_fir	FIR filter design	6
g_multirate	Multirate signal processing	7
g_iir	IIR filter design	8
g_adapt	Adaptive signal processing	9

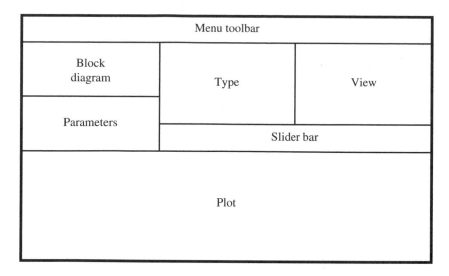

Figure B.3 Screen Template for Chapter GUI Modules

The *View* window includes radio button controls that allow the user to select the information to be graphically displayed in the *Plot* window that appears along the bottom half of the screen. The choices depend on the type of signal processing operation being performed. Below the *Type* and *View* windows is a horizonal *Slider bar* control that provides a very simple way for the user to directly modify a scalar design parameter, such as the sampling rate, the number of samples, the number of bits, or the filter order. The *Parameters*, *Type*, and *View* windows also contain checkbox controls that allow the user to turn options on and off such as dB display, signal clipping, and additive noise.

Along the top of the screen is a *Menu* toolbar that includes options specific to each GUI module. All modules have certain standard options including *Save*, *Caliper*, *Print*, *Help*, and *Exit*. When results are saved in a MAT-file with the *Save* option, they can be reloaded using the same or different GUI modules. All GUI modules save data using a single standardized format, namely,

$$a, b, x, y, fs$$

In this way, data exported from one GUI module can be imported into another GUI module using the User-defined input option. The *Caliper* option allows the user to measure any point on the current plot using crosshairs and the mouse. The *Print* option generates a hard copy of the current plot, while the *Help* option displays help on how to effectively use the GUI module. Full page sample screens of each of the chapter GUI modules can be found at the end of this appendix.

● ● ● ● ● ● ● ● ● ● ● ● ● ● ● ●

B.4 FDSP Toolbox Functions

The use of the GUI modules is very convenient, but it is not as flexible as having users write their own MATLAB scripts to perform signal processing tasks. Algorithms developed in the text are implemented as FDSP toolbox functions. These functions

Name	Description
f_caliper	Measure points on plot using mouse crosshairs
f_clip	Clip value to an interval, check calling arguments
f_getsound	Record signal from the PC microphone
f_labels	Label graphical output
f_prompt	Prompt for a scalar in a specified range
f_randg	Gaussian random matrix
f_randu	Uniformly distributed random matrix
f_torow	Convert vector to a row vector
f_tocol	Convert vector to a column vector
f_wait	Pause to examine displayed output

fall into two broad categories, main-program support functions and chapter functions. Instructions for usage of any of the FDSP functions and GUI modules can be obtained by using the *helpwin* command with a command-line argument.

```
helpwin fdsp          % Help for all FDSP toolbox functions
helpwin f_dsp         % Help for the FDSP driver module
helpwin g_xxx         % Help for GUI module g_xxx
```

Once the name of a function or module is determined, more detailed help on that item can be obtained by selecting that link. Alternatively, if the name of the function of interest is known, a *Help* window for that function can be obtained by using the function name as the command-line argument.

```
helpwin f_xxx         % Help for FDSP toolbox function f_xxx
```

The *Help* option in the FDSP driver module in Figure B.2 also provides documentation on all of the chapter GUI modules and the FDSP functions. The MATLAB *lookfor* command can be used to locate the names of functions containing a key word in their first comment line.

The main program support functions consist of general low-level utility functions that are designed to simplify the process of writing MATLAB scripts by performing some routine tasks. These functions are summarized in Table B.2.

The second group of toolbox functions are implementations of algorithms developed in the chapters. Specialized functions are developed in those instances where corresponding built-in MATLAB functions are not available as part of the Student Version of MATLAB. In order to minimize the expense to the student, it is assumed that no supplementary MATLAB toolboxes are available. Instead, these functions are provided in the FDSP toolbox.

Summaries of the FDSP functions, organized by chapters, can be found in Table B.3. To learn more about the usage of any of these functions, simply type *helpwin* followed by the function name.

continued on the next page

Name	Description
Chapter 7	
f_decimate	Integer sampling rate decimator
f_interpolate	Integer sampling rate interpolator
f_rateconv	Rational sampling rate converter
Chapter 8	
f_iirres	Design resonator IIR filter
f_iirnotch	Design notch IIR filter
f_iircomb	Design comb IIR filter
f_iirinv	Design inverse comb IIR filter
f_butters	Design analog Butterworth lowpass filter
f_cheby1s	Design analog Chebyshev-I lowpass filter
f_cheby2s	Design analog Chebyshev-II lowpass filter
f_elliptics	Design analog elliptic lowpass filter
f_low2lows	Lowpass-to-lowpass analog frequency transformation
f_low2highs	Lowpass-to-highpass analog frequency transformation
f_low2bps	Lowpass-to-bandpass analog frequency transformation
f_low2bss	Lowpass-to-bandstop analog frequency transformation
f_bilin	Bilinear analog-to-digital filter transformation
f_butterz	Design Butterworth IIR filter
f_cheby1z	Design Chebyshev-I IIR filter
f_cheby2z	Design Chebyshev-II IIR filter
f_ellipticz	Design elliptic IIR filter
f_string	Plucked-string IIR filter output
f_reverb	Reverb IIR filter output
Chapter 9	
f_lms	Least mean square (LMS) method
f_normlms	Normalized LMS method
f_corrlms	Correlation LMS method
f_leaklms	Leaky LMS method
f_fxlms	Filtered-x LMS active noise control
f_signal	Signal-synthesis active noise control
f_rls	Recursive least squares (RLS) method
f_rbfw	Radial basis function (RBF) system identification
f_rbf0	Zero-order RBF network evaluation
f_state	Evaluate state of nonlinear discrete-time system

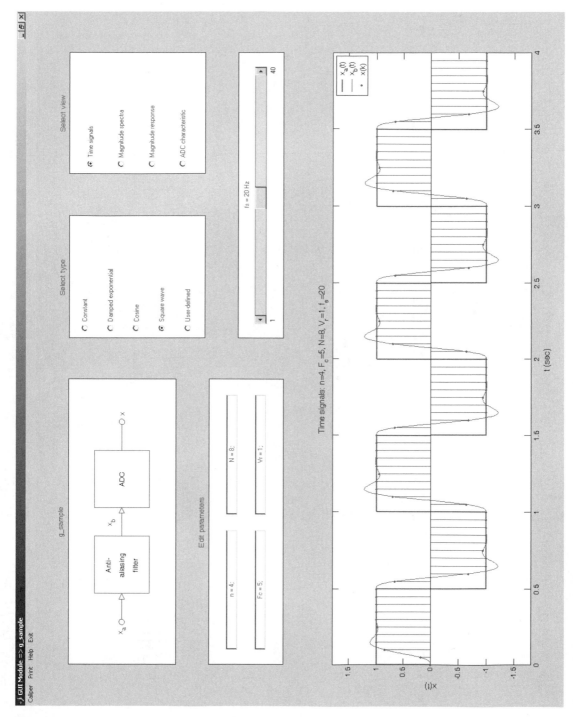

Figure B.4 GUI Module *g_sample*, Chapter 1

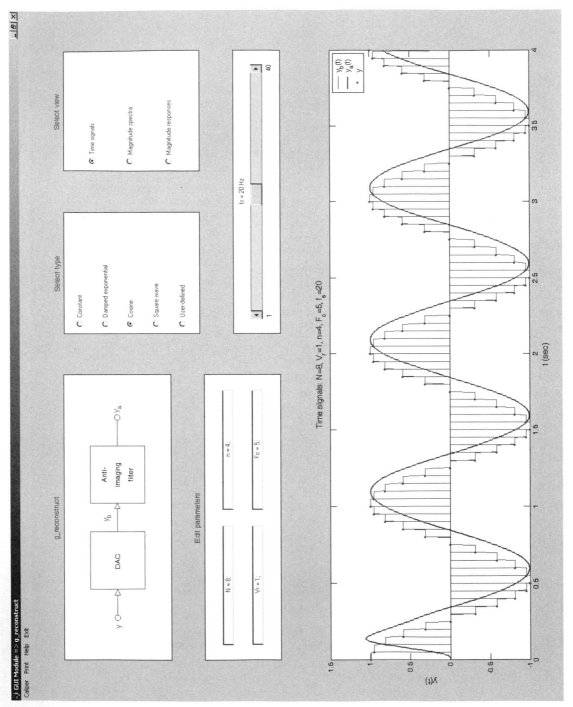

Figure B.5 GUI Module *g_reconstruct*, Chapter 1

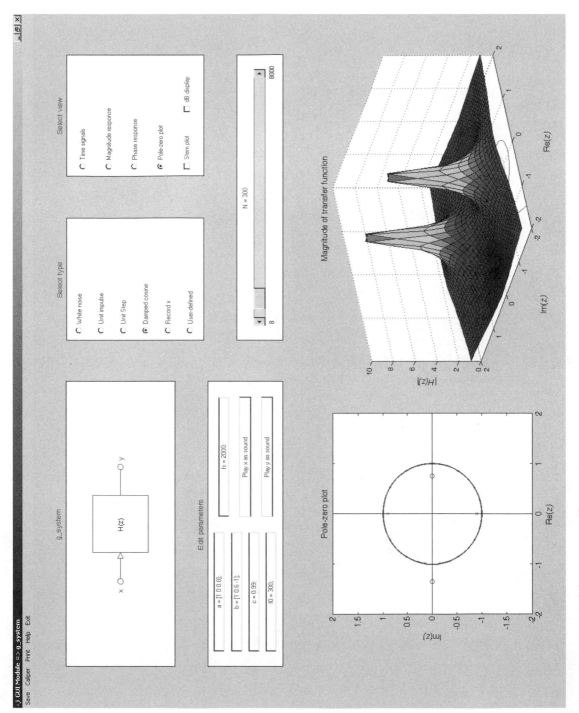

Figure B.6   GUI Module *g_system*, Chapter 2

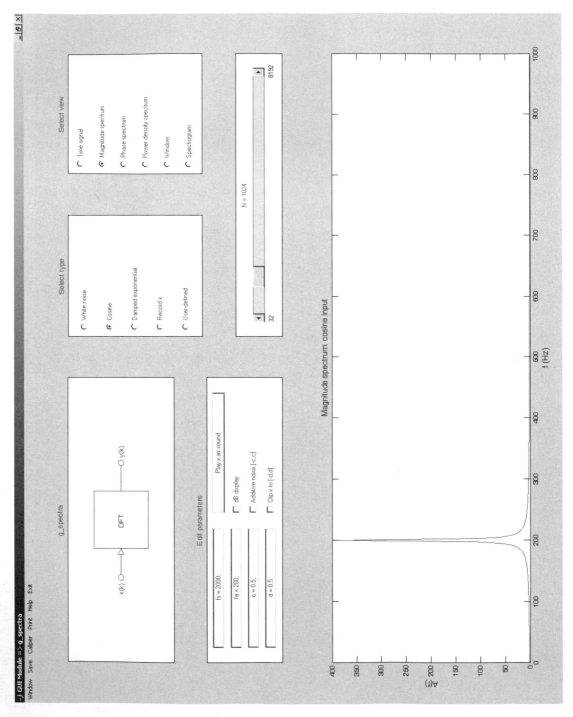

Figure B.7 GUI Module *g_spectra*, Chapter 3

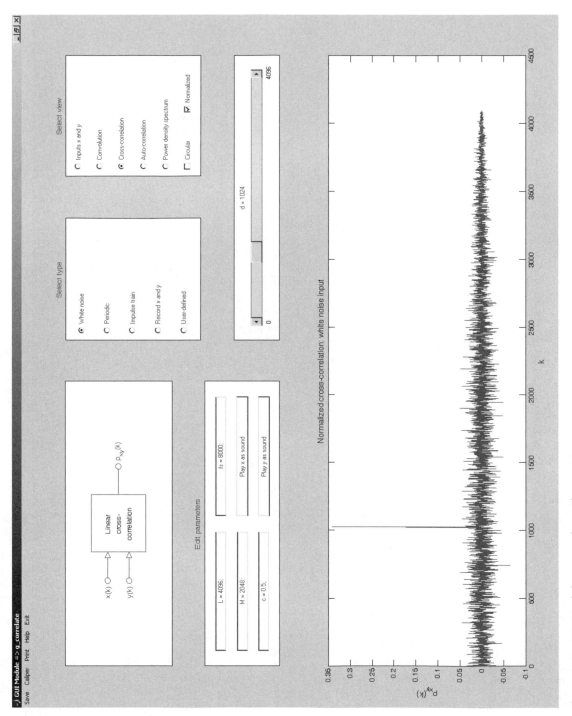

Figure B.8 GUI Module *g_correlate*, Chapter 4

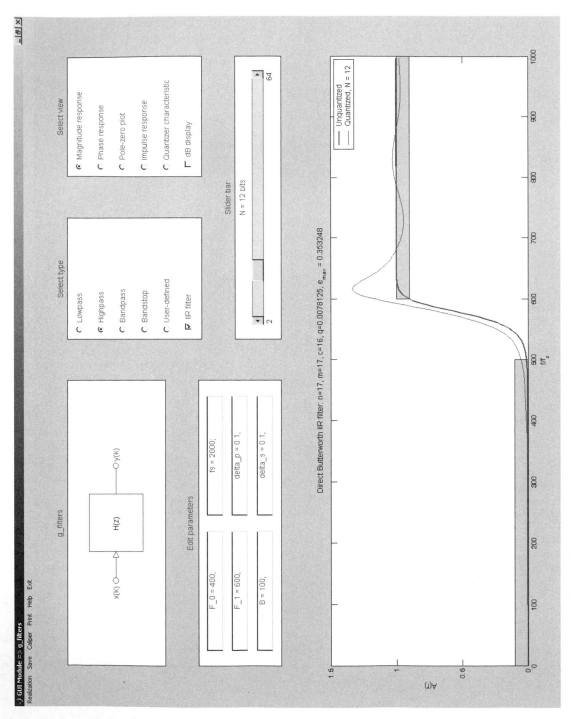

Figure B.9   GUI Module *g_filters*, Chapter 5

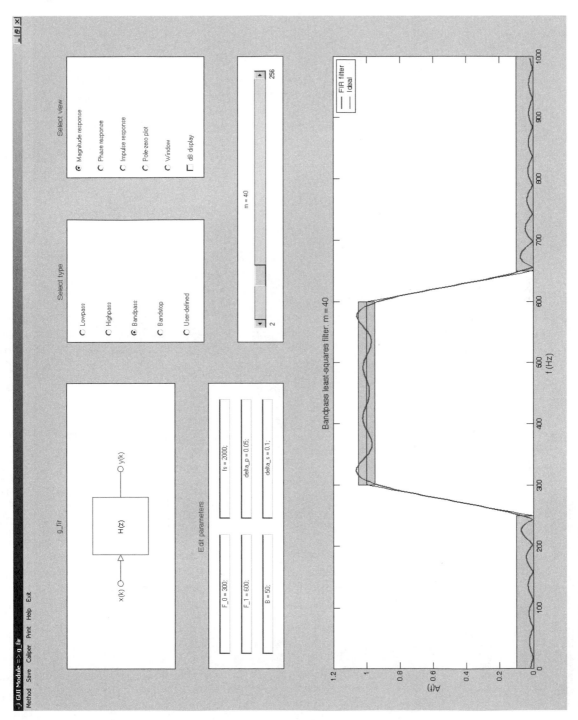

Figure B.10 GUI Module *g_fir*, Chapter 6

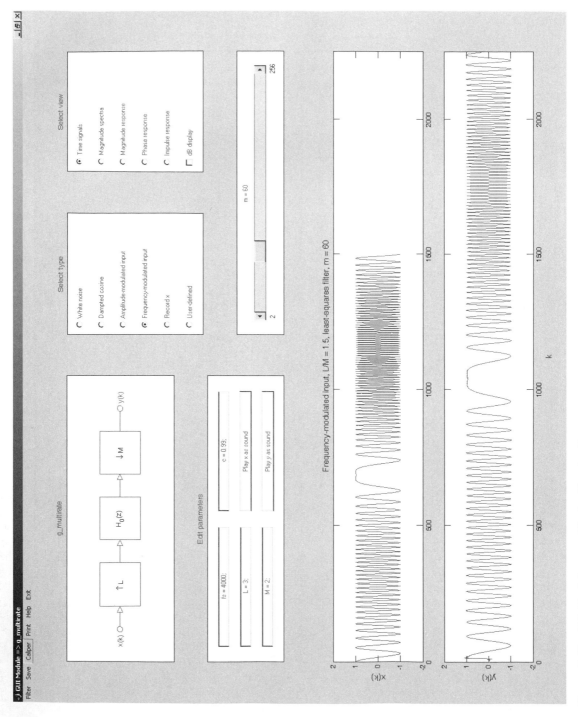

Figure B.11   GUI Module *g_multirate*, Chapter 7

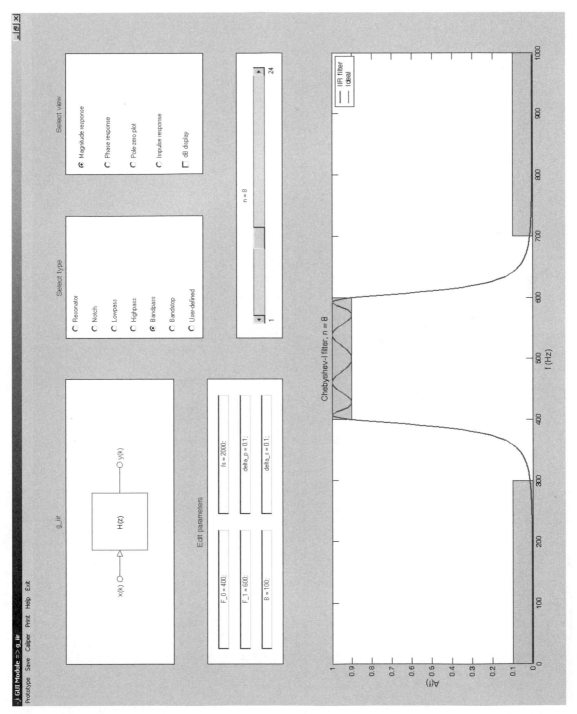

Figure B.12 GUI Module *g_iir*, Chapter 8

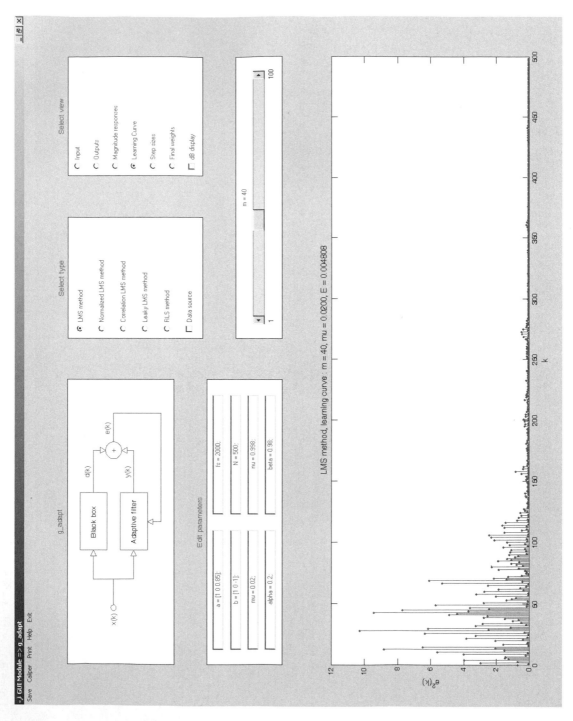

Figure B.13 GUI Module *g_adapt*, Chapter 9

Transform Tables

C.1 Fourier Series

Complex Form:

$$x_a(t + T) = x_a(t)$$

$$x_a(t) = \sum_{k=-\infty}^{\infty} c_k \exp\left(\frac{j2\pi kt}{T}\right)$$

$$c_k = \frac{1}{T} \int_{-T/2}^{T/2} x_a(t) \exp\left(\frac{-j2\pi kt}{T}\right)$$

Trigonometric Form:

$$x_a(t) = \frac{a_0}{2} + \sum_{k=1}^{\infty} a_k \cos\left(\frac{2\pi kt}{T}\right) + b_k \sin\left(\frac{2\pi kt}{T}\right)$$

$$a_k = \frac{2}{T} \int_{-T/2}^{T/2} x_a(t) \cos\left(\frac{2\pi kt}{T}\right) dt = 2\text{Re}\{c_k\}$$

$$b_k = \frac{2}{T} \int_{-T/2}^{T/2} x_a(t) \sin\left(\frac{2\pi kt}{T}\right) dt = -2\text{Im}\{c_k\}$$

Cosine Form:

$$x_a(t) = \frac{d_0}{2} + \sum_{k=1}^{\infty} d_k \cos\left(\frac{2\pi kt}{T} + \theta_k\right)$$

$$d_k = \sqrt{a_k^2 + b_k^2} = 2|c_k|$$

$$\theta_k = \tan^{-1}\left(\frac{-b_k}{a_k}\right) = \tan^{-1}\left(\frac{-\text{Im}\{c_k\}}{\text{Re}\{c_k\}}\right)$$

Table C.1: ▶
Fourier Series Pairs

Description	$x_a(t)$	Fourier series
Odd square wave	$\text{sgn}\left[\sin\left(\frac{2\pi t}{T}\right)\right]$	$\frac{4}{\pi} \sum_{k=1}^{\infty} \frac{1}{2k-1} \sin\left[\frac{2\pi(2k-1)t}{T}\right]$
Even square wave	$\text{sgn}\left[\cos\left(\frac{2\pi t}{T}\right)\right]$	$\frac{4}{\pi} \sum_{k=1}^{\infty} \frac{(-1)^{k-1}}{2k-1} \cos\left[\frac{2\pi(2k-1)t}{T}\right]$
Impulse train	$\delta_T(t)$	$\frac{1}{T} + \frac{2}{T} \sum_{k=1}^{\infty} \cos\left(\frac{2\pi kt}{T}\right)$
Even pulse train	$u_a\left[\cos\left(\frac{2\pi t}{T}\right) - \cos\left(\frac{2\pi \tau}{T}\right)\right]$	$\frac{2\tau}{T} + \frac{4\tau}{T} \sum_{k=1}^{\infty} \text{sinc}\left(\frac{2\pi k\tau}{T}\right) \cos\left(\frac{2\pi kt}{T}\right)$
Rectified sine wave	$\left|\sin\left(\frac{2\pi t}{T}\right)\right|$	$\frac{2}{\pi} - \frac{4}{\pi} \sum_{k=1}^{\infty} \frac{1}{4k^2-1} \cos\left(\frac{4\pi kt}{T}\right)$
Sawtooth wave	$\text{mod}(t, T)$	$\frac{1}{2} - \frac{1}{\pi} \sum_{k=1}^{\infty} \frac{1}{k} \sin\left(\frac{2\pi kt}{T}\right)$

• • • • • • • • • • • • • • • • •

C.2 Fourier Transform

$$X_a(f) \overset{\Delta}{=} \int_{-\infty}^{\infty} x_a(t) \exp(-j2\pi f t) dt, \quad f \in R$$

Table C.2: ▶
Fourier Transform Pairs

Entry	$x_a(t)$	$X_a(f)$	Description
1	$\exp(-ct)u_a(t)$	$\dfrac{1}{c + j2\pi f}$	Causal exponential
2	$\exp(-c|t|)$	$\dfrac{2c}{c^2 + 4\pi^2 f^2}$	Double exponential
3	$\exp[-(ct)^2]$	$\dfrac{\sqrt{\pi} \exp[-(\pi f/c)^2]}{c}$	Gaussian
4	$\exp(-ct)\cos(2\pi f_0 t)u_a(t)$	$\dfrac{c + j2\pi f}{(c + j2\pi f)^2 + (2\pi f_0)^2}$	Damped cosine
5	$\exp(-ct)\sin(2\pi f_0 t)u_a(t)$	$\dfrac{2\pi f_0}{(c + j2\pi f)^2 + (2\pi f_0)^2}$	Damped sine
6	$p_T(t) \overset{\Delta}{=} u_a(t + T) - u_a(t - T)$	$2\tau \mathrm{sinc}(2\pi T f)$	Pulse
7	$2B\mathrm{sinc}(2\pi Bt)$	$p_B(f)$	Sinc function
8	$\delta_a(t)$	1	Unit impulse
9	$u_a(t)$	$\dfrac{\delta_a(f)}{2} + \dfrac{1}{j2\pi f}$	Unit step
10	c	$c\delta_a(f)$	Constant
11	$\mathrm{sgn}(t)$	$\dfrac{1}{j\pi f}$	Signum function
12	$\cos(2\pi f_0 t)$	$\dfrac{\delta_a(f + f_0) + \delta_a(f - f_0)}{2}$	Cosine
13	$\sin(2\pi f_0 t)$	$\dfrac{j[\delta_a(f + f_0) - \delta_a(f - f_0)]}{2}$	Sine

Property	$x_a(t)$	$X_a(f)$
Linearity	$ax_1(t) + bx_2(t)$	$aX_1(f) + bX_2(f)$
Time scale	$x_a(at)$	$\dfrac{1}{\|a\|} X_a\left(\dfrac{f}{a}\right)$
Reflection	$x_a(-t)$	$X_a(-f)$
Duality	$X_a(t)$	$x_a(-f)$
Complex conjugate	$x_a^*(t)$	$X_a^*(-f)$
Time shift	$x_a(t - T)$	$\exp(-j2\pi fT)X_a(f)$
Frequency shift	$\exp(j2\pi f_0 t)x_a(t)$	$X_a(f - f_0)$
Time differentiation	$\dfrac{d^k x_a(t)}{dt^k}$	$(j2\pi f)^k X_a(f)$
Frequency differentiation	$t^k x_a(t)$	$\left(\dfrac{1}{2\pi}\right)^k \dfrac{d^k X_a(f)}{df^k}$
Time convolution	$\displaystyle\int_{-\infty}^{\infty} x_1(\tau)x_2(t - \tau)d\tau$	$X_1(f)X_2(f)$
Frequency convolution	$x_1(t)x_2(t)$	$\displaystyle\int_{-\infty}^{\infty} X_1(\alpha)X_2(f - \alpha)d\alpha$
Cross-correlation	$\displaystyle\int_{-\infty}^{\infty} x_1(\tau)x_2^*(t + \tau)d\tau$	$X_1(f)X_2^*(f)$

● ● ● ● ● ● ● ● ● ● ● ● ● ● ● ● ● ● ●
C.3 **Laplace Transform**

$$X_a(s) \overset{\Delta}{=} \int_0^\infty x_a(t) \exp(-st)dt, \quad s \in C$$

Table C.4: ▶
Laplace Transform
Pairs

Entry	$x_a(t)$	$X_a(s)$	Description
1	$\delta_a(t)$	1	Unit impulse
2	$u_a(t)$	$\dfrac{1}{s}$	Unit step
3	$t^m u_a(t)$	$\dfrac{m!}{s^{m+1}}$	Polynomial
4	$\exp(-ct)u_a(t)$	$\dfrac{1}{s+c}$	Exponential
5	$\exp(-ct)t^m u_a(t)$	$\dfrac{m!}{(s+c)^{m+1}}$	Damped polynomial
6	$\sin(2\pi f_0 t)u_a(t)$	$\dfrac{2\pi f_0}{s^2 + (2\pi f_0)^2}$	Sine
7	$\cos(2\pi f_0 t)u_a(t)$	$\dfrac{s}{s^2 + (2\pi f_0)^2}$	Cosine
8	$\exp(-ct)\sin(2\pi f_0 t)u_a(t)$	$\dfrac{2\pi f_0}{(s+c)^2 + (2\pi f_0)^2}$	Damped sine
9	$\exp(-ct)\cos(2\pi f_0 t)u_a(t)$	$\dfrac{s+c}{(s+c)^2 + (2\pi f_0)^2}$	Damped cosine
10	$t\sin(2\pi f_0 t)u_a(t)$	$\dfrac{4\pi f_0}{[s^2 + (2\pi f_0)^2]^2}$	Polynomial sine
11	$t\cos(2\pi f_0 t)u_a(t)$	$\dfrac{s^2 - (2\pi f_0)^2}{[s^2 + (2\pi f_0)^2]^2}$	Polynomial cosine

Property	$x_a(t)$	$X_a(f)$
Linearity	$ax_1(t) + bx_2(t)$	$aX_1(s) + bX_2(s)$
Time scale	$x_a(at), a > 0$	$\dfrac{1}{a}X_a\left(\dfrac{s}{a}\right)$
Time multiplication	$tx_a(t)$	$-\dfrac{dX_a(s)}{ds}$
Time division	$\dfrac{x_a(t)}{t}$	$\displaystyle\int_s^\infty X_a(\sigma)d\sigma$
Time shift	$x_a(t - T)u_a(t - T)$	$\exp(-sT)X_a(s)$
Frequency shift	$\exp(-at)x_a(t)$	$X_a(s + a)$
Derivative	$\dfrac{dx_a(t)}{dt}$	$sX_a(s) - x_a(0^+)$
Integral	$\displaystyle\int_0^t x_a(\tau)d\tau$	$\dfrac{X_a(s)}{s}$
Differentiation	$\dfrac{d^k x_a(t)}{dt^k}$	$s^k X_a(s) - \displaystyle\sum_{i=0}^{k-1} s^{k-i-1}\dfrac{d^i x_a(0^+)}{dt^i}$
Convolution	$\int_0^t x_1(\tau)x_2(t - \tau)d\tau$	$X_1(s)X_2(s)$
Initial value	$x_a(0^+)$	$\displaystyle\lim_{s\to\infty} sX_a(s)$
Final value	$\displaystyle\lim_{t\to\infty} x_a(t)$, stable	$\displaystyle\lim_{s\to 0} sX_a(s)$

C.4 Z-transform

$$X(z) \triangleq \sum_{k=0}^{\infty} x(k)z^{-k}, \quad z \in C$$

Table C.6: ▶
Z-transform Pairs

Entry	$x(k)$	$X(z)$	Description
1	$\delta(k)$	1	Unit impulse
2	$u(k)$	$\dfrac{z}{z-1}$	Unit step
3	$ku(k)$	$\dfrac{z}{(z-1)^2}$	Unit ramp
4	$k^2u(k)$	$\dfrac{z(z+1)}{(z-1)^3}$	Unit parabola
5	$a^k u(k)$	$\dfrac{z}{z-a}$	Exponential
6	$ka^k u(k)$	$\dfrac{az}{(z-a)^2}$	Linear exponential
7	$k^2 a^k u(k)$	$\dfrac{az(z+a)}{(z-a)^3}$	Quadratic exponential
8	$\sin(bk)u(k)$	$\dfrac{z\sin(b)}{z^2 - 2z\cos(b) + 1}$	Sine
9	$\cos(bk)u(k)$	$\dfrac{z[z - \cos(b)]}{z^2 - 2z\cos(b) + 1}$	Cosine
10	$a^k \sin(bk)u(k)$	$\dfrac{az\sin(b)}{z^2 - 2az\cos(b) + a^2}$	Damped sine
11	$a^k \cos(bk)u(k)$	$\dfrac{z[z - a\cos(b)]}{z^2 - 2az\cos(b) + a^2}$	Damped cosine

Property	$x(k)$	$X(z)$
Linearity	$ax_1(k) + bx_2(k)$	$aX_1(z) + bX_2(z)$
Delay	$x(k - M)u(k - M)$	$z^{-M}X(z)$
Time multiplication	$kx(k)$	$-z\dfrac{dX(z)}{dz}$
Z-scale	$a^k x(k)$	$X\left(\dfrac{z}{a}\right)$
Initial value	$x(0)$	$\lim\limits_{z \to \infty} X(z)$
Final value	$\lim\limits_{k \to \infty} x(k)$, stable	$\lim\limits_{z \to 1}(z - 1)X(z)$
Linear convolution	$x(k) \star y(k)$	$X(z)Y(z)$
Linear cross-correlation	$r_{xy}(k)$	$X(z)Y^*(z)$

C.5 Discrete-time Fourier Transform (DTFT)

$$X(f) = \sum_{k=0}^{\infty} x(k)\exp(-jk2\pi fT), \quad f \in R$$

Property	$x(k)$	$X(f)$
Linearity	$ax(k) + by(k)$	$aX(f) + bY(f)$
Periodic	$x(k)$	$X(f + f_s) = X(f)$
Symmetry	$x(k)$, real	$X(-f) = X^*(f)$
Time shift	$x(k - M)u(k - M)$	$\exp(-jM2\pi fT)X(f)$
Frequency shift	$\exp(jk2\pi f_0 T)x(k)$	$X(f - f_0)$
Convolution	$x(k) \star y(k)$	$X(f)Y(f)$
Parseval's theorem	$x(k)$	$\sum\limits_{k=0}^{\infty} |x(k)|^2 = \dfrac{1}{f_s} \int_{-f_s/2}^{f_s/2} |X(f)|^2 df$

C.6 Discrete Fourier Transform (DFT)

$$X(i) \stackrel{\Delta}{=} \sum_{k=0}^{N-1} x(k) \exp\left(\frac{-jki2\pi}{N}\right), \quad 0 \le i < N$$

Table C.9: ▶
Discrete Fourier
Transform Properties

Property	$x(k)$	$X(i)$
Linearity	$ax(k) + by(k)$	$aX(i) + bY(i)$
Periodic	$x(k)$	$X(i + N) = X(i)$
Symmetry	$x(k)$, real	$X(N - i) = X^*(i)$
Circular shift	$x_p(k - M)$	$\exp\left(\frac{-jMi2\pi}{N}\right) X(i)$
Reflection	$x_p(-k)$, real	$X^*(i)$
Circular convolution	$x(k) \circ y(k)$	$X(i)Y(i)$
Circular cross-correlation	$c_{xy}(k)$	$X(i)Y^*(i)$
Parseval's theorem	$x(k)$	$\sum_{k=0}^{N-1} |x(k)|^2 = \frac{1}{N} \sum_{i=0}^{N-1} |X(i)|^2$

Mathematical Identities

D.1 Complex Numbers

$$j = \sqrt{-1}$$

$$z = x + jy$$

$$z^* = x - jy$$

$$z + z^* = 2\mathrm{Re}(z) = 2x$$

$$z - z^* = j2\mathrm{Im}(z) = j2y$$

$$zz^* = |z|^2 = x^2 + y^2$$

$$z = A\exp(j\phi)$$

$$A = \sqrt{x^2 + y^2}$$

$$\phi = \tan^{-1}\left(\frac{y}{x}\right)$$

D.2 Euler's Identity

$$\exp(\pm j\phi) = \cos(\phi) \pm j\sin(\phi)$$

$$\cos(\phi) = \frac{\exp(j\phi) + \exp(-j\phi)}{2}$$

$$\sin(\phi) = \frac{\exp(j\phi) - \exp(-j\phi)}{j2}$$

$$\exp(\pm j\pi/2) = \pm j$$

● ● ● ● ● ● ● ● ● ● ● ● ● ● ● ●

D.3 Trigonometric Identities

$$\cos^2(a) + \sin^2(a) = 1$$

$$\cos(a \pm b) = \cos(a)\cos(b) \mp \sin(a)\sin(b)$$

$$\sin(a \pm b) = \sin(a)\cos(b) \pm \cos(a)\sin(b)$$

$$\cos(2a) = \cos^2(a) - \sin^2(a)$$

$$\sin(2a) = 2\sin(a)\cos(a)$$

$$\cos^2(a) = \frac{1 + \cos(2a)}{2}$$

$$\sin^2(a) = \frac{1 - \cos(2a)}{2}$$

$$\cos(a)\cos(b) = \frac{\cos(a + b) + \cos(a - b)}{2}$$

$$\sin(a)\sin(b) = \frac{\cos(a - b) - \cos(a + b)}{2}$$

$$\sin(a)\cos(b) = \frac{\sin(a + b) + \sin(a - b)}{2}$$

● ● ● ● ● ● ● ● ● ● ● ● ● ● ● ●

D.4 Inequalities

$$|a + b| \le |a| + |b|$$

$$\|x + y\| \le \|x\| + \|y\|$$

$$|x^T y| \le \|x\| \cdot \|y\|$$

$$\left\| \int_a^b x(t)dt \right\| \le \int_a^b \|x(t)\|dt$$

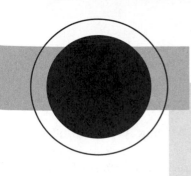

References and Further Reading

[1] Ahmed, N., and Natarajan, T., *Discrete-Time Signals and Systems*, Reston, VA: Reston, 1983.

[2] Al-Ajlouni, A. F., Schilling, R. J., and Harris, S. L., "Identification of Nonlinear Discrete-time Systems using Raised-cosine Radial Basis Function Networks." *Int. J. Systems Science*, Vol. 35, No. 4, pp. 211–221, 2004.

[3] Bartlett, M. S., "Smoothing Periodograms from Time Series with Continuous Spectra." *Nature*, Vol. 161, pp. 686–687, May 1948.

[4] Bellanger, M., *Adaptive Digital Filters and Signal Analysis*. New York: Marcel Dekker, 1987.

[5] Bequette, B. W., *Process Control: Modeling, Design, and Simulation*. Upper Saddle River, NJ: Prentice Hall, 2003.

[6] Burrrus, C. S., and Guo, H., *Introduction to Wavelets and Wavelet Transforms: A Primer*. Upper Saddle River, NJ: Prentice Hall, 1997.

[7] Candy, J. C., and Temes, G. C., *Oversampling Delta-Sigma Data Converters*. New York, NY: IEEE Press, 1992.

[8] Chapman, S. J., *MATLAB Programming for Engineers, Second Edition*. Pacific Grove, CA: Brooks/Cole, 2002.

[9] Chatfield, C., *The Analysis of Time Series*. London: Chapman and Hall, 1980.

[10] Constantinides, A. G., "Spectral Transformations for Digital Filters." *Proc. Inst. Elec. Engr.*, Vol. 117, pp. 1585–1590, 1970.

[11] Cook, T. A., *The Curves of Life*. Dover: 1979.

[12] Cooley, J. W., and Tukey, R. W., "An Algorithm for Machine Computation of Complex Fourier Series." *Mathematics of Computation*, Vol. 19, pp. 297–301, 1965.

[13] Crochiere, R. E., and Rabiner, L. R., "Optimum FIR Digital Filter Implementations for Decimation, Interpolation, and Narrow-band Filtering." *IEEE Trans. Acoustics, Speech, and Signal Processing*. Vol. 23, No. 5, pp. 444–456, 1975.

[14] Crochiere, R. E., and Rabiner, L. R., "Further Considerations in the Design of Decimators and Interpolators," *IEEE Trans. Acoustics, Speech, and Signal Processing*. Vol. 24, pp. 296–311, 1976.

[15] Dorf, R. C., and Svoboda, J. A., *Introduction to Electric Circuits*. New York: Wiley, 2000.

[16] Durbin, J., "Efficient Estimation of Parameters in Moving Average Model." *Biometrika*, Vol. 46, pp. 306–316, 1959.

[17] Dwight, H. B., *Tables of Integrals and other Mathematical Data, Fourth Edition*. New York: MacMillan, 1961.

[18] Elliott, S. J., Stothers, I. M., and Nelson, P. A., "A Multiple Error LMS Algorithm and Its Application to the Active Control of Sound and Vibration." *IEEE Trans. Acoustics, Speech, and Signal Processing*, Vol. ASSP-35, pp. 1423–1434, 1987.

[19] Franklin, G. E., Powell, J. D., and Workman, M. L., *Digital Control of Dynamic Systems, Second Edition*. Reading, MA: Addison-Wesley Publishing, 1990.

[20] Gerald, C. F., and Wheatley, P. O., *Applied Numerical Analysis, Fourth Edition*. Reading, MA: Addison-Wesley, 1989.

[21] Gitlin, R. D., Meadors, H. C., and Weinstein, S. B., "The Tap-Leakage Algorithm: An Algorithm for the Stable Operation of a Digitally Implemented, Fractional Adaptive Spaced Equalizer." *Bell System Tech. J.*, Vol. 61, pp. 1817–1839, 1982.

[22] Grover, D., and Deller J. R., *Digital Signal Processing and the Microcontroller*. Upper Saddle River, NJ: Prentice Hall, 1999.

[23] Hanselman, D., and Littlefield, B., *Mastering MATLAB 6*. Upper Saddle River, NJ: Prentice Hall, 2001.

[24] Hassibi, B. A., Sayed, H., and Kailath, T., "H^∞ Optimality of the LMS Algorithm." *IEEE Trans. on Signal Processing*, Vol. 44, pp. 267–280, 1996.

[25] Haykin, S., *Adaptive Filter Theory, Fourth Edition*. Upper Saddle River, NJ: Prentice Hall, 2002.

[26] Hernandez, E., and Arkun, Y., "Stability of Nonlinear Polynomial ARMA Models and Their Inverse," *Int. J. Control*, Vol. 63, No. 5, pp. 885–906, 1996.

[27] Ifeachor, E. C., and Jervis, B. W., *Digital Signal Processing: A Practical Approach, Second Edition*. Harlow, UK: Prentice Hall, 2002.

[28] Ingle, V. K., and Proakis, J. G., *Digital Signal Processing Using MATLAB*. Pacific Grove, CA: Brooks/Cole, 2000.

[29] Jackson, L. B., *Digital Filters and Signal Processing, Third Edition*. Boston: Kluwer Academic Publishers, 1996.

[30] Jaffe, D. A., and Smith, J. O., "Extensions of the Karplus-Strong Plucked-string Algorithm." *Computer Music Journal*, Vol. 7, No. 2, pp. 56–69, 1983.

[31] Jameco Electronics Catalog, 1355 Shoreway Road, Belmont, CA, 94002-4100, 2004.

[32] Jansson, P. A., *Deconvolution*. Academic Press, 1997.

[33] Kailath, T., "Estimating Filters for Linear Time-Invariant Channels." *Quarterly Progress Rep. 58*, MIT Research Laboratory for Electronics, Cambridge, MA, pp. 185–197, 1960.

[34] Kaiser, J. F., "Digital Filters." *System Analysis by Digital Computer*, Chap. 7, F. F. Kuo and J. F. Kaiser, Eds., New York: Wiley, 1966.

[35] Kaiser, J. F., "Nonrecursive Digital Filter Design Using the I_0-sinh Window Function." *Proc. 1974 IEEE Int. Symp. on Circuits and Systems*, San Francisco, CA, pp. 20–23, April 1974.

[36] Kuo, S. M., and Gan, W.-S., *Digital Signal Processing: Architectures, Implementations, and Applications*. Upper Saddle River, NJ: Pearson Prentice Hall, 2005.

[37] Kuo, S. M., and Morgan, D. R., *Active Noise Control Systems: Algorithms and DSP Implementations*. New York: Wiley, 1996.

[38] Lam, H. Y.-F., *Analog and Digital Filters*. Englewood Cliffs, NJ: Prentice Hall, 1979.

[39] Levinson, N., "The Wiener RMS Criterion in Filter Design and Prediction." *J. Math. Phys.*, Vol. 25, pp. 261–278, 1947.

[40] Ljung, L., Morf, M., and Falconer, D., "Fast Calculation of Gain Matrices for Recursive Estimation Schemes." *Int. J. Control*, Vol. 27, pp 1–19, 1978.

[41] Ludeman, L. C., *Fundamentals of Digital Signal Processing*. New York: Harper and Row, 1986.

[42] Markel J. D., and Gray, A. H., Jr., *Linear Prediction of Speech*. New York: Springer-Verlag, 1976.

[43] Marwan, N., "Make Install Tool for MATLAB." www.agnld.uni-potsdam.de/marwan/6.download/ whitepaper_makeinstall.html, Potsdam, Germany, 2003.

[44] McGillem, C. D., and Cooper, G. R., *Continuous and Discrete Signal and System Analysis*. New York: Holt, Rhinehart and Winston, 1974.

[45] Mitra, S. K., *Digital Signal Processing: A Computer-Based Approach, Second Edition*. Boston: McGraw-Hill Irwin, 2001.

[46] Moorer, J. A., "About the Reverberation Business." *Computer Music Journal*, Vol. 3, No. 2, pp. 13–28, 1979.

[47] Moustakides, G. V., "Study of the Transient Phase of the Forgetting Factor RLS." *IEEE Trans. Signal Processing*, Vol. 45, pp. 2468–2476, 1997.

[48] Nilsson, J. W., *Electric Circuits*. Reading, MA: Addison-Wesley, 1982.

[49] Noble, B., *Applied Linear Algebra*. Englewood Cliffs, NJ: Prentice Hall, 1969.

[50] Oppenheim, A. V., Schafer, R. W., and Buck, J. R., *Discrete-Time Signal Processing*. Upper Saddle River, NJ: Prentice Hall, 1999.

[51] Papamichalis, P., "Digital Signal Processing Applications with the TMS320 Family. Theory, Algorithms, and Implementations," Vol. 3, Dallas, TX: Texas Instruments, 1990.

[52] Park, S. K., and Miller, K. W., "Random Number Generators: Good Ones are Hard to Find." *Communications of the ACM*, Vol. 31, pp. 1192–1201, 1988.

[53] Parks, T. W., and Burrus, C. S., *Digital Filter Design*. New York: Wiley, 1987.

[54] Parks, T. W., and McClellan, J. H., "Chebyshev Approximation for Nonrecursive Digital Filters with Linear Phase." *IEEE Trans. Circuit Theory*, Vol. CT-19, pp. 189–194, Mar. 1972.

[55] Parks, T. W., and McClellan, J. H., "A Program for the Design of Linear Phase Finite Impulse Response Filters." *IEEE Trans. Audio Electroacoustics*, Vol. AU-20, No. 3, pp. 195–199, Aug. 1972.

[56] Porat, B., *A Course in Digital Signal Processing*. New York: Wiley, 1997.

[57] Proakis, J. G., and Manolakis, D. G., *Digital Signal Processing: Principles, Algorithms, and Applications, Second Edition*. New York: Macmillan Publishing, 1992.

[58] Rabiner, L. R., and Crochiere, R. E., "A Novel Implementation for Narrowband FIR Digital Filters" *IEEE Trans. Acoustics, Speech, and Signal Processing*, Vol. 23, No. 5, pp. 457–464, 1975.

[59] Rabiner, L. R., Gold, B., and McGonegal, C. A., "An Approach to the Approximation Problem for Nonrecursive Digital Filters." *IEEE Trans. Audio and Electroacoustic*, Vol. AU-18, pp. 83–106, June 1970.

[60] Rabiner, L. R., McClellan, J. H., and Parks, T. W., "FIR Digital Filter Design Techniques using Weighted Chebyshev Approximation." *Proc. IEEE*, Vol. 63, pp. 595–610, 1975.

[61] Rabiner, L. R., and Schafer, R. W., *Digital Processing of Speech Signals*. Englewood Cliffs, NJ: Prentice Hall, 1978.

[62] Remez, E. Y., "General Computational Methods of Chebyshev Approximations." *Atomic Energy Translation 4491*, Kiev, USSR, 1957.

[63] Roads, C., Pope, S. T., Piccialli, A., and DePolki, G., Editors, *Musical Signal Processing*. Lisse, Netherlands: Swets & Zeitlinger, 1997.

[64] Schilling, R. J., Al-Ajlouni, A., Carroll, J. J., and Harris, S. L., "Active Control of Narrowband Acoustic Noise of Unknown Frequency using a Phase-Locked Loop." *Int. J. Systems Science*, Vol. 29, No. 3, pp. 287–295, 1998.

[65] Schilling, R. J., Carroll, J. J., and Al-Ajlouni, A., "Approximation of Nonlinear Systems with Radial Basis Function Neural Networks." *IEEE Trans. Neural Networks*, Vol. 12, No. 1, pp. 1–15, 2001.

[66] Schilling, R. J., and Harris, S. L., *Applied Numerical Methods for Engineers Using MATLAB and C*. Pacific Grove, CA: Brooks/Cole, 2000.

[67] Schilling, R. J., and Lee, H., *Engineering Analysis: A Vector Space Approach*. New York: Wiley, 1988.

[68] Shan, T. J., and Kailath, T., "Adaptive Algorithms with an Automatic Gain Control Feature." *IEEE Trans. Circuits and Systems*, Vol. CAS-35, pp. 122–127, 1988.

[69] Shannon, C. E., "Communication in the Process of Noise." *Proc. IRE*, pp. 10–21, Jan. 1949.

[70] Slock, T. T. M., "On the Convergence Behavior of the LMS and the Normalized LMS Algorithms." *IEEE Trans. Signal Processing*, Vol. 41, pp. 2811–2825, 1993.

[71] Steiglitz, K., *A Digital Signal Processing Primer with Applications to Digital Audio and Computer Music*. Menlo Park, CA: Addison-Wesley, 1996.

[72] Strum, R. E., and Kirk, D. E., *First Principles of Discrete Systems and Digital Signal Processing*. Reading, MA: Addison-Wesley, 1988.

[73] Treichler, J. R., Johnson, C. R., Jr., and Larimoore, M. G., *Theory and Design of Adaptive Filters*. Upper Saddle River, NJ: Prentice Hall, 2001.

[74] Tretter, S. A., *Introduction to Discrete-Time Signal Processing*. New York: Wiley, 1976.

[75] Warnaka, G. E., Poole, L. A., and Tichy, J., "Active Acoustic Attenuators." U.S. Patent 4,473906, Sept. 25, 1984.

[76] Wasserman, P. D., *Neural Computing: Theory and Practice*. New York: Van Nostrand Reinhold, 1989.

[77] Webb, A., and Shannon, S., "Shape-Adaptive Radial Basis Functions," *IEEE Trans. Neural Networks*, Vol. 9, Nov. 1998.

[78] Weiner, N., and Paley, R. E. A. C., *Fourier Transforms in the Complex Domain*. Providence, RI: American Mathematical Society, 1934.

[79] Welch, P. D., "The Use of Fast Fourier Transform for the Estimation of Power Spectra: A Method Based on Time Averaging over Short Modified Periodograms." *IEEE Trans. Audio and Electroacoustics*, Vol. AU-15, pp. 70–73, June 1967.

[80] Widrow, B., and Hoff, M. E., Jr., "Adaptive Switching Circuits." *IRE WESCON Conv. Rec., Part 4*, pp. 96–104, 1960.

[81] Widrow, B., and Stearns, S. D., *Adaptive Signal Processing*. Englewood Cliffs, NJ: Prentice Hall, 1985.

[82] Wilkinson, J. H., *Rounding Error in Algebraic Processes*. Englewood Cliffs, NJ: Prentice Hall, 1963.

[83] Woodbury, M., "Inverting Modified Matrices." *Mem. Rep. 42*, Princeton, NJ: Statistical Research Group, Princeton University, 1950.

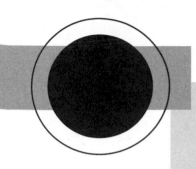

Index